Direct and Inverse Spectral Problems for Ordinary Differential and Functional-Differential Operators

Direct and Inverse Spectral Problems for Ordinary Differential and Functional-Differential Operators

Editor

Natalia Bondarenko

MDPI

Basel • Beijing • Wuhan • Barcelona • Belgrade • Novi Sad • Cluj • Manchester

Editor
Natalia Bondarenko
Samara National Research
University
Samara, Russia

Editorial Office
MDPI AG
Grosspeteranlage 5
4052 Basel, Switzerland

This is a reprint of articles from the Special Issue published online in the open access journal *Mathematics* (ISSN 2227-7390) (available at: https://www.mdpi.com/si/mathematics/direct_and_inverse_spectral_problems_for_ordinary_differential_and_functional_differential_operators).

For citation purposes, cite each article independently as indicated on the article page online and as indicated below:

Lastname, A.A.; Lastname, B.B. Article Title. *Journal Name* **Year**, *Volume Number*, Page Range.

ISBN 978-3-7258-1595-1 (Hbk)
ISBN 978-3-7258-1596-8 (PDF)
doi.org/10.3390/books978-3-7258-1596-8

Contents

About the Editor

Natalia Bondarenko

Natalia Bondarenko graduated from the Faculty of Mechanics and Mathematics of Saratov National Research State University in 2010. In 2011, she received her Candidate of Science (PhD) degree in mathematics and physics. In 2022, Natalia Bondarenko received her Doctor of Science degree from Lomonosov Moscow State University. Now, she works as a full professor at the Department of Applied Mathematics and Physics of Samara National Research University (Samara, Russia). In addition, Natalia Bondarenko currently participates in research projects at Saratov National Research State University, Peoples' Friendship University of Russia, and Lomonosov Moscow State University. Her research is focused on the theory of inverse spectral problems and related topics. In 2021, Natalia Bondarenko received the Research Excellence Award Russia from Elsevier for her outstanding contribution to the field of Mathematics at the national and international levels.

Preface

This Special Issue is devoted to the spectral theory of ordinary differential and functional–differential operators, as well as their applications and related fields. The initial goal of organizing this Special Issue was to invite researchers specializing in direct and inverse problems of spectral analysis to present their scientific results on these topics. Direct spectral problems consist in studying the properties of spectral characteristics such as asymptotical formulas for eigenvalues and eigenfunctions, trace formulas, completeness and basicity of root functions, etc. Inverse spectral problems consist in the recovery of operators from their spectral characteristics. In recent years, the theory of inverse problems has been actively developing not only for differential operators, but also for integro-differential operators, functional–differential operators with delays, and those with the so-called frozen arguments. Besides these topics, the Special Issue has attracted high-quality research papers on a variety of related fields, such as theory of partial differential equations and of pseudo-differential equations with fractional derivatives, asymptotical analysis for solutions of differential equations, spectral theory for abstract operators in Hilbert spaces, inverse nodal problems, numerical methods for solving spectral and scattering problems, etc. I hope that this Special Issue will be interesting to specialists in differential equations and mathematical physics, as well as in applications of spectral theory.

I would like to take this opportunity to thank everyone who contributed to the success of this Special Issue, namely, the authors for their high-quality scientific papers, the reviewers for their valuable comments, the editors for conducting peer-review process, and the MDPI staff for their organizational and technical assistance. I would especially like to thank the Section Manager Editor, Ms. Caitlynn Tong, for her continuous help, patience, attention, and support during the work on this Special Issue.

Natalia Bondarenko
Editor

mathematics MDPI

Article

Reconstruction of Higher-Order Differential Operators by Their Spectral Data

Natalia P. Bondarenko [1,2]

[1] Department of Applied Mathematics and Physics, Samara National Research University, Moskovskoye Shosse 34, Samara 443086, Russia; bondarenkonp@info.sgu.ru
[2] Department of Mechanics and Mathematics, Saratov State University, Astrakhanskaya 83, Saratov 410012, Russia

Abstract: This paper is concerned with inverse spectral problems for higher-order ($n > 2$) ordinary differential operators. We develop an approach to the reconstruction from the spectral data for a wide range of differential operators with either regular or distribution coefficients. Our approach is based on the reduction of an inverse problem to a linear equation in the Banach space of bounded infinite sequences. This equation is derived in a general form that can be applied to various classes of differential operators. The unique solvability of the linear main equation is also proved. By using the solution of the main equation, we derive reconstruction formulas for the differential expression coefficients in the form of series and prove the convergence of these series for several classes of operators. The results of this paper can be used for the constructive solution of inverse spectral problems and for the investigation of their solvability and stability.

Keywords: inverse spectral problems; higher-order differential operators; distribution coefficients; constructive solution; method of spectral mappings

MSC: 34A55; 34B09; 34B05; 34E05; 46F10

Citation: Bondarenko, N.P. Reconstruction of Higher-Order Differential Operators by Their Spectral Data. *Mathematics* **2022**, *10*, 3882. https://doi.org/10.3390/math10203882

Academic Editor: Vasily Novozhilov

Received: 13 September 2022
Accepted: 17 October 2022
Published: 19 October 2022

1. Introduction

This paper is concerned with the inverse spectral theory for operators generated by the differential expression

$$\ell_n(y) := y^{(n)} + \sum_{k=0}^{\lfloor n/2\rfloor-1} (\tau_{2k}(x)y^{(k)})^{(k)} + \sum_{k=0}^{\lfloor (n-1)/2\rfloor-1} \left((\tau_{2k+1}(x)y^{(k)})^{(k+1)} + (\tau_{2k+1}(x)y^{(k+1)})^{(k)}\right), \; x \in (0,1), \quad (1)$$

where the notation $\lfloor a\rfloor$ means rounding down, and the functions $\{\tau_\nu\}_{\nu=0}^{n-2}$ can be either integrable or distributional. Various aspects of spectral theory for such operators and related issues have been intensively studied in recent years (see, e.g., [1–9]). However, the general theory of inverse spectral problems for (1) with arbitrary $n > 2$ has not been created yet. This paper aims to develop an approach to the reconstruction of the coefficients $\{\tau_\nu\}_{\nu=0}^{n-2}$ from the spectral data for a wide class of differential operators.

1.1. Historical Background

Inverse problems of spectral analysis consist in the recovery of differential operators from their spectral information. Such problems arise in practice when one needs to determine certain physical parameters of a system from some measured data or to construct a model with desired properties. The majority of physical applications are concerned with linear differential operators of form (1) with $n = 2, 3, 4$.

For $n = 2$, expression (1) turns into the Sturm–Liouville (Schrödinger) operator

$$-\ell_2(y) = -y'' + q(x)y, \tag{2}$$

which models string vibrations in classical mechanics, electron motion in quantum mechanics, and is widely used in other branches of science and engineering. The third-order linear differential operators arise in the inverse problem method for integration of the nonlinear Boussinesq equation (see [10,11]), in mechanical problems of modeling the thin membrane flow of viscous liquid and elastic beam vibrations (see [12] and references therein). Inverse spectral problems for the fourth-order linear differential operators attract much attention from scholars because of their applications in mechanics and geophysics (see [13–20] and references therein).

The classical results of the inverse problem theory were obtained for the Sturm–Liouville operator (2) with integrable potential $q(x)$ in the 1950s by Marchenko, Levitan, and their followers (see [21,22]). They developed the transformation operator method, which reduces the nonlinear inverse Sturm–Liouville spectral problem to the linear Fredholm integral equation of the second kind. However, the transformation operator method appeared to be ineffective for the higher-order differential operators

$$y^{(n)} + \sum_{k=0}^{n-2} p_k(x) y^{(k)}, \quad n > 2. \tag{3}$$

Note that the differential expression (1) can be transformed into (3) in the case of sufficiently smooth coefficients $\{\tau_\nu\}_{\nu=0}^{n-2}$.

Thus, the development of inverse spectral theory for the higher-order operators (3) required new approaches. Relying on the ideas of Leibenson [23,24], Yurko created the method of spectral mappings. This method allowed him to construct inverse problem solutions for the higher-order differential operators (3) with regular (integrable) coefficients on the half-line $x > 0$ and on a finite interval $x \in (0, T)$ (see [25,26]). The case of Bessel-type singularities also was considered [27,28]. Later on, the ideas of the method of spectral mappings were applied to a wide range of inverse spectral problems, e.g., to inverse problems for the first-order differential systems [29], for differential operators on graphs [30], and for quadratic differential pencils [31]. This method is based on the theory of analytic functions and mainly on the contour integration in the complex plane of the spectral parameter. The method of spectral mappings reduces a nonlinear inverse problem to a linear equation in a suitable Banach space. This space is constructed in different ways for different operator classes. In particular, for differential operators on a finite interval, the main equation is usually derived in the space m of infinite bounded sequences. It is also worth mentioning that an approach to inverse scattering problems for higher-order differential operators (3) on the full line was developed by Beals et al. [32,33].

During the last 20 years, the inverse problems have been actively investigated for the second-order differential operators with distributional potentials (see, e.g., [34–43]). In particular, Hryniv and Mykytyuk [34–36] transferred the transformation operator method to the Sturm–Liouville operators (2) with potential $q(x)$ of class $W_2^{-1}(0,1)$ and so generalized the basic results of inverse problem theory to this class of operators. Note that the space W_2^{-1} contains the Dirac δ-function and the Coulumb potential $\frac{1}{x}$, which are used for modeling particle interactions in quantum mechanics [44]. The method of spectral mappings has been extended to the Sturm–Liouville operators with potentials of W_2^{-1} in [37,43,45]. This opens the possibility of constructing the inverse spectral theory for higher-order differential operators with distribution coefficients. However, till now, only the first steps have been taken in this direction. In [9,46], the uniqueness of recovering the higher-order differential operators with distribution coefficients on a finite interval and on the half-line has been studied. The goals of this paper are to derive the linear main equation of the inverse problem to prove its unique solvability and to obtain reconstruction formulas for the coefficients $\{\tau_\nu\}_{\nu=0}^{n-2}$ of various classes.

1.2. Problem Statement and Methods

Our treatment of the differential expression (1) is based on *the regularization approach.* Namely, we assume the differential equation

$$\ell_n(y) = \lambda y, \quad x \in (0,1), \tag{4}$$

where λ is the spectral parameter, can be equivalently transformed into the first-order system

$$Y'(x) = (F(x) + \Lambda)Y(x), \quad x \in (0,1), \tag{5}$$

where $Y(x)$ is a column vector function of size n, Λ is the $(n \times n)$-matrix whose entry at the position $(n,1)$ equals λ and all the other entries are zero, and $F(x) = [f_{k,j}(x)]_{k,j=1}^n$ is a matrix function with the following properties:

$$\begin{array}{llll} f_{k,j}(x) \equiv 0, & k+1<j, & f_{k,k+1}(x) \equiv 1, & k = \overline{1,n-1}, \\ f_{k,k} \in L_2(0,1), & k = \overline{1,n}, & f_{k,j} \in L_1(0,1), & k > j, \quad \mathrm{trace}(F(x)) = 0. \end{array} \tag{6}$$

We denote the class of $(n \times n)$ matrix functions satisfying (6) by $\mathfrak{F}_n$.

By using any matrix $F \in \mathfrak{F}_n$, one can define the quasi-derivatives

$$y^{[0]} := y, \quad y^{[k]} = (y^{[k-1]})' - \sum_{j=1}^{k} f_{k,j} y^{[j-1]}, \quad k = \overline{1,n}, \tag{7}$$

and the domain

$$\mathcal{D}_F = \{y \colon y^{[k]} \in AC[0,1],\, k = \overline{0,n-1}\}.$$

Definition 1. *A matrix function $F(x) \in \mathfrak{F}_n$ is called an associated matrix of the differential expression $\ell_n(y)$ if $\ell_n(y) = y^{[n]}$ for any $y \in \mathcal{D}_F$. We call a function y a solution of Equation (4) if $y \in \mathcal{D}_F$ and $y^{[n]} = \lambda y$, $x \in (0,1)$.*

For a function $y \in \mathcal{D}_F$, introduce the notation $\vec{y}(x) = \mathrm{col}(y^{[0]}(x), y^{[1]}(x), \dots, y^{[n-1]}(x))$. Obviously, y is a solution of Equation (4) if and only if $Y = \vec{y}$ satisfies (5).

The associated matrices for various classes of differential expressions $\ell_n(y)$ have been constructed, e.g., in [1,3,46–48] (see also Sections 4.3–4.5 of this paper). For example, for the differential expression $\ell_2(y) = y'' - \tau_0 y$, $\tau_0 \in W_2^{-1}(0,1)$, that is, $\tau_0 = \sigma_0'$, $\sigma_0 \in L_2(0,1)$, the associated matrix has the form (see [49]):

$$F(x) = \begin{bmatrix} \sigma_0(x) & 1 \\ -\sigma_0^2(x) & -\sigma_0(x) \end{bmatrix}.$$

For the regular case $\tau_\nu \in L_1(0,1)$, $\nu = \overline{0,n-2}$, the construction of associated matrix $F(x)$ is well-known (see [50] and Section 4.4 of this paper). The regularization of even-order ($n = 2m$) differential operators (1) with distribution coefficients $\tau_{2k+j} \in W_2^{-(m-k-j)}(0,1)$, $k = \overline{0,m-1}$, $j = 0,1$, has been obtained by Mirzoev and Shkalikov [1]. Later on, the case of odd-order n was considered in [47]. Vladimirov [51] suggested a more general construction which, in particular, includes both cases [1,47]. It is worth mentioning that, in [1,47,51], the differential expressions of more general form than (1) were studied, with the coefficients at $y^{(n)}$ and $y^{(n-1)}$ not necessarily equal 1 and 0, respectively. However, in this paper, we confine ourselves to the form (1), which is natural for studying the inverse problems [9,46].

In this paper, we assume that $\ell_n(y)$ is any differential expression that has an associated matrix in terms of Definition 1. We do not impose any additional restrictions on $\{\tau_\nu\}_{\nu=0}^{n-2}$, since we are interested in formulating the abstract results which can be applied to various classes of differential operators. Certain restrictions on $\{\tau_\nu\}_{\nu=0}^{n-2}$ are imposed below when necessary.

Let us proceed to the inverse problem formulation. Suppose that we have a differential expression of form (1) and an associated matrix $F(x) = [f_{k,j}]_{k,j=1}^{n}$. By using the corresponding quasi-derivatives (7), define the linear forms

$$\mathcal{U}_{s,a}(y) := y^{[p_{s,a}]}(a) + \sum_{j=1}^{p_{s,a}} u_{s,j,a} y^{[j-1]}(a), \quad s = \overline{1,n}, \quad a = 0, 1, \tag{8}$$

where $p_{s,a} \in \{0, \ldots, n-1\}$, $p_{s,a} \neq p_{k,a}$ for $s \neq k$, and $u_{s,j,a}$ are some complex numbers. In addition, introduce the matrices $U_a = [u_{s,j,a}]_{s,j=1}^{n}$, $u_{s,j,a} := \delta_{j,p_{s,a}+1}$ for $j > p_{s,a}$, $a = 0, 1$. Here, and below, $\delta_{j,k}$ is the Kronecker delta. We call the triple $(F(x), U_0, U_1)$ by the problem $\mathcal{L}$. Below, we introduce various characteristics related to the problem $\mathcal{L}$.

Denote by $\{C_k(x, \lambda)\}_{k=1}^{n}$ the solutions of Equation (4) satisfying the initial conditions

$$\mathcal{U}_{s,0}(C_k) = \delta_{s,k}, \quad s = \overline{1,n}. \tag{9}$$

Equivalently, the $(n \times n)$-matrix function $C(x, \lambda) := [\vec{C}_k(x, \lambda)]_{k=1}^{n}$ is the solution of the system (5) with the initial condition $C(0, \lambda) = U_0^{-1}$. Therefore, the solutions $\{C_k(x, \lambda)\}_{k=1}^{n}$ are uniquely defined. Moreover, their quasi-derivatives $C_k^{[j]}(x, \lambda)$ are entire in λ for each fixed $x \in [0, 1]$, $k = \overline{1,n}$, $j = \overline{0, n-1}$.

It has been proved in ([9], Section 4) that, for all $\lambda \in \mathbb{C}$ except for a countable set, Equation (4) has the so-called *Weyl solutions* $\{\Phi_k(x, \lambda)\}_{k=1}^{n}$ satisfying the boundary conditions

$$\mathcal{U}_{s,0}(\Phi_k) = \delta_{s,k}, \quad s = \overline{1,k}, \qquad \mathcal{U}_{s,1}(\Phi_k) = 0, \quad s = \overline{k+1,n}, \tag{10}$$

Define the matrix function $\Phi(x, \lambda) = [\vec{\Phi}_k(x, \lambda)]_{k=1}^{n}$. The columns of the matrices $C(x, \lambda)$ and $\Phi(x, \lambda)$ form fundamental solution systems of (5). Consequently, the following relation holds:

$$\Phi(x, \lambda) = C(x, \lambda) M(\lambda) \tag{11}$$

where the matrix function $M(\lambda)$ is called *the Weyl matrix* of the problem $\mathcal{L}$ (see [9]).

The notion of Weyl matrix generalizes the notion of Weyl function for the second-order operators (see [21,26]). Weyl functions and their generalizations play an important role in the inverse spectral theory for various classes of differential operators. In particular, Yurko [25–28] has used the Weyl matrix as the main spectral characteristics for the reconstruction of the higher-order differential operators (3) with regular coefficients. The analogous inverse problem for the differential expression of form (1) can be formulated as follows.

Problem 1. *Given the Weyl matrix $M(\lambda)$, find the coefficients $\{\tau_\nu\}_{\nu=0}^{n-2}$.*

The uniqueness of Problem 1's solution has been proved in [9] for the Mirzoev–Shkalikov case: $n = 2m$, $\tau_{2k+j} \in W_2^{-(m-k-j)}(0, 1)$ and $n = 2m + 1$, $\tau_{2k+j} \in W_1^{-(m-k-j)}(0, 1)$, $j = 0, 1$. In [46], the uniqueness of recovering the boundary condition coefficients from the Weyl matrix has been studied.

It has been shown in ([9], Section 4) that the Weyl matrix $M(\lambda) = [M_{j,k}(\lambda)]_{j,k=1}^{n}$ is unit lower-triangular, and its nontrivial entries have the form

$$M_{j,k}(\lambda) = -\frac{\Delta_{j,k}(\lambda)}{\Delta_{k,k}(\lambda)}, \quad 1 \le k < j \le n, \tag{12}$$

where $\Delta_{k,k}(\lambda) := \det[\mathcal{U}_{s,1}(C_r)]_{s,r=k+1}^{n}$ and $\Delta_{j,k}(\lambda)$ is obtained from $\Delta_{k,k}(\lambda)$ by the replacement of C_j by C_k. The functions $C_r^{[s]}(1, \lambda)$, $r = \overline{1,n}$, $s = \overline{0, n-1}$ are entire analytic in λ, and so are the functions $\Delta_{j,k}(\lambda)$, $1 \le k \le j \le n$. Hence, $M(\lambda)$ is meromorphic in λ, and the poles of the k-th column of $M(\lambda)$ coincide with the zeros of $\Delta_{k,k}(\lambda)$. At the same time, the zeros of the entire functions $\Delta_{j,k}(\lambda)$, $1 \le k \le j \le n$ coincide with the eigenvalues of some boundary value problems for Equation (4), and the inverse problem by the Weyl matrix (Problem 1) is related to the inverse problem by $\frac{n(n+1)}{2}$ spectra (see [9] for details).

We say that the problem $\mathcal{L}$ belongs to the class W if all the zeros of $\Delta_{k,k}(\lambda)$ are simple for $k = \overline{1, n-1}$. Then, in view of (12), the poles of $M(\lambda)$ are simple. In general, the function $\Delta_{k,k}(\lambda)$ can have at most a finite number of multiple zeros. The latter case can be treated by developing the methods of Buterin et al. [52,53], who considered the non-self-adjoint Sturm–Liouville operators ($n = 2$) with regular potentials. However, the case of multiple zeros is much more technically complicated, so, in this paper, we always assume that $\mathcal{L} \in W$.

Denote by Λ the set of the Weyl matrix poles. Consider the Laurent series

$$M(\lambda) = \frac{M_{\langle -1 \rangle}(\lambda_0)}{\lambda - \lambda_0} + M_{\langle 0 \rangle}(\lambda_0) + M_{\langle 1 \rangle}(\lambda_0)(\lambda - \lambda_0) + \dots, \quad \lambda_0 \in \Lambda.$$

Denote

$$\mathcal{N}(\lambda_0) := [M_{\langle 0 \rangle}(\lambda_0)]^{-1} M_{\langle -1 \rangle}(\lambda_0), \quad \lambda_0 \in \Lambda, \tag{13}$$

We call the collection $\{\lambda_0, \mathcal{N}(\lambda_0)\}_{\lambda_0 \in \Lambda}$ *the spectral data* of the problem $\mathcal{L}$. Obviously, the spectral data are uniquely specified by the Weyl matrix $M(\lambda)$, so Problem 1 can be reduced to the following problem:

Problem 2. *Given the spectral data $\{\lambda_0, \mathcal{N}(\lambda_0)\}_{\lambda_0 \in \Lambda}$, find the coefficients $\{\tau_\nu\}_{\nu=0}^{n-2}$.*

It is more convenient to study the reconstruction question for Problem 2. It is worth mentioning that, in fact, the Weyl matrix and the spectral data can be constructed according to the above definitions for any matrix function $F(x)$ of class $\mathfrak{F}_n$, not necessarily associated with any differential expression of form (1). However, in general, the matrix $F(x)$ is not uniquely specified by the Weyl matrix (see Example 4.5 in [46]). Therefore, in this paper, the solution of Problem 2 is divided into the two steps:

$$\{\lambda_0, \mathcal{N}(\lambda_0)\}_{\lambda_0 \in \Lambda} \overset{(1)}{\to} \{\Phi_k(x,\lambda)\}_{k=1}^n \overset{(2)}{\to} \{\tau_\nu\}_{\nu=0}^{n-2}.$$

The recovery of the Weyl solutions $\{\Phi_k(x,\lambda)\}_{k=1}^n$ from the spectral data is studied for a matrix $F(x)$ of general form, and then reconstruction formulas are derived for $\{\tau_\nu\}_{\nu=0}^{n-2}$ of certain classes.

For a fixed $F \in \mathfrak{F}_n$, we define the quasi-derivatives (7), the expression $\ell_n(y) := y^{[n]}$, the problem $\mathcal{L} = (F(x), U_0, U_1)$, its spectral data $\{\lambda_0, \mathcal{N}(\lambda_0)\}_{\lambda_0 \in \Lambda}$ as above, and focus on the following auxiliary problem.

Problem 3. *Given the spectral data $\{\lambda_0, \mathcal{N}(\lambda_0)\}_{\lambda_0 \in \Lambda}$, find the Weyl solutions $\{\Phi_k(x,\lambda)\}_{k=1}^n$.*

Let us briefly describe the method of solution. Along with $\mathcal{L}$, we consider another problem $\tilde{\mathcal{L}} = (\tilde{F}(x), \tilde{U}_0, \tilde{U}_1)$ of the same form but with different coefficients. Similarly to $\Phi(x,\lambda)$, define $\tilde{\Phi}(x,\lambda)$ for $\tilde{\mathcal{L}}$. An important role in our analysis is played by *the matrix of spectral mappings*:

$$\mathcal{P}(x,\lambda) = \Phi(x,\lambda)[\tilde{\Phi}(x,\lambda)]^{-1}.$$

For each fixed $x \in [0,1]$, the matrix function $\mathcal{P}(x,\lambda)$ is meromorpic in λ with poles at the eigenvalues $\Lambda \cup \tilde{\Lambda}$. The method is based on the integration of some functions by a special family of contours enclosing these eigenvalues. Applying the Residue theorem, we derive an infinite system of linear equations. Furthermore, that system is transformed into a linear equation in the Banach space m of infinite bounded sequences. The main equation of the inverse problem has the form

$$(\mathbf{I} - \tilde{R}(x))\psi(x) = \tilde{\psi}(x), \quad x \in [0,1],$$

where, for each fixed $x \in [0,1]$, $\psi(x)$ and $\tilde{\psi}(x)$ are elements of m, $\tilde{R}(x)$ is a linear compact operator in m, and $\mathbf{I}$ is the unit operator. The element $\tilde{\psi}(x)$ and the operator $\tilde{R}(x)$ are constructed by the model problem $\tilde{\mathcal{L}}$ and by the spectral data $\{\lambda_0, \mathcal{N}(\lambda_0)\}_{\lambda_0 \in \Lambda}$, $\{\tilde{\lambda}_0, \tilde{\mathcal{N}}(\tilde{\lambda}_0)\}_{\tilde{\lambda}_0 \in \tilde{\Lambda}}$ of the two problems $\mathcal{L}$, $\tilde{\mathcal{L}}$, respectively, while the unknown element $\psi(x)$ is related to the desired functions $\{\Phi_k(x,\lambda)\}_{k=1}^n$. We prove that the operator $(\mathbf{I} - \tilde{R}(x))$ has the bounded inverse, and so the main equation is uniquely solvable (see Theorem 1). This implies the uniqueness of the solution for Problem 3. Using the main equation, we obtain a constructive procedure for solving Problem 3 (see Algorithm 1). These results can be applied to a wide range of differential operators (1) with associated matrices of class $\mathfrak{F}_n$.

Furthermore, by using the solution of the main equation, we derive reconstruction formulas for $\{\tau_\nu\}_{\nu=0}^{n-2}$. We describe the general idea and then apply it to the certain classes of operators:

(i) $n = 3$, $\tau_1 \in L_2(0,1)$, $\tau_0 \in W_2^{-1}(0,1)$.
(ii) n is even, $\tau_\nu \in L_2(0,1)$, $\nu = \overline{0, n-2}$.
(iii) n is even, $\tau_\nu \in W_2^{-1}(0,1)$, $\nu = \overline{0, n-2}$.

We obtain the uniqueness theorems and constructive algorithms for solving Problem 2 for the cases (i)–(iii). Note that, although the functions τ_ν in the case (ii) are regular, this case has less smoothness than the one considered by Yurko [26].

The reconstruction formulas have the form of series, and the main difficulties in our analysis are related to studying the convergence of those series. These difficulties increase for the case of nonsmooth and/or distribution coefficients. In order to prove the series convergence, we use the Birkhoff-type solutions constructed by Savchuk and Shkalikov [2] and the precise asymptotic formulas for the spectral data obtained in [54]. For the cases (ii) and (iii), we reconstruct the functions τ_ν step-by-step for $\nu = n-2, n-3, \dots, 1, 0$. The similar approach can be used in the case of odd n, which requires technical modifications.

By using the reconstruction formulas, one can develop numerical methods for solving inverse spectral problems (see [55] for the second-order case). However, this issue requires an additional work. In this paper, we obtain theoretical algorithms, which in the future can be used for the investigation of existence and stability of the inverse problem solution.

It is worth mentioning that our method of inverse problem solution is the first one for higher-order differential operators with distribution coefficients. The obtained main equation and reconstruction formulas generalize the results of [45] for the Sturm–Liouville operators with distribution potential. The other methods which applied to the second-order operators (see, e.g., [34,39]), to the best of the author's knowledge, appear to be ineffective for higher orders.

The paper is organized as follows. In Section 2, we provide preliminaries and study the properties of the spectral data. Section 3 is devoted to the contour integration and to the derivation of the main equation of the inverse problem in a Banach space. The unique solvability of the main equation is also proved. As a result, an algorithm for solving the auxiliary Problem 3 is obtained for arbitrary $F \in \mathfrak{F}_n$. In Section 4, we derive the reconstruction formulas for the coefficients $\{\tau_\nu\}_{\nu=0}^{n-2}$ and study the convergence of the obtained series. Section 5 contains a brief summary of the main results.

2. Preliminaries

Throughout the paper, we use the following **notations**.

1. I is the $(n \times n)$ unit matrix, e_k is the k-th column of I, $k = \overline{1, n}$.
2. The sign T denotes the matrix transpose.
3. $\delta_{k,j} = \begin{cases} 1, & k = j, \\ 0, & k \neq j. \end{cases}$
4. $J := [(-1)^{k+1}\delta_{k,n-j+1}]_{k,j=1}^n$, $J_a := [(-1)^{p^\star_{k,a}}\delta_{k,n-j+1}]_{k,j=1}^n$, where $p^\star_{k,a} := n - 1 - p_{k,a}$, $a = 0, 1$.

5. If for $\lambda \to \lambda_0$
$$A(\lambda) = \sum_{k=-q}^{p} a_k(\lambda - \lambda_0)^k + o((\lambda - \lambda_0)^p),$$
then
$$[A(\lambda)]^{\langle k \rangle}_{|\lambda=\lambda_0} = A_{\langle k \rangle}(\lambda_0) := a_k.$$
6. The notations $\lfloor x \rfloor$ and $\lceil x \rceil$ are used for rounding a real number x down and up, respectively.
7. The binomial coefficients are denoted by $C_n^k = \dfrac{n!}{k!(n-k)!}$.
8. Along with $\mathcal{L}$, we consider the problems $\tilde{\mathcal{L}}$, $\mathcal{L}^\star$, $\tilde{\mathcal{L}}^\star$ of the same form but with different coefficients. We agree that, if a symbol γ denotes an object related to $\mathcal{L}$, then the symbols $\tilde{\gamma}$, $\gamma^\star$, $\tilde{\gamma}^\star$ denote the analogous objects related to $\tilde{\mathcal{L}}$, $\mathcal{L}^\star$, $\tilde{\mathcal{L}}^\star$, respectively. Note that the quasi-derivatives for the problems $\tilde{\mathcal{L}}$, $\mathcal{L}^\star$, $\tilde{\mathcal{L}}^\star$ are defined by using the matrices $\tilde{F}(x)$, $F^\star(x)$, $\tilde{F}^\star(x)$, respectively, which may be different from $F(x)$.
9. The notation $y^{[k]}$ is used for quasi-derivatives defined by (7) (or analogously by using the entries of $\tilde{F}(x)$, $F^\star(x)$, or $\tilde{F}^\star(x)$). The notation $\vec{y}(x)$ is used for the column vector of the quasi-derivatives $y^{[0]}(x), y^{[1]}(x), \dots, y^{[n-1]}(x)$.
10. In estimates, the symbol C is used for various positive constants independent of x, l, k, etc.
11. $a \overset{if\ (condition)}{\times} b = \begin{cases} ab, & \text{if (condition) holds,} \\ a, & \text{otherwise.} \end{cases}$

In Section 2.1, we define an auxiliary problem $\mathcal{L}^\star = (F^\star(x), U_0^\star, U_1^\star)$ and study its properties. In Section 2.2, the properties of the spectral data $\{\lambda_0, \mathcal{N}(\lambda_0)\}_{\lambda_0 \in \Lambda}$ are investigated.

2.1. Problems $\mathcal{L}$ and $\mathcal{L}^\star$

For a matrix $F \in \mathfrak{F}_n$, define the matrix $F^\star(x) = [f^\star_{k,j}(x)]_{k,j=1}^n$ as follows:
$$f^\star_{k,j}(x) := (-1)^{k+j+1} f_{n-j+1,n-k+1}(x). \tag{14}$$

Obviously, $F^\star \in \mathfrak{F}_n$.

Let $F(x)$ be a fixed matrix function of class $\mathfrak{F}_n$. Suppose that $y \in \mathcal{D}_F$ and $z \in \mathcal{D}_{F^\star}$; the quasi-derivatives for y are defined via (7) by using the elements of $F(x)$, and the quasi-derivatives for z are defined as
$$z^{[0]} := z, \quad z^{[k]} = (z^{[k-1]})' - \sum_{j=1}^{k} f^\star_{k,j} z^{[j-1]}, \quad k = \overline{1,n}, \tag{15}$$
and
$$\mathcal{D}_{F^\star} := \{z \colon z^{[k]} \in AC[0,1],\, k = \overline{0, n-1}\}.$$

Define
$$\ell_n(y) := y^{[n]}, \quad \ell_n^\star(z) := (-1)^n z^{[n]}, \quad \langle z, y \rangle := \sum_{j=0}^{n-1} (-1)^j z^{[j]} y^{[n-j-1]}.$$

Lemma 1. *The following relation holds:*
$$\frac{d}{dx}\langle z, y \rangle = z\ell_n(y) - y\ell_n^\star(z). \tag{16}$$

Proof. Differentiation implies

$$\frac{d}{dx}\langle z,y\rangle = \sum_{j=0}^{n-1}(-1)^j (z^{[j]})' y^{[n-j-1]} + \sum_{j=0}^{n-1}(-1)^j z^{[j]} (y^{[n-j-1]})'. \tag{17}$$

From (7) and (15), we obtain

$$(z^{[j]})' = z^{[j+1]} + \sum_{s=1}^{j+1} f^{\star}_{j+1,s} z^{[s-1]}, \quad (y^{[n-j-1]})' = y^{[n-j]} + \sum_{s=1}^{n-j} f_{n-j,s} y^{[s-1]}.$$

Substituting the latter relations into (17), we obtain

$$\frac{d}{dx}\langle z,y\rangle = \sum_{j=0}^{n-1}(-1)^j y^{[n-j]} z^{[j]} + \sum_{j=0}^{n-1}(-1)^j \sum_{s=1}^{n-j} f_{n-j,s} y^{[s-1]} z^{[j]}$$
$$+ \sum_{j=0}^{n-1}(-1)^j y^{[n-j-1]} z^{[j+1]} + \sum_{j=0}^{n-1}(-1)^j \sum_{s=1}^{j+1} f^{\star}_{j+1,s} y^{[n-j-1]} z^{[s-1]}.$$

Note that

$$\sum_{j=0}^{n-1}(-1)^j y^{[n-j]} z^{[j]} + \sum_{j=0}^{n-1}(-1)^j y^{[n-j-1]} z^{[j+1]} = y^{[n]} z + (-1)^{n-1} y z^{[n]},$$
$$\sum_{j=0}^{n-1}(-1)^j \sum_{s=1}^{n-j} f_{n-j,s} y^{[s-1]} z^{[j]} = \sum_{1\le s\le j\le n} (-1)^{s+1} f_{n-s+1,n-j+1} y^{[n-j]} z^{[s-1]},$$
$$\sum_{j=0}^{n-1}(-1)^j \sum_{s=1}^{j+1} f^{\star}_{j+1,s} y^{[n-j-1]} z^{[s-1]} = \sum_{1\le s\le j\le n} (-1)^{j+1} f^{\star}_{j,s} y^{[n-j]} z^{[s-1]}.$$

Taking (14) into account, we arrive at (16). □

If y and z satisfy the relations $\ell_n(y) = \lambda y$ and $\ell^{\star}_n(z) = \mu z$, respectively, then (16) readily implies

$$\frac{d}{dx}\langle z,y\rangle = (\lambda-\mu) yz. \tag{18}$$

Define $\vec{y}(x) = \mathrm{col}(y^{[0]}(x), y^{[1]}(x), \dots, y^{[n-1]}(x))$ and $\vec{z}(x) = \mathrm{col}(z^{[0]}(x), z^{[1]}(x), \dots, z^{[n-1]}(x))$ by using the corresponding quasi-derivatives (7) and (15), and the matrix $J := [(-1)^{k+1}\delta_{k,n-j+1}]_{k,j=1}^{n}$. Then,

$$\langle z,y\rangle_{|x=a} = [\vec{z}(a)]^T J \vec{y}(a). \tag{19}$$

For $a = 0,1$, let $U_a = [u_{s,j,a}]_{s,j=1}^{n}$ be an $(n\times n)$ matrix such that $u_{s,j,a} = \delta_{j,p_{s,a}+1}$ for $j > p_{s,a}$, where $p_{s,a} \in \{0,\dots,n-1\}$, and $p_{s,a} \ne p_{k,a}$ for $s \ne k$. The matrices U_a define the linear forms $\mathcal{U}_{s,a}$ via (8).

Along with U_a, consider the matrices

$$U^{\star}_a := [J_a^{-1} U_a^{-1} J]^T, \quad a = 0,1, \tag{20}$$

where $J_a = [(-1)^{p^{\star}_{k,a}} \delta_{k,n-j+1}]_{k,j=1}^{n}$, $p^{\star}_{k,a} := n-1-p_{n-k+1,a}$. The matrices $U^{\star}_a$, $a = 0,1$, generate the linear forms

$$\mathcal{U}^{\star}_{s,a}(z) = z^{[p^{\star}_{s,a}]}(a) + \sum_{j=1}^{p^{\star}_{s,a}} u^{\star}_{s,j,a} z^{[j-1]}(a), \quad s = \overline{1,n}, \quad a = 0,1.$$

The matrices $U_a^\star$ are chosen is such a way that the following relation holds:

$$\langle z, y \rangle_{|x=a} = \sum_{s=1}^{n} (-1)^{p^\star_{s,a}} \mathcal{U}^\star_{s,a}(z) \mathcal{U}_{n-s+1,a}(y) \tag{21}$$

for any $y \in \mathcal{D}_F$, $z \in \mathcal{D}_{F^\star}$. Indeed, the right-hand side of (21) can be represented in the matrix form

$$[U_a^\star \vec{z}(a)]^T J_a U_a \vec{y}(a),$$

Taking (19) and (20) into account, we arrive at (21).

Consider the problems $\mathcal{L} = (F(x), U_0, U_1)$ and $\mathcal{L}^\star = (F^\star(x), U_0^\star, U_1^\star)$. For $\mathcal{L}$, the matrix functions $C(x,\lambda)$, $\Phi(x,\lambda)$, and $M(\lambda)$ were defined in the Introduction. For $\mathcal{L}^\star$, similarly denote by $\{C_k^\star(x,\lambda)\}_{k=1}^n$ and $\{\Phi_k^\star(x,\lambda)\}_{k=1}^n$ the solutions of equation $\ell_n^\star(z) = \lambda z$, $x \in (0,1)$, satisfying the conditions

$$\begin{aligned} \mathcal{U}^\star_{s,0}(C_k^\star) = \delta_{s,k}, \quad s = \overline{1,n}, \\ \mathcal{U}^\star_{s,0}(\Phi_k^\star) = \delta_{s,k}, \quad s = \overline{1,k}, \qquad \mathcal{U}^\star_{s,1}(\Phi_k^\star) = 0, \quad s = \overline{k+1,n}. \end{aligned} \tag{22}$$

Put $C^\star(x,\lambda) := [\vec{C}_k^\star(x,\lambda)]_{k=1}^n$, $\Phi^\star(x,\lambda) := [\vec{\Phi}_k^\star(x,\lambda)]_{k=1}^n$. Then, the relation

$$\Phi^\star(x,\lambda) = C^\star(x,\lambda) M^\star(\lambda) \tag{23}$$

holds, where $M^\star(\lambda)$ is the Weyl matrix of the problem $\mathcal{L}^\star$.

Lemma 2. *The following relations hold:*

$$[M^\star(\lambda)]^T J_0 M(\lambda) = J_0, \tag{24}$$

$$[\Phi^\star(x,\lambda)]^T J \Phi(x,\lambda) = J_0. \tag{25}$$

Proof. The initial conditions (9) are equivalent to $U_0 C(0,\lambda) = I$. Using (11), we obtain $M(\lambda) = U_0 \Phi(0,\lambda)$. Similarly, $M^\star(\lambda) = U_0^\star \Phi^\star(0,\lambda)$. Hence,

$$A(\lambda) := [M^\star(\lambda)]^T J_0 M(\lambda) = [U_0^\star \Phi^\star(0,\lambda)]^T J_0 U_0 \Phi(0,\lambda), \quad A(\lambda) = [A_{k,j}(\lambda)]_{k,j=1}^n,$$

$$A_{k,j}(\lambda) = [U_0^\star \vec{\Phi}_k^\star(0,\lambda)]^T J_0 U_0 \vec{\Phi}_j(0,\lambda) = \sum_{s=1}^{n} (-1)^{p^\star_{s,0}} \mathcal{U}^\star_{s,0}(\Phi_k^\star) \mathcal{U}_{n-s+1,0}(\Phi_j). \tag{26}$$

On the one hand, using (10), (22), and (26), we obtain $A_{k,j}(\lambda) = 0$ if $k + j > n + 1$ and $A_{k,j}(\lambda) = (-1)^{p^\star_{k,0}}$ if $k + j = n + 1$. On the other hand, (21) and (26) imply $A_{k,j}(\lambda) = \langle \Phi_k^\star, \Phi_j \rangle_{|x=0}$. It follows from (18) that $\langle \Phi_k^\star, \Phi_j \rangle$ does not depend on x. Consequently,

$$\langle \Phi_k^\star, \Phi_j \rangle_{|x=0} = \langle \Phi_k^\star, \Phi_j \rangle_{|x=1} = \sum_{s=1}^{n} (-1)^{p^\star_{s,1}} \mathcal{U}^\star_{s,1}(\Phi_k^\star) \mathcal{U}_{n-s+1,1}(\Phi_j).$$

Using the boundary conditions (10) and (22) at $x = 1$, we conclude that $A_{k,j}(\lambda) = 0$ if $k + j < n + 1$. Thus, $A(\lambda) = J_0$ and (24) is proved.

Using the relation $A_{k,j}(\lambda) = \langle \Phi_k^\star, \Phi_j \rangle$ for $k, j = \overline{1,n}$ and (19), we obtain

$$A(\lambda) = [\Phi^\star(x,\lambda)]^T J \Phi(x,\lambda).$$

This implies (25). □

2.2. Spectral Data

Consider the Weyl matrix $M(\lambda)$ of the problem $\mathcal{L} = (F(x), U_0, U_1)$, where $F \in \mathfrak{F}_n$. Recall that the poles of the k-th column of $M(\lambda)$ coincide with the zeros of $\Delta_{k,k}(\lambda) =$

$\det[\mathcal{U}_{s,1}(C_r)]_{s,r=k+1}^n$. One can easily show that the zeros of $\Delta_{k,k}(\lambda)$ coincide with the eigenvalues of the following boundary value problem $\mathcal{L}_k$:

$$\ell_n(y) = \lambda y, \quad x \in (0,1), \qquad \mathcal{U}_{s,0}(y) = 0, \quad s = \overline{1,k}, \qquad \mathcal{U}_{s,1}(y) = 0, \quad s = \overline{k+1,n}.$$

By virtue of Theorem 1.1 in [54], the spectrum of $\mathcal{L}_k$ is a countable set of eigenvalues $\Lambda_k := \{\lambda_{l,k}\}_{l\geq 1}$ having the following asymptotics (counting with multiplicities):

$$\lambda_{l,k} = (-1)^{n-k}\left(\frac{\pi}{\sin\frac{\pi k}{n}}(l + \chi_k + \varkappa_{l,k})\right)^n, \tag{27}$$

where $\{\varkappa_{l,k}\} \in l_2$ and χ_k are constants which depend only on n, k, and $\{p_{s,a}\}$. Hence, for a fixed $k \in \{1,\ldots,n-1\}$ and sufficiently large l, the eigenvalues $\lambda_{l,k}$ are simple.

Assume that $\mathcal{L} \in W$, that is, all the zeros of $\Delta_{k,k}(\lambda)$ are simple for $k = \overline{1,n-1}$. Then, in view of (12) and (24), the poles of $M(\lambda)$ and $M^\star(\lambda)$ are simple. It follows from (11) and (23) that the matrix functions $\Phi(x,\lambda)$ and $\Phi^\star(x,\lambda)$ for each fixed $x \in [0,1]$ also have only simple poles.

Denote $\Lambda := \bigcup_{k=1}^{n-1}\Lambda_k$. Similarly to $\mathcal{N}(\lambda_0)$, denote

$$\mathcal{N}^\star(\lambda_0) := [M^\star_{\langle 0\rangle}(\lambda_0)]^{-1}M^\star_{\langle -1\rangle}(\lambda_0), \quad \lambda_0 \in \Lambda. \tag{28}$$

For $\lambda_0 \notin \Lambda$, we mean that $\mathcal{N}(\lambda_0) = \mathcal{N}^\star(\lambda_0) = 0$.

Let us study some properties of the matrices $\mathcal{N}(\lambda_0)$ and $\mathcal{N}^\star(\lambda_0)$. Denote by $\phi(x,\lambda)$ the first row of the matrix function $\Phi(x,\lambda)$: $\phi(x,\lambda) = e_1^T\Phi(x,\lambda) = [\Phi_k(x,\lambda)]_{k=1}^n$.

Lemma 3. *The following relations hold for each $\lambda_0 \in \Lambda$: $\mathcal{N}^2(\lambda_0) = 0$,*

$$[\mathcal{N}^\star(\lambda_0)]^T = -J_0\mathcal{N}(\lambda_0)J_0^{-1}, \tag{29}$$

$$\Phi_{\langle -1\rangle}(x,\lambda_0) = \Phi_{\langle 0\rangle}(x,\lambda_0)\mathcal{N}(\lambda_0), \quad \Phi^\star_{\langle -1\rangle}(x,\lambda_0) = \Phi^\star_{\langle 0\rangle}(x,\lambda_0)\mathcal{N}^\star(\lambda_0), \tag{30}$$

$$\ell_n(\phi_{\langle 0\rangle}(x,\lambda_0)) = \lambda_0\phi_{\langle 0\rangle}(x,\lambda_0) + \phi_{\langle 0\rangle}(x,\lambda_0)\mathcal{N}(\lambda_0). \tag{31}$$

Proof. The relation (24) implies

$$[M(\lambda)]^{-1} = J_0^{-1}[M^\star(\lambda)]^TJ_0, \tag{32}$$

$$M(\lambda)J_0^{-1}[M^\star(\lambda)]^T = J_0^{-1}. \tag{33}$$

It follows from (33) that

$$M_{\langle -1\rangle}(\lambda_0)J_0^{-1}[M^\star_{\langle -1\rangle}(\lambda_0)]^T = 0, \tag{34}$$

$$M_{\langle 0\rangle}(\lambda_0)J_0^{-1}[M^\star_{\langle -1\rangle}(\lambda_0)]^T + M_{\langle -1\rangle}(\lambda_0)J_0^{-1}[M^\star_{\langle 0\rangle}(\lambda_0)]^T = 0. \tag{35}$$

Using (13), (28), and (35), we obtain (29). Multiplying (29) by $\mathcal{N}(\lambda_0)$ and using (34), we derive

$$\mathcal{N}(\lambda_0)J_0^{-1}[\mathcal{N}^\star(\lambda_0)]^T = -\mathcal{N}^2(\lambda_0)J_0^{-1} = 0.$$

Hence $\mathcal{N}^2(\lambda_0) = 0$.

Using (11) and (32), we obtain

$$C(x,\lambda) = \Phi(x,\lambda)[M(\lambda)]^{-1} = \Phi(x,\lambda)J_0^{-1}[M^\star(\lambda)]^TJ_0.$$

Since $C(x,\lambda)$ is entire in λ for each fixed $x \in [0,1]$, then we obtain

$$\Phi_{\langle 0\rangle}(x,\lambda_0)J_0^{-1}[M^\star_{\langle -1\rangle}(\lambda_0)]^TJ_0 + \Phi_{\langle -1\rangle}(x,\lambda_0)J_0^{-1}[M^\star_{\langle 0\rangle}(\lambda_0)]^TJ_0 = 0, \quad \lambda_0 \in \Lambda. \tag{36}$$

Using (36) and (28), we derive

$$\Phi_{\langle 0\rangle}(x,\lambda_0)J_0^{-1}[\mathcal{N}^\star(\lambda_0)]^T J_0 + \Phi_{\langle -1\rangle}(x,\lambda_0) = 0.$$

Taking (29) into account, we arrive at the first relation in (30). The second one is similar.

It follows from the relation $\ell_n(\phi(x,\lambda)) = \lambda\phi(x,\lambda)$ that

$$\ell_n(\phi_{\langle -1\rangle}(x,\lambda_0)) = \lambda_0\phi_{\langle -1\rangle}(x,\lambda_0),$$
$$\ell_n(\phi_{\langle 0\rangle}(x,\lambda_0)) = \lambda_0\phi_{\langle 0\rangle}(x,\lambda_0) + \phi_{\langle -1\rangle}(x,\lambda_0).$$

Using (30), we arrive at (31). □

Consider the entries of the matrix $\mathcal{N}(\lambda_0) = [\mathcal{N}_{k,j}(\lambda_0)]_{k,j=1}^n$. Since $M(\lambda)$ is unit lower-triangular, we have $\mathcal{N}_{k,j}(\lambda_0) = 0$ for all $k \le j$, $\lambda_0 \in \Lambda$. The structural properties of $\mathcal{N}(\lambda_0)$ are described by the following lemma.

Lemma 4. *(i) If $\lambda_0 \notin \Lambda_k$, then $\mathcal{N}_{s,j}(\lambda_0) = 0$, $s = \overline{k+1,n}$, $j = \overline{1,k}$.*

(ii) If $\lambda_0 \in \Lambda_s$ for $s = \overline{\nu+1,k-1}$, $\lambda_0 \notin \Lambda_\nu$, $\lambda_0 \notin \Lambda_k$, $1 \le \nu+1 < k \le n$, then $\mathcal{N}_{k,\nu+1}(\lambda_0) \ne 0$. (Here $\Lambda_0 = \Lambda_n = \varnothing$).

Proof. This lemma is proved similarly to Lemma 2.3.1 in [26], so we outline the proof briefly. If $\lambda_0 \notin \Lambda_k$, then $\Phi_{k,\langle -1\rangle}(x,\lambda_0) = 0$. On the other hand, it follows from (30) that

$$\Phi_{k,\langle -1\rangle}(x,\lambda_0) = \sum_{s=k+1}^{n} \mathcal{N}_{s,k}(\lambda_0)\Phi_{s,\langle 0\rangle}(x,\lambda_0).$$

Applying the linear forms $\mathcal{U}_{s,0}$ to this relation for $s = \overline{k+1,n}$, we conclude that $\mathcal{N}_{s,k}(\lambda_0) = 0$, $s = \overline{k+1,n}$. Thus, the assertion (i) is proved for $j = k$. The proof for $j = k-1,\dots,2,1$ can be obtained by induction.

In order to prove (ii), we suppose that $\Delta_{\nu,\nu}(\lambda_0) \ne 0$, $\Delta_{s,s}(\lambda_0) = 0$ for $s = \overline{\nu+1,k-1}$. Then, it can be shown that $\mathcal{U}_{s,1}(\Phi_{s,\langle 0\rangle}(x,\lambda_0)) \ne 0$, $s = \overline{\nu+2,k-1}$ and $\Phi_{\nu+1,\langle -1\rangle}(x,\lambda_0) \not\equiv 0$. Suppose that $\mathcal{N}_{k,\nu+1}(\lambda_0) = 0$. Consequently, (30) implies

$$\Phi_{\nu+1,\langle -1\rangle}(x,\lambda_0) = \sum_{s=\nu+2}^{k-1} \mathcal{N}_{s,\nu+1}(\lambda_0)\Phi_{s,\langle 0\rangle}(x,\lambda_0).$$

Applying the linear forms $\mathcal{U}_{s,1}$ for $s = \overline{\nu+2,k-1}$, we conclude that $N_{s,\nu+1}(\lambda_0) = 0$, $s = \overline{\nu+2,k-1}$, and so $\Phi_{\nu+1,\langle -1\rangle}(x,\lambda_0) \equiv 0$. This contradiction yields (ii). □

In view of the asymptotics (27), we have $\lambda_{l,k} \ne \lambda_{r,k+1}$ for sufficiently large l and r. Therefore, Lemma 4 implies the following corollary.

Corollary 1. *For sufficiently large $|\lambda_0|$, $\lambda_0 \in \Lambda$, all the entries of $\mathcal{N}(\lambda_0)$ equal zero except $\mathcal{N}_{k+1,k}(\lambda_0)$, $k = \overline{1,n-1}$.*

Define *the weight numbers* $\beta_{l,k} := \mathcal{N}_{k+1,k}(\lambda_{l,k})$. It is worth considering $\beta_{l,k}$ only for sufficiently large l. It follows from (13) and (12) that

$$\beta_{l,k} = M_{k+1,k,\langle -1\rangle}(\lambda_{l,k}) = -\frac{\Delta_{k+1,k}(\lambda_{l,k})}{\frac{d}{d\lambda}\Delta_{k,k}(\lambda_{l,k})}.$$

Consequently, Theorem 6.2 from [54] yields the asymptotics

$$\beta_{l,k} = l^{n-1+p_{k+1,0}-p_{k,0}}(\beta_k^0 + \varkappa_{l,k}^0), \quad \{\varkappa_{l,k}^0\} \in l_2, \quad k = \overline{1,n-1}, \tag{37}$$

where the constants β_k^0 depend only on n, k, and $\{p_{s,a}\}$.

3. Main Equation

This section is devoted to the constructive solution of the auxiliary Problem 3, that is, to the recovery of the Weyl solutions $\{\Phi_k(x,\lambda)\}_{k=1}^n$ from the spectral data $\{\lambda_0, \mathcal{N}(\lambda_0)\}_{\lambda_0 \in \Lambda}$. We consider this problem for $\mathcal{L} = (F(x), U_0, U_1) \in W$ with an arbitrary $F \in \mathfrak{F}_n$. Thus, the results of this section can be applied to a wide class of differential expressions (1) with the associated matrix of $\mathfrak{F}_n$.

Along with $\mathcal{L}$, we consider another problem $\tilde{\mathcal{L}} = (\tilde{F}(x), \tilde{U}_0, \tilde{U}_1)$ of the same form but with different coefficients. Assume that $\tilde{F} \in \mathfrak{F}_n$, $p_{s,a} = \tilde{p}_{s,a}$, $s = \overline{1,n}$, $a = 0,1$. The quasi-derivatives for $\tilde{\mathcal{L}}$ are defined by the matrix $\tilde{F}(x)$, so they are different from the quasi-derivatives of the problem $\mathcal{L}$. The problem $\tilde{\mathcal{L}}^\star$ is defined similarly to $\mathcal{L}^\star$. For simplicity, we assume that $\tilde{\mathcal{L}} \in W$. The case $\tilde{\mathcal{L}} \in W$ requires technical modifications (see Remark 1). Denote $\mathcal{I} := \Lambda \cup \tilde{\Lambda}$.

In Section 3.1, we reduce the studied problem to the infinite system (68) of linear equations with respect to some entries of $\phi_{\langle 0 \rangle}(x,\lambda_0)$, $\lambda_0 \in \mathcal{I}$. Our technique is based on the contour integration in the λ-plane and on the Residue theorem. In Section 3.2, the system (68) is transformed into the main Equation (80) in the Banach space m of infinite bounded sequences. The unique solvability of the main equation is proved. Finally, we arrive at the constructive Algorithm 1 for finding $\{\Phi_k(x,\lambda)\}_{k=1}^n$ by the spectral data. This algorithm is used in the next section for solving the inverse spectral problem.

3.1. Contour Integration

In order to formulate and prove the main lemma of this subsection (Lemma 6), we first need some preliminaries. Introduce the notations

$$D(x,\mu,\lambda) := (\lambda-\mu)^{-1}[\Phi(x,\mu)]^{-1}\Phi(x,\lambda), \quad \tilde{D}(x,\mu,\lambda) := (\lambda-\mu)^{-1}[\tilde{\Phi}(x,\mu)]^{-1}\tilde{\Phi}(x,\lambda), \tag{38}$$

$$D_{\langle \alpha \rangle}(x,\lambda_0,\lambda) := [D(x,\mu,\lambda)]^{\langle \alpha \rangle}_{|\mu=\lambda_0}, \quad \alpha \in \mathbb{Z}. \tag{39}$$

and similarly define $\tilde{D}_{\langle \alpha \rangle}(x,\lambda_0,\lambda)$.

Lemma 5. *The following relations hold:*

$$D_{\langle -1 \rangle}(x,\lambda_0,\lambda) = -\mathcal{N}(\lambda_0) D_{\langle 0 \rangle}(x,\lambda_0,\lambda), \tag{40}$$

$$[D(x,\mu,\lambda)]^{\langle -1 \rangle}_{|\lambda=\lambda_0} = [D(x,\mu,\lambda)]^{\langle 0 \rangle}_{|\lambda=\lambda_0}\mathcal{N}(\lambda_0), \tag{41}$$

$$[(\lambda-\lambda_0)I + \mathcal{N}(\lambda_0)] D_{\langle 0 \rangle}(x,\lambda_0,\lambda) = J_0^{-1}\langle [\phi^\star_{\langle 0 \rangle}(x,\lambda_0)]^T, \phi(x,\lambda)\rangle, \tag{42}$$

$$D'(x,\mu,\lambda) = J_0^{-1}[\phi^\star(x,\mu)]^T \phi(x,\lambda). \tag{43}$$

Proof. Using (25) and (38), we obtain

$$D(x,\mu,\lambda) = (\lambda-\mu)^{-1} J_0^{-1}[\Phi^\star(x,\mu)]^T J \Phi(x,\lambda). \tag{44}$$

It follows from (44) and (39) that

$$D_{\langle -1 \rangle}(x,\lambda_0,\lambda) = (\lambda-\lambda_0)^{-1} J_0^{-1}[\Phi^\star_{\langle -1 \rangle}(x,\lambda_0)]^T J\Phi(x,\lambda), \tag{45}$$

$$D_{\langle 0 \rangle}(x,\lambda_0,\lambda) = (\lambda-\lambda_0)^{-1} J_0^{-1}[\Phi^\star_{\langle 0 \rangle}(x,\lambda_0)]^T J\Phi(x,\lambda) + (\lambda-\lambda_0)^{-2} J_0^{-1}[\Phi^\star_{\langle -1 \rangle}(x,\lambda_0)]^T J\Phi(x,\lambda). \tag{46}$$

Using (45) and (46) together with Lemma 3, we derive (40). The relation (41) is proved similarly.

It follows from (19) that

$$[\Phi^\star(x,\mu)]^T J\Phi(x,\lambda) = \langle [\phi^\star(x,\mu)]^T, \phi(x,\lambda)\rangle. \tag{47}$$

Using (45), (46), and (47), we obtain

$$(\lambda - \lambda_0) D_{\langle 0 \rangle}(x, \lambda_0, \lambda) = J_0^{-1} \langle [\phi^\star_{\langle 0 \rangle}(x, \mu)]^T, \phi(x, \lambda) \rangle + D_{\langle -1 \rangle}(x, \lambda_0, \lambda).$$

Taking (40) into account, we arrive at (42).

In order to prove (43), we combine (44), (47), and (18):

$$D'(x, \mu, \lambda) = (\lambda - \mu)^{-1} J_0^{-1} \frac{d}{dx} \langle [\phi^\star(x, \mu)]^T, \phi(x, \lambda) \rangle = J_0^{-1} [\phi^\star(x, \mu)]^T \phi(x, \lambda).$$

□

Put $\hat{\mathcal{N}}(\lambda_0) := \mathcal{N}(\lambda_0) - \tilde{\mathcal{N}}(\lambda_0)$. Below, in this section, we suppose that $x \in [0, 1]$ is fixed.

Lemma 6. *The following relations hold:*

$$\phi(x, \lambda) = \tilde{\phi}(x, \lambda) + \sum_{\lambda_0 \in \mathcal{I}} \phi_{\langle 0 \rangle}(x, \lambda_0) \hat{\mathcal{N}}(\lambda_0) \tilde{D}_{\langle 0 \rangle}(x, \lambda_0, \lambda), \tag{48}$$

$$D(x, \mu, \lambda) - \tilde{D}(x, \mu, \lambda) = \sum_{\lambda_0 \in \mathcal{I}} [D(x, \mu, \xi)]^{\langle 0 \rangle}_{\xi = \lambda_0} \hat{\mathcal{N}}(\lambda_0) \tilde{D}_{\langle 0 \rangle}(x, \lambda_0, \lambda), \tag{49}$$

where the series converge in the sense

$$\sum_{\lambda_0 \in \mathcal{I}} = \lim_{R \to \infty} \sum_{\lambda_0 \in \mathcal{I}_R}, \quad \mathcal{I}_R := \{\lambda \in \mathcal{I} \colon |\lambda| < R\},$$

uniformly by λ, μ on compact sets of $(\mathbb{C} \setminus \mathcal{I})$.

Proof. In this proof, a crucial role is played by the matrix of spectral mappings

$$\mathcal{P}(x, \lambda) = \Phi(x, \lambda) [\tilde{\Phi}(x, \lambda)]^{-1}. \tag{50}$$

It follows from (25) and (50) that

$$\mathcal{P}(x, \lambda) = \Phi(x, \lambda) J_0^{-1} [\tilde{\Phi}^\star(x, \lambda)]^T J. \tag{51}$$

The proof consists of three steps.

STEP 1. REGIONS AND CONTOURS. Choose a circle $\mathcal{C}_* := \{\lambda \in \mathbb{C} \colon |\lambda| < \lambda_*\}$ of sufficiently large radius λ_*. Choose the $\sqrt[n]{\lambda}$ branch so that $\arg(\sqrt[n]{\lambda}) \in (-\frac{\pi}{2n}, \frac{3\pi}{2n})$. Then, it follows from the asymptotics (27) that the roots $\rho_0 := \sqrt[n]{\lambda_0}$ of the eigenvalues $\lambda_0 \in (\mathcal{I} \setminus \mathcal{C}_*)$ lie in the two strips

$$\mathcal{S}_j := \{\rho \colon \operatorname{Re}(\epsilon_j \rho) > 0, \ |\operatorname{Im}(\epsilon_j \rho)| < c\}, \quad \epsilon_j := \exp(-2\pi i j / n), \quad j = 0, 1, \tag{52}$$

for an appropriate choice of the constant c. More precisely, $\sqrt[n]{\lambda_{l,k}} \in \mathcal{S}_0$ if $(n - k)$ is even and $\sqrt[n]{\lambda_{l,k}} \in \mathcal{S}_1$ otherwise. For $j = 0, 1$, denote by Ξ_j the image of $\mathcal{S}_j$ in the λ-plane under the mapping $\lambda = \rho^n$. Put $\Xi := \Xi_0 \cup \Xi_1 \cup \mathcal{C}_*$. Clearly, $\mathcal{I} \subset \Xi$.

Furthermore, fix a sufficiently small $\delta > 0$ and define the regions

$$\mathcal{S}_{j,\delta} := \{\rho \in \mathcal{S}_j \colon \exists \rho_0 \in \mathcal{S}_j \cap \mathcal{I} \text{ s.t. } |\rho - \rho_0| < \delta\}, \quad j = 0, 1.$$

For $j = 0, 1$, denote by $\Xi_{j,\delta}$ the image of $\mathcal{S}_j$ in the λ-plane under the mapping $\lambda = \rho^n$. Put

$$\mathcal{H}_\delta := \mathbb{C} \setminus (\Xi_{1,\delta} \cup \Xi_{2,\delta} \cup \mathcal{C}_*).$$

Let $\lambda = \rho^n$, $\Theta(\rho) := \mathrm{diag}\{1, \rho, \dots, \rho^{n-1}\}$. It can be shown in the standard way (see, e.g., the relation (2.1.37) in [26], and the proof of Theorem 2 in [9]) that

$$\mathcal{P}(x, \lambda) = \Theta(\rho)(I + o(1))[\Theta(\rho)]^{-1}, \quad |\lambda| \to \infty, \tag{53}$$

uniformly with respect to $\lambda \in \mathcal{H}_\delta$.

For sufficiently large values of $R > 0$, define the regions (see Figure 1):

$$\Xi_R := \{\lambda \in \Xi \colon |\lambda| < R\}, \quad \Xi_R^\pm := \{\lambda \colon |\lambda| < R,\, \lambda \notin \Xi,\, \pm \mathrm{Im}\lambda > 0\},$$

and their boundaries $\gamma_R := \partial \Xi_R$, $\gamma_R^\pm := \partial \Xi_R^\pm$ with the counter-clockwise circuit. Below, we consider only such radii R that $\gamma_R \subset \mathcal{H}_\delta$.

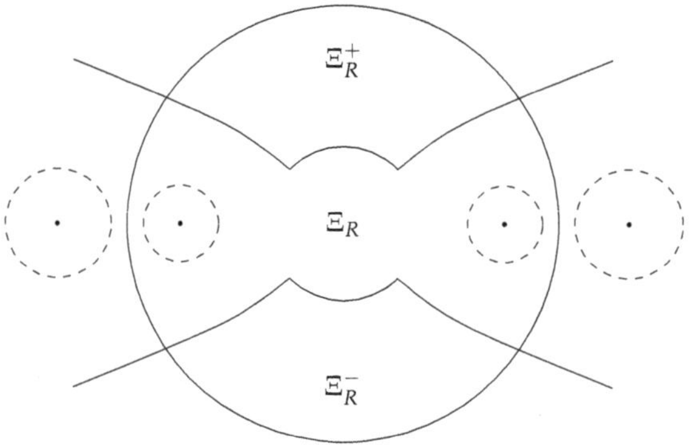

Figure 1. Contours.

STEP 2. CONTOUR INTEGRATION. In view of (51), the matrix function $\mathcal{P}(x, \lambda)$ is meromorpic in λ with the poles $\mathcal{I}$. Hence, $\mathcal{P}(x, \lambda)$ is analytic in $\Xi_R^\pm$. Let $\mathcal{P}_1(x, \lambda)$ be the first row of $\mathcal{P}(x, \lambda)$. The Cauchy formula implies

$$\mathcal{P}_1(x, \lambda) - e_1^T = -\frac{1}{2\pi i} \oint_{\gamma_R^\pm} \frac{\mathcal{P}_1(x, \xi) - e_1^T}{\lambda - \xi} \, d\xi, \quad \lambda \in \Xi_R^\pm,$$

$$\frac{\mathcal{P}(x, \lambda) - \mathcal{P}(x, \mu)}{\lambda - \mu} = -\frac{1}{2\pi i} \oint_{\gamma_R^\pm} \frac{\mathcal{P}(x, \xi)}{(\lambda - \xi)(\xi - \mu)} \, d\xi, \quad \lambda, \mu \in \Xi_R^\pm.$$

Consequently,

$$\mathcal{P}_1(x, \lambda) = e_1^T + \frac{1}{2\pi i} \oint_{\gamma_R} \frac{\mathcal{P}_1(x, \xi)}{\lambda - \xi} \, d\xi - \frac{1}{2\pi i} \oint_{|\xi| = R} \frac{\mathcal{P}_1(x, \xi) - e_1^T}{\lambda - \xi} \, d\xi, \tag{54}$$

$$\frac{\mathcal{P}(x, \lambda) - \mathcal{P}(x, \mu)}{\lambda - \mu} = \frac{1}{2\pi i} \oint_{\gamma_R} \frac{\mathcal{P}(x, \xi)}{(\lambda - \xi)(\xi - \mu)} \, d\xi - \frac{1}{2\pi i} \oint_{|\xi| = R} \frac{\mathcal{P}(x, \xi)}{(\lambda - \xi)(\xi - \mu)} \, d\xi. \tag{55}$$

Using (38), (50), (54), and (55), we derive

$$\phi(x, \lambda) = \mathcal{P}_1(x, \lambda) \tilde{\Phi}(x, \lambda) = \tilde{\phi}(x, \lambda) + \frac{1}{2\pi i} \oint_{\gamma_R} \frac{\mathcal{P}_1(x, \xi) \tilde{\Phi}(x, \lambda)}{\lambda - \xi} \, d\xi + \varepsilon_R^1(x, \lambda), \tag{56}$$

$$D(x, \mu, \lambda) - \tilde{D}(x, \mu, \lambda) = \frac{[\Phi(x, \mu)]^{-1} (\mathcal{P}(x, \lambda) - \mathcal{P}(x, \mu)) \tilde{\Phi}(x, \lambda)}{\lambda - \mu}$$

$$= \frac{1}{2\pi \mathrm{i}} \oint\limits_{\gamma_R} \frac{[\Phi(x,\mu)]^{-1}\Phi(x,\xi)}{\xi-\mu} \frac{[\tilde{\Phi}(x,\xi)]^{-1}\tilde{\Phi}(x,\lambda)}{\lambda-\xi}\, d\xi + \varepsilon_R^2(x,\mu,\lambda)$$
$$= \frac{1}{2\pi \mathrm{i}} \oint\limits_{\gamma_R} D(x,\mu,\xi)\tilde{D}(x,\xi,\lambda)\, d\xi + \varepsilon_R^2(x,\mu,\lambda), \tag{57}$$

where

$$\varepsilon_R^1(x,\lambda) := -\frac{1}{2\pi \mathrm{i}} \oint\limits_{|\xi|=R} \frac{(\mathcal{P}_1(x,\xi) - e_1^T)\tilde{\Phi}(x,\lambda)}{\lambda-\xi}\, d\xi,$$
$$\varepsilon_R^2(x,\mu,\lambda) := -\frac{1}{2\pi \mathrm{i}} \oint\limits_{|\xi|=R} \frac{[\Phi(x,\mu)]^{-1}\mathcal{P}(x,\xi)\tilde{\Phi}(x,\lambda)}{(\lambda-\xi)(\xi-\mu)}\, d\xi.$$

It follows from (53) that

$$\lim_{\substack{R\to\infty \\ \gamma_R\subset\mathcal{H}_\delta}} \varepsilon_R^1(x,\lambda) = 0, \quad \lim_{\substack{R\to\infty \\ \gamma_R\subset\mathcal{H}_\delta}} \varepsilon_R^2(x,\mu,\lambda) = 0. \tag{58}$$

STEP 3. RESIDUES. Using the first row of (51):

$$\mathcal{P}_1(x,\lambda) = \phi(x,\lambda) J_0^{-1} [\tilde{\Phi}^\star(x,\lambda)]^T J$$

and the Residue theorem, we obtain

$$\frac{1}{2\pi \mathrm{i}} \oint\limits_{\gamma_R} \frac{\mathcal{P}_1(x,\xi)\tilde{\Phi}(x,\lambda)}{\lambda-\xi}\, d\xi = \sum_{\lambda_0\in\mathcal{I}_R} \operatorname*{Res}_{\xi=\lambda_0} \phi(x,\xi)\tilde{D}(x,\xi,\lambda). \tag{59}$$

Using (56), (58), and (59), we obtain

$$\phi(x,\lambda) = \tilde{\phi}(x,\lambda) + \sum_{\lambda_0\in\mathcal{I}} (\phi_{\langle -1\rangle}(x,\lambda_0)\tilde{D}_{\langle 0\rangle}(x,\lambda_0,\lambda) + \phi_{\langle 0\rangle}(x,\lambda_0)\tilde{D}_{\langle -1\rangle}(x,\lambda_0,\lambda)). \tag{60}$$

It follows from (30) that

$$\phi_{\langle -1\rangle}(x,\lambda_0) = \phi_{\langle 0\rangle}(x,\lambda_0)\mathcal{N}(\lambda_0). \tag{61}$$

Substituting (40) for $\tilde{D}_{\langle -1\rangle}(x,\lambda_0,\lambda)$ and (61) into (60), we derive the relation (48).

It remains to prove (49). Using Lemma 5, we derive

$$\begin{aligned} \operatorname*{Res}_{\xi=\lambda_0} D(x,\mu,\xi)\tilde{D}(x,\xi,\lambda) &= [D(x,\mu,\xi)]_{|\xi=\lambda_0}^{\langle -1\rangle}\tilde{D}_{\langle 0\rangle}(x,\lambda_0,\lambda) + [D(x,\mu,\xi)]_{|\xi=\lambda_0}^{\langle 0\rangle}\tilde{D}_{\langle -1\rangle}(x,\lambda_0,\lambda) \\ &= [D(x,\mu,\xi)]_{|\xi=\lambda_0}^{\langle 0\rangle}\hat{\mathcal{N}}(\lambda_0)\tilde{D}_{\langle 0\rangle}(x,\lambda_0,\lambda). \end{aligned} \tag{62}$$

Combining (57), (58), and (62) all together and applying the Residue theorem, we arrive at (49).

Now, (48) and (49) are proved only for $\lambda,\mu\in(\mathbb{C}\setminus\Xi)$. Using analytic continuation, we conclude that these relations hold for $\lambda,\mu\in(\mathbb{C}\setminus\mathcal{I})$. □

Our next goal is to obtain an infinite system of linear equations with respect to some entries of $\phi_{\langle 0\rangle}(\lambda_0)$, $\lambda_0\in\mathcal{I}$. Introduce the ordered set

$$V := \{(l,k,\varepsilon) \colon l\ge 1,\ k\in\{1,\dots,n-1\},\ \varepsilon\in\{0,1\}\}.$$

For $v=(l,k,\varepsilon)$, $v_0=(l_0,k_0,\varepsilon_0)$ and $v,v_0\in V$, we mean that $v<v_0$ if $l<l_0$ or ($l=l_0$ and $k<k_0$) or ($l=l_0$, $k=k_0$ and $\varepsilon<\varepsilon_0$). Denote

$$\lambda_{l,k,0} := \lambda_{l,k}, \quad \lambda_{l,k,1} := \tilde{\lambda}_{l,k}, \quad \mathcal{N}_0(\lambda_0) := \mathcal{N}(\lambda_0), \quad \mathcal{N}_1(\lambda_0) := \tilde{\mathcal{N}}(\lambda_0), \tag{63}$$

$$\varphi_{l,k,\varepsilon}(x) := \Phi_{k+1,\langle 0\rangle}(x, \lambda_{l,k,\varepsilon}), \quad \tilde{\varphi}_{l,k,\varepsilon}(x) := \tilde{\Phi}_{k+1,\langle 0\rangle}(x, \lambda_{l,k,\varepsilon}), \tag{64}$$

$$\tilde{P}_{l,k,\varepsilon}(x, \lambda) := e_{k+1}^T \mathcal{N}_\varepsilon(\lambda_{l,k,\varepsilon}) \tilde{D}_{\langle 0\rangle}(x, \lambda_{l,k,\varepsilon}, \lambda), \tag{65}$$

$$\tilde{G}_{(l,k,\varepsilon),(l_0,k_0,\varepsilon_0)}(x) := [\tilde{P}_{l,k,\varepsilon}(x, \lambda)]^{\langle 0\rangle}_{\lambda=\lambda_{l_0,k_0,\varepsilon_0}} e_{k_0+1}, \tag{66}$$

and similarly define $P_{l,k,\varepsilon}(x, \lambda)$, $G_{(l,k,\varepsilon),(l_0,k_0,\varepsilon_0)}(x)$. Using these notations, we obtain the following corollary of Lemma 6.

Corollary 2. *The following relations hold:*

$$\phi(x, \lambda) = \tilde{\phi}(x, \lambda) + \sum_{(l,k,\varepsilon)\in V} (-1)^\varepsilon \varphi_{l,k,\varepsilon}(x) \tilde{P}_{l,k,\varepsilon}(x, \lambda), \tag{67}$$

$$\varphi_{l_0,k_0,\varepsilon_0}(x) = \tilde{\varphi}_{l_0,k_0,\varepsilon_0}(x) + \sum_{(l,k,\varepsilon)\in V} (-1)^\varepsilon \varphi_{l,k,\varepsilon}(x) \tilde{G}_{(l,k,\varepsilon),(l_0,k_0,\varepsilon_0)}(x), \tag{68}$$

$$G_{(l_0,k_0,\varepsilon_0),(l_1,k_1,\varepsilon_1)}(x) - \tilde{G}_{(l_0,k_0,\varepsilon_0),(l_1,k_1,\varepsilon_1)}(x) = \sum_{(l,k,\varepsilon)\in V} (-1)^\varepsilon G_{(l_0,k_0,\varepsilon_0),(l,k,\varepsilon)}(x) \tilde{G}_{(l,k,\varepsilon),(l_1,k_1,\varepsilon_1)}(x), \tag{69}$$

where $x \in [0, 1]$, $(l_0, k_0, \varepsilon_0), (l_1, k_1, \varepsilon_1) \in V$.

Proof. Taking Lemma 4 on the structure of $\mathcal{N}(\lambda_0)$ and $\tilde{\mathcal{N}}(\lambda_0)$ into account, we rewrite (48) in the form

$$\phi(x, \lambda) = \tilde{\phi}(x, \lambda) + \sum_{(l,k,\varepsilon)\in V} (-1)^\varepsilon \Phi_{k+1,\langle 0\rangle}(x, \lambda_{l,k,\varepsilon}) e_{k+1}^T \mathcal{N}_\varepsilon(\lambda_{l,k,\varepsilon}) \tilde{D}_{\langle 0\rangle}(x, \lambda_{l,k,\varepsilon}, \lambda).$$

Using (64) and (65), we arrive at (67). Taking the $(k_0 + 1)$-th entry in the relation (67), putting $\lambda = \lambda_{l_0,k_0,\varepsilon_0}$, and using (64) and (66), we readily obtain (68).

Analogously, we represent (49) as follows:

$$D(x, \mu, \lambda) - \tilde{D}(x, \mu, \lambda) = \sum_{(l,k,\varepsilon)\in V} (-1)^\varepsilon [D(x, \mu, \xi)]^{\langle 0\rangle}_{\xi=\lambda_{l,k,\varepsilon}} e_{k+1} e_{k+1}^T \mathcal{N}_\varepsilon(\lambda_{l,k,\varepsilon}) \tilde{D}_{\langle 0\rangle}(x, \lambda_{l,k,\varepsilon}, \lambda).$$

Passing from $D(x, \mu, \lambda)$ and $\tilde{D}(x, \mu, \lambda)$ to $P_{l_0,k_0,\varepsilon_0}(x, \lambda)$ and $\tilde{P}_{l_0,k_0,\varepsilon_0}(x, \lambda)$, respectively, we derive

$$P_{l_0,k_0,\varepsilon_0}(x, \lambda) - \tilde{P}_{l_0,k_0,\varepsilon_0}(x, \lambda) = \sum_{(l,k,\varepsilon)\in V} (-1)^\varepsilon [P_{l_0,k_0,\varepsilon_0}(x, \xi)]^{\langle 0\rangle}_{\xi=\lambda_{l,k,\varepsilon}} e_{k+1} \tilde{P}_{l,k,\varepsilon}(x, \lambda).$$

Using (66) and the analogous relation for $G_{(l,k,\varepsilon),(l_0,k_0,\varepsilon_0)}(x)$, we finally arrive (69). □

The relations (68) can be considered as an infinite linear system with respect to $\varphi_{l,k,\varepsilon}(x)$, $(l, k, \varepsilon) \in V$. However, it is inconvenient to use (68) as the main equation system for the inverse problem, because the series in (68) converges only "with brackets":

$$\sum_{(l,k,\varepsilon)\in V} = \sum_{(l,k)} \left(\sum_{\varepsilon=0,1} (\dots) \right).$$

Therefore, in the next section, we transform the system (68) to a linear equation in a suitable Banach space. The relation (69) is used to prove the unique solvability of the main equation.

Remark 1. *If $\tilde{\mathcal{L}} \notin W$, that is, the poles of $\tilde{M}(\lambda)$ are not necessarily simple, then this influences the calculation of the residues in (59). Consequently, we obtain the following relation instead of (48):*

$$\phi(x, \lambda) = \tilde{\phi}(x, \lambda) + \sum_{\lambda_0\in\mathcal{I}} \Big[\phi_{\langle 0\rangle}(x, \lambda_0)(\mathcal{N}(\lambda_0)\tilde{D}_{\langle 0\rangle}(x, \lambda_0, \lambda) + \tilde{D}_{\langle -1\rangle}(x, \lambda_0, \lambda))$$

$$+\sum_{k=1}^{m_{\lambda_0}-1}\phi_{\langle k\rangle}(x,\lambda_0)\tilde{D}_{\langle-(k+1)\rangle}(x,\lambda_0,\lambda)\Bigg], \tag{70}$$

where m_{λ_0} is the multiplicity of $\lambda_0 \in \tilde{\Lambda}$. Using (70), one can derive an infinite system analogous to (68), containing not only entries of the vectors $\phi_{\langle 0\rangle}(x,\lambda_0)$ but also of $\phi_{\langle k\rangle}(x,\lambda_0)$ for $k=\overline{1,m_{\lambda_0}-1}$.

3.2. Linear Equation in a Banach Space

Define the numbers $\{\xi_l\}$, which characterize "the difference" of the two spectral data sets $\{\lambda_0,\mathcal{N}(\lambda_0)\}_{\lambda_0\in\Lambda}$ and $\{\tilde{\lambda}_0,\tilde{\mathcal{N}}(\tilde{\lambda}_0)\}_{\tilde{\lambda}_0\in\tilde{\Lambda}}$:

$$\xi_l := \sum_{k=1}^{n-1}\left(|\lambda_{l,k}-\tilde{\lambda}_{l,k}| + \sum_{j=k+1}^{n}|\mathcal{N}_{j,k}(\lambda_{l,k})-\tilde{\mathcal{N}}_{j,k}(\tilde{\lambda}_{l,k})|l^{p_{k,0}-p_{k+1,0}}\right)l^{1-n}, \quad l\ge 1. \tag{71}$$

Taking Corollary 1 into account, we reduce (71) to the following form for all sufficiently large values of l:

$$\xi_l = \sum_{k=1}^{n-1}(|\lambda_{l,k}-\tilde{\lambda}_{l,k}| + |\beta_{l,k}-\tilde{\beta}_{l,k}|l^{p_{k,0}-p_{k+1,0}})l^{1-n}. \tag{72}$$

Relation (72), together with the asymptotics (27) and (37), implies $\{\xi_l\}\in l_2$.

Lemma 7. *The following estimates hold for $(l,k,\varepsilon),(l_0,k_0,\varepsilon_0)\in V$:*

$$|\varphi_{l,k,\varepsilon}(x)|\le Cw_{l,k}(x), \quad |\varphi_{l,k,0}(x)-\varphi_{l,k,1}(x)|\le Cw_{l,k}(x)\xi_l,$$

$$|G_{(l,k,\varepsilon),(l_0,k_0,\varepsilon_0)}(x)|\le \frac{C}{|l-l_0|+1}\cdot\frac{w_{l_0,k_0}(x)}{w_{l,k}(x)},$$

$$|G_{(l,k,0),(l_0,k_0,\varepsilon_0)}(x)-G_{(l,k,1),(l_0,k_0,\varepsilon_0)}(x)|\le \frac{C\xi_l}{|l-l_0|+1}\cdot\frac{w_{l_0,k_0}(x)}{w_{l,k}(x)},$$

$$|G_{(l,k,\varepsilon),(l_0,k_0,0)}(x)-G_{(l,k,\varepsilon),(l_0,k_0,1)}(x)|\le \frac{C\xi_{l_0}}{|l-l_0|+1}\cdot\frac{w_{l_0,k_0}(x)}{w_{l,k}(x)},$$

$$|G_{(l,k,0),(l_0,k_0,0)}(x)-G_{(l,k,0),(l_0,k_0,1)}(x)-G_{(l,k,1),(l_0,k_0,0)}(x)+G_{(l,k,1),(l_0,k_0,1)}(x)|\le \frac{C\xi_l\xi_{l_0}}{|l-l_0|+1}\cdot\frac{w_{l_0,k_0}(x)}{w_{l,k}(x)},$$

where

$$w_{l,k}(x) := l^{-p_{k+1,0}}\exp(-xl\cot(k\pi/n)),$$

and the constant C does not depend on $x,l,\varepsilon,k,l_0,\varepsilon_0,k_0$.

The proof of Lemma 7 repeats the technique of ([26], Section 2.3.3), so we omit it. The similar estimates are valid for $\tilde{\varphi}_{l,k,\varepsilon}(x)$ and $\tilde{G}_{(l_0,k_0,\varepsilon_0),(l,k,\varepsilon)}(x)$.

Put $\theta_l := \xi_l^{-1}$ if $\xi_l\ne 0$ and $\theta_l=0$ otherwise. Introduce the notations

$$\begin{bmatrix}\psi_{l,k,0}(x)\\ \psi_{l,k,1}(x)\end{bmatrix} := w_{l,k}^{-1}(x)\begin{bmatrix}\theta_l & -\theta_l\\ 0 & 1\end{bmatrix}\begin{bmatrix}\varphi_{l,k,0}(x)\\ \varphi_{l,k,1}(x)\end{bmatrix}, \tag{73}$$

$$\begin{bmatrix}R_{(l_0,k_0,0),(l,k,0)}(x) & R_{(l_0,k_0,0),(l,k,1)}(x)\\ R_{(l_0,k_0,1),(l,k,0)}(x) & R_{(l_0,k_0,1),(l,k,1)}(x)\end{bmatrix} :=$$
$$\frac{w_{l,k}(x)}{w_{l_0,k_0}(x)}\begin{bmatrix}\theta_{l_0} & -\theta_{l_0}\\ 0 & 1\end{bmatrix}\begin{bmatrix}G_{(l,k,0),(l_0,k_0,0)}(x) & G_{(l,k,1),(l_0,k_0,0)}(x)\\ G_{(l,k,0),(l_0,k_0,1)}(x) & G_{(l,k,1),(l_0,k_0,1)}(x)\end{bmatrix}\begin{bmatrix}\xi_l & 1\\ 0 & -1\end{bmatrix}. \tag{74}$$

For brevity, put $\psi_v(x) := \psi_{l,k,\varepsilon}(x)$, $R_{v_0,v}(x) := R_{(l_0,k_0,\varepsilon_0),(l,k,\varepsilon)}(x)$, $v = (l,k,\varepsilon)$, $v_0 = (l_0,k_0,\varepsilon_0)$, $v, v_0 \in V$. The functions $\tilde{\psi}_v(x)$ and $\tilde{R}_{v_0,v}(x)$ are defined analogously.

Using (68) and (69), and the above notations, we obtain

$$\psi_{v_0}(x) = \tilde{\psi}_{v_0}(x) + \sum_{v \in V} \tilde{R}_{v_0,v}(x)\psi_v(x), \quad v_0 \in V, \tag{75}$$

$$R_{v_1,v_0}(x) - \tilde{R}_{v_1,v_0}(x) = \sum_{v \in V} \tilde{R}_{v_1,v}(x) R_{v,v_0}(x), \quad v_1, v_0 \in V. \tag{76}$$

Lemma 7 yields the estimates

$$|\psi_v(x)| \le C, \quad |R_{v_0,v}(x)| \le \frac{C\xi_l}{|l - l_0| + 1}, \quad v, v_0 \in V, \tag{77}$$

and the similar estimates for $\tilde{\psi}_v(x)$, $\tilde{R}_{v_0,v}(x)$. Consequently, the Cauchy—Bunyakovsky-Schwarz inequality

$$\sum_l \frac{\xi_l}{|l - l_0| + 1} \le \left(\sum_l \xi_l^2\right)^{1/2} \left(\sum_l \frac{1}{(|l - l_0| + 1)^2}\right)^{1/2} < \infty, \tag{78}$$

implies the absolute convergence of the series in (75) and (76).

Consider the Banach space m of bounded infinite sequences $\alpha = [\alpha_v]_{v \in V}$ with the norm $\|\alpha\|_m = \sum_{v \in V} |\alpha_v|$. Obviously, $\psi(x), \tilde{\psi}(x) \in m$ for each fixed $x \in [0,1]$. Define the linear operator $R(x) = [R_{v_0,v}(x)]_{v_0,v \in V}$ acting on an element $\alpha = [\alpha_v]_{v \in V} \in m$ by the following rule:

$$[R(x)\alpha]_{v_0} = \sum_{v \in V} R_{v_0,v}(x)\alpha_v, \quad v_0 \in V. \tag{79}$$

The operator $\tilde{R}(x) = [\tilde{R}_{v_0,v}(x)]_{v_0,v \in V}$ is defined similarly. It follows from (77) and (78) that the operators $R(x)$, $\tilde{R}(x)$ are bounded from m to m for each fixed $x \in [0,1]$. Denote by $\mathbf{I}$ the unit operator in m.

Using the introduced notations, we obtain the following theorem on the main equation and its unique solvability.

Theorem 1. *For each fixed $x \in [0,1]$, the linear operator $R(x)$ is compact in m and can be approximated by finite-rank operators: $R(x) = \lim_{N \to \infty} R^N(x)$. The same properties are valid for $\tilde{R}(x)$. Furthermore, the following relation holds*

$$(\mathbf{I} - \tilde{R}(x))\psi(x) = \tilde{\psi}(x), \quad x \in [0,1], \tag{80}$$

which is called the main equation of the inverse problem. The operator $(\mathbf{I} + \tilde{R}(x))$ has a bounded inverse of form

$$(\mathbf{I} - \tilde{R}(x))^{-1} = \mathbf{I} + R(x). \tag{81}$$

Thus, the main Equation (80) is uniquely solvable in m for each fixed $x \in [0,1]$.

Proof. For $N \in \mathbb{N}$, define the index set $V^N := \{v = (l,k,\varepsilon) \in V : l \le N\}$ and the finite-rank operator $R^N(x)$:

$$[R^N(x)\alpha]_{v_0} = \sum_{v \in V^N} R_{v_0,v}(x)\alpha_v. \tag{82}$$

Using (77)–(82), we show that

$$\|R(x) - R^N(x)\|_{m \to m} = \sup_{v_0 \in V} \sum_{v \in (V \setminus V^N)} |R_{v_0,v}(x)| \le \sup_{l_0} \sum_{l \ge N} \frac{C\xi_l}{|l - l_0| + 1} \to 0, \quad N \to \infty.$$

Hence, the operator $R(x)$ is compact.

According to our notations, the relations (75) and (76) take the form (80) and

$$R(x) - \tilde{R}(x) = \tilde{R}(x)R(x),$$

respectively. The latter relation implies (81), which completes the proof. □

Thus, we arrive at the following algorithm for solving Problem 3.

Algorithm 1: Suppose that the spectral data $\{\lambda_0, \mathcal{N}(\lambda_0)\}_{\lambda_0 \in \Lambda}$ of the problem $\mathcal{L} \in W$ are given. We have to find the Weyl solutions $\{\Phi_k(x,\lambda)\}_{k=1}^n$.

1. Choose an arbitrary model problem $\tilde{\mathcal{L}} \in W$ with $\tilde{p}_{s,a} = p_{s,a}$, $s = \overline{1,n}$, $a = 0,1$. In particular, one can take $\tilde{F}(x) = [\delta_{k+1,j}]_{k,j=1}^n$, $\tilde{U}_a = [\delta_{j,p_{s,a}+1}]_{s,j=1}^n$.
2. For the problem $\tilde{\mathcal{L}}$, find the matrix function $\tilde{\Phi}(x,\lambda)$ and then $\tilde{D}(x,\mu,\lambda)$ by (38).
3. Using $\tilde{\Phi}(x,\lambda)$, $\tilde{D}(x,\mu,\lambda)$, the spectral data $\{\lambda_0, \mathcal{N}(\lambda_0)\}_{\lambda_0 \in \Lambda}$, $\{\tilde{\lambda}_0, \tilde{\mathcal{N}}(\tilde{\lambda}_0)\}_{\tilde{\lambda}_0 \in \tilde{\Lambda}}$, and the notations (63), find $\tilde{\varphi}_{l,k,\varepsilon}(x)$, $\tilde{P}_{l,k,\varepsilon}(x,\lambda)$, and $\tilde{G}_{(l,k,\varepsilon),(l_0,k_0,\varepsilon_0)}$ for $(l,k,\varepsilon), (l_0,k_0,\varepsilon_0) \in V$ via (64), (65), and (66), respectively.
4. Construct the infinite sequence $\tilde{\psi}(x)$ and the operator $\tilde{R}(x)$ by using (73) and (74) (with tilde), respectively.
5. Find $\psi(x)$ by solving the main Equation (80).
6. Find $\{\varphi_{l,k,\varepsilon}(x)\}_{(l,k,\varepsilon)\in V}$ from (73):

$$\begin{bmatrix} \varphi_{l,k,0}(x) \\ \varphi_{l,k,1}(x) \end{bmatrix} = w_{l,k}(x) \begin{bmatrix} \xi_l & 1 \\ 0 & 1 \end{bmatrix} \begin{bmatrix} \psi_{l,k,0}(x) \\ \psi_{l,k,1}(x) \end{bmatrix}$$

7. Construct $\phi(x,\lambda) = [\Phi_k(x,\lambda)]_{k=1}^n$ by (67).

4. Reconstruction Formulas

In this section, we use the solution $\psi(x)$ of the main Equation (80) to obtain the solution of Problem 2 for some classes of differential operators. We derive the reconstruction formulas in the form of series for the coefficients $\{\tau_\nu\}_{\nu=0}^{n-2}$ of the differential expression (1).

In Section 4.1, the general approach to obtaining reconstruction formulas is described. However, for certain classes of the coefficients $\{\tau_\nu\}_{\nu=0}^{n-2}$, the convergence of the obtained series has to be studied in the corresponding spaces. Therefore, in Section 4.2, we prove an auxiliary lemma on the series convergence. In Sections 4.3–4.5, we study the three classes of operators:

(i) $n = 3$, $\tau_0 \in W_2^{-1}(0,1)$, $\tau_1 \in L_2(0,1)$;
(ii) n is even, $\tau_\nu \in L_2(0,1)$, $\nu = \overline{0, n-2}$;
(iii) n is even, $\tau_\nu \in W_2^{-1}(0,1)$, $\nu = \overline{0, n-2}$.

For each case, we provide the uniqueness theorem of the inverse problem solution in an appropriate statement, obtain reconstruction formulas, and prove the convergence of the series, and so obtain constructive algorithms for solving Problem 2. For the cases (ii) and (iii), we recover the coefficients $\tau_{n-2}, \tau_{n-3}, \dots, \tau_1, \tau_0$ one-by-one in order to achieve the convergence estimates for the corresponding series. The even order in (ii) and (iii) is considered for definiteness. Similar ideas can be applied to the odd-order differential operators. For simplicity, in all the three cases, we choose such boundary conditions that their coefficients cannot be uniquely recovered from the spectral data and so do not consider their reconstruction. However, for other types of boundary conditions, the recovery of their coefficients also can be studied similarly to the regular case (see Lemma 2.3.7 in [26]).

Let us introduce some notations used throughout this section. Note that the collection $\{\lambda_{l,k,\varepsilon}\}_{(l,k,\varepsilon)\in V}$ may contain multiple eigenvalues for a fixed $\varepsilon \in \{0,1\}$: $\lambda_{l,k,\varepsilon} = \lambda_{l_0,k_0,\varepsilon}$, $(l,k) \neq (l_0,k_0)$. In order to exclude such values, we define the set

$$V' := \{(l,k,\varepsilon) \in V \colon \nexists (l_0,k_0,\varepsilon) \in V \text{ s.t. } (l_0,k_0) < (l,k) \text{ and } \lambda_{l_0,k_0,\varepsilon} = \lambda_{l,k,\varepsilon}\}.$$

In this section, we use the following notations for an index $v = (l, k, \varepsilon) \in V'$:

$$\lambda_v := \lambda_{l,k,\varepsilon}, \quad \phi_v(x) := \phi_{\langle 0 \rangle}(x, \lambda_v), \quad \tilde{P}_v(x, \lambda) := (-1)^\varepsilon \mathcal{N}_\varepsilon(\lambda_v) \tilde{D}_{\langle 0 \rangle}(x, \lambda_v, \lambda), \tag{83}$$

$$c_v := (-1)^\varepsilon \mathcal{N}_\varepsilon(\lambda_v) J_0^{-1}, \quad \tilde{g}_v(x) := [\tilde{\phi}^\star_{\langle 0 \rangle}(x, \lambda_v)]^T. \tag{84}$$

Additionally, define the scalar functions

$$\tilde{\eta}_{l,k,\varepsilon}(x) := (-1)^\varepsilon e_{k+1}^T \mathcal{N}_\varepsilon(\lambda_{l,k,\varepsilon}) J_0^{-1} [\tilde{\phi}^\star_{\langle 0 \rangle}(x, \lambda_{l,k,\varepsilon})]^T, \quad v \in V. \tag{85}$$

4.1. General Approach

In terms of the notations (83), the relation (48) can be rewritten as

$$\phi(x, \lambda) = \tilde{\phi}(x, \lambda) + \sum_{v \in V'} \phi_v(x) \tilde{P}_v(x, \lambda).$$

Formal calculations show that

$$\ell_n(\phi(x, \lambda)) = \ell_n(\tilde{\phi}(x, \lambda)) + \sum_{v \in V'} \ell_n(\phi_v(x) \tilde{P}_v(x, \lambda)).$$

Recall that

$$\ell_n(\phi(x, \lambda)) = \lambda \phi(x, \lambda), \quad \tilde{\ell}_n(\tilde{\phi}(x, \lambda)) = \lambda \tilde{\phi}(x, \lambda),$$

and, by virtue of (31),

$$\ell_n(\phi_v(x)) = \lambda_v \phi_v(x) + \phi_v(x) \mathcal{N}_0(\lambda_v).$$

Define $\hat{\ell}_n(y) := \ell_n(y) - \tilde{\ell}_n(y)$. Consequently,

$$\begin{aligned} \lambda(\phi(x, \lambda) - \tilde{\phi}(x, \lambda)) - \sum_{v \in V'} \ell_n(\phi_v(x)) \tilde{P}_v(x, \lambda) &= \sum_{v \in V'} \phi_v(x) [(\lambda - \lambda_v) I - \mathcal{N}_0(\lambda_v)] \tilde{P}_v(x, \lambda) \\ &= \hat{\ell}_n(\tilde{\phi}(x, \lambda)) + \sum_{v \in V'} \ell_n(\phi_v(x) \tilde{P}_v(x, \lambda)) - \sum_{v \in V'} \ell_n(\phi_v(x)) \tilde{P}_v(x, \lambda). \end{aligned} \tag{86}$$

Using (83) and (42), we derive

$$\begin{aligned} [(\lambda - \lambda_v) - \mathcal{N}_0(\lambda_v)] \tilde{P}_v(x, \lambda) = &(-1)^\varepsilon \mathcal{N}_\varepsilon(\lambda_v) J_0^{-1} \langle [\tilde{\phi}^\star_v(x)]^T, \tilde{\phi}(x, \lambda) \rangle \\ &+ (-1)^{\varepsilon+1} [\mathcal{N}_\varepsilon(\lambda_v) \mathcal{N}_1(\lambda_v) + \mathcal{N}_0(\lambda_v) \mathcal{N}_\varepsilon(\lambda_v)] \tilde{D}_{\langle 0 \rangle}(x, \lambda_0, \lambda). \end{aligned}$$

The summation yields

$$\sum_{v \in V'} \phi_v(x) [(\lambda - \lambda_v) I - \mathcal{N}_0(\lambda_v)] \tilde{P}_v(x, \lambda) = \sum_{v \in V'} \phi_v(x) c_v \langle \tilde{g}_v(x), \tilde{\phi}(x, \lambda) \rangle, \tag{87}$$

where c_v and $\tilde{g}_v(x)$ are defined by (84). Combining (86) and (87) together, we obtain

$$\sum_{v \in V'} \phi_v(x) c_v \langle \tilde{g}_v(x), \tilde{\phi}(x, \lambda) \rangle = \hat{\ell}_n(\tilde{\phi}(x, \lambda)) + \sum_{v \in V'} \ell_n(\phi_v(x) \tilde{P}_v(x, \lambda)) - \sum_{v \in V'} \ell_n(\phi_v(x)) \tilde{P}_v(x, \lambda). \tag{88}$$

Suppose that the differential expression $y^{[n]} = \ell_n(y)$ has the form (1). Then, $\ell_n(y)$ can be formally represented as

$$\ell_n(y) = y^{(n)} + \sum_{s=0}^{n-2} p_s(x) y^{(s)}, \tag{89}$$

where

$$p_s = \sum_{k=\lceil s/2 \rceil}^{\min\{s, \lfloor n/2 \rfloor - 1\}} C_k^{s-k} [\tau_{2k}^{(2k-s)} + \tau_{2k+1}^{(2k-s+1)}] + \sum_{k=\lceil (s-1)/2 \rceil}^{\min\{s, \lfloor (n-1)/2 \rfloor\} - 1} 2 C_k^{s-k-1} \tau_{2k+1}^{(2k+1-s)}. \tag{90}$$

(We assume that $\tau_{n-1}(x) \equiv 0$). Suppose that $\tilde{\ell}_n(y)$ has a form similar to (89) with the coefficients $\tilde{p}_s(x)$, so

$$\hat{\ell}_n(y) := \sum_{s=0}^{n-2} \hat{p}_s(x) y^{(s)}, \quad \hat{p}_s := p_s - \tilde{p}_s. \tag{91}$$

Using (89), we derive

$$\ell_n(\phi_v \tilde{P}_v) = \ell_n(\phi_v)\tilde{P}_v + \sum_{k=1}^{n} C_n^k \sum_{v \in V'} \phi_v^{(n-k)} \tilde{P}_v^{(k)} + \sum_{k=1}^{n-2} p_k \sum_{r=1}^{k} C_k^r \sum_{v \in V'} \phi_v^{(k-r)} \tilde{P}_v^{(r)}. \tag{92}$$

The relations (43) and (83) imply

$$\tilde{P}_v'(x, \lambda) = c_v \tilde{g}_v(x) \tilde{\phi}(x, \lambda). \tag{93}$$

Substituting (92) into (93) and grouping the terms at $\tilde{\phi}^{(s)}(x, \lambda)$, we obtain

$$\ell_n(\phi_v \tilde{P}_v) - \ell_n(\phi_v)\tilde{P}_v = \sum_{s=0}^{n-1} t_{n,s} \tilde{\phi}^{(s)} + \sum_{s=0}^{n-3} \sum_{k=s+1}^{n-2} p_k t_{k,s} \tilde{\phi}^{(s)}, \tag{94}$$

where

$$t_{k,s}(x) := \sum_{r=s}^{k-1} C_k^{r+1} C_r^s T_{k-r-1,r-s}(x), \qquad T_{j_1,j_2}(x) := \sum_{v \in V'} \phi_v^{(j_1)}(x) c_v \tilde{g}_v^{(j_2)}(x). \tag{95}$$

Combining (88), (91), and (94) all together, we arrive at the relation

$$\begin{aligned}\sum_{v \in V'} \phi_v(x) c_v \langle \tilde{g}_v(x), \tilde{\phi}(x, \lambda) \rangle = \sum_{s=0}^{n-2} \hat{p}_s(x) \tilde{\phi}^{(s)}(x, \lambda) + \sum_{s=0}^{n-1} t_{n,s}(x) \tilde{\phi}^{(s)}(x, \lambda) \\ + \sum_{s=0}^{n-3} \sum_{k=s+1}^{n-2} p_k(x) t_{k,s}(x) \tilde{\phi}^{(s)}(x, \lambda)\end{aligned} \tag{96}$$

For definiteness, suppose that $\tilde{p}_s(x) = 0$, $s = \overline{0, n-2}$. Then, $y^{[s]} = y^{(s)}$, $s = \overline{0, n}$, for the problem $\tilde{\mathcal{L}}$, and so

$$\langle \tilde{g}_v(x), \tilde{\phi}(x, \lambda) \rangle = \sum_{s=0}^{n-1} (-1)^{n-s-1} \tilde{g}_v^{(n-s-1)}(x) \tilde{\phi}^{(s)}(x, \lambda).$$

Therefore, combining the terms at $\tilde{\phi}^{(s)}(x, \lambda)$, we obtain the formulas for finding the coefficients

$$p_s = (-1)^{n-s-1} \sum_{v \in V'} \phi_v(x) c_v \tilde{g}_v^{(n-s-1)}(x) - t_{n,s}(x) - \sum_{k=s+1}^{n-2} p_k(x) t_{k,s}(x), \tag{97}$$

where $s = n-2, n-1, \dots, 1, 0$. These formulas coincide with the ones for the regular case (see ([26]), Lemma 2.3.7).

Using the relations (90) and (97), one can find τ_ν for $\nu = n-2, n-3, \dots, 1, 0$. However, the Formulas (97) have been obtained by formal calculations. They can be used for reconstruction if the coefficients $\{\tau_\nu\}_{\nu=0}^{n-2}$ are so smooth that the series in (95) and (97) converge. If the coefficients $\{\tau_\nu\}_{\nu=0}^{n-2}$ are nonsmooth or even distributional, then the convergence of the series is a nontrivial question, which should be investigated separately for different classes of operators. For some classes, this question is considered in Sections 4.3–4.5.

4.2. Series Convergence

In this subsection, we prove the following auxiliary lemma.

Lemma 8. *Suppose that $j_1, j_2 \in \{0, 1, \ldots, n-1\}$ and $\{l^{(j_1+j_2)}\xi_l\} \in l_2$. Then, there exist constants $\{A_v\}_{v \in V'}$, such that the series*

$$\sum_{v \in V'} (\phi_v^{[j_1]}(x) c_v \tilde{g}_v^{[j_2]}(x) - A_v) \tag{98}$$

converges in $L_2(0,1)$. Moreover, if $\{l^{(j_1+j_2)}\xi_l\} \in l_1$, then the series

$$\sum_{v \in V'} \phi_v^{[j_1]}(x) c_v \tilde{g}_v^{[j_2]}(x) \tag{99}$$

converges absolutely and uniformly on $[0,1]$.

Here, and below, the quasi-derivatives for $\phi_v(x)$ are generated by the matrix $F(x)$ and for $\tilde{g}_v(x)$, by $\tilde{F}^\star(x)$. In order to prove Lemma 8, we need to formulate preliminary propositions.

Consider the sector $\Gamma_1 = \{\rho \in \mathbb{C} \colon 0 < \arg \rho < \frac{\pi}{n}\}$. Denote by $\{\omega_k\}_{k=1}^n$ the roots of the equation $\omega^n = 1$, numbered so that

$$\operatorname{Re}(\rho\omega_1) < \operatorname{Re}(\rho\omega_2) < \cdots < \operatorname{Re}(\rho\omega_n), \quad \rho \in \Gamma_1.$$

In addition, define the extended sector

$$\Gamma_{1,h} := \{\rho \in \mathbb{C} \colon \rho + h\exp(\tfrac{i\pi}{2n}) \in \Gamma_1\}, \quad h > 0.$$

In the proof of Lemma 8, we need the following proposition on the Birkhoff-type solutions of Equation (4) with certain asymptotic behavior as $|\rho| \to \infty$.

Proposition 1 ([2]). *For some $\rho^* > 0$, Equation (4) has a fundamental system of solutions $\{y_k(x,\rho)\}_{k=1}^n$ whose quasi-derivatives $y_k^{[j]}(x,\rho)$, $k = \overline{1,n}$, $j = \overline{0,n-1}$ are continuous for $x \in [0,1]$, $\rho \in \overline{\Gamma}_{1,h}$, $|\rho| \geq \rho^*$, analytic in $\rho \in \Gamma_{1,h}$, $|\rho| > \rho^*$ for each fixed $x \in [0,1]$, and satisfy the relation*

$$y_k^{[j]}(x,\rho) = (\rho\omega_k)^j \exp(\rho\omega_k x)(1 + \zeta_{jk}(x,\rho)),$$

where

$$\max_{j,k,x} |\zeta_{jk}(x,\rho)| \leq C(\Upsilon(\rho) + |\rho|^{-1}), \quad \rho \in \overline{\Gamma}_{1,h}, \ |\rho| \geq \rho^*,$$

and $\Upsilon(\rho)$ fulfills the condition $\{\Upsilon(\rho_l)\} \in l_2$ for any noncondensing sequence $\{\rho_l\} \subset \Gamma_{1,h}$.

Consider the strip $\mathcal{S}_0$ defined by (52). Clearly, for a suitable choice of h and c, we have $\mathcal{S}_0 \subset \Gamma_{1,h}$ and $\lambda_{l,k,\varepsilon} = \rho_{l,k,\varepsilon}^n$, $\rho_{l,k,\varepsilon} \in \mathcal{S}_0$ for even $(n-k)$ and for sufficiently large l. Furthermore, in this section, we confine ourselves to considering even $(n-k)$, since the case of odd $(n-k)$ is similar.

Proposition 2. *Suppose that $k \in \{1, 2, \ldots, n-1\}$ and $(n-k)$ is even. Then, the Weyl solution can be expanded as*

$$\Phi_{k+1}(x,\lambda) = \sum_{s=1}^n b_{s,k+1}(\rho) y_s(x,\rho), \quad \lambda = \rho^n, \quad \rho \in \mathcal{S}_0,$$

where the coefficients $b_{s,k+1}(\rho)$ are analytic in $\rho \in \mathcal{S}_0$, $|\rho| \geq \rho^*$ and fulfill the estimate

$$b_{s,k+1}(\rho) = O\left(\rho^{-p_{k+1,0}} \overset{if\ s>k+1}{\times} \exp(\rho(\omega_{k+1} - \omega_s))\right). \tag{100}$$

Proof. The properties of the coefficient $b_{s,k+1}(\rho)$ follow from the certain formulas for these coefficients obtained in the proof of Lemma 3 in [9]. □

Proposition 3. *Let z be a nonzero complex with $\operatorname{Re} z \le 0$, and let $\{\varkappa_l\}_{l\ge 1} \in l_2$. Then, the series $\sum\limits_{l\ge 1} \varkappa_l \exp(zlx)$ converges in $L_2(0,1)$.*

Proof of Lemma 8. Let $j_1, j_2 \in \{0, 1, \dots, n-1\}$ be fixed. In order to prove the convergence of the series (98) and (99), it is sufficient to consider their terms for $v = (l, k, \varepsilon)$ with sufficiently large l. For technical simplicity, let us assume that $\lambda_{l_1,k_1,\varepsilon} \ne \lambda_{l_2,k_2,\varepsilon}$ for any sufficiently large l_1, l_2, such that $l_1 \ne l_2$. In view of Corollary 1, we have

$$\sum_{v:\, l \text{ is fixed}} \varphi_v^{[j_1]}(x) c_v \tilde{g}_v^{[j_2]}(x) = \sum_{k=1}^{n-1} (-1)^{n-1-p_{k,0}} \mathscr{Z}_{l,k}(x), \tag{101}$$
$$\mathscr{Z}_{l,k}(x) := \sum_{\varepsilon=0,1} (-1)^{\varepsilon} \beta_{l,k,\varepsilon} \varphi_{l,k,\varepsilon}^{[j_1]}(x) \tilde{\varphi}_{l,n-k,\varepsilon}^{\star[j_2]}(x),$$

where

$$\varphi_{l,k,\varepsilon}^{[j_1]}(x) = \Phi_{k+1}^{[j_1]}(x, \lambda_{l,k,\varepsilon}), \quad \tilde{\varphi}_{l,n-k,\varepsilon}^{\star[j_2]}(x) = \tilde{\Phi}_{n-k+1}^{\star[j_2]}(x, \lambda_{l,k,\varepsilon}), \quad \beta_{l,k,0} := \beta_{l,k}, \quad \beta_{l,k,1} := \tilde{\beta}_{l,k},$$

Fix $k \in \{1, 2, \dots, n-1\}$ such that $(n-k)$ is even. Then, by Proposition 2, we have

$$\Phi_{k+1}^{[j_1]}(x, \lambda_{l,k,\varepsilon}) = \sum_{s_1=1}^{n} b_{s_1,k+1}(\rho_{l,k,\varepsilon}) y_{s_1}^{[j_1]}(x, \rho_{l,k,\varepsilon}),$$
$$\tilde{\Phi}_{n-k+1}^{\star[j_2]}(x, \lambda_{l,k,\varepsilon}) = \sum_{s_2=1}^{n} \tilde{b}_{n-s_2+1,n-k+1}^{\star}(\rho_{l,k,\varepsilon}) \tilde{y}_{n-s_2+1}^{\star[j_2]}(x, \rho_{l,k,\varepsilon}). \tag{102}$$

Using the above relations and Proposition 1, we obtain

$$\mathscr{Z}_{l,k}(x) = \sum_{s_1=1}^{n} \sum_{s_2=1}^{n} Z_{l,k,s_1,s_2}(x),$$
$$Z_{l,k,s_1,s_2}(x) = \sum_{\varepsilon=0,1} \alpha_{l,k,s_1,s_2,\varepsilon} \exp(\rho_{l,k,\varepsilon}(\omega_{s_1} - \omega_{s_2})x)(1 + \zeta_{s_1,j_1}(x, \rho_{l,k,\varepsilon}))(1 + \tilde{\zeta}_{n-s_2+1,j_2}^{\star}(x, \rho_{l,k,\varepsilon})),$$
$$\alpha_{l,k,s_1,s_2,\varepsilon} := \beta_{l,k,\varepsilon} b_{s_1,k+1}(\rho_{l,k,\varepsilon}) b_{n-s_2+1,n-k+1}^{\star}(\rho_{l,k,\varepsilon}) (\omega_{s_1})^{j_1} (-\omega_{s_2})^{j_2} \rho_{l,k,\varepsilon}^{j_1+j_2}. \tag{103}$$

Consider the sums

$$Z_{l,k,s_1,s_2}(x) = Z_{l,k,s_1,s_2}^1(x) + Z_{l,k,s_1,s_2}^2(x) + Z_{l,k,s_1,s_2}^3(x) + Z_{l,k,s_1,s_2}^4(x),$$
$$Z_{l,k,s_1,s_2}^1(x) := \sum_{\varepsilon=0,1} \alpha_{l,k,s_1,s_2,\varepsilon} \exp(\rho_{l,k,\varepsilon}(\omega_{s_1} - \omega_{s_2})x),$$
$$Z_{l,k,s_1,s_2}^2(x) := \sum_{\varepsilon=0,1} \alpha_{l,k,s_1,s_2,\varepsilon} \exp(\rho_{l,k,\varepsilon}(\omega_{s_1} - \omega_{s_2})x) \zeta_{s_1,j_1}(x, \rho_{l,k,\varepsilon}),$$
$$Z_{l,k,s_1,s_2}^3(x) := \sum_{\varepsilon=0,1} \alpha_{l,k,s_1,s_2,\varepsilon} \exp(\rho_{l,k,\varepsilon}(\omega_{s_1} - \omega_{s_2})x) \tilde{\zeta}_{n-s_2+1,j_2}^{\star}(x, \rho_{l,k,\varepsilon}),$$
$$Z_{l,k,s_1,s_2}^4(x) := \sum_{\varepsilon=0,1} \alpha_{l,k,s_1,s_2,\varepsilon} \exp(\rho_{l,k,\varepsilon}(\omega_{s_1} - \omega_{s_2})x) \zeta_{s_1,j_1}(x, \rho_{l,k,\varepsilon}) \tilde{\zeta}_{n-s_2+1,j_2}^{\star}(x, \rho_{l,k,\varepsilon}).$$

Thus, it is sufficient to study the convergence of the series $\sum\limits_{l\ge l_0} Z_{l,k,s_1,s_2}^{\nu}(x)$ for fixed k, s_1, s_2, and $\nu = \overline{1,4}$.

The asymptotics (27) and (37) imply

$$|\rho_{l,k,\varepsilon}| \le Cl, \quad |\beta_{l,k,\varepsilon}| \le Cl^{n-1+p_{k+1,0}-p_{k,0}}. \tag{104}$$

Using (103) together with the estimates (100) and (104), we obtain

$$|\alpha_{l,k,s_1,s_2,\varepsilon}| \le Cl^{j_1+j_2} \overset{if\ s_1>k+1}{\times} \exp(\mathrm{Re}\,(\omega_{k+1}-\omega_{s_1})r_k l) \overset{if\ s_2<k}{\times} \exp(\mathrm{Re}\,(\omega_{s_2}-\omega_k)r_k l),$$

where $r_k := \frac{\pi}{\sin\frac{\pi k}{n}}$. The relation (72) yields

$$|\rho_{l,k,0}-\rho_{l,k,1}| \le C\xi_l, \quad |\beta_{l,k,0}-\beta_{l,k,1}| \le C\xi_l l^{n-1+p_{k+1,0}-p_{k,0}}.$$

Since the functions $b_{s,k+1}(\rho)$ are analytic and satisfy (100), we obtain

$$|b_{s,k+1}(\rho_{l,k,0})-b_{s,k+1}(\rho_{l,k,1})| \le C\xi_l l^{-p_{k+1,0}} \overset{if\ s>k+1}{\times} \exp(\mathrm{Re}\,(\omega_s-\omega_k)r_k l).$$

It follows from (103) that

$$\begin{aligned}\alpha_{l,k,s_1,s_2,0}-\alpha_{l,k,s_1,s_2,1} = {} & (\beta_{l,k,0}-\beta_{l,k,1})b_{s_1,k+1}(\rho_{l,k,0})b^\star_{n-s_2+1,n-k+1}(\rho_{l,k,0})(\omega_{s_1})^{j_1}(-\omega_{s_2})^{j_2}\rho_{l,k,0}^{j_1+j_2}\\ &+\beta_{l,k,1}(b_{s_1,k+1}(\rho_{l,k,0})-b_{s_1,k+1}(\rho_{l,k,1}))b^\star_{n-s_2+1,n-k+1}(\rho_{l,k,0})(\omega_{s_1})^{j_1}(-\omega_{s_2})^{j_2}\rho_{l,k,0}^{j_1+j_2}\\ &+\beta_{l,k,1}b_{s_1,k+1}(\rho_{l,k,1})(b^\star_{n-s_2+1,n-k+1}(\rho_{l,k,0})-b^\star_{n-s_2+1,n-k+1}(\rho_{l,k,1}))(\omega_{s_1})^{j_1}(-\omega_{s_2})^{j_2}\rho_{l,k,0}^{j_1+j_2}\\ &+\beta_{l,k,1}b_{s_1,k+1}(\rho_{l,k,1})b^\star_{n-s_2+1,n-k+1}(\rho_{l,k,1})(\omega_{s_1})^{j_1}(-\omega_{s_2})^{j_2}(\rho_{l,k,0}^{j_1+j_2}-\rho_{l,k,1}^{j_1+j_2}).\end{aligned}$$

Consequently, we estimate

$$\begin{aligned}|\alpha_{l,k,s_1,s_2,0}-\alpha_{l,k,s_1,s_2,1}| \le Cl^{j_1+j_2}\xi_l & \overset{if\ s_1>k+1}{\times} \exp(\mathrm{Re}\,(\omega_{k+1}-\omega_{s_1})r_k l)\\ & \overset{if\ s_2<k}{\times} \exp(\mathrm{Re}\,(\omega_{s_2}-\omega_k)r_k l).\end{aligned}$$

Suppose that $\{l^{j_1+j_2}\xi_l\} \in l_2$. Consider the cases:

1. If $s_1 = s_2 \notin \{k,k+1\}$, then the terms of the series $\sum\limits_{l\ge l_0} Z^1_{l,k,s_1,s_2}(x)$ decay exponentially, so the series converges absolutely.
2. If $s_1 = s_2 \in \{k,k+1\}$, then the series $\sum\limits_{l\ge l_0}(\alpha_{l,k,s_1,s_2,0}-\alpha_{l,k,s_1,s_2,1})$ does not necessarily converge.
3. If $s_1 \ne s_2$, then

$$\begin{aligned}Z^1_{l,k,s_1,s_2}(x) = {} & ((\alpha_{l,k,s_1,s_2,0}-\alpha_{l,k,s_1,s_2,1})\\ &+\alpha_{l,k,s_1,s_2,1}[(\rho_{l,k,0}-\rho_{l,k,1})(\omega_{s_1}-\omega_{s_2})x + O(\xi_l^2)])\exp(\rho_{l,k,0}(\omega_{s_1}-\omega_{s_2})x).\end{aligned}$$

Consequently, the series $\sum\limits_{l\ge l_0} Z^1_{l,k,s_1,s_2}(x)$ converges in $L_2(0,1)$ by virtue of Proposition 3.

Using Proposition 1, we show that

$$\begin{aligned}&|\zeta_{s_1,j_1}(x,\rho_{l,k,\varepsilon})| \le C(\Upsilon(\rho_{l,k,\varepsilon})+l^{-1}),\\ &|\zeta_{s_1,j_1}(x,\rho_{l,k,0})-\zeta_{s_1,j_1}(x,\rho_{l,k,1})| \le C\xi_l(\Upsilon(\rho^*_{l,k,0})+l^{-1}),\end{aligned}$$

where $\Upsilon(\rho^*_{l,k,0}) = \max\limits_{|\rho-\rho_{l,k,0}|\le\delta}\Upsilon(\rho)$. Note that $\{\Upsilon(\rho^*_{l,k,0})\} \in l_2$. Consequently, the series $\sum\limits_{l\ge l_0} Z^2_{l,k,s_1,s_2}(x)$ converges absolutely and uniformly on $[0,1]$. The proof for Z^3 and Z^4 is analogous. Thus, the regularized series $\sum\limits_{l\ge l_0}(\mathscr{X}_{l,k}(x)-A_{l,k})$ converges in $L_2(0,1)$ with the constants

$$A_{l,k} = \sum_{s=k,k+1}(\alpha_{l,k,s,s,0}-\alpha_{l,k,s,s,1}).$$

Using the arguments above, we obtain the estimate

$$|\mathscr{Z}_{l,k}(x)| \le C l^{j_1+j_2}\xi_l.$$

Hence, in the case $\{l^{j_1+j_2}\xi_l\} \in l_1$, the series $\sum\limits_{l \ge l_0} \mathscr{Z}_{l,k}(x)$ converges absolutely and uniformly with respect to $x \in [0,1]$. Taking (101) into account, we arrive at the assertion of the lemma. □

4.3. Case $n = 3$

Consider the differential expression

$$\ell_3(y) = y^{(3)} + (\tau_1(x)y)' + \tau_1(x)y' + \tau_0(x)y, \quad x \in (0,1),$$

where $\tau_1 \in L_2(0,1)$ and $\tau_0 \in W_2^{-1}(0,1)$, that is, $\tau_0 = \sigma_0'$, $\sigma_0 \in L_2(0,1)$. The associated matrix has the form (see, e.g., [47]):

$$F(x) = \begin{bmatrix} 0 & 1 & 0 \\ -(\sigma_0 + \tau_1) & 0 & 1 \\ 0 & (\sigma_0 - \tau_1) & 0 \end{bmatrix}, \tag{105}$$

so, $y^{[1]} = y'$, $y^{[2]} = y'' + (\sigma_0 + \tau_1)y$, $y^{[3]} = \ell_3(y)$.

Suppose that $p_{s,0} = s - 1$, $p_{s,1} = 3 - s$, $s = \overline{1,3}$, in the linear forms (8). Using the technique of [54], we obtain the eigenvalue asymptotics

$$\lambda_{l,k} = (-1)^{k+1}\left(\frac{2\pi}{\sqrt{3}}\left(l + \frac{1}{6} + \frac{(-1)^k}{\pi^2 l}\int_0^1 \tau_1(t)\,dt + \frac{\varkappa_{l,k}}{l}\right)\right)^3, \quad \{\varkappa_{l,k}\} \in l_2, \quad l \ge 1,\ k = 1,2. \tag{106}$$

Assume that $\mathcal{L} \in W$. It can be easily shown that, if $\Lambda_1 \cap \Lambda_2 = \varnothing$, then the spectral data $\{\lambda_0, \mathcal{N}(\lambda_0)\}_{\lambda_0 \in \Lambda}$ do not depend on the boundary condition coefficients $u_{s,j,a}$. Therefore, let us assume that $U_0 = I$, $U_1 = [\delta_{k,4-j}]_{k,j=1}^3$. Consider the following inverse problem.

Consider the problems $\mathcal{L} = (F(x), U_0, U_1) \in W$ and $\tilde{\mathcal{L}} = (\tilde{F}(x), U_0, U_1) \in W$, where $\tilde{F}(x)$ is the matrix function associated with the differential expression $\tilde{\ell}_3(y)$ having the coefficients $\tilde{\tau}_1 \in L_2(0,1)$ and $\tilde{\tau}_0 = \tilde{\sigma}_0' \in W_2^{-1}(0,1)$. Under the above assumptions, the following uniqueness theorem for solution of Problem 2 is valid.

Theorem 2. *If $\Lambda = \tilde{\Lambda}$ and $\mathcal{N}(\lambda_0) = \tilde{\mathcal{N}}(\lambda_0)$ for all $\lambda_0 \in \Lambda$, then $\tau_1(x) = \tilde{\tau}_1(x)$ and $\sigma_0(x) = \tilde{\sigma}_0(x) + const$ a.e. on $(0,1)$. Thus, the spectral data $\{\lambda_0, \mathcal{N}(\lambda_0)\}_{\lambda_0 \in \Lambda}$ uniquely specify $\tau_1 \in L_2(0,1)$ and $\tau_0 \in W_2^{-1}(0,1)$.*

In order to prove Theorem 2, we need the following auxiliary lemma, which is valid for n not necessarily equal to 3.

Lemma 9. *If $\mathcal{L}, \tilde{\mathcal{L}} \in W$, $\Lambda = \tilde{\Lambda}$ and $\mathcal{N}(\lambda_0) = \tilde{\mathcal{N}}(\lambda_0)$ for all $\lambda_0 \in \Lambda$, then the matrix of spectral mappings $\mathcal{P}(x, \lambda)$ defined by (50) does not depend on λ.*

Proof. It follows from (25) and (50) that

$$\mathcal{P}(x, \lambda) = \Phi(x, \lambda) J_0^{-1} [\tilde{\Phi}^\star(x, \lambda)]^T J.$$

Using (29) and (30), we derive for $\lambda_0 \in \Lambda$:

$$\begin{aligned} \mathcal{P}_{\langle -2 \rangle}(x, \lambda) J^{-1} &= \Phi_{\langle -1 \rangle}(x, \lambda_0) J_0^{-1} [\tilde{\Phi}^\star_{\langle -1 \rangle}(x, \lambda_0)]^T \\ &= \Phi_{\langle 0 \rangle}(x, \lambda_0) \mathcal{N}(\lambda_0) J_0^{-1} [\mathcal{N}^*(\lambda_0)]^T [\tilde{\Phi}^\star_{\langle 0 \rangle}(x, \lambda_0)]^T = 0, \\ \mathcal{P}_{\langle -1 \rangle}(x, \lambda) J^{-1} &= \Phi_{\langle -1 \rangle}(x, \lambda_0) J_0^{-1} [\tilde{\Phi}^\star_{\langle 0 \rangle}(x, \lambda_0)]^T + \Phi_{\langle 0 \rangle}(x, \lambda_0) J_0^{-1} [\tilde{\Phi}^\star_{\langle -1 \rangle}(x, \lambda_0)]^T \end{aligned}$$

$$= \Phi_{\langle 0 \rangle}(x, \lambda_0)(\mathcal{N}(\lambda_0)J_0^{-1} + J_0^{-1}[\mathcal{N}^\star(\lambda_0)]^T)[\tilde{\Phi}^\star_{\langle 0 \rangle}(x, \lambda_0)]^T = 0.$$

Hence, $\mathcal{P}(x, \lambda)$ is entire in λ. Using the asymptotics (53) and Liouville's theorem, we conclude that $\mathcal{P}(x, \lambda) \equiv \mathcal{P}(x)$, $x \in [0, 1]$. □

Proof of Theorem 2. This proof is similar to the proof of Theorem 2 in [9], so we outline it briefly. By Lemma 9, $\mathcal{P}(x, \lambda) \equiv \mathcal{P}(x)$. Furthermore, $\mathcal{P}(x)$ is a unit lower-triangular matrix. One can easily show that

$$\mathcal{P}'(x) + \mathcal{P}(x)\tilde{F}(x) = F(x)\mathcal{P}(x), \quad x \in (0, 1), \tag{107}$$

where the matrix functions $F(x)$ and $\tilde{F}(x)$ have the form (105). In the element-wise form, (107) implies $\mathcal{P}_{2,1} = \mathcal{P}_{3,2} = \mathcal{P}'_{3,1} = 0$, $\mathcal{P}_{3,1} = \hat{\sigma}_0 \pm \hat{\tau}_1$. Hence, $\hat{\tau}_1 = 0$, $\hat{\sigma}_0 = const$ in $L_2(0, 1)$, which concludes the proof. □

Now, suppose that the spectral data $\{\lambda_0, \mathcal{N}(\lambda_0)\}_{\lambda_0 \in \Lambda}$ of the problem $\mathcal{L} = (F(x), U_0, U_1)$ are given. Using the asymptotics (106), one can find the number $\tilde{\tau}_1 := \int_0^1 \tau_1(t)\,dt$. Put

$$\tilde{F}(x) = \begin{bmatrix} 0 & 1 & 0 \\ -\tilde{\tau}_1 & 0 & 1 \\ 0 & -\tilde{\tau}_1 & 0 \end{bmatrix}, \tag{108}$$

and $\tilde{\mathcal{L}} := (\tilde{F}(x), U_0, U_1)$. Clearly, $\tilde{F}^\star(x) = \tilde{F}(x)$. Consequently, in our case,

$$\langle \tilde{g}_v, \tilde{\phi} \rangle = \tilde{g}_v'' \tilde{\phi} - \tilde{g}_v' \tilde{\phi}' + \tilde{g}_v \tilde{\phi}'' + 2\tilde{\tau}_1 \tilde{g}_v \tilde{\phi}.$$

Hence, the relation (88) takes the form

$$\begin{aligned} &T_{0,0}\tilde{\phi}'' - T_{0,1}\tilde{\phi}' + (T_{0,2} + 2\tilde{\tau}_1 T_{0,0})\tilde{\phi} \\ &\quad = \hat{\tau}_1 \tilde{\phi}' + (\hat{\tau}_1' + \hat{\tau}_0)\tilde{\phi} + T_{0,0}\tilde{\phi}'' + (3T_{1,0} + 2T_{0,1})\tilde{\phi}' + (3T_{2,0} + 3T_{1,1} + T_{0,2} + 2\tau_1 T_{0,0})\tilde{\phi}, \end{aligned}$$

where T_{j_1, j_2} were defined in (95). Grouping the terms at $\tilde{\phi}'(x, \lambda)$ and $\tilde{\phi}(x, \lambda)$, we derive the formulas

$$\begin{aligned} \tau_1 &= \tilde{\tau}_1 - \frac{3}{2} \sum_{v \in V'} (\phi_v' c_v \tilde{g}_v + \phi_v c_v \tilde{g}_v'), \\ \tau_0 &= -\hat{\tau}_1' - 3\frac{d}{dx}\left(\sum_{v \in V'} \phi_v' c_v \tilde{g}_v \right) - 2\hat{\tau}_1 \sum_{v \in V'} \phi_v c_v \tilde{g}_v. \end{aligned}$$

By virtue of Corollary 1.3 and Theorem 6.4 from [54] and (72), we have $\{l\xi_l\} \in l_2$. Applying Lemma 8 to prove the series convergence in suitable spaces and using the notations (85), we arrive at the following reconstruction formulas for τ_1 and τ_0.

Theorem 3. *Let $\mathcal{L}$ and $\tilde{\mathcal{L}}$ be the problems defined above in this section. The following relations hold:*

$$\tau_1 = \tilde{\tau}_1 - \frac{3}{2} \sum_{(l,k,\varepsilon) \in V} (\varphi_{l,k,\varepsilon}' \tilde{\eta}_{l,k,\varepsilon} + \varphi_{l,k,\varepsilon} \tilde{\eta}_{l,k,\varepsilon}'), \tag{109}$$

$$\tau_0 = -\hat{\tau}_1' - 3\frac{d}{dx}\left(\sum_{(l,k,\varepsilon) \in V} \varphi_{l,k,\varepsilon}' \tilde{\eta}_{l,k,\varepsilon} \right) - 2\hat{\tau}_1 \sum_{(l,k,\varepsilon) \in V} \varphi_{l,k,\varepsilon} \tilde{\eta}_{l,k,\varepsilon}. \tag{110}$$

The series in (109) converges in $L_2(0, 1)$. In (110), the series in brackets converges in $L_2(0, 1)$ with regularization, and the second series converges absolutely and uniformly with respect to $x \in [0, 1]$, so the right-hand side of (110) belongs to $W_2^{-1}(0, 1)$.

Following the proof of Lemma 8, one can easily show that the regularization constants A_v for the series in (109) equal zero. The regularization constants in (110) are omitted because of the differentiation. Finally, we arrive at the following Algorithm 2 for solving Problem 2.

Algorithm 2: Suppose that the spectral data $\{\lambda_0, \mathcal{N}(\lambda_0)\}_{\lambda_0 \in \Lambda}$ of the problem $\mathcal{L} = \mathcal{L}(F(x), U_0, U_1) \in W$ are given. Here, $F(x)$ is defined by (105), $U_0 = I$, $U_1 = [\delta_{k,4-j}]_{k,j=1}^3$. We have to find τ_1 and τ_0.

1. Find $\tilde{\tau}_1 = \int_0^1 \tau_1(x)\,dx$ from the eigenvalue asymptotics (106).
2. Take the model problem $\tilde{\mathcal{L}} = \mathcal{L}(\tilde{F}(x), U_0, U_1)$, where $\tilde{F}(x)$ is defined by (108).
3. Implement the steps 2–6 of Algorithm 1 to obtain $\{\varphi_{l,k,\varepsilon}(x)\}_{(l,k,\varepsilon)\in V}$.
4. Using the problem $\tilde{\mathcal{L}}$ and the spectral data $\{\lambda_0, \mathcal{N}(\lambda_0)\}_{\lambda_0\in\Lambda}$, $\{\tilde{\lambda}_0, \tilde{\mathcal{N}}(\tilde{\lambda}_0)\}_{\tilde{\lambda}_0\in\tilde{\Lambda}}$, construct the functions $\{\tilde{\eta}_{l,k,\varepsilon}(x)\}_{(l,k,\varepsilon)\in V}$ by (85).
5. Construct $\tau_1(x)$ and $\tau_0(x)$ by (109) and (110), respectively.

4.4. *Case of Even n, $\tau_\nu \in L_2(0,1)$*

Consider the differential expression (1) with even n and $\tau_\nu \in L_2(0,1)$, $\nu = \overline{0, n-2}$. The associated matrix $F(x) = [f_{k,j}(x)]_{k,j=1}^n$ is given by the relations

$$f_{n-k,k+1} = -\tau_{2k}, \quad k = \overline{0, \lfloor n/2 \rfloor - 1},$$
$$f_{n-k-1,k+1} = f_{n-k,k+2} = -\tau_{2k+1}, \quad k = \overline{0, \lfloor n/2 \rfloor - 2},$$

and all the other elements are defined by $f_{k,j} = \delta_{k,j-1}$. For instance,

$$\ell_6(y) = y^{(6)} + (\tau_4 y'')'' + [(\tau_3 y'')' + (\tau_3 y')''] + (\tau_2 y')' + [(\tau_1 y)' + \tau_1 y'] + \tau_0 y,$$

and the corresponding associated matrix is

$$F(x) = \begin{bmatrix} 0 & 1 & 0 & 0 & 0 & 0 \\ 0 & 0 & 1 & 0 & 0 & 0 \\ 0 & 0 & 0 & 1 & 0 & 0 \\ 0 & -\tau_3 & -\tau_4 & 0 & 1 & 0 \\ -\tau_1 & -\tau_2 & -\tau_3 & 0 & 0 & 1 \\ -\tau_0 & -\tau_1 & 0 & 0 & 0 & 0 \end{bmatrix}.$$

Suppose that $U_0 = I$, $U_1 = [\delta_{k,n-j+1}]_{k,j=1}^n$, $\mathcal{L} = (F(x), U_0, U_1) \in W$ and $\tilde{\mathcal{L}} = (\tilde{F}(x), U_0, U_1)$, where $\tilde{F}(x)$ is constructed in the same way as $F(x)$ by different coefficients $\tilde{\tau}_\nu \in L_2(0,1)$, $\nu = \overline{0, n-2}$. The following uniqueness theorem is proved similarly to Theorem 2.

Theorem 4. *If $\Lambda = \tilde{\Lambda}$ and $\mathcal{N}(\lambda_0) = \tilde{\mathcal{N}}(\lambda_0)$ for all $\lambda_0 \in \Lambda$, then $\tau_\nu(x) = \tilde{\tau}_\nu(x)$ a.e. on $(0,1)$, $\nu = \overline{0, n-2}$. Thus, the spectral data $\{\lambda_0, \mathcal{N}(\lambda_0)\}_{\lambda_0\in\Lambda}$ uniquely specify $\tau_\nu \in L_2(0,1)$, $\nu = \overline{0, n-2}$.*

Furthermore, we need the following proposition, which is an immediate corollary of Theorems 1.2 and 6.4 from [54] for the problems $\mathcal{L}$, $\tilde{\mathcal{L}}$ defined above in this subsection and the sequence $\{\xi_l\}$ defined by (71) (see also Example 5.2 in [54]).

Proposition 4 ([54]). *Suppose that $\nu_0 \in \{1, 2, \dots, n-1\}$, $\tau_\nu(x) = \tilde{\tau}_\nu(x)$ a.e. on $(0,1)$ for $\nu = \overline{\nu_0, n-2}$, and $\int_0^1 \hat{\tau}_{\nu_0-1}(x)\,dx = 0$. Then, $\{l^{n-\nu_0}\xi_l\} \in l_2$.*

We construct the solution of Problem 2 step-by-step.

STEP 1. Take the model problem $\tilde{\mathcal{L}} = \tilde{\mathcal{L}}^{(1)} := (\tilde{F}^{(1)}(x), U_0, U_1)$, where $\tilde{F}^{(1)}(x)$ is the associated matrix for the differential expression $\tilde{l}_n^{(1)}(y)$ with the coefficients $\tilde{\tau}_{n-2} := \int_0^1 \tau_{n-2}(x)\,dx$, $\tilde{\tau}_\nu := 0$, $\nu = \overline{0, n-3}$. The coefficient $\int_0^1 \tau_{n-2}(x)\,dx$ can be found from the eigenvalue asymptotics similarly to the case in Section 4.3. Using the terms of (88) at $\tilde{\phi}^{(n-2)}(x,\lambda)$, we derive the reconstruction formula

$$\tau_{n-2} = \tilde{\tau}_{n-2} - t_{n,n-2} - T_{0,1} = \tilde{\tau}_{n-2} - n \sum_{v\in V'} (\phi_v' c_v \tilde{g}_v + \phi_v c_v \tilde{g}_v').$$

By virtue of Proposition 4, $\{l\xi_l\} \in l_2$. Therefore, Lemma 8 implies that the obtained series converges in $L_2(0,1)$ with the regularization constants $A_v = 0$.

STEP 2. Take the model problem $\tilde{\mathcal{L}} = \tilde{\mathcal{L}}^{(2)} := (\tilde{F}^{(2)}(x), U_0, U_1)$, where $\tilde{F}^{(2)}(x)$ is the associated matrix for the differential expression $\tilde{l}_n^{(2)}(y)$ with the coefficients $\tilde{\tau}_{n-2} := \tau_{n-2}$, $\tilde{\tau}_{n-3} := \int_0^1 \tau_{n-3}(x)\,dx$, $\tilde{\tau}_\nu := 0$, $\nu = \overline{0, n-4}$. The coefficient $\int_0^1 \tau_{n-3}(x)\,dx$ can be found from the eigenvalue asymptotics. Using the terms of (88) at $\tilde{\phi}^{(n-2)}(x,\lambda)$, we show that $T_{0,0}'(x) = 0$. One can easily show that $T_{0,0}(0) = 0$, so $T_{0,0}(x) \equiv 0$. Consequently, grouping the terms of (88) at $\tilde{\phi}^{(n-3)}(x,\lambda)$, we obtain

$$\begin{aligned} 2\tau_{n-3} &= 2\tilde{\tau}_{n-3} - t_{n,n-3} + T_{0,2} \\ &= 2\tilde{\tau}_{n-3} - \sum_{v\in V'} \left(\tfrac{n(n-1)}{2} \phi_v'' c_v \tilde{g}_v + n(n-2)\phi_v' c_v \tilde{g}_v' + [\tfrac{(n-1)(n-2)}{2} - 1]\phi_v c_v \tilde{g}_v'' \right). \end{aligned}$$

By virtue of Proposition 4, $\{l^2\xi_l\} \in l_2$. Lemma 8 implies that the series converges in $L_2(0,1)$ with the zero regularization constants.

STEP s. Take the model problem $\tilde{\mathcal{L}} = \tilde{\mathcal{L}}^{(s)} := (\tilde{F}^{(s)}(x), U_0, U_1)$, where $\tilde{F}^{(s)}(x)$ is the associated matrix for the differential expression $\tilde{l}_n^{(s)}(y)$ with

$$\tilde{\tau}_\nu := \tau_\nu,\ \nu = \overline{n-s, n-2}, \quad \tilde{\tau}_{n-s-1} := \int_0^1 \tau_{n-s-1}(x)\,dx, \quad \tilde{\tau}_\nu := 0,\ \nu = \overline{0, n-s-2}. \quad (111)$$

For this model problem, we have $T_{j_1,j_2}(x) \equiv 0$ for all $j_1 + j_2 \le s-2$. Grouping the terms of (88) at $\tilde{\phi}^{(n-s-1)}(x,\lambda)$, we obtain

$$\begin{aligned} \tau_{n-s-1} &= \tilde{\tau}_{n-s-1} - (t_{n,n-s-1} + (-1)^{s+1} T_{0,s}) \overset{if\ s\ is\ even}{\times} \tfrac{1}{2} \\ &= \tilde{\tau}_{n-s-1} - \sum_{v\in V'} \left(\sum_{r=n-s}^{n} C_n^r C_{r-1}^{n-s-1} \phi_v^{(n-r)} c_v \tilde{g}_v^{(r-n+s)} + (-1)^{s+1} \phi_v c_v \tilde{g}_v^{(s)} \right) \overset{if\ s\ is\ even}{\times} \tfrac{1}{2} \\ &= \tilde{\tau}_{n-s-1} - \sum_{v\in V'} \left(\sum_{r=n-s}^{n} C_n^r C_{r-1}^{n-s-1} \phi_v^{[n-r]} c_v \tilde{g}_v^{[r-n+s]} + (-1)^{s+1} \phi_v c_v \tilde{g}_v^{[s]} \right) \overset{if\ s\ is\ even}{\times} \tfrac{1}{2} \end{aligned} \quad (112)$$

Proposition 4 implies that $\{l^s\xi_l\} \in l_2$. Therefore, it follows from Lemma 8 that the series in (112) converges in $L_2(0,1)$. The regularization constants equal zero because

$$\sum_{r=n-s}^{n} C_n^r C_{r-1}^{n-s-1} (-1)^r + (-1)^{s+1} = 0.$$

Note that all functions $\{\tau_\nu\}$ necessary for computation of the quasi-derivatives $\phi_v^{[n-r]}$ in (112) are computed at the previous steps, so the formula (112) can be used for finding τ_{n-s-1}. In terms of the notations (85), the relation (112) can be written as follows:

$$\tau_{n-s-1} = \tilde{\tau}_{n-s-1} - \sum_{(l,k,\varepsilon)\in V} \left(\sum_{r=n-s}^{n} C_n^r C_{r-1}^{n-s-1} \varphi_{l,k,\varepsilon}^{[n-r]} \tilde{\eta}_{l,k,\varepsilon}^{[r-n+s]} + (-1)^{s+1} \varphi_{l,k,\varepsilon} \tilde{\eta}_{l,k,\varepsilon}^{[s]} \right) \overset{if\ s\ is\ even}{\times} \tfrac{1}{2} \quad (113)$$

Thus, we obtain the following Algorithm 3 for solving Problem 2 in the considered case.

Algorithm 3: Suppose that the spectral data $\{\lambda_0, \mathcal{N}(\lambda_0)\}_{\lambda_0 \in \Lambda}$ of the problem $\mathcal{L} = (F(x), U_0, U_1) \in W$ are given. Here, $F(x)$ is the matrix associated with the differential expression $\ell_n(y)$, n is even, $\tau_\nu \in L_2(0,1)$, $\nu = \overline{0, n-2}$, $U_0 = I$, and $U_1 = [\delta_{k,n-j+1}]_{k,j=1}^n$. We have to find $\{\tau_\nu\}_{\nu=0}^{n-2}$. For simplicity, assume that the values $\int_0^1 \tau_\nu(x)\,dx$ are known. In fact, they can be found from the eigenvalue asymptotics.

For $s = 1, 2, \dots, n-1$, we find τ_{n-s-1} implementing the following steps:

1. Take the model problem $\tilde{\mathcal{L}} = \tilde{\mathcal{L}}^{(s)} = (\tilde{F}^{(s)}, U_0, U_1)$ induced by the differential expression $\tilde{\ell}_n^{(s)}(y)$ with the coefficients (111).
2. Implement steps 2–6 of Algorithm 1 to find $\{\varphi_{l,k,\varepsilon}(x)\}_{(l,k,\varepsilon) \in V}$.
3. Using the problem $\tilde{\mathcal{L}}$ and the spectral data $\{\lambda_0, \mathcal{N}(\lambda_0)\}_{\lambda_0 \in \Lambda}$, $\{\tilde{\lambda}_0, \tilde{\mathcal{N}}(\tilde{\lambda}_0)\}_{\tilde{\lambda}_0 \in \tilde{\Lambda}}$, construct the functions $\{\tilde{\eta}_{l,k,\varepsilon}(x)\}_{(l,k,\varepsilon) \in V}$ by (85).
4. Construct $\tau_{n-s-1}(x)$ by (113).

4.5. Case of Even n, $\tau_\nu \in W_2^{-1}(0,1)$

Suppose that n is even and $\tau_\nu \in W_2^{-1}(0,1)$ in (1) for $\nu = \overline{0, n-2}$, that is, $\tau_\nu = \sigma_\nu'$, where $\sigma_\nu \in L_2(0,1)$, and the derivative is understood in the sense of distributions. Put $m := \lfloor n/2 \rfloor$, and define the matrix function

$$Q(x) = [q_{r,j}(x)]_{r,j=0}^m := \sum_{\nu=0}^{n-2} (-1)^{\lfloor (\nu-1)/2 \rfloor} \chi_\nu \sigma_\nu(x),$$

where $\chi_\nu := [\chi_{\nu;r,j}]_{r,j=0}^m$,

$$\chi_{2k;k,k+1} = \chi_{2k;k+1,k} = 1, \quad \chi_{2k+1;k,k+2} = -\chi_{2k+1;k+2,k} = 1,$$

and all the other entries $\chi_{\nu;r,j}$ equal zero. The associated matrix $F(x) = [f_{k,j}(x)]_{k,j=1}^n$ for $\ell_n(y)$ is defined as follows (see [46] for details):

$$f_{m,j} := (-1)^{m+1} q_{j-1,m}, \; j = \overline{1,m}, \qquad f_{k,m+1} := (-1)^{k+1} q_{m,2m-k}, \; k = \overline{m+1, 2m},$$
$$f_{k,j} := (-1)^{k+1} q_{j-1,2m-k} + (-1)^{m+k} q_{j-1,m} q_{m,2m-k}, \quad k = \overline{m+1, 2m}, \; j = \overline{1,m},$$

and $f_{k,j} = \delta_{k,j-1}$ for all the other indices. Clearly, $F \in \mathfrak{F}_n$. For example, if $n = 4$, then

$$Q(x) = \begin{bmatrix} 0 & -\sigma_0 & \sigma_1 \\ -\sigma_0 & 0 & \sigma_2 \\ -\sigma_1 & \sigma_2 & 0 \end{bmatrix}, \quad F(x) = \begin{bmatrix} 0 & 1 & 0 & 0 \\ -\sigma_1 & -\sigma_2 & 1 & 0 \\ -\sigma_0 - \sigma_1\sigma_2 & -\sigma_2^2 & \sigma_2 & 1 \\ -\sigma_1^2 & \sigma_0 - \sigma_1\sigma_2 & \sigma_1 & 0 \end{bmatrix}$$

Consider Problem 2 for $\mathcal{L} = (F(x), U_0, U_1)$, $U_0 = I$, $U_1 = [\delta_{k,n-j+1}]_{k,j=1}^n$. Let $\tilde{\mathcal{L}} = (\tilde{F}(x), U_0, U_1)$, where $\tilde{F}(x)$ is the associated matrix for the differential expression $\tilde{l}_n(y)$ with the coefficients $\tilde{\tau}_\nu = \tilde{\sigma}_\nu' \in W_2^{-1}(0,1)$, $\nu = \overline{0, n-2}$. The following uniqueness theorem is proved analogously to Theorem 2.

Theorem 5. *If $\Lambda = \tilde{\Lambda}$ and $\mathcal{N}(\lambda_0) = \tilde{\mathcal{N}}(\lambda_0)$ for all $\lambda_0 \in \Lambda$, then $\sigma_\nu(x) = \tilde{\sigma}_\nu(x) + const$ a.e. on $(0,1)$ for $\nu = \overline{0, n-2}$. Thus, the spectral data $\{\lambda_0, \mathcal{N}(\lambda_0)\}_{\lambda_0 \in \Lambda}$ uniquely specify $\tau_\nu \in W_2^{-1}(0,1)$, $\nu = \overline{0, n-2}$.*

The functions $\{\sigma_\nu\}_{\nu=0}^{n-2}$ are specified uniquely up to a constant, so for simplicity we assume that $\int_0^1 \sigma_\nu(x)\,dx = 0$, $\nu = \overline{0, n-2}$.

Theorems 1.2 and 6.4 of [54] (see also Example 5.3 in [54]) readily imply the following proposition for the problems $\mathcal{L}$ and $\tilde{\mathcal{L}}$ of the considered form and the sequence $\{\xi_l\}$ defined by (71).

Proposition 5 ([54]). *Suppose that $\nu_0 \in \{1, 2, \dots, n-1\}$ and $\sigma_\nu(x) = \tilde{\sigma}_\nu(x)$ a.e. on $(0,1)$ for $\nu = \overline{\nu_0, n-2}$. Then, $\{l^{n-\nu_0-1}\xi_l\} \in l_2$.*

The algorithm of recovering the coefficients $\{\tau_\nu\}_{\nu=0}^{n-2}$ from the spectral data is similar to Algorithm 3. At STEP s, we take the model problem $\tilde{\mathcal{L}} = \tilde{\mathcal{L}}^{(s)}$ induced by the coefficients $\tilde{\sigma}_\nu := \sigma_\nu$, $\nu = \overline{n-s, n-2}$, and $\tilde{\sigma}_\nu := 0$, $\nu = \overline{0, n-s-1}$. Note that the series in (112) has the form

$$a_0 T_{s,0} + a_1 T_{s-1,1} + \ldots + a_s T_{0,s} = (b_0 T_{s-1,0} + b_1 T_{s-2,1} + \ldots + b_{s-1} T_{0,s-1})',$$

where

$$a_j := C_n^{s-j} C_{n-s+j-1}^{j} \overset{if\ j=s}{+} (-1)^{s+1}, \quad \sum_{j=0}^{s} a_j = 0,$$

$$b_j := \sum_{i=0}^{j} (-1)^{j-i} a_i, \quad j = \overline{0, s-1}.$$

Using this idea, we derive

$$\tau_{n-s-1} = -\frac{d}{dx} \sum_{v \in V'} \left(\sum_{j=0}^{s-1} b_j \phi_v^{[s-j-1]} c_v \tilde{g}_v^{[j]} \right) \overset{if\ s\ is\ even}{\times} \frac{1}{2}. \tag{114}$$

In view of Proposition 5, we have $\{l^{s-1}\xi_l\} \in l_2$. Hence, by virtue of Lemma 8, the series in (114) converges in $L_2(0,1)$ with some regularization constants. Because of the differentiation, we omit these constants. Thus, Formula (114) induces a function of $W_2^{-1}(0,1)$, and σ_{n-s-1} can be found uniquely up to a constant. This constant is chosen so that $\int_0^1 \sigma_{n-s-1}(x)\,dx = 0$. Taking $s = 1, 2, \dots, n-1$, we step-by-step construct all the coefficients $\tau_{n-2}, \tau_{n-3}, \dots, \tau_1, \tau_0$.

Note that the algorithms of this section are valid for $\tilde{\mathcal{L}} \in W$. However, the case $\tilde{\mathcal{L}} \notin W$ requires only technical modifications due to Remark 1, which do not influence the convergence of the series.

5. Conclusions

Let us briefly summarize the results of this paper. We studied the inverse spectral problem, which consists in recovering the coefficients $\{\tau_\nu\}_{\nu=0}^{n-2}$ of the differential expression (1) from the spectral data $\{\lambda_0, \mathcal{N}(\lambda_0)\}_{\lambda_0 \in \Lambda}$. An approach to the constructive solution of the inverse problem is developed. Our approach can be applied to a wide class of differential expressions $\ell_n(y)$, which admit regularization in terms of associated matrix.

The inverse problem solution consists of the two steps. First, we consider the auxiliary problem of finding the Weyl solutions $\{\Phi_k(x,\lambda)\}_{k=1}^{n}$ by using the spectral data. This problem is reduced to the linear Equation (80) in the Banach space m of bounded infinite sequences. Theorem 1 on the unique solvability of the main Equation (80) is proved. Second, by using the solution of the main equation, we derive reconstruction formulas for the coefficients $\{\tau_\nu\}_{\nu=0}^{n-2}$ and investigate the convergence of resulting series.

Let us mention the most important **advantages** of our approach:

1. The obtained results can be applied to a wide range of differential operators of arbitrary order with either integrable or distributional coefficients of various classes.

2. Our approach does not require self-adjointness.
3. Our method is constructive.
4. The results of this paper can be used for studying the existence and stability of the inverse problem solution as well as for developing numerical methods.

Funding: This work was supported by Grant 21-71-10001 of the Russian Science Foundation. https://rscf.ru/en/project/21-71-10001/ (accessed on 18 October 2022).

Data Availability Statement: Not applicable.

Conflicts of Interest: The author declares that this paper has no conflict of interest.

References

1. Mirzoev, K.A.; Shkalikov, A.A. Differential operators of even order with distribution coefficients. *Math. Notes* **2016**, *99*, 779–784. [CrossRef]
2. Savchuk, A.M.; Shkalikov, A.A. Asymptotic analysis of solutions of ordinary differential equations with distribution coefficients. *Sb. Math.* **2020**, *211*, 1623–1659. [CrossRef]
3. Konechnaja, N.N.; Mirzoev, K.A. The leading term of the asymptotics of solutions of linear differential equations with first-order distribution coefficients. *Math. Notes* **2019**, *106*, 81–88. [CrossRef]
4. Behncke, H.; Hinton, D.B. Spectral theory of higher order differential operators by examples. *J. Spectr. Theory* **2013**, *3*, 361–398. [CrossRef] [PubMed]
5. Papanicolaou, V.G. Some results on ordinary differential operators with periodic coefficients. *Complex Anal. Oper. Theory* **2016**, *10*, 1227–1265. [CrossRef]
6. Bao, Q.; Sun, J.; Hao, X.; Zettl, A. Characterization of self-adjoint domains for regular even order C-symmetric differential operators. *Electron. J. Qual. Theory Differ. Eq.* **2019**, *62*, 1–17.
7. Perera, U.; Böckmann, C. Solutions of Sturm-Liouville problems. *Mathematics* **2020**, *8*, 2074. [CrossRef]
8. Gal'kovskii, E.D.; Nazarov, A.I. A trace formula for higher order ordinary differential operators. *Sb. Math.* **2021**, *212*, 676–697. [CrossRef]
9. Bondarenko, N.P. Inverse spectral problems for arbitrary-order differential operators with distribution coefficients. *Mathematics* **2021**, *9*, 2989. [CrossRef]
10. Deift, P.; Tomei, C.; Trubowitz, E. Inverse scattering and the Boussinesq equation. *Comm. Pure Appl. Math.* **1982**, *35*, 567–628. [CrossRef]
11. McKean, H. Boussinesq's equation on the circle, Comm. *Pure Appl. Math.* **1981**, *34*, 599–691. [CrossRef]
12. Braeutigam, I.N.; Polyakov, D.M. On the asymptotics of eigenvalues of a third-order differential operator. *St. Petersburg Math. J.* **2020**, *31*, 585–606. [CrossRef]
13. Barcilon, V. On the solution of inverse eigenvalue problems of high orders. *Geophys. J. R. Astr. Soc.* **1974**, *39*, 143–154. [CrossRef]
14. McLaughlin, J.R. An inverse eigenvalue problem of order four. *SIAM J. Math. Anal.* **1976**, *7*, 646–661. [CrossRef]
15. Papanicolaou, V.G.; Kravvaritisz, D. An inverse spectral problem for the Euler-Bernoulli equation for the vibrating beam. *Inverse Probl.* **1997**, *13*, 1083–1092. [CrossRef]
16. Caudill, L.F.; Perry, P.A.; Schueller, A.W. Isospectral sets for fourth-order ordinary differential operators. *SIAM J. Math. Anal.* **1998**, *29*, 935–966. [CrossRef]
17. Gladwell, G.M.L. *Inverse Problems in Vibration*, 2nd ed.; Solid Mechanics and Its Applications; Springer: Dordrecht, The Netherlands, 2005; Volume 119.
18. Morassi, A. Exact construction of beams with a finite number of given natural frequencies. *J. Vibr. Cont.* **2015**, *21*, 591–600. [CrossRef]
19. Badanin, A.; Korotyaev, E. Inverse problems and sharp eigenvalue asymptotics for Euler-Bernoulli operators. *Inverse Probl.* **2015**, *31*, 055004. [CrossRef]
20. Jiang, X.; Li, X.; Xu, X. Numerical algorithms for inverse Sturm-Liouville problems. *Numer. Algorithms* **2022**, *89*, 1287–1309. [CrossRef]
21. Marchenko, V.A. *Sturm-Liouville Operators and Their Applications*; Birkhauser: Basel, Switzerland, 1986.
22. Levitan, B.M. *Inverse Sturm-Liouville Problems*; VNU Science Press: Utrecht, The Netherlands, 1987.
23. Leibenson, Z.L. The inverse problem of spectral analysis for higher-order ordinary differential operators. *Tr. Moskov. Mat. Obshch.* **1966**, *15*, 70–144; English Translated in *Trans. Moscow Math. Soc.* **1966**, *15*, 70–144.
24. Leibenson, Z.L. Spectral expansions of transformations of systems of boundary value problems. *Tr. Moskov. Mat. Obshch.* **1971**, *25*, 15–58; English Translated in *Trans. Moscow Math. Soc.* **1971**, *25*,, 15–58.
25. Yurko, V.A. Recovery of nonselfadjoint differential operators on the half-line from the Weyl matrix. *Sb. Math.* **1992**, *72*, 413–438. [CrossRef]
26. Yurko, V.A. *Method of Spectral Mappings in the Inverse Problem Theory, Inverse and Ill-Posed Problems Series;* VNU Science: Utrecht, The Netherlands, 2002.
27. Yurko, V.A. On higher-order differential operators with a singular point. *Inverse Probl.* **1993**, *9*, 495–502. [CrossRef]

28. Yurko, V.A. On higher-order differential operators with a regular singularity. *Sb. Math.* **1995**, *186*, 901–928. [CrossRef]
29. Yurko, V. Inverse spectral problems for differential systems on a finite interval. *Results Math.* **2005**, *48*, 371–386. [CrossRef]
30. Yurko, V.A. Inverse spectral problems for differential operators on spatial networks. *Russ. Math. Surv.* **2016**, *71*, 539–584. [CrossRef]
31. Buterin, S.A.; Yurko, V.A. Inverse problems for second-order differential pencils with Dirichlet boundary conditions. *J. Inv. Ill-Pos. Probl.* **2012**, *20*, 855–881. [CrossRef]
32. Beals, R. The inverse problem for ordinary differential operators on the line. *Am. J. Math.* **1985**, *107*, 281–366. [CrossRef]
33. Beals, R.; Deift, P.; Tomei, C. *Direct and Inverse Scattering on the Line, Mathematical Surveys and Monographs*; AMS: Providence, RI, USA, 1988; Volume 28.
34. Hryniv, R.O.; Mykytyuk, Y.V. Inverse spectral problems for Sturm-Liouville operators with singular potentials. *Inverse Probl.* **2003**, *19*, 665–684. [CrossRef]
35. Hryniv, R.O.; Mykytyuk, Y.V. Inverse spectral problems for Sturm-Liouville operators with singular potentials, II. Reconstruction by two spectra. *N.-Holl. Math. Stud.* **2004**, *197*, 97–114.
36. Hryniv, R.O.; Mykytyuk, Y.V. Half-inverse spectral problems for Sturm-Liouville operators with singular potentials. *Inverse Probl.* **2004**, *20*, 1423–1444. [CrossRef]
37. Freiling, G.; Ignatiev, M.Y.; Yurko, V.A. An inverse spectral problem for Sturm-Liouville operators with singular potentials on star-type graph. *Proc. Symp. Pure Math.* **2008**, *77*, 397–408.
38. Mykytyuk, Ya.V.; Trush, N.S. Inverse spectral problems for Sturm-Liouville operators with matrix-valued potentials. *Inverse Probl.* **2009**, *26*, 015009. [CrossRef]
39. Savchuk, A.M.; Shkalikov, A.A. Inverse problems for Sturm-Liouville operators with potentials in Sobolev spaces: Uniform stability. *Funct. Anal. Appl.* **2010**, *44*, 270–285. [CrossRef]
40. Hryniv, R.; Pronska, N. Inverse spectral problems for energy-dependent Sturm-Liouville equations. *Inverse Probl.* **2012**, *28*, 085008. [CrossRef]
41. Eckhardt, J.; Gesztesy, F.; Nichols, R.; Teschl, G. Supersymmetry and Schrödinger-type operators with distributional matrix-valued potentials. *J. Spectr. Theory* **2014**, *4*, 715–768. [CrossRef]
42. Guliyev, N.J. Schrödinger operators with distributional potentials and boundary conditions dependent on the eigenvalue parameter. *J. Math. Phys.* **2019**, *60*, 063501. [CrossRef]
43. Bondarenko, N.P. Inverse problem solution and spectral data characterization for the matrix Sturm-Liouville operator with singular potential. *Anal. Math. Phys.* **2021**, *11*, 145. [CrossRef]
44. Albeverio, S.; Gesztesy, F.; Hoegh-Krohn, R.; Holden, H. *Solvable Models in Quantum Mechanics*, 2nd ed.; AMS Chelsea Publishing: Providence, RI, USA, 2005.
45. Bondarenko, N.P. Solving an inverse problem for the Sturm-Liouville operator with singular potential by Yurko's method. *Tamkang J. Math.* **2021**, *52*, 125–154. [CrossRef]
46. Bondarenko, N.P. Linear differential operators with distribution coefficients of various singularity orders. *arXiv* **2022**, arXiv:2204.02052.
47. Mirzoev, K.A.; Shkalikov, A.A. Ordinary differential operators of odd order with distribution coefficients. *arXiv* **2019**, arXiv:1912.03660.
48. Valeev, N.F.; Nazirova, E.A.; Sultanaev, Y.T. On a method for studying the asymptotics of solutions of odd-order differential equations with oscillating coefficients. *Math. Notes* **2021**, *109*, 980–985. [CrossRef]
49. Savchuk, A.M.; Shkalikov, A.A. Sturm-Liouville operators with singular potentials. *Math. Notes* **1999**, *66*, 741–753. [CrossRef]
50. Everitt, W.N.; Marcus, L. *Boundary Value Problems and Symplectic Algebra for Ordinary Differential and Quasi-Differential Operators*; Mathematical Surveys and Monographs; AMS: Providence, RI, USA, 1999; Volume 61.
51. Vladimirov, A.A. On one approach to definition of singular differential operators. *arXiv* **2017**, arXiv:1701.08017.
52. Buterin, S.A. On inverse spectral problem for non-selfadjoint Sturm-Liouville operator on a finite interval. *J. Math. Anal. Appl.* **2007**, *335*, 739–749. [CrossRef]
53. Buterin, S.A.; Shieh, C.-T.; Yurko, V.A. Inverse spectral problems for non-selfadjoint second-order differential operators with Dirichlet boundary conditions. *Bound. Value Probl.* **2013**, *2013*, 180. [CrossRef]
54. Bondarenko, N.P. Spectral data asymptotics for the higher-order differential operators with distribution coefficients. *arXiv* **2022**, arXiv:2208.12956.
55. Ignatiev, M.; Yurko, V. Numerical methods for solving inverse Sturm-Liouville problems. *Results Math.* **2008**, *52*, 63–74. [CrossRef]

mathematics

MDPI

Review

Partial Inverse Sturm-Liouville Problems

Natalia P. Bondarenko [1,2,3]

[1] Department of Mechanics and Mathematics, Saratov State University, Astrakhanskaya 83, Saratov 410012, Russia; bondarenkonp@info.sgu.ru
[2] Department of Applied Mathematics and Physics, Samara National Research University, Moskovskoye Shosse 34, Samara 443086, Russia
[3] S.M. Nikolskii Mathematical Institute, Peoples' Friendship University of Russia (RUDN University), 6 Miklukho-Maklaya Street, Moscow 117198, Russia

Abstract: This paper presents a review of both classical and modern results pertaining to partial inverse spectral problems for differential operators. Such problems consist in the recovery of differential expression coefficients in some part of the domain (a finite interval or a geometric graph) from spectral characteristics, while the coefficients in the remaining part of the domain are known a priori. Usually, partial inverse problems require less spectral data than complete inverse problems. In this review, we pay considerable attention to partial inverse problems on graphs and to the unified approach based on the reduction of partial inverse problems to Sturm-Liouville problems with entire analytic functions in a boundary condition. We not only describe the results of selected studies but also compare them with each other and establish interconnections.

Keywords: inverse spectral problems; Sturm-Liouville operator; differential operators on graphs; Hochstadt-Lieberman problem; half-inverse problem

MSC: 34A55; 34B09; 34B07; 34B24; 34B45; 34K08; 34K29

Citation: Bondarenko, N.P. Partial Inverse Sturm-Liouville Problems. *Mathematics* **2023**, *11*, 2408. https://doi.org/10.3390/math11102408

Academic Editor: Luigi Rodino

Received: 18 April 2023
Revised: 17 May 2023
Accepted: 18 May 2023
Published: 22 May 2023

1. Introduction

This paper contains an overview of results pertaining to partial inverse spectral problems for ordinary differential operators. Such problems consist in the recovery of differential expression coefficients on some part of the domain (a finite interval or a geometric graph) from spectral characteristics, while the coefficients on the remaining part of the domain are known a priori. Usually, partial inverse problems require fewer spectral data than complete inverse problems. In the literature, partial inverse problems are also called half-inverse problems, Hochstadt–Lieberman-type problems, inverse problems with mixed data, and incomplete inverse problems.

We begin with some classical results regarding complete inverse spectral problems. The greatest success in inverse spectral theory has been achieved for the second-order Sturm-Liouville (one-dimensional Schrödinger) equation (see the monographs [1–5] and the references therein):

$$-y'' + q(x)y = \lambda y, \quad x \in (0,1), \tag{1}$$

where the function $q(x)$ is usually called the potential, and λ is the spectral parameter. In 1946, Borg [6] proved that the potential $q(x)$ is uniquely specified by the two spectra $\{\lambda_{n,j}\}_{n\geq 1}$ and $j = 0, 1$ of the boundary value problems for Equation (1) subject to the boundary conditions

$$y(0) = y^{(j)}(1) = 0, \quad j = 0, 1.$$

In their seminal paper [7], Gelfand and Levitan developed a constructive method for solving the inverse Sturm-Liouville problem. This method allowed the authors to obtain the necessary and sufficient conditions of the inverse problem's solvability. Since then, inverse spectral theory has been developing all over the world for various classes of differential operators with applications in classical and quantum mechanics, geophysics, nanotechnology, acoustics, electronics, and other fields of science and engineering.

In 1978, Hochstadt and Lieberman [8] proved that, if the potential $q(x)$ of the Sturm-Liouville (Schrödinger) Equation (1) is known a priori on the half-interval $\left(\frac{1}{2}, 1\right)$, then, in contrast to the Borg problem, the spectrum $\{\lambda_{n,0}\}_{n\geq 1}$ alone is sufficient for the unique specification of $q(x)$ on $\left(0, \frac{1}{2}\right)$. Thus, knowledge of the potential on part of the interval reduces the amount of spectral data needed for the operator reconstruction. The Hochstadt–Lieberman problem was the first partial inverse problem. Later on, various generalizations of this problem were considered by Hald [9], Gestezy and Simon [10], Horváth [11,12], and other scholars. Constructive methods and solvability conditions for the Hochstadt–Lieberman problem have been obtained by Sakhnovich [13], Hryniv and Mykytyuk [14], Buterin [15,16], and Martinyuk and Pivovarchik [17,18].

In recent years, considerable attention has been paid by mathematicians and physicists to the inverse transmission eigenvalue problem, which has applications in acoustics. In [19], McLaughlin and Polyakov presented an inverse transmission problem statement, which generalized the Hochstadt–Lieberman problem. The investigation of the McLaughlin–Polyakov problem continued in [20–23] and other studies, offering a series of new results in the theory of partial inverse problems.

A variety of partial inverse problems arise for differential operators on geometrical graphs, also called quantum graphs. Such operators are used to model various processes in graph-like structures and networks in organic chemistry, mesoscopic physics, nanotechnology, hydrodynamics, waveguide theory, and other applications (see, e.g., the monographs [24,25] and the references therein). A basic introduction to quantum graphs can be found in [26]. There is an extensive literature on inverse spectral problems for differential operators on graphs (see the survey [27] on this topic). In this review, we focus on partial inverse problems. Such problems on graphs arise when differential operator coefficients (for example, Sturm-Liouville potentials) are known a priori for part of the graph. These coefficients can be obtained by either measurements or a reconstruction method. In the second case, the solution of partial inverse problems can be used as an auxiliary step in solving complete inverse problems on graphs.

The first results on partial inverse problems on graphs were obtained by Pivovarchik [28], Yurko [29], and Yang et al. [30–32]. However, the results of these papers were limited to uniqueness theorems for the Sturm-Liouville (Schrödinger) operators on graphs of an elementary structure (star-shaped graphs and simple graphs with loops). Later on, Bondarenko developed a constructive method to solve partial inverse problems on graphs of various types. Using this method, a number of new results have been obtained for differential operators and pencils on star-shaped graphs [33–36], simple graphs with cycles [37,38], tree graphs (graphs without cycles) [39], and even graphs of an arbitrary geometrical structure [40,41]. These results included not only uniqueness theorems but also constructive algorithms for the solution, solvability, and stability of partial inverse problems.

Following the investigation of partial inverse problems on graphs, a unified approach to various classes of partial inverse problems arose [41–44]. This approach was based on the reduction of a partial inverse problem on either an interval or a graph to an inverse problem for a differential operator on an "unknown" interval with entire analytic functions in one of the boundary conditions. In [41–44], an inverse problem theory was created for Sturm-Liouville operators with entire functions in a boundary condition. This theory included the necessary and sufficient conditions of uniqueness, constructive methods for a solution, global solvability, local solvability, and stability. These results have been applied to the Hochstadt–Lieberman problem, the inverse transmission eigenvalue problem, and partial inverse problems on graphs. Later on, this approach was developed in [45] for differential pencils and in [46] for Sturm-Liouville operators with polynomial boundary conditions.

In addition, it is worth mentioning that partial inverse problems have been considered for other types of operators, in particular, for integro-differential operators [47–49], functional-differential operators with a constant delay [50–52], higher-order differential operators [53,54], and matrix Sturm-Liouville operators [55,56].

The goal of this review was to summarize classical and recent work on partial inverse problems. Below, we describe some features of this review. Since the amount of literature on partial inverse problems is enormous, this review includes only the results of selected papers, which, in the author's opinion, could help the reader to form a general picture. Most attention was paid to partial inverse problems on geometrical graphs and the unified approach, which has been investigated by the author in recent years. However, we also paid attention to classical results and different modern directions of research. In view of the huge amount of information available, we focused on describing *the results* of the selected papers. Unfortunately, we could not provide a full description of *the methods* by which these results were obtained. Nevertheless, the reader can find more details in the referenced literature. In this review, we compare the results for different problems and establish connections between them.

The paper is organized as follows. In Section 2, we consider the Hochstadt–Lieberman problem and its generalizations on intervals. Section 2.1 is devoted to the uniqueness theorems, and Section 2.2 focuses on constructive methods and solvability conditions. Section 2.3 is concerned with the inverse transmission eigenvalue problem (mostly the McLaughlin–Polyakov problem). In Section 3, we describe the known results on partial inverse problems for differential operators on graphs. Star-shaped graphs are considered in Section 3.1, simple graphs with loops in Section 3.2, and graphs of a general structure in Section 3.3. Section 4 is concerned with the unified approach to various classes of partial inverse problems. In Section 4.1, the inverse spectral theory of the Sturm-Liouville problem with entire functions in a boundary condition is presented. In Section 4.2, this theory is applied to partial inverse problems. In Section 5, we consider partial inverse problems for classes of operators other than Sturm-Liouville operators and pencils. Section 6 contains the conclusions.

Here, we present a few remarks about notations. When describing the results, we mostly preserve the notations of the original papers. Therefore, the notations included throughout the review can have different meanings. The symbol λ usually denotes the spectral parameter, unless stated otherwise. In the formulations of the uniqueness theorems, along with one problem (e.g., problem L), we often consider another problem (e.g., $\tilde{L}$) of the same form but with different coefficients. If a symbol α denotes an object related to the problem without a tilde, then the symbol $\tilde{\alpha}$ denotes the analogous object related to the problem with a tilde. In addition, all the boundary value problems in this review were considered on finite intervals or compact graphs, so their spectra are countable sets of eigenvalues.

2. Hochstadt–Lieberman Problem and Generalizations

2.1. Uniqueness Theorems

Let us begin with the famous result of Hochstadt and Lieberman [8]. They considered the Sturm-Liouville problem

$$\left.\begin{aligned} -y'' + q(x)y = \lambda y, \quad x \in (0,1), \\ y(0)\cos\alpha + y'(0)\sin\alpha = 0, \quad y(1)\cos\beta + y'(1)\cos\beta = 0, \end{aligned}\right\} \tag{2}$$

where $q \in L_1(0,1)$ and $\alpha, \beta \in [0,\pi)$. The Hochstadt–Lieberman problem is formulated as follows:

Problem 1 ([8]). *Suppose that the potential $q(x)$ on $\left(\frac{1}{2},1\right)$ and the constants α and β are known a priori. Given the spectrum $\{\lambda_n\}_{n\ge1}$ of the problem (2), find $q(x)$ on $\left(0,\frac{1}{2}\right)$.*

Hochstadt and Lieberman proved the following uniqueness theorem for Problem 1:

Theorem 1 ([8]). *Let $\{\lambda_n\}_{n\geq 1}$ be the spectrum of the problem* (2), *and let $\{\tilde{\lambda}_n\}_{n\geq 1}$ be the spectrum of a similar problem with an integrable potential $\tilde{q}(x)$. Suppose that $q(x) = \tilde{q}(x)$ on $\left(\frac{1}{2}, 1\right)$, and $\lambda_n = \tilde{\lambda}_n$, $n \geq 1$. Then, $q(x) = \tilde{q}(x)$ a.e. on $(0, 1)$.*

Hald [9] generalized Hochstadt and Lieberman's findings to the Sturm-Liouville problem with discontinuity. In addition, Hald showed that the coefficient in the left boundary condition is also uniquely specified by the spectrum.

Theorem 2 ([9]). *Consider the eigenvalue problem*

$$-y'' + q(x)y = \lambda y, \quad x \in (0, \pi),$$

with the boundary conditions

$$y'(0) - hy(0) = y'(\pi) + Hy(\pi) = 0$$

and the jump conditions

$$y(d+) = ay(d-), \quad y'(d+) = a^{-1}y(d-) + by(d-),$$

where q is an integrable function; $0 < d < \frac{\pi}{2}$; $a > 0$; and $|a-1| + |b| > 0$. Let $\{\lambda_n\}_{n\geq 0}$ be the eigenvalues. Consider the eigenvalue problem with a, b, d, h, H, λ, and q replaced by $\tilde{a}, \tilde{b}, \tilde{d}, \tilde{h}, \tilde{H}, \tilde{\lambda}$, and $\tilde{q}$, respectively. If $\lambda_n = \tilde{\lambda}_n$, $H = \tilde{H}$, and $q = \tilde{q}$ a.e. on $\left(\frac{\pi}{2}, \pi\right)$, then $a = \tilde{a}$, $b = \tilde{b}$, $d = \tilde{d}$, $h = \tilde{h}$, and $q = \tilde{q}$ a.e. on $(0, \pi)$.

Gesztesy and Simon [10] investigated a case where the potential $q(x)$ on an interval $(0, a)$, $a > \frac{1}{2}$ is known. Then, the potential is uniquely determined by a fractional part of the spectrum.

Theorem 3 ([10]). *Let $\sigma(H)$ denote the spectrum of the operator $H = -\frac{d^2}{dx^2} + q$ in $L_2(0, 1)$ with the boundary conditions*

$$y'(0) + h_0 y(0) = 0, \quad y'(1) + h_1 y(1) = 0, \quad h_0, h_1 \in \mathbb{R}.$$

Then, q on $\left[0, \frac{\pi}{2} + \frac{\alpha}{2}\right]$ for some $\alpha \in (0, 1)$; h_0; and a subset $S \subseteq \sigma(H)$ satisfying

$$\#\{\lambda \in S \colon \lambda \leq \lambda_0\} \geq (1-\alpha)\#\{\lambda \in \sigma(H) \colon \lambda \leq \lambda_0\} + \frac{\alpha}{2} \tag{3}$$

for all sufficiently large $\lambda_0 \in \mathbb{R}$ uniquely determine h_1 and q on $[0, 1]$.

The uniqueness of recovering the Sturm-Liouville potential from parts of spectra described by conditions analogous to (3) was also investigated in [57–59].

Horváth [11] noticed that, to recover the potential $q(x)$, one can use eigenvalues of several spectra $\sigma_j = \sigma(q, \alpha_j, \beta)$, $j = 1, \ldots, N$ of the Sturm-Liouville problems

$$\left.\begin{aligned} -y'' + q(x)y = \lambda y, \quad x \in (0, \pi), \\ y(0)\cos\alpha_j + y'(0)\sin\alpha_j = 0, \quad y'(\pi)\cos\beta + y(\pi)\sin\beta = 0. \end{aligned}\right\} \tag{4}$$

The following main result of [11] generalized Theorem 3 of Gestezy and Simon.

Theorem 4 ([11]). *Suppose that $\lambda_n^{(j)} \in \sigma_j$ is known for $n \in S_j$ and let*

$$n_j(t) = \#\{n \in S_j \colon \lambda_n^{(j)} < t^2\}, \quad t \geq 0.$$

Let $0 \le a < \pi$, $0 \le \gamma \le 1$, *and suppose that there exist* $t_0 > 0$ *and* $\delta > 0$ *such that, for* $t \ge t_0$,

$$\sum_{j=1}^{N} n_j(t) \ge \begin{cases} 2(1-\frac{a}{\pi})\{\gamma[t+\frac{1}{2}] + (1-\gamma)([t]+\frac{1}{2})\} + O(t^{-\delta}), & \text{if } \sin\beta \ne 0, \\ 2(1-\frac{a}{\pi})\{\gamma[t+\frac{1}{2}] + (1-\gamma)([t]+\frac{1}{2})\} - 1 + O(t^{-\delta}), & \text{if } \sin\beta = 0. \end{cases}$$

Then, q *on* $(0,a)$ *and the eigenvalues* $\left\{\lambda_n^{(j)} : n \in S_j\right\}$, $j = 1,\dots,N$, *determine* q *a.e. on* $(0,\pi)$.

An analogous result was obtained in [11] for Dirac operators.

The disadvantage of Theorems 3 and 4 is that their conditions are sufficient but not necessary for the unique specification of the potential by part of the spectrum. In [12], Horváth obtained the necessary and sufficient conditions for the uniqueness of a solution for the following inverse problem in terms of closed exponential systems.

Problem 2 ([12]). *Given the eigenvalues* $\{\lambda_n\}_{n\ge1}$, *where each* λ_n *belongs to the spectrum* $\sigma(q,\alpha_n,\beta)$ *of the Sturm-Liouville problem* (4), *find the potential* q.

For definiteness, we provide the results of [12] for $\sin\beta = 0$.

Theorem 5 ([12]). *Let* $1 \le p \le \infty$, $q \in L_p(0,\pi)$, *and* $0 \le a < \pi$, *and let* $\lambda_n \in \sigma(q,\alpha_n,0)$ *be real numbers with* $\lambda_n \not\to -\infty$. *Then,* $\beta = 0$, q *on* $(0,a)$ *and the eigenvalues* λ_n *determine* q *in* $L_p(0,\pi)$, *and*

$$e(\Lambda) = \left\{e^{\pm 2i\mu x}, e^{\pm 2i\sqrt{\lambda_n}x} : n \ge 1\right\} \tag{5}$$

is closed in $L_p(a-\pi,\pi-a)$ *for* $\mu \ne \pm\sqrt{\lambda_n}$. *Note that, if the sequence* $e(\Lambda)$ *is closed for at least one* $\mu \ne \pm\sqrt{\lambda_n}$, *then it is closed for any such value of* μ.

Furthermore, in [12], Horváth noticed that Problem 2 was closely related to the reconstruction of the potential $q(x)$ from the values of the Weyl function at a countable set of points. Let $v(x,\lambda)$ denote the solution to the following initial value problem:

$$-v'' + q(x)v = \lambda v, \quad x \in (0,\pi), \quad v(\pi,\lambda) = \sin\beta, \quad v'(\pi,\lambda) = -\cos\beta.$$

Then, the Weyl function is defined as follows:

$$m_\beta(\lambda) = \frac{v'(0,\lambda)}{v(0,\lambda)}.$$

According to the classical results [1,6], the Weyl function $m_\beta(\lambda)$ uniquely specifies the potential $q(x)$. Horváth [12] obtained the following necessary and sufficient conditions for the uniqueness of the potential reconstruction using the values $\{m_\beta(\lambda_n)\}_{n\ge1}$.

Problem 3 ([12]). *Given the values* $\{m_\beta(\lambda_n)\}_{n\ge1}$, *find* q.

Theorem 6 ([12]). *Let* $1 \le p \le \infty$ *and* λ_n, $n \ge 1$ *be different arbitrary real numbers with* $\lambda_n \not\to -\infty$. *Let* $\beta = 0$, $q, \tilde{q} \in L_p(0,\pi)$ *and consider the Weyl functions* $m_0(\lambda)$, *and* $\tilde{m}_0(\lambda)$, *defined by* q *and* $\tilde{q}$, *respectively. Then, the relation*

$$m_0(\lambda_n) = \tilde{m}_0(\lambda_n), \quad n \ge 1 \tag{6}$$

implies that $m_0(\lambda) \equiv \tilde{m}_0(\lambda)$ *if and only if the system* $e(\Lambda)$ *defined by* (5) *is closed in* $L_p(-\pi,\pi)$.

Note that both sides of (6) are allowed to be infinite. Results analogous to Theorems 5 and 6 for the case $\sin\beta \ne 0$ can also be found in [12].

The results of Horváth [11,12] motivated the further study of Problem 2 and its analogs. In particular, Horváth and Kiss [60,61] investigated the stability of the problem. Horváth and Sáfár [62] obtained the necessary and sufficient conditions for the uniqueness of the potential reconstruction on a subinterval by some of the eigenvalues and norming constants.

Note that the Hochstadt–Lieberman problem and the abovementioned generalizations deal with cases in which the potential is known on the right-hand (left-hand) subinterval, as in Figure 1. Naturally, the following question arises: if the potential is known on either the middle subinterval, as in Figure 2, or the boundary subintervals, as in Figure 3, then is the potential on the remaining part of the interval uniquely specified by the spectrum or any other spectral data?

Figure 1. Hochstadt–Lieberman-type problems.

Figure 2. When $q(x)$ is known on the middle subinterval.

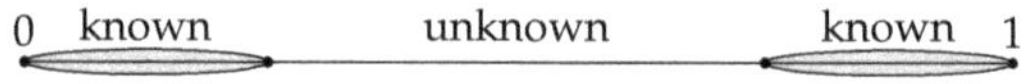

Figure 3. When $q(x)$ is unknown on the middle subinterval.

The question regarding the case in Figure 2 was answered by Guo and Wei [63]. Let us formulate their result. Let L denote the Sturm-Liouville operator $-y'' + q(x)y$ subject to the boundary conditions

$$y'(0) - hy(0) = 0, \quad y'(1) + Hy(1) = 0,$$

where $q \in L_1(0,1)$ is a real-valued function, and $h, H \in \mathbb{R}$. Let $\sigma(L) = \{\lambda_n\}_{n \ge 0}$ be the spectrum of L, and let $\psi(x,\lambda)$ be the solution of the Sturm-Liouville equation under the initial conditions $\psi(1,\lambda) = 1$, $\psi'(1,\lambda) = -H$. For a set $A = \{x_n\}_{n\ge 0}$ of positive reals, define

$$N_A(t) := \{n \in \mathbb{N} \cup \{0\} : x_n \le t\}.$$

Theorem 7 ([63])**.** *Let $[a_1, a_2] \subset [0,1]$ with $a_1 \le \frac{1}{2}$ and $a_1 + a_2 \ge 1$, where the two equalities do not occur simultaneously. Then, q on $[a_1, a_2]$ together with the subset S of $\sigma(L)$ and the interior spectral data $\dfrac{\psi(a_2, \lambda_n)}{\psi'(a_2, \lambda_n)}$ for $\lambda_n \in S'$, $S' \subset S$, where the subsets S and S' satisfy*

$$N_S(t) \ge 2a_1 N_\sigma(t) - a_1, \quad N_{S'}(t) \ge 2(1 - a_2) N_\sigma(t) + a_2 - 1$$

for all sufficiently large values of t, uniquely determine h, H, and q a.e. on $[0,1]$.

Thus, given the potential $q(x)$ on an interior subinterval $[a_1, a_2]$, some part of the spectrum together with additional spectral data related to the point a_2 uniquely specify the operator. It is interesting that, in Theorem 7, $[a_1, a_2]$ can be an arbitrarily small interval containing $\frac{1}{2}$.

The case in Figure 3, to the best of the author's knowledge, remains an open problem.

2.2. Solvability Conditions and Constructive Solution

The results of the previous subsection were concerned only with the uniqueness theorems. In this subsection, we consider constructive methods for solving the Hochstadt–Lieberman problem and the existence of its solution.

The first results in this direction were obtained by Sakhnovich [13]. He considered the Sturm-Liouville problem

$$-y'' + q(x)y = \lambda y, \quad x \in [0,1], \tag{7}$$
$$y(0) = y(1) = 0, \tag{8}$$

where the potential $q(x)$ is real-valued and continuous. Let $y(x, \lambda)$ denote the solution of Equation (7) satisfying the initial conditions $y(0, \lambda) = 0$, $y'(0, \lambda) = 1$. The main result of [13] was the following theorem, which provided sufficient conditions for solvability of the Hochstadt–Lieberman problem.

Theorem 8 ([13]). *Let the given functions $h(t)$, $t \in [0,1]$, and $p(x)$, $x \in [0, \frac{1}{2}]$ satisfy the following conditions:*
1. *The function $h(t)$ has a bounded derivative, and $h(0) = 0$.*
2. *The function $p(x)$ is bounded on the segment $[0, \frac{1}{2}]$.*
3. *The following inequality holds:*

$$\sup_{0 \le t \le 1} |h'(t)| + \frac{1}{4} \sup_{0 \le x \le \frac{1}{2}} |p(x)| < \frac{1}{2}. \tag{9}$$

Then, there exists a bounded function $q(x)$, $x \in [0,1]$, such that

$$q(x) = p(x), \quad x \in [0, \tfrac{1}{2}], \tag{10}$$

and the corresponding function $y(1, \lambda)$ has the form

$$y(1, \lambda) = \frac{\sin \sqrt{\lambda}}{\sqrt{\lambda}} + \int_0^1 \frac{\sin \sqrt{\lambda} t}{\sqrt{\lambda}} h(t)\, dt. \tag{11}$$

It is important to note that $y(1, \lambda)$ is the characteristic function of the problem (7)–(8), that is, the zeros of $y(1, \lambda)$ coincide with the eigenvalues $\{\lambda_n\}_{n \ge 1}$ of (7)–(8). Using the eigenvalues $\{\lambda_n\}_{n \ge 1}$, one can construct the function $y(1, \lambda)$ using Hadamard's factorization theorem and then find the function $h(t)$ satisfying (11) using the Fourier transform. Thus, Theorem 8 provides the sufficient conditions for the existence of the potential $q(x)$ that satisfies (10) and has the given spectrum $\{\lambda_n\}_{n \ge 1}$.

Theorem 8 is proved by a constructive method that finds the potential $q(x)$ from $p(x)$ and $h(x)$ via approximations. For the convergence of these approximations, the inequality (9) is crucial. Thus, the result of [13] has a local nature.

The necessary and sufficient conditions for the Hochstadt–Lieberman problem's solvability, to the best of the author's knowledge, were obtained for the first time by Hryniv and Mykytyuk [14]. They considered the Sturm-Liouville equation (Equation (1)) with the potential q of class $W_2^{-1}(0,1)$. In this case, it is convenient to write the Sturm-Liouville differential expression $-y'' + q(x)y$ as $\ell_\sigma(y) = -(y^{[1]})' - \sigma y'$, where $q = \sigma'$, $\sigma \in L_2(0,1)$, and $y^{[1]} := y' - \sigma y$ is the quasi-derivative. Let us use the notation $\operatorname{Re} L_2(0,a)$ for the class of real-valued functions of $L_2(0,a)$, $a > 0$. For $\sigma \in \operatorname{Re} L_2(0,1)$ and $h \in \mathbb{R}$, let $T_{\sigma,h}$ denote the operator in $L_2(0,1)$ that acts as $T_{\sigma,h} y = \ell_\sigma(y)$ on the domain

$$\operatorname{dom} T_{\sigma,h} = \{y \in W_1^1(0,1) \colon y^{[1]} \in W_1^1(0,1),\ \ell_\sigma(y) \in L_2(0,1),\ y^{[1]}(0) = 0,\ y^{[1]}(1) = hy(1)\}.$$

The operator $T_{\sigma,h}$ is self-adjoint, and its spectrum is a countable set of real simple eigenvalues $\{\lambda_n^2\}_{n \ge 0}$ satisfying $\{(\lambda_n - n)\}_{n \ge 0} \in l_2$. Using the shift $\sigma(x) := \sigma(x) + cx$, $h := h - c$, one can achieve the positivity $\lambda_n > 0$, $n \ge 0$. In [14], the following analog of the Hochstadt–Lieberman problem was considered:

Problem 4 ([14]). *Given a function $\sigma_0(x)$, $x \in (0, \frac{1}{2})$ and reals $\{\lambda_n\}_{n\geq 0}$, find a function $\sigma \in \mathrm{Re}\, L_2(0,1)$ and $h \in \mathbb{R}$ such that $\sigma(x) = \sigma_0(x)$ a.e. on $(0, \frac{1}{2})$ and the spectrum of $T_{\sigma,h}$ coincides with $\{\lambda_n^2\}_{n\geq 0}$.*

In order to formulate the results of [14], one needs some additional definitions. Let $\mathfrak{L}$ denote the set of all strictly increasing sequences $\Lambda = \{\lambda_n\}_{n\geq 0}$, in which $\lambda_n > 0$ and $\{(\lambda_n - \pi n)\}_{n\geq 0} \in l_2$. Let us fix an arbitrary $\Lambda = \{\lambda_n\} \in \mathfrak{L}$ and denote by Π_Λ the set of all real-valued functions $\psi \in L_2(0,1)$ of the form

$$\psi(x) = \sum_{n=0}^{\infty} (\alpha_n \cos(\lambda_n x) - \cos(\pi n x)) + \frac{1}{2}, \tag{12}$$

where $\{\alpha_n\}_{n\geq 0}$ is a sequence of positive numbers such that $\{(\alpha_n - 1)\}_{n\geq 0} \in l_2$.

For $\sigma_0 \in \mathrm{Re}\, L_2(0, \frac{1}{2})$, let $y_0(x,\lambda)$ denote the solution of the initial value problem

$$\ell_{\sigma_0}(y_0) = \lambda^2 y_0, \quad x \in (0, \tfrac{1}{2}), \quad y_0(0,\lambda) = 1, \quad y_0^{[1]}(0,\lambda) = 0.$$

Let $K(x,t)$ be the transformation operator kernel (see the details in [14]):

$$\cos \lambda x = y_0(x,\lambda) + \int_0^x K(x,t) y_0(t,\lambda)\, dt.$$

The necessary and sufficient conditions for the solvability of Problem 4 are provided by the following theorem:

Theorem 9 ([14]). *Assume that $\Lambda = \{\lambda_n\}_{n\geq 0} \in \mathfrak{L}$, $\sigma_0 \in \mathrm{Re}\, L_2(0, \frac{1}{2})$ and*

$$\phi_0(2x) := -\frac{1}{2}\sigma_0(x) + \int_0^x K^2(x,t)\, dt, \quad x \in (0, \tfrac{1}{2}).$$

1. *Problem 4 is solvable for the mixed spectral data (σ_0, Λ) if and only if $\phi_0 \in \Pi_\Lambda$.*
2. *If $\phi_0 \in \Pi_\Lambda$, then the solution of Problem 4 is unique, that is, there exists a unique $\sigma \in \mathrm{Re}\, L_2(0,1)$ and a unique $h \in \mathbb{R}$ such that σ is an extension of σ_0 and the spectrum of $T_{\sigma,h}$ coincides with $\Lambda^2 = \{\lambda_n^2\}_{n\geq 0}$.*

As corollaries of Theorem 9, Hryniv and Mykytuyk [14] also obtained some results for the case of the regular potential $q \in L_2(0,1)$. The proof of Theorem 9 is based on the transformation operator method (see [1,2]). Note that the numbers α_n in Expansion (12) for $\phi_0(x)$ equal the weight numbers $\|y_n\|^{-2}_{L_2(0,1)}$, where $\{y_n(x)\}_{n\geq 0}$ are the eigenfunctions of the operator $T_{\sigma,h}$ corresponding to the eigenvalues $\{\lambda_n^2\}_{n\geq 0}$. Hence, the requirement $\phi_0 \in \Pi_\Lambda$ means that the weight numbers are positive and have the asymptotics $\{(\alpha_n - 1)\}_{n\geq 0} \in l_2$, which is valid by necessity. Roughly speaking, the method of [14] reduced Problem 4 to the classical inverse problem using the spectral data $\{\lambda_n^2, \alpha_n\}_{n\geq 0}$ and required the necessary and sufficient conditions for the solvability of the latter problem. Such conditions in the case of a singular potential of class $W_2^{-1}(0,1)$ were obtained in [64].

Theorem 10 ([64]). *For the numbers $\{\lambda_n^2, \alpha_n\}_{n\geq 0}$ to be the spectral data of a positive Sturm-Liouville operator $T_{\sigma,h}$ with $\sigma \in \mathrm{Re}\, L_2(0,1)$ and $h \in \mathbb{R}$, it is necessary and sufficient that $\Lambda = \{\lambda_n\}_{n\geq 0} \in \mathfrak{L}$, $\alpha_n > 0$ for $n \geq 0$ and $\{(\alpha_n - 1)\}_{n\geq 0} \in l_2$.*

An alternative approach to the solution of the Hochstadt–Lieberman problem was developed in parallel by Buterin [15,16] and by Martinyuk and Pivovarchik [17]. In [15], Buterin considered the Sturm-Liouville problem

$$-y'' + q(x)y = \lambda y, \quad x \in (0,\pi), \tag{13}$$

$$y'(0) - hy(0) = 0, \quad y'(\pi) + Hy(\pi) = 0, \tag{14}$$

where $q \in L_1(0,\pi)$.

Let $S(x,\lambda)$, $\varphi(x,\lambda)$, and $\psi(x,\lambda)$ denote, respectively, the solutions of Equation (13) satisfying the initial conditions

$$S(0,\lambda)=0,\quad S'(0,\lambda)=1,\quad \varphi(0,\lambda)=1,\quad \varphi'(0,\lambda)=h,\quad \psi(\pi,\lambda)=1,\quad \psi'(\pi,\lambda)=-H.$$

Then, the eigenvalues $\{\lambda_n\}_{n\geq 0}$ of the boundary value problem (13)–(14) coincide with the zeros of the characteristic function $\Delta(\lambda)=\varphi'(\pi,\lambda)+H\varphi(\pi,\lambda)$. Thus,

$$\begin{gathered}\Delta^0(\lambda)=\psi(0,\lambda),\quad \Delta_1(\lambda)=-\psi'(\pi/2,\lambda),\quad \Delta_1^0(\lambda)=\psi(\pi/2,\lambda),\\ \Theta(\lambda)=\varphi(\pi/2,\lambda),\quad \Xi(\lambda)=\varphi'(\pi/2,\lambda).\end{gathered} \tag{15}$$

The main idea of [15] consists in the fact that, if the potential $q(x)$ on $(0,\pi/2)$ is known, then the functions $\Theta(\lambda)$ and $\Xi(\lambda)$ can be found. Thus, using the following relations between the characteristic functions,

$$\begin{aligned}\Delta_1^0(\lambda)&=\Delta^0(\lambda)\Theta(\lambda)-\Delta(\lambda)S(\pi/2,\lambda),\\ -\Delta_1(\lambda)&=\Delta^0(\lambda)\Xi(\lambda)-\Delta(\lambda)S'(\pi,\lambda),\end{aligned}$$

one can find $\Delta_1^0(\lambda)$ and $\Delta_1(\lambda)$ by the interpolation of entire functions and recover the potential $q(x)$ and the coefficient H from the Weyl function $M(\lambda)=-\dfrac{\Delta_1(\lambda)}{\Delta_1^0(\lambda)}$ for the interval $(\pi/2,\pi)$. Indeed, the zeros of the functions $\Delta_1(\lambda)$ and $\Delta_1^0(\lambda)$ are the two spectra of the Borg problem on this interval.

Let $\{\xi_n\}_{n\geq 0}$ and $\{\theta_n\}_{n\geq 0}$ denote the zeros of the entire functions $\Xi(\lambda)$ and $\Theta(\lambda)$, respectively. If these zeros are simple, the Hochstadt–Lieberman problem can be solved by the following constructive algorithm:

Method 1 ([15]). *Suppose that the spectrum $\{\lambda_n\}_{n\geq 0}$, the coefficient h, and the potential $q(x)$ on the interval $(0,\pi/2)$ are given. We must find $q(x)$ on $(\pi/2,\pi)$ and H.*

1. *Find $\Delta(\lambda)$ by the formula*

$$\Delta(\lambda)=\pi(\lambda_0-\lambda)\prod_{n=1}^{\infty}\frac{\lambda_n-\lambda}{n^2}.$$

2. *Construct the functions $\Theta(\lambda)$ and $\Xi(\lambda)$ using* (15) *and find their zeros θ_n, ξ_n, $n\geq 0$.*
3. *Calculate the numbers*

$$\begin{aligned}d(\xi_n)&=\Delta(\xi_n)S'(\pi/2,\xi_n)+\sqrt{\xi_n}\sin(\sqrt{\xi_n}\pi/2),\\ d_0(\theta_n)&=-\Delta(\theta_n)S(\pi/2,\theta_n)-\cos(\sqrt{\theta_n}\pi/2).\end{aligned}$$

4. *By interpolation, find the functions*

$$d(\lambda)=\sum_{n=0}^{\infty}d(\xi_n)\frac{\Xi(\lambda)}{(\lambda-\xi_n)\Xi'(\xi_n)},\quad d_0(\lambda)=\sum_{n=0}^{\infty}d_0(\theta_n)\frac{\Theta(\lambda)}{(\lambda-\theta_n)\Theta'(\theta_n)}.$$

5. *Let $\Delta_1(\lambda)=-\sqrt{\lambda}\sin(\sqrt{\lambda}\pi/2)+d(\lambda)$, $\Delta_1^0(\lambda)=\cos(\sqrt{\lambda}\pi/2)+d_1(\lambda)$.*
6. *Recover $q(x)$ on $(\pi/2,\pi)$ and H from the Weyl function $M(\lambda)=-\dfrac{\Delta_1^0(\lambda)}{\Delta_1(\lambda)}$.*

Note that Method 1 does not require the self-adjointness of the problem (13)–(14) and so works for complex-valued potentials $q(x)$, h, and H. The last step of Method 1 in the non-self-adjoint case can be implemented by an algorithm presented in [65]. Method 1 is also valid for multiple eigenvalues $\{\xi_n\}$ and $\{\theta_n\}$ after minor modifications.

In [16], Buterin generalized Method 1 to quadratic differential pencils of the form

$$y'' + (\rho^2 - 2\rho q_1(x) - q_0(x))y = 0, \quad x \in (0, \pi),$$
$$y'(0) - (h_1\rho + h_0)y(0) = 0, \quad y'(\pi) + (H_1\rho + H_0)y(\pi) = 0,$$

where ρ is the spectral parameter; $q_j(x) \in W_1^j[0,1]$ are complex-valued functions; $h_j, H_j \in \mathbb{C}$; $j = 0, 1$; $h_1 \neq \pm i$; and $H_1 \neq \pm i$. The half-inverse problem of [16] consists in the recovery of the coefficients q_0, q_1, H_0, and H_1 from the spectrum $\{\rho_n\}$, while q_0 and q_1 on $(0, \pi/2)$, h_0, and h_1 are known a priori. A similar problem for the quadratic differential pencil with another type of boundary conditions was investigated in [66].

An analogous approach was used by Martinyuk and Pivovarchik [17] to obtain the necessary and sufficient conditions of the Hochstadt–Lieberman problem's solvability. They considered the Sturm-Liouville problem

$$-y'' + q(x)y = \lambda y, \quad x \in (0, a), \quad y(0) = y(a) = 0$$

in the following equivalent form:

$$-y_j'' + q_j(x)y_j = \lambda^2 y_j, \quad x \in [0, a/2], \quad j = 1, 2, \tag{16}$$
$$y_j(0) = 0, \quad j = 1, 2, \quad y_1(a/2) = y_2(a/2), \quad y_1'(a/2) + y_2'(a/2) = 0. \tag{17}$$

The boundary value problem (16)–(17) can be treated as the Sturm-Liouville problem on a two-edge star-shaped graph with the standard matching conditions in the interior vertex (see Figure 4). In the Hochstadt–Lieberman problem, the potential q_1 on the first edge is known, and the potential q_2 on the second edge must be recovered from the spectrum $\{\lambda_k\}_{k \in \mathbb{Z}_0}$ of the boundary value problem (16)–(17), $\mathbb{Z}_0 := \mathbb{Z} \setminus \{0\}$.

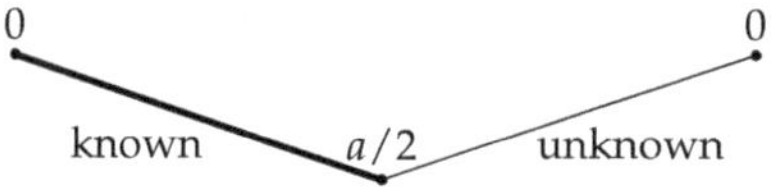

Figure 4. Two-edge graph.

The authors of [17] assumed that $q_j \in L_2(0, a/2)$ and denoted by $s_j(\lambda, x)$, $j = 1, 2$ the solutions of the corresponding Equation (16) satisfying the initial conditions $s_j(\lambda, 0) = 0$, $s_j'(\lambda, 0) = 1$.

Theorem 11 ([17]). *Let a real-valued function $q_1 \in L_2(0, a/2)$ be given, together with a set $\{\lambda_k\}_{k \in \mathbb{Z}_0}$ of numbers that satisfy the conditions:*

1. $\lambda_k = -\lambda_k$,
2. $-\infty < \lambda_1^2 < \lambda_2^2 < \cdots < \lambda_k^2 < \ldots$,
3. $\lambda_k = \dfrac{\pi k}{a} + \dfrac{K}{\pi k} + \dfrac{\beta_k}{k}$.

   Here, $K \in \mathbb{R}$, and $\{\beta_k\}_{k \in \mathbb{Z}_0} \in l_2$.

   If the function $\dfrac{s_2(\sqrt{\lambda}, a/2)}{s_2'(\sqrt{\lambda}, a/2)}$ belongs to the Nevanlinna class, then there exists a real-valued function $q_2(x) \in L_2[0, a/2]$ such that the spectrum of the problems (16) *and* (17) *generated by q_1 and q_2 coincides with $\{\lambda_k\}_{k \in \mathbb{Z}_0}$.*

It is supposed in Theorem 11 that the functions $s_2(\lambda, a/2)$ and $s_2'(\lambda, a/2)$ are recovered from q_1 and $\{\lambda_k\}_{k \in \mathbb{Z}_0}$ by a procedure analogous to Method 1. Note that the conditions of Theorem 11 are not only sufficient but also necessary. Indeed, the conditions 1–3 are the standard properties of Sturm-Liouville eigenvalues, and the function $\dfrac{s_2(\sqrt{\lambda}, a/2)}{s_2'(\sqrt{\lambda}, a/2)}$ is the Weyl function of the Sturm-Liouville problem on the second edge, which belongs to the Nevanlinna class by necessity. In fact, Martinyuk and Pivovarchik [17] reduced the Hochstadt–Lieberman problem to the classical inverse problem using the Weyl function on a subinterval corresponding to the second edge and then imposed an additional requirement of belonging to the Nevanlinna class. Analogous results for the Robin boundary conditions were obtained in [18].

Thus, both Theorems 9 and 11 pertaining to the necessary and sufficient conditions proposed by Hryniv and Mykytyuk [14] and by Martinyuk and Pivovarchik [17], respectively, contain some a posteriori conditions, which have to be checked after the implementation of several steps of a constructive procedure for solution. It seems that such conditions are unavoidable for Hochstadt–Lieberman-type problems.

Additionally, numerical techniques for solving the Hochstadt–Lieberman problem were developed by Rundell and Sacks [67]. An overview of some other work on the Hochstadt-Leiberman problems on an interval can be found in [68].

2.3. *McLaughlin–Polyakov Problem*

In this section, we consider the so-called transmission eigenvalue problem

$$-y'' + q(x)y = \lambda y, \quad x \in (0,1), \tag{18}$$

$$y(0) = 0, \quad y(1)\cos\rho a - y'(1)\frac{\sin\rho a}{\rho} = 0, \quad \rho = \sqrt{\lambda}, \tag{19}$$

where q is a real-valued potential of $L_2(0,1)$ and $a \geq 0$. The boundary condition at $x = 1$ has a non-linear and even non-polynomial dependence on the spectral parameter λ.

The problem (18)–(19) arise in connection with the investigation of the acoustic inverse scattering problem in a non-homogeneous medium (see [19]). *Transmission eigenvalues* are the eigenvalues k^2 of the boundary value problem

$$\begin{array}{ll} \Delta u + k^2 n(x) u = 0, & x \in B_1, \\ \Delta v + k^2 v = 0, & x \in B_1, \\ u(x) = v(x), & x \in \partial B_1, \\ \frac{\partial}{\partial r}(u(x) - v(x)) = 0, & x \in \partial B_1, \end{array} \tag{20}$$

where B_1 is the ball in $\mathbb{R}^3$ of radius 1 centered at the origin, ∂B_1 is its boundary, $n(x) > 0$ is the refractive index, and $\frac{\partial}{\partial r}$ is the normal derivative. The inverse transmission eigenvalue problems consist in the recovery of the function $n(x)$ (related to the speed of sound in acoustics) from the transmission eigenvalues. In spherically symmetric cases, the problem (20) can be reduced to the one-dimensional form (18)–(19) using the separation of variables and subsequent transforms (see [19]).

Difficulties in the investigation of the problem (18)–(19) are caused by the non-regularity of its boundary conditions in the Birkhoff and Stone senses (see [69]). Therefore, the transmission problem involves more complex spectral behavior than the classical Sturm-Liouville problems.

McLaughlin and Polyakov [19] showed that, for $a \neq 1$, the transmission eigenvalue problem has a subspectrum $\{\lambda_n\}_{n\geq 1}$ with the asymptotics

$$\sqrt{\lambda_n} = \frac{\pi n}{1-a} + \frac{\omega_0}{\pi n} + \frac{\varkappa_n}{n}, \quad \omega_0 := \frac{1}{2}\int_0^1 q(x)dx, \quad n \geq 1, \tag{21}$$

Furthermore, $\lambda_n \in \mathbb{R}$ for a sufficiently large n. Buterin and Yang [70] suggested that an eigenvalue sequence $\{\lambda_n\}_{n\geq 1}$ possessing these properties should be called *an almost real subspectrum*. Note that a finite number of the first eigenvalues in an almost real subspectrum may be complex and/or multiple. Since the potential $q(x)$ is real-valued, then, without a loss of generality, we can assume that an almost real subspectrum is symmetrical with respect to the real axis, that is, values λ and $\overline{\lambda}$ are only contained in the sequence

$\{\lambda_n\}_{n\ge1}$ simultaneously and have the same multiplicity. An almost real subspectrum can be non-unique. The results of this section are valid for any almost real subspectrum.

In [19], an investigation of the following inverse transmission eigenvalue problem was initiated.

Problem 5 ([19]). *Given an almost real subspectrum $\{\lambda_n\}_{n\ge1}$ and the potential $q(x)$ on the interval $(\alpha, 1)$, where $\alpha := \min\{|a-1|/2, 1\}$, find $q(x)$ on $(0, \alpha)$.*

We call Problem 5 *the McLaughlin–Polyakov problem*. Obviously, if $a = 0$, then the McLauglin-Polyakov problem coincides with the Hochstadt–Lieberman problem, and an almost real subspectrum coincides with the whole spectrum. McLaughin and Polyakov [19] proved the uniqueness theorem for the solution of Problem 5.

Theorem 12 ([19]). *Suppose that $a \ge 0$, $a \ne 1$. If $\lambda_n = \tilde\lambda_n$ for $n \ge 1$ and $q(x) = \tilde q(x)$ a.e. on $(\alpha, 1)$, then $q(x) = \tilde q(x)$ a.e. on $(0, \alpha)$.*

Note that, for $a \ge 3$, an almost real subspectrum uniquely specifies the potential on the whole interval $(0, 1)$, and some part of an almost real subspectrum is sufficient for $a > 3$. The investigation of Problem 5 was continued by McLauglin et al. in [20] for $a \ge 3$ and in [21] for $a \in (0,1) \cup (1,3)$. The authors of [20,21] developed numerical methods for reconstructing the potential based on the ideas of Rundell and Sacks [67]. However, they did not study the existence and stability of the solution.

In [22], Bondarenko and Buterin proved the following theorem on the local solvability and stability of the McLaughlin–Polyakov problem:

Theorem 13 ([22]). *Fix $a \in [0,1) \cup (1,3]$. For any real-valued potential $q \in L_2(0,1)$, there exists $\delta > 0$ such that, for any sequence $\{\tilde\lambda_n\}_{n\ge1}$ symmetric with respect to the real axis and an arbitrary real-valued function $q_1 \in L_2(\alpha, 1)$ satisfying*

$$\int_\alpha^1 q_1(x)\,dx = \int_\alpha^1 q(x)\,dx,$$

the closeness

$$\Lambda := \sqrt{\sum_{n=1}^{\infty} |\lambda_n - \tilde\lambda_n|^2} \le \delta, \quad Q := \|q - q_1\|_{L_2(\alpha,1)} \le \delta$$

implies the existence of a unique function $\tilde q(x) \in L_2(0,1)$ such that $\tilde q(x) = q_1(x)$ a.e. on $(\alpha, 1)$ and $\{\tilde\lambda_n\}_{n\ge1}$ is an almost real subspectrum of the boundary value problems (18) *and* (19) *with $q(x)$ replaced by $\tilde q(x)$. Moreover, the following estimate holds:*

$$\|q - \tilde q\|_{L_2(0,\alpha)} \le C(\Lambda + Q),$$

where C does not depend on $\{\tilde\lambda_n\}_{n\ge1}$ and $q_1(x)$.

Theorem 13 represents the first existence result for the inverse transmission eigenvalue problem. Furthermore, for $a = 0$, it provides the first full stability result for the Hochstadt–Lieberman problem, in which perturbations of both the spectrum and the potential on $(1/2, 1)$ are taken into account. In addition, Theorem 13 implies the minimality of the McLaughlin–Polyakov data in the case $a \in [0,1) \cup (1,3]$. For $a > 3$, the stability does not hold, since Problem 5 is overdetermined. The proof of Theorem 13 is constructive. Later on, by relying on the ideas of [22], a unified approach to partial inverse problems was developed (see Section 4).

In [23], the methods of Bondarenko and Buterin [22] were used to obtain further stability results for Problem 5. It is worth mentioning that inverse transmission eigenvalue problems were studied using statements other than the McLaughlin–Polyakov statement in [69–74] and other papers.

3. Partial Inverse Problems on Graphs

In this section, we consider generalizations of the Hochstadt–Lieberman problem on metric graphs. We treat the boundary value problems on graphs as differential systems. The geometrical graph structure is used only for defining the matching conditions. In interior graph vertices, the problems in this section mostly feature the standard matching conditions, which, from a physical viewpoint, express Kirchoff's law in electrical circuits, the balance of tension in elastic string networks, etc.

3.1. Star-Shaped Graphs

The majority of results on partial inverse problems for metric graphs have been obtained for star-shaped graphs. We start with the complete inverse problem statement for the Sturm-Liouville equations on such a graph.

Let G be a star-shaped graph containing $m \geq 3$ edges $\{e_j\}_{j=1}^m$ of equal lengths π. Each edge e_j joins the internal vertex v_0 with the boundary vertex v_j. For each edge e_j, we introduce the parameter $x_j \in [0, \pi]$. The value $x_j = 0$ corresponds to the boundary vertex v_j, and the value $x_j = \pi$ corresponds to the internal vertex v_0 (see Figure 5).

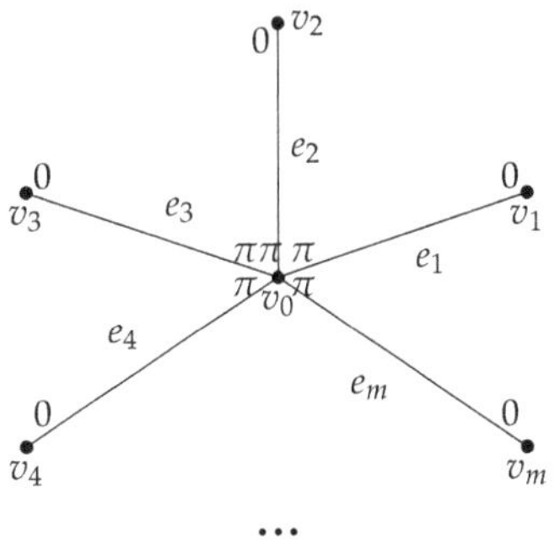

Figure 5. Star-shaped graph.

On the graph G, consider the Sturm-Liouville equations

$$-y_j''(x_j) + q_j(x_j)y_j(x_j) = \lambda y_j(x_j), \quad x_j \in (0, \pi), \quad j = \overline{1, m}, \tag{22}$$

with real-valued potentials $q_j \in L_2(0, \pi)$, $j = \overline{1, m}$, and the standard matching conditions at the internal vertex:

$$y_1(\pi) = y_j(\pi), \quad j = \overline{2, m}, \qquad \sum_{j=1}^{m} y_j'(\pi) = 0. \tag{23}$$

Let Λ and Λ_k, $k = \overline{1, m}$, denote the spectra of the boundary value problems L and L_k, $k = \overline{1, m}$, respectively, for Equation (22), subject to the matching conditions (23) and the following boundary conditions:

$$\begin{aligned} L: &\quad y_j(0) = 0, \quad j = \overline{1, m}, \\ L_k: &\quad y_k'(0) = 0, \quad y_j(0) = 0, \quad j = \overline{1, m} \setminus k. \end{aligned} \tag{24}$$

The spectra Λ and Λ_k, $k = \overline{1, m}$, are countable sets of real eigenvalues.

The complete inverse Sturm-Liouville problem on graph G is formulated as follows:

Problem 6 ([75]). *Given the spectra Λ and Λ_k, $k = \overline{1, m-1}$, find the potentials $\{q_j\}_{j=1}^m$.*

Problem 6 is a special case of the well-studied inverse spectral problems for Sturm-Liouville operators on trees (see [27,75]). In [75], the uniqueness of the inverse problem solution was proved, and a constructive algorithm for its solution based on the method of spectral mappings [4] was developed. Note that, for the recovery of the potentials on the whole graph, a sufficiently large amount of spectral data must be used (m spectra). To the best of the author's knowledge, the minimality of these data is an open question. In addition, the following question arises:

Can the amount of spectral data for reconstruction be reduced if the potentials on some edges are given a priori?

The first partial inverse problems on graphs were considered by Pivovarchik [28]. He studied the Sturm-Liouville problem (22)–(24) on a three-edge star graph ($m = 3$) with real-valued non-negative potentials $q_j \in L_2(0,\pi)$, $j = 1,2,3$. In addition, for $j = 1,2,3$, the spectrum of the Sturm-Liouville Equation (22) on the edge e_j subject to the boundary conditions $y_j(0) = y_j(\pi) = 0$ is denoted by $\mathfrak{S}_j$. The main results of [28] were concerned with the following inverse problem:

Problem 7 ([28]). *Given the spectra Λ and $\mathfrak{S}_j$, $j = 1,2,3$, find q_j for $j = 1,2,3$.*

A disadvantage of Problem 7 is that the spectra $\mathfrak{S}_j$, which carry information not from the whole graph but from the separate edges, are used for reconstruction. Nevertheless, as a corollary of the main results, the uniqueness of the solution was proved in [28] for the following partial inverse problem:

Problem 8 ([28]). *Given the potentials q_1 and q_2 and the spectrum Λ, find q_3.*

In fact, Problem 8 is overdetermined. Yang [30] showed that, for the unique recovery of the potential on one edge, the fractional part $\frac{2}{m}$ of the spectrum is sufficient if the potentials on all the other edges are supposed to be known. In [30], a Sturm-Liouville problem on a star-shaped graph G with general boundary conditions was considered:

$$y_j(0,\lambda)\cos\alpha_j + y_j'(0,\lambda)\sin\alpha_j = 0, \quad \alpha_j \in [0,\pi), \quad j = \overline{1,m}.$$

For simplicity, we formulate the results of Yang [30] for the Dirichlet boundary conditions (24).

Consider the Sturm-Liouville problem L presented by (22)–(24) with real-valued potentials of class $L_1(0,\pi)$. For each $j = \overline{1,m}$, let $S_j(x_j,\lambda)$ denote the solution of Equation (22) satisfying the initial conditions $S_j(0,\lambda) = 0$, $S_j'(0,\lambda) = 1$. The eigenvalues of L coincide with the zeros of the characteristic function

$$\Delta(\lambda) := \sum_{j=1}^{m} S_j'(\pi,\lambda) \prod_{\substack{k=1 \\ k\neq j}}^{m} S_k(\pi,\lambda) \tag{25}$$

and can be denoted as $\{\lambda_{nk}\}_{n\geq 1, k=\overline{1,m}}$ (counting with multiplicities), so that the following asymptotic relations hold:

$$\sqrt{\lambda_{n1}} = n - \frac{1}{2} + O\left(n^{-1}\right), \tag{26}$$

$$\sqrt{\lambda_{nk}} = n + O\left(n^{-1}\right), \quad k = \overline{2,m}. \tag{27}$$

The partial inverse problem of [30] is formulated as follows:

Problem 9 ([30]). *Suppose that the potentials $\{q_j\}_{j=2}^{m}$ are known a priori. Given a subspectrum $\Omega := \{\lambda_{nk}\}_{n\geq 1, k=1,2}$, find q_1.*

In view of symmetry, one can replace the potential q_1 with an arbitrary q_j, $j = \overline{2,m}$, and the eigenvalues $\{\lambda_{n2}\}_{n\geq 1}$ with a sequence $\{\lambda_{nk}\}_{n\geq 1}$ containing an arbitrary fixed $k = \overline{3,m}$ in the problem statement. Note that the subspectrum Ω can contain a finite number of multiple eigenvalues. Furthermore, Ω is not uniquely determined by the asymptotics (26) and (27), so any suitable subspectrum can be considered. Obviously, in the case $m = 2$, Ω is the whole spectrum, and Problem 9 turns into the Hochstadt–Lieberman problem.

In [30], the following uniqueness theorem for Problem 9 was proved:

Theorem 14 ([30]). *Let $\Omega = \{\lambda_{nk}\}_{n\geq 1, k=1,2}$ be a subspectrum of problem L satisfying the asymptotics* (26)–(27) *and the condition $\Omega \cap \mathfrak{S}_j = \varnothing$ for $j = \overline{2,m}$. If $q_j(x) = \tilde{q}_j(x)$ a.e. on $(0,\pi)$ for $j = \overline{2,m}$ and $\Omega = \tilde{\Omega}$, then $q_1(x) = \tilde{q}_1(x)$ a.e. on $(0,\pi)$.*

The condition $\Omega \cap \mathfrak{S}_j = \varnothing$ is crucial for the uniqueness. Suppose that this condition is violated, that is, there exist $j \in \{2,\dots,m\}$ and λ_0 such that $\lambda_0 \in \Omega \cap \mathfrak{S}_j$. Obviously, $\lambda_0 \in \mathfrak{S}_j$ implies $S_j(\pi,\lambda_0) = 0$. Taking (25) into account, we conclude that $S_i(\pi,\lambda_0) = 0$, and so $\lambda_0 \in \mathfrak{S}_i$ for some $i \neq j$. Thus, λ_0 is the eigenvalue of the two Dirichlet problems for separate edges e_i and e_j. If $i \neq 1$, this eigenvalue carries no information about the potential q_1. In [42], the validity of Theorem 14 was proved for complex-valued potentials $\{q_j\}_{j=1}^m$ and the condition $\Omega \cap \mathfrak{S}_j = \varnothing$, $j = \overline{2,m}$, replaced by the following less restrictive condition:

Condition 1. *For every $\lambda_{nk} \in \Omega$, there do not exist indices i and j such that $2 \leq i,j \leq m$, $i \neq j$ and $S_i(\pi,\lambda_{nk}) = S_j(\pi,\lambda_{nk}) = 0$.*

It is worth mentioning the paper by Yurko [29] in which uniqueness was studied for the following partial inverse problem.

Problem 10 ([29]). *Suppose that $\{q_j\}_{j=2}^m$ are known a priori and q_1 is known on the subinterval (b,π), where $b < \pi$. Given part of the spectrum Λ of the problem* (22)–(24)*, find q_1 on $(0,b)$.*

The investigation of Problem 9 was continued by Bondarenko [33]. In [33], a constructive algorithm for solution was developed, and the local solvability and stability were proved. In order to formulate the results of [33], we needed the following precise eigenvalue asymptotics:

Theorem 15 ([76]). *The eigenvalues $\{\lambda_{nk}\}_{n\geq 1, k=\overline{1,m}}$ (counting with multiplicities) of the boundary value problem L with real-valued potentials $q_j \in L_2(0,\pi)$, $j = \overline{1,m}$, can be numbered so that*

$$\sqrt{\lambda_{n1}} = n - \frac{1}{2} + \frac{\hat{\omega}}{\pi n} + \frac{\varkappa_{n1}}{n}, \tag{28}$$

$$\sqrt{\lambda_{nk}} = n + \frac{z_{k-1}}{\pi n} + \frac{\varkappa_{nk}}{n}, \quad k = \overline{2,m}, \tag{29}$$

where $\{\varkappa_{nk}\}_{n\in\mathbb{N}} \in l_2$, $k = \overline{1,m}$, $\hat{\omega} = \frac{1}{m}\sum\limits_{j=1}^{m} \omega_j$, $\omega_j = \frac{1}{2}\int\limits_0^{\pi} q_j(x)\,dx$ and $\{z_k\}_{k=1}^{m-1}$ are the roots of the characteristic polynomial

$$P(z) = \frac{d}{dz}\prod_{k=1}^{m}(z - \omega_k).$$

In [33], the following theorem regarding the local solvability and stability of Problem 9 was proved.

Theorem 16 ([33]). *Suppose that the boundary value problem L of the forms* (22) *and* (24) *with potentials $q_j \in L_2(0,\pi)$, $j = \overline{1,m}$ and its subspectrum $\{\lambda_{nk}\}_{n\geq 1, k=\overline{1,m}}$ satisfy the following assumptions:*

1. *All the eigenvalues* $\{\lambda_{nk}\}_{n\in\mathbb{N},\,k=1,2}$ *are distinct;*
2. $\lambda_{nk} > 0$, $n \in \mathbb{N}$, $k = 1,2$;
3. $S_j(\pi, \lambda_{nk}) \neq 0$, $j = \overline{1,m}$, $n \in \mathbb{N}$, $k = 1,2$;
4. $z_1 \neq \omega_j$, $j = \overline{1,m}$;
5. $S_1(\pi, 0) \neq 0$, $S_1'(\pi, 0) \neq 0$.

Then, there exists $\varepsilon_0 > 0$ *such that, for arbitrary real numbers* $\{\tilde{\lambda}_{nk}\}_{n\in\mathbb{N},\,k=1,2}$ *satisfying*

$$\left(\sum_{n=1}^{\infty} \sum_{k=1,2} \left(n(\lambda_{nk}^{1/2} - \tilde{\lambda}_{nk}^{1/2}) \right)^2 \right)^{1/2} < \varepsilon, \quad \varepsilon \le \varepsilon_0,$$

there exists a unique real function $q_1 \in L_2(0,\pi)$, *which is the solution of Problem 9 for* $\{\tilde{\lambda}_{nk}\}_{n\in\mathbb{N},\,k=1,2}$ *and* q_j, $j = \overline{2,m}$. *Moreover, the following estimate holds:*

$$\|q_1 - \tilde{q}_1\|_{L_2(0,\pi)} < C\varepsilon,$$

where the constant C depends only on L and ε_0.

Let us show that Theorem 16 implies the minimality of the spectral data of Problem 9. Suppose that problem L and the subspectrum $\Omega = \{\lambda_{nk}\}_{n\ge 1,\,k=1,2}$ satisfy the hypothesis of Theorem 16 and exclude one eigenvalue: $\Omega^- := \Omega \setminus \{\lambda_{11}\}$. Then, the subspectrum Ω^- does not uniquely specify q_1 if $\{q_j\}_{j=2}^m$ are fixed. Indeed, for any real number $\tilde{\lambda}_{11}$ sufficiently close to λ_{11}, Problem 9 with the data $\Omega^- \cup \{\tilde{\lambda}_{11}\}$ has a solution $\tilde{q}_1 \neq q_1$. Thus, there are two potentials, q_1 and $\tilde{q}_1$, corresponding to the subspectrum Ω^-.

In [34], the boundary value problem (22)–(23) on a star-shaped graph G was considered with complex-valued potentials $q_j \in L_2(0,\pi)$ and conditions of different types (Robin and Dirichlet) in the boundary vertices:

$$y_j'(0) - h_j y_j(0) = 0,\ j = \overline{1,p}, \quad y_j(0) = 0,\ j = \overline{p+1,m}, \tag{30}$$

where $1 \le p < m$ and $\{h_j\}_{j=1}^p$ are complex constants.

In [34], the following asymptotic formulas were obtained for the eigenvalues $\{\lambda_{nk}\}_{n\in\mathbb{N},\,k=\overline{1,m}}$ of the boundary value problem (22), (23), and (30):

$$\begin{aligned}
\sqrt{\lambda_{n1}} &= n - 1 + \frac{\alpha}{\pi} + \frac{\sigma}{\pi n} + \frac{\varkappa_{n1}}{n}, \quad \{\varkappa_{n1}\} \in l_2, \\
\sqrt{\lambda_{n2}} &= n - \frac{\alpha}{\pi} + \frac{\sigma}{\pi n} + \frac{\varkappa_{n2}}{n}, \quad \{\varkappa_{n2}\} \in l_2, \\
\sqrt{\lambda_{nk}} &= n - \frac{1}{2} + \frac{z_k}{\pi n} + \frac{\varkappa_{nk}}{n}, \quad k \in \mathcal{I}_3, \quad \varkappa_{n3} = o(1), \\
\sqrt{\lambda_{nk}} &= n + \frac{t_k}{\pi n} + \frac{\varkappa_{nk}}{n}, \quad k \in \mathcal{I}_4, \quad \varkappa_{n4} = o(1),
\end{aligned}$$

where α, σ, z_k, and t_k are certain constants, and $\mathcal{I}_3$ and $\mathcal{I}_4$ are fixed sets of indices such that $\mathcal{I}_3 \cup \mathcal{I}_4 = \overline{3,m}$, $\mathcal{I}_3 \cap \mathcal{I}_4 = \varnothing$, $|\mathcal{I}_3| = p - 1$, and $|\mathcal{I}_4| = m - p - 1$. To be precise, we assumed that $3 \in \mathcal{I}_3$ and $4 \in \mathcal{I}_4$ if these sets are nonempty.

The author of [34] was concerned with the following partial inverse problem for all possible cases depending on p and $1 \le k_1 < k_2 \le 4$:

Problem 11 ([34]). *Let the potentials* q_j, $j = \overline{2,m}$, *the coefficients* h_j, $j = \overline{2,p}$, *and the sequence* $\{\lambda_{nk}\}_{n\in\mathbb{N},\,k\in\{k_1,k_2\}}$ *of the eigenvalues of L be given. Find the potential* q_1 *and the coefficient* h_1.

The results of [34] included:

- Eigenvalue asymptotics;
- Uniqueness;
- A constructive solution.

The proof technique of [34] was derived from the study of the Riesz basis property for the sequences $\{\sin(n+\beta)t\}_{n\geq 1}$ and $\{1\} \cup \{\cos(n+\beta)t\}_{n\geq 1}$ [77].

In [35], the Sturm-Liouville problem $\mathcal{L}$ with singular potentials was considered on a star-shaped graph with different edge lengths $\{d_j\}_{j=1}^m$:

$$\ell_j y_j = -(y_j^{[1]})' - \sigma_j(x_j) y_j^{[1]} - \sigma_j^2(x_j) y_j, \quad x_j \in (0, d_j), \quad \sigma_j \in L_2(0, d_j), \quad j = \overline{1, m}, \tag{31}$$

with standard matching conditions

$$y_1(d_1) = y_j(d_j), \quad j = \overline{2, m}, \quad \sum_{j=1}^{m} y_j^{[1]}(d_j) = 0,$$

and Dirichlet boundary conditions

$$y_j(0) = 0, \quad j = \overline{1, m}.$$

Fix an integer p, $1 \leq p < m$. Let $\{\lambda_n\}_{n\in T}$, $T \subseteq \mathbb{N}$ be some subset of the spectrum.

Problem 12 ([35])**.** *Given the potentials $\{\sigma_j\}_{j=p+1}^m$, the subspectrum $\{\lambda_n\}_{n\in T}$, and the sequence $\{\omega_k\}_{k\geq 1}$, find $\{\sigma_j\}_{j=1}^p$.*

The numbers $\{\omega_k\}_{k\geq 1}$ are defined as follows. For $j = \overline{1, m}$, let $S_j(x_j, \lambda)$ be the solution of Equation (31) on the edge e_j satisfying the initial conditions $S_j(0, \lambda) = 0$, $S_j^{[1]}(0, \lambda) = 1$, and let $\{\lambda_{nj}\}_{n\geq 1}$ be the zeros of $S_j(d_j, \lambda)$. Since the function σ_j is real-valued, then the zeros of $\{\lambda_{nj}\}_{n\geq 1}$ are real and distinct as the eigenvalues of the corresponding operator.

Assume that the functions $S_j(d_j, \lambda)$ and $j = \overline{1, p}$ do not have any common zeros. Let $\{\mu_k\}_{k\geq 1}$ denote the union $\bigcup_{j=1}^{p} \{\lambda_{nj}\}_{n\geq 1}$ by arranging the numbers in increasing order: $\mu_k < \mu_{k+1}$, $k \in \mathbb{N}$. In view of our assumption, for every $k \in \mathbb{N}$, there exists exactly one index $j =: \omega_k \in \{1, \ldots, p\}$, such that $\mu_k \in \{\lambda_{nj}\}_{n\geq 1}$. The sequence $\{\omega_k\}_{k\geq 1}$ is used as additional data for the partial inverse problem.

Using a subspectrum $\{\lambda_n\}_{n\in T}$, it is possible to recover only the sum of the Weyl functions $\sum_{j=1}^{p} M_j(\lambda)$ for separate edges $\{e_j\}_{j=1}^p$. In order to find $M_j(\lambda)$ separately, one also needs $\{\omega_k\}_{k\geq 1}$.

Impose the assumptions:

(A_1) $S_j(d_j, \lambda_n) \neq 0$, $j = \overline{1, m}$, $n \in T$.

(A_2) The functions $S_j(d_j, \lambda)$ and $j = \overline{1, p}$ do not have any common zeros.

(A_3) $\lambda_n \neq \lambda_k$, $n \neq k$, $n, k \in T$.

(A_4) $\lambda_n > 0$, $n \in T$.

(A_5) $\lambda_{nj} > 0$, $n \in \mathbb{N}$, $j = \overline{1, p}$.

In [35], three approaches to uniqueness for the solution of Problem 12 were suggested. The first approach was based on the estimate of the infinite product

$$\Delta_T(\lambda) := \prod_{n\in T} \left(1 - \frac{\lambda}{\lambda_n}\right).$$

Theorem 17 ([35])**.** *Suppose that $\sigma_j = \tilde{\sigma}_j$ in $L_2(0, d_j)$ for $j = \overline{p+1, m}$, $\{\lambda_n\}_{n\in T} = \{\tilde{\lambda}_n\}_{n\in \tilde{T}}$, and $\omega_k = \tilde{\omega}_k$, $k \geq 1$. Moreover, let the assumptions (A_1)–(A_5) hold for both problems $\mathcal{L}$ and $\tilde{\mathcal{L}}$ and the corresponding subspectra, and let the estimate*

$$|\Delta_T(\lambda)| \geq C|\lambda|^{(1-2p)/2} \exp(2l|Im\sqrt{\lambda}|), \quad |\lambda| \geq \lambda^*, \quad \arg\lambda = \varphi, \tag{32}$$

be valid, where $\varphi \in (0, 2\pi)$ *and* $\lambda^* > 0$ *are fixed numbers,* $l := \sum_{j=1}^{p} d_j$*. Then,* $\sigma_j = \tilde{\sigma}_j$ *in* $L_2(0, d_j)$ *for* $j = \overline{1, p}$*.*

The second approach relied on the ideas of Gesztesy and Simon [10] and generalized Theorem 3.

Theorem 18 ([35])**.** *Suppose that* $\sigma_j = \tilde{\sigma}_j$ *in* $L_2(0, d_j)$ *for* $j = \overline{p+1, m}$*,* $\{\lambda_n\}_{n\in T} = \{\tilde{\lambda}_n\}_{n\in \tilde{T}}$*, and* $\omega_k = \tilde{\omega}_k$*,* $k \in \mathbb{N}$*. Moreover, let the assumptions* (A_1)*–*(A_5) *hold for both problems* $\mathcal{L}$ *and* $\tilde{\mathcal{L}}$ *and the corresponding subspectra, and for all sufficiently large* $t > 0$*, we have*

$$\#\{n \in T : \lambda_n < t\} \geq \alpha \#\{n \in \mathbb{N} : \lambda_n < t\} + \beta,$$

where

$$L = \sum_{j=1}^{m} d_j, \quad \alpha = \frac{2l}{L}, \quad \beta = \frac{1}{2}(\alpha(m-1) - 2p + 1).$$

Then, $\sigma_j = \tilde{\sigma}_j$ *in* $L_2(0, d_j)$ *for* $j = \overline{1, p}$*.*

The third approach of [35] was based on the construction of a special sequence of vector functions in the Hilbert space $L_2(0, l) \oplus L_2(0, l)$. The completeness of this sequence implies the uniqueness of the partial inverse problem solution.

3.2. Simple Graphs with Loops

The study of partial inverse problems on graphs with loops began with [31,32] for graph G presented in Figure 6. Graph G has the vertices $\{v_j\}_{j=0}^{r}$ and the edges $\{e_j\}_{j=1}^{r+r_1}$, where $e_j = [v_j, v_0]$ for $j = \overline{1, r}$ are boundary edges and e_j for $j = \overline{r+1, r_1}$ are loops.

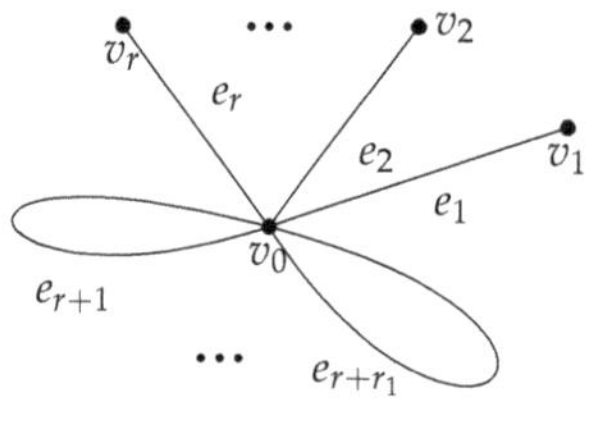

Figure 6. Graph with loops.

The Sturm-Liouville problem on G is given by the equations

$$-y_j'' + q_j(x) y_j = \lambda y_j, \quad x \in (0, 1), \quad j = \overline{1, r + r_1},$$

subject to the matching conditions in the internal vertex v_0:

$$y_j(1) = y_i(0), \quad j = \overline{1, r + r_1}, \quad i = \overline{r+1, r_1 + 1},$$
$$\sum_{j=1}^{r+r_1} y_j'(1) = \sum_{i=r+1}^{r+r_1} y_i'(0),$$

and the boundary conditions in the vertices v_j, $j = \overline{1, r}$. In [31], the Robin boundary conditions $y_j'(0) - h_j y_j(0) = 0$, $j = \overline{1, r}$ were considered and, in [32], the Dirichlet boundary conditions $y_j(0) = 0$, $j = \overline{1, r}$. The potentials $\{q_j\}_{j=1}^{r+r_1}$ were assumed to be real-valued and integrable.

In [31,32], the uniqueness theorems for the solution of the following partial inverse problem were proved:

Problem 13 ([31,32]). *Suppose that the potentials $\{q_j\}_{j=2}^{r+r_1}$ on $(0,1)$ and the potential q_1 on the subinterval $(b,1)$, $b<1$ are known a priori. Given a subspectrum, find q_1 on $(0,b)$.*

In [31], the constants $\{h_j\}_{j=2}^{r}$ of the boundary conditions were assumed to be known, and h_1 had to be recovered together with q_1 on $(0,b)$. Problem 13 was studied under a separation condition and the completeness condition of a cosine system. These conditions guaranteed the uniqueness of the solution.

Yang and Bondarenko [37] investigated a partial inverse problem on a lasso graph (see Figure 7).

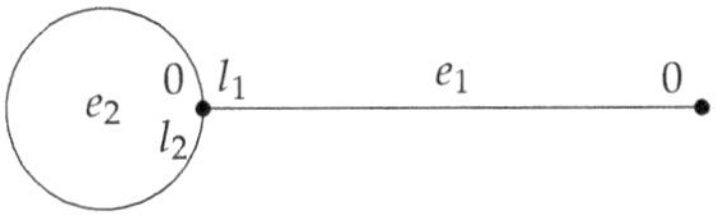

Figure 7. Lasso graph.

In [37], the following Sturm-Liouville problem on a lasso graph with singular potentials $q_j = W_2^{-1}(0,l_j)$, $q_j=\sigma_j'$, $j=1,2$ was considered:

$$\ell_j y_j = -(y_j^{[1]})' - \sigma_j(x_j)y_j^{[1]} - \sigma_j^2(x_j)y_j = \lambda y_j, \quad \sigma_j \in L_2(0,l_j), \quad j=1,2,$$
$$y_1(0)=0, \quad y_1(l_1)=y_2(0)=y_2(l_2), \quad y_1^{[1]}(l_1) - y_2^{[1]}(0) + y_2^{[1]}(l_1) = 0,$$

where $y_j^{[1]} = y_j' - \sigma_j y_j$, $j=1,2$, $l_1 = m \in \mathbb{N}$, $l_2=1$.

Problem 14 ([37]). *Given the function σ_1, the subspectrum Λ, and the signs Ω, find the function σ_2.*

The signs Ω in the problem statement are defined as follows. Let $S_2(x,\lambda)$ and $C_2(x,\lambda)$ be the solutions of equation $\ell_2 y_2 = \lambda y_2$ under the initial conditions $S_2(0,\lambda) = C_2^{[1]}(0,\lambda) = 0$, $S_2^{[1]}(0,\lambda) = C_2(0,\lambda) = 1$. Define

$$h(\lambda) := S_2(1,\lambda), \quad H(\lambda) := C_2(1,\lambda) - S_2^{[1]}(1,\lambda), \quad d(\lambda) := C_2(1,\lambda) + S_2^{[1]}(1,\lambda) - 2.$$

The zeros $\{\nu_n\}_{n\ge 1}$ of $h(\lambda)$ are the eigenvalues of the Dirichlet boundary value problem:

$$\ell_2 y_2 = \lambda y_2, \quad y_2(0) = y_2(1) = 0.$$

The zeros $\{\mu_n\}_{n\in\mathbb{Z}}$ of $d(\lambda)$ are the eigenvalues of the periodic problem:

$$\ell_2 y_2 = \lambda y_2, \quad y_2(0) = y_2(1), \quad y_2^{[1]}(0) = y_2^{[1]}(1).$$

Let $\omega_n := \operatorname{sign} H(\nu_n)$ and $\Omega := \{\omega_n\}_{n\ge 1}$. The partial inverse problem on the lasso graph (Problem 14) is reduced to the following periodic inverse problem on a finite interval:

Problem 15 ([37]). *Given the sequences $\{\nu_n\}_{n\ge1}$ and $\{\mu_n\}_{n\in\mathbb{Z}}$ and the sequence of signs Ω, construct the function σ_2.*

In [37], the solution of Problem 15 was obtained for the case of singular potentials $q \in W_2^{-1}(0,1)$. Thus, [37] contained the following results for the partial inverse problem:

- Eigenvalue asymptotics;
- Uniqueness;
- Algorithm;
- The solution of the inverse periodic problem with a singular potential.

Bondarenko and Shieh [38] studied partial inverse problems for a quadratic differential pencil on the graph with a cycle presented in Figure 8.

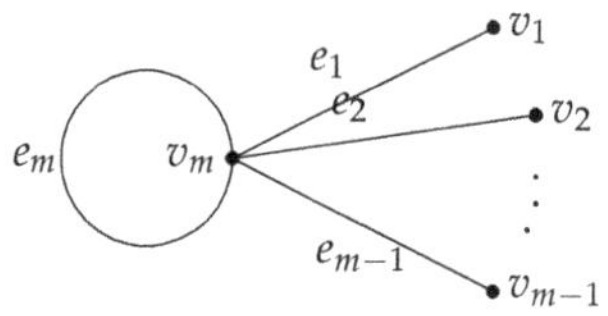

Figure 8. Graph with a cycle.

In [38], the following boundary value problem with nonlinear dependence on the spectral parameter λ was considered:

$$-y_j'' + q_j(x_j)y_j + 2\lambda p_j(x_j)y_j = \lambda^2 y_j, \quad x_j \in (0,\pi), \quad j = \overline{1,m},$$
$$y_j(0) = 0, \quad j = \overline{1,m-1},$$
$$y_m(0) = y_j(\pi), \quad j = \overline{1,m}, \qquad y_m'(0) = \sum_{j=1}^{m} y_j'(\pi),$$

where $p_j \in AC[0,\pi]$ and $q_j \in L_1(0,\pi)$, $j = \overline{1,m}$, are complex-valued functions. The following two partial inverse problems were studied.

Problem 16 ([38]). *Given the functions $\{p_j\}_{j=2}^m$ and $\{q_j\}_{j=2}^m$ and a subspectrum Λ, find p_1 and q_1.*

Problem 17 ([38]). *Let the functions $\{p_j\}_{j=1}^{m-1}$ and $\{q_j\}_{j=1}^{m-1}$, a subspectrum Λ, and the sequence of signs Ω be given. Find p_m and q_m.*

In Problem 17, Ω is the sequence of signs for the auxiliary periodic problem (see [38] for details). The results of [38] included:

- Eigenvalue asymptotics;
- Uniqueness;
- A constructive solution.

In [38], methods of working with multiple eigenvalues and vector-functional sequences containing exponents were developed. An important role in the proofs was played by the Riesz basicity of exponential systems $\{\exp(i\lambda_k t)\}$ in $L_2(-\pi,\pi)$. Later on, these methods were generalized to graphs of an arbitrary structure (see [40]).

3.3. Graphs of a General Structure

The analysis of partial inverse problems on star-shaped graphs and simple graphs with loops showed that such problems present specific features for each case. Therefore, it is difficult to obtain results for graphs of a general structure. Until now, only the following cases have been studied:

- The reconstruction of the potentials on an arbitrary tree graph (graph without cycles) by several spectra, while the potential on one edge is known a priori (see [78]).
- The reconstruction of the potential on one boundary edge of an arbitrary graph using part of the spectrum, while the potentials on all other edges are known a priori (see [40,41]).
- For a tree graph, the reconstruction of the potentials on a connected subtree from parts of several spectra, while the potentials on the remaining edges are known a priori (see [39]).

In this section, we briefly describe these results.

Proceeding to the statement of the Sturm-Liouville problem on a graph of a general structure, let $\mathcal{G}$ be a graph with a set of vertices $\mathcal{V}$ and edges $\{e_j\}_{j=1}^m$ with the corresponding lengths $\{T_j\}_{j=1}^m$. The graph may contain cycles, loops, and multiple edges. For each edge e_j, $j = \overline{1,m}$, introduce the parameter $x_j \in [0, T_j]$. Let us denote the ends of e_j as w_{2j-1} and w_{2j} so that $x_j = 0$ corresponds to w_{2j-1} and $x_j = T_j$ to w_{2j}. Every vertex v of the graph $\mathcal{G}$ can be considered as the equivalence class of all the ends w_j incident to this vertex: $v = \{w_{j_1}, w_{j_2}, \dots, w_{j_r}\}$. The number of elements in this class is called *the degree* of the vertex. We assume that the graph $\mathcal{G}$ does not have vertices of degree 2. Otherwise, the two edges incident to such vertices could be merged into one edge. The vertices of degree 1 are called *the boundary vertices*, and the others are called *the internal vertices*. An edge incident to a boundary vertex is called *a boundary edge*. Let $\partial\mathcal{G}$ and $\operatorname{int}\mathcal{G}$ denote the sets of the boundary vertices and the internal vertices, respectively, $\mathcal{V} = \partial\mathcal{G} \cup \operatorname{int}\mathcal{G}$.

A *function* on the graph $\mathcal{G}$ is a vector function $y = [y_j]_{j=1}^m$ with components $y_j = y_j(x_j)$, $x_j \in [0, T_j]$. A function y belongs to a class $\mathscr{A}(\mathcal{G})$ if $y_j \in \mathscr{A}[0, T_j]$ for $j = \overline{1,m}$, where $\mathscr{A} = L_1$, AC, etc. In order to define matching and boundary conditions, one needs the following notations:

$$\begin{array}{ll} y_{|w_{2j-1}} = y_j(0), & y_{|w_{2j}} = y_j(T_j), \\ y'_{|w_{2j-1}} = -y'_j(0), & y'_{|w_{2j}} = y'_j(T_j), \end{array} \quad j = \overline{1,m}.$$

For $v \in \partial\mathcal{G}$, we write $y(v)$ and $y'(v)$ for $y_{|w_k}$ and $y'_{|w_k}$, respectively, where $w_k \in v$.

Bondarenko and Shieh [78] considered the Sturm-Liouville equations

$$-y_j'' + q_j(x_j)y_j = \lambda y_j, \quad x_j \in (0, T_j), \quad j = \overline{1,m}, \tag{33}$$

on a tree graph $\mathcal{G}$ (a graph without cycles) with the potential $q = [q_j]_{j=1}^m \in L_1(\mathcal{G})$ and the standard matching conditions

$$\left.\begin{array}{lll} y_{|w_j} = y_{|w_k}, & w_j, w_k \in v & \text{(continuity conditions)} \\ \sum\limits_{w_j \in v} y'_{|w_j} = 0 & & \text{(Kirchhoff's condition)} \end{array}\right\} \quad v \in \operatorname{int}\mathcal{G}. \tag{34}$$

Let L_0 and L_k, $v_k \in \partial\mathcal{G}$, be the boundary value problems for the system (33) with the matching conditions (34) and the following conditions in the boundary vertices:

$$\begin{aligned} &L_0\colon y(v_i) = 0, \quad v_i \in \partial\mathcal{G}, \\ &L_k\colon y'(v_k) = 0, \quad y(v_i) = 0, \quad v_i \in \partial\mathcal{G}\backslash\{v_k\}. \end{aligned}$$

The problems L_k have discrete spectra, which are the countable sets of eigenvalues $\Lambda_k = \{\lambda_{ks}\}_{s\geq 1}$, $k = 0$ or $v_k \in \partial\mathcal{G}$.

Fix a vertex $v_r \in \partial\mathcal{G}$ as *a root* of the tree $\mathcal{G}$. Let e_r be the edge incident to v_r. Then, the uniqueness theorem for the *complete* inverse problem on the tree is formulated as follows:

Theorem 19 ([75]). *The spectra Λ_0 and Λ_k, $k \in \partial\mathcal{G} \setminus \{v_r\}$, uniquely determine the potential q on the whole tree $\mathcal{G}$.*

Thus, if the number of boundary vertices is b, then b spectra are required for the recovery of the potentials. Bondarenko and Shieh [78] assumed that the potential q_f is known a priori on one edge e_f and proved that the remaining potentials can be uniquely recovered from $(b-1)$ spectra. If an internal edge e_f is removed, then the tree $\mathcal{G}$ splits into two parts. Let us denote them by P_1 and P_2 and their sets of boundary vertices by ∂P_1 and ∂P_2, respectively.

Theorem 20 ([78]). *Let the potential q_f on an edge e_f ($f \neq r$) be known.*

1. *If e_f is a boundary edge, the spectra Λ_0 and Λ_k, $v_k \in \partial\mathcal{G} \setminus \{v_f, v_r\}$, uniquely determine the potential q on the whole graph $\mathcal{G}$.*
2. *If e_f is an internal edge, the spectra Λ_0 and Λ_k $v_k \in \partial\mathcal{G} \setminus \{v_{r1}, v_{r2}\}$, where $v_{r1} \in \partial P_1$ and $v_{r2} \in \partial P_2$, uniquely determine the potential q on the whole graph $\mathcal{G}$.*

Theorem 20 was proved by a constructive method, developing from the ideas of [75]. The case of the internal edge e_f is the most difficult. It is crucial that the two end vertices of the internal edge e_f have degrees of at least 3. Consequently, Theorem 20 cannot be applied to an interval with the potential given on a middle subinterval.

Bondarenko [39] investigated another type of partial inverse problem on a tree $\mathcal{G}$. The edge lengths T_j in [39] were assumed to be equal π. The Sturm-Liouville equation presented in Equation (33) was considered with the singular potentials $q_j \in W_2^{-1}(0,\pi)$, $j = \overline{1,m}$. Therefore, Equation (33) was represented in the form

$$-(y_j^{[1]})' - \sigma_j(x_j)y_j^{[1]} - \sigma_j^2(x_j)y_j = \lambda y_j, \quad x \in (0, T_j), \quad j = \overline{1,m}, \tag{35}$$

where $q_j = \sigma_j'$, $\sigma_j \in L_2(0,T_j)$, $y_j^{[1]} = y_j' - \sigma_j y_j$, $j = \overline{1,m}$.

Let $\{\gamma_j\}_{j=1}^m$ be some real constants. In order to define the matching conditions, one uses the following notations:

$$\begin{array}{ll} y_{|w_{2j-1}} = y_j(0), & y_{|w_{2j}} = y_j(T_j), \\ y^{[1]}_{|w_{2j-1}} = -y_j^{[1]}(0), & y^{[1]}_{|w_{2j}} = y_j^{[1]}(T_j) + \gamma_j y_j(T_j), \end{array} \quad j = \overline{1,m}. \tag{36}$$

For $v \in \partial\mathcal{G}$, $y(v)$ and $y^{[1]}(v)$ are written for $y_{|w_k}$ and $y^{[1]}_{|w_k}$, respectively, where $w_k \in v$. Let us divide the set of the boundary vertices into two disjoint subsets:

$$\partial\mathcal{G} = \mathcal{V}^D \cup \mathcal{V}^N, \quad \mathcal{V}^D \cap \mathcal{V}^N = \varnothing.$$

Thus, in [39], the boundary value problem L for the Sturm-Liouville equation presented in Equation (35) was considered subject to the matching conditions

$$\left.\begin{array}{l} y_{|w_j} = y_{|w_k}, \quad w_j, w_k \in v \\ \sum\limits_{w_j \in v} y^{[1]}_{|w_j} = 0 \end{array}\right\} \quad v \in \operatorname{int}\mathcal{G} \tag{37}$$

and the boundary conditions

$$y(v) = 0, \quad v \in \mathcal{V}^D, \qquad y^{[1]}(v) = 0, \quad v \in \mathcal{V}^N. \tag{38}$$

Let the tree $\mathcal{G}$ be divided into two subtrees $\mathcal{G}_{known}$ and $\mathcal{G}_{unknown}$ with a common vertex $w \in \operatorname{int}\mathcal{G}$ (see Figure 9). Let E_{known} and $E_{unknown}$ denote the edge sets of $\mathcal{G}_{known}$ and $\mathcal{G}_{unknown}$, respectively. Let $\{v_k\}_{k=1}^b$ denote the boundary vertices $\partial G_{unknown} \setminus \{w\}$. For each $k = \overline{1,b}$, let L_k denote the boundary value problem (35), (37) with the boundary conditions (38) for $v \in \partial\mathcal{G} \setminus \{v_k\}$ and $y(v_k) = 0$ if $v_k \in \mathcal{V}^N$, or $y^{[1]}(v_k) = 0$ if $v_k \in \mathcal{V}^D$. In other words, if the problem L has the Dirichlet boundary condition in v_k, then L_k has the Neumann boundary condition, and vice versa.

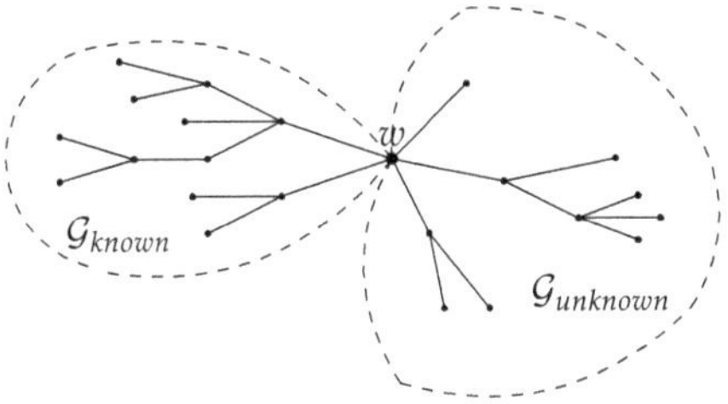

Figure 9. Tree graph.

Problem 18 ([39]). *Suppose that the functions σ_j on the edges $e_j \in E_{known}$ and the constants $\{\gamma_j\}_{j=1}^m$ are known a priori. Given some subspectra of the problems L and L_k for $k = \overline{1, b-1}$, find σ_j for all $e_j \in E_{unknown}$.*

Note that, in view of Theorem 19, the full spectra of L and L_k for $k = \overline{1, b-1}$ determine the potentials on the whole tree $\mathcal{G}$. Strictly speaking, Yurko [75] proved Theorem 19 for regular potentials $q_j \in L_2(0, T_j)$. For singular potentials $q_j \in W_2^{-1}(0, T_j)$, similar results were obtained by Vasiliev [79].

In [39], it was shown that, if the potentials σ_j are known on E_{known}, then only part of the spectra can be used for reconstruction. Sufficient conditions for the uniqueness were formulated in terms of completeness for some special vector functional sequences, which were constructed using the known functions σ_j and the given subspectra. Furthermore, the uniqueness conditions in terms of the eigenvalue asymptotics were obtained. In addition, in [39], a constructive algorithm for solving Problem 18 was developed. This algorithm allows one to reduce the partial inverse problem to a complete inverse problem for the "unknown" subtree.

Proceeding to general graphs containing cycles, for such graphs, partial inverse problems were investigated only for case in which a potential is unknown on one edge. Let $\mathcal{G}$ be a graph of an arbitrary structure with arbitrary edge lengths $\{T_j\}_{j=1}^m$. In [40], Sturm-Liouville differential equations with quadratic dependence on the spectral parameter λ were considered on the graph $\mathcal{G}$:

$$-y_j''(x_j) + (q_j(x_j) + 2\lambda p_j(x_j) - \lambda^2) y_j(x_j) = \lambda y_j(x_j), \quad x_j \in (0, T_j), \quad j = \overline{1, m}, \tag{39}$$

where $y = [y_j]_{j=1}^m$, $p = [p_j]_{j=1}^m$, and $q = [q_j]_{j=1}^m$ are complex-valued functions on $\mathcal{G}$, $y \in W_2^2(\mathcal{G})$, $p \in AC(\mathcal{G})$, $q \in L_1(\mathcal{G})$.

Let γ_{jk} be some complex numbers, defined for the ends $w_j \in v$, $v \in \operatorname{int}\mathcal{G}$, $k = \overline{1, 4}$, $\gamma_{jk} \neq 0$ for $k = 1, 2$. Define the linear forms

$$U_j(y) := y'_{|w_j} + (\lambda\gamma_{j3} + \gamma_{j4}) y_{|w_j}.$$

Thus, in [40], the differential pencil L given by Equation (39) subject to the following conditions was considered:

$$\gamma_{j1} y_{|w_j} = \gamma_{k1} y_{|w_k}, \quad w_j, w_k \in v, \quad v \in \operatorname{int}\mathcal{G},$$
$$\sum_{w_j \in v} \gamma_{j2} U_j(y) = 0, \quad v \in \operatorname{int}\mathcal{G},$$
$$y_{|w_j} = 0, \quad w_j \in v, \quad v \in \partial\mathcal{G}.$$

For certainty, we assume that e_1 is a boundary edge.

Problem 19 ([40]). *Suppose that $\{p_j\}_{j=2}^m$, $\{q_j\}_{j=2}^m$, and $\{\gamma_{jk}\}$ are known a priori. Given a subspectrum Λ' of the pencil L, find p_1 and q_1 (see Figure 10).*

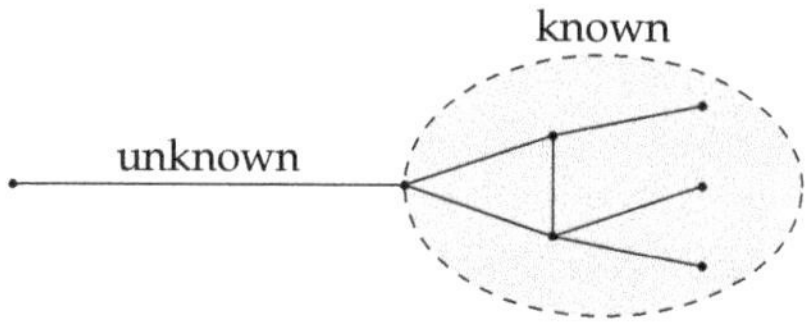

Figure 10. Graph of an arbitrary structure.

In [40], Problem 19 was studied under a separation condition, which generalized Condition 1 for a star-shaped graph. For a general graph, the separation condition had a complicated technical formulation, so we omit it here. The results of [40] for Problem 19 included:

- Uniqueness in the general case;
- A constructive solution for rationally dependent edge lengths.

In particular, the recovery of the coefficients p_1 and q_1 from the whole spectrum Λ of the pencil L was investigated. The characteristic function of L satisfies the following asymptotic relation:

$$\Delta(\lambda) = \lambda^r(\Delta_0(\lambda) + O\Big(|\lambda|^{-1}\exp(M|\mathrm{Im}\,\lambda|)\Big)), \quad |\lambda| \to \infty,$$

where $r \in \mathbb{Z}$, $M = \sum\limits_{j=1}^{m} T_j$, and $\Delta_0(\lambda)$ is a polynomial of $\cos(\lambda T_j)$ and $\sin(\lambda T_j)$, $j = \overline{1, m}$, $\Delta_0(\lambda) = O(\exp(M|\mathrm{Im}\,\lambda|))$. In [40], the following regularity condition was imposed:

$$|\Delta_0(i\tau)| \geq C\exp(M|\tau|), \quad \tau \in \mathbb{R}, \quad |\tau| \geq \tau^*, \tag{40}$$

for some $\tau^* > 0$. Under the conditions in (40) and the separation condition, the spectrum Λ uniquely specifies p_1 and q_1 if $T_1 < M/2$ or $T_1 = M/2, r = -2$ (see Theorem 2 in [40] for details). In other words, the length of the "unknown" edge has to be less than or equal to the total length of the graph for the unique determination of the pencil coefficients on this edge by the spectrum.

In the case of rationally dependent edge lengths $T_j = \pi n_j$, $n_j \in \mathbb{N}$, $j = \overline{1, m}$, the spectrum Λ of the regular pencil L contains subsequences of eigenvalues $\{\lambda_{nk}\}_{n \in \mathbb{Z}}$, $k = \overline{1, s}$, satisfying the asymptotic relation

$$\lambda_{nk} = 2n + \beta_k + o(1), \quad |n| \to \infty.$$

It was shown in [40] that one can choose a certain number of such subsequences to uniquely recover p_1 and q_1. The constructive method of [40] developed the ideas of [33,38] and other papers. This method was based on the reduction of the partial inverse problem (Problem 19) to a complete inverse problem for the Sturm-Liouville quadratic pencil on the interval $(0, T_1)$.

The most complete results for a partial inverse problem on an arbitrary graph were obtained in [41] for the boundary value problem given by (35), (37), and (38):

$$\left.\begin{array}{c} -(y_j^{[1]})' - \sigma_j(x_j)y_j^{[1]} - \sigma_j^2(x_j)y_j = \lambda y_j, \quad x \in (0, T_j), \quad j = \overline{1, m}, \\ y_{|w_j} = y_{|w_k}, \quad w_j, w_k \in v, \quad \sum\limits_{w_j \in v} y_{|w_j}^{[1]} = 0, \quad v \in \mathrm{int}\,\mathcal{G}, \\ y(v) = 0, \quad v \in \mathcal{V}^D, \qquad y^{[1]}(v) = 0, \quad v \in \mathcal{V}^N, \end{array}\right\} \tag{41}$$

where $y_{|w_k}$ are defined by (36), with $\gamma_j = 0$. The edge lengths were assumed to be rationally dependent: $T_j = 2\pi n_j$, $n_j \in \mathbb{N}$, $j = \overline{1, m}$.

For certainty, assume that v_1 is a boundary vertex corresponding to the end $w_1 \sim x_1 = 0$ of the edge e_1 and $v_1 \in \mathcal{V}^D$.

Problem 20 ([41]). *Suppose that the functions $\{\sigma_j\}_{j=2}^m$ are known a priori. Given a subspectrum Λ, find σ_1 (see Figure 10).*

In [41], the following results were obtained for Problem 20:

- A uniqueness theorem;
- A constructive solution;
- Sufficient conditions for global solvability;
- Local solvability and stability.

Let us formulate a uniqueness theorem for Problem 20. For this purpose, one first needs to construct the characteristic function of the Sturm-Liouville problem on an arbitrary graph. For each fixed $j = \overline{1,m}$, introduce the solutions $C_j(x_j, \lambda)$ and $S_j(x_j, \lambda)$ of Equation (35) satisfying the initial conditions

$$C_j(0,\lambda) = S_j^{[1]}(0,\lambda) = 1, \quad C_j^{[1]}(0,\lambda) = S_j(0,\lambda) = 0.$$

Every solution $[y_j]_{j=1}^m$ of system (35) can be represented in the form

$$y_j(x_j,\lambda) = a_j(\lambda)C_j(x_j,\lambda) + b_j(\lambda)S_j(x_j,\lambda), \quad j = \overline{1,m}, \tag{42}$$

with some coefficients $a_j(\lambda)$ and $b_j(\lambda)$ independent of x. Substituting (42) into (37) and (38), one obtains the system of linear equations $\mathscr{S}$ with respect to $a_j(\lambda)$ and $b_j(\lambda)$, $j = \overline{1,m}$. The determinant $\Delta(\lambda)$ of this system is the characteristic function of L, that is, the spectrum of the problem (41) coincides with the zeros of $\Delta(\lambda)$.

The characteristic function can be represented in the form

$$\Delta(\lambda) = S_1(T_1,\lambda)\Delta^K(\lambda) + S_1^{[1]}(T_1,\lambda)\Delta^\Pi(\lambda), \tag{43}$$

where $\Delta^K(\lambda)$ and $\Delta^\Pi(\lambda)$ are the determinants of the linear systems obtained from $\mathscr{S}$ by the replacements $S_1(T_1,\lambda) \mapsto 1$, $S_1^{[1]}(T_1,\lambda) \mapsto 0$ and $S_1(T_1,\lambda) \mapsto 0$, $S_1^{[1]}(T_1,\lambda) \mapsto 1$, respectively. Note that our construction defines the functions $\Delta(\lambda)$, $\Delta^\Pi(\lambda)$, and $\Delta^K(\lambda)$ uniquely up to the sign, which depends on the order of equations and variables in the system $\mathscr{S}$. However, it is possible to fix such signs so that Formula (43) is valid. Clearly, the functions $\Delta(\lambda)$, $\Delta^K(\lambda)$, and $\Delta^\Pi(\lambda)$ are entire, and $\Delta^K(\lambda)$ and $\Delta^\Pi(\lambda)$ do not depend on σ_1.

The separation condition for Problem 20 reads as follows:

Condition 2. *For each $\lambda \in \Lambda$, $\Delta^\Pi(\lambda) \neq 0$ or $\Delta^K(\lambda) \neq 0$.*

Condition 2 is essential, since otherwise, if $\Delta^\Pi(\lambda) = \Delta^K(\lambda) = 0$, then (43) readily implies $\Delta(\lambda) = 0$, but such an eigenvalue λ is related to the "known" part of the graph (see Figure 10) and carries no information on σ_1.

The spectrum of the problem (41) consists of eigenvalue subsequences with the asymptotics

$$\sqrt{\lambda_{nk}} = n + r_k + \varkappa_{nk}, \quad \{\varkappa_{nk}\} \in l_2, \tag{44}$$

where $k = \overline{1,N}$, $N := 2\sum_{j=1}^m n_j$, $n_j = \frac{T_j}{2\pi}$, $n \in \mathbb{N}$ or $n \in N \cup \{0\}$ depending on k, and $\{r_k\}_{k=1}^N \subseteq [0,1)$. Furthermore, for each $r_k \neq 0$, there exists $r_s = 1 - r_k$. The numbers $\{r_k\}_{k=1}^N$ depend on the graph structure and not on $\{\sigma_j\}_{j=1}^m$. The asymptotics (44) are obtained by the reduction of the Sturm-Liouville problem to the matrix form.

Let us impose the following condition on the subspectrum Λ in Problem 20:

Condition 3. $\Lambda = \{\lambda_{nk}\}_{n\geq 0, k\in\mathcal{K}}$, *where λ_{nk} satisfies the asymptotics in* (44) *and a subset $\mathcal{K} \subseteq \{1,\dots,N\}$ fulfills the following conditions:*

1. *All the values* $\{r_k\}_{k\in\mathcal{K}}$ *from* (44) *are distinct.*
2. $r_k \notin \{0, \frac{1}{2}\}$, $k \in \mathcal{K}$.
3. *For each* $k \in \mathcal{K}$, *there exists* $s \in \mathcal{K}$ *such that* $r_k + r_s = 1$.
4. $|\mathcal{K}| = 4n_1$.

Note that Λ is non-uniquely determined by $\mathcal{K}$, and any finite number of eigenvalues in Λ can be chosen arbitrarily. In particular, Λ can contain a finite number of multiple eigenvalues. The condition $|\mathcal{K}| = 4n_1$ connects the length of the "unknown" edge $T_1 = 2\pi n_1$ with the "size" of the subspectrum, which is used for the reconstruction. The following theorem asserts the uniqueness of the solution to Problem 20.

Theorem 21 ([41]). *Let* Λ *be a subspectrum of the problem in* (41) *satisfying Conditions 2 and 3. If* $\sigma_j = \tilde{\sigma}_j$ *in* $L_2(0, T_j)$ *for* $j = \overline{2, m}$ *and* $\Lambda = \tilde{\Lambda}$ *(with respect to multiplicities), then* $\sigma_1 = \tilde{\sigma}_1$ *in* $L_2(0, T_1)$.

The proof of Theorem 21, the constructive solution, and the study of the solvability and stability of the partial inverse problem in [41] were based on the unified approach, which is described in the next section.

4. Unified Approach

In this section, we describe a unified approach to partial inverse problems on intervals and graphs that was developed in [41–44] and subsequent studies. This approach allows one to reduce a partial inverse problem to a complete inverse problem on an "unknown" part of an interval or graph. The central role in the reduction technique is played by a special vector functional sequence $\{v_n\}_{n\geq 0}$ in the Hilbert space $\mathcal{H} = L_2(0, l) \oplus L_2(0, l)$, where l is the length of an "unknown" subinterval. The completeness and the Riesz basis property of this sequence imply uniqueness and a constructive solution to the corresponding partial inverse problem, respectively. The unified approach also allows one to obtain the solvability conditions and stability of partial inverse problems.

The initial ideas behind this approach appeared in [22,33], studies of the inverse transmission eigenvalue problem and a partial inverse Sturm-Liouville problem on a star-shaped graph, respectively. Later on, Bondarenko [42,43] noticed that partial inverse problems for various classes of differential operators can be represented as Sturm-Liouville problems with entire analytic functions in one of the boundary conditions, and an inverse spectral theory for such problems was created. As corollaries of this general theory, both well-known and novel results for the Hochstast-Lieberman problem and its generalizations have been deduced.

In Section 4.1, we provide the inverse spectral theory of the Sturm-Liouville equation with entire functions in a boundary condition, mostly based on the results of [42,43]. In Section 4.2, applications to partial inverse problems are discussed.

4.1. Sturm-Liouville Problem with Entire Functions in a Boundary Condition

Consider the following Sturm-Liouville problem $R(q, f_1, f_2)$:

$$-y''(x) + q(x)y(x) = \lambda y(x), \quad x \in (0, \pi), \tag{45}$$

$$y(0) = 0, \quad f_1(\lambda)y'(\pi) + f_2(\lambda)y(\pi) = 0, \tag{46}$$

where $q(x)$ is a complex-valued potential of $L_2(0, \pi)$, and $f_1(\lambda)$ and $f_2(\lambda)$ are entire analytic functions of the spectral parameter λ.

Let $S(x, \lambda)$ denote the solution of Equation (45) satisfying the initial conditions $S(0, \lambda) = 0$, $S'(0, \lambda) = 1$. The spectrum of $R(q, f_1, f_2)$ consists of the eigenvalues, which coincide with the zeros of the entire characteristic function

$$\Delta(\lambda) := f_1(\lambda)S'(\pi, \lambda) + f_2(\lambda)S(\pi, \lambda). \tag{47}$$

Depending on the functions $f_1(\lambda)$ and $f_2(\lambda)$, the spectrum can be at most countable or coincide with the whole complex plane. If there is no additional information on $f_1(\lambda)$ and $f_2(\lambda)$, then one cannot study specific properties of the spectrum (e.g., eigenvalue asymptotics). However, one can consider the following inverse problem:

Problem 21 ([42,43]). *Suppose that the functions $f_1(\lambda)$ and $f_2(\lambda)$ are known a priori. Given a subspectrum $\{\lambda_n\}_{n\ge1}$ of the problem $R(q, f_1, f_2)$ and the number $\omega := \frac{1}{2}\int_0^\pi q(x)\,dx$, find the potential q.*

In [42,43], Problem 21 was studied under certain restrictions on $\{\lambda_n\}_{n\ge1}$ that guaranteed the uniqueness and existence of the solution, etc. Note that, in applications to partial inverse problems, the number ω can usually be found from the eigenvalue asymptotics. However, in general cases, it has to be given.

Introduce the notations

$$s(x,\lambda) = \sqrt{\lambda}\sin(\sqrt{\lambda}x), \quad c(x,\lambda) = \cos(\sqrt{\lambda}x).$$

Then, the sine-type solution $S(x,\lambda)$ can be represented in terms of the transformation operator:

$$S(x,\lambda) = \frac{s(x,\lambda)}{\lambda} + \frac{1}{\lambda}\int_0^x \mathscr{K}(x,t)s(t,\lambda)\,dt. \tag{48}$$

Let $\eta_1(\lambda) := S(\pi,\lambda)$ and $\eta_2(\lambda) := S'(\pi,\lambda)$. Applying differentiation and integration by parts in (48), one can easily obtain the following standard relations:

$$\eta_1(\lambda) = \frac{s(\pi,\lambda)}{\lambda} - \frac{\omega c(\pi,\lambda)}{\lambda} + \frac{1}{\lambda}\int_0^\pi K(t)c(t,\lambda)\,dt, \tag{49}$$

$$\eta_2(\lambda) = c(\pi,\lambda) + \frac{\omega s(\pi,\lambda)}{\lambda} + \frac{1}{\lambda}\int_0^\pi N(t)s(t,\lambda)\,dt, \tag{50}$$

where

$$K(t) = \mathscr{K}_t(\pi,t), \quad N(t) = \mathscr{K}_x(\pi,t), \quad K, N \in L_2(0,\pi). \tag{51}$$

The pair of functions $\{K, N\}$ from (49) and (50) is called *the Cauchy data* of the potential q. This name was chosen because the eigenvalue problem for Equation (45) with the boundary conditions $y(0) = y(\pi) = 0$ is related to the initial value (Cauchy) problem

$$u_{tt} - u_{xx} + q(x)u = 0, \quad 0 \le |t| \le x \le \pi,$$
$$u(\pi,t) = \mathscr{K}(\pi,t), \quad u_x(\pi,t) = \mathscr{K}_x(\pi,t), \quad -\pi \le t \le \pi,$$

where $\mathscr{K}(x,t) = -\mathscr{K}(x,-t)$ for $t < 0$. This problem has the unique solution $u(x,t) \equiv \mathscr{K}(x,t)$. The initial data of the Cauchy problem are the functions $\{\mathscr{K}(\pi,t), \mathscr{K}_x(\pi,t)\}$, which are related to $K(t)$ and $N(t)$ by (51).

The method of [42,43] was based on the reduction of Problem 21 to the following auxiliary inverse problem.

Problem 22. *Given the Cauchy data $\{K, N\}$, find the potential q.*

Problem 22 is equivalent to classical inverse spectral problems. Indeed, using the Cauchy data, one can construct $S(\pi,\lambda)$ and $S'(\pi,\lambda)$ via (49) and (50) and the Weyl function $M(\lambda) := -\dfrac{S'(\pi,\lambda)}{S(\pi,\lambda)}$, which uniquely specifies q (see, e.g., [4]). Thus, the uniqueness of Problem 22's solution follows from the classical result obtained by Borg [6]. Its constructive solution can be obtained by the standard methods (see [4]). Some numerical techniques for the reconstruction of the potential using the Cauchy data are described in [67].

For simplicity, we assume that $\lambda_n \neq 0$ for $n \geq 1$ and the eigenvalues $\{\lambda_n\}_{n\geq 1}$ are simple, that is, $\lambda_n \neq \lambda_m$ for $n \neq m$. In [42,43], results were provided for the general case of multiple eigenvalues.

Introduce the complex Hilbert space of vector functions

$$\mathcal{H} := L_2(0,\pi) \oplus L_2(0,\pi) = \{h = [h_1, h_2] \colon h_j \in L_2(0,\pi),\, j = 1,2\}$$

with the following scalar product and norm:

$$(g,h)_{\mathcal{H}} := \int_0^\pi (\overline{g_1(t)}h_1(t) + \overline{g_2(t)}h_2(t))\,dt, \quad \|h\|_{\mathcal{H}} = \sqrt{(h,h)_{\mathcal{H}}},$$
$$g,h \in \mathcal{H}, \quad g = [g_1,g_2], \quad h = [h_1,h_2].$$

Substituting the representations (49) and (50) into (47) and letting $\lambda = \lambda_n$, one derives the relation

$$(u, v_n)_{\mathcal{H}} = w_n \tag{52}$$

for $n \geq 1$, where

$$u(t) := [\overline{N(t)}, \overline{K(t)}], \quad v_n(t) = v(t,\lambda_n), \quad w_n = w(\lambda_n), \quad n \geq 1, \tag{53}$$

$$v(t,\lambda) := [f_1(\lambda)s(t,\lambda), f_2(\lambda)c(t,\lambda)], \tag{54}$$

$$w(\lambda) := -f_1(\lambda)(\lambda c(\pi,\lambda) + \omega s(\pi,\lambda)) - f_2(\lambda)(s(\pi,\lambda) - \omega c(\pi,\lambda)). \tag{55}$$

Since the function $S(\pi,\lambda)$ is analytical at $\lambda = 0$, one obtains an additional relation of the form (52) for $n = 0$ from (49) with

$$v_0(t) := [0,1], \quad w_0 := \omega. \tag{56}$$

In [42], the following conditions were introduced:

(COMPLETE)—the sequence $\{v_n\}_{n\geq 0}$ is complete in $\mathcal{H}$.

(BASIS)—the sequence $\{v_n\}_{n\geq 0}$ is an unconditional basis in $\mathcal{H}$.

Clearly, the condition (BASIS) implies (COMPLETE). It was shown in [42] that the condition (COMPLETE) is necessary and sufficient for the uniqueness of Problem 21's solution.

Theorem 22 ([42])**.** *Let $\{\lambda_n\}_{n\geq 1}$ and $\{\tilde{\lambda}_n\}_{n\geq 1}$ be subspectra of the problems $R(q,f_1,f_2)$ and $R(\tilde{q},f_1,f_2)$, respectively. Suppose that $R(q,f_1,f_2)$ and $\{\lambda_n\}_{n\geq 1}$ satisfy the condition* (COMPLETE)*, and let $\lambda_n = \tilde{\lambda}_n$, $n \geq 1$, $\omega = \tilde{\omega}$. Then, $q = \tilde{q}$ in $L_2(0,\pi)$.*

Theorem 23 ([42])**.** *Let $\{\lambda_n\}_{n\geq 1}$ be a subspectrum of the problem $R(q,f_1,f_2)$. Suppose that the sequence $\{v_n\}_{n\geq 0}$ is incomplete in $\mathcal{H}$. Then, there exists a complex-valued function $\tilde{q} \in L_2(0,\pi)$, $\tilde{q} \neq q$ such that $\omega = \tilde{\omega}$, and $\{\lambda_n\}_{n\geq 1}$ is a subspectrum of $R(\tilde{q},f_1,f_2)$.*

Under the condition (BASIS), the following constructive algorithm for solving Problem 21 was obtained in [42]:

Method 2 ([42])**.** *Let the functions $f_j(\lambda)$, $j = 1,2$, the subspectrum $\{\lambda_n\}_{n\geq 1}$, and the number ω be given. One must construct the potential q.*

1. *Using $f_j(\lambda)$, $j = 1,2$, $\{\lambda_n\}_{n\geq 1}$, and ω, construct the vector functions $\{v_n\}_{n\geq 0}$ and the numbers $\{w_n\}_{n\geq 0}$ using Formulas* (53)–(56)*.*
2. *For the basis $\{v_n\}_{n\geq 0}$, find the biorthonormal basis $\{v_n^*\}_{n\geq 0}$, that is, $(v_n, v_k^*)_{\mathcal{H}} = \delta_{nk}$, $n,k \geq 0$.*
3. *Construct the element $u \in \mathcal{H}$ satisfying* (52) *using the formula*

$$u = \sum_{n=0}^{\infty} \overline{w_n} v_n^*.$$

4. *Using the elements of* $u(t) = [\overline{N(t)}, \overline{K(t)}]$, *solve Problem 22 with the Cauchy data and find* q.

It is worth noting that, in the case of simple eigenvalues $\{\lambda_n\}_{n\geq 1}$, Problem 21 is a special case of Problem 2, which was studied by Horváth [12]. Indeed, the numbers $\{\lambda_n\}_{n\geq 1}$ can be treated as the eigenvalues of different boundary value problems for Equation (45) subject to the boundary conditions

$$y(0) = 0, \quad y'(\pi)\cos\beta_n + y(\pi)\sin\beta_n = 0, \quad \beta_n := \arctan\frac{f_2(\lambda_n)}{f_1(\lambda_n)}.$$

On the other hand, using the given data of Problem 21, one can easily find the values of the Weyl function in the points $\{\lambda_n\}_{n\geq 1}$:

$$M(\lambda_n) = -\frac{S'(\pi,\lambda_n)}{S(\pi,\lambda_n)} = \frac{f_2(\lambda_n)}{f_1(\lambda_n)}.$$

Thus, Problem 21 is closely related to the problem of potential reconstruction from the values $\{M(\lambda_n)\}_{n\geq 1}$ (see Problem 3). The uniqueness of this problem solution was studied by Horváth [12]. A constructive solution is provided by Method 2 with necessary modifications. Namely, in the definitions of v_n and w_n, one should replace $f_1(\lambda_n)$ with 1 and $f_2(\lambda_n)$ with $M(\lambda_n)$. Thus, to the best of the author's knowledge, a constructive algorithm for the recovery of the potential q from the values $\{M(\lambda_n)\}_{n\geq 1}$ of the Weyl function at a countable set of points was obtained for the first time in [42]. Effective numerical algorithms for this reconstruction were developed by Kravchenko and Torba [80]. The technique of [80] was based on the representations of the Sturm-Liouville equation solutions as Neumann series of Bessel functions. The methods of [80] can be applied to various classes of partial inverse problems.

Proceeding with the results of [42], in applications to partial inverse problems, it can be difficult to verify the conditions (COMPLETE) and (BASIS). Therefore, the following easily verified conditions are introduced (for convenience, let $\lambda_0 = 0$):

(COMPLETE C)—the sequence $\{\cos(\sqrt{\lambda_n}t)\}_{n\geq 0}$ is complete in $L_2(0,2\pi)$.

(BASIS C)—the sequence $\{\cos(\sqrt{\lambda_n}t)\}_{n\geq 0}$ is a Riesz in $L_2(0,2\pi)$.

(SEPARATION)—for every $n \geq 1$, there exists $f_1(\lambda_n) \neq 0$ or $f_2(\lambda_n) \neq 0$.

(ASYMPTOTICS)—$\operatorname{Im}\rho_n = O(1)$, $n \to \infty$, and $\{\rho_n^{-1}\}_{n\geq n_0} \in l_2$, where $\rho_n := \sqrt{\lambda_n}$, $\arg\rho_n \in \left[-\frac{\pi}{2}, \frac{\pi}{2}\right)$.

Theorem 24 ([42]).

1. (SEPARATION) *and* (COMPLETE C) *together imply* (COMPLETE).
2. (SEPARATION), (ASYMPTOTICS), *and* (BASIS C) *together imply* (BASIS).

Thus, one can replace the condition (COMPLETE) in Theorem 22 with (SEPARATION) and (COMPLETE C) and the condition (BASIS) in Method 2 with (SEPARATION), (ASYMPTOTICS), and (BASIS C). The results remain valid.

The investigation of Problem 21 was continued in [43], which studied the solvability and stability of the inverse problem. In particular, the following sufficient conditions for the global solvability of Problem 21 were obtained:

Theorem 25 ([43]). *Let* $f_j(\lambda)$, $j = 1,2$, *be entire functions, and let* $\{\lambda_n\}_{n\geq 1}$ *and* ω *be complex numbers such that the sequence* $\{v_n\}_{n\geq 0}$ *constructed by them satisfies the condition* (BASIS) *and* $\left\{\frac{w_n}{\|v_n\|_{\mathcal{H}}}\right\} \in l_2$. *Then, following Method 2, one can construct the functions* $K, N \in L_2(0,\pi)$. *If the zeros* $\{\theta_{nj}\}_{n\geq 1}$, $j = 1,2$, *of the corresponding functions* $\eta_j(\lambda)$, $j = 1,2$, *defined by* (49) *and* (50) *are real and interlace in the sense*

$$\theta_{n2} < \theta_{n1} < \theta_{n+1,2}, \quad n \in \mathbb{N}, \tag{57}$$

then there exists a unique real-valued function $q \in L_2(0,\pi)$ *such that the sequence* $\{\lambda_n\}_{n\ge1}$ *is a subspectrum of* $R(q, f_1, f_2)$ *and* $\frac{1}{2}\int_0^\pi q(x)\,dx = \omega$.

Note that the interlacing property (57) appears from the necessary and sufficient conditions for the solvability of the classical Borg problem:

Theorem 26 ([4]). *For sequences* $\{\theta_{nj}\}_{n\ge1}$, $j = 1,2$, *of real numbers to be the spectra of the corresponding problems* $L_j(q)$, $j = 1,2$, *for Sturm-Liouville Equation* (45) *with a real-valued potential* $q \in L_2(0,\pi)$ *subject to the boundary conditions* $y(0) = y^{(j-1)}(\pi) = 0$, *it is necessary and sufficient to have the asymptotics*

$$\sqrt{\theta_{nj}} = n - \frac{j-1}{2} + \frac{\omega}{\pi n} + \frac{\varkappa_{nj}}{n}, \quad n \in \mathbb{N}, \quad j = 1,2, \quad \{\varkappa_{nj}\} \in l_2,$$

and the interlacing property (57).

In fact, Problem 21 is reduced to the Borg problem by Method 2, and then the a posteriori condition (57) is imposed. Analogous a posteriori conditions appeared in the papers of Hryniv and Mykytyuk [14] and Martinyuk and Pivovarchik [17] for the Hochstadt–Lieberman problem (see Theorems 9 and 11). As already mentioned in Section 2.2, such conditions seem to be unavoidable for the solvability of partial inverse problems.

Furthermore, in [43], the local solvability and stability of Problem 21 were obtained. In order to formulate these results, one needs the following additional condition:

(ESTIMATES)—there exist constants $a_j > 0$, $j = 1,2,3$, and $\{\alpha_n\}_{n\ge1}$ such that

$$|f_j(\rho^2)| \le a_1|\rho_n|^{\alpha_n+j-1}, \quad j = 1,2, \quad |\rho - \rho_n| \le \frac{a_2}{|\rho_n|},$$
$$|w(\rho^2)| \le a_1|\rho_n|^{\alpha_n+1}, \quad |\rho - \rho_n| \le \frac{a_2}{|\rho_n|},$$
$$|f_1(\lambda_n)|^2 + |\lambda_n|^{-1}|f_2(\lambda_n)|^2 \ge a_3|\lambda_n|^{\alpha_n}, \quad n \ge 1.$$

Although these estimates look complicated, they naturally appear in applications involving partial inverse problems on graphs, the inverse transmission eigenvalue problem, etc.

Theorem 27 ([43]). *Let* $R(q, f_1, f_2)$ *be a fixed boundary value problem of the form* (45) *and* (46), *and let* $\{\lambda_n\}_{n\ge1}$ *be a fixed subspectrum of* $R(q, f_1, f_2)$. *Suppose that the conditions* (BASIS), (ASYMPTOTICS), *and* (ESTIMATES) *are fulfilled. Then, there exists* $\varepsilon > 0$ *(depending on* $R(q, f_1, f_2)$ *and* $\{\lambda_n\}_{n\ge1}$*) such that, for every complex sequence* $\{\tilde\lambda_n\}_{n\ge1}$ *satisfying the estimate*

$$\Xi := \left(\sum_{n=1}^{\infty} (|\rho_n| + 1)^{-2}|\rho_n - \tilde\rho_n|^2\right)^{1/2} \le \varepsilon, \quad \tilde\rho_n := \sqrt{\tilde\lambda_n}, \tag{58}$$

there exists a complex-valued function $\tilde q \in L_2(0,\pi)$ *such that* $\omega = \tilde\omega$, *and* $\{\tilde\lambda_n\}_{n\ge1}$ *is a subspectrum of the corresponding problem* $R(\tilde q, f_1, f_2)$. *Moreover,*

$$\|q - \tilde q\|_{L_2(0,\pi)} \le C\Xi, \tag{59}$$

where the constant C depends only on $R(q, f_1, f_2)$, $\{\lambda_n\}_{n\ge1}$ *and not on* $\{\tilde\lambda_n\}_{n\ge1}$.

Note that here, Theorem 27 was formulated for simple eigenvalues $\{\lambda_n\}_{n\ge1}$. However, in [43], it was proved for the general case of multiple eigenvalues. The multiplicities in the sequences $\{\lambda_n\}_{n\ge1}$ and $\{\tilde\lambda_n\}_{n\ge1}$ may be distinct, since the groups of multiple eigenvalues in $\{\lambda_n\}_{n\ge1}$ may split into smaller groups under a small perturbation. This effect was taken into account in [43]. The proof of Theorem 27 relies on Method 2 and the local solvability and stability of Problem 22 using the Cauchy data, which was proved in [42]. In addition, note that Theorem 27 contains no a posteriori conditions of the type (57).

Thus, for Problem 21, the following results have been obtained:

- The necessary and sufficient conditions of uniqueness;
- A constructive solution;
- Simple sufficient conditions for uniqueness and the algorithm;
- Sufficient conditions for the global solvability;
- Local solvability and stability.

Below, we discuss the studies on inverse problems with entire functions in the boundary conditions for other types of operators. The Sturm-Liouville problem analogous to (45) and (46) with the Robin boundary condition $y'(0) - hy(0) = 0$ was considered in [44]. However, in [44], proofs were provided only for simple eigenvalues. Moreover, in the proof of the local solvability and stability theorem, reduction to the Borg problem by the two spectra was used. Unfortunately, the application of the Borg theorem in [44] allows us to obtain only the stability estimate $\|q - \tilde{q}\| \le C\Xi^{1/p}$, where p is the maximal eigenvalue multiplicity in the Borg problem (see [44] for details). The reduction to the inverse problem using the Cauchy data allows us to obtain a better estimate (59) without the power $1/p$.

In [41], the inverse problem analogous to Problem 21 was studied for a singular potential $q \in W_2^{-1}(0, \pi)$, and the results were applied to a partial inverse problem on an arbitrary graph (see Section 3.3 for details).

Kuznetsova [45] studied the inverse problem for the differential pencil

$$-y'' + q(x)y + 2\lambda p(x)y = \lambda^2 y, \quad x \in (0, \pi),$$

$$y(0) = 0, \quad f_1(\lambda)y^{[1]}(\pi) + f_2(\lambda)y(\pi) = 0,$$

where $q \in W_2^{-1}(0, \pi)$, $p \in L_2(0, \pi)$, $y^{[1]} = y' - \sigma y$, $q = \sigma'$, $\sigma \in L_2(0, \pi)$. The results of [45] included:

- Uniqueness;
- A constructive solution;
- Simple sufficient conditions for uniqueness and the algorithm;
- Application to Hochstadt–Lieberman-type problems.

Bondarenko and Chitorkin [46] investigated the inverse problem for the Sturm-Liouville equation (45) subject to the boundary conditions

$$p_1(\lambda)y'(0) + p_2(\lambda)y(0) = 0, \quad f_1(\lambda)y'(\pi) + f_2(\lambda)y(\pi) = 0,$$

where $p_1(\lambda)$ and $p_2(\lambda)$ are relative prime polynomials of the spectral parameter λ, and $f_1(\lambda)$ and $f_2(\lambda)$ are entire functions. In [46], the uniqueness of the inverse problem solution was studied, and the results were applied to Hochstadt–Lieberman-type problems with polynomial dependence on λ not only in the boundary conditions but also in the discontinuity conditions inside the interval.

4.2. Applications to Partial Inverse Problems

In this subsection, we show how partial inverse problems can be reduced to Problem 21 with entire functions in the boundary conditions. As examples, we consider the following partial inverse problems:

- The Hochstadt–Lieberman problem (Problem 1);
- The McLaughlin–Polyakov problem (Problem 5);
- A partial inverse problem on a star-shaped graph (Problem 9);
- A partial inverse problem on a graph of an arbitrary structure (Problem 20).

We start with the application to the Hochstadt–Lieberman problem, which is described in [42]. Consider the following eigenvalue problem:

$$-y''(x) + q(x)y(x) = \lambda y(x), \quad x \in (0, 2\pi), \tag{60}$$

$$y(0) = y(2\pi) = 0, \tag{61}$$

with a complex-valued potential $q \in L_2(0, 2\pi)$. Let $\{\lambda_n\}_{n\geq 1}$ denote the eigenvalues of the problems presented in (60) and (61), counted with their multiplicities and numbered according to their asymptotics

$$\sqrt{\lambda_n} = \frac{n}{2} + \frac{\omega_{2\pi}}{\pi n} + o\left(n^{-1}\right), \quad n \to \infty, \tag{62}$$

where $\omega_{2\pi} := \frac{1}{2}\int_0^{2\pi} q(x)\,dx$. The Hochstadt–Lieberman problem in this case is formulated as follows:

Problem 23 ([42]). *Suppose that the potential $q(x)$ is known a priori for $x \in (\pi, 2\pi)$. Given the spectrum $\{\lambda_n\}_{n\geq 1}$ (counting with multiplicities), find the potential $q(x)$ for $x \in (0, \pi)$.*

Let us show that Problem 23 can be reduced to Problem 21 with entire functions in the boundary condition. Let $S(x, \lambda)$ and $\psi(x, \lambda)$ denote the solution of Equation (60) satisfying the initial conditions

$$S(0, \lambda) = 0, \quad S'(0, \lambda) = 1, \quad \psi(2\pi, \lambda) = 0, \quad \psi'(2\pi, \lambda) = -1.$$

The eigenvalues of (60) and (61) coincide with the zeros of the characteristic function

$$\Delta(\lambda) = \psi(\pi, \lambda)S'(\pi, \lambda) - \psi'(\pi, \lambda)S(\pi, \lambda). \tag{63}$$

Comparing (63) with (47), one can conclude that the eigenvalue problems presented in (60) and (61) are equivalent to the problem $R(q, f_1, f_2)$ given by (45) and (46) with

$$f_1(\lambda) := \psi(\pi, \lambda), \quad f_2(\lambda) := -\psi'(\pi, \lambda). \tag{64}$$

Note that these functions $f_j(\lambda), j = 1, 2$, are entire in the λ-plane and can be constructed by the known part of the potential $q(x)$, $x \in (\pi, 2\pi)$. The constant ω can also be found using the given data of Problem 23 by the formula

$$\omega = \omega_{2\pi} - \frac{1}{2}\int_{\pi}^{2\pi} q(x)\,dx,$$

where $\omega_{2\pi}$ can be determined from the asymptotics in (62). Thus, Problem 23 is reduced to Problem 21.

Suppose that the eigenvalues $\{\lambda_n\}_{n\geq 1}$ of the problem (60)–(61) are simple. Then, one can easily show that the conditions (Basis C), (Separation), (Asymptotics), and (Estimates) of the previous subsection hold. Therefore, Theorems 22 and 24 imply the following corollary:

Corollary 1 ([42]). *Let $\{\lambda_n\}_{n\geq 1}$ and $\{\tilde{\lambda}_n\}_{n\geq 1}$ be the spectra of the boundary value problems of the form (60) and (61) with potentials q and $\tilde{q}$, respectively. Suppose that $q(x) = \tilde{q}(x)$ a.e. on $(\pi, 2\pi)$ and $\lambda_n = \tilde{\lambda}_n$ for all $n \geq 1$. Then, $q(x) = \tilde{q}(x)$ a.e. on $(0, \pi)$. In other words, the solution of Problem 23 is unique. This solution can be found using Method 2, taking (64) into account.*

Obviously, the uniqueness of Corollary 1 is similar to the Hochstadt–Lieberman theorem (Theorem 1) for complex-valued potentials. Method 2 generalizes the algorithms of Buterin [16,65] (see Method 1) and Martinyuk and Pivovarchik [17] for solving the Hochstadt–Lieberman problem.

Theorem 27 implies the following corollary on the local solvability and stability of the Hochstadt–Lieberman problem:

Corollary 2. *For any complex-valued function $q \in L_2(0, 2\pi)$, there exists $\varepsilon > 0$ such that, for any complex sequence $\{\tilde{\lambda}_n\}_{n \geq 1}$ close to the spectrum $\{\lambda_n\}_{n \geq 1}$ of the problem* (60)–(61) *in the sense* (58)*, there exists a complex-valued function $\tilde{q} \in L_2(0, 2\pi)$ such that $q(x) = \tilde{q}(x)$ a.e. on $(\pi, 2\pi)$ and $\{\tilde{\lambda}_n\}_{n \geq 1}$ is the spectrum of the problem* (60)–(61) *with the potential $\tilde{q}$. Moreover, $\|q - \tilde{q}\|_{L_2(0,\pi)} \leq C\Xi$, where the constant C depends only on q.*

It is worth noting that, since the potential $q(x)$ in (60) is complex-valued, a finite number of eigenvalues can be multiple. In this case, Corollaries 1 and 2 remain valid, and Method 2 is also valid with necessary technical modifications (see [42,43] for details). Therefore, to the best of the author's knowledge, Theorem 27 provides the first results on the local solvability and stability of the Hochstadt–Lieberman problem in the general case of a complex-valued potential with eigenvalues that are not necessarily simple. Theorem 25 can also be transferred to the Hochstadt–Lieberman problem.

An analogous reduction can be applied to Hochstadt–Lieberman-type inverse problems with the discontinuity conditions

$$y(d+) = ay(d-), \quad y'(d+) = a^{-1}y(d-) + by(d-),$$

and/or polynomial dependence on the spectral parameter in the boundary conditions (see, e.g., [9,81]). If all the discontinuities and the polynomial dependence lie on the "known" part of the interval, then such a partial inverse problem can be similarly reduced to Problem 21 for (45) and (46). The opposite case requires a separate investigation, which can be implemented analogously.

Proceeding to the McLaughlin–Polyakov problem (Problem 5), the reduction of this problem to Problem 21 was briefly described in [43]. We present it here in more detail.

Suppose that $a \in [0, 1) \cup (1, 3]$. Let $y_j(x, \lambda)$, $j = 1, 2$, denote the solutions of Equation (18) satisfying the initial conditions

$$y_1(1, \lambda) = y_2'(1, \lambda) = 0, \quad -y_1'(1, \lambda) = y_2(1, \lambda) = 1.$$

Obviously, the function

$$\xi(x, \lambda) := y_2(x, \lambda)\frac{\sin \rho a}{\rho} - y_1(x, \lambda) \cos \rho a. \tag{65}$$

for each $\lambda \in \mathbb{C}$ is the only solution (up to a constant multiplier) of Equation (18) satisfying the boundary condition (19) at $x = 1$. Therefore, for every eigenvalue λ_n of the boundary value problem (18)–(19), the corresponding eigenfunction has the form $S(x, \lambda_n) = c_n \xi(x, \lambda_n)$, where c_n is a constant. Consequently, the transmission eigenvalues coincide with the zeros of the characteristic function

$$\Delta(\lambda) := \begin{vmatrix} S(x, \lambda) & \xi(x, \lambda) \\ S'(x, \lambda) & \xi'(x, \lambda) \end{vmatrix}.$$

For $x = \alpha$, we have

$$\Delta(\lambda) = S(\alpha, \lambda)\xi'(\alpha, \lambda) - S'(\alpha, \lambda)\xi(\alpha, \lambda).$$

Comparing this relation with (47), one can conclude that the transmission eigenvalue problem can be represented as a Sturm-Liouville problem on the interval $(0, \alpha)$ with the entire functions

$$f_1(\lambda) := -\xi(\alpha, \lambda), \quad f_2(\lambda) := \xi'(\alpha, \lambda) \tag{66}$$

in the right-hand boundary condition. The only difference from the problem (45)–(46) is the interval length α instead of π. With this technical difference in mind, the McLaughlin–Polyakov problem is equivalent to Problem 21 with the functions $f_j(\lambda)$, $j = 1,2$, defined by (66) and with an almost real subspectrum $\{\lambda_n\}_{n\geq 1}$. The number $\omega = \frac{1}{2}\int_0^\alpha q(x)\,dx$ can be found using the asymptotics (21) and the known potential q on the subinterval $(\alpha, 1)$:

$$\omega = \lim_{n\to+\infty}((1-a)\sqrt{\lambda_n} - \pi n)\pi n - \frac{1}{2}\int_\alpha^1 q(x)\,dx.$$

It can be shown that, in the case of the simple subspectrum $\{\lambda_n\}_{n\geq 1}$, the conditions (BASIS C), (SEPARATION), (ASYMPTOTICS), and (ESTIMATES) of Section 4.1 hold. For the case of multiple eigenvalues, all the results are valid with some technical modifications. Consequently, the uniqueness theorem of McLaughlin and Polyakov (Theorem 12) can be easily deduced as a corollary of Theorems 22 and 24. The solution of the McLaughlin–Polyakov problem can be found using Method 2, taking the relation (66) into account and replacing π with α. Theorem 25 implies the following corollary on the global solvability of the McLaughlin–Polyakov problem:

Corollary 3. *Let numbers $a \in [0,1) \cup (1,3]$, $\omega \in \mathbb{R}$, and a real-valued function $\tilde{q} \in L_2(\alpha, 1)$ be fixed. For a sequence $\{\lambda_n\}_{n\geq 1}$ to be an almost real subspectrum of the transmission eigenvalue problem* (18)–(19) *with a potential $q \in L_2(0,1)$ such that $q(x) = \tilde{q}(x)$ a.e. on $(\alpha, 1)$ and $\frac{1}{2}\int_0^1 q(x)\,dx = \omega_0$, the following conditions are necessary and sufficient:*

1. *$\{\lambda_n\}_{n\geq 1}$ satisfies the asymptotics* (21).
2. *The zeros $\{\theta_{nj}\}_{n\geq 1, j=1,2}$ of the functions $\eta_j(\lambda)$, $j = 1,2$, defined by* (49) *and* (50) *using the functions $K, N \in L_2(0,\alpha)$, which are constructed by Method 2, are real and interlace in the sense of* (57).

Note that the global solvability of the inverse transmission eigenvalue problem was also investigated by Buterin et al. [69]. However, in [69], another problem statement was considered. The potential was not assumed to be known a priori on the subinterval $(\alpha, 1)$.

In addition, one can apply Theorem 27 to obtain the local solvability and stability of the McLaughlin–Polyakov problem. However, this result would be weaker than that of Theorem 13 proposed by Bondarenko and Buterin [22], because Theorem 27 does not allow one to take perturbations of the potential $q(x)$ on $(\alpha, 1)$ into account.

It is worth mentioning that the transmission eigenvalue problem (18)–(19) can be represented as the following boundary value problem on the three-edge graph in Figure 11:

$$\begin{gathered}
-y_j''(x_j) + q_j(x_j)y_j(x_j) = \lambda y_j(x_j), \quad x_j \in (0, T_j), \quad j = 1,2,3, \\
y_1(0) = 0, \quad y_1(T_1) = y_2(0), \quad y_1'(T_1) = y_2'(0), \\
y_2(T_2) = y_3(0), \quad y_2'(T_2) = -y_3'(0), \quad y_3(T_3) = 0, \\
T_1 := \alpha, \quad T_2 := 1-\alpha, \quad T_3 := a, \quad q_1(x) := q(x), \quad q_2(x) := q(x+\alpha), \quad q_3(x) := 0.
\end{gathered}$$

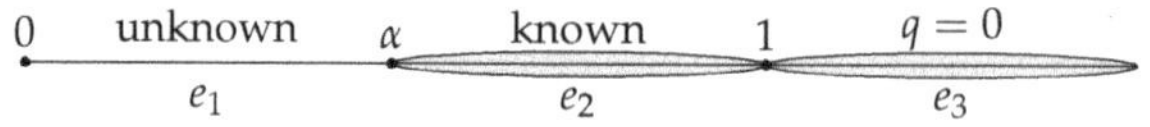

Figure 11. Graph representation of the transmission eigenvalue problem.

In order to model the condition $y'(1)\cos\rho a - y'(1)\dfrac{\sin\rho a}{\rho} = 0$, one can add a dummy edge of length a with a zero potential. Note that the matching conditions at the vertex joining e_2 and e_3 are non-standard and irregular. Nevertheless, the methods for partial inverse problems on graphs can also be used for the McLaughlin–Polyakov problem.

Next, consider Problem 9 for the Sturm-Liouville problem L of the form (22)–(24) on a star-shaped graph. In contrast to [33], we suppose that the potentials $\{q_j\}_{j=1}^m$ are complex-valued. Recall that the characteristic function of problem L is given by Formula (25):

$$\Delta(\lambda) := \sum_{j=1}^{m} S_j'(\pi,\lambda) \prod_{\substack{k=1 \\ k\neq j}}^{m} S_k(\pi,\lambda).$$

Comparing (25) with (47), one can easily see that the eigenvalue problem L on the star-shaped graph is equivalent to the problem (45) and (46) with $q = q_1$ and with the following entire functions in the boundary condition:

$$f_1(\lambda) := \prod_{k=2}^{m} S_k(\pi,\lambda), \quad f_2(\lambda) := \sum_{j=2}^{m} S_j'(\pi,\lambda) \prod_{\substack{k=2 \\ k\neq j}}^{m} S_k(\pi,\lambda).$$

Suppose that a subspectrum $\Omega = \{\lambda_{nk}\}_{n\geq 1, k=1,2}$ satisfying the asymptotics (28) and (29) and Condition 1 is given together with the potentials $\{q_j\}_{j=2}^m$. Then, Condition 1 implies the separation condition $f_1(\lambda_{nk}) \neq 0$ or $f_2(\lambda_{nk}) \neq 0$ for $n \geq 1$, $k = 1,2$. The number $\omega = \omega_1$ can be found from the asymptotics (28). The functions $f_1(\lambda)$ and $f_2(\lambda)$ can be constructed using the potentials $\{q_j\}_{j=2}^m$. Thus, Problem 9 is reduced to Problem 21 by the subspectrum Ω. In [43], the results of Section 4.1 were applied to this problem, and so the results of [30,33] were generalized to the case of complex-valued potentials. Certain other conditions of [30,33] were weakened. In particular, the local solvability and stability theorem (generalizing Theorem 16) was proved in the following form:

Theorem 28 ([43]). *Let $\{q_j\}_{j=1}^m$ be fixed complex-valued functions of $L_2(0,\pi)$, and let $\{\lambda_{nk}\}_{n\geq 1, k=1,2}$ be eigenvalues of the problem L satisfying the asymptotic relations* (28) *and* (29)*. Suppose that Condition 1 holds and $z_2 \neq \omega_j$, $j = \overline{2,m}$. Then, there exists $\varepsilon > 0$ (depending on $\{q_j\}_{j=1}^m$ and $\{\lambda_{nk}\}_{n\geq 1, k=1,2}$) such that, for any sequence $\{\tilde{\lambda}_{nk}\}_{n\geq 1, k=1,2}$ satisfying the estimate*

$$\Xi := \left(\sum_{n=1}^{\infty} \sum_{k=1}^{2} (|\lambda_{nk}| + 1) |\sqrt{\lambda_{nk}} - \sqrt{\tilde{\lambda}_{nk}}|^2 \right)^{1/2} \leq \varepsilon,$$

there exists a unique complex-valued function $\tilde{q}_1 \in L_2(0,\pi)$ such that $\{\tilde{\lambda}_{nk}\}_{n\geq 1, k=1,2}$ is a subspectrum of the problem $\tilde{L}$ with $\tilde{q}_1$ instead of q_1. Moreover, $\|q_1 - \tilde{q}_1\|_{L_2(0,\pi)} \leq C\Xi$, where the constant C depends only on $\{q_j\}_{j=1}^m$ and $\{\lambda_{nk}\}_{n\geq 1, k=1,2}$.

An analogous reduction was applied to Problem 20 on an arbitrary graph with an unknown potential on a boundary edge in [41]. The characteristic function for the corresponding boundary value problem in (41) is given by Formula (43):

$$\Delta(\lambda) = S_1(T_1,\lambda)\Delta^K(\lambda) + S_1^{[1]}(T_1,\lambda)\Delta^\Pi(\lambda).$$

Consequently, the problem in (41) can be represented in the form

$$\left. \begin{aligned} -(y^{[1]}(x))' - \sigma(x)y^{[1]}(x) - \sigma^2(x)y(x) = \lambda y(x), \quad x \in (0,T), \\ y(0) = 0, \quad f_1(\lambda)y^{[1]}(T) + f_2(\lambda)y(T) = 0, \end{aligned} \right\} \tag{67}$$

where $\sigma := \sigma_1$, $y^{[1]} = y' - \sigma y$, $T := T_1$, $f_1(\lambda) := \Delta^\Pi(\lambda)$, $f_2(\lambda) := \Delta^K(\lambda)$.

In [41], an inverse spectral theory for the problem (67) was created analogously to the theory in Section 4.1. Consequently, the results for the partial inverse problem on an arbitrary graph (Problem 20), which were described in Section 3.3, were obtained.

5. Other Types of Operators

This section deals with partial inverse problems for other classes of operators different from Sturm-Liouville differential operators and pencils. Namely, we consider the known results for the following types of operators:

- Integro-differential operators;
- Functional differential operators with a constant delay;
- Higher-order differential operators;
- Matrix Sturm-Liouville operators.

The most complete results in this direction have been obtained for integro-differential operators with an integral term in the form of convolution. Wang and Wei [47] studied a partial inverse problem for the integro-differential equation

$$-y'' + q(x)y + \int_0^x M(x-t)y(t)\,dt = \lambda y, \quad x \in (0, \pi), \tag{68}$$

with the Robin boundary conditions

$$y'(0) - hy(0) = 0, \quad y'(\pi) + Hy(\pi) = 0, \tag{69}$$

where $q(x)$ and $M(x)$ are real-valued functions of $L_2(0, \pi)$, and h and H are real constants. The spectrum of the problem (68)–(69) is denoted by $\sigma(L) = \{\lambda_n\}_{n \geq 0}$.

The following Gestezy–Simon-type uniqueness theorem was proved for the problem (68)–(69):

Theorem 29 ([47]). *Suppose that $a \in [0, \pi)$, $h = \tilde{h}$, $M(x) = \tilde{M}(x)$ a.e. on $(0, a)$, and $q(x) = \tilde{q}(x)$ on $(0, \pi)$. Then, for any $\varepsilon > 0$, if a subspectrum $S \subseteq \sigma(L) \cap \sigma(\tilde{L})$ satisfies*

$$\#\{\lambda_n \in S \colon |\lambda_n| \leq t\} \geq \left(1 - \frac{a}{\pi}\right)\#\{\lambda_n \in \sigma(L) \colon |\lambda_n| \leq t\} + \frac{a}{2\pi} - \frac{1}{2} + \varepsilon,$$

where $t \geq t_0$, t_0 is a positive constant, then $H = \tilde{H}$ and $M(x) = \tilde{M}(x)$ a.e. on (a, π).

However, the results of [47] are limited to uniqueness. Later on, Buterin and Sat [48] studied not only uniqueness but also reconstruction and subspectrum characterization for an integro-differential operator half-inverse problem. In [48], the integro-differential Equation (68) was considered subject to the Dirichlet boundary conditions

$$y(0) = y(\pi) = 0. \tag{70}$$

The functions $q(x)$ and $(\pi - x)M(x)$ were assumed to be complex-valued and belong to $L_2(0, \pi)$.

Buterin and Sat [48] studied the following inverse problem:

Problem 24 ([48]). *Given the even subspectrum $\{\lambda_{2n}\}_{n \geq 1}$, find the function $M(x)$ on $(\pi/2, \pi)$, provided that $M(x)$ on $(0, \pi/2)$ and the potential $q(x)$ are known.*

Buterin and Sat also proved the following theorem, which provides the uniqueness of the solution and the even subspectrum characterization of Problem 24.

Theorem 30 ([48]). *Let arbitrary complex-valued functions $q(x) \in L_2(0, \pi)$ and $f(x) \in L_2(0, \pi/2)$ be given and fixed. Then, for any sequence of complex numbers $\{\mu_n\}_{n \geq 1}$ of the form*

$$\mu_n = \left(2n + \frac{A}{2n} + \frac{\varkappa_n}{n}\right)^2, \quad A = \frac{1}{2\pi}\int_0^\pi q(x)\,dx, \quad \{\varkappa_n\} \in l_2, \quad n \geq 1, \tag{71}$$

there exists a unique (up to a set of measure zero) function $M(x)$ *such that* $(\pi - x)M(x) \in L_2(0, \pi)$, $M(x) = f(x)$ *on* $(0, \pi/2)$, *and* $\{\mu_n\}_{n \geq 1}$ *is the even subspectrum (i.e.,* $\lambda_{2n} = \mu_n$*) of the boundary value problem* (68)–(70).

Moreover, Buterin and Sat [48] provided a constructive algorithm for solving Problem 24. The method of [48] was based on the technique created by Buterin for solving inverse problems for integro-differential operators (see [82] and the references therein).

The results of [48] showed the principal difference between differential and integro-differential operators. In half-inverse problems for integro-differential operators, the given mixed data (eigenvalues and operator coefficients on a subinterval) are independent of each other. In Problem 24, one can take arbitrary numbers satisfying the eigenvalue asymptotics (71) and an arbitrary function $M(x)$ on $(0, \pi/2)$ and reconstruct $M(x)$ on $(\pi/2, \pi)$. In Hochstadt–Lieberman-type problems for differential operators, the spectrum and the potential $q(x)$ on a subinterval are related to each other. This relationship implies hard-to-verify conditions in the characterization theorems (see, e.g., Theorems 9 and 11).

It is worth mentioning that Sat and Yilmaz [49] attempted to study a partial inverse problem of another kind for the integro-differential operator (68)–(70). Namely, they assumed that the kernel $M(x)$ is known on $(0, \pi)$ and the potential $q(x)$ is known on the half-interval $(\pi/2, \pi)$ and investigated the recovery of $q(x)$ on the interval $(0, \pi/2)$ from the spectrum. However, the results of [49] were wrong, and the proofs contained mistakes. Namely, the estimate $O\left(\frac{1}{\rho^2}\right)$ after Formula (2.10) in [49] was incorrect. Therefore, the problem of recovering $q(x)$ on a subinterval while $M(x)$ is known remains open.

Bondarenko and Yurko [50] studied the following partial inverse problem for a Sturm-Liouville-type operator with a constant delay. Let $\{\lambda_{n,j}\}_{n \geq 1}$, $j = 0, 1$, denote the eigenvalues of the corresponding boundary value problems

$$-y''(x) + q(x)y(x - a) = \lambda y(x), \quad 0 < x < \pi, \tag{72}$$

$$y(0) = y^{(j)}(\pi) = 0, \tag{73}$$

where $a \in \left[\frac{\pi}{3}, \frac{\pi}{2}\right)$, $q(x)$ is a complex-valued potential of $L_2(0, \pi)$, and $q(x) = 0$ a.e. on $(0, a)$.

Problem 25 ([50]). *Assume that* $q(x)$ *is known a priori for* $x \in \left[\frac{3a}{2}, \pi - \frac{a}{2}\right]$. *Given subspectra* $\{\lambda_{n_k,j}\}_{k \geq 1}$, $j = 0, 1$, *find* $q(x)$ *on* (a, π) *(see Figure 12).*

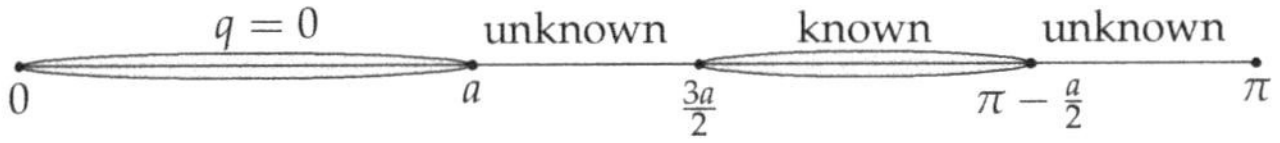

Figure 12. Partial inverse problem with delay.

Note that Problem 25 is different from the Hochstadt–Lieberman problem, since the potential $q(x)$ is given on an interior subinterval. However, for differential operators with a constant delay, the statement of Problem 25 appears to be natural.

Bondarenko and Yurko [50] proved the following uniqueness theorem and obtained a constructive algorithm for finding the solution of Problem 25.

Theorem 31 ([50]). *Suppose that the sequences* $\{\cos n_k x\}_{k \geq 0}$ $(n_0 := 0)$ *and* $\left\{\sin\left(n_k - \frac{1}{2}\right)x\right\}_{k \geq 1}$ *are complete in* $L_2(0, \pi - a)$, $q(x) = \tilde{q}(x)$ *a.e. on* $\left[\frac{3a}{2}, \pi - \frac{a}{2}\right]$ *and* $\lambda_{n_k,j} = \tilde{\lambda}_{n_k,j}$, $k \geq 1$, $j = 0, 1$. *Then,* $q(x) = \tilde{q}(x)$ *a.e. on* (a, π).

Djurić and Vladičić [51] considered the boundary value problem (72)–(73) in the case $a \in \left(\frac{\pi}{3}, \frac{2\pi}{5}\right)$ and noticed that the two full spectra $\{\lambda_{n,j}\}_{n \geq 1}$, $j = 0, 1$, uniquely specify the potential on not only the boundary subintervals $(a, \frac{3a}{2})$ and $(\pi - \frac{a}{2}, \pi)$, but also the interior subinterval $(\pi - a, 2a)$ (see Figure 13). In this case, knowledge of the potential on the subintervals $\left(\frac{3a}{2}, \pi - a\right)$ and $(\pi - a, 2a)$ is unnecessary.

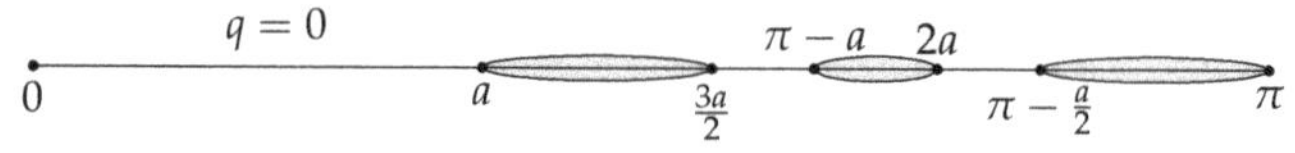

Figure 13. The potential recovered by Djurić and Vladičić.

Theorem 32 ([51]). *The spectra $\{\lambda_{n,j}\}_{n \geq 1}$, $j = 0, 1$, uniquely determine the potential $q(x)$ on the set $\left(a, \frac{3a}{2}\right) \cup (\pi - a, 2a) \cup \left(\pi - \frac{a}{2}, \pi\right)$.*

Moreover, the following uniqueness theorem for a partial inverse problem was proved.

Theorem 33 ([51]). *Assume that the potential is known on the set $\left(\frac{3a}{2}, \frac{\pi}{2} + \frac{a}{4}\right)$ as well as the integral $\int_{\pi/2+a/4}^{\pi-a} q(x)\,dx$. Then, the spectra $\{\lambda_{n,j}\}_{n \geq 1}$, $j = 0, 1$, uniquely determine the potential $q(x)$ on (a, π).*

In [52], Buterin et al. conducted a comprehensive study of inverse spectral problems for quadratic differential pencils with delays of the form

$$y''(x) + \rho^2 y(x) = q_0(x) y_0(x - a_0) + 2\rho q_1(x) y_1(x - a_1), \quad x \in (0, \pi), \tag{74}$$

where ρ is the spectral parameter, $a_0 \in \left[\frac{\pi}{3}, \pi\right)$, $a_1 \in \left[\frac{\pi}{2}, \pi\right)$, $a_0 + a_1 \geq \pi$, $q_\nu \in W_2^\nu[a_\nu, \pi]$, $q_\nu(x) = 0$ on $(0, a_\nu)$, and $\int_{a_1}^{\pi} q_1(x)\,dx = 0$. Let $\{\rho_{n,j}\}$ denote the spectra of the boundary value problems for Equation (74) subject to the boundary conditions $y(0) = y^{(j)}(\pi) = 0$, $j = 0, 1$. In particular, Buterin et al. [52] generalized Theorem 32 to the pencil in (74).

Theorem 34 ([52]). *Let both spectra $\{\rho_{n,j}\}$, $j = 0, 1$, be specified. Then, the function $q_0(x)$ is uniquely determined a.e. on $\left(a, \frac{3a_0}{2}\right) \cup (\pi - a_0, 2a_0) \cup \left(\pi - \frac{a_0}{2}, \pi\right)$, while $q_1(x)$ is uniquely determined on $[a_1, \pi]$.*

Theorem 33 was also generalized (see [52] for details).

Next, let us consider the higher-order differential equation

$$y^{(n)} + \sum_{k=0}^{n-2} p_k(x) y^{(k)} = \lambda y, \quad n > 2, \quad x \in (0, T), \tag{75}$$

on a finite interval ($T < \infty$) and the half-line ($T = \infty$). The general theory of inverse spectral problems for Equation (75) was created by Yurko [53]. In Section 4 of [53], Yurko considered partial inverse problems that consisted in the recovery of part of the coefficients $\{p_{\varkappa_j}\}_{j=1}^N$ ($\varkappa = \{\varkappa_j\}_{j=1}^N \subseteq \{0, 1, \dots, n-2\}$) from the Weyl functions $\{\mathfrak{M}_i(\lambda)\}_{i=1}^N$, which were defined using suitable boundary conditions (see [53] for details). The other coefficients $\{p_k\}_{k \notin \varkappa}$ were assumed to be known a priori and integrable on either the finite or infinite interval $(0, T)$. The unknown coefficients $\{p_k\}_{k \in \varkappa}$ were assumed to be piece-wise analytic functions. The partial inverse problem was considered under a specific information condition, which guaranteed its unique solvability. The solution was constructed by the method of standard models.

Recently, Chen et al. [54] attempted to study the Hochstadt–Lieberman-type inverse problem for the fourth-order differential equation

$$y^{(4)} + q(x)y = \lambda^4 y, \quad x \in (0,1), \quad q \in L_1(0,1), \tag{76}$$

subject to the boundary conditions

$$y(0) = y'(0) = 0, \quad y(1) = y'(1) = 0. \tag{77}$$

The inverse problem of [54] consists in the recovery of the potential $q(x)$ on the half-interval $(1/2, 1)$ from the eigenvalues $\{\lambda_k\}$ of (76) and (77), while the potential $q(x)$ on $(0, 1/2)$ is known a priori. However, the main result of [54] (Theorem 1.1) was wrong. In particular, the authors of [54] asserted that, for any sequence $\{\lambda_k\}_{k\in\mathbb{Z}\setminus\{0\}}$ satisfying the conditions $\lambda_{-k} = \lambda_k$, $0 < \lambda_1^4 \le \lambda_2^4 \le \dots \le \lambda_N^4 < \lambda_{N+1}^4 < \dots$, and the asymptotics

$$\lambda_k = \left(k - \frac{1}{2}\right)\pi + \beta_k, \quad \{\beta_k\} \in l_2, \tag{78}$$

there exists a corresponding potential q of class L_1. However, the asymptotics (78) are not precise. For example, Polyakov [83] recently obtained more precise eigenvalue asymptotics, implying that not every sequence satisfying (78) together with the other conditions of [54] can be a spectrum of the problem (76) and (77) with potential $q \in L_1(0,1)$. This was not the only mistake of [54]. Furthermore, it is surprising that, in the Hochstadt–Lieberman-type theorem in [54], the eigenvalues $\{\lambda_k\}$ and the potential $q(x)$ on $(0, 1/2)$ are not related to each other. For the second-order case, there is such a relationship (see, e.g., Theorems 9 and 11). Nevertheless, the problem stated in [54] is a challenging issue for future investigation.

Malamud [55,56] proved the following analog of the Hochstadt–Lieberman theorem for the matrix Sturm-Liouville equations

$$-y'' + Q(x)y = \lambda^2 y, \quad -\tilde{y}'' + \tilde{Q}(x)\tilde{y} = \lambda^2 \tilde{y}, \quad x \in (0,1), \tag{79}$$

where $Q(x)$ and $\tilde{Q}(x)$ are $(n \times n)$ matrix functions. Let I_n denote the $(n \times n)$ unit matrix.

Theorem 35 ([55,56]). *Let the entries of $Q(x)$ and $\tilde{Q}(x)$ be complex-valued functions of $L_1(0,1)$, and let $Q(x) = \tilde{Q}(x)$ for a.a. $x \in [1/2, 1]$. Let $Y(x,\lambda)$ and $\tilde{Y}(x,\lambda)$ be the $(n \times n)$ matrix solutions of the initial value problems*

$$Y(0,\lambda) = \tilde{Y}(0,\lambda) = I_n, \quad Y'(0,\lambda) = H_1, \, \tilde{Y}'(0,\lambda) = \tilde{H}_1$$

for the first and second equations in (79)*, respectively. If*

$$Y'(1,\lambda) + H_2 Y(1,\lambda) = \tilde{Y}'(1,\lambda) + H_2 \tilde{Y}(1,\lambda) = 0, \quad \lambda \in \mathbb{C},$$

for some $(n \times n)$ complex matrix H_2, then $H_1 = \tilde{H}_1$ and $Q(x) = \tilde{Q}(x)$ for a.a. $x \in [0,1]$.

Theorem 35 shows that the monodromy matrix $Y'(1,\lambda) + H_1 Y(1,\lambda)$ uniquely determines the matrix potential $Q(x)$ on the half-interval $[0, 1/2]$ if $Q(x)$ is known on $[1/2, 1]$.

6. Conclusions

In this review, we considered selected results on partial inverse spectral problems for differential operators.

The most complete results were obtained for the Hochstadt–Lieberman problem. Several constructive methods were developed that allowed researchers to obtain numerical algorithms for solutions and the necessary and sufficient conditions for the solvability of half-inverse problems. The uniqueness of the inverse problem solution was studied fairly completely for cases in which the potential is known a priori on a subinterval $(0, a)$. Some results have also been obtained for the known potential on an interior subinterval

$(a, b) \subset (0, 1)$. However, cases in which the potential is unknown on an interior subinterval and is known on some boundary subintervals remain open.

For differential operators on geometrical graphs, the most simple situation occurs when the potential is unknown only on a boundary edge or even on part of a boundary edge. Such partial inverse problems can be reduced to inverse problems on an interval with entire functions in a boundary condition using the unified approach. These entire functions are constructed by the operator coefficients on the "known" part of the graph. Therefore, for this kind of problems, uniqueness, constructive solutions, global solvability, local solvability, and stability have been obtained even on graphs of an arbitrary geometrical structure. Analogous ideas can be applied to cases in which the potential is unknown on a boundary subgraph. Cases in which the potential is known on some interior edges of the graph have also been considered. For the unknown potential on an interior part of a graph, the question is open, as with the case of the interval.

In addition, there have been several attempts to study partial inverse problems for non-local operators, higher-order differential operators, and differential systems. However, the results of these studies are fragmentary, and they do not form a general picture. Some ideas are easily transferred from Hochstadt–Lieberman problems for differential operators to other types of operators. However, for functional differential operators with a delay, higher-order differential operators, and other types of operators, fundamentally new problem statements appear to be natural and, consequently, different methods are required for their investigation.

In conclusion, we formulated several open problems.

Problem 26. *Determine the potential $q(x)$ of the Sturm-Liouville equation $-y'' + q(x)y = \lambda y$ on an interior subinterval $(a, b) \subset (0, 1)$ from fewer spectral data than are used for the complete inverse problem, while $q(x)$ is known on $(0, 1) \setminus (a, b)$ (see Figure 3).*

Problem 27. *Investigate the solvability and stability of the inverse Sturm-Liouville partial inverse problem on the interval $(0, 1)$ in the case of a known potential on an interior subinterval (see Figure 2) using the spectral data of Guo and Wei [63] or any other spectral data.*

Problem 28. *Study partial Sturm-Liouville inverse problems on graphs in case where the potentials are known on an interior subgraph. Determine the spectral data that are sufficient for the unique reconstruction of the potentials on the whole graph. This problem is open even for simple graphs (star-shaped graphs, lasso graphs, and trees).*

Problem 29. *Investigate the solvability and stability of partial inverse problems on graphs for cases in which the potentials are known on a boundary part of the graph. These issues have been studied only for an unknown potential on one edge.*

Problem 30. *Construct an inverse problem theory for the Sturm-Liouville equation with entire analytical functions in one of the boundary conditions* (45)–(46) *and discontinuity conditions of the form*

$$y(d+) = ay(d-), \quad y'(d+) = a^{-1}y(d-) + by(d-)$$

at one or several points inside the interval. Note that the investigation of this problem will open up the possibility of studying a wide class of partial inverse problems with discontinuities. Inverse Sturm-Liouville problems with discontinuities in interior points appear in electronics when constructing the parameters of heterogeneous electric lines with desirable technical charateristics [84] and in geophysical models of the Earth's oscillations [9].

Problem 31. *Study the reconstruction of the potential $q(x)$ of the integro-differential equation*

$$-y'' + q(x)y + \int_0^x M(x - t)y(t)\,dt = \lambda y$$

on a half-interval from spectral data under the assumption that $M(x)$ is known.

Problem 32. *Suppose that the coefficients $\{p_k\}_{k=0}^{n-2}$ of the higher-order differential equation*

$$y^{(n)} + \sum_{k=0}^{n-2} p_k(x)y^{(k)} = \lambda y, \quad n > 2, \quad x \in (0,1),$$

are known on the half-interval $(0,1/2)$. How many spectral data are sufficient for the unique specification of these coefficients on $(1/2,1)$? In particular, one can study this half-inverse problem for the fourth-order differential equation

$$y^{(4)} - (p(x)y')' + q(x)y = \lambda y. \tag{80}$$

Note that this equation is important for mechanical applications, since the Euler-Bernoulli equation $(a(x)u'')'' = \mu b(x)u$, which describes beam vibrations, can be reduced to the form of (80) *(see [85]).*

Thus, the theory of partial inverse spectral problems still poses many challenges.

Funding: This work was supported by grant 21-71-10001 of the Russian Science Foundation, https://rscf.ru/en/project/21-71-10001/, accessed on 18 April 2023 .

Data Availability Statement: Not applicable.

Acknowledgments: The author is grateful to Sergey Buterin for his valuable comments.

Conflicts of Interest: The author declares no conflict of interests.

References

1. Marchenko, V.A. *Sturm-Liouville Operators and Their Applications*; Birkhäuser: Basel, Switzerland, 1986.
2. Levitan, B.M. *Inverse Sturm-Liouville Problems*; VNU Science Press: Utrecht, The Netherlands, 1987.
3. Pöschel, J.; Trubowitz, E. *Inverse Spectral Theory*; Academic Press: New York, NY, USA, 1987.
4. Freiling, G.; Yurko, V. *Inverse Sturm-Liouville Problems and Their Applications*; Nova Science Publishers: Huntington, NY, USA, 2001.
5. Kravchenko, V.V. *Direct and Inverse Sturm-Liouville Problems*; Birkhäuser: Cham, Switzerland, 2020.
6. Borg, G. Eine Umkehrung der Sturm-Liouvilleschen Eigenwertaufgabe: Bestimmung der Differentialgleichung durch die Eigenwerte. *Acta Math.* **1946**, *78*, 1–96. (In German) [CrossRef]
7. Gel'fand, I.M.; Levitan, B.M. On the determination of a differential equation from its spectral function. *Izv. Akad. Nauk SSSR Ser. Mat.* **1951**, *15*, 309–360. (In Russian)
8. Hochstadt, H.; Lieberman, B. An inverse Sturm-Liouville problem with mixed given data. *SIAM J. Appl. Math.* **1978**, *34*, 676–680. [CrossRef]
9. Hald, O. Discontinuous inverse eigenvalue problem. *Commun. Pure Appl. Math.* **1984**, *37*, 539–577. [CrossRef]
10. Gesztesy, F.; Simon, B. Inverse spectral analysis with partial information on the potential, II. The case of discrete spectrum. *Trans. AMS* **2000**, *352*, 2765–2787. [CrossRef]
11. Horváth, M. On the inverse spectral theory of Schrödinger and Dirac operators. *Trans. AMS* **2001**, *353*, 4155–4171. [CrossRef]
12. Horváth, M. Inverse spectral problems and closed exponential systems. *Ann. Math.* **2005**, *162*, 885–918. [CrossRef]
13. Sakhnovich, L. Half-inverse problems on the finite interval. *Inverse Probl.* **2001**, *17*, 527–532. [CrossRef]
14. Hryniv, R.O.; Mykytyuk, Y.V. Half-inverse spectral problems for Sturm-Liouville operators with singular potentials. *Inverse Probl.* **2004**, *20*, 1423–1444. [CrossRef]
15. Buterin, S.A. On a constructive solution of the incomplete inverse Sturm-Liouville problem. *Math. Mekhanika Saratov State Univ.* **2009**, *11*, 8–12. (In Russian)
16. Buterin, S.A. On half inverse problem for differential pencils with the spectral parameter in boundary conditions. *Tamkang J. Math.* **2011**, *42*, 355–364 [CrossRef]
17. Martinyuk, O.; Pivovarchik, V. On the Hochstadt-Lieberman theorem. *Inverse Probl.* **2010**, *26*, 035011. [CrossRef]
18. Pivovarchik, V. On the Hald-Gesztesy-Simon theorem. *Integr. Equ. Oper. Theor.* **2012**, *73*, 383–393. [CrossRef]
19. McLaughlin, J.R.; Polyakov, P.L. On the uniqueness of a spherically symmetric speed of sound from transmission eigenvalues. *J. Diff. Equ.* **1994**, *107*, 351–382. [CrossRef]
20. McLaughlin, J.R.; Polyakov, P.L.; Sacks, P.E. Reconstruction of a spherically symmetric speed of sound. *SIAM J. Appl. Math.* **1994**, *54*, 1203–1223. [CrossRef]

21. McLaughlin, J.R.; Sacks, P.E.; Somasundaram, M. Inverse scattering in acoustic media using interior transmission eigenvalues. In *Inverse Problems in Wave Propagation*; Chavent, G., Papanicolaou, G., Sacks, P., Symes, W., Eds.; Springer: New York, NY, USA, 1997; pp. 357–374.
22. Bondarenko, N.; Buterin, S. On a local solvability and stability of the inverse transmission eigenvalue problem. *Inverse Probl.* **2017**, *33*, 115010. [CrossRef]
23. Xu, X.-C.; Ma, L.-J.; Yang, C.-F. On the stability of the inverse transmission eigenvalue problem from the data of McLaughlin and Polyakov. *J. Diff. Equ.* **2022**, *316*, 222–248. [CrossRef]
24. Berkolaiko, G.; Kuchment, P. *Introduction to Quantum Graphs*; American Mathematical Society: Providence, RI, USA, 2013.
25. Pokorny, Y.V.; Penkin, O. M.; Pryadiev, V.L.; Borovskikh, A.V.; Lazarev, K.P.; Shabrov, S.A. *Differential Equations on Geometrical Graphs*; Fizmatlit: Moscow, Russia, 2005. (In Russian)
26. Kuchment, P. Quantum graphs. Some basic structures. *Waves Random Media* **2004**, *14*, S107–S128. [CrossRef]
27. Yurko, V.A. Inverse spectral problems for differential operators on spatial networks. *Russ. Math. Surv.* **2016**, *71*, 539–584. [CrossRef]
28. Pivovarchik, V.N. Inverse problem for the Sturm-Liouville equation on a simple graph. *SIAM J. Math. Anal.* **2000**, *32*, 801–819. [CrossRef]
29. Yurko, V. Inverse nodal problems for Sturm-Liouville operators on star-type graphs. *J. Inverse Ill-Posed Probl.* **2008**, *16*, 715–722. [CrossRef]
30. Yang, C.-F. Inverse spectral problems for the Sturm-Liouville operator on a *d*-star graph. *J. Math. Anal. Appl.* **2010**, *365*, 742–749. [CrossRef]
31. Yang, C.-F. Inverse problems for the differential operator on a graph with cycles. *J. Math. Anal. Appl.* **2017**, *445*, 1548–1562. [CrossRef]
32. Yang, C.-F.; Wang, F. Inverse problems on a graph with loops. *J. Inverse Ill-Posed Probl.* **2017**, *25*, 373–380. [CrossRef]
33. Bondarenko, N.P. A partial inverse problem for the Sturm-Liouville operator on a star-shaped graph. *Anal. Math. Phys.* **2018**, *8*, 155–168. [CrossRef]
34. Bondarenko, N.P. Partial inverse problems for the Sturm-Liouville operator on a star-shaped graph with mixed boundary conditions. *J. Inverse Ill-Posed Probl.* **2018**, *26*, 1–12. [CrossRef]
35. Bondarenko, N.P.; Yang, C.-F. Partial inverse problems for the Sturm-Liouville operator on a star-shaped graph with different edge lengths. *Results Math.* **2018**, *73*, 56. [CrossRef]
36. Wang, Y.P.; Shieh, C.-T. Inverse problems for Sturm-Liouville operators on a star-shaped graph with mixed spectral data. *Appl. Anal.* **2020**, *99*, 2371–2380. [CrossRef]
37. Yang, C.-F.; Bondarenko, N.P. A partial inverse problem for the Sturm-Liouville operator on the lasso-graph. *Inverse Probl. Imaging* **2019**, *13*, 69–79. [CrossRef]
38. Bondarenko, N.P.; Shieh, C.-T. Partial inverse problems for quadratic differential pencils on a graph with a loop. *J. Inverse Ill-Posed Probl.* **2020**, *28*, 449–463. [CrossRef]
39. Bondarenko, N.P. An inverse problem for Sturm-Liouville operators on trees with partial information given on the potentials. *Math. Meth. Appl. Sci.* **2019**, *42*, 1512–1528. [CrossRef]
40. Bondarenko, N.P. Inverse problem for the differential pencil on an arbitrary graph with partial information given on the coefficients. *Anal. Math. Phys.* **2019**, *9*, 1393–1409. [CrossRef]
41. Bondarenko, N.P. A partial inverse Sturm-Liouville problem on an arbitrary graph. *Math. Meth. Appl. Sci.* **2021**, *44*, 6896–6910. [CrossRef]
42. Bondarenko, N.P. Inverse Sturm-Liouville problem with analytical functions in the boundary condition. *Open Math.* **2020**, *18*, 512–528. [CrossRef]
43. Bondarenko, N.P. Solvability and stability of the inverse Sturm-Liouville problem with analytical functions in the boundary condition. *Math. Meth. Appl. Sci.* **2020**, *43*, 7009–7021. [CrossRef]
44. Yang, C.-F.; Bondarenko, N.P.; Xu, X.-C. An inverse problem for the Sturm-Liouville pencil with arbitrary entire functions in the boundary condition. *Inverse Probl. Imaging* **2020**, *14*, 153–169. [CrossRef]
45. Kuznetsova, M.A. On recovering quadratic pencils with singular coefficients and entire functions in the boundary conditions. *Math. Meth. Appl. Sci.* **2023**, *46*, 5086–5098. [CrossRef]
46. Bondarenko, N.P.; Chitorkin, E.E. Inverse Sturm-Liouville problem with spectral parameter in the boundary conditions. *Mathematics* **2023**, *11*, 1138. [CrossRef]
47. Wang, Y.; Wei, G. The uniqueness for Sturm-Liouville problems with aftereffect. *Acta Math. Sci.* **2012**, *32A*, 1171–1178. (In Chinese)
48. Buterin, S.A.; Sat, M. On the half inverse spectral problem for an integro-differential operator. *Inverse Probl. Sci. Eng.* **2017**, *25*, 1508–1518. [CrossRef]
49. Sat, M.; Yilmaz, E. A Hochstadt-Lieberman theorem for integro-differential operator. *Inter. J. Pure Appl. Math.* **2013**, *88*, 413–423. [CrossRef]
50. Bondarenko, N.P.; Yurko, V.A. Partial inverse problems for the Sturm-Liouville equation with deviating argument. *Math. Meth. Appl. Sci.* **2018**, *41*, 8350–8354. [CrossRef]
51. Djurić, N.; Vladičić, V. Incomplete inverse problem for Sturm-Liouville type differential equation with constant delay. *Results Math.* **2019**, *74*, 161. [CrossRef]

52. Buterin, S.A.; Malyugina, M.A.; Shieh, C.-T. An inverse spectral problem for second-order functional-differential pencils with two delays. *Appl. Math. Comput.* **2021**, *411*, 126475. [CrossRef]
53. Yurko, V.A. Inverse problems of spectral analysis for differential operators and their applications. *J. Math. Sci.* **2000**, *98*, 319–426. [CrossRef]
54. Chen, L.; Shi, G.; Yan, J. On the Hochstadt-Lieberman theorem for the fourth-order binomial operator. *J. Math. Phys.* **2023**, *64*, 043503. [CrossRef]
55. Malamud, M.M. Uniqueness questions in inverse problems for systems of differential equations on a finite interval. *Trans. Mosc. Math. Soc.* **1999**, *60*, 173–224.
56. Malamud, M.M. Uniqueness of the matrix Sturm-Liouville equation given a part of the monodromy matrix, and Borg type results. In *Sturm-Liouville Theory*; Birkhäuser: Basel, Switzerland, 2005; pp. 237–270.
57. del Rio, R.; Gesztesy, F.; Simon, B. Inverse spectral analysis with partial information on the potential, III. Updating boundary conditions. *Intl. Math. Res. Not.* **1997**, *15*, 75–758.
58. Amour, L.; Raoux, T. Inverse spectral results for Schrödinger operators on the unit interval with potentials in l^p spaces. *Inverse Probl.* **2007**, *23*, 2367–2373. [CrossRef]
59. Amour, L.; Faupin, J.; Raoux, T. Inverse spectral results for Schrödinger operator on the unit interval with partial information given on the potentials. *J. Math. Phys.* **2009**, *50*, 033505. [CrossRef]
60. Horváth, M.; Kiss, M. Stability of direct and inverse eigenvalue problems for Schrödinger operators on finite intervals. *Int. Math. Res. Not.* **2010**, *2010*, 2022–2063. [CrossRef]
61. Horváth, M.; Kiss, M. Stability of direct and inverse eigenvalue problems: The case of complex potentials. *Inverse Probl.* **2011**, *27*, 095007. [CrossRef]
62. Horváth, M.; Sáfár, O. Inverse eigenvalue problems. *J. Math. Phys.* **2016**, *57*, 112102. [CrossRef]
63. Guo, Y.; Wei, G. Inverse Sturm-Liouville problems with the potential known on an interior subinterval. *Appl. Anal.* **2015**, *94*, 1025–1031. [CrossRef]
64. Hryniv, R.O.; Mykytyuk, Y.V. Inverse spectral problems for Sturm-Liouville operators with singular potentials. *Inverse Probl.* **2003**, *19*, 665–684. [CrossRef]
65. Buterin, S.A. On inverse spectral problem for non-selfadjoint Sturm-Liouville operator on a finite interval. *J. Math. Anal. Appl.* **2007**, *335*, 739–749. [CrossRef]
66. Buterin, S.A.; Shieh, C.-T. Incomplete inverse spectral and nodal problems for differential pencils. *Results Math.* **2012**, *62*, 167–179. [CrossRef]
67. Rundell, W.; Sacks, E. Reconstruction techniques for classical inverse Sturm-Liouville problems. *Math. Comput.* **1992**, *58*, 161–183. [CrossRef]
68. Hatinoğlu, B. Mixed data in inverse spectral problems for the Schrödinger operators. *J. Spec. Theor.* **2021**, *11*, 281–322. [CrossRef]
69. Buterin, S.A.; Choque-Rivero, A.E.; Kuznetsova, M.A. On a regularization approach to the inverse transmission eigenvalue problem. *Inverse Probl.* **2020**, *36*, 105002. [CrossRef]
70. Buterin, S.A.; Yang, C.-F. On an inverse transmission problem from complex eigenvalues. *Results Math.* **2017**, *71*, 859–866. [CrossRef]
71. Cakoni, F.; Colton, D.; Monk, P. On the use of transmission eigenvalues to estimate the index of refraction from far field data. *Inverse Probl.* **2007**, *23*, 507–522. [CrossRef]
72. Aktosun, T.; Gintides, D.; Papanicolaou, V.G. The uniqueness in the inverse problem for transmission eigenvalues for the spherically symmetric variable-speed wave equation. *Inverse Probl.* **2011**, *27*, 115004. [CrossRef]
73. Wei, G.; Xu, H.-K. Inverse spectral analysis for the transmission eigenvalue problem. *Inverse Probl.* **2013**, *29*, 115012. [CrossRef]
74. Gintides, D.; Pallikarakis, N. The inverse transmission eigenvalue problem for a discontinuous refractive index. *Inverse Probl.* **2017**, *33*, 055006. [CrossRef]
75. Yurko, V. Inverse spectral problems for Sturm-Liouville operators on graphs. *Inverse Probl.* **2005**, *21*, 1075–1086. [CrossRef]
76. Pivovarchik, V. Inverse problem for the Sturm-Liouville equation on a star-shaped graph. *Math. Nachr.* **2007**, *280*, 1595–1619. [CrossRef]
77. Sedletskii, A.M. Nonharmonic analysis. *J. Math. Sci.* **2003**, *116*, 3551–3619. [CrossRef]
78. Bondarenko, N.; Shieh, C.-T. Partial inverse problems for Sturm-Liouville operators on trees. *Proc. R. Soc. Edinb. Sect. A Math.* **2017**, *147*, 917–933. [CrossRef]
79. Vasiliev, S.V. An inverse spectral problem for Sturm-Liouville operators with singular potentials on arbitrary compact graphs. *Tamkang J. Math.* **2019**, *50*, 293–305. [CrossRef]
80. Kravchenko, V.V.; Torba, S.M. A practical method for recovering Sturm-Liouville problems from the Weyl function. *Inverse Probl.* **2021**, *37*, 065011. [CrossRef]
81. Guliyev, N.J. Essentially isospectral transformations and their applications. *Ann. Mat. Pura Appl.* **2020**, *199*, 1621–1648. [CrossRef]
82. Buterin, S. Uniform full stability of recovering convolutional perturbation of the Sturm-Liouville operator from the spectrum. *J. Diff. Equ.* **2021**, *282*, 67–103. [CrossRef]
83. Polyakov, D.M. On the spectral properties of a fourt-order self-adjoint operator. *Diff. Equ.* **2023**, *59*, 168–173. [CrossRef]

84. Meschanov, V.P.; Feldstein, A.L. *Automatic Design of Directional Couplers*; Sviaz: Moscow, Russia, 1980. (In Russian)
85. Gladwell, G.M.L. Inverse Problems in Vibration. In *Solid Mechanics and Its Applications*, 2nd ed.; Springer: Dordrecht, The Netherlands, 2005; Volume 119.

mathematics MDPI

Article

Geometric Approximation of Point Interactions in Three-Dimensional Domains

Denis Ivanovich Borisov [1,2,3]

[1] Institute of Mathematics, Ufa Federal Research Center, Russian Academy of Sciences, Ufa 450008, Russia; borisovdi@yandex.ru
[2] Institute of Mathematics, Informatics and Robotics, Bashkir State University, Ufa 450076, Russia
[3] Nikol'skii Mathematical Institute, Peoples Friendship University of Russia (RUDN University), Moscow 117198, Russia

Abstract: In this paper, we study a three-dimensional second-order elliptic operator with a point interaction in an arbitrary domain. The operator is supposed to be self-adjoint. We cut out a small cavity around the center of the interaction and consider an operator in such perforated domain with the Robin condition on the boundary of the cavity. Our main result states that once the coefficient in this Robin condition is appropriately chosen, the operator in the perforated domain converges to that with the point interaction in the norm resolvent sense. We also succeed in establishing order-sharp estimates for the convergence rate.

Keywords: point interaction; small cavity; Robin condition; norm resolvent convergence; convergence rate

MSC: 46L87

Citation: Borisov, D.I. Geometric Approximation of Point Interactions in Three-Dimensional Domains. *Mathematics* **2024**, *12*, 1031. https://doi.org/10.3390/math12071031

Academic Editor: Natalia Bondarenko

Received: 19 February 2024
Revised: 27 March 2024
Accepted: 27 March 2024
Published: 29 March 2024

1. Introduction

Operators with singular point interactions are a popular model in modern mathematical physics, which have attracted a lot of attention. They have been used to model physical systems, in which an interaction is supported in a small area [1]. While for one-dimensional operators such operators look rather simple, the two- and three-dimensional cases are more delicate. In the pioneering work [2], Berezin and Faddeev provided a method of dealing with such cases. After that, there appeared many works devoted to operators with point interactions. Here, we mention only a famous monograph [3] and refer to many references provided therein.

One of the directions of studying operators with point interactions is a corresponding perturbation theory. Namely, there is a natural question of how to approximate such operators by the ones with regular coefficients in the norm resolvent sense. A usual method is to use operators with regular coefficients and to suppose that some of these coefficients are located in a small area and are large in the area. The results of such kind are discussed in much detail in [3]; see also [4,5].

In our recent works [6,7], we suggested a completely new alternative approach to approximating two-dimensional operators with point interactions via an appropriate geometric perturbation. In [6], we considered differential operators with a fixed differential expression and the perturbation was a small cavity about the center of the interaction, which was cut out from the domain. On the boundary of the cavity, a special Robin boundary condition was imposed. The coefficient in this condition was large and depended on a small parameter, which governed the size of the cavity. Once the cavity shrank to the center of the interaction, we showed that, in the sense of the norm resolvent convergence, the perturbed operator converges to an operator with a point interaction and the latter is determined by the shape of the cavity and the coefficient in the Robin condition. However, it turned out that in this way, we could approximate not all values of the coupling constant, and the admissible values of such coupling constant should satisfy a certain upper bound. At the same time, an important feature of our result is that it was established for an operator with a general differential expression and not just for the Laplacian, which has been treated in many previous works. To the best of our knowledge, a general definition of operators with point interaction on manifolds with arbitrary differential expressions was given for the first time in a very recent work [8].

In [7], we succeeded in dealing with non-self-adjoint operators, but the boundary condition on the boundary of the cavity was non-local. Such non-locality as well as non-self-adjointness allowed us to omit the aforementioned upper bound from [6] for the admissible values of the coupling constant.

It should be noted that small cavities are a very classical example in singular perturbation theory. The case of classical boundary conditions has been studied many times. Here, we mention only some books [9–12] as well as many references therein. Typical results show a convergence of the solutions for given right hand sides and the convergence is either weak or strong in appropriate Sobolev spaces. Once the right-hand sides in a problem are smooth enough, it is also possible to construct asymptotic expansions for the solutions, and this has been performed in many situations in a series of works. We also mention some recent results on norm resolvent convergence for problems in perforated domains [13–19]. However, in all these works, the boundary conditions were not too singular and could not produce point interactions in the limit.

In this present paper, we extend the approach of [6,7] to the three-dimensional case. Namely, we consider an arbitrary second-order differential operator in an arbitrary three-dimensional domain with varying coefficients. As in [6], we suppose that this operator is self-adjoint. Then, we add a point interaction to this operator and show how to approximate it by cutting out a small cavity. On the boundary of this cavity, we, again, impose a Robin condition with an appropriately scaled coefficient. Then, we show that once the coupling constant satisfies an appropriate upper bound, the operator on the domain with the cavity approximates the operator with the point interaction in the resolvent sense. Moreover, we succeed in providing estimates for the convergence rate and show that they are order-sharp. The established norm resolvent convergence implies the convergence of the spectrum and of the associated spectral projections.

Our technique generally follows the lines of [6,7]. However, the three-dimensional case turns out to be much more difficult. The main difficulty is due to the completely different behavior of the fundamental solution of the Laplace operator in comparison with the two-dimensional case. Such difference destroys certain crucial local estimates from [6,7], and this is why, instead, we have to analyze a special Steklov problem corresponding to the considered cavity. Such analysis turns out to be an independent problem, which we solve in Section 4.1, and nothing like this is needed in the two-dimensional case.

2. Problem and Results

In the three-dimensional space $\mathbb{R}^3$, we choose an arbitrary non-empty domain, which is either bounded or unbounded, and we denote this domain by Ω. The situation in which Ω coincides with the entire space is possible. Once the boundary of the domain Ω is non-empty, we suppose that its smoothness is C^2. We use x_0 to denote an arbitrary fixed point in Ω, while ω is a bounded simply connected domain in $\mathbb{R}^3$ containing the origin; the boundary of ω is C^3-smooth. We introduce a small cavity around the point x_0 as $\omega_\varepsilon := \{x : (x - x_0)\varepsilon^{-1} \in \omega\}$, where $x = (x_1, x_2, x_3)$ are the Cartesian coordinates in $\mathbb{R}^3$ and ε is a small positive parameter.

Let $A_{ij} = A_{ij}(x)$, $A_j = A_j(x)$, and $A_0 = A_0(x)$ be real functions defined on the closure $\overline{\Omega}$ possessing the following smoothness: $A_{ij} \in C^4(\overline{\Omega})$, $A_j \in C^3(\overline{\Omega})$, $A_0 \in C^2(\overline{\Omega})$. The functions A_{ij} obey the standard ellipticity condition

$$A_{ij} = A_{ji}, \qquad \sum_{i,j=1}^{3} A_{ij}(x)\xi_i\xi_j \geqslant c_0(\xi_1^2 + \xi_2^2 + \xi_3^2)$$

for all $\xi_i \in \mathbb{R}$ and $x \in \overline{\Omega}$ with a fixed positive constant c_0 independent of x and ξ.

We consider a self-adjoint, scalar second-order differential operator $\mathcal{H}_\varepsilon$ with the differential expression

$$\hat{\mathcal{H}} := -\sum_{i,j=1}^{3} \frac{\partial}{\partial x_i} A_{ij} \frac{\partial}{\partial x_j} + \mathrm{i} \sum_{j=1}^{3} \left(A_j \frac{\partial}{\partial x_j} + \frac{\partial}{\partial x_j} A_j \right) + A_0$$

in $\Omega_\varepsilon := \Omega \setminus \overline{\omega_\varepsilon}$ subject to boundary conditions

$$\mathcal{B}u = 0 \quad \text{on} \quad \partial\Omega, \tag{1}$$

$$\frac{\partial u}{\partial \mathrm{n}} + \alpha(x,\varepsilon)u = 0 \quad \text{on} \quad \partial\omega_\varepsilon, \tag{2}$$

where

$$\alpha(x,\varepsilon) := \alpha_0(x - x_0) + \alpha_1\big((x - x_0)\varepsilon^{-1}\big), \tag{3}$$

$$\frac{\partial}{\partial \mathrm{n}} := \sum_{i,j=1}^{3} A_{ij}\nu_i \frac{\partial}{\partial x_i} - \mathrm{i} \sum_{j=1}^{3} \nu_j A_j,$$

and $\nu = (\nu_1, \nu_2, \nu_3)$ stands for the unit normal on $\partial\omega_\varepsilon$ directed inside ω_ε. $\mathcal{B}$ denotes an arbitrary boundary operator. The only restriction for this operator is that it should obey implicit assumptions, which we impose in what follows. Particular examples for the operator $\mathcal{B}$ are the ones corresponding to the Dirichlet, Neumann, Robin, or quasi-periodic boundary conditions. If $\partial\Omega$ is empty, then boundary condition (1) is not needed. The function α_0 is introduced as

$$\alpha_0(x) := -|\mathrm{A}_0^{-\frac{1}{2}} x| \nu \cdot \mathrm{A}_0 \nabla_x |\mathrm{A}_0^{-\frac{1}{2}} x|^{-1} = \frac{\nu \cdot x}{|\mathrm{A}_0^{-\frac{1}{2}} x|^2}, \tag{4}$$

where ν is the unit normal on $\partial\omega$ directed inside ω and $\mathrm{A}_0 := \mathrm{A}(0)$,

$$\mathrm{A}(x) := \begin{pmatrix} A_{11}(x) & A_{12}(x) & A_{13}(x) \\ A_{21}(x) & A_{22}(x) & A_{23}(x) \\ A_{31}(x) & A_{32}(x) & A_{33}(x) \end{pmatrix}.$$

The function $\alpha_1 = \alpha_1(s)$ is supposed to be real and continuous on $\partial\omega$, and it will be fixed later.

This paper aims to study the behavior of the resolvent of the operator $\mathcal{H}_\varepsilon$ for a small ε. Before formulating our main result, we need to introduce additional notation. $B_r(a)$ denotes the open ball of radius r centered at a point a. The definition of the cavity ω_ε implies the chain of inclusions

$$\omega_\varepsilon \subset B_{R_1\varepsilon}(x_0) \subset B_{2R_1\varepsilon}(x_0) \subset B_{R_2}(x_0) \subset B_{2R_2}(x_0) \subset \Omega_0 \subset \Omega$$

with some fixed positive constants R_1, R_2 independent of ε.

Let $\mathcal{H}_\Omega$ be the operator in $L_2(\Omega)$ with the differential expression $\hat{\mathcal{H}}$ subject to boundary condition (1); the associated sesquilinear form is denoted by $\mathfrak{h}_\Omega$. We make the following assumptions on the operator $\mathcal{H}_\Omega$ and its form $\mathfrak{h}_\Omega$, which are, in fact, implicit assumptions for the coefficients A_{ij}, A_j, and A_0 and for the boundary operator $\mathcal{B}$. The operator $\mathcal{H}_\Omega$ is self-adjoint and is lower semi-bounded, while the form $\mathfrak{h}_\Omega$ is closed and symmetric, and its domain $\mathfrak{D}(\mathfrak{h}_\Omega)$ is a subspace of $W_2^1(\Omega)$. The domain Ω contains a subdomain Ω_0 such that $x_0 \in \Omega_0$ and the restriction of each function from the domain $\mathfrak{D}(\mathcal{H}_\Omega)$ to Ω_0 belongs to $W_2^2(\Omega_0)$. The estimate

$$\mathfrak{h}_\Omega(u,u) - \mathfrak{h}_{\Omega_0}(u,u) + c_1\|u\|^2_{L_2(\Omega\setminus\Omega_0)} \geqslant c_2\|u\|^2_{W_2^1(\Omega\setminus\Omega_0)} \tag{5}$$

holds for all $u \in \mathfrak{D}(\mathfrak{h}_\Omega)$ with constants c_1, c_2 independent of u, and the constant c_2 is strictly positive. Given an arbitrary subdomain $\tilde\Omega \subset \Omega$ on $W_2^1(\tilde\Omega)$, we introduce an auxiliary form:

$$\begin{aligned}\mathfrak{h}_{\tilde\Omega}(u,v) := {} & \sum_{i,j=1}^{3}\left(A_{ij}\frac{\partial u}{\partial x_j}, \frac{\partial v}{\partial x_i}\right)_{L_2(\tilde\Omega)} + \mathrm{i}\sum_{j=1}^{3}\left(\frac{\partial u}{\partial x_j}, A_j v\right)_{L_2(\tilde\Omega)} \\ & - \mathrm{i}\sum_{j=1}^{3}\left(A_j u, \frac{\partial v}{\partial x_j}\right)_{L_2(\tilde\Omega)} + (A_0 u, v)_{L_2(\tilde\Omega)}.\end{aligned}$$

We suppose that for bounded subdomains $\tilde\Omega$ such that $\partial\tilde\Omega \cap \partial\Omega = \emptyset$, this auxiliary form satisfies the lower bound

$$\mathfrak{h}_{\tilde\Omega}(u,u) + c_1\|u\|^2_{L_2(\tilde\Omega)} \geqslant c_2\|u\|^2_{W_2^1(\tilde\Omega)} \tag{6}$$

with the constants c_1, c_2 from (5).

Rigorously, we introduce the operator $\mathcal{H}_\varepsilon$ in terms of the operator $\mathcal{H}_\Omega$ in the same way as in the two-dimensional case in [6]. Namely, we first introduce an auxiliary infinitely differentiable cut-off function χ with values in $[0,1]$ equal to the ones in $B_{2R_2}(x_0)$ and vanishing outside Ω_0. Then, $\mathcal{H}_\varepsilon$ is the operator in $L_2(\Omega_\varepsilon)$ with the differential expression $\hat{\mathcal{H}}$ on the domain $\mathfrak{D}(\mathcal{H}_\varepsilon)$, which consists of the functions u satisfying condition (2) and

$$(1-\chi)u \in \mathfrak{D}(\mathcal{H}_\Omega), \qquad \chi u \in W_2^2(\Omega_0 \setminus \omega_\varepsilon).$$

On this domain, the operator $\mathcal{H}_\varepsilon$ acts as follows:

$$\mathcal{H}_\varepsilon u := \mathcal{H}_\Omega(1-\chi)u + \hat{\mathcal{H}}\chi u.$$

It is proven in Section 3 in Lemma 3 that the boundary value problem

$$(\hat{\mathcal{H}} + c_1)G = 0 \quad \text{in} \quad \Omega\setminus\{x_0\}, \qquad \mathcal{B}G = 0 \quad \text{on} \quad \partial\Omega, \tag{7}$$

where c_1 is the constant from (5) and (6), possesses a unique solution in the space $W_2^2(\Omega \setminus B_\delta(x_0)) \cap C^2(\overline{\Omega_0} \setminus \{x_0\})$ for some $\delta > 0$ with the differentiable asymptotic at x_0:

$$\begin{aligned}& G(x) = G_{-1}(x-x_0) + G_0(x-x_0) + a_0 + O(|x-x_0|), \qquad x \to x_0, \qquad (8)\\ & G_{-1}(x) := |\mathrm{A}_0^{-\frac{1}{2}}x|^{-1},\\ & G_0(x) := \sum_{i,j=1}^{3} a_{ij}(x)\frac{\partial^2}{\partial x_i \partial x_j}|\mathrm{A}_0^{-\frac{1}{2}}x| + \sum_{i,j,k=1}^{3} a_{ijk}\frac{\partial^3}{\partial x_i\partial x_j\partial x_k}|\mathrm{A}_0^{-\frac{1}{2}}x|^3 + \sum_{j=1}^{3} a_j\frac{\partial}{\partial x_j}|\mathrm{A}_0^{-\frac{1}{2}}x|,\end{aligned}$$

where a_{ij} are homogeneous polynomials of order 1 with real coefficients; a_{ijk}, a_0 are real constants; and a_j are complex constants. We denote

$$\beta_0 := \sum_{j=1}^{3} \int\limits_{\partial\omega} x_j G_{-1}(x)\nu \cdot \frac{\partial \mathrm{A}}{\partial x_j}(x_0)\nabla G_{-1}(x)\,ds + \sum_{j=1}^{3} \int\limits_{\partial\omega} G_{-1}(x)\nu \cdot \mathrm{A}_0 \nabla \operatorname{Re} G_0(x)\,ds \\ - \sum_{j=1}^{3} \int\limits_{\partial\omega} \operatorname{Re} G_0(x)\nu \cdot \mathrm{A}_0 \nabla G_{-1}(x)\,ds. \tag{9}$$

It is shown in Lemma 3 that this constant is real.

We consider an auxiliary eigenvalue problem

$$\begin{aligned} &\operatorname{div}_\xi \mathrm{A}_0 \nabla_\xi \psi = 0 \quad \text{in} \quad \mathbb{R}^3 \setminus \omega, \qquad \lambda\nu \cdot \mathrm{A}_0 \nabla_\xi \psi + \alpha_0 \psi = 0 \quad \text{on} \quad \partial\omega, \\ &\psi(\xi) = C|\mathrm{A}_0^{-\frac{1}{2}}\xi|^{-1} + O(|\mathrm{A}_0^{-\frac{1}{2}}\xi|^{-2}), \qquad \xi \to \infty, \end{aligned} \tag{10}$$

where C is some constant depending on the choice of the function ψ. We show in Section 4.1 that this problem has at most countably many eigenvalues, each of these eigenvalues is real, and the greatest eigenvalue is equal to 1 and is simple. κ denotes the distance from 1 to the next closest eigenvalue of problem (10).

We let

$$\beta := a_0 - \frac{1}{4\pi(\det \mathrm{A}_0)^{\frac{1}{4}}} \left(\beta_0 + \int\limits_{\partial\omega} \alpha_1(s) G_{-1}^2(x)\,ds \right) \tag{11}$$

and assume that $\beta \neq a_0$. We also impose the condition

$$\beta_0 + \int\limits_{\partial\omega} \alpha_1(s) G_{-1}^2(x)\,ds < \kappa \|G\|_{L_2(\Omega)}^2. \tag{12}$$

$\mathcal{H}_{0,\beta}$ denotes the operator in $L_2(\Omega)$ with the differential expression $\hat{\mathcal{H}}$ and a point interaction at the point x_0. The domain of this operator and its action read as follows:

$$\mathfrak{D}(\mathcal{H}_{0,\beta}) := \left\{ u = u(x) : u(x) = v(x) + (\beta - a)^{-1} v(x_0) G(x),\ v \in \mathfrak{D}(\mathcal{H}_\Omega) \right\} \tag{13}$$

$$\mathcal{H}_{0,\beta} u = \mathcal{H}_\Omega v s. - c_1 (\beta - a)^{-1} v(x_0) G. \tag{14}$$

Here, the constant c_1 comes from (5) and (6), $\|\cdot\|_{X \to Y}$ denotes the norm of a bounded operator acting from a Hilbert space X into a Hilbert space Y, while $\sigma(\cdot)$ stands for a spectrum of an operator.

Our main result is as follows.

Theorem 1. *The operators $\mathcal{H}_\varepsilon$ and $\mathcal{H}_{0,\beta}$ are self-adjoint and satisfy the estimates*

$$\|(\mathcal{H}_\varepsilon - \lambda)^{-1} - (\mathcal{H}_{0,\beta} - \lambda)^{-1}\|_{L_2(\Omega) \to L_2(\Omega_\varepsilon)} \leqslant C\varepsilon^{\frac{1}{2}}, \tag{15}$$

$$\|\chi_{\tilde{\Omega}}((\mathcal{H}_\varepsilon - \lambda)^{-1} - (\mathcal{H}_{0,\beta} - \lambda)^{-1})\|_{L_2(\Omega) \to \mathfrak{D}(\mathfrak{h}_\Omega)} \leqslant C\varepsilon^{\frac{1}{2}}. \tag{16}$$

Here, $\tilde{\Omega}$ is an arbitrary fixed subdomain of Ω, the closure of which does not contain the point x_0, while $\chi_{\tilde{\Omega}}$ is an infinitely differentiable cut-off function equal to one on $\tilde{\Omega}$ and vanishing outside some larger fixed domain, the closure of which also does not contain the point x_0. The symbol C denotes positive constants independent of ε but depending on λ and additionally on the choice of $\tilde{\Omega}$ in (16). These estimates are order-sharp.

The convergence of the resolvents established in the above theorem implies the convergence of the spectrum and spectral projections. Such convergence can be established by a literal reproduction of the proof of Theorem 2.2 in [6]. This gives our second main result; in the following theorem, $\sigma(\,\cdot\,)$ denotes the spectrum of an operator.

Theorem 2. *The spectrum of the operator $\mathcal{H}_\varepsilon$ converges to that of $\mathcal{H}_{0,\beta}$ as $\varepsilon \to +0$. Namely, if $\lambda \notin \sigma(\mathcal{H}_{0,\beta})$, then $\lambda \notin \sigma(\mathcal{H}_\varepsilon)$ provided ε is small enough. If $\lambda \in \sigma(\mathcal{H}_{0,\beta})$; then, there exists a point $\lambda_\varepsilon \in \sigma(\mathcal{H}_\varepsilon)$ such that $\lambda_\varepsilon \to \lambda$ as $\varepsilon \to +0$. For any $\varrho_1, \varrho_2 \notin œ(\mathcal{H}_{0,\beta})$, $\varrho_1 < \varrho_2$, the spectral projection of $\mathcal{H}_\varepsilon$ corresponding to the segment $[\varrho_1, \varrho_2]$ converges to the spectral projection of $\mathcal{H}_{0,\beta}$ corresponding to the same segment in the sense of the norm $\|\cdot\|_{L_2(\Omega)\to L_2(\Omega_\varepsilon)}$.*

For each fixed segment $J := [\varrho_1, \varrho_2]$ of the real line, the inclusion

$$\sigma(\mathcal{H}_\varepsilon) \cap J \subset \{\lambda \in J : \operatorname{dist}(\lambda, \sigma(\mathcal{H}_{0,\beta}) \cap J) \leqslant C\varepsilon^{\frac{1}{2}}\}$$

holds, where C is a fixed constant independent of ε but depending on Q. If λ_0 is an isolated eigenvalue of $\mathcal{H}_{0,\beta}$ of multiplicity n, there exist exactly n eigenvalues of the operator $\mathcal{H}_\varepsilon$, counting multiplicities, which converge to λ_0 as $\varepsilon \to +0$. The total projection $\mathcal{P}_\varepsilon$ associated with these perturbed eigenvalues and the projection $\mathcal{P}_{0,\beta}$ onto the eigenspace associated with λ_0 satisfy estimates similar to (15) and (16).

Let us briefly discuss our problem and the main results. First of all, we stress that the operators we consider are rather general, namely, they have general differential expressions with variable coefficients and these coefficients can have a rather arbitrary behavior outside the domain Ω_0. Namely, once it is possible to define properly the operator $\mathcal{H}_\Omega$, our scheme works, and we can introduce the operators $\mathcal{H}_\varepsilon$ and $\mathcal{H}_{0,\beta}$. Such approach worked perfectly for two-dimensional operators in [6,7], and, here, we extend it to three-dimensional operators.

Our first main result, Theorem 1, states that a general three-dimensional operator with a point interaction can be approximated by cutting out a small hole around the center of the point interaction and by imposing a special Robin condition on its boundary. This condition is given by (1), and in view of the definition of the function α_0 in (4), we immediately see that

$$\alpha(x,\varepsilon) = \varepsilon^{-1}\alpha_0\left(\frac{x-x_0}{\varepsilon}\right) + \alpha_1\left(\frac{x-x_0}{\varepsilon}\right),$$

which means that the coefficient in this Robin condition grows as ε tends to zero. Under an appropriate choice of the function α_1, Theorem 1 states the convergence of the resolvent of $\mathcal{H}_\varepsilon$ to that of $\mathcal{H}_{0,\beta}$ in the operator norm $\|\cdot\|_{L_2(\Omega)\to L_2(\Omega_\varepsilon)}$ (see (15)). The convergence rate is $O(\varepsilon^{\frac{1}{2}})$, which is shown to be order-sharp. The second convergence expressed in estimate (16) means that once we consider the restriction of the resolvent of the operator $\mathcal{H}_\varepsilon$ to a subdomain of Ω separated from the point x_0, then the convergence also holds a stronger $\|\cdot\|_{L_2(\Omega)\to\mathfrak{D}(\mathfrak{h}_\Omega)}$-norm. The mentioned subdomain is controlled by the cut-off function. We also stress that both estimates (15) and (16) are order-sharp, and in Section 6, we adduce examples proving this statement. We also note that the norm $\|(\mathcal{H}_\varepsilon - \lambda)^{-1} - (\mathcal{H}_{0,\beta} - \lambda)^{-1}\|_{L_2(\Omega)\to W_2^1(\Omega_\varepsilon)}$ does not go to zero as $\varepsilon \to 0$ since an example from Section 6 shows that we only have

$$\|(\mathcal{H}_\varepsilon - \lambda)^{-1} - (\mathcal{H}_{0,\beta} - \lambda)^{-1}\|_{L_2(\Omega)\to W_2^1(\Omega_\varepsilon)} = O(1), \qquad \varepsilon \to 0. \tag{17}$$

The constant β describing the point interaction in the operator $\mathcal{H}_{0,\beta}$ cannot take all values on the real line because of assumption (12). This condition is, in fact, an upper bound for β, and it involves the constant κ, which is an implicit characteristic of the cavity ω. At the same time, we a priori know that $\kappa > 0$, and to obey (12), it is sufficient to suppose that

$$\beta_0 + \int\limits_{\partial\omega} \alpha_1(s) G_{-1}^2(x)\, ds < 0.$$

A more gentle sufficient condition for (12) could be given once we have a lower bound for κ expressed in some geometric characteristics of the boundary $\partial\omega$. Unfortunately, we fail in trying to find such lower bound. A possible way of getting it could be based on using a nice formula for the eigenvalues of (10), which we establish in this work (see (53)), and trying to obtain an appropriate minimax principle on its base. However, we fail in trying to find an appropriate set of functions over which we can take such minimax. We also mention that in the two-dimensional case, for self-adjoint operators, we have an upper bound for admissible values of β (see [6]).

Comparing our results with the ones established in [6,7] for the two-dimensional case, we mention the following important differences. The first of them is that the convergence rates in estimates (15) and (16) are now powers of ε, while in [6,7], similar rates are powers of $|\ln\varepsilon|^{-1}$. This means that for the three-dimensional operators, our approximation is better. A deep reason explaining this situation is a difference between the fundamental solutions of the Laplace operator in two and three dimensions. Due to the same reason, we to modify quite essentially a part of our proof for the three-dimensional operator, and this is the second important difference. Namely, one of the key ingredients is a lower-semiboundedness of the form associated with the perturbed operator, and we do need an explicit lower bound for this form. In the two-dimensional case, such lower bound is based on certain local estimates similar to the ones in Lemma 5 below. In the three-dimensional case, these local estimates are not enough, and we have to analyze an auxiliary Steklov problem; see Section 4.1 below.

Once we have the resolvent convergence stated in Theorem 1, it is possible to prove the convergence of the spectrum and the associated spectral projections. This can be performed by a literal reproduction of the proof of a similar theorem from [6], and it leads us to Theorem 2. This is why we do not provide the proof of Theorem 2 in this paper.

3. Auxiliary Statements

Here, we establish several lemmas, which are important ingredients in the proof of Theorem 1.

Lemma 1. *The identities*

$$\int\limits_{\partial\omega}\frac{\alpha_0(s)}{|\mathrm{A}_0^{-\frac{1}{2}}x|}\,\mathrm{d}s=\int\limits_{\partial\omega}\frac{\nu\cdot x}{|\mathrm{A}_0^{-\frac{1}{2}}x|^3}\,ds=4\pi\big(\det\mathrm{A}_0\big)^{\frac{1}{4}},\tag{18}$$

$$\int\limits_{\partial\omega}\frac{\alpha_0(s)}{|\mathrm{A}_0^{-\frac{1}{2}}x|^2}\,\mathrm{d}s=-\int\limits_{\mathbb{R}^3\setminus\omega}\frac{dx}{|\mathrm{A}_0^{-\frac{1}{2}}x|^4}\tag{19}$$

hold true.

Proof. We begin with an obvious equation:

$$\operatorname{div}\mathrm{A}_0\nabla_x|\mathrm{A}_0^{-\frac{1}{2}}x|^{-1}=0,\qquad x\in\mathbb{R}^3\setminus\{0\}.$$

We integrate by parts this equation over $\omega\setminus\{x:|x|<\delta\}$ with a sufficiently small δ:

$$\begin{aligned}0=&-\int\limits_{\partial\omega}\nu\cdot\mathrm{A}_0\nabla_x|\mathrm{A}_0^{-\frac{1}{2}}x|^{-1}\,ds+\int\limits_{\{x:\,|x|=\delta\}}\frac{x}{|x|}\cdot\mathrm{A}_0\nabla_x|\mathrm{A}_0^{-\frac{1}{2}}x|^{-1}\,ds\\ =&\int\limits_{\partial\omega}\frac{\nu\cdot x}{|\mathrm{A}_0^{-\frac{1}{2}}x|^3}\,ds-\int\limits_{\{x:\,|x|=\delta\}}\frac{|x|}{|\mathrm{A}_0^{-\frac{1}{2}}x|^3}\,ds=\int\limits_{\partial\omega}\frac{\nu\cdot x}{|\mathrm{A}_0^{-\frac{1}{2}}x|^3}\,ds-\int\limits_{\{x:\,|x|=1\}}\frac{ds}{|\mathrm{A}_0^{-\frac{1}{2}}x|^3}.\end{aligned}\tag{20}$$

Since the matrix A_0 is positive definite and Hermitian, there exists an orthogonal matrix reducing A_0 to its diagonal form. Performing the change of the variables with this matrix in the latter integral, then passing to the spherical coordinates, and denoting with Λ_j the eigenvalues of the matrix $\mathrm{A}_0^{-\frac{1}{2}}$, we obtain the following:

$$\begin{aligned}
\int\limits_{\{x:\,|x|=1\}} \frac{1}{|\mathrm{A}_0^{-\frac{1}{2}}x|^3}\,ds &= \int\limits_{\{x:\,|x|=1\}} \left(\sum_{j=1}^{3} \Lambda_j x_j^2\right)^{-\frac{3}{2}} ds \\
&= \int\limits_0^{2\pi} d\phi \int\limits_{-\frac{\pi}{2}}^{\frac{\pi}{2}} \frac{\cos\vartheta\, d\vartheta}{\left((\Lambda_1\cos^2\phi + \Lambda_2\sin^2\phi)\cos^2\vartheta + \Lambda_3\sin^2\vartheta\right)^{\frac{3}{2}}} \\
&= \int\limits_0^{2\pi} \frac{\sin\vartheta}{(\Lambda_1\cos^2\phi + \Lambda_2\sin^2\phi)\left((\Lambda_1\cos^2\phi + \Lambda_2\sin^2\phi)\cos^2\vartheta + \Lambda_3\sin^2\vartheta\right)^{\frac{1}{2}}}\Bigg|_{\vartheta=-\frac{\pi}{2}}^{\vartheta=\frac{\pi}{2}} d\phi \\
&= \frac{2}{\sqrt{\Lambda_3}} \int\limits_0^{2\pi} \frac{d\phi}{\Lambda_1\cos^2\phi + \Lambda_2\sin^2\phi} = \frac{8}{\sqrt{\Lambda_3}} \int\limits_0^{\frac{\pi}{2}} \frac{d\phi}{\Lambda_1\cos^2\phi + \Lambda_2\sin^2\phi} \\
&= \frac{8}{\sqrt{\Lambda_1\Lambda_2\Lambda_3}} \arctan\frac{\sqrt{\Lambda_2}}{\sqrt{\Lambda_1}}\tan\phi\Bigg|_{\phi=0}^{\phi=\frac{\pi}{2}} = 4\pi(\det \mathrm{A}_0)^{\frac{1}{4}}.
\end{aligned}$$

This formula and (20) imply an identity (18).

Similar to the above calculations, we integrate by parts as follows:

$$\begin{aligned}
0 &= \lim_{R\to+\infty} \int\limits_{B_R(0)\setminus\omega} |\mathrm{A}_0^{-\frac{1}{2}}x|^{-1} \operatorname{div}_x \mathrm{A}_0 |\mathrm{A}_0^{-\frac{1}{2}}x|^{-1}\,dx \\
&= \int\limits_{\partial\omega} |\mathrm{A}_0^{-\frac{1}{2}}x|^{-1}\nu\cdot \mathrm{A}_0\nabla_x|\mathrm{A}_0^{-\frac{1}{2}}x|^{-1}\,ds - \int\limits_{\mathbb{R}^3\setminus\omega} \mathrm{A}_0\nabla_x|\mathrm{A}_0^{-\frac{1}{2}}x|^{-1}\cdot\nabla_x|\mathrm{A}_0^{-\frac{1}{2}}x|^{-1}\,dx \\
&= -\int\limits_{\partial\omega} \frac{\alpha_0(s)}{|\mathrm{A}_0^{-\frac{1}{2}}x|^2}\,\mathrm{d}s - \int\limits_{\mathbb{R}^3\setminus\omega} \frac{dx}{|\mathrm{A}_0^{-\frac{1}{2}}x|^4},
\end{aligned}$$

and this proves (19). The proof is complete. □

With $\mathbb{Z}_+$ we denote the set of non-negative integral numbers, that is, $\mathbb{Z}_+ := \mathbb{N}\cup\{0\}$.

Lemma 2. *For each $m\in\mathbb{Z}_+$, each polynomial $P = P(x)$, and each multi-index $\gamma\in\mathbb{Z}_+^3$, the equation*

$$\sum_{i,j=1}^{3} A_{ij}(x_0)\frac{\partial^2 u(x)}{\partial x_i \partial x_j} = P(x-x_0)\frac{\partial^\gamma}{\partial x^\gamma}|\mathrm{A}_0^{-\frac{1}{2}}(x-x_0)|^{2m-1}, \qquad x\in\mathbb{R}^3\setminus\{x_0\},$$

possesses a solution of the form

$$u(x) = \sum_{\substack{\theta\in\mathbb{Z}_+^3\\ |\theta|\leqslant\deg P}} Q_\theta(x-x_0)\frac{\partial^{\gamma+\theta}}{\partial x^{\gamma+\theta}}|\mathrm{A}_0^{-\frac{1}{2}}(x-x_0)|^{2m+1+2|\theta|}, \tag{21}$$

where Q_θ are some polynomials with degrees obeying the inequality

$$\deg Q_\theta \leqslant \deg P - |\theta|.$$

Proof. It is sufficient to study only the case when A_0 coincides with the unit matrix E and $x_0 = 0$, since the general case is reduced to the one mentioned by the linear change $y = \mathrm{A}_0^{-\frac{1}{2}}(x-x_0)$. This is why we provide only the proof of the particular mentioned case.

We prove the lemma by induction in the degree of the polynomial P. We first consider the case $\deg P = 0$, that is, P is a constant. Then, it is straightforward to confirm that the equation

$$\Delta u = P\frac{\partial^\gamma}{\partial x^\gamma}|x|^{2m-1}, \qquad x \in \mathbb{R}^3 \setminus \{0\}, \tag{22}$$

possesses a solution

$$u(x) = \frac{P}{(2m+1)(2m+2)}\frac{\partial^\gamma}{\partial x^\gamma}|x|^{2m+1}.$$

Suppose that Equation (22) possesses a solution of the form in (21), with $\mathrm{A}_0 = \mathrm{E}$ and $x_0 = 0$ for all γ and all polynomials P with $\deg P \leqslant k$ for some $k \in \mathbb{Z}_+$. We then consider Equation (22) with a polynomial P such that $\deg P = k+1$ and seek its solution as

$$u(x) = \frac{P(x)}{(2m+1)(2m+2)}\frac{\partial^\gamma}{\partial x^\gamma}|x|^{2m+1} - \tilde{u}(x) \tag{23}$$

and for $\tilde{u}$, we then obtain the equation

$$\Delta u = \frac{2}{(2m+1)(2m+2)}\sum_{i=1}^{3}\frac{\partial P}{\partial x_i}\frac{\partial}{\partial x_i}\frac{\partial^\gamma}{\partial x^\gamma}|x|^{2m+1} + \Delta P\frac{\partial^\gamma}{\partial x^\gamma}|x|^{2m+1}.$$

The degrees of the polynomials $\frac{\partial P}{\partial x_i}$ and ΔP are at most $\deg P - 1$ and $\deg P - 2$, respectively, and by the induction assumption, the above equation possesses a solution of the form in (21), namely,

$$\tilde{u} = \sum_{\substack{\theta\in\mathbb{Z}_+^3\\ |\theta|\leqslant k}} Q_\theta(x)\frac{\partial^{\gamma+\theta}}{\partial x^{\gamma+\theta}}|x|^{2m+3+2|\theta|}$$

with some polynomials Q_θ of degrees $\deg Q_\theta \leqslant k - |\theta|$. Substituting this formula into (23), we arrive at (21) for $\deg P = k+1$. The proof is complete. $\square$

Estimates (5) and (6) show that the spectrum of the self-adjoint operator $\mathcal{H}_\Omega$ is a subset of the half-line $[c_2 - c_1, \infty)$. Then, the positivity of the constant c_2 implies that the resolvent $(\mathcal{H}_\Omega + c_1)^{-1}$ is well-defined.

We introduce an auxiliary sesquilinear form

$$\begin{aligned}\mathfrak{g}_\varepsilon(u,v) :=& \mathfrak{h}_\Omega((1-\chi)u,(1-\chi)v) + \mathfrak{h}_{\Omega_0\setminus\omega_\varepsilon}(\chi u,(1-\chi)v)\\ &+ \mathfrak{h}_{\Omega_0\setminus\omega_\varepsilon}((1-\chi)u,\chi v) + \mathfrak{h}_{\Omega_0\setminus\omega_\varepsilon}(\chi u,\chi v)\end{aligned} \tag{24}$$

on the domain

$$\mathfrak{D}(\mathfrak{g}_\varepsilon) := \Big\{u:\ (1-\chi)u \in \mathfrak{D}(\mathfrak{h}_\Omega),\ \chi u \in W_2^1(\Omega_0\setminus\omega_\varepsilon)\Big\}. \tag{25}$$

It is clear that this form is symmetric.

Lemma 3. *The boundary value problem (7) possesses a unique solution in* $W_2^2(\Omega\setminus B_{2R_2}(x_0)) \cap C^2(\overline{\Omega_0}\setminus\{x_0\})$ *with a differentiable asymptotic (8). The identity*

$$\left(\frac{\partial G}{\partial \mathrm{n}} + \alpha_0 G, G\right)_{L_2(\partial\omega_\varepsilon)} = \beta_0 + O(\varepsilon) \tag{26}$$

holds true, where β_0 *is introduced in (9), and this constant is real.*

Proof. We expand the coefficients A_{ij}, A_j, and A_0 of the differential expression $\hat{\mathcal{H}}$ by the Taylor formula about the point x_0, and using Lemma 2, we see that there exists a function $G_1(x)$ of the form

$$G_1(x) = G_0(x) + \sum_{j,\theta} P_{j,\theta}(x) \frac{\partial^\theta}{\partial x^\theta} |\mathrm{A}_0^{-\frac{1}{2}}(x)|^j, \tag{27}$$

where the sum is finite and is taken over $j \in \mathbb{Z}_+$ and $\theta \in \mathbb{Z}_+^3$, and $P_{j,\theta}$ are some polynomials, where

$$\sum_{j,\theta} P_{j,\theta}(x) \frac{\partial^\theta}{\partial x^\theta} |\mathrm{A}_0^{-\frac{1}{2}}(x)|^j = O(1), \qquad x \to 0, \tag{28}$$

such that

$$(\hat{\mathcal{H}} + c_1)(G_{-1}(x - x_0) + G_1(x - x_0)) = F_0(x),$$

where F_0 is continuous; the Lipschitz in $\overline{B_{2R_2}(x_0)}$ is infinitely differentiable in $B_{2R_2}(x_0) \setminus \{x_0\}$, and

$$F_0(x) = O(|x - x_0|), \qquad x \to x_0.$$

We seek the solution to the boundary value problem (7), (24) as

$$G(x) = G_2(x) + G_3(x), \qquad G_2(x) := \big(G_{-1}(x - x_0) + G_1(x - x_0)\big)\chi(x), \tag{29}$$

and the unknown function G_2 should solve the equation

$$(\mathcal{H}_\Omega + c_1)G_3 = F_2, \qquad F_2 := -\chi F_0 + F_1, \tag{30}$$

and F_1 is a certain polynomial expression of the derivatives of G_0 and χ up to the second order. This yields $F_2 \in L_2(\Omega) \cap C^\gamma(\overline{\Omega_0})$ for each $\gamma \in (0,1)$.

Since the point $-c_1$ is outside the resolvent set of the operator $\mathcal{H}_\Omega$, Equation (30) is uniquely solvable in $\mathfrak{D}(\mathcal{H}_\Omega)$. The standard Schauder estimates [20] imply that this solution belongs to $C^{2+\gamma}(\overline{\Omega_0})$. Therefore, the function G_3 satisfies the Taylor formula

$$G_3(x) = a_0 + O(|x - x_0|), \qquad x \to x_0,$$

with some constant a_0. Now, recovering the function G by Formula (29), we conclude that problem (7), (24) is uniquely solvable in $W_2^2(\Omega \setminus B_{2R_2}(x_0)) \cap C^2(\overline{\Omega_0} \setminus \{x_0\})$, and the solution satisfies asymptotics (8).

We confirm that the constant a_0 is real. We proceed as in the proof of Lemma 3.2 in [6], namely, as in Equations (3.9)–(3.12) in [6], and we obtain the following:

$$\mathfrak{h}_\Omega(G_3, G_3) + c_1 \|G_3\|^2_{L_2(\Omega)} - \big((\hat{\mathcal{H}} + c_1)G_2, G_2\big)_{L_2(\Omega)} = -\big((\hat{\mathcal{H}} + c_1)G_2, G\big)_{L_2(\Omega)}$$

and, denoting $\Omega^\delta := \Omega \setminus B_\delta(x_0)$,

$$\begin{aligned}
&((\mathcal{H}_\Omega + c_1)G_2, G_2)_{L_2(\Omega)} - \big((\hat{\mathcal{H}} + c_1)G_2, G\big)_{L_2(\Omega)} \\
&= \lim_{\delta \to +0} \left(\sum_{i,j=1}^{3} \left(A_{ij} \frac{\partial G_2}{\partial x_j}, \frac{\partial G_2}{\partial x_i} \right)_{L_2(\Omega^\delta)} - 2\,\mathrm{Im} \sum_{j=1}^{2} \left(A_j \frac{\partial G_2}{\partial x_j}, G_2 \right)_{L_2(\Omega^\delta)} \right. \\
&\quad \left. + ((A_0 + c_1)G_2, G_2)_{L_2(\Omega^\delta)} - \int_{\partial B_\delta(x_0)} G \overline{\frac{\partial G_2}{\partial \mathrm{n}}}\, ds \right).
\end{aligned} \tag{31}$$

We let

$$b_\delta(x) := \sum_{j=1}^{3} \nu_j A_j(x_0), \qquad x \in \partial B_\delta(x_0),$$
$$G_4(x) := G_{-1}(x-x_0) + \operatorname{Re} G_0(x-x_0), \qquad G_5(x) := \operatorname{Im} G_0(x-x_0),$$

where $x_{0,i}$ are the coordinates of the point x_0, and we observe that the functions b_δ and G_5 are odd with respect to each of the variables $x_i - x_{0,i}$, $i = 1,2,3$. Then, it follows from asymptotics (8) and Formulas (27)–(29) that, as $\delta \to +0$,

$$\begin{aligned}
\int\limits_{\partial B_\delta(x_0)} G \overline{\frac{\partial G_2}{\partial \mathrm{n}}}\, ds =& \int\limits_{\partial B_\delta(x_0)} (G_2 + a_0) \overline{\frac{\partial G_2}{\partial \mathrm{n}}}\, ds \\
=& -\frac{1}{\delta} \int\limits_{\partial B_\delta(x_0)} (G_2 + a_0) \sum_{i,j=1}^{3} A_{ij}(x_i - x_{i,0}) \frac{\partial \overline{G_2}}{\partial x_i}\, ds \\
& - \mathrm{i} \int\limits_{\partial B_\delta(x_0)} b_\delta |\mathrm{A}_0^{-\frac{1}{2}}(x-x_0)|^{-2}\, ds + o(1) \\
=& -\frac{a_0}{\delta} \int\limits_{\partial B_\delta(x_0)} \sum_{i,j=1}^{3} A_{ij}(x_0)(x_i - x_{i,0}) \frac{\partial}{\partial x_i} G_{-1}(x-x_0)\, ds \\
& -\frac{1}{\delta} \int\limits_{\partial B_\delta(x_0)} G_4 \sum_{i,j=1}^{3} A_{ij}(x)(x_i - x_{i,0}) \frac{\partial}{\partial x_i} G_{-1}(x-x_0)\, ds \\
& -\frac{\mathrm{i}}{\delta} \int\limits_{\partial B_\delta(x_0)} G_5 \sum_{i,j=1}^{3} A_{ij}(x_0)(x_i - x_{i,0}) \frac{\partial}{\partial x_i} G_{-1}(x-x_0)\, ds + o(1) \\
=& 4\pi a_0 (\det \mathrm{A}_0)^{\frac{1}{4}} - \frac{1}{\delta} \int\limits_{\partial B_\delta(x_0)} G_4 \sum_{i,j=1}^{3} A_{ij}(x)(x_i - x_{i,0}) \frac{\partial}{\partial x_i} G_{-1}(x-x_0)\, ds + o(1).
\end{aligned}$$

We substitute this identity into (31), and this leads us to a formula for a_0, which implies that this constant is real.

We proceed to proving (26). We first represent the function G as $G = \chi G + (1-\chi)G$, and we see that the function $(1-\chi)G$ is the solution of the equation

$$(\mathcal{H}_\Omega + c_1)(1-\chi)G = -(\hat{\mathcal{H}} + c_1)\chi G. \tag{32}$$

Hence,

$$\begin{aligned}
\mathfrak{h}_\Omega((1-\chi)G,(1-\chi)G) + c_1\|(1-\chi)G\|_{L_2(\Omega)}^2 =& -\big((\hat{\mathcal{H}} + c_1)\chi G, (1-\chi)G\big)_{L_2(\Omega_0)} \\
=& -\mathfrak{h}_{\Omega_0\setminus\omega_\varepsilon}\big(\chi G, (1-\chi)G\big)_{L_2(\Omega_0)} \\
& - c_1\big(\chi G, (1-\chi)G\big)_{L_2(\Omega_0)}.
\end{aligned} \tag{33}$$

We then consider Equation (32) pointwise in Ω_ε, multiply it by χG in $L_2(\Omega_0)$, and integrate it once by parts. This gives the following:

$$\begin{aligned}
&\big((\mathcal{H}_\Omega + c_1)(1-\chi)G, \chi G\big)_{L_2(\Omega_0\setminus\omega_\varepsilon)} = -\big((\hat{\mathcal{H}} + c_1)\chi G, \chi G\big)_{L_2(\Omega_0\setminus\omega_\varepsilon)}, \\
&\mathfrak{h}_{\Omega_0\setminus\omega_\varepsilon}((1-\chi)G, \chi G) = -\mathfrak{h}_{\Omega_0\setminus\omega_\varepsilon}(\chi G, \chi G)_{L_2(\Omega_0\setminus\omega_\varepsilon)} - \left(\frac{\partial G}{\partial \mathrm{n}}, G\right)_{L_2(\partial\omega_\varepsilon)}.
\end{aligned} \tag{34}$$

Summing this identity with (33) and taking into consideration definition (24) of the form $\mathfrak{g}_\varepsilon$, we find that

$$\mathfrak{g}_\varepsilon(G,G)+\|G\|^2_{L_2(\Omega_\varepsilon)}=\left(\frac{\partial G}{\partial \mathrm{n}},G\right)_{L_2(\partial\omega_\varepsilon)}. \tag{35}$$

In view of asymptotics (8) and definition (4) of the function α_0, we then see that

$$\mathfrak{g}_\varepsilon(G,G)+\|G\|^2_{L_2(\Omega_\varepsilon)}+(\alpha_0 G,G)_{L_2(\partial\omega_\varepsilon)}=\left(\frac{\partial G}{\partial \mathrm{n}}+\alpha_0 G,G\right)_{L_2(\partial\omega_\varepsilon)}=\tilde{\beta}_0+O(\varepsilon),$$

where

$$\begin{aligned}\tilde{\beta}_0:=&\sum_{j=1}^{3}\int\limits_{\partial\omega} x_j G_{-1}(x)\nu\cdot\frac{\partial \mathrm{A}}{\partial x_j}(x_0)\nabla G_{-1}(x)\,ds+\sum_{j=1}^{3}\int\limits_{\partial\omega} G_{-1}(x)\nu\cdot \mathrm{A}_0\nabla G_0(x)\,ds\\ &-\sum_{j=1}^{3}\int\limits_{\partial\omega}\overline{G_0(x)}\nu\cdot \mathrm{A}_0\nabla G_{-1}(x)\,ds-\mathrm{i}\sum_{j=1}^{3}\nu_j A_j(x_0)\int\limits_{\partial\omega}G_{-1}^2(x)\,ds.\end{aligned}$$

Since the initialexpression in the above formulas is real due to Formula (35), the same is true for the constant $\tilde{\beta}_0$, and identity (26) holds true. Moreover, since the constant $\tilde{\beta}_0$ is real, we immediately see that $\tilde{\beta}_0=\operatorname{Re}\tilde{\beta}_0=\beta_0$, and this completes the proof. $\square$

We let $\Pi_\varepsilon:=B_{2R_2}(x_0)\setminus\omega_\varepsilon$.

Lemma 4. *These estimates hold:*

$$\|v\|^2_{L_2(\partial\omega_\varepsilon)}\leqslant C\varepsilon\|v\|^2_{W_2^1(\Pi_\varepsilon)},\qquad v\in W_2^1(\Pi_\varepsilon), \tag{36}$$

$$\|v\|^2_{L_2(\omega_\varepsilon)}\leqslant C\varepsilon^2\|v\|^2_{W_2^1(B_{2R_2}(x_0))},\qquad v\in W_2^1(B_{2R_2}(x_0)), \tag{37}$$

where C is a fixed constant independent of ε and v.

Estimates (36) and (37) are proven in Lemmas 2.1 and 2.2 in [14]. Using these estimates and reproducing the proof of Lemmas 3.4 and 3.5 in [6] with obvious minor changes, we arrive at the following statement.

Lemma 5. *For all $v\in W_2^1(\Pi_\varepsilon)$ satisfying the condition*

$$\int\limits_{\partial\omega_\varepsilon} v\,ds=0 \tag{38}$$

the inequality

$$\|v\|^2_{L_2(\partial\omega_\varepsilon)}\leqslant C\varepsilon\|\nabla v\|^2_{L_2(\Pi_\varepsilon)} \tag{39}$$

holds, where C is a constant independent of ε and v. If, in addition, the function v is defined on the entire ball $B_{2R_2}(x_0)$ and $v\in W_2^2(B_{2R_2}(x_0))$, then

$$\|v\|^2_{L_2(\partial\omega_\varepsilon)}\leqslant C\varepsilon^3\|v\|^2_{W_2^2(B_{2R_2}(x_0))}, \tag{40}$$

where C is a constant independent of ε and v.

For all $\varphi\in C^1(\partial\omega)$ and all $v\in W_2^2(B_{2R_2}(x_0))$, the inequality

$$\left|\varepsilon^{-2}\int\limits_{\partial\omega_\varepsilon}\varphi\left(\frac{x-x_0}{\varepsilon}\right)v(x)\,\mathrm{d}s-c(\varphi)v(x_0)\right|\leqslant C\varepsilon^{\frac{1}{2}}\|v\|_{W_2^2(B_{2R_2}(x_0))},\quad c(\varphi):=\int\limits_{\partial\omega}\varphi(x)\,\mathrm{d}s, \tag{41}$$

holds true, where C is a constant independent of ε and v.

Proof. For each function $v \in W_2^1(\Pi_\varepsilon)$, we let $\tilde{v}(\xi) := v(x_0 + \varepsilon\xi)$. The latter function is an element of $W_2^1(B_{R_1}(0) \setminus \omega)$, and

$$\int\limits_{\partial\omega} \tilde{v}\, ds = 0.$$

Hence,

$$\|\tilde{v}\|^2_{L_2(\partial\omega)} \leqslant C\|\nabla_\xi \tilde{v}\|^2_{L_2(B_{R_1}(0)\setminus\omega)},$$

where C is a fixed constant independent of $\tilde{v}$. Rewriting the obtained inequality in terms of the function v, we obtain

$$\|v\|^2_{L_2(\partial\omega_\varepsilon)} \leqslant C\varepsilon\|\nabla v\|^2_{L_2(B_{R_1\varepsilon}(x_0)\setminus\omega_\varepsilon)}, \tag{42}$$

where C is a constant independent of ε and v. This proves (39). If, in addition, $v \in W_2^2(B_{2R_2}(x_0))$, then we apply estimate (37) with v replaced by its derivatives and ω_ε replaced by $B_{R_1\varepsilon}(x_0)$ to the right hand side of (42), and this leads us to (40).

We proceed to proving (41). The boundary value problem

$$\Delta_\xi Y = 0 \quad \text{in} \quad \omega \setminus \{0\}, \qquad \frac{\partial Y}{\partial \nu} = \varphi \quad \text{on} \quad \partial\omega, \qquad Y(\xi) = -\frac{c(\varphi)}{4\pi|\xi|} + O(1)$$

is solvable and possesses a unique solution, such that

$$\int\limits_{\omega} Y(\xi)\, d\xi = 0.$$

By the standardSchauder estimates, the function $Y + \frac{c(\varphi)}{4\pi|\cdot|}$ belongs to $C^{2+\gamma}(\overline{\omega})$ for each $\gamma \in (0,1)$.

Let $v \in C^2(B_{2R_2}(x_0))$, then the function $\tilde{v}(\xi) := v(x_0 + \varepsilon\xi)$ is an element of $C^2(\overline{\omega})$. Using the above definition of the function and integrating by parts, we easily find that

$$0 = \lim_{r\to+0} \int\limits_{\omega\setminus B_r(0)} \tilde{v}\Delta_\xi Y\, ds = -\int\limits_{\partial\omega} \tilde{v}\varphi\, ds + c(\varphi)\tilde{v}(0) + \int\limits_{\partial\omega} Y\frac{\partial\tilde{v}}{\partial\nu}\, ds + \int\limits_{\omega} Y\Delta_\xi\tilde{v}\, ds.$$

Returning back tothe function v, we obtain

$$\varepsilon^{-2}\int\limits_{\partial\omega_\varepsilon} v\varphi\Big(\frac{\cdot - x_0}{\varepsilon}\Big)\, ds - c(\varphi)v(x_0) = \varepsilon^{-1}\int\limits_{\partial\omega_\varepsilon} Y\Big(\frac{\cdot - x_0}{\varepsilon}\Big)\frac{\partial v}{\partial\nu}\, ds + \varepsilon^{-1}\int\limits_{\omega} Y\Big(\frac{\cdot - x_0}{\varepsilon}\Big)\Delta v\, ds.$$

Using the aforementionedsmoothness of the function Y and estimating the right-hand side of the obtained identity, in view of (36), we obtain

$$\begin{aligned}\left|\varepsilon^{-2}\int\limits_{\partial\omega_\varepsilon} v\varphi\Big(\frac{\cdot - x_0}{\varepsilon}\Big)\, ds - c(\varphi)v(x_0)\right| \leqslant & C\varepsilon^{-1}\left\|\frac{\partial v}{\partial\nu}\right\|_{L_2(\partial\omega_\varepsilon)}\left\|Y\Big(\frac{\cdot - x_0}{\varepsilon}\Big)\right\|_{L_2(\partial\omega_\varepsilon)} \\ & + C\varepsilon^{-1}\|v\|_{W_2^2(\omega_\varepsilon)}\left\|Y\Big(\frac{\cdot - x_0}{\varepsilon}\Big)\right\|_{L_2(\omega_\varepsilon)} \\ \leqslant & C\varepsilon^{\frac{1}{2}}\|v\|_{W_2^2(B_{2R_2}(x_0))},\end{aligned}$$

where the Cs are some constants independent of ε and v. Since the space $C^2(\overline{B_{2R_2}(x_0)})$ is dense in $W_2^2(B_{2R_2}(x_0))$, the above estimate also holds for all $v \in W_2^2(B_{2R_2}(x_0))$, and we arrive at (41). The proof is complete. □

4. Lower Semi-Boundedness and Self-Adjointness

In this section, we establish the self-adjointness of the operators $\mathcal{H}_\varepsilon$ and $\mathcal{H}_{0,\beta}$. In addition, we show that the operator $\mathcal{H}_\varepsilon$ is lower semi-bounded, and this is a key ingredient in proving estimates (15) and (16).

We introduce a sesquilinear form

$$\mathfrak{h}_\varepsilon(u,v) := \mathfrak{g}_\varepsilon(u,v) + (\alpha u, v)_{L_2(\partial\omega_\varepsilon)} \tag{43}$$

on the domain $\mathfrak{D}(\mathfrak{h}_\varepsilon) := \mathfrak{D}(\mathfrak{g}_\varepsilon)$, and we recall that the form $\mathfrak{g}_\varepsilon$ and its domain are introduced in (24) and (25). The form $\mathfrak{h}_\varepsilon$ is symmetric. Literally reproducing Equations (4.4)–(4.7) from [6], we see that the form $\mathfrak{h}_\varepsilon$ is associated with the operator $\mathcal{H}_\varepsilon$. Proceeding, then, as in inequalities (4.16)–(4.18) from [6], we also obtain

$$\mathfrak{g}_\varepsilon(u,u) + c_1\|u\|^2_{L_2(\Omega_\varepsilon)} \geqslant c_2\|u\|^2_{W_2^1(\Omega_\varepsilon)} \tag{44}$$

for all $u \in \mathfrak{D}(\mathfrak{g}_\varepsilon)$.

The proof of the self-adjointness of the operator $\mathcal{H}_\varepsilon$ is based on the lower semi-boundedness of its form $\mathfrak{h}_\varepsilon$. In order to prove the latter, we need to study an auxiliary operator similar to a Neumann-to-Dirichlet map and an associated Steklov problem.

4.1. Auxiliary Operator

We first establish the closedness of the form $\mathfrak{g}_\varepsilon$.

Lemma 6. *The form $\mathfrak{g}_\varepsilon$ is closed.*

Proof. We recall that, by our assumptions, the form $\mathfrak{h}_\Omega$ is closed. Let $u_n \in \mathfrak{D}(\mathfrak{g}_\varepsilon)$ be a sequence such that $\mathfrak{g}_\varepsilon(u_n - u_m) \to 0$ as $n,m \to +\infty$ and $u_n \to u$ in $L_2(\Omega_\varepsilon)$. Then, by inequality (44), we immediately conclude that u is an element of $W_2^1(\Omega_\varepsilon)$ and $u_n \to u$ in the norm of this space. Hence,

$$\begin{aligned}\mathfrak{h}_{\Omega_0\setminus\omega_\varepsilon}(\chi(u_n-u),(1-\chi)(u_n-u)) + \mathfrak{h}_{\Omega_0\setminus\omega_\varepsilon}((1-\chi)(u_n-u),\chi(u_n-u))\\ + \mathfrak{h}_{\Omega_0\setminus\omega_\varepsilon}(\chi(u_n-u),\chi(u_n-u)) \to 0, \qquad n\to+\infty,\end{aligned}$$

and, therefore, by definition (43) of the form $\mathfrak{g}_\varepsilon$, we see that

$$\mathfrak{h}_\Omega((1-\chi)(u_n-u),(1-\chi)(u_n-u)) \to 0 \quad \text{as} \quad n\to+\infty.$$

The closedness of the form $\mathfrak{h}_\Omega$ then implies that $(1-\chi)u \in \mathfrak{D}(\mathfrak{g}_\varepsilon)$, and by the definition of the cut-off function, we conclude that $u = (1-\chi)u + \chi u \in \mathfrak{D}(\mathfrak{g}_\varepsilon)$ and $\mathfrak{h}_\Omega(u_n - u, u_n - u) \to 0$ as $n \to +\infty$. The proof is complete. □

We equip the linear space $\mathfrak{D}(\mathfrak{g}_\varepsilon)$ with the scalar product

$$(\cdot\,,\cdot)_{\mathfrak{g}_\varepsilon} := \mathfrak{g}_\varepsilon(\cdot\,,\cdot) + c_1(\cdot\,,\cdot)_{L_2(\mathbb{R}^3\setminus\omega)}$$

and owing to the symmetricity and closedness of the form, as well as to inequality (44), this makes the space $\mathfrak{D}(\mathfrak{g}_\varepsilon)$ a Hilbert one. Since by (44) we have $W_2^1(\Omega_\varepsilon) \subseteq \mathfrak{D}(\mathfrak{g}_\varepsilon)$, each $u \in \mathfrak{D}(\mathfrak{g}_\varepsilon)$ possesses a trace on $\partial\omega_\varepsilon$. The operator, which maps $u \in \mathfrak{D}(\mathfrak{g}_\varepsilon)$ into its trace on $\partial\omega_\varepsilon$, is well-defined as a bounded one from $\mathfrak{D}(\mathfrak{g}_\varepsilon)$ into $L_2(\partial\omega_\varepsilon)$; we denote this operator by $\mathcal{T}_\varepsilon$. In view of inequalities (36) and (44) and the compactness of the trace operator from $W_2^1(\Pi_\varepsilon)$ into $L_2(\partial\omega_\varepsilon)$, the operator $\mathcal{T}_\varepsilon : \mathfrak{D}(\mathfrak{g}_\varepsilon) \to L_2(\partial\omega_\varepsilon)$ is compact and satisfies the estimate

$$\|\mathcal{T}_\varepsilon\|_{\mathfrak{D}(\mathfrak{g}_\varepsilon)\to L_2(\partial\omega_\varepsilon)} \leqslant C\varepsilon^{\frac{1}{2}}, \tag{45}$$

where C is a constant independent of ε.

For each $\phi \in \mathfrak{D}(\mathfrak{g}_\varepsilon)$, we consider the boundary value problem

$$(\hat{\mathcal{H}} + c_1)u = 0 \quad \text{in} \quad \Omega_\varepsilon, \qquad \mathcal{B}u = 0 \quad \text{on} \quad \partial\Omega, \qquad \frac{\partial u}{\partial \mathrm{n}} = -\alpha\phi \quad \text{on} \quad \partial\omega_\varepsilon. \tag{46}$$

The solution isunderstood in the generalized sense, namely, a solution is a function $u \in \mathfrak{D}(\mathfrak{g}_\varepsilon)$ such that

$$(u, v)_{\mathfrak{g}_\varepsilon} + (\alpha\phi, v)_{L_2(\partial\omega_\varepsilon)} = 0 \quad \text{for all} \quad v \in \mathfrak{D}(\mathfrak{g}_\varepsilon). \tag{47}$$

Since $\mathfrak{g}_\varepsilon$ is the scalar product on the Hilbert space $\mathfrak{D}(\mathfrak{g}_\varepsilon)$, boundary value problem (46) is uniquely solvable for each $\phi \in L_2(\partial\omega_\varepsilon)$. By $\mathcal{A}_\varepsilon^0$, we denote the operator mapping ϕ into the solution of problem (46). This operator is bounded as acting from $L_2(\partial\omega_\varepsilon)$ into $\mathfrak{D}(\mathfrak{g}_\varepsilon)$. Moreover, by estimates (36), (44) we easily find that

$$\|u\|_{\mathfrak{g}_\varepsilon}^2 = -(\alpha\phi, u)_{L_2(\partial\omega_\varepsilon)} \leqslant C\varepsilon^{-\frac{1}{2}}\|\phi\|_{L_2(\partial\omega\varepsilon)}\|u\|_{W_2^1(\Pi_\varepsilon)} \leqslant C\varepsilon^{-\frac{1}{2}}\|\phi\|_{L_2(\partial\omega\varepsilon)}\|u\|_{\mathfrak{g}_\varepsilon}, \tag{48}$$

where the Cs are constants independent of ε, u, and ϕ. Hence,

$$\|\mathcal{A}_\varepsilon^0\|_{L_2(\partial\omega_\varepsilon)\to\mathfrak{D}(\mathfrak{g}_\varepsilon)} \leqslant C\varepsilon^{-\frac{1}{2}}, \tag{49}$$

where C is a constant independent of ε. It also follows from the symmetricity of the form $\mathfrak{g}_\varepsilon$ and the identity (47) that the operator $\mathcal{A}_\varepsilon := \mathcal{A}_\varepsilon^0\mathcal{T}_\varepsilon$ acting on $\mathfrak{D}(\mathfrak{g}_\varepsilon)$ is self-adjoint. Since the operator $\mathcal{T}_\varepsilon$ is compact, the same is true for $\mathcal{A}_\varepsilon$. Estimates (45) and (49) imply that

$$\|\mathcal{A}_\varepsilon\|_{\mathfrak{D}(\mathfrak{g}_\varepsilon)\to\mathfrak{D}(\mathfrak{g}_\varepsilon)} \leqslant C,$$

where C is a constant independent of ε. The spectrum of the operator $\mathcal{A}_\varepsilon$ consists of discrete eigenvalues, which can accumulate only at zero, and the latter is the only possible point of the essential spectrum.

It is possible to construct an asymptotic expansion for the operator $\mathcal{A}_\varepsilon$ as $\varepsilon \to 0$ on the base of the classical method of matching asymptotic expansions similarly to Chapter III in [10] and Chapter II, Section 2.3.4 in [12]. The application of this technique shows that

$$\|\mathcal{A}_\varepsilon - \mathcal{L}_\varepsilon\|_{\mathfrak{D}(\mathfrak{g}_\varepsilon)\to\mathfrak{D}(\mathfrak{g}_\varepsilon)} \to 0, \qquad \varepsilon \to 0, \qquad \mathcal{L}_\varepsilon := \zeta_\varepsilon\mathcal{S}_\varepsilon^{-1}\mathcal{L}\mathcal{S}_\varepsilon\mathcal{T}_\varepsilon + \varepsilon(1-\zeta_\varepsilon)G\mathcal{C}\mathcal{S}_\varepsilon\mathcal{T}_\varepsilon, \tag{50}$$

$$(\mathcal{S}_\varepsilon u)(\xi) := u(\varepsilon\xi), \qquad \xi \in \mathbb{R}^3 \setminus \omega, \qquad \mathcal{C}u := \frac{1}{4\pi(\det \mathrm{A}_0)^{\frac{1}{4}}}\int\limits_{\partial\omega} \alpha_0(\xi)u(\xi)\,ds.$$

Here, $\zeta_\varepsilon(x) = \zeta(|x|\varepsilon^{-\frac{1}{2}})$, and $\zeta = \zeta(t)$ is an infinitely differentiable cut-off function equal to one as $t < 1$ and vanishing as $t > 2$. By $\mathcal{L}$, we denote an operator from $L_2(\partial\omega)$ into $W_2^1(B_{R_1}(0) \setminus \omega) \cap C^\infty(\mathbb{R}^3 \setminus \omega)$ mapping each function $\phi \in L_2(\partial\omega)$ into the unique solution $U = U(\xi)$ of the boundary value problem

$$\operatorname{div}_\xi \mathrm{A}_0\nabla_\xi U = 0 \quad \text{in} \quad \mathbb{R}^3 \setminus \omega, \qquad \nu \cdot \mathrm{A}_0\nabla_\xi U + \alpha_0\phi = 0 \quad \text{on} \quad \partial\omega,$$
$$U(\xi) = \mathcal{C}(\phi)|\mathrm{A}_0^{-\frac{1}{2}}\xi|^{-1} + O(|\mathrm{A}_0^{-\frac{1}{2}}\xi|^{-2}), \qquad \xi \to \infty,$$

and the above asymptotic for U is differentiable. Since the operator $\mathcal{T}_\varepsilon$ is compact and $\mathcal{C}$ is a linear functional, it follows from the definition of the operator $\mathcal{L}_\varepsilon$ in (50) that this operator is compact. Hence, its spectrum consists of eigenvalues of finite multiplicities, which can accumulate only at zero, and the latter is the only possible point of the continuous spectrum.

Let $\mathcal{T} : W_2^1(B_{R_1}(0) \setminus \omega) \to L_2(\partial\omega)$ be the operator of taking the trace on $\partial\omega$; this operator is obviously compact. Then, the operator $\mathcal{T}\mathcal{L} : L_2(\partial\omega) \to L_2(\partial\omega)$ is compact as well. The eigenvalues of this operator coincide with those of the operator $\mathcal{L}_\varepsilon$, counting the multiplicities. Indeed, let $\lambda \neq 0$ be an eigenvalue of the operator $\mathcal{T}\mathcal{L}$. This means that there exists a non-trivial solution of boundary value problem (10). Hence, λ is an eigenvalue of the operator $\mathcal{L}_\varepsilon$, and the associated eigenfunction is $\zeta_\varepsilon \mathcal{S}_\varepsilon^{-1}\psi + \varepsilon\lambda^{-1}(1 - \zeta_\varepsilon)GC(\psi)$. Furthermore, vice versa, let $\lambda \neq 0$ be an eigenvalue of the operator $\mathcal{L}_\varepsilon$ and ψ be an associated eigenfunction. Then, we consider the eigenvalue equation $\mathcal{L}_\varepsilon\psi = \lambda\psi$ as the identity for two functions defined for $x \in B_{\frac{1}{\varepsilon^2}} \setminus \omega_\varepsilon$, and we see immediately that ψ solves problem (10) and, hence, λ is an eigenvalue of the operator $\mathcal{T}\mathcal{L}$.

Since the operator $\mathcal{T}\mathcal{L}$ is compact, its spectrum consists of discrete eigenvalues, which can accumulate only at zero, and the latter is the only possible point of the essential spectrum. Since the eigenvalues of the operator $\mathcal{T}\mathcal{L}$ coincide with those of the operator $\mathcal{L}_\varepsilon$, in view of the convergence in (50), we conclude that the eigenvalues of $\mathcal{T}\mathcal{L}$ are the limits of the eigenvalues of $\mathcal{L}_\varepsilon$ as $\varepsilon \to 0$, counting the multiplicities, and, hence, the eigenvalues of $\mathcal{T}\mathcal{L}$ are real.

Let $\lambda \neq 0$ be an eigenvalue of the operator $\mathcal{T}\mathcal{L}$, then problem (10) possesses a non-trivial solution. We multiply the equation in (10) by ψ in $L_2(\mathbb{R}^3 \setminus \omega)$ and integrate once by parts using the boundary condition in (10). This gives

$$\lambda = -\frac{(\alpha_0\psi, \psi)_{L_2(\partial\omega)}}{(\mathrm{A}_0\nabla_\xi\psi, \nabla_\xi\psi)_{L_2(\mathbb{R}^3\setminus\omega)}}. \tag{51}$$

We represent ψ as

$$\psi(\xi) = E(\xi)\Psi(x), \qquad E(\xi) := |\mathrm{A}_0^{-\frac{1}{2}}\xi|^{-1}. \tag{52}$$

Since the function E is non-zero on $\mathbb{R}^3 \setminus \omega$, the above representation for ψ is well-defined, and in view of the asymptotic at infinity in problem (10), the function $\Psi(\xi)$ possesses the following differentiable asymptotic at infinity:

$$\Psi(\xi) = \lambda^{-1}C(\psi) + O(|\mathrm{A}_0^{-\frac{1}{2}}\xi|^{-1}), \qquad \xi \to \infty.$$

We substitute representation (52) into the denominator of (51) and integrate by parts using the definition of E and α_0:

$$\begin{aligned}(\mathrm{A}_0\nabla_\xi\psi, \nabla_\xi\psi)_{L_2(\mathbb{R}^3\setminus\omega)} =&(\mathrm{A}_0E\nabla_\xi\Psi, E\nabla_\xi\Psi)_{L_2(\mathbb{R}^3\setminus\omega)} + (\mathrm{A}_0\Psi\nabla_\xi E, \Psi\nabla_\xi E)_{L_2(\mathbb{R}^3\setminus\omega)}\\ &+ (\mathrm{A}_0\Psi\nabla_\xi E, E\nabla_\xi\Psi)_{L_2(\mathbb{R}^3\setminus\omega)} + (\mathrm{A}_0E\nabla_\xi\Psi, \Psi\nabla_\xi E)_{L_2(\mathbb{R}^3\setminus\omega)}\\ =&(\mathrm{A}_0E\nabla_\xi\Psi, E\nabla_\xi\Psi)_{L_2(\mathbb{R}^3\setminus\omega)} + \int\limits_{\partial\omega} \overline{\Psi}E\nu\cdot\mathrm{A}_0\Psi\nabla_\xi E\,ds\\ &- \int\limits_{\mathbb{R}^3\setminus\omega} E\,\mathrm{div}\,\mathrm{A}_0|\Psi|^2\nabla_\xi E\,d\xi + (\mathrm{A}_0\Psi\nabla_\xi E, E\nabla_\xi\Psi)_{L_2(\mathbb{R}^3\setminus\omega)}\\ &+ (\mathrm{A}_0E\nabla_\xi\Psi, \Psi\nabla_\xi E)_{L_2(\mathbb{R}^3\setminus\omega)}\\ =&(\mathrm{A}_0E\nabla_\xi\Psi, E\nabla_\xi\Psi)_{L_2(\mathbb{R}^3\setminus\omega)} - (\alpha_0E\Psi, E\Psi)_{L_2(\partial\omega)},\end{aligned}$$

and the final expression is positive since the same is true for the initial scalar product $(\mathrm{A}_0\nabla_\xi\psi, \nabla_\xi\psi)_{L_2(\mathbb{R}^3\setminus\omega)}$. Substituting these identities and representation (52) into (51), we obtain

$$\lambda = \frac{-(\alpha_0E\Psi, E\Psi)_{L_2(\partial\omega)}}{(\mathrm{A}_0E\nabla_\xi\Psi, E\nabla_\xi\Psi)_{L_2(\mathbb{R}^3\setminus\omega)} - (\alpha_0E\Psi, E\Psi)_{L_2(\partial\omega)}}. \tag{53}$$

Since the denominatorof the obtained quotient is positive, we immediately conclude that $\lambda < 1$ once Ψ is not identically one. It is straightforward to confirm that $\lambda = 1$ is an eigenvalue of the operator $\mathcal{T}\mathcal{L}$, and the corresponding non-trivial solution of problem (10) is $\psi = E$. Identity (53), then, implies that $\lambda = 1$ is a simple eigenvalue of the operator $\mathcal{T}\mathcal{L}$.

In view of the established facts on the eigenvalues of the operator $\mathcal{T}\mathcal{L}$ and the convergence in (50), the greatest eigenvalue of the operator $\mathcal{A}_\varepsilon$ is simple and converges to 1 as $\varepsilon \to 0$. We denote the next eigenvalue of the operator $\mathcal{A}_\varepsilon$ by $\tilde{\lambda}_\varepsilon$. This eigenvalue converges to the next eigenvalue of the operator $\mathcal{T}\mathcal{L}$, which is strictly less than one. Let

ψ_ε be a normalized eigenfunction, in $\mathfrak{D}(\mathfrak{g}_\varepsilon)$, of the operator $\mathcal{A}_\varepsilon$ associated with its greatest eigenvalue. Then, by the minimax principle applied to the operator $\mathcal{A}_\varepsilon$, we find

$$\frac{(\mathcal{A}_\varepsilon u, u)_{\mathfrak{g}_\varepsilon}}{\|u\|^2_{\mathfrak{g}_\varepsilon}} \leqslant \tilde{\lambda}_\varepsilon \quad \text{for all} \quad u \in \mathfrak{D}(\mathfrak{g}_\varepsilon) \quad \text{such that} \quad (u, \psi_\varepsilon)_{\mathfrak{g}_\varepsilon} = 0. \tag{54}$$

In view of identity (47), we can rewrite this inequality as

$$-(\mathcal{A}_\varepsilon u, u)_{\mathfrak{g}_\varepsilon} = (\alpha u, u)_{L_2(\partial\omega_\varepsilon)} \geqslant -\tilde{\lambda}_\varepsilon \|u\|^2_{\mathfrak{g}_\varepsilon} \tag{55}$$

for all $u \in \mathfrak{D}(\mathfrak{g}_\varepsilon)$ obeying the orthogonality condition from (54).

We also need asymptotics for the eigenvalue λ_ε and the associated eigenfunction ψ_ε; let us find them. It follows from problem (7), (24); the definition (4) of the function α_0; and Lemma 3 that the function G solves the following boundary value problem

$$\begin{aligned} &(\hat{\mathcal{H}} + c_1)G = 0 \quad \text{in} \quad \Omega_\varepsilon, \qquad \mathcal{B}G = 0 \quad \text{on} \quad \partial\Omega, \\ &\frac{\partial G}{\partial \mathrm{n}} = -\alpha G + \varepsilon^{-1} h_\varepsilon \quad \text{on} \quad \partial\omega_\varepsilon, \end{aligned} \tag{56}$$

where h_ε is a continuous function in $\partial\omega_\varepsilon$ bounded uniformly in the spatial variables in $\partial\omega_\varepsilon$ and the small parameter ε. U_ε denotes the solution of the problem

$$(\hat{\mathcal{H}} + c_1)U_\varepsilon = 0 \quad \text{in} \quad \Omega_\varepsilon, \qquad \mathcal{B}U_\varepsilon = 0 \quad \text{on} \quad \partial\Omega, \qquad \frac{\partial U_\varepsilon}{\partial \mathrm{n}} = h_\varepsilon \quad \text{on} \quad \partial\omega_\varepsilon, \tag{57}$$

and in view of the uniform boundedness of h, similarly to (48), we immediately obtain the following:

$$\|U_\varepsilon\|_{\mathfrak{g}_\varepsilon} = O(\varepsilon^{\frac{3}{2}}). \tag{58}$$

Lemma 3 also implies that

$$\|G\|_{L_2(\Omega_\varepsilon)} = \|G\|_{L_2(\Omega)} + O(\varepsilon^{\frac{1}{2}}), \qquad \|G\|_{\mathfrak{g}_\varepsilon} = \|G^2_{-1}\|_{L_2(\mathbb{R}^3\setminus\omega)} \varepsilon^{-\frac{1}{2}} + O(\varepsilon^{\frac{1}{2}}|\ln\varepsilon|), \tag{59}$$

where C is a positive constant independent of ε.

Comparing problems (46) and (56), we see that $G = \mathcal{A}_\varepsilon G + \varepsilon^{-1} U_\varepsilon$ and, hence,

$$G_\varepsilon = \mathcal{A}_\varepsilon G_\varepsilon + U_\varepsilon, \qquad G_\varepsilon := \frac{G}{\|G\|_{\mathfrak{g}_\varepsilon}}, \qquad V_\varepsilon := \frac{\varepsilon^{-1} U_\varepsilon}{\|G\|_{\mathfrak{g}_\varepsilon}}, \qquad \|V_\varepsilon\|_{\mathfrak{g}_\varepsilon} = O(\varepsilon). \tag{60}$$

We apply the resolvent $(\mathcal{A}_\varepsilon - 1)^{-1}$ to the obtained equation and employ standard results on the behavior of the resolvents of the self-adjoint operators near the isolated eigenvalues (see Chapter V, Section 3.5 in [21]). This gives the identity

$$G_\varepsilon = \frac{(V_\varepsilon, \psi_\varepsilon)_{\mathfrak{g}_\varepsilon}}{1 - \lambda_\varepsilon} \psi_\varepsilon + \mathcal{R}_\varepsilon V_\varepsilon, \tag{61}$$

where $\mathcal{R}_\varepsilon$ is the reduced resolvent at the point 1, and this is an operator in $\mathfrak{D}(\mathfrak{g}_\varepsilon)$ bounded uniformly in ε and acting into the orthogonal complement to ψ_ε in $\mathfrak{D}(\mathfrak{g}_\varepsilon)$. Hence,

$$\|\mathcal{R}_\varepsilon V_\varepsilon\|_{\mathfrak{g}_\varepsilon} = O(\varepsilon). \tag{62}$$

This estimate and identity (61) imply that

$$\|G_\varepsilon - c_\varepsilon \psi_\varepsilon\|_{\mathfrak{g}_\varepsilon} = O(\varepsilon), \qquad c_\varepsilon := \frac{(V_\varepsilon, \psi_\varepsilon)_{\mathfrak{g}_\varepsilon}}{1 - \lambda_\varepsilon} = (G_\varepsilon, \psi_\varepsilon)_{\mathfrak{g}_\varepsilon}. \tag{63}$$

Calculating the scalarproduct in $\mathfrak{D}(\mathfrak{g}_\varepsilon)$ of both sides of identity (61) with G_ε, in view of identity (62), we immediately see that

$$1 = |c_\varepsilon|^2 + O(\varepsilon). \tag{64}$$

Calculating the scalarproduct in $\mathfrak{D}(\mathfrak{g}_\varepsilon)$ of both sides of identity (61) with $c_\varepsilon \psi_\varepsilon$, by (58), (62), and (63), the definition of V_ε in (60), and the normalization of G_ε and ψ_ε in $\mathfrak{D}(\mathfrak{g}_\varepsilon)$, we see that

$$\lambda_\varepsilon - 1 = -\frac{1}{\varepsilon\|G\|_{\mathfrak{g}_\varepsilon}} \frac{(U_\varepsilon, c_\varepsilon\psi_\varepsilon)_{\mathfrak{g}_\varepsilon}}{(G_\varepsilon, c_\varepsilon\psi_\varepsilon)_{\mathfrak{g}_\varepsilon}} = -\frac{(U_\varepsilon, G)_{\mathfrak{g}_\varepsilon}}{\varepsilon\|G\|^2_{\mathfrak{g}_\varepsilon}}(1 + O(\varepsilon)). \tag{65}$$

Let us find the scalar product $(U_\varepsilon, G)_{\mathfrak{g}_\varepsilon}$. In order to do this, we write the definition of the generalized solution of problems (56), (57) with G as the test function:

$$(G, G)_{\mathfrak{g}_\varepsilon} + (\alpha G, G)_{L_2(\partial\omega_\varepsilon)} - \varepsilon^{-1}(h_\varepsilon, G)_{L_2(\partial\omega_\varepsilon)} = 0, \qquad (U_\varepsilon, G)_{\mathfrak{g}_\varepsilon} - (h_\varepsilon, G)_{\partial\omega_\varepsilon} = 0.$$

Hence,

$$\varepsilon^{-1}(U_\varepsilon, G)_{\mathfrak{g}_\varepsilon} = \varepsilon^{-1}(h_\varepsilon, G)_{L_2(\partial\omega_\varepsilon)} = \|G\|^2_{\mathfrak{g}_\varepsilon} + (\alpha G, G)_{L_2(\partial\omega_\varepsilon)}.$$

It also follows from asymptotics (8) and the definition of the function α that

$$\begin{aligned}(\alpha G, G)_{L_2(\partial\omega_\varepsilon)} =&\varepsilon^{-1}(\alpha_0 G_{-1}, G_{-1})_{L_2(\partial\omega)} + 2\,\mathrm{Re}(\alpha_0 G_0, G_{-1})_{L_2(\partial\omega)} \\ &+ (\alpha_1 G_{-1}, G_{-1})_{L_2(\partial\omega)} + O(\varepsilon).\end{aligned}$$

This identity and (11), (26) yield

$$\varepsilon^{-1}(U_\varepsilon, G)_{\mathfrak{g}_\varepsilon} = \beta_0 + (\alpha_1 G_{-1}, G_{-1})_{L_2(\partial\omega)} + O(\varepsilon) = -4\pi(\det \mathrm{A}_0)^{\frac{1}{4}}(\beta - a_0) + O(\varepsilon), \quad \varepsilon \to 0.$$

The obtained formula and (59), (19) allow us to rewrite (65) as

$$\lambda_\varepsilon - 1 = \varepsilon \frac{4\pi(\beta - a_0)(\det \mathrm{A}_0)^{\frac{1}{4}}}{\|G^2_{-1}\|^2_{L_2(\mathbb{R}^3\setminus\omega)}} + O(\varepsilon^2|\ln\varepsilon|).$$

4.2. *Lower Semi-Boundedness*

In this subsection, we prove the lower-semiboundedness of the form $\mathfrak{h}_\varepsilon$. We represent each function $u \in \mathfrak{D}(\mathfrak{g}_\varepsilon)$ as

$$u = u^\perp + (u, \psi_\varepsilon)_{\mathfrak{D}(\mathfrak{g}_\varepsilon)}\psi_\varepsilon, \qquad (u^\perp, \psi_\varepsilon)_{\mathfrak{D}(\mathfrak{g}_\varepsilon)} = 0.$$

Then, by (55), for all $u \in \mathfrak{D}(\mathfrak{g}_\varepsilon)$, we have the following:

$$\begin{aligned}\mathfrak{h}_\varepsilon(u,u) + c_1\|u\|^2_{L_2(\Omega_\varepsilon)} =&(u - \mathcal{A}_\varepsilon u, u)_{\mathfrak{g}_\varepsilon} \\ =&(u^\perp - \mathcal{A}_\varepsilon u^\perp, u^\perp)_{\mathfrak{g}_\varepsilon} + (1-\lambda_\varepsilon)|(u,\psi_\varepsilon)_{\mathfrak{D}(\mathfrak{g}_\varepsilon)}|^2 \\ \geqslant&(1-\tilde\lambda_\varepsilon)\|u\|^2_{\mathfrak{g}_\varepsilon} + (1-\lambda_\varepsilon)|(u,\psi_\varepsilon)_{\mathfrak{D}(\mathfrak{g}_\varepsilon)}|^2 \\ =&(\kappa + c_3(\varepsilon))\|u\|^2_{\mathfrak{g}_\varepsilon} + (1-\lambda_\varepsilon)|(u,\psi_\varepsilon)_{\mathfrak{D}(\mathfrak{g}_\varepsilon)}|^2.\end{aligned}$$

As it is established in the previous section, the eigenvalue $\tilde\lambda_\varepsilon$ converges to the second eigenvalue of the operator $\mathcal{TL}$, and this is why $1 - \tilde\lambda_\varepsilon = \kappa + c_3(\varepsilon)$, where $c_3(\varepsilon) \to 0$ as $\varepsilon \to +0$. This allows us to the rewrite the above estimate as

$$\mathfrak{h}_\varepsilon(u,u) + c_1\|u\|^2_{L_2(\Omega_\varepsilon)} \geqslant (\kappa + c_3(\varepsilon))\|u\|^2_{\mathfrak{g}_\varepsilon} + (1-\lambda_\varepsilon)|(u,\psi_\varepsilon)_{\mathfrak{D}(\mathfrak{g}_\varepsilon)}|^2. \tag{66}$$

For each $u \in \mathfrak{D}(\mathfrak{g}_\varepsilon)$, the function

$$u^\perp := u - (u, \psi_\varepsilon)_{\mathfrak{g}_\varepsilon}\psi_\varepsilon \tag{67}$$

satisfies the orthogonality condition in (54), and in view of (66), we have

$$
\begin{aligned}
\mathfrak{h}_\varepsilon(u,u)+c_1\|u\|^2_{L_2(\Omega_\varepsilon)} =&(u-\mathcal{A}_\varepsilon u,u)_{\mathfrak{g}_\varepsilon}=(u^\perp-\mathcal{A}_\varepsilon u^\perp,u^\perp)_{\mathfrak{g}_\varepsilon}+(1-\lambda_\varepsilon)|(u,\psi_\varepsilon)_{\mathfrak{g}_\varepsilon}|^2\\
\geqslant&(\kappa+c_3(\varepsilon))\|u^\perp\|^2_{\mathfrak{g}_\varepsilon}+(1-\lambda_\varepsilon)|(u,\psi_\varepsilon)_{\mathfrak{g}_\varepsilon}|^2\\
\geqslant&(\kappa+c_3(\varepsilon))\|u^\perp\|^2_{\mathfrak{g}_\varepsilon}\\
&-\varepsilon\left(\frac{4\pi(\beta-a_0)(\det \mathrm{A}_0)^{\frac{1}{4}}}{\|G_{-1}^2\|^2_{L_2(\mathbb{R}^3\setminus\omega)}}+C\varepsilon|\ln\varepsilon|\right)|(u,\psi_\varepsilon)_{\mathfrak{g}_\varepsilon}|^2
\end{aligned}
\tag{68}
$$

where C is some constant independent of ε and u.

For further purposes, it is more convenient to introduce another representation similar to (67): we let

$$
\begin{aligned}
&u_\perp:=u-(u,G_\varepsilon)_{\mathfrak{g}_\varepsilon}G_\varepsilon, \qquad \psi_{\varepsilon,\perp}:=\psi_\varepsilon-(\psi_\varepsilon,G_\varepsilon)_{\mathfrak{g}_\varepsilon}G_\varepsilon,\\
&(u_\perp,G_\varepsilon)_{\mathfrak{g}_\varepsilon}=(\psi_{\varepsilon,\perp},G_\varepsilon)_{\mathfrak{g}_\varepsilon}=0.
\end{aligned}
\tag{69}
$$

Comparing the above definition of $\psi_{\varepsilon,\perp}$ with (63), (64), we immediately see that

$$
\|\psi_{\varepsilon,\perp}\|_{\mathfrak{g}_\varepsilon}=O(\varepsilon), \qquad \psi_{\varepsilon,\perp}=\psi_\varepsilon-\overline{c_\varepsilon}G_\varepsilon.
\tag{70}
$$

It follows from (67), (69) that

$$
u^\perp=u_\perp+(u,G_\varepsilon-c_\varepsilon\psi_\varepsilon)_{\mathfrak{g}_\varepsilon}G_\varepsilon-(u,\psi_\varepsilon)_{\mathfrak{g}_\varepsilon}\psi_{\varepsilon,\perp}
$$

and in view of the orthogonality conditions in (69), we find

$$
\|u^\perp\|^2_{\mathfrak{g}_\varepsilon}=\|u_\perp-(u,\psi_\varepsilon)_{\mathfrak{g}_\varepsilon}\psi_{\varepsilon,\perp}\|^2_{\mathfrak{g}_\varepsilon}+|(u,G_\varepsilon-c_\varepsilon\psi_\varepsilon)_{\mathfrak{g}_\varepsilon}|^2.
\tag{71}
$$

By the Cauchy–Schwarz inequality and (70), we obtain

$$
\begin{aligned}
\|u_\perp-(u,\psi_\varepsilon)_{\mathfrak{g}_\varepsilon}\psi_{\varepsilon,\perp}\|^2_{\mathfrak{g}_\varepsilon}\geqslant&\|u_\perp\|^2_{\mathfrak{g}_\varepsilon}(1-|\ln\varepsilon|^{-1})-(|\ln\varepsilon|-1)|(u,\psi_\varepsilon)_{\mathfrak{g}_\varepsilon}|^2\|\psi_{\varepsilon,\perp}\|^2_{\mathfrak{g}_\varepsilon}\\
\geqslant&\|u_\perp\|^2_{\mathfrak{g}_\varepsilon}(1-|\ln\varepsilon|^{-1})-C\varepsilon^2|\ln\varepsilon||(u,\psi_\varepsilon)_{\mathfrak{g}_\varepsilon}|^2,
\end{aligned}
\tag{72}
$$

where the Cs are some positive constants independent of ε and u. It also follows from the Cauchy–Schwarz inequality and (63), (64) that

$$
\begin{aligned}
|(u,\psi_\varepsilon)_{\mathfrak{g}_\varepsilon}|^2=&|c_\varepsilon|^{-2}|(u,c_\varepsilon\psi_\varepsilon)_{\mathfrak{g}_\varepsilon}|^2=|c_\varepsilon|^{-2}|(u,G_\varepsilon)_{\mathfrak{g}_\varepsilon}-(u,G_\varepsilon-c_\varepsilon\psi_\varepsilon)_{\mathfrak{g}_\varepsilon}|^2\\
\geqslant&(1-|\ln\varepsilon|^{-1})|(u,G_\varepsilon)_{\mathfrak{g}_\varepsilon}|^2-(|\ln\varepsilon|-1)|(u,G_\varepsilon-c_\varepsilon\psi_\varepsilon)_{\mathfrak{g}_\varepsilon}|^2,\\
|(u,\psi_\varepsilon)_{\mathfrak{g}_\varepsilon}|^2\geqslant& 2|(u,G_\varepsilon)_{\mathfrak{g}_\varepsilon}|^2+2|(u,G_\varepsilon-c_\varepsilon\psi_\varepsilon)_{\mathfrak{g}_\varepsilon}|^2.
\end{aligned}
\tag{73}
$$

Substituting the latter inequality and (72) into (71), we obtain the following:

$$
\|u^\perp\|^2_{\mathfrak{g}_\varepsilon}\geqslant\|u_\perp\|^2_{\mathfrak{g}_\varepsilon}(1-|\ln\varepsilon|^{-1})-C\varepsilon^2|\ln\varepsilon||(u,G_\varepsilon)_{\mathfrak{g}_\varepsilon}|^2+\frac{1}{2}|(u,G_\varepsilon-c_\varepsilon\psi_\varepsilon)_{\mathfrak{g}_\varepsilon}|^2
$$

with some constant C independent of ε and u. This estimate and (73) allow us to rewrite (68) as

$$
\begin{aligned}
\mathfrak{h}_\varepsilon(u,u)+c_1\|u\|^2_{L_2(\Omega_\varepsilon)}\geqslant&(\kappa+c_3(\varepsilon))\|u_\perp\|^2_{\mathfrak{g}_\varepsilon}(1-|\ln\varepsilon|^{-1})\\
&-\varepsilon\left(\frac{4\pi(\beta-a_0)(\det \mathrm{A}_0)^{\frac{1}{4}}}{\|G_{-1}^2\|^2_{L_2(\mathbb{R}^3\setminus\omega)}}+C|\ln\varepsilon|^{-1}\right)|(u,G_\varepsilon)_{\mathfrak{g}_\varepsilon}|^2
\end{aligned}
\tag{74}
$$

where C is a constant independent of ε and u.

By the Cauchy–Schwarz inequality and (59), (67), we find that

$$\begin{aligned}\|u\|^2_{L_2(\Omega_\varepsilon)} =&\|u_\perp\|^2_{L_2(\Omega_\varepsilon)} + 2\,\mathrm{Re}(u, G_\varepsilon)_{\mathfrak{g}_\varepsilon}(G_\varepsilon, u_\perp)_{L_2(\Omega_\varepsilon)} + |(u, G_\varepsilon)_{\mathfrak{g}_\varepsilon}|^2\|G_\varepsilon\|^2_{L_2(\Omega_\varepsilon)}\\ \geqslant& -\eta\|u_\perp\|^2_{L_2(\Omega_\varepsilon)} + \frac{\eta\|G_\varepsilon\|^2_{L_2(\Omega_\varepsilon)}}{1+\eta}|(u, G_\varepsilon)_{\mathfrak{g}_\varepsilon}|^2\\ \geqslant& -\eta\|u_\perp\|^2_{L_2(\Omega_\varepsilon)} + \frac{\varepsilon\eta\|G\|^2_{L_2(\Omega)}(1-C\varepsilon^{\frac{1}{2}})}{(1+\eta)\|G^2_{-1}\|^2_{L_2(\mathbb{R}^3\setminus\omega)}}|(u, G_\varepsilon)_{\mathfrak{g}_\varepsilon}|^2\end{aligned}$$

for an arbitrary $\eta \in (0,1)$ with some constant C independent of ε, u, and η. By (74), for an arbitrary $c_4 > 0$, we then obtain

$$\begin{aligned}\mathfrak{h}_\varepsilon(u,u) + (c_1+c_4)\|u\|^2_{L_2(\Omega_\varepsilon)} \geqslant& \left(\kappa + c_3(\varepsilon) - c_4\eta(1-|\ln\varepsilon|^{-1})\right)\|u_\perp\|^2_{\mathfrak{g}_\varepsilon}\\ &+ \frac{\varepsilon\|G\|^2_{L_2(\Omega)}}{\|G^2_{-1}\|^2_{L_2(\mathbb{R}^3\setminus\omega)}}\left(\frac{c_4\eta}{(1+\eta)} - \frac{4\pi(\beta-a_0)(\det \mathrm{A}_0)^{\frac{1}{4}}}{\|G\|^2_{L_2(\Omega)}} - C|\ln\varepsilon|^{-1}\right)|(u, G_\varepsilon)_{\mathfrak{g}_\varepsilon}|^2.\end{aligned}$$

Hence, choosing η small enough and c_4 large enough, in view of condition (12), we conclude on the existence of the constants η and c_4 such that

$$\mathfrak{h}_\varepsilon(u,u) + (c_1+c_4)\|u\|^2_{L_2(\Omega_\varepsilon)} \geqslant c_5\|u_\perp\|^2_{\mathfrak{g}_\varepsilon} + c_5\varepsilon|(u, G_\varepsilon)_{\mathfrak{g}_\varepsilon}|^2 \tag{75}$$

for all $u \in \mathfrak{D}(\mathfrak{g}_\varepsilon)$ with a fixed positive constant c_5 independent of ε and u.

4.3. Self-Adjointness

We proceed to proving the self-adjointness of the operators $\mathcal{H}_\varepsilon$ and $\mathcal{H}_{0,\beta}$. We begin with the operator $\mathcal{H}_\varepsilon$. Since the form $\mathfrak{h}_\varepsilon$ is symmetric and lower-semi-bounded and is associated with the operator $\mathcal{H}_\varepsilon$, it is sufficient to show that it is closed, and then this will imply the self-adjointness of the operator $\mathcal{H}_\varepsilon$.

We choose an arbitrary sequence $u_n \in \mathfrak{D}(\mathfrak{h}_\varepsilon)$ such that

$$\|u_n - u\|_{L_2(\Omega_\varepsilon)} \to 0, \qquad \mathfrak{h}_\varepsilon(u_n-u_m, u_n-u_m) \to 0 \quad \text{as} \quad n,m\to\infty \tag{76}$$

for some $u \in L_2(\Omega_\varepsilon)$. We also observe that since

$$\|v\|^2_{\mathfrak{g}_\varepsilon} = \|v_\perp\|^2_{\mathfrak{g}_\varepsilon} + |(v, G_\varepsilon)_{\mathfrak{g}_\varepsilon}|^2$$

for each $v \in \mathfrak{D}(\mathfrak{g}_\varepsilon)$, then it follows from (75) that

$$\mathfrak{h}_\varepsilon(v,v) + (c_1+c_4)\|v\|^2_{L_2(\Omega_\varepsilon)} \geqslant c_5\varepsilon\|v\|^2_{\mathfrak{g}_\varepsilon}.$$

This estimate and (76) yield

$$\|u_n - u_m\|_{\mathfrak{g}_\varepsilon} \to 0 \quad \text{as} \quad n,m\to\infty.$$

Since the space $\mathfrak{D}(\mathfrak{g}_\varepsilon)$ is Hilbert and is a subspace of $L_2(\Omega_\varepsilon)$, the sequence u_n converges in $\mathfrak{D}(\mathfrak{g}_\varepsilon)$, and the limit is necessarily u. Hence, $u \in \mathfrak{D}(\mathfrak{g}_\varepsilon)$ and $\|u_n - u\|_{\mathfrak{g}_\varepsilon} \to 0$ as $n\to\infty$. By estimates (36) and (44), we also see that $\big(\alpha(u_n-u), (u_n-u)\big)_{L_2(\partial\omega_\varepsilon)} \to 0$ as $n\to+\infty$. Therefore, $\mathfrak{h}_\varepsilon(u_n-u) \to 0$ as $n\to+\infty$, and the form $\mathfrak{h}_\varepsilon$ is closed. This yields the self-adjointness of the operator $\mathcal{H}_\varepsilon$.

We proceed to the operator $\mathcal{H}_{0,\beta}$. We consider the adjoint operator $\mathcal{H}^*_{0,\beta}$, and by the definition of an adjoint operator, the domain of $\mathcal{H}^*_{0,\beta}$ consists of all functions $v \in L_2(\Omega)$, for which there exists a function $g \in L_2(\Omega)$ obeying the identity

$$(\mathcal{H}_{0,\beta}u, v)_{L_2(\Omega)} = (u, g)_{L_2(\Omega)} \quad \text{for all} \quad u \in \mathfrak{D}(\mathcal{H}_{0,\beta}), \qquad \mathcal{H}^*_{0,\beta}v = g.$$

Substituting the representation in (13) for the functions from the domain of the operator $\mathcal{H}_{0,\beta}$ into the above identity, we obtain

$$(\mathcal{H}_\Omega u_0, v)_{L_2(\Omega)} - (\beta - a)^{-1} u_0(x_0)(G, c_1 v + g)_{L_2(\Omega)} = (u_0, g)_{L_2(\Omega)}, \qquad u_0 \in \mathfrak{D}(\mathcal{H}_\Omega). \quad (77)$$

Similarly to (32)–(34), we confirm that

$$((\mathcal{H}_\Omega + c_1)u_0, G)_{L_2(\Omega\setminus B_\delta(x_0))} = -\left(\frac{\partial u_0}{\partial \mathrm{n}}, G\right)_{L_2(\partial B_\delta(x_0))} + \left(u_0, \frac{\partial G}{\partial \mathrm{n}}\right)_{L_2(\partial B_\delta(x_0))} \quad (78)$$

Since $u_0 \in W_2^2(\Omega)$, by (18), (36), and (41), we find that

$$\left(\frac{\partial u_0}{\partial \mathrm{n}}, G\right)_{L_2(\partial B_\delta(x_0))} \to 0, \qquad \left(u_0, \frac{\partial G}{\partial \mathrm{n}}\right)_{L_2(\partial B_\delta(x_0))} \to -4\pi(\det \mathrm{A}_0)^{\frac{1}{4}} u_0(x_0)$$

Passing, then, to the limit in (78), we obtain

$$((\mathcal{H}_\Omega + c_1)u_0, G)_{L_2(\Omega)} = -4\pi(\det \mathrm{A}_0)^{\frac{1}{4}} u_0(x_0). \quad (79)$$

This allows us to rewrite (77) as

$$(\mathcal{H}_\Omega u_0, v)_{L_2(\Omega)} - (\beta - a)^{-1}\kappa((\mathcal{H}_\Omega + c_1)u_0, G)_{L_2(\Omega)} = (u_0, g)_{L_2(\Omega)},$$
$$\rho := -\frac{\pi\sqrt{\det \mathrm{A}_0}(G, c_1 v + g)_{L_2(\Omega)}}{4},$$

which yields

$$(\mathcal{H}_\Omega u_0, v - (\beta - a)^{-1}\overline{\rho} G)_{L_2(\Omega)} = (u_0, g + (\beta - a)^{-1} c_1 \overline{\rho} G)_{L_2(\Omega)}.$$

Due to the self-adjointness of the operator $\mathcal{H}_\Omega$, we then obtain the identities

$$w := v - (\beta - a)^{-1}\overline{\rho} G \in \mathfrak{D}(\mathcal{H}_\Omega), \qquad \mathcal{H}_\Omega w = g + (\beta - a)^{-1} c_1 \overline{\rho} G. \quad (80)$$

Applying identity (79) with u_0 replaced by w, we find that

$$-4\pi(\det \mathrm{A}_0)^{\frac{1}{4}} w(x_0) = ((\mathcal{H}_\Omega + c_1)w, G)_{L_2(\Omega)} = (g + c_1 v, G)_{L_2(\Omega)} = -4\pi(\det \mathrm{A}_0)^{\frac{1}{4}}\overline{\rho}.$$

This identity and (80) imply that

$$v = w + (\beta - a)^{-1} w(x_0) G, \qquad \mathcal{H}_{0,\beta}^* w = g = \mathcal{H}_\Omega w - (\beta - a)^{-1} c_1 w(x_0)$$

and, hence, $\mathcal{H}_{0,\beta}^* = \mathcal{H}_{0,\beta}$.

In the next section, we also need the following auxiliary lemma, the proof of which literally reproduces that of Lemma 4.3 in [6].

Lemma 7. *Let $f \in L_2(\Omega)$, $\operatorname{Im}\lambda \neq 0$, $u := (\mathcal{H}_{0,\beta} - \lambda)^{-1} f$. Then, the function u satisfies the representation*

$$u(x) = v(x) + (\beta - a_0)^{-1} v(x_0) G(x), \qquad v \in \mathfrak{D}(\mathcal{H}_\Omega), \quad (81)$$

and the estimate

$$\mathfrak{h}_\Omega(v, v) + c_1\|v\|_{L_2(\Omega)}^2 + \|v_0\|_{W_2^2(B_{2R_2}(x_0))}^2 + |v_0(x_0)|^2 \leqslant C(\lambda)\|f\|_{L_2(\Omega)}^2 \quad (82)$$

holds, where $C(\lambda)$ is a constant independent of f.

5. Resolvent Convergence

In this section, we prove estimates (15) and (16). The operators $\mathcal{H}_\varepsilon$ and $\mathcal{H}_{0,\beta}$ are self-adjoint, and this is why their resolvents are well-defined for λ with a non-zero imaginary part. We arbitrarily fix such λ and a function $f \in L_2(\Omega)$, and we let

$$u_0 := (\mathcal{H}_{0,\beta} - \lambda)^{-1} f, \qquad u_\varepsilon := (\mathcal{H}_\varepsilon - \lambda)^{-1} f, \qquad v_\varepsilon := u_\varepsilon - u_0. \tag{83}$$

The function v_ε is an element of $W_2^2(\Omega_\varepsilon)$ and solves the boundary value problem

$$(\hat{\mathcal{H}} - \lambda) v_\varepsilon = 0 \quad \text{in} \quad \Omega_\varepsilon, \qquad \mathcal{B} v_\varepsilon = 0 \quad \text{on} \quad \partial\Omega, \qquad \frac{\partial v_\varepsilon}{\partial n} = -\alpha v_\varepsilon + p_\varepsilon \quad \text{on} \quad \partial\omega_\varepsilon,$$

where

$$p_\varepsilon := \left(\frac{\partial}{\partial n} + \alpha \right) u_0.$$

The associated integral identity with v_ε as the test function reads as

$$\mathfrak{h}_\varepsilon(v_\varepsilon, v_\varepsilon) - \lambda \|v_\varepsilon\|_{L_2(\Omega_\varepsilon)}^2 = (p_\varepsilon, v_\varepsilon)_{L_2(\partial\omega_\varepsilon)}. \tag{84}$$

Our next step is to estimate the right hand of this identity.

Since $u_0 \in \mathfrak{D}(\mathcal{H}_{0,\beta})$, it satisfies representation (81) with $v = v_0$ and estimate (82), while by (14), for the function f, we have

$$f = (\mathcal{H}_{0,\beta} - \lambda) u_0 = (\mathcal{H}_\Omega - \lambda) v_0 - (\beta - a_0)^{-1} (\lambda + c_1) v_0(x_0) G.$$

Following (69), we let

$$v_{\varepsilon,\perp} := v_\varepsilon - (v_\varepsilon, G_\varepsilon)_{\mathfrak{g}_\varepsilon} G_\varepsilon, \qquad (v_{\varepsilon,\perp}, G_\varepsilon)_{\mathfrak{g}_\varepsilon} = 0, \qquad v_\varepsilon = v_{\varepsilon,\perp} + (v_\varepsilon, G_\varepsilon)_{\mathfrak{g}_\varepsilon} G_\varepsilon. \tag{85}$$

Then, we represent the function p_ε as

$$\begin{aligned} &p_\varepsilon = p_{\varepsilon,1} + p_{\varepsilon,2} + p_{\varepsilon,3} + p_{\varepsilon,4}, \qquad p_{\varepsilon,1} := \frac{\partial v_0}{\partial n}, \\ &p_{\varepsilon,2} := (v_0 - \langle v_0 \rangle_{\partial\omega_\varepsilon}) \alpha, \qquad p_{\varepsilon,3} := (\langle v_0 \rangle_{\partial\omega_\varepsilon} - v_0(x_0)) \alpha, \\ &p_{\varepsilon,4} := v_0(x_0)(\beta - a_0)^{-1} \left(\frac{\partial G}{\partial n} + \alpha G \right) + \alpha v_0(x_0), \end{aligned} \tag{86}$$

where

$$\langle v_0 \rangle_{\partial\omega_\varepsilon} := \frac{1}{\varepsilon^2 \operatorname{mes} \partial\omega} \int\limits_{\partial\omega_\varepsilon} v_0(x)\, ds$$

and $\operatorname{mes} \partial\omega$ is the area of $\partial\omega$.

By estimates (36) and (82), we immediately obtain

$$\left| (p_{\varepsilon,1}, v_\varepsilon)_{L_2(\partial\omega_\varepsilon)} \right| \leqslant C\varepsilon \|v_0\|_{W_2^2(\Omega_0)} \|v_\varepsilon\|_{W_2^1(\Omega_0)} \leqslant C\varepsilon \|f\|_{L_2(\Omega)} \left(\|v_{\varepsilon,\perp}\|_{\mathfrak{g}_\varepsilon} + |(v_\varepsilon, G_\varepsilon)_{\mathfrak{g}_\varepsilon}| \right); \tag{87}$$

here and till the end of this section, C denotes various constants independent of f, u_0, u_ε, v_ε, ε, and spatial variables.

The function $v_0 - \langle v_0 \rangle_{\partial\omega_\varepsilon}$ satisfies condition (38) and belongs to $W_2^2(B_{2R_2}(x_0))$. This is why, by (36), (40), (82), and the definition of α, we obtain the following:

$$\begin{aligned} \left| (p_{\varepsilon,2}, v_\varepsilon)_{L_2(\partial\omega_\varepsilon)} \right| &\leqslant C\varepsilon^2 \|v_0\|_{W_2^2(B_{2R_2}(x_0))} \|v_\varepsilon\|_{W_2^1(\Omega_0)} \\ &\leqslant C\varepsilon^2 \|f\|_{L_2(\Omega)} \left(\|v_{\varepsilon,\perp}\|_{\mathfrak{g}_\varepsilon} + |(v_\varepsilon, G_\varepsilon)_{\mathfrak{g}_\varepsilon}| \right). \end{aligned} \tag{88}$$

Applying estimate (41) with $\phi = 1$ to the function v_0 and estimates (36) and (82), we obtain the following:

$$\begin{aligned}\big|(p_{\varepsilon,3}, v_\varepsilon)_{L_2(\partial\omega_\varepsilon)}\big| \leqslant & C|(v_0 - \langle v_0\rangle_{\partial\omega_\varepsilon})| \|v_\varepsilon\|_{L_2(\partial\omega_\varepsilon)} \leqslant C\varepsilon \|f\|_{L_2(\Omega)} \|v_\varepsilon\|_{\mathfrak{g}_\varepsilon} \\ \leqslant & C\varepsilon \|f\|_{L_2(\Omega)} \|v_\varepsilon\|_{\mathfrak{g}_\varepsilon} \big(\|v_{\varepsilon,\perp}\|_{\mathfrak{g}_\varepsilon} + |(v_\varepsilon, G_\varepsilon)_{\mathfrak{g}_\varepsilon}|\big).\end{aligned} \tag{89}$$

In view of the definition of $v_{\varepsilon,\perp}$ in (85), we have

$$(p_{\varepsilon,4}, v_\varepsilon)_{L_2(\partial\omega_\varepsilon)} = (p_{\varepsilon,4}, v_{\varepsilon,\perp})_{L_2(\partial\omega_\varepsilon)} + (v_\varepsilon, G_\varepsilon)_{\mathfrak{g}_\varepsilon} \frac{(p_{\varepsilon,4}, G)_{L_2(\partial\omega_\varepsilon)}}{\|G\|_{\mathfrak{g}_\varepsilon}}. \tag{90}$$

Using, then, the definition of the function α in (3), asymptotics for $\|G\|_{\mathfrak{g}_\varepsilon}$ in (59), estimate (36) applied for $v_{\varepsilon,\perp}$, inequality (82) for v_0, the boundary condition on $\partial\omega_\varepsilon$ in (56), and the uniform boundedness of the function, we find that

$$\big|(p_{\varepsilon,4}, v_{\varepsilon,\perp})_{L_2(\partial\omega_\varepsilon)}\big| \leqslant C\varepsilon^{\frac{1}{2}} \|p_{\varepsilon,4}\|_{L_2(\partial\omega_\varepsilon)} \|v_{\varepsilon,\perp}\|_{\mathfrak{g}_\varepsilon} \leqslant C\varepsilon^{\frac{1}{2}} \|f\|_{L_2(\Omega)} \|v_{\varepsilon,\perp}\|_{\mathfrak{g}_\varepsilon}. \tag{91}$$

Employing asymptotics(8) and (26), condition (11), and estimate (82), we find that

$$\big|(p_{\varepsilon,4}, G)_{L_2(\partial\omega_\varepsilon)}\big| \leqslant C\varepsilon \|f\|_{L_2(\Omega)}$$

and, hence, in view of (59), (90), and (91),

$$\big|(p_{\varepsilon,4}, v_\varepsilon)_{L_2(\partial\omega_\varepsilon)}\big| \leqslant C\varepsilon^{\frac{1}{2}} \|f\|_{L_2(\Omega)} \|v_{\varepsilon,\perp}\|_{\mathfrak{g}_\varepsilon} + C\varepsilon^{\frac{3}{2}} \|f\|_{L_2(\Omega)} |(v_\varepsilon, G_\varepsilon)_{\mathfrak{g}_\varepsilon}|.$$

Summing up this estimate and (87)–(89), in view of (86), we obtain

$$\big|(p_\varepsilon, v_\varepsilon)_{L_2(\partial\omega_\varepsilon)}\big| \leqslant C\varepsilon^{\frac{1}{2}} \|f\|_{L_2(\Omega)} \|v_{\varepsilon,\perp}\|_{\mathfrak{g}_\varepsilon} + C\varepsilon \|f\|_{L_2(\Omega)} |(v_\varepsilon, G_\varepsilon)_{\mathfrak{g}_\varepsilon}|.$$

We take the imaginary part of identity (84) and use the above estimate:

$$\|v_\varepsilon\|^2_{L_2(\Omega_\varepsilon)} \leqslant C\varepsilon^{\frac{1}{2}} \|f\|_{L_2(\Omega)} \|v_{\varepsilon,\perp}\|_{\mathfrak{g}_\varepsilon} + C\varepsilon \|f\|_{L_2(\Omega)} |(v_\varepsilon, G_\varepsilon)_{\mathfrak{g}_\varepsilon}|.$$

Then, we take the imaginary part of identity (84) and employ the above inequality and (75):

$$\|v_{\varepsilon,\perp}\|_{\mathfrak{g}_\varepsilon} + \varepsilon^{\frac{1}{2}} |(v_\varepsilon, G_\varepsilon)_{\mathfrak{g}_\varepsilon}| \leqslant C\varepsilon^{\frac{1}{2}} \|f\|_{L_2(\Omega)}.$$

This implies that

$$\|v_{\varepsilon,\perp}\|_{\mathfrak{g}_\varepsilon} \leqslant C\varepsilon^{\frac{1}{2}} \|f\|_{L_2(\Omega)}, \qquad |(v_\varepsilon, G_\varepsilon)_{\mathfrak{g}_\varepsilon}| \leqslant C \|f\|_{L_2(\Omega)}. \tag{92}$$

Inequality (15) follows from the above estimates, (83), (85), and (59). It is also easy to see that for an arbitrary domain $\tilde{\Omega}$ described in the formulation of the theorem, we have

$$\|\chi_{\tilde{\Omega}} G\|_{\mathfrak{D}(\mathfrak{h}_\Omega)} \leqslant C.$$

Using this estimate and (92), we arrive at (16).

6. Order Sharpness

In this section, we show that estimates (15) and (16) are order-sharp by providing an appropriate example. We let

$$\begin{aligned} &\Omega := B_1(0), \qquad x_0 := 0, \qquad \Omega_0 := B_{\frac{1}{2}}(0), \\ &\hat{\mathcal{H}} := -\Delta, \qquad c_1 := 0, \qquad \mathcal{B}u = u, \end{aligned}$$

Then, A_0 is the unit matrix and, assuming that α_1 is a constant function,

$$G(x) = |x|^{-1} - 1, \qquad \alpha_0(x) = -|x|^{-1},$$
$$a_0 = -1, \qquad \beta_0 = 0, \qquad \alpha_1 = -\beta - 1.$$

We choose u_0 as

$$u_0(x) := v_0(x) + (\beta+1)^{-1}G(x), \qquad v_0(x) := w(|x|\varepsilon^{-1}),$$

where $w = w(\xi)$ is an infinitely differentiable even function on $\mathbb{R}$, vanishing outside $[-2,2]$ and obeying the conditions

$$w(\xi) \equiv 1 \quad \text{on} \quad [-1,1], \qquad f_0(\xi) := w''(\xi) + 2\xi^{-1}w'(\xi) \not\equiv 0 \quad \text{on} \quad [-2,2]\setminus[-1,1]. \tag{93}$$

The function v_0 obviously belongs to $W_2^2(\Omega)$ and vanishes on $\partial\Omega$. The function u_0 solves the equation

$$(\mathcal{H}_{0,\beta} + \mathrm{i})u_0 = f,$$
$$f(x) := -\varepsilon^{-2} f_0(|x|\varepsilon^{-1}) + \mathrm{i}w(|x|\varepsilon^{-1}) + \mathrm{i}(\beta+1)^{-1}G(x).$$

In view of the assumption of f_0 in (93), we immediately see that

$$\|f\|^2_{L_2(\Omega)} \geqslant C\varepsilon^{-4}\|f_0(|\cdot|\varepsilon^{-1})\|^2_{L_2(B_{2\varepsilon}(0))} - C \geqslant C\varepsilon^{-1},$$

where the Cs are some positive constants independent of ε. Using the first assumption in (93), it is also straightforward to confirm that

$$\begin{aligned}\left(\frac{\partial}{\partial\nu} + \alpha_0 + \alpha_1\right)u_0 =& \varepsilon^{-1}(\alpha_1 - \beta)(\beta+1)^{-1} + \alpha_1\beta(\beta+1)^{-1} \\ =& -\varepsilon^{-1}(2 - (\beta+1)^{-1}) - \beta.\end{aligned} \tag{94}$$

The function

$$Q(x) := \frac{1}{|x|}\sinh\frac{1+\mathrm{i}}{\sqrt{2}}(|x|-1)$$

solves the problem

$$\begin{aligned}&(-\Delta + \mathrm{i})Q = 0 \quad \text{in} \quad x \in \Omega\setminus\{0\}, \qquad Q = 0 \quad \text{on} \quad \partial\Omega, \\ &\left(\frac{\partial}{\partial\nu} + \alpha_0 + \alpha_1\right) = -\varepsilon^{-1}\frac{1+\mathrm{i}}{\sqrt{2}}\left(\cosh\frac{1+\mathrm{i}}{\sqrt{2}}(1-\varepsilon) + (2+\beta)\sinh\frac{1+\mathrm{i}}{\sqrt{2}}(1-\varepsilon)\right).\end{aligned} \tag{95}$$

We also observe that

$$\|Q\|_{L_2(\Omega_\varepsilon)} \geqslant C, \qquad \|Q\|_{W_2^1(\Omega_\varepsilon)} \geqslant C\varepsilon^{-\frac{1}{2}}, \qquad \|Q\|_{W_2^1(\Omega\setminus B_r(0))} \geqslant C(r), \tag{96}$$

where C and $C(r)$ are some fixed positive constants independent of ε.

Using problem (95) and identity (94), we easily see that the corresponding function $u_\varepsilon = (\mathcal{H}_\varepsilon + \mathrm{i})^{-1}f$ reads as $u_\varepsilon = u_0 - c_\varepsilon Q$, where

$$\begin{aligned}c_\varepsilon :=& \frac{\sqrt{2}}{1+\mathrm{i}}\frac{2 - (\beta+1)^{-1} + \varepsilon\beta}{\left(\cosh\frac{1+\mathrm{i}}{\sqrt{2}}(1-\varepsilon) + (2+\beta)\sinh\frac{1+\mathrm{i}}{\sqrt{2}}(1-\varepsilon)\right)} \\ =& \frac{\sqrt{2}}{1+\mathrm{i}}\frac{2 - (\beta+1)^{-1}}{\left(\cosh\frac{1+\mathrm{i}}{\sqrt{2}} + (2+\beta)\sinh\frac{1+\mathrm{i}}{\sqrt{2}}\right)} + O(\varepsilon).\end{aligned}$$

Hence, in view of (96),

$$\frac{\|u_\varepsilon - u_0\|_{L_2(\Omega\setminus B_r(0))}}{\|f\|_{L_2(\Omega)}} \geqslant C\varepsilon^{\frac{1}{2}},$$
$$\frac{\|u_\varepsilon - u_0\|_{L_2(\Omega_\varepsilon)}}{\|f\|_{L_2(\Omega)}} \geqslant C(r)\varepsilon^{\frac{1}{2}},$$
$$\frac{\|u_\varepsilon - u_0\|_{W_2^1(\Omega_\varepsilon)}}{\|f\|_{L_2(\Omega)}} \geqslant C, \tag{97}$$

where C and $C(r)$ are some fixed constants independent of ε. The first estimate shows that estimate (15) is order-sharp, while the second estimate does the same for (16). Estimate (97) ensures that estimate (17) is order-sharp. The proof of Theorem 1 is complete.

Funding: This work was supported by the Program of Developing Scientific and Educational Volga Region Mathematical Center (agreement no. 075-02-2024-1444).

Data Availability Statement: No new data were created or analyzed in this study. Data sharing is not applicable to this article.

Conflicts of Interest: The author declare no conflicts of interest.

References

1. Fermi, E. Sul moto dei neutroni nelle sostanze idrogenate. *Ric. Sci.* **1936**, *7*, 13–52.
2. Berezin, F.A.; Faddeev, L.D. A remark on Schrödinger's equation with a singular potential. *Sov. Math. Dokl.* **1961**, *2*, 372–375.
3. Albeverio, S.; Gesztesy, F.; Høegh-Krohn, R.; Holden, H. *Solvable Models in Quantum Mechanics*, 2nd ed.; AMS Chelsea Publishing: Providence, RI, USA, 2005.
4. Albeverio, S.; Nizhnik, L. Approximation of general zero-range potentials. *Ukrainian Math. J.* **2000**, *52*, 582–589. [CrossRef]
5. Exner, P.; Neidhardt, H.; Zagrebnov, V.A. Potential approximations to δ': An inverse Klauder phenomenon with norm-resolvent convergence. *Commun. Math. Phys.* **2001**, *224*, 593–612. [CrossRef]
6. Borisov, D.I.; Exner, P. Approximation of point interactions by geometric perturbations in two-dimensional domains. *Bull. Math. Sci.* **2023**, *13*, 2250003. [CrossRef]
7. Borisov, D.I. Geometric approximation of point interactions in two-dimensional domains for non-self-adjoint operators. *Mathematics* **2023**, *11*, 947. [CrossRef]
8. Savin, A. The Friedrichs extension of elliptic operators with conditions on submanifolds of arbitrary dimension. *Mathematics* **2023**, *12*, 418. [CrossRef]
9. Díaz, J.I.; Gómez-Castro, D.; Shaposhnikova, T.A. *Nonlinear Reaction-Diffusion Processes for Nanocomposites: Anomalous Improved Homogenization*; De Gruyter: Berlin, Germany, 2021.
10. Il'in, A.M. *Matching of Asymptotic Expansions of Solutions of Boundary Value Problems*; American Mathematical Society: Providence, RI, USA, 1992.
11. Marchenko, V.A.; Khruslov, Y.E. *Boundary Value Problems in Domains with a Fine-Grained Boundary*; Naukova Dumka: Kiev, Ukraine, 1974. (In Russian)
12. Maz'ya, V.G.; Nazarov, S.A.; Plamenevskii, B.A. *Asymptotic Theory of Elliptic Boundary Value Problems in Singularly Perturbed Domains*; Birkhäuser: Basel, Switzerland, 2000; Volumes I and II.
13. Suslina, T.A. Spectral approach to homogenization of elliptic operators in a perforated space. *Rev. Math. Phys.* **2018**, *30*, 1840016. [CrossRef]
14. Borisov, D.I.; Mukhametrakhimova, A.I. On norm resolvent convergence for elliptic operators in multi-dimensional domains with small holes. *J. Math. Sci.* **2018**, *232*, 283–298. [CrossRef]
15. Zhikov, V.V. Spectral method in homogenization theory. *Proc. Steklov Inst. Math.* **2005**, *250*, 85–94.
16. Khrabustovskyi, A.; Plum, M. Operator estimates for homogenization of the Robin Laplacian in a perforated domain. *J. Diff. Equats.* **2022**, *338*, 474–517. [CrossRef]
17. Borisov, D.I. Operator estimates for non-periodically perforated domains with Dirichlet and nonlinear Robin conditions: Strange term. *Math. Model. Appl. Sci.* **2024**, *47*, 4122–4164. [CrossRef]
18. Borisov, D.I.; Kříž, J. Operator estimates for non-periodically perforated domains with Dirichlet and nonlinear Robin conditions: Vanishing limit. *Anal. Math. Phys.* **2023**, *13*, 5. [CrossRef]
19. Anné, C.; Post, O. Wildly perturbed manifolds: Norm resolvent and spectral convergence. *J. Spectr. Theory* **2021**, *11*, 229–279. [CrossRef]

20. Gilbarg, D.; Trudinger, N. *Elliptic Partial Differential Equations of Second Order*; Springer: New York, NY, USA, 1983.
21. Kato, T. *Perturbation Theory for Linear Operators*; Springer: Berlin, Germany, 1976.

mathematics

MDPI

Article

Spectrum of One-Dimensional Potential Perturbed by a Small Convolution Operator: General Structure

D. I. Borisov [1,*], A. L. Piatnitski [2,3] and E. A. Zhizhina [2,3]

1 Institute of Mathematics, Ufa Federal Research Center, Russian Academy of Sciences, Chernyshevsky str. 112, Ufa 450008, Russia
2 Faculty of Engineering Science and Technology, UiT The Arctic University of Norway, P.O. Box 385, 8505 Narvik, Norway; apiatnitski@gmail.com (A.L.P.); ejj@iitp.ru (E.A.Z.)
3 Institute for Information Transmission Problem (Kharkevich Institute) of RAS, Bolshoy Karetny per. 19, Build. 1, Moscow 127051, Russia
* Correspondence: borisovdi@yandex.ru

Abstract: We consider an operator of multiplication by a complex-valued potential in $L_2(\mathbb{R})$, to which we add a convolution operator multiplied by a small parameter. The convolution kernel is supposed to be an element of $L_1(\mathbb{R})$, while the potential is a Fourier image of some function from the same space. The considered operator is not supposed to be self-adjoint. We find the essential spectrum of such an operator in an explicit form. We show that the entire spectrum is located in a thin neighbourhood of the spectrum of the multiplication operator. Our main result states that in some fixed neighbourhood of a typical part of the spectrum of the non-perturbed operator, there are no eigenvalues and no points of the residual spectrum of the perturbed one. As a consequence, we conclude that the point and residual spectrum can emerge only in vicinities of certain thresholds in the spectrum of the non-perturbed operator. We also provide simple sufficient conditions ensuring that the considered operator has no residual spectrum at all.

Keywords: convolution operator; potential; perturbation; spectrum; emerging eigenvalues

MSC: 47G10; 47A55; 47A10

Citation: Borisov, D.I.; Piatnitski, A.L.; Zhizhina, E.A. Spectrum of One-Dimensional Potential Perturbed by a Small Convolution Operator: General Structure. *Mathematics* **2023**, *11*, 4042. https://doi.org/10.3390/math11194042

Academic Editor: Natalia Bondarenko

Received: 26 August 2023
Revised: 13 September 2023
Accepted: 14 September 2023
Published: 23 September 2023

1. Introduction

Over the last 20 years, there has been growing interest in non-local operators since they arise in various applications. Among such operators, there are convolution operators with integrable kernels. They appear in population dynamics, ecological problems and porous media theory. One of the interesting models of a nonlocal operator is a convolution operator perturbed by a potential, i.e., an operator

$$(\mathcal{L}u)(x) = \int\limits_{\mathbb{R}^d} a(x-y)u(y)\,dy + V(x)u(x) \quad \text{in} \quad L_2(\mathbb{R}^d). \tag{1}$$

While the spectra of the convolution operator and of the operator of multiplication by the potential can be found and characterized very easily, the description of the spectrum of their sum is a very non-trivial problem. At the same time, the spectral properties of such sums are not only of pure mathematical interest, but are important also for many applications. For instance, such operators arise in the mathematical theory of population dynamics and it is important to know whether a given operator of the form (1) possesses positive eigenvalues; such questions were studied in [1–4].

A more general problem regards the spectral properties of Schrödinger type operators, which are perturbations of a given pseudo-differential operator by a potential; see [5–8] and the references therein. The assumptions made in the cited papers ensured that the essential spectrum of the perturbed operator coincides with that of the unperturbed pseudo-differential operator. The main results described the existence of the discrete spectrum and Cwikel–Lieb–Rozenblum-type inequalities. A similar result was obtained in [9] for perturbations of a rather general class of Schrödinger type operators defined on a σ-compact metric space. In [10], various bounds were obtained for the number of negative eigenvalues produced by a perturbation of an operator $\mathcal{H}_0$ under the assumption that the Markov process with generator $-\mathcal{H}_0$ is recurrent.

In our recent works [11,12], we studied spectral properties of operator (1) assuming that it was self-adjoint. The essential spectrum was found explicitly. We established several sufficient conditions ensuring the existence of the discrete spectrum and obtained upper and lower bounds for the number of points of the discrete spectrum. We also provided sufficient conditions guaranteeing that the considered operator had infinitely many discrete eigenvalues accumulating to the thresholds of the essential spectrum. The structure of such sufficient conditions was quite different from similar well-known sufficient conditions for differential operators perturbed by localized potentials. The reason is that in the latter case, the unperturbed differential operator is unbounded and is perturbed by a bounded multiplication operator. In the case of the operator in (1), both the convolution operator and multiplication are equipollent and this essentially changes the spectral properties in comparison with the classical model of perturbed elliptic differential operators.

It is well known that a small localized perturbation of a differential operator with a non-empty essential spectrum can create eigenvalues emerging from certain thresholds in this essential spectrum. There are hundreds of works, in which such bifurcation was investigated for various models. Not trying to mention all such works, we cite only a few very classical ones, where this phenomenon was first rigorously studied [13–16]. In view of such results for differential operators, a natural and reasonable continuation of our studies in [11,12] is to consider similar the issue for operators (1), i.e., to study the operator

$$(\mathcal{L}^{\varepsilon}u)(x) = \int\limits_{\mathbb{R}^d} a(x-y)u(y)\,dy + \varepsilon V(x)u(x)$$

on $L_2(\mathbb{R}^d)$, where ε is a small parameter. Here, again, the unperturbed operator and the perturbed one are equipollent and we naturally expect that the mechanisms of the eigenvalue's emergence from the essential spectrum can be rather different from ones for differential operators. This is indeed the case; for instance, using the Fourier transform, we can replace the operator $\mathcal{L}^{\varepsilon}$ with a unitary equivalent one, in which the original convolution operator is replaced by the multiplication operator, while the potential generates a convolution operator with a small coupling constant:

$$(\hat{\mathcal{L}}^{\varepsilon}u)(x) = \hat{a}(x)u(x) + \varepsilon \int\limits_{\mathbb{R}^d} \hat{V}(x-y)u(y)\,dy.$$

Exactly this operator in the one-dimensional case ($d = 1$) is the main object of the study in the present work. We succeed in dropping the condition of self-adjointness of the operator and treating a general operator with a complex-valued potential and a general convolution kernel. For such a general non-self-adjoint operator, we explicitly find its essential spectrum; it turns out to be the union of the ranges of the potential and of the Fourier image of the convolution kernel. Then, we show that the entire spectrum is located in a thin neighbourhood of the spectrum of the unperturbed multiplication operator. Our most nontrivial result states that in some fixed neighbourhood of a typical part of the spectrum of the unperturbed operator, there are no eigenvalues and no residual spectrum. As a consequence, we conclude that the eigenvalues and the residual spectrum can emerge only in vicinities of certain thresholds in the essential spectrum of the unperturbed operator. We also provide simple sufficient conditions ensuring that the considered operator has no residual spectrum at all, and not only in the aforementioned vicinities.

The issue of the existence and behaviour of possible eigenvalues and the residual spectrum emerging from the aforementioned threshold is an interesting problem that deserves an independent study. We shall present such a study in our next paper, which is being prepared now.

2. Problem and Main Results

Let $V = V(x)$ and $a = a(x)$ be measurable complex-valued functions defined on $\mathbb{R}$. On the space $L_1(\mathbb{R})$, we introduce a Fourier transform by the formula

$$\mathcal{F}[u](x) := \int\limits_{\mathbb{R}^d} u(\xi)e^{-\mathrm{i}x\cdot\xi}\,d\xi$$

and then extend it to $L_2(\mathbb{R})$. We assume that the function a belongs to $L_1(\mathbb{R})$, while the function V is an image of some function $\hat{V} \in L_1(\mathbb{R}^d)$, i.e., $V = \mathcal{F}[\hat{V}]$. We let $\hat{a}(\xi) := \mathcal{F}[a](\xi)$.

The paper is devoted to studying an operator in $L_2(\mathbb{R})$ defined by the formula

$$\mathcal{L}^\varepsilon := \mathcal{L}_V + \varepsilon\mathcal{L}_{a\star}, \qquad (\mathcal{L}_{a\star}u)(x) := \int\limits_{\mathbb{R}^d} a(x-y)u(y)\,dy, \qquad (\mathcal{L}_V u)(x) := V(x)u(x),$$

where ε is a small positive parameter. This operator is bounded in $L_2(\mathbb{R})$; this fact can be easily proved by literally reproducing the proof of Lemma 4.1 in [11]. Our main aim is to describe the behaviour of the spectrum of this operator for sufficiently small ε.

Since the functions a and V are complex-valued, the operator $\mathcal{L}^\varepsilon$ is non-self-adjoint. In this paper, we follow a usual classification of the spectrum of a non-self-adjoint operator. Namely, the spectrum $\sigma(\,\cdot\,)$ of a given operator is introduced as a complement to its resolvent set. The point spectrum $\sigma_{\mathrm{pnt}}(\,\cdot\,)$ is the set of all eigenvalues. The essential spectrum $\sigma_{\mathrm{ess}}(\,\cdot\,)$ is defined in terms of the characteristic sequences, i.e., $\lambda \in \sigma_{\mathrm{ess}}(\mathcal{A})$ of a closed operator $\mathcal{A}$ in $L_2(\mathbb{R})$ if there exists a bounded non-compact sequence u_n in $L_2(\mathbb{R})$ such that $(\mathcal{A}-\lambda)u_n \to 0$ in $L_2(\mathbb{R})$ as $n \to \infty$. The residual spectrum $\sigma_{\mathrm{res}}(\,\cdot\,)$ is defined as

$$\sigma_{\mathrm{res}}(\,\cdot\,) := \sigma(\,\cdot\,) \setminus \big(\sigma_{\mathrm{pnt}}(\,\cdot\,) \cup \sigma_{\mathrm{ess}}(\,\cdot\,)\big).$$

We shall show in Section 4.3, see Lemma 8, that the residual spectrum is given by the formula

$$\sigma_{\mathrm{res}}(\mathcal{A}) = (\sigma_{\mathrm{pnt}}(\mathcal{A}^*))^\dagger \setminus \big(\sigma_{\mathrm{pnt}}(\mathcal{A}) \cup \sigma_{\mathrm{ess}}(\mathcal{A})\big), \tag{2}$$

where for an arbitrary set $S \subset \mathbb{C}$, the set $S^\dagger$ is obtained by the symmetric reflection with respect to the real axis, i.e., $S^\dagger := \{\overline{\lambda} :\ \lambda \in S\}$.

We first describe the essential spectrum of the operator $\mathcal{L}^\varepsilon$. In order to do this, we introduce two curves in the complex plane as the ranges of the functions V and $\hat{a}$:

$$\Upsilon := \{V(x) :\, x \in \mathbb{R}\}, \qquad \gamma := \{\hat{a}(x) :\, x \in \mathbb{R}\}.$$

Theorem 1. *The spectrum of the operator $\mathcal{L}^\varepsilon$ is located in a small neighbourhood of Υ, namely,*

$$\sigma(\mathcal{L}^\varepsilon) \subseteq \big\{\lambda \in \mathbb{C} :\ \mathrm{dist}(\lambda, \Upsilon) \leqslant \varepsilon\|a\|_{L_1(\mathbb{R})}\big\}. \tag{3}$$

For all ε the essential spectrum of the operator $\mathcal{L}^\varepsilon$ is given by the identity

$$\sigma_{\mathrm{ess}}(\mathcal{L}^\varepsilon) = \Upsilon \cup \varepsilon\gamma. \tag{4}$$

The sets Υ and γ are continuous closed curves in the complex plane that contain the origin.

Apart of the essential spectrum described in Theorem 1, the operator $\mathcal{L}^\varepsilon$ can also have point and residual spectra. Our second main result states that the eigenvalues of the operator $\mathcal{L}^\varepsilon$ and its residual spectrum can exist only in the vicinities of certain thresholds on the curve Υ and they are absent in certain neighbourhoods of finite pieces of this curve. In order to state such a result, we classify all points $x_0 \in \mathbb{R}$ by a behaviour of the function V in their vicinities. Namely, given two pairs $\alpha = (\alpha_-, \alpha_+)$ and $\beta = (\beta_-, \beta_+)$ with $\alpha_\pm \in \mathbb{C} \setminus \{0\}$ and $\beta_\pm \in (0, +\infty)$, a point $x_0 \in \mathbb{R}$ is called a (β, α) threshold if there exists a ρ-neighbourhood of the point x_0 such that

$$V(x) - V_0 = \alpha_\pm |x - x_0|^{\beta_\pm} v_\pm(x) \quad \text{as} \quad 0 \leqslant \pm(x - x_0) \leqslant \rho, \tag{5}$$

where $v_- \in C^2[x_0 - \rho, x_0]$, $v_+ \in C^2[x_0, x_0 + \rho]$ are some complex-valued functions such that

$$v_\pm(x_0) = 1, \qquad |v_-'(x)| \leqslant C \quad \text{on} \quad [x_0 - \rho, x_0], \qquad |v_+'(x)| \leqslant C \quad \text{on} \quad [x_0, x_0 + \rho], \tag{6}$$

where C is some constant independent of x.

A point $x_0 \in \mathbb{R}$ is called regular if there exists a ρ-neighbourhood of the point x_0 such that

$$V \in C^2[x_0 - \rho, x_0 + \rho], \qquad V'(x_0) \neq 0. \tag{7}$$

Let S be a connected close piece of the curve Υ not containing the origin. We assume that this piece is the image of finitely many disjoint segment $J_j := [b_j^-, b_j^+]$ on the real axis, i.e.,

$$S = \{V(x) : x \in J\}, \qquad V(x) \notin S \quad \text{as} \quad x \notin J := \bigcup_{j=1}^{n} J_j, \tag{8}$$

where $n \in \mathbb{N}$ and $b_j^\pm \in \mathbb{R}$ are fixed numbers and $b_j^- < b_j^+$. For $\delta > 0$, we let

$$S^\delta := \{\lambda \in \mathbb{C} : \operatorname{dist}(\lambda, S) \leqslant \delta\}.$$

By $B_r(y)$, we denote an open ball in the complex plane of a radius r centred at a point y.

Now, we are in a position to formulate our second main result.

Theorem 2. *Let S be a connected close piece of the curve Υ not containing the origin and obeying (8), each segment J_j contains only regular points and finitely many (β, α) thresholds, and for each of such thresholds, we have $\beta_\pm < 1$. Suppose that there exists a natural m such that for each $\lambda \in S$, each of the segment J_j contains at most m points x such that $V(x) = \lambda$. Suppose also that the generalize derivative a' exists and*

$$a \in L_1(\mathbb{R}) \cap W_2^1(\mathbb{R}), \qquad \operatorname*{esssup}_{\substack{(x,y)\in\mathbb{R}^2 \\ 0<|x-y|<1}} \frac{|a'(x) - a'(y)|}{|x - y|^\theta} < \infty, \tag{9}$$

where $\theta \in (0, 1]$ is some fixed number. Then, there exists a sufficiently small $\delta > 0$ such that for all sufficiently small ε, the closed δ-neighbourhood S^δ of the set S intersects neither with the point spectrum of the operator $\mathcal{L}^\varepsilon$, nor with its residual spectrum, i.e.,

$$\sigma_{\mathrm{pnt}}(\mathcal{L}^\varepsilon) \cap S^\delta = \emptyset, \qquad \sigma_{\mathrm{res}}(\mathcal{L}^\varepsilon) \cap S^\delta = \emptyset.$$

Our third result concerns the residual spectrum. It is well known that such a spectrum is always absent for self-adjoint operators. In view of the absence of the residual spectrum in the set S^δ stated in Theorem 2, there arises a natural question on sufficient conditions ensuring the absence of the residual spectrum for the operator $\mathcal{L}^\varepsilon$. The answer to this question is our third main result formulated in the following theorem.

Theorem 3. *Assume that one of the following conditions holds:*

$$V(x) = \overline{V(x)}, \qquad a(x) = \overline{a(-x)}, \tag{10}$$

or

$$V(\tau x + \varrho) = V(x), \qquad a(-\tau x) = a(x), \qquad x \in \mathbb{R}, \tag{11}$$

for some $\varrho \in \mathbb{R}$ and $\tau \in \{-1, +1\}$. Then, the residual spectrum of the operator $\mathcal{L}^\varepsilon$ is empty for all ε.

Let us briefly discuss the problem and the main results. The main feature of our operator $\mathcal{L}^\varepsilon$ is its non-self-adjointness, and in the general situation, both functions V and a are complex-valued. The convolution operator is multiplied by the small parameter and our operator $\mathcal{L}^\varepsilon$ is to be treated as a perturbation of the multiplication operator by a small convolution operator. As mentioned in the introduction, by applying the Fourier transform to the operator $\mathcal{L}^\varepsilon$, we can reduce it to a unitarily equivalent operator, in which the convolution and the potential parts interchange; then, we obtain a convolution operator perturbed by a small potential. The results of this work serve as a first step in studying how such a small perturbation deforms the spectrum of the unperturbed operator.

Our first result, Theorem 1, describes explicitly the location of the essential spectrum of the operator $\mathcal{L}^\varepsilon$. It turns out to be the union of the essential spectra of the unperturbed multiplication operator $\mathcal{L}_V$ and of the perturbed operator $\varepsilon\mathcal{L}_{a\star}$. These parts of the essential spectrum are the curves Υ and $\varepsilon\gamma$. The latter curve is small and is located in the vicinity of the origin. The spectrum of the operator $\mathcal{L}^\varepsilon$ also satisfies inclusion (3), which means that this spectrum is located in a thin tubular neighbourhood of the limiting spectrum Υ.

Our most nontrivial result is Theorem 2. It states that in a typical situation, there are fixed neighbourhoods of finite pieces of the curve Υ, which contain no point and residual spectra of the operator $\mathcal{L}^\varepsilon$. The choice of such finite pieces is characterized by the presence of (β, α) thresholds, and these pieces are to be generated by regular point and finitely many (β, α) thresholds with $\beta_\pm < 1$. The latter condition means that the function V approaches such threshold with a not very high rate; see (5). The fact that there should be finitely many such thresholds is important and is employed essentially in the proof of Theorem 2. Another important point is that the considered piece of the curve Υ should not pass the origin; the presence of an additional curve $\varepsilon\gamma$ of the essential spectrum seems to play a nontrivial role in the existence of the discrete and residual spectrum in the vicinity of the origin. Assumption (9) is also essentially employed in the proof, and what can happen once they are violated is an interesting open question. We conjecture that violation of these conditions can dramatically change the spectral picture for the operator $\mathcal{L}^\varepsilon$.

We also observe that the second condition in (9) means that the first generalized derivative a' is Hölder-continuous almost everywhere, and this can be guaranteed by assuming that the second generalized derivative a'' exists and belongs to $L_p(\mathbb{R})$ with some $p \in (1, +\infty)$ including the case $p = +\infty$. Indeed, if the second derivative is an element of $L_\infty(\mathbb{R})$, then the second condition in (9) is satisfied with $\theta = 1$, while for $1 < p < +\infty$, it is implied by the Hölder inequality:

$$|a'(x) - a'(y)| = \left| \int\limits_x^y a''(t)\,dt \right| \leqslant |x-y|^{1-\frac{1}{p}} \|a'\|_{L_p(\mathbb{R})}.$$

An important consequence of Theorem 2 is that the eigenvalues and the points of the residual spectrum can arise only in the vicinity of (β, α) thresholds, when at least one of the numbers β_+ and β_- exceeds or equal to 1; in the case $\beta_+ = \beta_- = 1$, we should additionally assume that $\alpha_+ \neq -\alpha_-$ to avoid the case of a regular point. This means that typically, the spectrum of the operator $\mathcal{L}^\varepsilon$ is as follows: there is the essential spectrum described in Theorem 1, and along the curve Υ, there are no eigenvalues and residual spectrum except vicinities of the origin and (β, α) thresholds with $\beta_+ \geqslant 1$ or/and $\beta_- \geqslant 1$. In such vicinities, the eigenvalues can indeed emerge; see an example in our recent work [12]. However, the study of possible emerging eigenvalues in the general situation is a non-trivial problem, which we postpone for our next paper.

Theorem 3 addresses one more question on the absence of the residual spectrum for the operator $\mathcal{L}^\varepsilon$. In contrast to Theorem 2, here we aim to find cases where the residual spectrum is completely absent rather than only in some neighbourhoods of some pieces of

Υ. Condition (10) guarantees that the operator $\mathcal{L}^\varepsilon$ is self-adjoint. Condition (11) is more delicate and, in fact, it means that the operator $\mathcal{L}^\varepsilon$ is $\mathcal{PT}$-symmetric, namely,

$$\mathcal{PT}(\mathcal{L}^\varepsilon)^* = \mathcal{L}^\varepsilon\mathcal{PT}. \tag{12}$$

Here $\mathcal{T}$ is the operator of the complex conjugation, i.e., $\mathcal{T}u = \overline{u}$. The symbol $\mathcal{P}$ is an operator acting as

$$(\mathcal{P}u)(x) = u(\tau x + \varrho). \tag{13}$$

We also observe that once condition (12) holds for some other operator $\mathcal{P}$, it also ensures the absence of the residual spectrum for the operator $\mathcal{L}^\varepsilon$. Indeed, if λ and ϕ are an eigenvalue and an associated eigenfunction of the adjoint operator $(\mathcal{L}^\varepsilon)^*$, then

$$\mathcal{L}^\varepsilon\mathcal{PT}\phi = \mathcal{PT}(\mathcal{L}^\varepsilon)^*\phi = \overline{\lambda}\mathcal{PT}\phi. \tag{14}$$

Hence, $\overline{\lambda}$ is an eigenvalue of the operator $\mathcal{L}^\varepsilon$, and by Formula (2), we see that the residual spectrum of the operator $\mathcal{L}^\varepsilon$ is empty.

3. Location of Spectrum and Essential Spectrum

In this section, we prove Theorem 1. We begin with checking identity (3). The spectrum of the operator $\mathcal{L}_V$ obviously coincides with Υ. As $\lambda \notin \Upsilon$, the inverse operator $(\mathcal{L}_V - \lambda)^{-1}$ is the multiplication by $(V - \lambda)^{-1}$ and it is easy to see that the norm of the operator $(\mathcal{L}_V - \lambda)^{-1}$ satisfies the estimate

$$\|(\mathcal{L}_V - \lambda)^{-1}\| \leqslant \frac{1}{\operatorname{dist}(\lambda, \Upsilon)}. \tag{15}$$

For $\lambda \notin \Upsilon$, we consider the resolvent equation

$$(\mathcal{L}^\varepsilon - \lambda)u = f$$

with an arbitrary $f \in L_2(\Omega)$, and we rewrite it as

$$u + \varepsilon(\mathcal{L}_V - \lambda)^{-1}\mathcal{L}_{a\star}u = (\mathcal{L}_V - \lambda)^{-1}f. \tag{16}$$

By $\|\cdot\|_{X\to Y}$, we denote the norm of a bounded operator acting from a Banach space X into a Banach space Y. As it was shown in the proof of Lemma 4.1 in [11], once $a \in L_1(\mathbb{R})$, the operator $\mathcal{L}_{a\star}$ is bounded in $L_2(\mathbb{R})$ and

$$\|\mathcal{L}_{a\star}\|_{L_2(\mathbb{R})\to L_2(\mathbb{R})} \leqslant \|a\|_{L_1(\mathbb{R})}. \tag{17}$$

This estimate and (15) yield that as

$$\varepsilon\|(\mathcal{L}_V - \lambda)^{-1}\mathcal{L}_{a\star}\| \leqslant \varepsilon\frac{\|a\|_{L_1(\mathbb{R})}}{\operatorname{dist}(\lambda, \Upsilon)} < 1,$$

the inverse operator $\big(\mathcal{I} + \varepsilon(\mathcal{L}_V - \lambda)^{-1}\mathcal{L}_{a\star}\big)^{-1}$ is well defined, where $\mathcal{I}$ is the identity operator. This allows us to solve Equation (16) and to find the resolvent of the operator $\mathcal{L}^\varepsilon$:

$$(\mathcal{L}^\varepsilon - \lambda)^{-1} = \big(\mathcal{I} + \varepsilon(\mathcal{L}_V - \lambda)^{-1}\mathcal{L}_{a\star}\big)^{-1}(\mathcal{L}_V - \lambda)^{-1} \qquad \text{as} \qquad \varepsilon\|a\|_{L_1(\mathbb{R})} < \operatorname{dist}(\lambda, \Upsilon).$$

Hence, each point in the spectrum of the operator $\mathcal{L}^\varepsilon$ satisfies the inequality $\operatorname{dist}(\lambda, \Upsilon) \leqslant \varepsilon\|a\|_{L_1(\mathbb{R})}$ and this proves inclusion (3).

In order to prove identity (4), we adapt the proof of Theorem 2.1 from [11] and below, we reproduce the main milestones from the cited work. It follows from our assumptions on a and $\hat{V}$ that the functions V and $\hat{a}$ are bounded and continuous on $\mathbb{R}$ and decay at infinity. We also observe the following unitary equivalence:

$$\left(\frac{1}{(2\pi)^{\frac{d}{2}}}\mathcal{F}\right)\mathcal{L}_{a\star}\left(\frac{1}{(2\pi)^{\frac{d}{2}}}\mathcal{F}\right)^{-1}=\mathcal{L}_{\hat{a}},\qquad\left(\frac{1}{(2\pi)^{\frac{d}{2}}}\mathcal{F}\right)\mathcal{L}_{V}\left(\frac{1}{(2\pi)^{\frac{d}{2}}}\mathcal{F}\right)^{-1}=\mathcal{L}_{\hat{V}\star}.\tag{18}$$

Hence,

$$\begin{aligned}\sigma(\mathcal{L}_{\varepsilon\hat{a}})=\sigma_{\mathrm{ess}}(\mathcal{L}_{\varepsilon\hat{a}})=\sigma(\mathcal{L}_{\varepsilon a\star})=\sigma_{\mathrm{ess}}(\mathcal{L}_{\varepsilon a\star})=\varepsilon\gamma,\\ \sigma(\mathcal{L}_{V})=\sigma_{\mathrm{ess}}(\mathcal{L}_{V})=\sigma(\mathcal{L}_{\hat{V}\star})=\sigma_{\mathrm{ess}}(\mathcal{L}_{\hat{V}\star})=\Upsilon.\end{aligned}\tag{19}$$

We are going to prove the inclusion

$$\Upsilon\cup\varepsilon\gamma\subseteq\sigma_{\mathrm{ess}}(\mathcal{L}^{\varepsilon}).\tag{20}$$

We let

$$\varphi_n(x):=\begin{cases}(2n)^{\frac{1}{2}} & \text{as}\quad |x|<\dfrac{1}{n},\\ 0 & \text{as}\quad |x|>\dfrac{1}{n}\end{cases}$$

for all natural n. For an arbitrary $\lambda\in\Upsilon$, there exists $x_0\in\mathbb{R}$ such that $V(x_0)=\lambda$. The sequence $\varphi_n(x-x_0)$, normalized and non-compact in $L_2(\mathbb{R})$, is obviously a characteristic one of the operator $\mathcal{L}_V$ at the point λ. We also have:

$$\begin{aligned}\|\mathcal{L}_{a\star}\varphi_n(\,\cdot\,-x_0)\|^2_{L_2(\mathbb{R})}\leqslant&2n\int\limits_{\mathbb{R}}dx\left(\int\limits_{x_0-\frac{1}{n}}^{x_0+\frac{1}{n}}|a(x-y)|\,dy\right)^2=2n\int\limits_{\mathbb{R}}dx\left(\int\limits_{x-\frac{1}{n}}^{x+\frac{1}{n}}|a(y)|\,dy\right)^2\\ \leqslant&2n\left(\sup_{x\in\mathbb{R}}\int\limits_{x-\frac{1}{n}}^{x+\frac{1}{n}}|a(y)|\,dy\right)\int\limits_{\mathbb{R}^d}dx\int\limits_{x-\frac{1}{n}}^{x+\frac{1}{n}}|a(y)|\,dy\\ =&2n\left(\sup_{x\in\mathbb{R}}\int\limits_{x-\frac{1}{n}}^{x+\frac{1}{n}}|a(y)|\,dy\right)\int\limits_{\mathbb{R}^d}dy|a(y)|\int\limits_{y-\frac{1}{n}}^{y+\frac{1}{n}}dx\\ =&\|a\|_{L_1(\mathbb{R})}\left(\sup_{x\in\mathbb{R}}\int\limits_{x-\frac{1}{n}}^{x+\frac{1}{n}}|a(y)|\,dy\right)\to0,\quad n\to\infty,\end{aligned}$$

where the latter convergence is due to the absolute continuity of the Lebesgue integral. Hence, $\varphi_n(x-x_0)$ is a characteristic sequence of the operator $\mathcal{L}^{\varepsilon}$ at λ and

$$\sigma_{\mathrm{ess}}(\mathcal{L}_V)\subseteq\sigma_{\mathrm{ess}}(\mathcal{L}^{\varepsilon}).\tag{21}$$

By unitary equivalence (18) and identity (19), we similarly obtain $\sigma_{\mathrm{ess}}(\mathcal{L}_{\varepsilon a\star})\subseteq\sigma_{\mathrm{ess}}(\mathcal{L})$, and in view of (21), this proves (20).

It remains to show that

$$\sigma_{\mathrm{ess}}(\mathcal{L}^{\varepsilon})\setminus(\Upsilon\cup\varepsilon\gamma)=\varnothing.$$

If $\lambda\in\sigma_{\mathrm{ess}}(\mathcal{L}^{\varepsilon})\setminus(\Upsilon\cup\varepsilon\gamma)$, there exists a bounded non-compact sequence $u_n\in L_2(\mathbb{R})$ such that

$$f_n:=(\mathcal{L}-\lambda)u_n\to0,\qquad n\to\infty.\tag{22}$$

Since $\lambda\notin\big(\sigma_{\mathrm{ess}}(\mathcal{L}_V)\cup\sigma_{\mathrm{ess}}(\mathcal{L}_{\varepsilon a\star})\big)$, in view of (19), the resolvents $(\mathcal{L}_V-\lambda)^{-1}$ and $(\mathcal{L}_{\varepsilon a\star}-\lambda)^{-1}$ are well defined and bounded. Then, we rewrite (22) as

$$\frac{1}{V-\lambda}\mathcal{L}_{\varepsilon a\star}u_n+u_n=\frac{f_n}{V-\lambda}\to0,\qquad n\to+\infty,\qquad V(x)\neq\lambda,\quad x\in\mathbb{R},$$

and we get

$$(\mathcal{L}_{\varepsilon a\star} - \lambda)u_n + V_1\mathcal{L}_{\varepsilon a\star}u_n = \frac{\lambda}{V-\lambda}f_n, \qquad V_1 := \frac{V}{V-\lambda}, \qquad \frac{1}{V-\lambda} = -\frac{1}{\lambda} + \frac{V_1}{\lambda}$$

where we have used zero as in $\sigma_{\text{ess}}(\mathcal{L}_V)$ and, therefore, $\lambda \neq 0$. Applying, then, the resolvent $(\mathcal{L}_{a\star} - \lambda)^{-1}$ to the obtained identity, we finally find:

$$u_n = (\mathcal{L}_{\varepsilon a\star} - \lambda)^{-1}\left(\frac{\lambda}{V-\lambda}f_n - V_1\mathcal{L}_{\varepsilon a\star}u_n\right).$$

Since the function V decays as infinity, the same holds for V_1. This ensures the compactness of the operator $V_1\mathcal{L}_{\varepsilon a\star}$ in $L_2(\mathbb{R})$ and, hence, by the above identity, the sequence u_n is compact, which is impossible. The proof is complete.

4. Absence of Point and Residual Spectrum

In this section, we prove Theorem 2. The proof consists of three main parts and we present them as separate subsections. After the proof of Theorem 2, we provide the proof of Theorem 3.

4.1. Absence of Embedded Eigenvalues

By our assumptions, the segment J_j contains only regular points and possibly finitely many (β,α) thresholds. We denote the latter thresholds by $x^{(j,i)}$, $i = 1,\ldots,m_j$, $j = 1,\ldots,n$, while the symbols $\beta_\pm^{(j,i)}$ and $\alpha_\pm^{(j,i)}$ stand for the corresponding values of $\beta_\pm$ and $\alpha_\pm$. The mentioned structure of the segment J_j implies that the function V is continuous on each of the segments J_j and is continuously differentiable on the same segments except the (β,α) thresholds. It also follows from the definition of the (β,α) thresholds and the regular points that

$$|V'(x)| \geqslant c_0 > 0, \qquad x \in J_j \setminus \{x^{(j,i)}, i = 1,\ldots,m_i\}, \quad j = 1,\ldots,n, \tag{23}$$

where c_0 is a fixed constant independent of x. As x approaches one of the thresholds $x^{(j,i)}$, the derivative V' blows up in the sense $|V'(x)| \to +\infty$ as $x \to x^{(j,i)}$.

It follows from (8) that there exists a small fixed δ_0 such that

$$V(x) \notin S \quad \text{as} \quad x \in [b_j^- - \delta_0, b_j^-) \cup (b_j^+, b_j^+ + \delta_0], \quad j = 1,\ldots,n,$$

and, by (24),

$$\operatorname{dist}(V(x), S) \geqslant c_1|x - b_j^\pm| \quad \text{as} \quad 0 < \pm(x - b_j^\mp) < \delta_0, \quad j = 1,\ldots,n. \tag{24}$$

with some fixed positive constant c_1 independent of x and j. We can additionally choose δ_0 small enough so that for all $j = 1,\ldots,n$, the intervals $[b_j^- - \delta_0, b_j^-) \cup (b_j^+, b_j^+ + \delta_0]$ contain only regular points and, if necessary, reducing the constant c_0, we can extend estimate (23) to $\tilde{J}_j$, namely,

$$|V'(x)| \geqslant c_0 > 0, \qquad x \in \tilde{J}_j \setminus \{x^{(j,i)}, i = 1,\ldots,m_i\}, \quad j = 1,\ldots,n. \tag{25}$$

Since S is a closed connected piece of the curve Υ, there exist two small fixed positive numbers δ_1 and c_2 such that

$$\operatorname{dist}(V(x), S^{\delta_1}) \geqslant c_2 \qquad \text{as} \qquad x \notin \tilde{J} := \bigcup_{j=1}^{n} \tilde{J}_j, \qquad \tilde{J}_j := [b_j^- - \delta_0, b_j^+ + \delta_0]. \tag{26}$$

We consider the eigenvalue equation for the operator $\mathcal{L}^\varepsilon$ with the spectral parameter ranging in S^{δ_1}:

$$(V - \lambda)\psi + \varepsilon\mathcal{L}_{a\star}\psi = 0. \tag{27}$$

Given an arbitrary measurable set $X \subseteq \mathbb{R}$, by $\mathcal{P}_X$, we denote the operator of restriction to X. This operator is considered as acting from $L_2(\mathbb{R})$ into $L_2(X)$ by the rule $(\mathcal{P}_X\psi)(x) := \psi(x)$, $x \in X$. Representing the real axis as $\mathbb{R} = \tilde{J} \cup (\mathbb{R} \setminus \tilde{J})$ and using an obvious decomposition $L_2(\mathbb{R}) = L_2(\tilde{J}) \oplus L_2(\mathbb{R} \setminus \tilde{J})$, we denote

$$\psi_{\tilde{J}} := \mathcal{P}_{\tilde{J}}\psi, \qquad \psi_{\mathbb{R}\setminus\tilde{J}} := \mathcal{P}_{\mathbb{R}\setminus\tilde{J}}\psi$$

and equivalently rewrite Equation (27) as a pair of two equations

$$\begin{aligned} &(V-\lambda)\psi_{\tilde{J}} + \varepsilon\mathcal{P}_{\tilde{J}}\mathcal{M}_{\tilde{J}}\psi_{\tilde{J}} + \varepsilon\mathcal{P}_{\tilde{J}}\mathcal{M}_{\mathbb{R}\setminus\tilde{J}}\psi_{\mathbb{R}\setminus\tilde{J}} = 0, \\ &(V-\lambda)\psi_{\mathbb{R}\setminus\tilde{J}} + \varepsilon\mathcal{P}_{\mathbb{R}\setminus\tilde{J}}\mathcal{M}_{\mathbb{R}\setminus\tilde{J}}\psi_{\mathbb{R}\setminus\tilde{J}} + \varepsilon\mathcal{P}_{\mathbb{R}\setminus\tilde{J}}\mathcal{M}_{\tilde{J}}\psi_{\tilde{J}} = 0, \end{aligned} \tag{28}$$

where for an arbitrary measurable set $X \subseteq \mathbb{R}$, the symbol $\mathcal{M}_X$ denotes a convolution operator acting from $L_2(X)$ into $L_2(\mathbb{R})$ by the rule

$$(\mathcal{M}_X\psi)(x) := \int\limits_X a(x-y)\psi(y)\,dy, \qquad x \in \mathbb{R}. \tag{29}$$

The first equation in (28) is to be treated as that in $L_2(\tilde{J})$, while the other equation is that in $L_2(\mathbb{R} \setminus \tilde{J})$.

Owing to (26), the norm of the operator of multiplication by $(V-\lambda)^{-1}$ in $L_2(\mathbb{R} \setminus \tilde{J})$ is bounded uniformly in $\lambda \in S^{\delta_1}$ by the constant c_2^{-1}. Applying this operator to the second equation in (28), we obtain an equivalent equation

$$\big(\mathcal{I}_{\mathbb{R}\setminus\tilde{J}} + \varepsilon(V-\lambda)^{-1}\mathcal{P}_{\mathbb{R}\setminus\tilde{J}}\mathcal{M}_{\mathbb{R}\setminus\tilde{J}}\big)\psi_{\mathbb{R}\setminus\tilde{J}} + \varepsilon(V-\lambda)^{-1}\mathcal{P}_{\mathbb{R}\setminus\tilde{J}}\mathcal{M}_{\tilde{J}}\psi_{\tilde{J}} = 0, \tag{30}$$

where $\mathcal{I}_{\mathbb{R}\setminus\tilde{J}}$ is the identity operator in $L_2(\mathbb{R} \setminus \tilde{J})$ and by estimate (17) we immediately see that $(V-\lambda)^{-1}\mathcal{P}_{\mathbb{R}\setminus\tilde{J}}\mathcal{M}_{\mathbb{R}\setminus\tilde{J}}$ is a bounded operator in $L_2(\mathbb{R} \setminus \tilde{J})$, and $(V-\lambda)^{-1}\mathcal{P}_{\mathbb{R}\setminus\tilde{J}}\mathcal{M}_{\tilde{J}}$ is a bounded operator from $L_2(\tilde{J})$ into $L_2(\mathbb{R} \setminus \tilde{J})$; both operators are bounded uniformly in $\lambda \in S^{\delta_1}$. Hence, for sufficiently small ε, the operator $\mathcal{I}_{\mathbb{R}\setminus\tilde{J}} + \varepsilon(V-\lambda)^{-1}\mathcal{P}_{\mathbb{R}\setminus\tilde{J}}\mathcal{M}_{\mathbb{R}\setminus\tilde{J}}$ is invertible for each $\lambda \in S^{\delta_1}$ and the inverse operator

$$\mathcal{Q}(\varepsilon,\lambda) := \big(\mathcal{I}_{\mathbb{R}\setminus\tilde{J}} + \varepsilon(V-\lambda)^{-1}\mathcal{P}_{\mathbb{R}\setminus\tilde{J}}\mathcal{M}_{\mathbb{R}\setminus\tilde{J}}\big)^{-1}$$

is bounded uniformly in ε and $\lambda \in S^{\delta_1}$ as an operator in $L_2(\mathbb{R} \setminus \tilde{J})$. Applying this operator to Equation (30), we immediately find $\psi_{\mathbb{R}\setminus\tilde{J}}$:

$$\psi_{\mathbb{R}\setminus\tilde{J}} = -\varepsilon\mathcal{Q}(\varepsilon,\lambda)(V-\lambda)^{-1}\mathcal{P}_{\mathbb{R}\setminus\tilde{J}}\mathcal{M}_{\tilde{J}}\psi_{\tilde{J}}, \tag{31}$$

and the operator $\mathcal{Q}(\varepsilon,\lambda)(V-\lambda)^{-1}\mathcal{P}_{\mathbb{R}\setminus\tilde{J}}\mathcal{M}_{\tilde{J}}$ from $L_2(\tilde{J})$ into $L_2(\mathbb{R} \setminus \tilde{J})$ is bounded uniformly in ε and $\lambda \in S^{\delta_1}$. Substituting this formula into the first equation in (28), we arrive at a single equation for $\psi_{\tilde{J}}$:

$$(V-\lambda)\psi_{\tilde{J}} + \varepsilon\mathcal{P}_{\tilde{J}}\mathcal{M}_{\tilde{J}}\psi_{\tilde{J}} - \varepsilon^2\mathcal{P}_{\tilde{J}}\mathcal{M}_{\mathbb{R}\setminus\tilde{J}}\mathcal{Q}(\varepsilon,\lambda)(V-\lambda)^{-1}\mathcal{P}_{\mathbb{R}\setminus\tilde{J}}\mathcal{M}_{\tilde{J}}\psi_{\tilde{J}} = 0. \tag{32}$$

We observe that the second and the third terms in the above equation can be rewritten as

$$\varepsilon\mathcal{P}_{\tilde{J}}\mathcal{M}_{\tilde{J}}\psi_{\tilde{J}} - \varepsilon^2\mathcal{P}_{\tilde{J}}\mathcal{M}_{\mathbb{R}\setminus\tilde{J}}\mathcal{Q}(\varepsilon,\lambda)(V-\lambda)^{-1}\mathcal{P}_{\mathbb{R}\setminus\tilde{J}}\mathcal{M}_{\tilde{J}}\psi_{\tilde{J}} = \varepsilon\mathcal{P}_{\tilde{J}}\mathcal{L}_{a\star}\mathcal{A}(\varepsilon,\lambda)\psi_{\tilde{J}},$$

where $\mathcal{A}$ is an operator from $L_2(\tilde{J})$ into $L_2(\mathbb{R})$ defined by the formula

$$\mathcal{A}(\varepsilon,\lambda)\psi_{\tilde{J}} := \begin{cases} \psi_{\tilde{J}} & \text{on} \quad \tilde{J}, \\ -\varepsilon\mathcal{Q}(\varepsilon,\lambda)(V-\lambda)^{-1}\mathcal{P}_{\mathbb{R}\setminus\tilde{J}}\mathcal{M}_{\tilde{J}}\psi_{\tilde{J}} & \text{on} \quad \mathbb{R}\setminus\tilde{J}. \end{cases} \tag{33}$$

This operator is bounded uniformly in ε and $\lambda \in S^{\delta_1}$, namely,

$$\|\mathcal{A}(\varepsilon,\lambda)\|_{L_2(\tilde{J})\to L_2(\mathbb{R})} \leqslant c_3, \tag{34}$$

where c_3 is a constant independent of ε and λ. Hence, Equation (32) becomes

$$(V-\lambda)\psi_{\tilde{J}} + \varepsilon\mathcal{P}_{\tilde{J}}\mathcal{L}_{a\star}\mathcal{A}(\varepsilon,\lambda)\psi_{\tilde{J}} = 0. \tag{35}$$

Our main aim is to prove that there exists a fixed positive $\delta \in (0,\delta_1]$ such that for $\lambda \in S^\delta$, Equation (35) can have only trivial solutions. First, we are going to show that such a statement holds for λ located on the curve $\Upsilon \cap S^{\delta_1}$; such a curve obviously contains S.

We arbitrarily choose $\lambda \in \Upsilon \cap S^{\delta_1}$ and let $z^{(j,i)}$ be all points of the segment $\tilde{J}_j$ such that $V(z^{(j,i)}) = \lambda$. Here, the superscript j ranges in some subset of $\{1,\ldots,n\}$ and i ranges from 1 to some natural number depending on j. Let us show that the total number of points $z^{(j,i)}$ in each segment $\tilde{J}_j$ is bounded by some constant $\tilde{m} \geqslant m$ independent of j and λ provided δ_1 and δ_0 are chosen small enough. Indeed, according to our assumptions, the total number of the points $z^{(j,i)}$ located in the segment J_j is bounded by m and we only need to estimate the total number of such points located in $\tilde{J}_j \setminus J_j$. If λ is such that one of the corresponding points $z^{(j,i)}$ is located in $[b_j^- - \delta_0, b_j^-)$ or in $(b_j^+, b_j^+ + \delta_0]$ for some j, then each of the mentioned intervals can contain at most one point $z^{(j,i)}$. This will be ensured by the inequality

$$V(x) \neq V(y) \quad \text{as} \quad x \neq y, \qquad x,y \in [b_j^- - \delta_0, b_j^-) \quad \text{or} \quad x,y \in (b_j^+, b_j^+ + \delta_0], \tag{36}$$

which we are going to prove. The point b_j^+ can be regular or a (β,α) threshold, and in both cases, owing to (5) and (7), for $x \in (b_j^+, b_j^+ + \delta_0]$ the function V can be represented as

$$V(x) = \alpha_0(x-b_j^+)^{\beta_0}v_0(x), \qquad v_0 \in C^2[b_j^+, b_j^+ + \delta_0],$$

provided δ_0 is small enough. Here, α_0 is some non-zero complex number, $\beta_0 \in (0,1]$ is some real number and v_0 is some complex-valued function such that $v_0(b_j^+) = 1$. Choosing $x,y \in (b_j^+, b_j^+ + \delta_0]$ arbitrarily, we have

$$\begin{aligned} V^{\frac{1}{\beta_0}}(x) - V^{\frac{1}{\beta_0}}(y) =& \alpha_0^{\frac{1}{\beta_0}}\left((x-b_j^+)v_0^{\frac{1}{\beta_0}}(x) - (y-b_j^+)v_0^{\frac{1}{\beta_0}}(y)\right) \\ =& \alpha_0^{\frac{1}{\beta_0}}\left((x-y)v_0^{\frac{1}{\beta_0}}(x) + (y-b_j^+)\left(v_0^{\frac{1}{\beta_0}}(x) - v_0^{\frac{1}{\beta_0}}(y)\right)\right). \end{aligned}$$

Applying the Lagrange rule, we obtain:

$$V^{\frac{1}{\beta_0}}(x) - V^{\frac{1}{\beta_0}}(y) = \alpha_0^{\frac{1}{\beta_0}}(x-y)\left(v_0^{\frac{1}{\beta_0}}(x) + (y-b_j^+)\tilde{v}_0(x,y)\right), \tag{37}$$

where $\tilde{v}_0(x,y)$ is some function obeying the uniform estimate

$$|\tilde{v}_0(x,y)| \leqslant \frac{1}{\beta_0}\|v_0\|_{C[b_j^+,b_j^+ + \delta_0]}^{\frac{1}{\beta_0}-1}\|v_0'\|_{C[b_j^+,b_j^+ + \delta_0]}.$$

Since each segment $\tilde{J}_j$ can contain only finitely many (β,α) thresholds and all other points are regular, the right-hand side of this inequality can be estimated from the above by some constant independent of j. Hence, in view of the identity $v_0(b_j^+) = 1$, the expression in the brackets on the right-hand side of (37) is close to 1 and can not vanish once we choose a small enough δ_0. This confirms inequality (36).

Let δ_2 be a fixed positive number such that the intervals $U^{(j,i)} := \tilde{J}_j \cap (z^{(j,i)} - \delta_2, z^{(j,i)} + \delta_2)$ are disjoint and each of these intervals contains no (β,α) thresholds except possibly that at $z^{(j,i)}$. Assume that $z^{(j,i)}$ is a regular point and let x range outside $U^{(j,i)}$, but still in some bigger neighbourhood of $z^{(j,i)}$. By the Lagrange rule, we then have

$$V(x) - \lambda = V(x) - V(z^{(j,i)}) = (x - z^{(j,i)})\left(\operatorname{Re} V'(x_{\mathrm{r}}^{(j,i)}) + \mathrm{i}\operatorname{Im} V'(x_{\mathrm{i}}^{(j,i)})\right),$$

where $x_{\mathrm{r}}^{(j,i)}$ and $x_{\mathrm{i}}^{(j,i)}$ are some points between x and $z^{(j,i)}$. By inequality (25), we see that for such x, the inequality holds:

$$|V(x)-\lambda| \geqslant c_2|x-z^{(j,i)}|. \tag{38}$$

If $z^{(j,i)}$ is a (β,α) threshold, we choose δ_2 small enough, so that in the interval $U^{(j,i)}$, representation (5) holds true. This representation implies immediately that

$$|V(x)-\lambda| \geqslant c_4|x-z^{(j,i)}|$$

again for x outside $U^{(j,i)}$, but still in some bigger neighbourhood of $z^{(j,i)}$; here, c_4 is a fixed positive constant independent of x, j and i. This estimate and (38) imply the existence of a positive constant c_5 depending on δ_2 but independent of the choice of λ such that

$$|V(x)-\lambda| \geqslant c_5 > 0 \qquad \text{as} \qquad x\in\tilde{J}\setminus U, \qquad U:=\bigcup_{j,i} U^{(j,i)}. \tag{39}$$

By $\chi^{(j,i)}=\chi^{(j,i)}(x)$, we denote the characteristic functions of the intervals $U^{(j,i)}$, while M_0 is the set of the superscripts (j,i) such that either the point $z^{(j,i)}$ is regular or it is a (β,α) threshold with at least one of $\beta_\pm$ obeying $\beta_\pm\in[\frac{1}{2},1]$. We return back to Equation (35) with $\lambda\in\Upsilon\cap S^{\delta_1}$ and let $\psi_{\tilde{J}}$ be its solution in $L_2(\tilde{J})$. Since the function $V-\lambda$ vanishes only at the corresponding points $z^{(j,i)}$, which form a set of zero measures, we can rewrite this equation as

$$\psi_{\tilde{J}}+\frac{\varepsilon}{V-\lambda}\mathcal{P}_{\tilde{J}}\mathcal{L}_{a\star}\mathcal{A}(\varepsilon,\lambda)\psi_{\tilde{J}}=0. \tag{40}$$

The second term in this equation can be represented as follows:

$$\frac{1}{V-\lambda}\mathcal{P}_{\tilde{J}}\mathcal{L}_{a\star}\mathcal{A}(\varepsilon,\lambda)\psi_{\tilde{J}}=\mathcal{B}_0(\varepsilon,\lambda)\psi_{\tilde{J}}+\mathcal{B}_1(\varepsilon,\lambda)\psi_{\tilde{J}}, \qquad \mathcal{B}_1(\varepsilon,\lambda)\psi_{\tilde{J}}:=\frac{1}{V-\lambda}\mathcal{B}_2(\varepsilon,\lambda)\psi_{\tilde{J}},$$

where

$$(\mathcal{B}_0(\varepsilon,\lambda)\psi_{\tilde{J}})(x):=\frac{1}{V(x)-\lambda}\sum_{(j,i)\in M_0}\chi^{(j,i)}(x)\int_{\mathbb{R}} a(z^{(j,i)}-y)\big(\mathcal{A}(\varepsilon,\lambda)\psi_{\tilde{J}}\big)(y)\,dy, \qquad x\in J, \tag{41}$$

$$(\mathcal{B}_2(\varepsilon,\lambda)\psi_{\tilde{J}})(x):=\sum_{(j,i)\in M_0}\int_{\mathbb{R}}\big(a(x-y)-a(z^{(j,i)}-y)\chi^{(j,i)}(x)\big)\big(\mathcal{A}(\varepsilon,\lambda)\psi_{\tilde{J}}\big)(y)\,dy, \qquad x\in J.$$

Let us show that $\mathcal{B}_1(\varepsilon,\lambda)$ is a bounded operator in $L_2(\tilde{J})$ and, moreover, its norm is bounded uniformly in $\lambda\in\Upsilon\cap S^{\delta_1}$. Indeed, as $x\in\tilde{J}\setminus U$, the function $(\mathcal{B}_1(\varepsilon,\lambda)\psi_{\tilde{J}})(x)$ reads as

$$(\mathcal{B}_1(\varepsilon,\lambda)\psi_{\tilde{J}})(x)=\frac{1}{V(x)-\lambda}\int_{\mathbb{R}} a(x-y)\big(\mathcal{A}(\varepsilon,\lambda)\psi_{\tilde{J}}\big)(y)\,dy.$$

Estimates (17), (34) and (39) then imply

$$\|\mathcal{B}_1(\varepsilon,\lambda)\psi_{\tilde{J}}\|_{L_2(\tilde{J}\setminus U)}\leqslant c_5^{-1}\|a\|_{L_1(\mathbb{R})}\|\mathcal{A}\psi_{\tilde{J}}\|_{L_2(\tilde{J})}\leqslant c_3c_5^{-1}\|a\|_{L_1(\mathbb{R})}\|\psi_{\tilde{J}}\|_{L_2(\tilde{J})}. \tag{42}$$

As $x\in U^{(j,i)}$, $(j,i)\in M_0$, the function $(\mathcal{B}_1(\varepsilon,\lambda)\psi_{\tilde{J}})(x)$ is given by the formula

$$\begin{aligned}(\mathcal{B}_2(\varepsilon,\lambda)\psi_{\tilde{J}})(x)=&\frac{1}{V(x)-\lambda}\int_{\mathbb{R}}\big(a(x-y)-a(z^{(j,i)}-y)\big)\big(\mathcal{A}(\varepsilon,\lambda)\psi_{\tilde{J}}\big)(y)\,dy\\ =&\frac{1}{V(x)-\lambda}\int_{\mathbb{R}} dy\,\big(\mathcal{A}(\varepsilon,\lambda)\psi_{\tilde{J}}\big)(y)\int_0^{x-z^{(j,i)}} a'(t+z^{(j,i)}-y)\,dt.\end{aligned} \tag{43}$$

Using, then, the definition of the regular points and (β,α) thresholds and estimate (25), by the Cauchy–Schwarz inequality and the uniform boundedness of the operator $\mathcal{A}$, we obtain:

$$\begin{aligned}
\big|(\mathcal{B}_1(\varepsilon,\lambda)\psi_{\tilde{J}})(x)\big|^2 \leqslant& \frac{C}{|x-z^{(j,i)}|^2}\left(\int\limits_{\mathbb{R}} dy\,\big|(\mathcal{A}(\varepsilon,\lambda)\psi_{\tilde{J}})(y)\big| \int\limits_{-|x-z^{(j,i)}|}^{|x-z^{(j,i)}|} \big|a'(t+z^{(j,i)}-y)\big|\,dt\right)^2 \\
\leqslant& \frac{C}{|x-z^{(j,i)}|^2}\|\mathcal{A}(\varepsilon,\lambda)\psi_{\tilde{J}}\|^2_{L_2(\mathbb{R})}\int\limits_{\mathbb{R}} dy\left(\int\limits_{-|x-z^{(j,i)}|}^{|x-z^{(j,i)}|} \big|a'(t+z^{(j,i)}-y)\big|\,dt\right)^2 \\
\leqslant& \frac{C}{|x-z^{(j,i)}|}\|\mathcal{A}(\varepsilon,\lambda)\psi_{\tilde{J}}\|^2_{L_2(\mathbb{R})}\int\limits_{\mathbb{R}} dy \int\limits_{-|x-z^{(j,i)}|}^{|x-z^{(j,i)}|} \big|a'(t+z^{(j,i)}-y)\big|^2\,dt \\
=& \frac{C}{|x-z^{(j,i)}|}\|\mathcal{A}(\varepsilon,\lambda)\psi_{\tilde{J}}\|^2_{L_2(\mathbb{R})}\int\limits_{-|x-z^{(j,i)}|}^{|x-z^{(j,i)}|} dt\int\limits_{\mathbb{R}} \big|a'(t+z^{(j,i)}-y)\big|^2\,dy \\
\leqslant& C\|a'\|_{L_2(\mathbb{R})}\|\psi_{\tilde{J}}\|^2_{L_2(\tilde{J})},
\end{aligned} \tag{44}$$

where the symbol C stands for various constants independent of x, $\lambda \in \Upsilon$ and $\psi_{\tilde{J}}$. Integrating the obtained estimate over $U^{(j,i)}$ and summing up the result over $(j,i) \in M_0$, we finally arrive at the inequality

$$\big\|\mathcal{B}_1(\varepsilon,\lambda)\psi_{\tilde{J}}\big\|_{L_2(U)} \leqslant c_6\|\psi_{\tilde{J}}\|_{L_2(\tilde{J})},$$

where c_6 is a constant independent of $\lambda \in \Upsilon \cap S^{\delta_1}$ and $\psi_{\tilde{J}}$. This inequality and (42) imply that the operator $\mathcal{B}_1$ is bounded in $L_2(\tilde{J})$ and its norm is bounded uniformly in $\lambda \in \Upsilon \cap S^{\delta_1}$.

Let us study the function $\mathcal{B}_0\psi_{\tilde{J}}$ defined in (41). If $\psi_{\tilde{J}}$ is a solution of Equation (40) in the space $L_2(\tilde{J})$, then the function $\mathcal{B}_1\psi_{\tilde{J}}$ is also an element of this space and, hence, $\mathcal{B}_0(\varepsilon,\lambda)\psi_{\tilde{J}}$ is necessarily in $L_2(\tilde{J})$. At the same time, as $x \in U^{(j,i)}$, this function reads

$$(\mathcal{B}_0(\varepsilon,\lambda)\psi_{\tilde{J}})(x) = \frac{1}{V(x)-\lambda}\int\limits_{\mathbb{R}} a(z^{(j,i)}-y)\,(\mathcal{A}(\varepsilon,\lambda)\psi_{\tilde{J}})(y)\,dy \tag{45}$$

and the integral is independent of x. The function $(V(x)-\lambda)^{-1}$ has a singularity at the point $z^{(j,i)}$ and since $z^{(j,i)}$ is either a regular point or a (β,α) threshold with at least one of $\beta_\pm$ not less than $\frac{1}{2}$, this function is not an element of $L_2(U^{(j,i)})$. Hence, the only possibility is that the integral in (45) necessarily vanishes. Then, $\mathcal{B}_0\psi_{\tilde{J}} = 0$ and Equation (40) becomes

$$(\mathcal{I}_{\tilde{J}} + \varepsilon\mathcal{B}_1(\varepsilon,\lambda))\psi_{\tilde{J}} = 0,$$

where $\mathcal{I}_{\tilde{J}}$ is the identity mapping in $L_2(\tilde{J})$. Since the operator $\mathcal{B}_1$ is bounded uniformly in λ, for sufficiently small ε, the operator $\mathcal{I} + \varepsilon\mathcal{B}_1(\varepsilon,\lambda)$ is boundedly invertible and the above equation can have only the trivial solution. Therefore, Equations (35) and (40) also have only the trivial solution as $\lambda \in \Upsilon \cap S^{\delta_1}$.

4.2. Reduction to System of Linear Algebraic Equations

We proceed to proving the existence of a small fixed positive $\delta \leqslant \delta_1$ such that the set $S^\delta \setminus \Upsilon$ contains no eigenvalues of the operator $\mathcal{L}^\varepsilon$. Namely, we are going to show that for $\lambda \in S^\delta \setminus \Upsilon$, Equation (35) possesses only the trivial solution. In this subsection, we make the first important step in studying this equation, i.e., we reduce it to a system of linear algebraic equations.

We choose a sufficiently small $\delta_3 \leqslant \min\left\{\frac{\delta_1}{2}, 1\right\}$ and introduce a finite covering of the curve S by open balls $B_{\delta_3}(P_k)$ with centers at some points $P_k \in S$, $k = 1,\ldots,N$, where

$N \in \mathbb{N}$ is the number of the balls. By our assumptions, for each k, the point P_k is the image of finitely many points in the segment J_j and, hence, the piece of curve $B_{2\delta_3}(P_k) \cap \Upsilon$ is the image of finitely many segments in $\tilde{J}_j$, namely,

$$B_{2\delta_3}(P_k) \cap \Upsilon = \bigcup_{j=1}^{n} \bigcup_{i=1}^{N_k^{(j)}} \left\{ V(x) : x \in I_k^{(j,i)} \right\}, \qquad P_k = V(Y_k^{(j,i)}), \qquad Y_k^{(j,i)} \in I_k^{(j,i)},$$

where $I_k^{(j,i)} \subset \tilde{J}_j$ are some open intervals, $N_k^{(j)}$ are some given natural numbers, and $Y_k^{(j,i)}$ are some points. Owing to inequality (25) and the assumed smoothness of the function V, by choosing a small enough δ_3, we can gain the following properties:

P1. The intervals $I_k^{(j,i)}$ are disjoint for different i, their lengths satisfy the estimate $|I_k^{(j,i)}| < 1$ and all possible thresholds in the interval J_j are among the points $Y_k^{(j,i)}$;

P2. The end points of the intervals $I_k^{(j,i)}$ do not coincide with the (β, α) thresholds located in the segment J_j, each of the intervals $I_k^{(j,i)}$ contains at most one (β, α) threshold and the distance from this threshold to other intervals $I_k^{(j,i)}$ is at least $c_7\delta_3$, where $c_7 > 0$ is a constant independent of δ_3, k, j, i; the image of each end point of each interval $I_k^{(j,i)}$ is located on $\partial B_{2\delta_3}(P_k)$;

P3. If some interval $I_k^{(j,i)}$ contains a (β, α) threshold, then the corresponding identity (5) holds true for the entire interval.

In what follows, given a curve and a point in the complex plane, we say that this point is projected onto this curve orthogonally to some non-zero complex number if this projection is made along the straight line orthogonal to the vector connecting the origin and this non-zero complex number. We suppose an extra two properties of δ_3.

P4. If a given interval $I_k^{(j,i)}$ contains only regular points, then for all $\lambda \in B_{\delta_3}(Y_k^{(j,i)}) \setminus \Upsilon$, there exists a unique projection of $\frac{\lambda}{V'(Y_k^{(j,i)})}$ onto the curve $\Gamma_k^{(j,i)} := \{V(x) : x \in I_k^{(j,i)}\}$ orthogonally to the number $V'(Y_k^{(j,i)})$ and the inequality holds:

$$\operatorname{Re} \frac{V'(x)}{V'(Y_k^{(j,i)})} \geqslant \frac{1}{2} \quad \text{for all } x \in I_k^{(j,i)}; \tag{46}$$

P5. If a given interval $I_k^{(j,i)}$ contains a (β, α) threshold at $Y_k^{(j,i)} \in I_k^{(j,i)}$ with corresponding $\alpha_\pm = \alpha_{\pm,k}^{(j,i)}$, then for all $\lambda \in B_{\delta_3}(Y_k^{(j,i)}) \setminus \Upsilon$ such that $\operatorname{Re} \frac{\lambda - P_k}{\alpha_{\pm,k}^{(j,i)}} > 0$ there exists a unique projection of $\frac{\lambda}{\alpha_{k,\pm}^{(j,i)}}$ onto the curve

$$\Gamma_{k,\pm}^{(j,i)} := \{V(x) : x \in I_{k,\pm}^{(j,i)}\}, \quad \text{where} \quad I_{k,\pm}^{(j,i)} := I_k^{(j,i)} \cap \{x : \pm(x - Y_k^{(j,i)}) > 0\}$$

orthogonally to the number $\alpha_{k,\pm}^{(j,i)}$; the functions $v_\pm = v_{k,\pm}^{(j,i)}$ from (5) corresponding to the (β, α) threshold at $Y_k^{(j,i)}$ satisfy the estimates

$$\begin{aligned} &v_{k,\pm}^{(j,i)} \geqslant \frac{1}{2}, \quad |v_{k,\pm}^{(j,i)}(x)| \leqslant 2, \quad |\operatorname{Im} v_{k,\pm}^{(j,i)}(x)| \leqslant \tan\frac{\pi\beta_0}{2} \operatorname{Re} v_{k,\pm}^{(j,i)}(x) \quad \text{as} \quad x \in I_{k,+}^{(j,i)}, \\ &|I_{k,\pm}^{(j,i)}| \|(v_{k,\pm}^{(j,i)})'\|_{C(\overline{I_{k,+}^{(j,i)}})} \leqslant \frac{1}{4^{1+\frac{1}{\beta_0}}}, \end{aligned} \tag{47}$$

where

$$\beta_0 := \frac{1}{2} \min_{k,j,i} \{\beta_{k,+}^{(j,i)}; \beta_{k,-}^{(j,i)}\}. \tag{48}$$

We observe that the definition of intervals $I_k^{(j,i)}$ implies immediately that

$$|V(x)-\lambda|\geqslant \delta_3 \quad \text{as} \quad \lambda\in B_{\delta_3}(P_k), \quad x\in\tilde{J}\setminus I_k^{(j)}, \quad j=1,\ldots,n, \qquad I_k^{(j)}:=\bigcup_{i=1}^{N_k^{(j)}} I_k^{(j,i)}. \tag{49}$$

Property P4 can be equivalently formulated as follows: there exists a unique solution to the equation

$$\operatorname{Re}\frac{V(Z)-P_k}{V'(Y_k^{(j,i)})}=\operatorname{Re}\frac{\lambda-P_k}{V'(Y_k^{(j,i)})} \tag{50}$$

for all $\lambda\in B_{\delta_3}(Y_k^{(j,i)})\setminus\Upsilon$. In view of the definition of a regular point, this equation is uniquely solvable, since for Z close to $Y_k^{(j,i)}$ the quotient on the left hand side behaves as

$$\frac{V(Z)-P_k}{V'(Y_k^{(j,i)})}=Z-Y_k^{(j,i)}+O((Z-Y_k^{(j,i)})^2).$$

The latter identity also ensures the possibility of satisfying (46). We denote the unique solution of (50) by $Z_k^{(j,i)}=Z_k^{(j,i)}(\lambda)$.

Property P5 can be also equivalently formulated as follows: there exists a unique solution to the equation

$$\operatorname{Re}\frac{V(Z_\pm)-P_k}{\alpha_{\pm,k}^{(j,i)}}=\operatorname{Re}\frac{\lambda-P_k}{\alpha_{\pm,k}^{(j,i)}} \tag{51}$$

for all $\lambda\in B_{\delta_3}(Y_k^{(j,i)})\setminus\Upsilon$ obeying an additional condition $\operatorname{Re}\frac{\lambda-P_k}{\alpha_{\pm,k}^{(j,i)}}>0$. These equations are again locally uniquely solvable owing to the definition of (β,α) threshold, which also ensures (47). We denote the solutions of (51) by $Z_{\pm,k}^{(j,i)}=Z_{\pm,k}^{(j,i)}(\lambda)$. We also let

$$Z_{\pm,k}^{(j,i)}(\lambda):=Y_k^{j,i} \qquad \text{as} \qquad \operatorname{Re}\frac{\lambda-P_k}{\alpha_{\pm,k}^{(j,i)}}\leqslant 0. \tag{52}$$

In what follows, we consider Equation (35) for $\lambda\in E_{k,\delta_3}$, where

$$E_{k,\delta_3}:=B_{\delta_3}(P_k)\setminus\Upsilon. \tag{53}$$

We rewrite this equation in form (40) and then we represent the second term in the latter equation as

$$\psi_{\tilde{J}}+\varepsilon\mathcal{B}_{3,k}(\varepsilon,\lambda)\psi_{\tilde{J}}+\varepsilon\mathcal{B}_{4,k}(\varepsilon,\lambda)\psi_{\tilde{J}}=0, \tag{54}$$

$$\mathcal{B}_{3,k}(\varepsilon,\lambda):=\sum_{j=1}^{n}\sum_{i=1}^{N_k^{(j)}}\frac{\xi_k^{(j,i)}}{V-\lambda}\mathcal{P}_{\tilde{J}}\mathcal{L}_{a\star}\mathcal{A}(\varepsilon,\lambda),$$

$$\mathcal{B}_{4,k}(\varepsilon,\lambda):=\sum_{j=1}^{n}\sum_{i=1}^{N_k^{(j)}}\frac{1-\xi_k^{(j,i)}}{V-\lambda}\mathcal{P}_{\tilde{J}}\mathcal{L}_{a\star}\mathcal{A}(\varepsilon,\lambda),$$

where $\xi_k^{(j,i)}$ are the characteristic functions of the intervals $I_k^{(j,i)}$. It follows immediately from the definitions of the operators $\mathcal{B}_{4,k}$ and the function $\xi_k^{(j,i)}$ and estimates (17), (34) and (49) that

$$\|\mathcal{B}_{4,k}\|_{L_2(\tilde{J})\to L_2(\tilde{J})}\leqslant\frac{c_8}{\delta_3}, \tag{55}$$

where c_8 is a constant independent of λ, k, δ_3.

We proceed to studying the operators $\mathcal{B}_{3,k}(\varepsilon,\lambda)$. Let M_1 be the set of all superscripts (j,i) such that the intervals $I_k^{(j,i)}$, $(j,i)\in M_1$, contain only regular points, while M_2 is the set of all superscripts (j,i) such that the intervals $I_k^{(j,i)}$, $(j,i) \in M_2$, possesses a (β,α) threshold at $Y_k^{(j,i)} \in I_k^{(j,i)}$. Bearing in mind Properties P4 and P5, we represent the operator $\mathcal{B}_{3,k}$ as a sum

$$\mathcal{B}_{3,k}(\varepsilon,\lambda) = \mathcal{B}_{5,k}(\varepsilon,\lambda) + \mathcal{B}_{6,k}(\varepsilon,\lambda)\mathcal{A}(\varepsilon,\lambda), \tag{56}$$

where $\mathcal{B}_{5,k}(\varepsilon,\lambda)$ and $\mathcal{B}_{6,k}(\varepsilon,\lambda)$ are operators in $L_2(\tilde{J})$ defined by the formulas

$$\begin{aligned}\mathcal{B}_{5,k}(\varepsilon,\lambda) :=& \sum_{(j,i)\in M_1} \phi_k^{(j,i)} \ell(Z_k^{(j,i)}(\lambda),\varepsilon,\lambda) + \sum_{(j,i)\in M_2} \phi_{k,+}^{(j,i)} \ell(Z_{k,+}^{(j,i)}(\lambda),\varepsilon,\lambda) \\ &+ \sum_{(j,i)\in M_2} \phi_{k,-}^{(j,i)} \ell(Z_{k,-}^{(j,i)}(\lambda),\varepsilon,\lambda), \\ \mathcal{B}_{6,k}(\varepsilon,\lambda) :=& \sum_{(j,i)\in M_1} \mathcal{B}_{6,k}^{(j,i)}(\varepsilon,\lambda) + \sum_{(j,i)\in M_2} \mathcal{B}_{6,k,+}^{(j,i)}(\varepsilon,\lambda) + \sum_{(j,i)\in M_2} \mathcal{B}_{6,k,-}^{(j,i)}(\varepsilon,\lambda),\end{aligned} \tag{57}$$

where

$$\begin{aligned}&\phi_k^{(j,i)} := \frac{\xi_k^{(j,i)}}{V-\lambda}, \qquad \phi_{k,\pm}^{(j,i)} := \frac{\xi_{k,\pm}^{(j,i)}}{V-\lambda}, \\ &(\mathcal{B}_{6,k}^{(j,i)}(\varepsilon,\lambda)\psi)(x) := \xi_k^{(j,i)}(x) \int_{\mathbb{R}} \frac{a(x-y) - a(Z_k^{(j,i)}(\lambda)-y)}{V(x)-\lambda} \psi(y)\,dy, \\ &(\mathcal{B}_{6,k,\pm}^{(j,i)}(\varepsilon,\lambda)\psi)(x) := \xi_k^{(j,i)}(x) \int_{\mathbb{R}} \frac{a(x-y) - a(Z_{k,\pm}^{(j,i)}(\lambda)-y)}{V(x)-\lambda} \psi(y)\,dy,\end{aligned} \tag{58}$$

$\xi_{k,\pm}^{(j,i)}$ are the characteristic functions of the intervals $I_{k,\pm}^{(j,i)}$, and $\ell(z,\varepsilon,\lambda)$, $z \in \mathbb{R}$, is a bounded linear functional on $L_2(\tilde{J})$ defined as

$$\ell(z,\varepsilon,\lambda)\psi_{\tilde{J}} := \int_{\mathbb{R}} a(z-y)(\mathcal{A}(\varepsilon,\lambda)\psi_{\tilde{J}})(y)\,dy.$$

In order to study the properties of the operators $\mathcal{B}_{5,k}(\varepsilon,\lambda)$ and $\mathcal{B}_{6,k}(\varepsilon,\lambda)$, we shall need the following lemma.

Lemma 1. *There exists $\delta_4 > 0$ independent of k such that for all λßnE_{k,δ_3}, all k and all $\delta_3 \leqslant \delta_4$ the estimates hold:*

$$\begin{aligned}&|V(x)-\lambda| \geqslant c_9|x - Z_k^{(j,i)}(\lambda)| \quad &&\text{as} \quad x \in I_k^{(j,i)}, \quad (j,i) \in M_1, \\ &|V(x)-\lambda| \geqslant c_9|x - Z_{k,\pm}^{(j,i)}(\lambda)| \quad &&\text{as} \quad x \in I_{k,\pm}^{(j,i)}, \quad (j,i) \in M_2,\end{aligned} \tag{59}$$

where c_9 is a positive constant independent of δ_3, x, λ, k, j and i.

Proof. We first consider the case $(j,i) \in M_1$. By Equation (50), estimate (25) and the Lagrange rule, we have:

$$\begin{aligned}|V(x)-\lambda| =& |V'(Y_k^{(j,i)})| \left| \frac{V(x)-P_k}{V'(Y_k^{(j,i)})} - \frac{\lambda - P_k}{V'(Y_k^{(j,i)})} \right| \\ \geqslant& |V'(Y_k^{(j,i)})| \left| \operatorname{Re} \frac{V(x)-P_k}{V'(Y_k^{(j,i)})} - \operatorname{Re} \frac{\lambda - P_k}{V'(Y_k^{(j,i)})} \right| \\ =& |V'(Y_k^{(j,i)})| \left| \operatorname{Re} \frac{V(x)-P_k}{V'(Y_k^{(j,i)})} - \operatorname{Re} \frac{V(Z_k^{(j,i)})-P_k}{V'(Y_k^{(j,i)})} \right| \geqslant \frac{c_0}{2}|x - Z_k^{(j,i)}|.\end{aligned} \tag{60}$$

We proceed to the case $(j,i) \in M_2$. We shall prove the second inequality in (59) only for $x \in I_{k,+}^{(j,i)}$; the case of the interval $I_{k,-}^{(j,i)}$ can be treated in the same way. In the considered case, the interval $I_k^{(j,i)}$ contains a (β,α) threshold at some internal point $Y_k^{(j,i)}$. We first suppose that $\operatorname{Re}\lambda\beta_+^{-1} \leqslant 0$. In view of (5) and (52), we have:

$$|V(x)-\lambda| \geqslant |\alpha_{k,+}^{(j,i)}| \left|\operatorname{Re}\frac{V(x)-\lambda}{\alpha_{k,+}^{(j,i)}}\right| \geqslant C|x-Y_k^{(j,i)}|^{\beta_{\pm,k}^{(j,i)}} \geqslant C|x-Y_k^{(j,i)}|, \tag{61}$$

where C is a constant independent of k, j, i and λ. This proves the second inequality in (59) as $\operatorname{Re}\lambda\beta_+^{-1} \leqslant 0$.

Suppose that $\operatorname{Re}\lambda\beta_+^{-1} > 0$. Then, we argue similarly to (60):

$$\begin{aligned}
|V(x)-\lambda| =& |\alpha_{+,k}^{(j,i)}| \left|\frac{V(x)-P_k}{\alpha_{+,k}^{(j,i)}} - \frac{\lambda-P_k}{\alpha_{+,k}^{(j,i)}}\right| \\
\geqslant& |\alpha_{+,k}^{(j,i)}| \left|\operatorname{Re}\frac{V(x)-P_k}{\alpha_{+,k}^{(j,i)}} - \operatorname{Re}\frac{V(Z_{+,k}^{(j,i)})-P_k}{\alpha_{+,k}^{(j,i)}}\right| \\
\geqslant& |\alpha_{+,k}^{(j,i)}| \left|\left((x-Y_k^{(j,i)})^{\beta_{k,+}^{(j,i)}} \operatorname{Re} v_{k,+}^{(j,i)}(x)\right)'\Big|_{x=\zeta}\right| |x-Z_{k,+}^{(j,i)}| \\
=& |\alpha_{+,k}^{(j,i)}| \left|\beta_{k,+}^{(j,i)} \operatorname{Re} v_{k,+}^{(j,i)}(\zeta) + (\zeta - Y_k^{(j,i)}) \operatorname{Re}(v_{k,+}^{(j,i)})'(\zeta)\right| \frac{|x-Z_{k,+}^{(j,i)}|}{|\zeta-Y_k^{(j,i)}|^{1-\beta_{k,+}^{(j,i)}}},
\end{aligned}$$

where ζ is some point between x and $Z_{k,+}^{(j,i)}$. It follows from the first and fourth inequalities in (47) and (48) that

$$\left|\beta_{k,+}^{(j,i)} \operatorname{Re} v_{k,+}^{(j,i)}(\zeta) + (\zeta - Y_k^{(j,i)}) \operatorname{Re}(v_{k,+}^{(j,i)})'(\zeta)\right| \geqslant \frac{\beta_{k,+}^{(j,i)}}{2} - |I_{k,+}^{(j,i)}| \|(v_{k,+}^{(j,i)})'\|_{C(\overline{I_{k,+}^{(j,i)}})} \geqslant \beta_0 - \frac{1}{4^{1+\frac{1}{\beta_0}}} > \frac{\beta_0}{2}.$$

This inequality and the inequality $|I_{k,+}^{(j,i)}| < |I_k^{(j,i)}| < 1$, see Property P1, allows us to continue the above estimating:

$$|V(x)-\lambda| \geqslant \frac{\beta_0|\alpha_{+,k}^{(j,i)}|}{2} \frac{|x-Z_{k,+}^{(j,i)}|}{|\zeta-Y_k^{(j,i)}|^{1-\beta_{k,+}^{(j,i)}}} \geqslant \frac{\beta_0|\alpha_{+,k}^{(j,i)}|}{2|I_{k,+}^{(j,i)}|^{1-\beta_{k,+}^{(j,i)}}} |x-Z_{k,+}^{(j,i)}| \geqslant \frac{\beta_0|\alpha_{+,k}^{(j,i)}|}{2}|x-Z_{k,+}^{(j,i)}|.$$

The proof is complete. □

Using this lemma and arguing as in (43) and (44), we easily see that the operators $\mathcal{B}_{6,k}(\varepsilon,\lambda)$ are bounded uniformly in ε and $\lambda \in E_{k,\delta_3}$ once $\delta_3 \leqslant \delta_4$, namely,

$$\|\mathcal{B}_{6,k}(\varepsilon,\lambda)\|_{L_2(\tilde{J})\to L_2(\tilde{J})} \leqslant c_{10}, \tag{62}$$

where c_{10} is a constant independent of ε and λ. This inequality and (55), (34) yield that the operator

$$\mathcal{G}(\varepsilon,\lambda) := \left(\mathcal{I}_J + \varepsilon\mathcal{B}_{4,k} + \varepsilon\mathcal{B}_{6,k}\mathcal{A}(\varepsilon,\lambda)\right)^{-1}$$

is well defined and bounded in $L_2(\tilde{J})$ provided

$$\varepsilon \leqslant \frac{\delta_3}{2(c_8 + c_{10}c_3\delta_3)}, \qquad \lambda \in E_{k,\delta_3}, \qquad 0 < \delta_3 \leqslant \delta_4$$

and for such values of ε, δ_3 and λ, it satisfies the estimate

$$\|\mathcal{G}(\varepsilon,\lambda)\|_{L_2(\tilde{J})\to L_2(\tilde{J})} \leqslant 2.$$

We substitute identity (56) into Equation (54) and then apply the operator $\mathcal{G}(\varepsilon,\lambda)$ to the resulting relation and use the definition of the operator $\mathcal{B}_{5,k}$. This implies one more equation:

$$\begin{aligned}\psi_{\bar{J}}+\varepsilon\sum_{(j,i)\in M_1}\Phi_k^{(j,i)}(\varepsilon,\lambda)\ell\big(Z_k^{(j,i)}(\lambda),\varepsilon,\lambda\big)\psi_{\bar{J}}+\varepsilon\sum_{(j,i)\in M_2}\Phi_{k,+}^{(j,i)}(\varepsilon,\lambda)\ell\big(Z_{k,+}^{(j,i)}(\lambda),\varepsilon,\lambda\big)\psi_{\bar{J}}\\+\varepsilon\sum_{(j,i)\in M_2}\Phi_{k,-}^{(j,i)}(\varepsilon,\lambda)\ell\big(Z_{k,-}^{(j,i)}(\lambda),\varepsilon,\lambda\big)\psi_{\bar{J}}=0,\end{aligned}\tag{63}$$

where
$$\Phi_k^{(j,i)}(\varepsilon,\lambda):=\mathcal{G}(\varepsilon,\lambda)\phi_k^{(j,i)},\qquad \Phi_{k,\pm}^{(j,i)}(\varepsilon,\lambda):=\mathcal{G}(\varepsilon,\lambda)\phi_{k,\pm}^{(j,i)}.\tag{64}$$

We arbitrarily choose $p\in\{1,\ldots,n\}$ and $i\in\{1,\ldots,N_k^{(p)}\}$ and if $(p,q)\in M_1$, we apply the functional $\ell\big(Z_k^{(p,q)}(\lambda),\varepsilon,\lambda\big)$ to Equation (63), while for $(p,q)\in M_2$ we apply the functionals $\ell\big(Z_{k,\pm}^{(p,q)}(\lambda),\varepsilon,\lambda\big)$ to the same equation. This gives the following identities:

$$\begin{aligned}&\ell\big(Z_k^{(p,q)}(\lambda),\varepsilon,\lambda\big)\psi_{\bar{J}}+\varepsilon\sum_{(j,i)\in M_1}A_k^{(p,q,j,i)}(\varepsilon,\lambda)\ell\big(Z_k^{(j,i)}(\lambda),\varepsilon,\lambda\big)\psi_{\bar{J}}\\&\quad+\varepsilon\sum_{(j,i)\in M_2}A_{k,+}^{(p,q,j,i)}(\varepsilon,\lambda)\ell\big(Z_{k,+}^{(j,i)}(\lambda),\varepsilon,\lambda\big)\psi_{\bar{J}}\\&\quad+\varepsilon\sum_{(j,i)\in M_2}A_{k,-}^{(p,q,j,i)}(\varepsilon,\lambda)\ell\big(Z_{k,-}^{(j,i)}(\lambda),\varepsilon,\lambda\big)\psi_{\bar{J}}=0,\qquad (p,q)\in M_1,\\&\ell\big(Z_{k,+}^{(p,q)}(\lambda),\varepsilon,\lambda\big)\psi_{\bar{J}}+\varepsilon\sum_{(j,i)\in M_1}A_{k,+}^{(p,q,j,i)}(\varepsilon,\lambda)\ell\big(Z_k^{(j,i)}(\lambda),\varepsilon,\lambda\big)\psi_{\bar{J}}\\&\quad+\varepsilon\sum_{(j,i)\in M_2}A_{k,+,+}^{(p,q,j,i)}(\varepsilon,\lambda)\ell\big(Z_{k,+}^{(j,i)}(\lambda),\varepsilon,\lambda\big)\psi_{\bar{J}}\\&\quad+\varepsilon\sum_{(j,i)\in M_2}A_{k,+,-}^{(p,q,j,i)}(\varepsilon,\lambda)\ell\big(Z_{k,-}^{(j,i)}(\lambda),\varepsilon,\lambda\big)\psi_{\bar{J}}=0,\qquad (p,q)\in M_2,\\&\ell\big(Z_{k,-}^{(p,q)}(\lambda),\varepsilon,\lambda\big)\psi_{\bar{J}}+\varepsilon\sum_{(j,i)\in M_1}A_{k,-}^{(p,q,j,i)}(\varepsilon,\lambda)\ell\big(Z_k^{(j,i)}(\lambda),\varepsilon,\lambda\big)\psi_{\bar{J}}\\&\quad+\varepsilon\sum_{(j,i)\in M_2}A_{k,-,+}^{(p,q,j,i)}(\varepsilon,\lambda)\ell\big(Z_{k,+}^{(j,i)}(\lambda),\varepsilon,\lambda\big)\psi_{\bar{J}}\\&\quad+\varepsilon\sum_{(j,i)\in M_2}A_{k,-,-}^{(p,q,j,i)}(\varepsilon,\lambda)\ell\big(Z_{k,-}^{(j,i)}(\lambda),\varepsilon,\lambda\big)\psi_{\bar{J}}=0,\qquad (p,q)\in M_2,\end{aligned}\tag{65}$$

where

$$A_k^{(p,q,j,i)}(\varepsilon,\lambda):=\ell\big(Z_k^{(p,q)}(\lambda),\varepsilon,\lambda\big)\Phi_k^{(j,i)}(\varepsilon,\lambda),\qquad A_{k,\pm}^{(p,q,j,i)}(\varepsilon,\lambda):=\ell\big(Z_k^{(p,q)}(\lambda),\varepsilon,\lambda\big)\Phi_{k,\pm}^{(j,i)}(\varepsilon,\lambda)\tag{66}$$

as $(p,q)\in M_1$ and

$$A_{k,\pm}^{(p,q,j,i)}(\varepsilon,\lambda):=\ell\big(Z_{k,\pm}^{(p,q)}(\lambda),\varepsilon,\lambda\big)\Phi_k^{(j,i)}(\varepsilon,\lambda),\qquad A_{k,\flat,\natural}^{(p,q,j,i)}(\varepsilon,\lambda):=\ell\big(Z_{k,\flat}^{(p,q)}(\lambda),\varepsilon,\lambda\big)\Phi_{k,\natural}^{(j,i)}(\varepsilon,\lambda)\tag{67}$$

as $(p,q)\in M_2$, where the symbols $\flat$ and $\natural$ are to be independently replaced by '+' or '−'. Identity (65) is a system of linear equations for the numbers $\ell\big(Z_k^{(j,i)}(\lambda),\varepsilon,\lambda\big)$ and $\ell\big(Z_{k,\pm}^{(j,i)}(\lambda),\varepsilon,\lambda\big)$. Once we find these numbers, we can recover the function $\psi_{\bar{J}}$ for Equation (63). If system (65) has only the trivial solution, this immediately implies that $\psi_{\bar{J}}$ vanishes identically, and by Formula (31), Equation (27) can have only the trivial solution; hence, the set E_{k,δ_3} contains no eigenvalues of the operator $\mathcal{L}^{\varepsilon}$. In order to prove that system (65) has only the trivial solution, it is sufficient to show that all functions $A_{k,\pm}^{(p,q,j,i)}(\varepsilon,\lambda)$ and $A_{k,\flat,\natural}^{(p,q,j,i)}(\varepsilon,\lambda)$ are bounded uniformly in ε and λ. The proof of this fact is our next important step.

4.3. Trivial Solution and Absence of the Spectrum

In this subsection, we prove the uniform boundedness of the functions $A_{k,\pm}^{(p,q,j,i)}(\varepsilon,\lambda)$ and $A_{k,\flat,\natural}^{(p,q,j,i)}(\varepsilon,\lambda)$ and this will allow us to complete the proof of Theorem 2. We first rewrite Formula (64) for the functions $\Phi_k^{(j,i)}(\varepsilon,\lambda)$ and $\Phi_{k,\pm}^{(j,i)}(\varepsilon,\lambda)$ as

$$\begin{aligned}\Phi_k^{(j,i)}(\varepsilon,\lambda)&=\phi_k^{(j,i)}-\varepsilon\mathcal{G}(\varepsilon,\lambda)\big(\mathcal{B}_{4,k}(\varepsilon,\lambda)+\mathcal{B}_{6,k}(\varepsilon,\lambda)\big)\phi_k^{(j,i)},\\ \Phi_{k,\pm}^{(j,i)}(\varepsilon,\lambda)&=\phi_{k,\pm}^{(j,i)}-\varepsilon\mathcal{G}(\varepsilon,\lambda)\big(\mathcal{B}_{4,k}(\varepsilon,\lambda)+\mathcal{B}_{6,k}(\varepsilon,\lambda)\big)\phi_{k,\pm}^{(j,i)}.\end{aligned}\tag{68}$$

The prove of the uniform boundedness of $A_k^{(p,q,j,i)}(\varepsilon,\lambda)$, $A_{k,\pm}^{(p,q,j,i)}(\varepsilon,\lambda)$, $A_{k,\pm}^{(p,q,j,i)}(\varepsilon,\lambda)$, $A_{k,\flat,\natural}^{(p,q,j,i)}(\varepsilon,\lambda)$ is based on a series of the following lemmas.

Lemma 2. *There exists $\delta_5>0$ such that as $\delta_3\leqslant\delta_5$, for all $\lambda\in E_{k,\delta_3}$ and $(j,i)\in M_1$ the estimates hold*

$$\left|\int_{\mathbb{R}}\phi_k^{(j,i)}(x)\,dx\right|\leqslant\frac{c_{11}}{\delta_3},$$

where c_{11} is a constant independent of k, j, i, δ_3 and λ.

Proof. We begin with representing the considered integral as

$$\begin{aligned}\int_{\mathbb{R}}\phi_k^{(j,i)}(x)\,dx&=\int_{I_k^{(j,i)}}\frac{dx}{V(x)-\lambda}\\ &=\frac{1}{V'(Z_k^{(j,i)})}\int_{I_k^{(j,i)}}\frac{V'(x)}{V(x)-\lambda}\,dx+\frac{1}{V'(Z_k^{(j,i)})}\int_{I_k^{(j,i)}}\frac{V'(Z_k^{(j,i)})-V'(x)}{V(x)-\lambda}\,dx.\end{aligned}\tag{69}$$

The first integral in the right hand side of the above representation can be immediately rewritten as

$$\int_{I_k^{(j,i)}}\frac{V'(x)}{V(x)-\lambda}\,dx=\int_{\Gamma_k^{(j,i)}}\frac{dt}{t-\lambda}.$$

The above integral over the curve $\Gamma_k^{(j,i)}$ is holomorphic in $\lambda\in B_{\delta_3}(P_k)\setminus\Upsilon$. As λ is such that $\operatorname{dist}(\lambda,\Upsilon)\geqslant\frac{\delta_3}{2}$, we have an obvious estimate

$$\left|\int_{\Gamma_k^{(j,i)}}\frac{dt}{t-\lambda}\right|\leqslant\frac{C}{\delta_3},\tag{70}$$

where C is a constant independent of λ, k, j, i and δ_3. We also easily find that

$$\frac{d}{d\lambda}\int_{\Gamma_k^{(j,i)}}\frac{dt}{t-\lambda}=\int_{\Gamma_k^{(j,i)}}\frac{dt}{(t-\lambda)^2}=\frac{1}{\partial_-\Gamma_k^{(j,i)}-\lambda}-\frac{1}{\partial_+\Gamma_k^{(j,i)}-\lambda},\tag{71}$$

where $\partial_\pm\Gamma_k^{(j,i)}$ are the end-points of the curve $\Gamma_k^{(j,i)}$. Definition (53) of the set E_{k,δ_3} ensures that

$$\frac{1}{|\partial_\pm\Gamma_k^{(j,i)}-\lambda|}\geqslant\frac{1}{\delta_3}.$$

Having this estimate and (70) in mind and integrating (71) with respect to λ, in view of (25), we immediately find

$$\left|\frac{1}{V'(Z_k^{(j,i)})}\int\limits_{I_k^{(j,i)}}\frac{V'(x)}{V(x)-\lambda}\,dx\right|\leqslant\frac{C}{\delta_3},\tag{72}$$

where C is a constant independent of λ, k, j, i and δ_3.

In order to estimate the second integral in the right hand side of (69), we employ estimate (59) and the Lagrange rule:

$$\left|\frac{1}{V'(Z_k^{(j,i)})}\int\limits_{I_k^{(j,i)}}\frac{V'(Z_k^{(j,i)})-V'(x)}{V(x)-\lambda}\,dx\right|\leqslant\frac{C}{|V'(Z_k^{(j,i)})|}\sup_{t\in I_k^{(j,i)}}|V''(t)|,\tag{73}$$

where C is a constant independent of λ, k, j, i and δ_3. According to the definition of the regular points, the function V is twice continuously differentiable on J_j except for (β,α) thresholds, which are denoted, we recall, by $x^{(j,i)}$. In the vicinity of the latter points, the first and the second derivatives of the function V have singularities of orders $O(|x-x^{(j,i)}|^{\beta_\pm^{(j,i)}-1})$ and $O(|x-x^{(j,i)}|^{\beta_\pm^{(j,i)}-2})$. According to Property P2, the minimal distance from the interval $I_k^{(j,i)}$ to the nearest (β,α) threshold is at least $c_7\delta_3$, and since the total number of the thresholds is finite, we conclude on the existence of $\delta_5>0$ such that for $\delta_3\leqslant\delta_5$ the estimate

$$\frac{\sup\limits_{t\in I_k^{(j,i)}}|V''(t)|}{|V'(Z_k^{(j,i)})|}\leqslant\frac{C}{\delta_3}$$

holds true, where C is a constant independent of δ_3, k, j, i. Substituting this estimate into (73), we obtain:

$$\left|\frac{1}{V'(Z_k^{(j,i)})}\int\limits_{I_k^{(j,i)}}\frac{V'(Z_k^{(j,i)})-V'(x)}{V(x)-\lambda}\,dx\right|\leqslant\frac{C}{\delta_3},$$

where C is a constant independent of δ_3, k, j, i. This estimate and (72) yield the desired estimate from the statement of the lemma. The proof is complete. □

Lemma 3. *For all $\lambda\in E_{k,\delta_3}$ and $(j,i)\in M_2$ the estimates hold*

$$\left|\int\limits_{\mathbb{R}}\phi_{k,\pm}^{(j,i)}(x)\,dx\right|\leqslant c_{12},\tag{74}$$

where c_{12} is a constant independent of k, j, i, and λ but depending on δ_3.

Proof. We provide the proof only for the integral with $\phi_{k,+}^{(j,i)}$; the other case can be treated in the same way. We first suppose that

$$\operatorname{Re}\frac{\lambda-P_k}{\alpha_{+,k}^{(j,i)}}\leqslant 0.$$

Then, by (61) and the assumed inequality $\beta_{k,+}^{(j,i)}<1$ we immediately obtain:

$$\left|\int\limits_{\mathbb{R}}\frac{\xi_{k,\pm}^{(j,i)}(x)}{V(x)-\lambda}\,dx\right|\leqslant C\int\limits_{\Gamma_{k,+}^{(j,i)}}\frac{dx}{(x-Y_k^{(j,i)})^{\beta_{k,+}^{j,i}}}\leqslant C,$$

where by C we denote some constants independent of λ, k, j, i and δ_3.

Suppose now that

$$\operatorname{Re}\eta > 0, \qquad \text{where} \qquad \eta := \frac{\lambda - P_k}{\alpha_{+,k}^{(j,i)}}. \tag{75}$$

Owing to the third inequality in (47) and (48) the function

$$w(x) := (x - Y_k^{(j,i)})\big(v_{k,+}^{(j,i)}(x)\big)^{\frac{1}{\beta_{k,+}^{(j,i)}}}$$

is well defined and

$$w^{\beta_{k,+}^{(j,i)}}(x) = (x - Y_k^{(j,i)})^{\beta_{k,+}^{(j,i)}} v_{k,+}^{(j,i)}(x) = \frac{V(x) - P_k}{\alpha_{k,+}^{(j,i)}}. \tag{76}$$

The assumed smoothness of $v_{k,+}^{(j,i)}$, see (5) and (6) yields that

$$w \in C^2(\overline{I_{k,+}^{(j,i)}}), \qquad \|w\|_{C^2(\overline{I_{k,+}^{(j,i)}})} \leqslant C, \tag{77}$$

where C is a constant independent of k, j, i. The first, second and fourth inequalities in (47) and identity (5) imply that for $x \in \overline{I_{k,+}^{(j,i)}}$, the estimates hold:

$$\begin{aligned}
\left(\frac{|x - Y_{k,+}^{(j,i)}|}{\beta_{k,+}^{(j,i)}} |v_{k,+}^{(j,i)}(x)|^{\frac{1}{\beta_{k,+}^{(j,i)}} - 1} |(v_{k,+}^{(j,i)})'(x)|\right)^{\beta_{k,+}^{(j,i)}} &\leqslant |(v_{k,+}^{(j,i)})'(x)|^{\beta_{k,+}^{(j,i)}} |v_{k,+}^{(j,i)}(x)| |I_{k,+}^{(j,i)}|^{\beta_{k,+}^{(j,i)}} \\
&\leqslant 2 |(v_{k,+}^{(j,i)})'(x)|^{\beta_{k,+}^{(j,i)}} |I_{k,+}^{(j,i)}|^{\beta_{k,+}^{(j,i)}} \\
&\leqslant \frac{2}{4^{\frac{\beta_{k,+}^{(j,i)}}{\beta_0} + \beta_{k,+}^{(j,i)}}} \leqslant \frac{1}{4^{\beta_{k,+}^{(j,i)}}} < \frac{1}{2^{\beta_{k,+}^{(j,i)}}} \leqslant |v_{k,+}^{(j,i)}(x)|^{\beta_{k,+}^{(j,i)}}.
\end{aligned}$$

Hence,

$$|w'(x)| \geqslant C, \qquad x \in \overline{I_{k,+}^{(j,i)}},$$

where C is a positive constant independent of x, k, j and i. We denote

$$\tilde{\Gamma} := \{w(x) : x \in I_{k,+}^{(j,i)}\}, \qquad \tilde{\alpha} := \alpha_{k,+}^{(j,i)}, \qquad \tilde{\beta} := \beta_{k,+}^{(j,i)}.$$

We rewrite the considered integral as follows:

$$\int_{\mathbb{R}} \phi_{k,\pm}^{(j,i)}(x)\,dx = \int_{I_{k,+}^{(j,i)}} \frac{dx}{V(x) - \lambda} = \frac{1}{w'(Z_{k,+}^{(j,i)})} \int_{I_{k,+}^{(j,i)}} \frac{w'(x)\,dx}{V(x) - \lambda} + \int_{I_{k,+}^{(j,i)}} \frac{w'(Z_{k,+}^{(j,i)}) - w'(x)}{V(x) - \lambda}\,dx.$$

Using, then, identity (76) and making the change in variable $t = w(x)$ in the first integral in the right hand side of the above identity, we obtain:

$$\int_{\mathbb{R}} \phi_{k,\pm}^{(j,i)}(x)\,dx = \frac{1}{\tilde{\alpha} w'(Z_{k,+}^{(j,i)})} \int_{\tilde{\Gamma}} \frac{dt}{t^{\tilde{\beta}} - \eta} + \int_{I_{k,+}^{(j,i)}} \frac{w'(Z_{k,+}^{(j,i)}) - w'(x)}{V(x) - \lambda}\,dx. \tag{78}$$

Owing to the above established smoothness of the function w, see (77), and the second inequality in (59), by applying the Lagrangue rule, we immediately estimate the second integral in the right hand side of the above identity:

$$\left|\int\limits_{I_{k,+}^{(j,i)}} \frac{w'(Z_{k,+}^{(j,i)}) - w'(x)}{V(x)-\lambda}\,dx\right| \leqslant C \int\limits_{I_{k,+}^{(j,i)}} \frac{|x - Z_{k,+}^{(j,i)}|}{|V(x)-\lambda|}\,dx \leqslant C, \tag{79}$$

where the symbol C denotes various constants independent of δ_3, λ, k, j and i.

Let us estimate the first integral in the right hand side of (78). Suppose that the point η is located above the curve $\tilde{\Gamma}$. Then, we choose the branch of the analytic function $z^{\tilde{\beta}}$ with the cut along the positive imaginary semi-axis and the argument of z ranging in $(-\frac{3\pi}{2}, \frac{\pi}{2}]$. Let $\tilde{z}$ be the end-point of the curve $\tilde{\Gamma}$ not coinciding with the origin. In the complex plane, we introduce extra two curves:

$$\tilde{\Gamma}_1 := \{z:\, z = e^{-\frac{\pi \mathrm{i}}{\tilde{\beta}}} s,\ s \in (0, |\tilde{z}|)\}, \qquad \tilde{\Gamma}_2 := \{z:\, |z| = |\tilde{z}|,\ \arg z \in (-\pi, \arg \tilde{z})\}.$$

Then, the closure of the union of these two curves and $\tilde{\Gamma}$ is a closed contour, and by the Cauchy integral theorem, we obtain:

$$\int\limits_{\tilde{\Gamma}} \frac{dt}{t^{\tilde{\beta}} - \eta} = -\int\limits_{\tilde{\Gamma}_1} \frac{dt}{t^{\tilde{\beta}} - \eta} - \int\limits_{\tilde{\Gamma}_2} \frac{dt}{t^{\tilde{\beta}} - \eta} = e^{-\frac{\pi \mathrm{i}}{\tilde{\beta}}} \int\limits_0^{|\tilde{z}|} \frac{ds}{s^{\tilde{\beta}} + \eta} - \int\limits_{\tilde{\Gamma}_2} \frac{dt}{t^{\tilde{\beta}} - \eta}. \tag{80}$$

Since $\lambda \in E_{k,\delta_3}$, it follows from the definition of η in (75) and Property P2 that

$$|\eta| \leqslant \frac{\delta_3}{|\tilde{\alpha}|}, \qquad |\tilde{z}|^{\tilde{\beta}} = \frac{2\delta_3}{|\tilde{\alpha}|}.$$

Hence, $|t^{\tilde{\beta}} - \eta| \geqslant \frac{\delta_3}{|\tilde{\alpha}|}$ on the curve $\tilde{\Gamma}_2$ and

$$\left|\int\limits_{\tilde{\Gamma}_2} \frac{dt}{t^{\tilde{\beta}} - \eta}\right| \leqslant \frac{2\pi|\tilde{\alpha}|}{\delta_3}. \tag{81}$$

Since $\operatorname{Re}\eta > 0$ by (75), the first integral in the right hand side of the above identity can be immediately estimated as

$$\left|e^{-\frac{\pi \mathrm{i}}{\tilde{\beta}}} \int\limits_0^{|\tilde{z}|} \frac{ds}{s^{\tilde{\beta}} + \eta}\right| \leqslant \int\limits_0^{|\tilde{z}|} \frac{ds}{s^{\tilde{\beta}}} = \frac{1}{1-\tilde{\beta}} \frac{1}{|\tilde{z}|^{1-\tilde{\beta}}} = \frac{1}{1-\tilde{\beta}} \frac{|\tilde{\alpha}|^{\frac{1}{\tilde{\beta}}-1}}{(2\delta_3)^{\frac{1}{\tilde{\beta}}-1}} \leqslant \frac{C}{\delta_3^{\frac{1}{\beta_0}-1}},$$

where C is a constant independent of δ_3, k, j and i. This estimate (80) and (81), (80) prove the uniform boundedness of the first integral in the right hand side of (78), and in view of (79), we arrive at estimate (74) for $\phi_{k,+}^{(j,i)}$. The proof is complete. □

Lemma 4. *The function a is an element of* $C(\overline{\tilde{J}})$.

Proof. Since $a \in W_2^1(\mathbb{R})$, by the standard embedding theorems, we conclude that $a \in C(\mathbb{R})$ and this proves the lemma. □

Lemma 5. *As* $\delta_3 \leqslant \min\{\delta_4, \delta_5\}$, *for* $\lambda \in E_{k,\delta_3}$ *the estimates hold:*

$$\|\mathcal{L}_{a\star}\phi_k^{(j,i)}\|_{L_\infty(\tilde{J}_k)} \leqslant c_{13}, \qquad \|\mathcal{L}_{a\star}\phi_{k,\pm}^{(j,i)}\|_{L_\infty(\tilde{J}_k)} \leqslant c_{13}, \tag{82}$$

$$\|\mathcal{L}_{a\star}\phi_k^{(j,i)}\|_{L_2(\mathbb{R})} \leqslant c_{13}, \qquad \|\mathcal{L}_{a\star}\phi_{k,\pm}^{(j,i)}\|_{L_2(\mathbb{R})} \leqslant c_{13}, \tag{83}$$

where c_{13} *is a constant independent of* λ, k, j, i *but depending on* δ_3.

Proof. We fix k and some (j,i) in the corresponding set M_1 and represent the function $\mathcal{L}_{a\star}\phi_k^{(j,i)}$ as

$$(\mathcal{L}_{a\star}\phi_{k,\pm}^{(j,i)})(x) = a(x - Z_k^{(j,i)}) \int\limits_{I_k^{(j,i)}} \phi_k^{(j,i)}(y)\,dy + \int\limits_{I_k^{(j,i)}} \frac{a(x-y) - a(x - Z_k^{(j,i)})}{V(y) - \lambda}\,dy. \tag{84}$$

By Lemmas 2 and 4, we immediately estimate the first integral in the right hand side of the above identity:

$$\begin{aligned} &\left| a(x - Z_k^{(j,i)}) \int\limits_{I_k^{(j,i)}} \phi_k^{(j,i)}(y)\,dy \right| \leqslant \frac{c_{11}}{\delta_3} \|a\|_{C(\bar{J})}, \\ &\left\| a(\,\cdot\, - Z_k^{(j,i)}) \int\limits_{I_k^{(j,i)}} \phi_k^{(j,i)}(y)\,dy \right\|_{L_2(\mathbb{R})} \leqslant \frac{c_{11}}{\delta_3} \|a\|_{L_2(\mathbb{R})}. \end{aligned} \tag{85}$$

To estimate the second integral in the right hand side of (84), we employ a representation similar to (43):

$$\int\limits_{I_k^{(j,i)}} \frac{a(x-y) - a(x - Z_k^{(j,i)})}{V(y) - \lambda}\,dy = \int\limits_{I_k^{(j,i)}} \frac{dy}{V(y) - \lambda} \int\limits_0^{y - Z_k^{(j,i)}} a'(x - Z_k^{(j,i)} - t)\,dt$$

and use then the Cauchy–Schwarz inequality and the first estimate from (59):

$$\begin{aligned} \left| \int\limits_{I_k^{(j,i)}} \frac{a(x-y) - a(x - Z_k^{(j,i)})}{V(y) - \lambda}\,dy \right| &\leqslant \frac{1}{c_9} \int\limits_{I_k^{(j,i)}} \frac{1}{|y - Z_k^{(j,i)}|^{\frac{1}{2}}} \left| \int\limits_{-|y - Z_k^{(j,i)}|}^{|y - Z_k^{(j,i)}|} \left| a'(x - Z_k^{(j,i)} - t) \right|^2 dt \right|^{\frac{1}{2}} dy \\ &\leqslant \frac{\|a'\|_{L_2(\mathbb{R})}}{c_9} \int\limits_{I_k^{(j,i)}} \frac{dy}{|y - Z_k^{(j,i)}|^{\frac{1}{2}}} \leqslant C, \end{aligned}$$

$$\begin{aligned} \left\| \int\limits_{I_k^{(j,i)}} \frac{a(\cdot - y) - a(\,\cdot\, - Z_k^{(j,i)})}{V(y) - \lambda}\,dy \right\|_{L_2(\mathbb{R})}^2 &\leqslant C \int\limits_{\mathbb{R}} dx \int\limits_{I_k^{(j,i)}} \frac{dy}{|y - Z_k^{(j,i)}|} \int\limits_{-|y - Z_k^{(j,i)}|}^{|y - Z_k^{(j,i)}|} \left| a'(x - Z_k^{(j,i)} - t) \right|^2 dt \\ &\leqslant C \|a'\|_{L_2(\mathbb{R})}^2 \int\limits_{I_k^{(j,i)}} \frac{dy}{|y - Z_k^{(j,i)}|} \int\limits_{-|y - Z_k^{(j,i)}|}^{|y - Z_k^{(j,i)}|} dt \leqslant C, \end{aligned}$$

where by C we denote various constants independent of λ, k, j and i. These estimates (84) and (85) prove the first inequalities in (82) and (83).

The proof of the second inequalities in (82) and (83) follows the same lines. Namely, in (84), we just replace $I_k^{(j,i)}$, $Z_k^{(j,i)}$, $\phi_k^{(j,i)}$ by $I_{k,+}^{(j,i)}$, $Z_{k,+}^{(j,i)}$, $\phi_{k,+}^{(j,i)}$. Then, a corresponding analogue of inequality (85) is implied by Lemmas 3 and 4, while estimating the second integral literally reproduces the above argument. The proof is complete. □

Lemma 6. *As $\delta_3 \leqslant \min\{\delta_4, \delta_5\}$, for $\lambda \in E_{k,\delta_3}$ the estimates hold:*

$$\|\mathcal{B}_{6,k}(\varepsilon,\lambda)\mathcal{A}(\varepsilon,\lambda)\phi_k^{(j,i)}\|_{L_2(\bar{J})} \leqslant c_{14}, \qquad \|\mathcal{B}_{6,k}(\varepsilon,\lambda)\mathcal{A}(\varepsilon,\lambda)\phi_{k,\pm}^{(j,i)}\|_{L_2(\bar{J})} \leqslant c_{14},$$

where c_{14} is a constant independent of λ, k, j, i but depending on δ_3.

Proof. In view of the definition of the operator $\mathcal{B}_{6,k}$ in (57), it is sufficient to prove the uniform boundedness of the norms

$$\begin{aligned}
&\|\mathcal{B}_{6,k}^{(p,q)}(\varepsilon,\lambda)\mathcal{A}(\varepsilon,\lambda)\phi_k^{(j,i)}\|_{L_2(\tilde{J})}, && \|\mathcal{B}_{6,k}^{(p,q)}(\varepsilon,\lambda)\mathcal{A}(\varepsilon,\lambda)\phi_{k,\pm}^{(j,i)}\|_{L_2(\tilde{J})}, && (p,q)\in M_1,\\
&\|\mathcal{B}_{6,k,\flat}^{(p,q)}(\varepsilon,\lambda)\mathcal{A}(\varepsilon,\lambda)\phi_k^{(j,i)}\|_{L_2(\tilde{J})}, && \|\mathcal{B}_{6,k,\flat}^{(p,q)}(\varepsilon,\lambda)\mathcal{A}(\varepsilon,\lambda)\phi_{k,\pm}^{(j,i)}\|_{L_2(\tilde{J})}, && (p,q)\in M_2, \quad \flat\in\{+,-\}.
\end{aligned} \tag{86}$$

Bearing in mind the definition of the operators $\mathcal{B}_{6,k}^{(p,q)}$ and $\mathcal{B}_{6,k,\pm}^{(p,q)}$ in (58), the definition of the operator $\mathcal{A}$ in (33), and inequalities (26) and (62), we can estimate the first of the above norms as follows:

$$\begin{aligned}
\|\mathcal{B}_{6,k}^{(p,q)}(\varepsilon,\lambda)\mathcal{A}(\varepsilon,\lambda)\phi_k^{(j,i)}\|_{L_2(\tilde{J})} \leqslant &\|\mathcal{B}_{6,k}^{(p,q)}(\varepsilon,\lambda)\phi_k^{(j,i)}\|_{L_2(\tilde{J})} + \|\mathcal{B}_{6,k}^{(p,q)}(\varepsilon,\lambda)\mathcal{P}_{\mathbb{R}\setminus\tilde{J}}\mathcal{A}(\varepsilon,\lambda)\phi_k^{(j,i)}\|_{L_2(\tilde{J})}\\
\leqslant &\|\mathcal{B}_{6,k}^{(p,q)}(\varepsilon,\lambda)\phi_k^{(j,i)}\|_{L_2(\tilde{J})} + C\|\mathcal{M}_{\tilde{J}}\phi_k^{(j,i)}\|_{L_2(\mathbb{R}\setminus\tilde{J})},
\end{aligned}$$

where C is a constant independent of λ, ε, k, j, i and δ_3. In the same way, we can estimate other norms in (86) and, hence, it is sufficient to prove the uniform boundedness only for the norms

$$\begin{aligned}
&\|\mathcal{B}_{6,k}^{(p,q)}(\varepsilon,\lambda)\phi_k^{(j,i)}\|_{L_2(\tilde{J})}, && \|\mathcal{B}_{6,k}^{(p,q)}(\varepsilon,\lambda)\phi_{k,\pm}^{(j,i)}\|_{L_2(\tilde{J})}, && (p,q)\in M_1,\\
&\|\mathcal{B}_{6,k,\flat}^{(p,q)}(\varepsilon,\lambda)\phi_k^{(j,i)}\|_{L_2(\tilde{J})}, && \|\mathcal{B}_{6,k,\flat}^{(p,q)}(\varepsilon,\lambda)\phi_{k,\pm}^{(j,i)}\|_{L_2(\tilde{J})}, && (p,q)\in M_2, \quad \flat\in\{+,-\},\\
&\|\mathcal{M}_{\tilde{J}}\phi_k^{(j,i)}\|_{L_2(\mathbb{R}\setminus\tilde{J})}, && \|\mathcal{M}_{\tilde{J}}\phi_{k,\pm}^{(j,i)}\|_{L_2(\mathbb{R}\setminus\tilde{J})}.
\end{aligned} \tag{87}$$

The uniform boundedness of the latter two norms follows immediately from (83) and definition (29) of the operator $\mathcal{M}_{\tilde{J}}$.

According to the definition of the operators $\mathcal{B}_{6,k}^{(j,i)}$ in (58), the identity holds:

$$\begin{aligned}
(\mathcal{B}_{6,k}^{(p,q)}(\varepsilon,\lambda)\phi_k^{(j,i)})(x) =&\xi_k^{(j,i)}(x)\int\limits_{I_k^{(j,i)}} \frac{a(x-y)-a(Z_k^{(j,i)}(\lambda)-y)}{(V(x)-\lambda)(V(y)-\lambda)}\,dy\\
=&\xi_k^{(j,i)}(x)\frac{a(x-Z_k^{(j,i)}(\lambda))-a(0)}{(V(x)-\lambda)}\int\limits_{I_k^{(j,i)}}\frac{dy}{V(y)-\lambda}\\
&+\xi_k^{(j,i)}(x)\int\limits_{I_k^{(j,i)}}\frac{a(x-y)-a(x-Z_k^{(j,i)}(\lambda))-a(Z_k^{(j,i)}(\lambda)-y)+a(0)}{(V(x)-\lambda)(V(y)-\lambda)}\,dy\\
=&\xi_k^{(j,i)}(x)\frac{a(x-Z_k^{(j,i)}(\lambda))-a(0)}{(V(x)-\lambda)}\int\limits_{I_k^{(j,i)}}\frac{dy}{V(y)-\lambda}\\
&+\xi_k^{(j,i)}(x)\int\limits_{I_k^{(j,i)}}\frac{dy}{(V(x)-\lambda)(V(y)-\lambda)}\int\limits_0^{x-Z_k^{(j,i)}}\big(a'(t+Z_k^{(j,i)}-y)-a'(t)\big)\,dt.
\end{aligned}$$

Since $\xi_k^{(j,i)}$ is the characteristic function of a bounded interval $I_k^{(j,i)}$ and $a\in L_1(\tilde{J})$ by Lemma 4, in view of Lemma 2, we immediately conclude that the first term in the right hand side of the above identity is an element of $L_2(\mathbb{R})$ and it is bounded uniformly in λ, k, j, i in the norm of this space. The norm of the second term is estimated by using (59) and the second condition in (9):

$$\int_{\mathbb{R}} \left| \xi_k^{(j,i)}(x) \int_{I_k^{(j,i)}} \frac{dy}{(V(x)-\lambda)(V(y)-\lambda)} \int_0^{x-Z_k^{(j,i)}} \left(a'(t+Z_k^{(j,i)}-y)-a'(t)\right) dt \right|^2 dx$$

$$\leqslant C \int_{I_k^{(j,i)}} dx \left| \int_{I_k^{(j,i)}} \frac{dy}{|x-Z_k^{(j,i)}||y-Z_k^{(j,i)}|} \int_{-|x-Z_k^{(j,i)}|}^{|x-Z_k^{(j,i)}|} \left| a'(t+Z_k^{(j,i)}-y)-a'(t) \right| dt \right|^2$$

$$\leqslant C \int_{I_k^{(j,i)}} dx \left| \int_{I_k^{(j,i)}} \frac{dy}{|x-Z_k^{(j,i)}||y-Z_k^{(j,i)}|} \int_{-|x-Z_k^{(j,i)}|}^{|x-Z_k^{(j,i)}|} |Z_k^{(j,i)}-y|^\theta \, dt \right|^2$$

$$\leqslant C \int_{I_k^{(j,i)}} dx \left| \int_{I_k^{(j,i)}} \frac{dy}{|y-Z_k^{(j,i)}|^{1-\theta}} \right|^2 \leqslant C,$$

where the symbol C stands for various constants independent of λ, k, j, i. Hence, the functions $\mathcal{B}_{6,k}^{(p,q)}(\varepsilon,\lambda)\phi_k^{(j,i)}$ are bounded in $L_2(\mathbb{R})$ uniformly in λ, k, j, i. Similar boundedness for remaining functions in (87) is established in the same way, and one should just use the second estimate from (59) and Lemma 3. The proof is complete. □

Lemma 7. *As $\delta_3 \leqslant \min\{\delta_4, \delta_5\}$, for $\lambda \in E_{k,\delta_3}$ the estimates hold:*

$$\|\mathcal{B}_{4,k}(\varepsilon,\lambda)\mathcal{A}(\varepsilon,\lambda)\phi_k^{(j,i)}\|_{L_2(\mathcal{J})} \leqslant c_{15}, \qquad \|\mathcal{B}_{4,k}(\varepsilon,\lambda)\mathcal{A}(\varepsilon,\lambda)\phi_{k,\pm}^{(j,i)}\|_{L_2(\mathcal{J})} \leqslant c_{15},$$

where c_{15} is a constant independent of λ, k, j, i but depending on δ_3.

Proof. Denoting

$$\mathcal{B}_{4,k}^{(p,q)} := \frac{1-\xi_k^{(p,q)}}{V-\lambda}\mathcal{P}_{\mathcal{J}}\mathcal{L}_{a\star},$$

we observe that

$$\mathcal{B}_{4,k} = \sum_{p=1}^{n}\sum_{q=1}^{N_k^{(j)}} \mathcal{B}_{4,k}^{(j,i)}\mathcal{A}(\varepsilon,\lambda). \tag{88}$$

Then, using inequality (49) and the definition of the operator $\mathcal{A}(\varepsilon,\lambda)$, we obtain:

$$\|\mathcal{B}_{4,k}^{(p,q)}\mathcal{A}(\varepsilon,\lambda)\phi_k^{(j,i)}\|_{L_2(\mathcal{J})} \leqslant C\|\mathcal{L}_{a\star}\phi_k^{(j,i)}\|_{L_2(\mathbb{R})}, \qquad \|\mathcal{B}_{4,k}^{(p,q)}\mathcal{A}(\varepsilon,\lambda)\phi_{k,\pm}^{(j,i)}\|_{L_2(\mathcal{J})} \leqslant C\|\mathcal{L}_{a\star}\phi_k^{(j,i)}\|_{L_2(\mathbb{R})},$$

where the symbol C denotes some constants independent of λ, k, j and i. Applying, then, estimates (83), we see that the norms in the above inequality are uniformly bounded, and together with, (88) this completes the proof. □

We substitute Formula (68) into (66) and (67) and apply Lemmas 6 and 7 and estimate (82). This yields the desired uniform boundedness of the functions $A_k^{(p,q,j,i)}(\varepsilon,\lambda)$, $A_{k,\pm}^{(p,q,j,i)}(\varepsilon,\lambda)$, $A_{k,\pm}^{(p,q,j,i)}(\varepsilon,\lambda)$, $A_{k,\flat,\natural}^{(p,q,j,i)}(\varepsilon,\lambda)$ with some fixed sufficiently small δ_3. All these functions are bounded by some constant c_{16} independent of ε, $\lambda \in E_{k,\delta_3}$, k, j, i. Hence, there exists $\varepsilon_0 > 0$ independent of k, λ, j, i such that as $\varepsilon < \varepsilon_0$, system (65) possesses only trivial solution simultaneously for all k. Therefore, there exists $\delta > 0$ such that the set S^δ contains no eigenvalues of the operator $\mathcal{L}^\varepsilon$.

In order to prove the absence of the residual spectrum, we first need to establish Formula (2). By $\mathrm{Ker}(\cdot)$ and $\mathrm{Ran}(\cdot)$, we denote the kernel and the range of a given closed operator.

Lemma 8. *Identity (2) is true.*

Proof. Given a closed operator $\mathcal{A}$ in a Hilbert space H, let $\lambda \notin \sigma_{\text{ess}}(\mathcal{A}) \cup \sigma_{\text{pnt}}(\mathcal{A})$. Then, $\text{Ker}(\mathcal{A} - \lambda) = \{0\}$ and hence, the inverse operator $(\mathcal{A} - \lambda)^{-1}$ is well defined on the range $\text{Ran}(\mathcal{A} - \lambda)$. This inverse operator is bounded. Indeed, if this operator was unbounded, this would mean the existence of a sequence $u_n \in \mathfrak{D}(\mathcal{A})$ such that $\|u_n\|_H = 1$ and $\|(\mathcal{A} - \lambda)u_n\|_H \to 0$ as $n \to \infty$. Since $\lambda \notin \sigma_{\text{ess}}(A)$, the sequence $\{u_n\}$ is compact, and choosing a subsequence if needed, we can suppose that u_n converges to some u_* in H. Then, by the closedness of the operator $\mathcal{A}$ and the normalization of u_n, we immediately conclude that $\|u_*\|_H = 1$ and $(\mathcal{A} - \lambda)u_* = 0$, i.e., u_* is an eigenfunction of $\mathcal{A}$ associated with its eigenvalue λ. This is impossible, since $\lambda \notin \sigma_{\text{pnt}}(\mathcal{A})$ and, therefore, the inverse operator $(\mathcal{A} - \lambda)^{-1}$ is bounded on the range $\text{Ran}(\mathcal{A} - \lambda)$. By [17] (Ch. 3, Sect. 2, Thm. 9), this yields that the range $\text{Ran}(\mathcal{A} - \lambda)$ is closed. Hence, as $\lambda \notin \sigma_{\text{ess}}(\mathcal{A}) \cup \sigma_{\text{pnt}}(\mathcal{A})$, it belongs to the spectrum $\sigma(\mathcal{A})$ if and only if $\overline{\text{Ran}(\mathcal{A} - \lambda)} = \text{Ran}(\mathcal{A} - \lambda) \neq H$, which is equivalent to $\text{Ker}(\mathcal{A}^* - \overline{\lambda}) \neq \{0\}$, i.e., if and only if $\overline{\lambda}$ is an eigenvalue of the adjoint operator $\mathcal{A}^*$. This completes the proof. □

In view of Formula (2), we observe that the adjoint operator for $\mathcal{L}^\varepsilon$ reads as

$$(\mathcal{L}^\varepsilon)^* = \mathcal{L}_{\overline{V}} + \varepsilon \mathcal{L}_{a^*\star}, \qquad a^*(z) := \overline{a(-z)}. \tag{89}$$

This adjoint operator is of the same structure as $\mathcal{L}^\varepsilon$ in particular, the essential spectrum of the operator $\mathcal{L}_{\overline{V}}$ is just the complex conjugation of the curve Υ, namely,

$$\sigma(\mathcal{L}_{\overline{V}}) = \sigma_{\text{ess}}(\mathcal{L}_{\overline{V}}) = \Upsilon^\dagger.$$

Then, we choose the complex conjugation of the piece S of this curve and we see that it also satisfies the assumptions of Theorem 2. The function a^* obeys Assumption (9). Then, lessening if needed the number δ, we conclude that the set $(S^\delta)^\dagger$ contains no eigenvalues of the operator $(\mathcal{L}^\varepsilon)^*$. Then, Formula (2) implies that the set S^δ also contains no points of the residual spectrum of the operator $\mathcal{L}^\varepsilon$ and this completes the proof of Theorem 2.

4.4. *Absence of Residual Spectrum*

In this subsection, we prove Theorem 3. We recall Formula (89) for the adjoint operator $\mathcal{L}^\varepsilon$ and immediately see that Condition (10) guarantees the self-adjointness of the operator $\mathcal{L}^\varepsilon$. This implies the absence of the residual spectrum.

Suppose that Condition (11) is obeyed. As it was stated in Section 2, see identities (12)–(14), it is sufficient to check the validity of $\mathcal{PT}$-symmetricity condition (12) with the operator $\mathcal{P}$ given in (13). This can be carried out by straightforward calculations for an arbitrary $\psi \in L_2(\mathbb{R})$ using conditions (11):

$$\begin{aligned}
(\mathcal{PT}(\mathcal{L}^\varepsilon)^*\psi)(x) =& \mathcal{PT}\left(\overline{V(x)}\phi(x) + \varepsilon \int_{\mathbb{R}} \overline{a(y-x)}\phi(y)\,dy\right) \\
=& V(\tau x + \varrho)\overline{\phi(\tau x + \varrho)} + \varepsilon \int_{\mathbb{R}} a(y - \tau x - \varrho)\overline{\phi(y)}\,dy \\
=& V(\tau x + \varrho)\overline{\phi(\tau x + \varrho)} + \varepsilon \int_{\mathbb{R}} a(\tau(y-x))\overline{\phi(\tau y + \varrho)}\,dy \\
=& V(x)(\mathcal{PT}\phi)(x) + \varepsilon \int_{\mathbb{R}} a(x-y)(\mathcal{PT}\phi)(y)\,dy = (\mathcal{L}^\varepsilon \mathcal{PT}\phi)(x).
\end{aligned}$$

This completes the proof.

Author Contributions: Conceptualization, D.I.B., A.L.P. and E.A.Z.; Methodology, D.I.B., A.L.P. and E.A.Z.; Validation, D.I.B.; Formal analysis, D.I.B., A.L.P. and E.A.Z.; Investigation, D.I.B., A.L.P. and E.A.Z.; Writing—original draft, D.I.B., A.L.P. and E.A.Z.; Writing—review & editing, D.I.B., A.L.P. and E.A.Z. All authors have read and agreed to the published version of the manuscript.

Funding: The research by D.I.B. is partially supported by the Ministry of Education of the Russian Federation under a state task (agreement No. 073-03-2023-010 of 26 January 2023). The work of the second and the third authors is partially supported by the Tromsø Research Foundation project "Pure Mathematics in Norway" and UiT Aurora project MASCOT. The research of the second author was supported by the Ministry of Science and Higher Education of the Russian Federation (megagrant No. 075-15-2022-1115).

Data Availability Statement: No datasets were generate during the study.

Conflicts of Interest: The authors declare no conflict of interest.

References

1. Berestycki, H.; Coville, J.; Vo, H.-H. On the definition and the properties of the principal eigenvalue of some nonlocal operators. *J. Funct. Anal.* **2016**, *271*, 2701–2751.
2. Kondratiev, Y.; Molchanov, S.; Pirogov, S.; Zhizhina, E. On ground state of some non local Schrodinger operator. *Appl. Anal.* **2017**, *96*, 1390–1400.
3. Kondratiev, Y.; Molchanov, S.; Piatnitski, A.; Zhizhina, E. Resolvent bounds for jump generators. *Appl. Anal.* **2018**, *97*, 323–336.
4. Kondratiev, Y.; Molchanov, S.; Vainberg, B. Spectral analysis of non-local Schrodinger operators. *J. Funct. Anal.* **2017**, *273*, 1020–1048.
5. Brüning, J.; Geyler, V.; Pankrashkin, K. On the discrete spectrum of spin–orbit Hamiltonians with singular interactions. *Russ. J. Math. Phys.* **2007**, *14*, 423–429.
6. Hainzl, C.; Seiringer, R. Asymptotic behaviour of eigenvalues of Schrödinger type operators with degenerate kinetic energy. *Math. Nachrichten* **2010**, *283*, 489–499.
7. Hoang, V.; Hundertmark, D.; Richter, J.; Vugalter, S. Quantitative bounds versus existence of weakly coupled bound states for Schrodinger type operators. *arXiv* **2017**, arXiv:1610.09891v2.
8. Pankrashkin, K. Variational principle for hamiltonians with degenerate bottom. In *Mathematical Results in Quantum Mechanics*; Beltita, I., Nenciu, G., Purice, R., Eds.; World Scientific: Singapore, 2008; pp. 231–240.
9. Molchanov, S.; Vainberg, B. On General Cwikel–Lieb–Rozenblum and Lieb–Thirring Inequalities. In *Around the Research of Vladimir Maz'ya. III. Analysis and Applications*; Laptev, A., Ed.; Springer: Berlin/Heidelberg, Germany, 2010; pp. 201–246.
10. Molchanov, S.; Vainberg, B. Bargmann type estimates for the counting function for general Schrödinger operators. *J. Math. Sci.* **2012**, *184*, 456–508.
11. Borisov, D.I.; Piatnitski, A.L.; Zhizhina, E.A. On spectra of convolution operators with potentials. *J. Math. Anal. Appl.* **2023**, 517, 126568.
12. Borisov, D.I.; Piatnitski, A.L.; Zhizhina, E.A. Spectrum of a convolution operator with potential. *Russ. Math. Surv.* **2022**, *77*, 546–548.
13. Klaus, M.; Simon, B. Coupling constant thresholds in nonrelativistic quantum mechanics, I. Short range two-body case. *Ann. Phys.* **1980**, *130*, 251–281.
14. Blankenbecler, R.; Goldberger, M.L.; Simon, B. The bound states of weakly coupled long-range one-dimensional quantum Hamiltonians. *Ann. Phys.* **1977**, *108*, 69–78.
15. Klaus, M. On the bound state of Schrdinger operators in one dimension. *Ann. Phys.* **1977**, *108*, 288–300.
16. Simon, B. The bound state of weakly coupled Schrödinger operators in one and two dimensions. *Ann. Phys.* **1976**, *97*, 279–288.
17. Birman, M.S.; Solomyak, M.Z. *Spectral Theory of Self-Adjoint Operators in Hilbert Space*; D. Reidel Publishing Co.: Dordrecht, The Netherlands, 1987.

mathematics

MDPI

Article

An Inverse Sturm–Liouville-Type Problem with Constant Delay and Non-Zero Initial Function

Sergey Buterin * and Sergey Vasilev

Department of Mathematics, Saratov State University, 410012 Saratov, Russia
* Correspondence: buterinsa@sgu.ru

Abstract: We suggest a new statement of the inverse spectral problem for Sturm–Liouville-type operators with constant delay. This inverse problem consists of recovering the coefficient (often referred to as potential) of the delayed term in the corresponding equation from the spectra of two boundary value problems with one common boundary condition. The previous studies, however, focus mostly on the case of zero initial function, i.e., they exploit the assumption that the potential vanishes on the corresponding subinterval. In the present paper, we waive that assumption in favor of a continuously matching initial function, which leads to the appearance of an additional term with a frozen argument in the equation. For the resulting new inverse problem, we pay special attention to the situation when one of the spectra is given only partially. Sufficient conditions and necessary conditions on the corresponding subspectrum for the unique determination of the potential are obtained, and a constructive procedure for solving the inverse problem is given. Moreover, we obtain the characterization of the spectra for the zero initial function and the Neumann common boundary condition, which is found to include an additional restriction as compared with the case of the Dirichlet common condition.

Keywords: Sturm–Liouville-type operator; functional-differential operator; constant delay; initial function; frozen argument; inverse spectral problem

MSC: 34A55; 34K29

Citation: Buterin, S.; Vasilev, S. An Inverse Sturm–Liouville-Type Problem with Constant Delay and Non-Zero Initial Function. *Mathematics* **2023**, *11*, 4746. https://doi.org/10.3390/math11234764

Academic Editor: Dongfang Li

Received: 26 October 2023
Revised: 20 November 2023
Accepted: 22 November 2023
Published: 25 November 2023

1. Introduction and Main Results

In recent years, there appeared a considerable interest in the inverse problem of recovering an integrable or a square-integrable potential $q(x)$ in the functional-differential equation

$$-y''(x) + q(x)y(x-a) = \lambda y(x), \quad 0 < x < \pi, \tag{1}$$

with constant delay $a \in (0, \pi)$ from the spectra of two boundary value problems for (1) with one common boundary condition (see [1–17] and references therein). For $a = 0$, this problem becomes the classical inverse Sturm–Liouville problem due to Borg [18,19], but the nonlocal case $a > 0$ requires other approaches. Moreover, it reveals some essentially different effects in the solution of the inverse problem than in the classical situation $a = 0$. For example, the solution of the inverse problem may be non-unique when $a \in (0, 2\pi/5)$ (see [12–14]).

Various equations with delay have been actively studied from the last century in connection with numerous applications (see, e.g., [20–26]). Such equations can be characterized by the possibility for the argument of the unknown function to go beyond its domain. For example, Equation (1) for $a > 0$ includes values of $y(x)$ for $x < 0$. In order to overcome this issue, one should specify an initial function, i.e., to impose $y(x) = f(x)$ for $x \in (-a, 0)$ with some known $f(x)$. In particular, one can put $q(x) = 0$ on $(0, a)$, which actually corresponds to specifying $f = 0$. We distinguish these two ways because rewriting Equation (1) in the form

$$-y''(x) + q^+(x)y(x-a) = \lambda y(x) - r(x), \quad 0 < x < \pi, \tag{2}$$

where $r(x) = q^-(x)f(x-a)$ and

$$q^-(x) = \begin{cases} q(x), & x \in (0,a), \\ 0, & x \in (a,\pi), \end{cases} \qquad q^+(x) = \begin{cases} 0, & x \in (0,a), \\ q(x), & x \in (a,\pi), \end{cases} \tag{3}$$

shows that $f \neq 0$ leads to a non-homogenous equation, while $f = 0$ deals with the corresponding homogenous one. Thus, for posing an eigenvalue problem, it is natural to choose the latter, i.e., to assume that $q(x) = 0$ on $(0,a)$. For this reason, the previous studies of inverse problems for (1) were focused mostly on this case, i.e., the reconstruction of $q(x)$ was actually carried out only for q^+, while q^- was a priori assumed to be zero.

A non-zero initial function f also may be appropriate for posing an eigenvalue problem, but it should be linearly dependent on the unknown function $y(x)$ on $[0,\pi]$ as, e.g.,

$$f(x) = y(0)g(x), \quad -a < x < 0. \tag{4}$$

This example corresponds to the classical theory [22] and ensures a continuous continuation of $y(x)$ to $[-a,0)$ whenever $g(x) \in C[-a,0]$ and $g(0) = 1$. Such continuation, however, is not always required (see, e.g., [25]). So, one can consider more general forms of an initial function such as $f(x) = Ly(x)$ with a linear operator L acting from $W_2^2[0,\pi]$ to $L_\infty(-a,0)$. Then, for keeping L in the frames of a perturbation, a natural requirement would be its relative compactness [27] with respect to the minimal operator of double differentiation. In particular, one can take $Ly(x) = F(y)g(x)$, where $F(y)$ is a linear functional relatively bounded to y'', e.g., $F(y) = y^{(j)}(b)$ for some $b \in [0,\pi]$ and $j \in \{0,1\}$. We will focus, however, on the special case (4).

An attempt to study the inverse problem for Equation (1) with a non-zero initial function $f(x)$ has been made in [16]. However, no dependence of $f(x)$ on $y(x)$ was assumed at all.

In the present paper, we refuse the usual assumption $q^- = 0$ but in favor of the initial function in the form (4). Then, Equation (1) can be rewritten with the so-called frozen argument:

$$-y''(x) + q^+(x)y(x-a) + p(x)y(0) = \lambda y(x), \quad 0 < x < \pi, \quad p(x) := q^-(x)g(x-a). \tag{5}$$

Since the functions $q^-(x)$ and $g(x-a)$ enter only in their product $p(x)$, they cannot be recovered simultaneously from any spectral information. Moreover, the reconstruction of $q^-(x)$ on any subinterval $(\alpha,\beta) \subset (0,a)$ can be possible only if $g(x) \neq 0$ a.e. on $(\alpha - a, \beta - a)$. For these reasons, we consider without loss of generality the canonical situation when $g(x) \equiv 1$.

For $j = 0,1$, $B_j(q)$ denotes the boundary value problem for Equation (1) with a complex-valued potential $q(x) \in L_2(0,\pi)$ under the boundary conditions

$$y'(0) = y^{(j)}(\pi) = 0$$

and under the initial-function condition

$$y(x) = y(0), \quad -a < x < 0. \tag{6}$$

Let $\{\lambda_{n,j}\}_{n\geq 0}$ be the spectrum of $B_j(q)$. Consider the following inverse problem.

Inverse Problem 1. *Given* $\{\lambda_{n,0}\}_{n\geq 0}$ *and* $\{\lambda_{n,1}\}_{n\geq 0}$, *find* $q(x)$.

The main results of the present paper (Theorems 1–3) are restricted to the case $a \geq \pi/2$. In accordance with [13,14], the solution of Inverse Problem 1 may be non-unique for $a \in (0, 2\pi/5)$, while the case $a \in [2\pi/5, \pi/2)$ will require an additional investigation. For future reference, however, we will mark those auxiliary assertions below whose proofs automatically extend to any wider ranges of a than just $a \in [\pi/2, \pi)$.

Everywhere below, one and the same symbol $\{\varkappa_n\}$ will denote *different* sequences in l_2. The following theorem gives basic necessary conditions for the solvability of Inverse Problem 1.

Theorem 1. *For $j = 0, 1$, the following asymptotics holds*

$$\lambda_{n,j} = \rho_{n,j}^2, \quad \rho_{n,j} = n + \frac{1-j}{2} + \frac{\omega}{\pi n} \cos\Big(n + \frac{1-j}{2}\Big)a + \frac{\varkappa_n}{n}, \quad \omega \in \mathbb{C}. \tag{7}$$

Here, the constant ω is determined by the formula

$$\omega = \frac{1}{2} \int_a^\pi q^+(x)\, dx. \tag{8}$$

Moreover, if the spectra $\{\lambda_{n,0}\}_{n\geq 0}$ and $\{\lambda_{n,1}\}_{n\geq 0}$ correspond to one and the same $q^-(x)$, then

$$i\theta_0(-ir) - \theta_1(-ir) = o(e^{(\pi-a)r}), \quad r \to +\infty, \tag{9}$$

where

$$\begin{aligned} \theta_0(\rho) &= \rho(\Delta_0(\rho^2) - \cos\rho\pi) - \omega \sin\rho(\pi - a), \\ \theta_1(\rho) &= \Delta_1(\rho^2) + \rho\sin\rho\pi - \omega\cos\rho(\pi - a), \end{aligned} \tag{10}$$

while the functions $\Delta_0(\lambda)$ and $\Delta_1(\lambda)$ are determined by the formulae

$$\Delta_0(\lambda) = \prod_{n=0}^\infty \frac{\lambda_{n,0} - \lambda}{(n+1/2)^2}, \quad \Delta_1(\lambda) = \pi(\lambda_{0,1} - \lambda) \prod_{n=1}^\infty \frac{\lambda_{n,1} - \lambda}{n^2}. \tag{11}$$

Condition (9) actually means that Inverse Problem 1 remains overdetermined as in the case $q^- = 0$ (see [6,15]). As will be seen below, it is sufficient to specify only one full spectrum and an appropriate part of the other one. For example, we also consider the following problem.

Inverse Problem 2. *Given $\{\lambda_{n_k,0}\}_{k\in\mathbb{N}}$ and $\{\lambda_{n,1}\}_{n\geq 0}$, find $q(x)$.*

Here, $\{n_k\}_{k\in\mathbb{N}}$ is an increasing sequence of non-negative integers. The next theorem gives sufficient conditions as well as necessary conditions on $\{n_k\}_{k\in\mathbb{N}}$ for the uniqueness of $q(x)$.

Theorem 2. *(i) If the system $\sigma_0 := \{\sin(n_k + 1/2)x\}_{k\in\mathbb{N}}$ is complete in $\mathcal{H} := L_2(0, \pi - a)$, then the potential $q(x)$ in Inverse Problem 2 is determined uniquely.*

(ii) Conversely, if the specification of $\{\lambda_{n_k,0}\}_{k\in\mathbb{N}}$ and $\{\lambda_{n,1}\}_{n\geq 0}$ uniquely determines $q(x)$, then the defect of σ_0 does not exceed 1, i.e., $\dim(\mathcal{H} \ominus \sigma_0) \leq 1$.

Since the system $\{\sin(n + 1/2)x\}_{n\geq 0}$ is complete in $L_2(0, \pi)$, this theorem, obviously, implies the unique determination of $q(x)$ by both complete spectra as in Inverse Problem 1.

The use of subspectra in the inverse problem with delay began in [6] for the zero initial function, where necessary and sufficient conditions were obtained on parts of both spectra to ensure the uniqueness of $q^+(x)$ in the case of the Dirichlet common condition at the origin.

We note that the gap between the sufficient and the necessary conditions in Theorem 2 is actually caused by imposing the common Neumann boundary condition. By the same reason, the conditions in Theorem 1 do not suffice for the solvability of Inverse Problem 1.

In the case of the Dirichlet common condition, necessary and sufficient conditions for the solvability of the corresponding inverse problem were obtained in [15] when $q^- = 0$. Here, we provide such conditions in the same case $q^- = 0$ but for the Neumann common condition, which brings to them an additional item. Specifically, the following theorem holds.

Theorem 3. *Arbitrary complex sequences $\{\lambda_{n,0}\}_{n\geq 0}$ and $\{\lambda_{n,1}\}_{n\geq 0}$ of the form (7) sharing one and the same $\omega \in \mathbb{C}$ are the spectra of the problems $B_0(q)$ and $B_1(q)$, respectively, with $q(x) = 0$ a.e. on $(0, a)$ if and only if the exponential types of the functions $\theta_0(\rho)$ and $\theta_1(\rho)$ determined by (10) and (11) do not exceed $\pi - a$ and the following relation is fulfilled:*

$$\lambda_{0,1} \prod_{n=1}^{\infty} \frac{\lambda_{n,1}}{n^2} = \frac{2\omega}{\pi}. \tag{12}$$

The latter relation is an additional characterizing condition, which does not exist in the Dirichlet case [15]. We note that the relevant difference between both cases was pointed out in [12] (see Remark 2 therein).

There are also various studies of recovering the operator with purely frozen argument

$$\ell y := -y''(x) + q(x)y(b), \quad y^{(\alpha)}(0) = y^{(\beta)}(\pi) = 0,$$

from its spectrum, where $b \in [0, \pi]$ and $\alpha, \beta \in \{0, 1\}$ (see [28–35] and references therein). In particular, its unique solvability depends on the value of b as well as on α and β. We note that both related to Inverse Problem 1 situations: $b = 0$, $\alpha = 1$, $\beta = 0$ and $b = 0$, $\alpha = \beta = 1$ belong to the so-called non-generate case, when the solution is unique (see, e.g., [28,29,32]).

We note that Theorem 2 also formally holds for $a = \pi$, which follows from Theorem 4.1 in [28] or Theorem 2 in [29]. In this case, $q^- = q$ and $q^+ = 0$. Then, Equation (1) under the initial-function condition (6) becomes an equation with purely frozen argument:

$$-y''(x) + q(x)y(0) = \lambda y(x), \quad 0 < x < \pi,$$

where $q(x)$ is uniquely determined by the single spectrum $\{\lambda_{n,1}\}_{n\geq 0}$.

The paper is organized as follows. In the next section, we construct transformation operators for a fundamental system of solutions of the homogeneous equation in (2), i.e., when $r(x) = 0$. In Section 3, Green's function of the Cauchy problem for the non-homogeneous Equation (2) under the zero initial conditions is constructed. In Section 4, we study the characteristic functions of the problems $B_j(q)$ and prove Theorem 1. Proofs of Theorems 2 and 3 are given in Section 5 along with a constructive procedure for solving the inverse problems. In the last section, we summarize the main innovations of the paper and discuss the results.

Throughout the paper, we agree that ρ is connected with λ by the relation $\rho^2 = \lambda$, while f' and $f^{(j)}$ denote the partial derivatives of a function f with respect to the *first* argument:

$$f'(x_1, \ldots, x_m) := \frac{d}{dx_1} f(x_1, \ldots, x_m), \quad f^{(j)}(x_1, \ldots, x_m) := \frac{d^j}{dx_1^j} f(x_1, \ldots, x_m).$$

2. Transformation Operators

Let $C(x, \lambda)$ and $S(x, \lambda)$ be solutions of the homogeneous equation in (2), i.e., the equation

$$-y''(x) + q^+(x)y(x - a) = \lambda y(x), \quad 0 < x < \pi, \tag{13}$$

under the initial conditions

$$C(0, \lambda) = S'(0, \lambda) = 1, \quad C'(0, \lambda) = S(0, \lambda) = 0.$$

They form a fundamental system of solutions of equation (13) (see, e.g., [14]).

In this section, we obtain representations for the functions $C(x,\lambda)$ and $S(x,\lambda)$ involving the so-called transformation operators, which connect them with the corresponding solutions of the simplest equation with the zero potential. Specifically, the following lemma holds.

Lemma 1. *Let $a \geq \pi/2$. The functions $S(x,\lambda)$ and $C(x,\lambda)$ admit the representations*

$$S(x,\lambda) = \frac{\sin\rho x}{\rho} + \int_a^x P(x,t)\frac{\sin\rho(x-t)}{\rho}\,dt, \tag{14}$$

$$C(x,\lambda) = \cos\rho x + \int_a^x K(x,t)\cos\rho(x-t)\,dt, \tag{15}$$

where (in accordance with our standing agreement) $\rho^2 = \lambda$ and

$$P(x,t) = \frac{1}{2}\int_{\frac{a+t}{2}}^{x+\frac{a-t}{2}} q^+(\tau)\,d\tau, \tag{16}$$

$$K(x,t) = \frac{1}{2}\int_a^{\frac{a+t}{2}} q^+(\tau)\,d\tau + \frac{1}{2}\int_a^{x+\frac{a-t}{2}} q^+(\tau)\,d\tau. \tag{17}$$

Proof. The assertion for $S(x,\lambda)$ is a particular case of Lemma 1 in [15]. So we will prove only (15) and (17). It is easy to see that the Cauchy problem for $C(x,\lambda)$ is equivalent to the integral equation

$$C(x,\lambda) = \cos\rho x + \int_a^x \frac{\sin\rho(x-t)}{\rho} q^+(t)C(t-a,\lambda)\,dt.$$

Taking into account that $a \geq \pi/2$, we calculate

$$\int_a^x \frac{\sin\rho(x-t)}{\rho} q^+(t)C(t-a,\lambda)\,dt = \int_a^x \frac{\sin\rho(x-t)}{\rho} q^+(t)\cos\rho(t-a)\,dt$$

$$= \int_a^x q^+(t)\cos\rho(t-a)\,dt \int_0^{x-t} \cos\rho\tau\,d\tau$$

$$= \frac{1}{2}\int_a^x q^+(t)\,dt \int_0^{x-t} \Big(\cos\rho(t-a+\tau) + \cos\rho(t-a-\tau)\Big)\,d\tau$$

$$= \frac{1}{2}\int_a^x q^+(t)\,dt \int_a^{2(x-t)+a} \cos\rho(x-\tau)\,d\tau = \frac{1}{2}\int_a^{2x-a} \cos\rho(x-t)\,dt \int_a^{x+\frac{a-t}{2}} q^+(\tau)\,d\tau$$

$$= \frac{1}{2}\int_a^x \Big(\int_a^{x+\frac{a-t}{2}} q^+(\tau)\,d\tau + \int_a^{\frac{a+t}{2}} q^+(\tau)\,d\tau\Big)\cos\rho(x-t)\,dt,$$

which finishes the proof. □

Remark 1. *While the imposed restriction $a \geq \pi/2$ is vital for (16) and (17), representations (14) and (15) also remain valid for all smaller $a \geq 0$ but with more complicated kernels. In particular, Lemma 1 in [15] gives an integral equation for $P(x,t)$ for all $a \in [0,\pi/2)$. Moreover, it extends representation (14) to quadratic pencils with two delays.*

The following corollary can be easily checked by direct calculations.

Corollary 1. *The following representations hold:*

$$C(x,\lambda) = \cos\rho x + \omega(x)\frac{\sin\rho(x-a)}{\rho} + \int_a^x K_0(x,t)\frac{\sin\rho(x-t)}{\rho}\,dt, \tag{18}$$

$$C'(x,\lambda) = -\rho \sin \rho x + \omega(x) \cos \rho(x-a) + \int_a^x K_1(x,t) \cos \rho(x-t)\, dt, \tag{19}$$

where

$$\omega(x) = \frac{1}{2}\int_a^x q^+(t)\, dt, \quad K_j(x,t) = \frac{1}{4}\Big(q^+\Big(\frac{a+t}{2}\Big) - (-1)^j q^+\Big(x + \frac{a-t}{2}\Big)\Big), \ j = 0,1. \tag{20}$$

3. Green's Function of the Cauchy Operator

Here, we obtain the solution $z(x,\lambda) = z(x,\lambda;r)$ of the Cauchy problem for the non-homogeneous Equation (2) with an arbitrary free term $r(x)$ under the zero initial conditions

$$z(0,\lambda) = z'(0,\lambda) = 0. \tag{21}$$

In the next section, we will need representations for $z(\pi,\lambda;q^-)$ and $z'(\pi,\lambda;q^-)$.

As in the local case $a=0$, the function $z(x,\lambda)$ is expected to have the form

$$z(x,\lambda) = \int_0^x G(x,t,\lambda) r(t)\, dt, \tag{22}$$

where $G(x,t,\lambda)$ is called Green's function. Let us obtain an explicit formula for it.

Lemma 2. *Let $a \in [0,\pi]$. Then,*

$$G(x,t,\lambda) = y_t(x-t), \quad t \le x \le \pi, \tag{23}$$

where the function $y_t(x)$ for each fixed $t \in [0,\pi)$ solves the Cauchy problem

$$-y_t''(x) + q_t(x) y_t(x-a) = \lambda y_t(x), \quad 0 < x < \pi - t, \quad y_t(0) = 0, \quad y_t'(0) = 1, \tag{24}$$

with

$$q_t(x) := \begin{cases} 0, & 0 < x < \min\{a, \pi - t\}, \\ q^+(x+t), & a < x < \pi - t. \end{cases} \tag{25}$$

Proof. Since the function $G(x,t,\lambda)$ is uniquely determined by the representation (22), one has the right to impose any restrictions on it that will finally lead to (22). In particular, it is natural to assume that $G(x,t,\lambda)$ is sufficiently smooth and obeys the conditions

$$G(x,x,\lambda) = 0, \quad G'(x,x,\lambda) = 1. \tag{26}$$

Then, substituting (22) into (2) and taking the arbitrariness of $r(x)$ into account, we obtain the relations

$$\begin{aligned} -G''(x,t,\lambda) &= \lambda G(x,t,\lambda), \quad 0 < t < x < a, \\ -G''(x,t,\lambda) + q^+(x) G(x-a,t,\lambda) &= \lambda G(x,t,\lambda), \quad 0 < t < x-a < \pi - a, \\ -G''(x,t,\lambda) &= \lambda G(x,t,\lambda), \quad 0 < x-a < t < x < \pi, \end{aligned}$$

which, in turn, along with (26) guarantee that (22) is a solution of the problem (2) and (21).

Substituting $x+t$ into the above three relations instead of x, we obtain

$$-G''(x+t,t,\lambda) = \lambda G(x+t,t,\lambda), \quad 0 < x < a-t < a, \tag{27}$$

$$-G''(x+t,t,\lambda) + q^+(x+t) G(x+t-a,t,\lambda) = \lambda G(x+t,t,\lambda), \quad a < x < \pi - t < \pi, \tag{28}$$

$$-G''(x+t,t,\lambda) = \lambda G(x+t,t,\lambda), \quad \max\{0, a-t\} < x < \min\{a, \pi - t\}. \tag{29}$$

Denote $y_t(x) := G(x+t,t,\lambda)$. Then, combining (27) and (29), one can rewrite:

$$-y_t''(x) = \lambda y_t(x), \quad 0 < x < \min\{a, \pi - t\},$$

while (28) takes the form

$$-y_t''(x) + q^+(x+t)y_t(x-a) = \lambda y_t(x), \quad a < x < \pi - t < \pi.$$

Using the designation (25) along with initial conditions (26), we arrive at (24).

Finally, note that after solving the Cauchy problem (24) by the standard approach (see, e.g., [14]), it is easy to see that $G(x,t,\lambda)$ is a continuous function with respect to all arguments. Hence, the integral in (22) exists and gives a solution to the Cauchy problem (2) and (21). □

Lemma 3. *Let $a \geq \pi/2$. Then, the following representations hold:*

$$G(x,t,\lambda) = \frac{\sin\rho(x-t)}{\rho}, \quad \max\{0, x-a\} \leq t \leq x \leq \pi, \tag{30}$$

and

$$G(x,t,\lambda) = \frac{\sin\rho(x-t)}{\rho} + \frac{1}{2}\int_{a+t}^{x} \frac{\sin\rho(x-\tau)}{\rho}\, d\tau \int_{\frac{a+t+\tau}{2}}^{x+\frac{a+t-\tau}{2}} q^+(\eta)\, d\eta \tag{31}$$

whenever $0 \leq t \leq x - a \leq \pi - a$.

Proof. By virtue of (24) and Lemma 1, we have the representation

$$y_t(x) = \frac{\sin\rho x}{\rho} + \frac{1}{2}\int_a^x \frac{\sin\rho(x-\tau)}{\rho}\, d\tau \int_{\frac{a+\tau}{2}}^{x+\frac{a-\tau}{2}} q_t(\eta)\, d\eta, \quad 0 \leq x \leq \pi - t,$$

which, in accordance with (23) and (25), leads to (30) and (31). □

By substituting (30) and (31) into (22) and changing the order of integration, we obtain

$$z(x,\lambda) = \int_0^x \Big(r(t) + \frac{1}{2}\int_0^{t-a} r(\tau)\, d\tau \int_{\frac{a+t+\tau}{2}}^{x+\frac{a+\tau-t}{2}} q^+(\eta)\, d\eta \Big) \frac{\sin\rho(x-t)}{\rho}\, dt, \quad 0 \leq x \leq \pi, \tag{32}$$

where $r(x) = 0$ for $x < 0$.

Further, differentiating (30) and (31) with respect to x, we arrive at the formulae

$$G'(x,t,\lambda) = \cos\rho(x-t), \quad \max\{0, x-a\} \leq t \leq x \leq \pi,$$

and

$$G'(x,t,\lambda) = \cos\rho(x-t) + \frac{1}{2}\int_{a+t}^{x} \Big(\int_{\frac{a+t+\tau}{2}}^{x} q^+(\eta)\, d\eta + \int_{x+\frac{a+t-\tau}{2}}^{x} q^+(\eta)\, d\eta \Big) \cos\rho(x-\tau)\, d\tau$$

as soon as $0 \leq t \leq x - a \leq \pi - a$. Substituting them into

$$z'(x,\lambda) = \int_0^x G'(x,t,\lambda) r(t)\, dt,$$

we analogously obtain the representation

$$z'(x,\lambda) = \int_0^x \Big(r(t) + \frac{1}{2}\int_0^{t-a} \Big(\int_{\frac{a+t+\tau}{2}}^{x} q^+(\eta)\, d\eta + \int_{x+\frac{a+\tau-t}{2}}^{x} q^+(\eta)\, d\eta \Big) r(\tau)\, d\tau \Big) \cos\rho(x-t)\, dt. \tag{33}$$

4. Characteristic Functions

Consider the entire functions

$$\Delta_j(\lambda) := C^{(j)}(\pi,\lambda) + z^{(j)}(\pi,\lambda;q^-), \quad j = 0,1. \tag{34}$$

The next lemma holds for any $a \in [0,\pi]$.

Lemma 4. *For $j = 0,1$, eigenvalues of the problem $B_j(q)$ coincide with zeros of $\Delta_j(\lambda)$.*

Proof. Since the sum $C(x,\lambda) + z(x,\lambda;q^-)$ cannot be identically zero, any zero of $\Delta_j(\lambda)$ is an eigenvalue of the problem $B_j(q)$, which, in turn, under our settings has the form

$$-y''(x) + q^+(x)y(x-a) + q^-(x)y(0) = \lambda y(x), \quad y'(0) = y^{(j)}(\pi) = 0. \tag{35}$$

Conversely, let λ be an eigenvalue of $B_j(q)$, and let $y(x)$ be the corresponding eigenfunction, i.e., a nontrivial solution of (35). Then, $y(0) \neq 0$ since, obviously, $y(x) \equiv 0$ otherwise. Without loss of generality, one can assume that $y(0) = 1$, which will imply $y(x) = C(x,\lambda) + z(x,\lambda;q^-)$ due to the uniqueness of solution of the Cauchy problem. Hence, $\Delta_j(\lambda) = y^{(j)}(\pi) = 0$. □

As usual, we call $\Delta_j(\lambda)$ the characteristic function of the problem $B_j(q)$. The following lemma based on the two preceding sections gives representations for both characteristic functions.

Lemma 5. *The characteristic functions admit the representations*

$$\Delta_0(\lambda) = \cos\rho\pi + \omega\frac{\sin\rho(\pi-a)}{\rho} + \int_0^\pi w_0(x)\frac{\sin\rho x}{\rho}\,dx, \quad w_0(x) \in L_2(0,\pi), \tag{36}$$

$$\Delta_1(\lambda) = -\rho\sin\rho\pi + \omega\cos\rho(\pi-a) + \int_0^\pi w_1(x)\cos\rho x\,dx, \quad w_1(x) \in L_2(0,\pi). \tag{37}$$

Moreover, the constant ω is determined by (8), and

$$w_0(\pi-x) = w_1(\pi-x) = q^-(x), \quad 0 < x < a, \tag{38}$$

while for $a < x < \pi$:

$$w_0(\pi-x) = \frac{1}{4}\Big(q^+\Big(\frac{a+x}{2}\Big) - q^+\Big(\pi+\frac{a-x}{2}\Big)\Big) + \frac{1}{2}\int_0^{x-a} q^-(t)\,dt\int_{\frac{a+x+t}{2}}^{\pi+\frac{a+t-x}{2}} q^+(\tau)\,d\tau, \tag{39}$$

$$w_1(\pi-x) = \frac{1}{4}\Big(q^+\Big(\frac{a+x}{2}\Big) + q^+\Big(\pi+\frac{a-x}{2}\Big)\Big)$$
$$+\frac{1}{2}\int_0^{x-a}\Big(\int_{\frac{a+x+t}{2}}^{\pi} q^+(\tau)\,d\tau + \int_{\pi+\frac{a+t-x}{2}}^{\pi} q^+(\tau)\,d\tau\Big)q^-(t)\,dt. \tag{40}$$

Proof. Substituting $x = \pi$ into (18) and (19) and using (8) and (20), we obtain

$$C(\pi,\lambda) = \cos\rho\pi + \omega\frac{\sin\rho(\pi-a)}{\rho} + \int_0^{\pi-a} u_0(x)\frac{\sin\rho x}{\rho}\,dx, \tag{41}$$

$$C'(\pi,\lambda) = -\rho\sin\rho\pi + \omega\cos\rho(\pi-a) + \int_0^{\pi-a} u_1(x)\cos\rho x\,dx, \tag{42}$$

where

$$u_j(\pi-x) = K_j(\pi,x) = \frac{1}{4}\Big(q^+\Big(\frac{a+x}{2}\Big) - (-1)^j q^+\Big(\pi+\frac{a-x}{2}\Big)\Big), \quad a < x < \pi, \ j = 0,1. \tag{43}$$

Further, substituting $r = q^-$ and $x = \pi$ into (32) and (33), we arrive at

$$z(\pi, \lambda; q^-) = \int_0^\pi v_0(x) \frac{\sin \rho x}{\rho}\, dx, \quad z'(\pi, \lambda; q^-) = \int_0^\pi v_1(x) \cos \rho x\, dx, \tag{44}$$

where

$$v_0(\pi - x) = v_1(\pi - x) = q^-(x), \quad 0 < x < a, \tag{45}$$

$$v_0(\pi - x) = \frac{1}{2} \int_0^{x-a} q^-(t)\, dt \int_{\frac{a+x+t}{2}}^{\pi + \frac{a+t-x}{2}} q^+(\tau)\, d\tau, \quad a < x < \pi, \tag{46}$$

$$v_1(\pi - x) = \frac{1}{2} \int_0^{x-a} \Big(\int_{\frac{a+x+t}{2}}^{\pi} q^+(\tau)\, d\tau + \int_{\pi + \frac{a+t-x}{2}}^{\pi} q^+(\tau)\, d\tau \Big) q^-(t)\, dt, \quad a < x < \pi. \tag{47}$$

According to (34), (41), (42), and (44), we obtain (36) and (37) with

$$w_j(x) = u_j(x) + v_j(x), \quad j = 0, 1, \tag{48}$$

where $u_0(x) = u_1(x) = 0$ on $(\pi - a, \pi)$. Finally, substituting (43) and (45)–(47) into (48), we arrive at (38)–(40). □

In the rest of this section, we provide auxiliary facts about arbitrary functions of the form (36) and (37) and give the proof of Theorem 1.

Lemmas 6–8 below are valid for any fixed $a \in [0, 2\pi]$. By the standard approach (see, e.g., [19,36]) involving Rouché's theorem, one can prove the following assertion.

Lemma 6. *For $j = 0, 1$, any $\Delta_j(\lambda)$ has infinitely many zeros $\{\lambda_{n,j}\}_{n \ge 0}$ of the form (7).*

The next assertion for $a = 0$ can be found in [19], but the proof does not depend on the value of a as soon as it ranges within $[0, 2\pi]$.

Lemma 7. *Any functions of the forms (36) and (37) are determined by their zeros uniquely. Moreover, the representations in (11) hold.*

Now, we are in position to give the proof of Theorem 1.

Proof of Theorem 1. The asymptotics (7) is a direct corollary of Lemmas 5 and 6. It remains to make note that, by virtue of (10), (36), and (37) along with Lemma 7, we have

$$i\theta_0(\rho) - \theta_1(\rho) = i \int_0^\pi w_0(x) \sin \rho x\, dx - \int_0^\pi w_1(x) \cos \rho x\, dx = \frac{\theta_+(\rho) - \theta_-(\rho)}{2},$$

where, according to (38),

$$\theta_+(\rho) = \int_0^{\pi - a} (w_0 - w_1)(x) \exp(i\rho x)\, dx, \quad \theta_-(\rho) = \int_0^\pi (w_0 + w_1)(x) \exp(-i\rho x)\, dx,$$

which implies (9). □

Statements analogous to the next lemma are often used for finding necessary and sufficient conditions for the solvability of inverse problems, i.e., a characterization of the spectral data (see Remark 2 in [36]). For its proof, we will follow a new simple idea suggested in [36].

Lemma 8. *For $j = 0, 1$, let $\{\lambda_{n,j}\}_{n \ge 0}$ be arbitrary complex sequences of the form (7). Then, the function $\Delta_j(\lambda)$ constructed by the corresponding formula in (11) has the form (36) or (37), respectively.*

Proof. Since the assertion of the lemma for $j = 0$ formally follows from Lemma 6 in [15], we focus on the case $j = 1$. Let a sequence $\{\lambda_{n,1}\}_{n\geq 0}$ of the form (7) be given. First, let all values $\lambda_{n,1}$ be distinct and $\lambda_{0,1} = 0$. Denote $\rho_{-n,1} := -\rho_{n,1}$ for $n \geq 1$. By virtue of Lemma 2 in [36], the system $\{\exp(i\rho_{n,1}x)\}_{n\in\mathbb{Z}}$ is a Riesz basis in $L_2(-\pi,\pi)$. Moreover, the asymptotics (7) implies $\{\theta(\rho_{n,1})\}_{n\in\mathbb{Z}} \in l_2$, where $\theta(\rho) := \rho\sin\rho\pi - \omega\cos\rho(\pi - a)$ and ω is as in (7). Hence, there exists a unique function $W_1(x) \in L_2(-\pi,\pi)$ obeying the relations

$$\theta(\rho_{n,1}) = \int_{-\pi}^{\pi} W_1(x)\exp(i\rho_{n,1}x)\,dx, \quad n \in \mathbb{Z}.$$

Obviously, $W_1(x)$ is even. Thus, $\lambda_{n,1} = (\rho_{n,1})^2$, $n \geq 0$, are zeros of the function $\Delta_1(\lambda)$ determined by (37) with $w_1(x) = 2W_1(x)$. By Lemma 6, $\Delta_1(\lambda)$ has no other zeros, while by Lemma 7, it admits the second representation in (11), which finishes the proof for a simple sequence $\{\lambda_{n,1}\}_{n\geq 0}$ containing a zero element.

For the general case, it is sufficient to note that multiplying $\Delta_1(\lambda)$ with any function

$$h(\lambda) := \prod_{n\in A} \frac{\lambda - \tilde{\lambda}_{n,1}}{\lambda - \lambda_{n,1}}, \quad A \subset \mathbb{N} \cup \{0\}, \quad \#A < \infty,$$

preserves the form (37) and changes only $w_1(x)$. Indeed, we have

$$h(\lambda)\Delta_1(\lambda) = -\rho\sin\rho\pi + \omega\cos\rho(\pi - a) + H(\lambda),$$

where

$$H(\lambda) = (1 - h(\lambda))\big(\rho\sin\rho\pi - \omega\cos\rho(\pi - a)\big) + h(\lambda)\int_0^{\pi} w_1(x)\cos\rho x\,dx.$$

The function $H(\lambda)$ is whole as soon as $\lambda_{n,1}$ are zeros of $\Delta_1(\lambda)$. Moreover, in the ρ-plane, we, obviously, have $H(\rho^2) \in L_2(-\infty,+\infty)$ and $H(\rho^2) = o(\exp(|\mathrm{Im}\,\rho|\pi))$ as $\rho \to \infty$. Thus, by virtue of the Paley–Wiener theorem (see, e.g., [37]), it has the form

$$H(\lambda) = \int_0^{\pi} \tilde{w}_1(x)\cos\rho x\,dx, \quad \tilde{w}_1(x) \in L_2(0,\pi),$$

which finishes the proof completely. □

Finally, let us give one more auxiliary assertion, which will be used in the proof of Theorem 2. Let $\{n_k\}_{k\in\mathbb{N}}$ be an increasing sequence of non-negative integers. Without loss of generality, assume that multiple elements in the subspectrum $\{\lambda_{n_k,0}\}_{k\in\mathbb{N}}$ are neighboring, i.e.,

$$\lambda_{n_k,0} = \lambda_{n_{k+1},0} = \ldots = \lambda_{n_{k+m_k-1},0},$$

where m_k is the multiplicity of the value $\lambda_{n_k,0}$ in this subspectrum. Put

$$\mathcal{S} := \{1\} \cup \{k : \lambda_{n_k,0} \neq \lambda_{n_{k-1},0},\ k \geq 2\}$$

and consider the functional system $\sigma := \{s_n(x)\}_{n\in\mathbb{N}}$, where

$$s_{k+\nu}(x) := \Big(n_k + \frac{1}{2}\Big)\frac{d^\nu}{d\lambda^\nu}\frac{\sin\rho x}{\rho}\Big|_{\lambda=\lambda_{n_k,0}}, \quad k \in \mathcal{S}, \quad \nu = \overline{0, m_k - 1}.$$

Lemma 9. *The system σ is a Riesz basis in $\mathcal{H}_b := L_2(0,b)$ if and only if so is the system $\sigma_0 = \{\sin(n_k + 1/2)x\}_{k\in\mathbb{N}}$. Moreover, they have equal defects, i.e.,* $\dim(\mathcal{H}_b \ominus \sigma_0) = \dim(\mathcal{H}_b \ominus \sigma)$.

Proof. Let there exist d linearly independent entire functions $h_\nu(\lambda)$, $\nu = \overline{1,d}$, of the form

$$h_\nu(\lambda) = \int_0^b f_\nu(x)\frac{\sin\rho x}{\rho}\,dx, \quad f_\nu(x) \in L_2(0,b), \tag{49}$$

whose zeros have the common part $\{(n_k + 1/2)^2\}_{k \in \mathbb{N}}$. In other words, the space $\mathcal{H}_b \ominus \sigma_0$ contains at least d linearly independent functions $f_\nu(x)$. Consider the meromorphic function

$$F(\lambda) := \prod_{k=1}^{\infty} \frac{\lambda_{n_k,0} - \lambda}{(n_k + 1/2)^2 - \lambda}.$$

Then, the entire (after removing singularities) function $\tilde{h}_\nu(\lambda) := F(\lambda) h_\nu(\lambda)$ has the form

$$\tilde{h}_\nu(\lambda) = \int_0^b \tilde{f}_\nu(x) \frac{\sin \rho x}{\rho} dx, \quad \tilde{f}_\nu(x) \in L_2(0, b). \tag{50}$$

Indeed, as in the proof of Lemma 2 in [36], one can show that $|F(\rho^2)| < C_\delta$ whenever

$$|\rho \pm (n_k + 1/2)| \geq \delta, \quad k \in \mathbb{N},$$

for each fixed $\delta > 0$. Hence, we have $|\rho \tilde{h}_\nu(\rho^2)| \leq C_\delta |\rho h_\nu(\rho^2)|$ for such ρ. Thus, according to (49), the function $\rho \tilde{h}_\nu(\rho^2)$ is square-integrable on the line $\rho = i\delta$, while the maximum modulus principle for analytic functions gives $\rho \tilde{h}_\nu(\rho^2) = o(\exp(|\operatorname{Im} \rho| b))$ as $\rho \to \infty$ in the entire plane. Using the Paley–Wiener theorem [37] and taking the oddness of $\rho \tilde{h}_\nu(\rho^2)$ into account, we obtain (50). Obviously, the functions $\tilde{h}_\nu(\lambda)$, $\nu = \overline{1,d}$, are linearly independent, and their zeros have the common part $\{\lambda_{n_k,0}\}_{k \in \mathbb{N}}$ with account of multiplicity. Therefore, $\dim(\mathcal{H} \ominus \sigma_0) \leq \dim(\mathcal{H} \ominus \sigma)$. Analogously, one can prove the inequality $\dim(\mathcal{H} \ominus \sigma_0) \geq \dim(\mathcal{H} \ominus \sigma)$.

We have proved the second assertion of the lemma, which means, in particular, that the systems σ and σ_0 can be complete in $\mathcal{H}_b$ only simultaneously. Hence, by virtue of Proposition 1.8.5 in [19], the simultaneous Riesz-basisness follows from their quadratical closeness

$$\sum_{k=1}^{\infty} \|s_k - s_k^0\|_{L_2(0,b)}^2 < \infty, \quad s_k^0(x) := \sin \gamma_k x, \quad \gamma_k := n_k + \frac{1}{2}.$$

The last inequality, in turn, is ensured by the estimate

$$s_k(x) - s_k^0(x) = \sin \rho_{n_k,0} x - \sin \gamma_k x + O\left(\frac{1}{k^2}\right)$$

$$= 2\cos \frac{(\rho_{n_k,0} + \gamma_k)x}{2} \sin \frac{(\rho_{n_k,0} - \gamma_k)x}{2} + O\left(\frac{1}{k^2}\right) = O\left(\frac{1}{k}\right), \quad k \to \infty,$$

which holds uniformly in $x \in [0, b]$. □

5. Solution of the Inverse Problems

When the functions $w_0(x)$ and $w_1(x)$ are specified, relations (38)–(40) can be considered as a nonlinear integral equation with respect to $q(x) = q^-(x) + q^+(x)$. The following lemma actually implies its unique solvability.

Lemma 10. *For any functions $w_0(x), w_1(x), q^-(x) \in L_2(0, \pi - a)$, the linear system consisting of (39) and (40) has a unique solution $q^+(x) \in L_2(a, \pi)$.*

Proof. Summing up equations (39) and (40) and then subtracting one from the other, we obtain

$$\left. \begin{array}{l} 2(w_1 + w_0)(\pi - x) = q^+\left(\frac{a+x}{2}\right) + 2 \int_0^{x-a} q^-(t)\, dt \int_{\frac{a+x+t}{2}}^{\pi} q^+(\tau)\, d\tau, \\[6pt] 2(w_1 - w_0)(\pi - x) = q^+\left(\pi + \frac{a-x}{2}\right) + 2 \int_0^{x-a} q^-(t)\, dt \int_{\pi + \frac{a-x-t}{2}}^{\pi} q^+(\tau)\, d\tau, \end{array} \right\} \quad a < x < \pi.$$

Changing the variable, we arrive at the relations

$$2(w_1 + w_0)(\pi + a - 2x) = q^+(x) + 2\int_0^{2(x-a)} q^-(t)\, dt \int_{x + \frac{t}{2}}^{\pi} q^+(\tau)\, d\tau, \quad a < x < \frac{a + \pi}{2},$$

$$2(w_1 - w_0)(2x - \pi - a) = q^+(x) + 2\int_0^{2(\pi-x)} q^-(t)\,dt \int_{x+\frac{t}{2}}^{\pi} q^+(\tau)\,d\tau, \quad \frac{a+\pi}{2} < x < \pi.$$

Then, changing the order of integration in the last two formulae, we obtain the system

$$2(w_1 + w_0)(\pi + a - 2x) = q^+(x) + 2\int_x^{2x-a} q^+(t)\,dt \int_0^{2(t-x)} q^-(\tau)\,d\tau$$
$$+2\int_{2x-a}^{\pi} q^+(t)\,dt \int_0^{2(x-a)} q^-(\tau)\,d\tau, \quad a < x < \frac{a+\pi}{2},$$
$$2(w_1 - w_0)(2x - \pi - a) = q^+(x) + 2\int_x^{\pi} q^+(t)\,dt \int_0^{2(t-x)} q^-(\tau)\,d\tau, \quad \frac{a+\pi}{2} < x < \pi.$$

Using the designations

$$W(x) := \begin{cases} 2(w_1 + w_0)(\pi + a - 2x), & a < x < \dfrac{a+\pi}{2}, \\ 2(w_1 - w_0)(2x - \pi - a), & \dfrac{a+\pi}{2} < x < \pi, \end{cases} \tag{51}$$

$$Q(x,t) := \begin{cases} 2\displaystyle\int_0^{2(x-a)} q^-(\tau)\,d\tau, & a < 2x - a < t < \pi, \\ 2\displaystyle\int_0^{2(t-x)} q^-(\tau)\,d\tau, & a < x < t < \min\{2x - a, \pi\}, \end{cases} \tag{52}$$

one can rewrite the latter system as a Volterra integral equation of the second kind:

$$W(x) = q^+(x) + \int_x^{\pi} Q(x,t) q^+(t)\,dt, \quad a < x < \pi, \tag{53}$$

which possesses a unique solution $q^+(x) \in L_2(a,\pi)$ (see, e.g., [38]). □

Proof of Theorem 2. First of all, note that due to (7), the value ω is always determined by specifying $\{\lambda_{n,1}\}_{n\geq 0}$ via the formula

$$\omega = \pi \lim_{k\to\infty} \tilde{n}_k \frac{\rho_{\tilde{n}_k,1} - \tilde{n}_k}{\cos \tilde{n}_k a}, \tag{54}$$

where the natural sequence $\{\tilde{n}_k\}$ is chosen so that $|\cos \tilde{n}_k a| \geq c > 0$. Alternatively, in accordance with (37), one can use the relation

$$\omega = \lim_{n\to\infty} \left(\Delta_1(\xi_n^2) + \xi_n \sin \xi_n \pi \right), \quad \xi_n = \frac{2\pi n}{\pi - a}, \tag{55}$$

where $\Delta_1(\lambda)$ is constructed by the second formula in (11).

(i) Let the system σ_0 be complete in $\mathcal{H}$. Since, according to Lemma 7, the characteristic function $\Delta_1(\lambda)$ is uniquely determined by its zeros, so is also $w_1(x)$ in (37). By virtue of (38), the function $w_0(x)$ coincides with $w_1(x)$ a.e. on $(\pi - a, \pi)$, i.e., it becomes known too.

By differentiating (36) $\nu = \overline{0, m_k - 1}$ times and substituting $\lambda = \lambda_{n_k,0}$ for $k \in \mathcal{S}$, we arrive at the relations

$$\beta_n = \int_0^{\pi - a} w_0(x) s_n(x)\,dx, \quad n \in \mathbb{N}, \tag{56}$$

where m_k, $\mathcal{S}$ and $s_n(x)$ were defined before Lemma 9 and

$$\beta_{k+\nu} = -(n_k + 1)\frac{d^\nu}{d\lambda^\nu}\left(\cos \rho\pi + \omega \frac{\sin \rho(\pi - a)}{\rho} + \gamma(\lambda) \right)\Big|_{\lambda=\lambda_{n_k,0}}, \quad k \in \mathcal{S}, \quad \nu = \overline{0, m_k - 1}, \tag{57}$$

$$\gamma(\lambda) = \int_{\pi-a}^{\pi} w_1(x) \frac{\sin \rho x}{\rho}\,dx. \tag{58}$$

Hence, by virtue of Lemma 9, the function $w_0(x)$ is determined uniquely also on $(0, \pi - a)$. Thus, it remains to recall representations (3) and (38), as well as to apply Lemma 10.

(ii) Assume that $q(x)$ is uniquely determined by $\{\lambda_{n_k,0}\}_{k\in\mathbb{N}}$ and $\{\lambda_{n,1}\}_{n\geq 0}$ and, to the contrary, that $\dim(\mathcal{H} \ominus \sigma_0) > 1$. Then, according to Lemma 9, we have $\dim(\mathcal{H} \ominus \sigma) > 1$, i.e., there exist at least two linearly independent functions $f_1(x), f_2(x) \in L_2(0, \pi - a)$ such that

$$\int_0^{\pi-a} f_\nu(x) s_n(x)\, dx = 0, \quad n \in \mathbb{N}, \quad \nu = 1, 2. \tag{59}$$

Let $\tilde{q}^+(x)$ be a solution of the integral equation

$$W(x) + \alpha_1 F_1(x) + \alpha_2 F_2(x) = \tilde{q}^+(x) + \int_x^\pi Q(x,t)\tilde{q}^+(t)\, dt, \quad a < x < \pi, \tag{60}$$

where $W(x)$ is defined in (51), while

$$F_\nu(x) := \begin{cases} 2f_\nu(\pi + a - 2x), & a < x < \dfrac{a+\pi}{2}, \\ -2f_\nu(2x - \pi - a), & \dfrac{a+\pi}{2} < x < \pi, \end{cases} \qquad \nu = 1, 2. \tag{61}$$

According to (53), we have $\tilde{q}^+(x) = q^+(x) + \alpha_1 g_1(x) + \alpha_2 g_2(x)$, where

$$g_\nu(x) = F_\nu(x) + \int_x^\pi Q_1(x,t) F_\nu(t)\, dt, \quad \nu = 1, 2,$$

while $Q_1(x,t)$ is the resolvent kernel for the kernel $Q(x,t)$. Choose α_1 and α_2 so that they do not vanish simultaneously and

$$\frac{1}{2}\int_a^\pi \tilde{q}^+(x)\, dx = \omega. \tag{62}$$

Since the functions $F_1(x)$ and $F_2(x)$ are linearly independent, so are $g_1(x)$ and $g_2(x)$. Hence, $\tilde{q}^+ \neq q^+$. Continue $\tilde{q}^+(x)$ to $(0, a)$ as zero and consider the function $\tilde{q}(x) = q^-(x) + \tilde{q}^+(x)$. By virtue of (62) and Lemma 5, the characteristic functions $\tilde{\Delta}_0(\lambda)$ and $\tilde{\Delta}_1(\lambda)$ of the problems $B_0(\tilde{q})$ and $B_1(\tilde{q})$, respectively, have the forms

$$\tilde{\Delta}_0(\lambda) = \cos\rho\pi + \omega\frac{\sin\rho(\pi - a)}{\rho} + \int_0^\pi \tilde{w}_0(x)\frac{\sin\rho x}{\rho}\, dx, \quad \tilde{w}_0(x) \in L_2(0,\pi),$$

$$\tilde{\Delta}_1(\lambda) = -\rho\sin\rho\pi + \omega\cos\rho(\pi - a) + \int_0^\pi \tilde{w}_1(x)\cos\rho x\, dx, \quad \tilde{w}_1(x) \in L_2(0,\pi),$$

and $\tilde{w}_j(x) = w_j(x)$ a.e. on $(\pi - a, \pi)$ for $j = 0, 1$. Moreover, analogously to (53), we have

$$\tilde{q}^+(x) + \int_x^\pi Q(x,t)\tilde{q}^+(t)\, dt = \begin{cases} 2(\tilde{w}_1 + \tilde{w}_0)(\pi + a - 2x), & a < x < \dfrac{a+\pi}{2}, \\ 2(\tilde{w}_1 - \tilde{w}_0)(2x - \pi - a), & \dfrac{a+\pi}{2} < x < \pi. \end{cases}$$

Comparing this with (51), (60), and (61), we obtain $\tilde{w}_1(x) = w_1(x)$ a.e. on $(0, \pi)$ and

$$\tilde{w}_0(x) = w_0(x) + \alpha_1 f_1(x) + \alpha_2 f_2(x) \ \text{ a.e. on } \ (0, \pi - a). \tag{63}$$

Hence, the spectra of $B_1(q)$ and $B_1(\tilde{q})$ coincide. Moreover, according to (56)–(59) and (63), the sequence $\{\lambda_{n_k,0}\}_{k\in\mathbb{N}}$ is a subsequence of zeros of $\tilde{\Delta}_0(\lambda)$. Hence, $\{\lambda_{n_k,0}\}_{k\in\mathbb{N}}$ is a subspectrum also of the problem $B_0(\tilde{q})$. Thus, we obtained another potential $\tilde{q} \neq q$ with the same spectral data $\{\lambda_{n_k,0}\}_{k\in\mathbb{N}}$ and $\{\lambda_{n,1}\}_{n\geq 0}$ as q has. This contradiction finishes the proof. □

Now, we are in a position to give a constructive procedure for solving Inverse Problem 1 (Algorithm 1).

Algorithm 1 Constructive procedure for solving Inverse Problem 1

Let the spectra $\{\lambda_{n,0}\}_{n\geq 0}$ and $\{\lambda_{n,1}\}_{n\geq 0}$ be given. Then:
(i) Construct the functions $\Delta_0(\lambda)$ and $\Delta_1(\lambda)$ by the formulae in (11);
(ii) Find the value ω by (54) or (55);
(iii) Calculate the functions $w_0(x)$ and $w_1(x)$ in (36) and (37) by inverting the corresponding Fourier transforms:

$$w_0(x) = \frac{2}{\pi}\sum_{n=1}^{\infty} a_n \sin nx, \quad w_1(x) = \frac{2}{\pi}\sum_{n=0}^{\infty} b_n \cos nx,$$

where

$$a_n = n(\Delta_0(n^2) - (-1)^n) + \omega(-1)^n \sin na, \quad n \geq 1, \quad b_n = \Delta_1(n^2) - \omega(-1)^n \cos na, \quad n \geq 0;$$

(iv) Find $q^-(x) \in L_2(0,a)$ by any relation in (38) and put $q^-(x) = 0$ for $x \in (a,\pi)$;
(v) Construct the functions $W(x)$ and $Q(x,t)$ by the formulae (51) and (52), respectively, and find $q^+(x) \in L_2(a,\pi)$ by solving the Volterra integral Equation (53);
(vi) Finally, construct $q(x) = q^-(x) + q^+(x)$, where $q^+(x) = 0$ on $(0,a)$.

This algorithm can be easily extended to Inverse Problem 2 if $\{\sin(n_k + 1/2)x\}_{k\in\mathbb{N}}$ is a Riesz basis in $L_2(0,\pi - a)$. Then, by virtue of Lemma 9, so is the system $\{s_n(x)\}_{n\in\mathbb{N}}$. Therefore, on step (iii), the function $\omega_0(x)$ can be constructed in accordance with (56) by the formula

$$w_0(x) = \sum_{n=1}^{\infty} \beta_n s_n^*(x), \quad 0 < x < \pi - a,$$

where the coefficients β_n are determined by relations (57) and (58), while $\{s_n^*(x)\}_{n\in\mathbb{N}}$ is the biorthogonal basis to the basis $\{s_n(x)\}_{n\in\mathbb{N}}$. It remains to note that, according to (38), the knowledge of $w_0(x)$ on $(\pi - a, \pi)$ is excessive since $w_1(x)$ has been found completely.

Proof of Theorem 3. Let us begin with the necessity part. According to (10), (36), and (37), we have

$$\theta_0(\rho) = \int_0^{\pi} w_0(x) \sin \rho x \, dx, \quad \theta_1(\rho) = \int_0^{\pi} w_1(x) \cos \rho x \, dx.$$

Hence, by virtue of (3) and (38), the exponential types of $\theta_0(\rho)$ and $\theta_1(\rho)$ do not exceed $\pi - a$. Finally, the relation (12) follows from Lemmas 5 and 7 after substituting $\lambda = 0$ into (37) and the second formula in (11). Indeed, according to (38) and (40), the assumption $q^- = 0$ implies

$$\int_0^{\pi} w_1(x)\, dx = \frac{1}{4}\int_a^{\pi} \left(q^+\left(\frac{a+x}{2}\right) + q^+\left(\pi + \frac{a-x}{2}\right)\right) dx = \frac{1}{2}\int_a^{\pi} q^+(x)\, dx = \omega. \quad (64)$$

For the sufficiency, we construct the functions $\Delta_0(\lambda)$ and $\Delta_1(\lambda)$ by the formulae in (11) using the given sequences $\{\lambda_{n,0}\}_{n\geq 0}$ and $\{\lambda_{n,1}\}_{n\geq 0}$. By virtue of Lemma 8, these functions have the forms (36) and (37), respectively, with some $w_0(x), w_1(x) \in L_2(0,\pi)$, which, in turn, vanish a.e. on $(\pi - a, \pi)$ by the first condition along with the Paley–Wiener theorem [37].

By virtue of Lemma 10, there exists a unique solution $q^+(x) \in L_2(a,\pi)$ of the system (39) and (40) with $q^-(x) = 0$. As in (64), we calculate

$$\tilde{\omega} := \int_0^{\pi - a} w_1(x)\, dx = \frac{1}{2}\int_a^{\pi} q^+(x)\, dx$$

and, hence,

$$\Delta_1(0) = \omega + \tilde{\omega}. \quad (65)$$

On the other hand, the second formula in (11) and condition (12) imply $\Delta_1(0) = 2\omega$, which, along with (65), gives $\tilde{\omega} = \omega$. Consider the problems $B_0(q)$ and $B_1(q)$ with the potential

$$q(x) = \begin{cases} 0, & x \in (0,a), \\ q^+(x), & x \in (a,\pi). \end{cases}$$

According to Lemma 5, $\Delta_0(\lambda)$ and $\Delta_1(\lambda)$ are their characteristic functions, respectively. Hence, $\{\lambda_{n,j}\}_{n\geq 0}$ is the spectrum of $B_j(q)$ for $j = 0, 1$. □

6. Conclusions and Discussing the Results

The paper thus connects two different directions in the inverse spectral theory, namely: for operators with constant delay [1–17] and for operators with a frozen argument [28–35], which have been developed independently before the present study. Such a fusion is naturally caused by replacing the standard assumption of the vanishing of the potential $q(x)$ on $(0,a)$ in equation (1) by imposing a continuously matching initial function (4). This leads to the appearance of a new term with a frozen argument at zero in Equation (5). Alternative forms of an initial function may give rise to considering also other equations with frozen argument

$$-y''(x) + q^+(x)y(x-a) + q^-(x)g(x-a)y^{(j)}(b) = \lambda y(x), \quad 0 < x < \pi, \tag{66}$$

where $g(x) \in L_\infty(0,a)$ and $b \in [0,\pi]$, while $j \in \{0,1\}$, or more general equations

$$-y''(x) + q^+(x)y(x-a) + q^-(x)Ly(x) = \lambda y(x), \quad 0 < x < \pi, \tag{67}$$

with some known linear operator $L : W_2^2[0,\pi] \to L_\infty(-a,0)$ under the reasonable assumption of the relative compactness with respect to the operator of double differentiation.

The usual restriction $q^-(x) = 0$ means that the two spectra must carry excessive information about the potential. For this reason, the reconstruction of $q(x)$ given only parts of the spectra was initiated in [6]. In particular, necessary and sufficient conditions for arbitrary subspectra guaranteeing the uniqueness of the potential were established. Later in [15], necessary and sufficient conditions for the solvability of the inverse problem from the complete spectra were obtained. Due to the overdetermination, these conditions besides the asymptotics also included some restrictions on the growth of certain entire functions constructed by the spectra.

Refusing the assumption $q^-(x) = 0$ would obviously lead to an increase of the required information for the unique recovery of $q(x)$. However, Theorem 2 shows that one of the spectra can still be specified partially. This effect is caused by the unique determination of the corresponding operator with the purely frozen argument, when $q^+(x) = 0$, from only one spectrum.

The proof of Theorem 2 gave Algorithm 1 for solving the inverse problem, which can be implemented numerically. We note that, in spite of the growing interest in recovering operators with constant delay, still no numerical results in this direction are known. For implementing Algorithm 1, one can adapt the numerical method suggested in [39] for integro-differential operators and involving approximations by entire functions of exponential type.

Author Contributions: Conceptualization, S.B. and S.V.; Methodology, S.B. and S.V.; Formal analysis, S.B. and S.V.; Investigation, S.B. and S.V.; Writing—original draft, S.B. and S.V.; Writing—review & editing, S.B. and S.V.; Supervision, S.B.; Project administration, S.B. All authors have read and agreed to the published version of the manuscript.

Funding: This research was supported by the Russian Science Foundation, Grant No. 22-21-00509, https://rscf.ru/project/22-21-00509/.

Data Availability Statement: The data presented in this study are openly available in arXiv at https://doi.org/10.48550/arXiv.2304.05487.

Acknowledgments: The authors are grateful to Maria Kuznetsova for reading the manuscript and making valuable comments as well as to anonymous referees for helpful remarks and recommendation to add Section 6.

Conflicts of Interest: The authors declare no conflict of interest.

References

1. Pikula M. Determination of a Sturm–Liouville-type differential operator with delay argument from two spectra. *Mat. Vestnik* **1991**, *43*, 159–171.
2. Freiling, G.; Yurko, V.A. Inverse problems for Sturm–Liouville differential operators with a constant delay. *Appl. Math. Lett.* **2012**, *25*, 1999–2004. [CrossRef]
3. Yang C.-F. Inverse nodal problems for the Sturm–Liouville operator with a constant delay. *J. Diff. Eqns.* **2014**, *257*, 1288–1306. [CrossRef]
4. Ignatiev M.Y. On an inverse Regge problem for the Sturm–Liouville operator with deviating argument. *J. Samara State Tech. Univ. Ser. Phys. Math. Sci.* **2018**, *22*, 203–211. [CrossRef]
5. Bondarenko, N.; Yurko, V. An inverse problem for Sturm–Liouville differential operators with deviating argument. *Appl. Math. Lett.* **2018**, *83*, 140–144. [CrossRef]
6. Buterin, S.A.; Yurko, V.A. An inverse spectral problem for Sturm–Liouville operators with a large delay. *Anal. Math. Phys.* **2019**, *9*, 17–27. [CrossRef]
7. Pikula, M.; Vladičić, V.; Vojvodić, B. Inverse spectral problems for Sturm–Liouville operators with a constant delay less than half the length of the interval and Robin boundary conditions. *Results Math.* **2019**, *74*, 45. [CrossRef]
8. Djuric, N.; Vladicic, V. Incomplete inverse problem for Sturm–Liouville type differential equation with constant delay. *Results Math.* **2019**, *74*, 161. [CrossRef]
9. Sat, M.; Shieh, C.-T. Inverse nodal problems for integro-differential operators with a constant delay. *J. Inverse Ill-Posed Probl.* **2019**, *27*, 501–509. [CrossRef]
10. Wang, Y.P.; Shieh, C.T.; Miao, H.Y. Reconstruction for Sturm–Liouville equations with a constant delay with twin-dense nodal subsets. *Inv. Probl. Sci. Eng.* **2019**, *27*, 608–617. [CrossRef]
11. Djurić N. Inverse problems for Sturm–Liouville-type operators with delay: Symmetric case. *Appl. Math. Sci.* **2020**, *14*, 505–510. [CrossRef]
12. Djurić, N.; Buterin, S. On an open question in recovering Sturm–Liouville-type operators with delay. *Appl. Math. Lett.* **2021**, *113*, 106862. [CrossRef]
13. Djurić, N.; Buterin, S. On non-uniqueness of recovering Sturm–Liouville operators with delay. Commun. *Nonlinear Sci. Numer.* Simulat. **2021**, *102*, 105900. [CrossRef]
14. Djurić, N.; Buterin, S. Iso-bispectral potentials for Sturm–Liouville-type operators with small delay. *Nonlin. Anal. Real World Appl.* **2022**, *63*, 103390. [CrossRef]
15. Buterin, S.A.; Malyugina, M.A.; Shieh, C.-T. An inverse spectral problem for second-order functional-differential pencils with two delays. *Appl. Math. Comput.* **2021**, *411*, 126475. [CrossRef]
16. Pikula, M.; Nedić, D.; Kal, co, I.; Kvesić, L. Inverse spectral boundary problem Sturm Liouville type with constant delay and non-zero initial function. *Math. Montisnigri* **2021**, *51*, 18–30. [CrossRef]
17. Wang, Y.P.; Keskin, B.; Shieh, C.-T. A partial inverse problem for non-self-adjoint Sturm–Liouville operators with a constant delay. *J. Inverse Ill-Posed Probl.* **2023**, *31*, 479–486. [CrossRef]
18. Borg G. Eine Umkehrung der Sturm–Liouvilleschen Eigenwertaufgabe. *Acta Math.* **1946**, *78*, 1–96. [CrossRef]
19. Freiling, G.; Yurko, V.A. *Inverse Sturm–Liouville Problems and Their Applications*; NOVA Science Publishers: New York, NY, USA, 2001.
20. Myshkis A.D. *Linear Differential Equations with a Delay Argument*; Nauka: Moscow, Russia, 1951.
21. Bellman, R.; Cooke, K.L. *Differential-Difference Equations*; R-374-PR; The RAND Corp.: Santa Monica, CA, USA , 1963.
22. Norkin S.B. *Second Order Differential Equations with a Delay Argument*; Nauka: Moscow, Russia, 1965.
23. Hale J. *Theory of Functional-Differential Equations*; Springer: New York, NY, USA, 1977.
24. Skubachevskii A.L. *Elliptic Functional Differential Equations and Applications*; Birkhäuser: Basel, Switzerland, 1997.
25. Azbelev, N.V.; Maksimov, V.P.; Rakhmatullina, L.F. *Introduction to the Theory of Functional Differential Equations: Methods and Applications*; Hindawi: New York, NY, USA, 2007.
26. Muravnik A.B. Nonlocal problems and functional-differential equations: Theoretical aspects and applications to mathematical modelling. *Math. Model. Nat. Phenom.* **2019**, *14*, 601. [CrossRef]
27. Kato T. *Perturbation Theory for Linear Operators*; Springer: Berlin/Heidelberg, Germany, 1980.
28. Bondarenko, N.P.; Buterin, S.A.; Vasiliev, S.V. An inverse spectral problem for Sturm–Liouville operators with frozen argument. *J. Math. Anal. Appl.* **2019**, *472*, 1028–1041. [CrossRef]
29. Buterin, S.; Kuznetsova, M. On the inverse problem for Sturm–Liouville-type operators with frozen argument: Rational case. *Comp. Appl. Math.* **2020**, *39*, 5. [CrossRef]
30. Wang, Y.P.; Zhang, M.; Zhao, W.; Wei, X. Reconstruction for Sturm–Liouville operators with frozen argument for irrational cases. *Appl. Math. Lett.* **2021**, *111*, 106590. [CrossRef]

31. Bondarenko N. Finite-difference approximation of the inverse Sturm–Liouville problem with frozen argument. *Appl. Math. Comput.* **2022**, *413*, 126653. [CrossRef]
32. Tsai, T.-M.; Liu, H.-F.; Buterin, S.; Chen, L.-H.; Shieh, C.-T. Sturm–Liouville-type operators with frozen argument and Chebyshev polynomials. *Math. Meth. Appl. Sci.* **2022**, *45*, 9635–9652. [CrossRef]
33. Dobosevych, O.; Hryniv, R. Reconstruction of differential operators with frozen argument. *Axioms* **2022**, *11*, 24. [CrossRef]
34. Kuznetsova M.A. Necessary and sufficient conditions for the spectra of the Sturm–Liouville operators with frozen argument. *Appl. Math. Lett.* **2022**, *131*, 108035. [CrossRef]
35. Kuznetsova M.A. Uniform stability of recovering Sturm-Liouville-type operators with frozen argument. *Results Math.* **2023**, *78*, 169. [CrossRef]
36. Buterin, S.A. On the uniform stability of recovering sine-type functions with asymptotically separated zeros. *Math. Notes* **2022**, *111*, 343–355. [CrossRef]
37. Levin B.Y. *Lectures on Entire Functions*; AMS: Providence, RI, USA, 1996.
38. Tricomi F.G. *Integral Equations*; Interscience Publishers, Inc.: New York, NY, USA, 1957.
39. Bondarenko, N.; Buterin, S. Numerical solution and stability of the inverse spectral problem for a convolution integro-differential operator. *Commun. Nonlinear Sci. Numer. Simulat.* **2020**, *89*, 105298. [CrossRef]

mathematics

MDPI

Article

A Class of Quasilinear Equations with Distributed Gerasimov–Caputo Derivatives

Vladimir E. Fedorov [1,2,*] and Nikolay V. Filin [1,2]

1 N.N. Krasovskii Institute of Mathematics and Mechanics of the Ural Branch of the Russian Academy of Sciences, 16, S.Kovalevskaya St., 620108 Yekaterinburg, Russia; filinnv@csu.ru
2 Mathematical Analysis Department, Mathematics Faculty, Chelyabinsk State University, 129, Kashirin Brothers St., 454001 Chelyabinsk, Russia
* Correspondence: kar@csu.ru; Tel.: +7-351-799-7235

Abstract: Quasilinear equations in Banach spaces with distributed Gerasimov–Caputo fractional derivatives, which are defined by the Riemann–Stieltjes integrals, and with a linear closed operator A, are studied. The issues of unique solvability of the Cauchy problem to such equations are considered. Under the Lipschitz continuity condition in phase variables and two types of continuity over all variables of a nonlinear operator in the equation, we obtain two versions on a theorem on the nonlocal existence of a unique solution. Two similar versions of local unique solvability of the Cauchy problem are proved under the local Lipschitz continuity condition for the nonlinear operator. The general results are used for the study of an initial boundary value problem for a generalization of the nonlinear phase field system of equations with distributed derivatives with respect to time.

Keywords: distributed fractional derivative; fractional differential equation; Cauchy problem; quasiliner equation; fixed point theorem; initial boundary value problem

MSC: 34G20; 35R11; 34A08; 47D99

Citation: Fedorov, V.E.; Filin, N.V. A Class of Quasilinear Equations with Distributed Gerasimov–Caputo Derivatives. *Mathematics* **2023**, *11*, 2472. https://doi.org/10.3390/math11112472

Academic Editor: Natalia Bondarenko

Received: 18 April 2023
Revised: 22 May 2023
Accepted: 23 May 2023
Published: 27 May 2023

1. Introduction

Various classes of fractional differential equations are the subjects of intensive research by many scientists in recent decades. Such equations are of interest both because of their increasing importance in applied investigations [1–4], and from the point of view of the development of theory [5–8]. A special class consists of equations with distributed derivatives (or so-called continual derivatives, mean derivatives), which, in partial, are applied to the research of some real phenomena and processes, when an order of a fractional derivative in a model depends on the process parameters: in the theory of viscoelastic media [9], in modeling dielectric induction and diffusion [10,11], in the kinetic theory [12], and in other scientific fields [13–16]. These works initiated other investigations of the equations with distributed derivatives from the point of view of the qualitative theory of differential equations [17–22].

The main aim of the present work is to investigate the Cauchy problem for a class of abstract quasilinear equations with distributed derivatives. Let $\mathcal{Z}$ be a Banach space, D^β be the fractional Gerasimov–Caputo derivative for $\beta > 0$ and the fractional Riemann–Liouville integral for $\beta \leq 0$, A be a linear closed densely defined in $\mathcal{Z}$ operator. Consider the Cauchy problem

$$D^k z(t_0) = z_k, \quad k = 0, 1, \dots, m-1, \tag{1}$$

for the quasilinear equation

$$\int_b^c D^\alpha z(t)d\mu(\alpha) = Az(t) + B\left(t, \int_{b_1}^{c_1} D^\alpha z(t)d\mu_1(\alpha), \ldots, \int_{b_n}^{c_n} D^\alpha z(t)d\mu_n(\alpha)\right), \tag{2}$$

where $b < c$, $m-1 < c \le m \in \mathbb{N}$, $\mu \in BV((b,c];\mathbb{C})$ (i.e., μ is a function of a bounded variation), c is a variation point of the measure $d\mu(\alpha)$, $b_l < c_l$, $c_1 \le c_2 \le \cdots \le c_n < c$, $\mu_l \in BV((b_l,c_l];\mathbb{C})$, c_l is a variation point of the measure $d\mu_l(\alpha)$, $l = 1,2,\ldots,n$. Equality (2) contains the Riemann–Stieltjes integrals.

Linear equations with a distributed order derivative

$$\int_b^c \omega(\alpha)D^\alpha z(t)d\alpha = Az(t) \tag{3}$$

were studied in works [23–25], where $\omega : (b,c) \to \mathbb{C}$ and A is a bounded operator, or a generator of an analytic resolving family of a fractional equation. For $b = 0$, $c \in (0,1]$, a criteria in terms of conditions on a linear closed operator A for the existence of an analytic resolving family of operators for Equation (3) were obtained in paper [26]. In the work [27], these criteria were generalized to the case $c > 1$ and a perturbations theorem on generators of analytic resolving operators families for (3) was obtained. Analogous results for the equation with a discretely distributed Gerasimov–Caputo derivative

$$\sum_{k=1}^n \omega_k D^{\alpha_k} z(t) = Az(t).$$

were obtained in [28]. All these results were generalized and combined in general formulations with the Riemann–Stieltjes integral in the definition of the distributed derivative [29]. Recall that an arbitrary function μ with a bounded variation has the form $\mu = \mu_c + \mu_d$, where μ_c is a continuous function with a bounded variation, and μ_d is a jumps function. Consequently, the left-hand side of (2) has the form

$$\int_b^c D^\alpha z(t)d\mu(\alpha) = \int_b^c \mu_c'(\alpha)D^\alpha z(t)d\alpha + \sum_{k=1}^n \omega_k D^{\alpha_k} z(t),$$

if there exists an appropriate derivative μ_c', α_k are points of jumps of the function μ_d, ω_k are values of jumps, $k = 1,2,\ldots,n$.

Each result in the listed works [23–29] on the linear homogeneous equation is accompanied by theorems on the solvability of the corresponding linear inhomogeneous equation. Here, such theorems are used for the study of the Cauchy problem (1) to quasilinear Equation (2). Note that the above-mentioned papers concern equations with distributed order derivatives in finite-dimensional spaces, or in the linear case, or with a bounded operator A in a Banach space (see [25]). In the present paper, we have studied for the first time a quasilinear equation with distributed derivatives and an unbounded A operator in an infinite-dimensional space.

In the second section of the present work, the main definitions and results on the solvability of the inhomogeneous equation are formulated. The third section contains the definition of special functional spaces, statements and proofs of their properties and properties of operators of distributed Gerasimov–Caputo fractional derivatives, which are acting in these spaces. In the fourth section, theorems on nonlocal solvability of Cauchy problem (1) and (2) are proved under the condition $B \in C([t_0,T] \times \mathcal{Z}^n; D_A)$, where D_A is the domain of A with its graph norm, or with $B \in C([t_0,T] \times \mathcal{Z}^n; \mathcal{Z})$, but in a slightly narrower functional space. In the fifth section, analogous results were obtained on the local unique solvability of problem (1) and (2) with $B \in C(U; D_A)$, or $B \in C(U; \mathcal{Z})$, where U is an open set in $\mathbb{R} \times \mathcal{Z}^n$. The last section contains an application of abstract results to the research of an initial boundary value problem for some generalization of the phase field system of equations with the distributed order Gerasimov–Caputo time-derivatives.

2. Linear Equation and Resolving Families

Let $\mathcal{Z}$ be a Banach space, denote for $\beta > 0$, $h : (t_0, \infty) \to \mathcal{Z}$ the Riemann–Liouville fractional integral of an order $\beta > 0$

$$J^\beta h(t) := \frac{1}{\Gamma(\beta)} \int_{t_0}^{t} (t-s)^{\beta-1} h(s)ds, \quad t > t_0.$$

Let $m - 1 < \alpha \leq m \in \mathbb{N}$, D^m be the derivative of the m-th order, then

$$D^\alpha h(t) := D^m J^{m-\alpha} \left(h(t) - \sum_{k=0}^{m-1} h^{(k)}(t_0) \frac{(t-t_0)^k}{k!} \right)$$

is the Gerasimov–Caputo derivative of the order α [1,2,30]. It will be assumed that $D^\alpha h(t) := J^{-\alpha} h(t)$ for $\alpha < 0$.

For a function $h : \mathbb{R}_+ \to \mathcal{Z}$, the Laplace transform is denoted by $\widehat{h}$ or $\mathrm{Lap}[h]$, if the expression for h is too large. For the Gerasimov–Caputo derivative of an order $\alpha \in (m-1, m]$, it is known the equality (see, e.g., [6])

$$\widehat{D^\alpha h}(\lambda) = \lambda^\alpha \widehat{h}(\lambda) - \sum_{k=0}^{m-1} h^{(k)}(0) \lambda^{\alpha-1-k}. \tag{4}$$

The notations $S_{\theta,a} := \{\mu \in \mathbb{C} : |\arg(\mu - a)| < \theta, \, \mu \neq a\}$ for $\theta \in [\pi/2, \pi]$, $a \in \mathbb{R}$, $\Sigma_\psi := \{t \in \mathbb{C} : |\arg t| < \psi, t \neq 0\}$ for $\psi \in (0, \pi/2]$ will be used later. Besides, the Banach space of all linear continuous operators from $\mathcal{Z}$ to $\mathcal{Z}$ will be denoted by $\mathcal{L}(\mathcal{Z})$, and denote the set of all linear closed operators, densely defined in $\mathcal{Z}$, acting in the space $\mathcal{Z}$, by $\mathcal{C}l(\mathcal{Z})$. The domain D_A of an operator $A \in \mathcal{C}l(\mathcal{Z})$ endows by its graph norm $\|\cdot\|_{D_A} := \|\cdot\|_{\mathcal{Z}} + \|A\cdot\|_{\mathcal{Z}}$. Hence, D_A is a Banach space with this norm due to the closedness of A.

Consider the Cauchy problem

$$D^k z(0) = z_k, \quad k = 0, 1, \ldots, m-1, \tag{5}$$

for the distributed order equation

$$\int_b^c D^\alpha z(t) d\mu(\alpha) = Az(t), \ t > 0, \tag{6}$$

where $b, c \in \mathbb{R}$, $b < c$, $m - 1 < c \leq m \in \mathbb{N}$, $\mu : (b, c] \to \mathbb{C}$ is a function with a bounded variation, briefly $\mu \in BV((b,c];\mathbb{C})$, c is a variation point of the measure $d\mu(\alpha)$. Equality (6) contains the Riemann–Stieltjes integral. A solution of problem (5) and (6) is a function $z \in C^{m-1}(\overline{\mathbb{R}}_+; \mathcal{Z}) \cap C(\mathbb{R}_+; D_A)$, such that $\int_b^c D^\alpha z(t) d\mu(\alpha) \in C(\mathbb{R}_+; \mathcal{Z})$ and equalities (5) and (6) for $t \in \mathbb{R}_+$ are fulfilled. Hereafter, $\overline{\mathbb{R}}_+ := \mathbb{R}_+ \cup \{0\}$.

Under the conditions of this section, consider the analytic on $S_{\pi,0}$ functions

$$W(\lambda) := \int_b^c \lambda^\alpha d\mu(\alpha) \quad W_k(\lambda) := \int_k^c \lambda^\alpha d\mu(\alpha), \ k = 0, 1, \ldots, m-1,$$

also defined by Riemann–Stieltjes integrals. Here and further, the main branch of the power function is considered.

Lemma 1 ([29]). *Let $b, c \in \mathbb{R}$, $b < c$, $m-1 < c \le m \in \mathbb{N}$, $\mu \in BV((b,c];\mathbb{C})$, c be a variation point of the measure $d\mu(\alpha)$. Then for $k, l = 0, 1, \dots, m-1$, $k > l$,*

$$\forall \varepsilon \in (0,c) \quad \exists C, \varrho > 0 \quad \forall \lambda \in S_{\pi,0} \setminus \{\lambda \in \mathbb{C} : |\lambda| < \varrho\} \quad |W_k(\lambda)| \ge C|\lambda|^{c-\varepsilon};$$

$$\forall \varepsilon \in (0,c) \quad \exists C, \varrho > 0 \quad \forall \lambda \in S_{\pi,0} \setminus \{\lambda \in \mathbb{C} : |\lambda| < \varrho\} \quad |W(\lambda)| \ge C|\lambda|^{c-\varepsilon};$$

$$\exists C, \varrho > 0 \quad \forall \lambda \in S_{\pi,0} \setminus \{\lambda \in \mathbb{C} : |\lambda| < \varrho\} \quad |W_k(\lambda) - W_l(\lambda)| \le C|\lambda|^{k};$$

$$\exists C, \varrho > 0 \quad \forall \lambda \in S_{\pi,0} \setminus \{\lambda \in \mathbb{C} : |\lambda| < \varrho\} \quad |W_k(\lambda) - W(\lambda)| \le C|\lambda|^{k}.$$

Definition 1 ([29]). *A family of operators $\{S_k(t) \in \mathcal{L}(\mathcal{Z}) : t \ge 0\}$, $k \in \{0, 1, \dots, m-1\}$, is called k-resolving for Equation* (6)*, if:*

(i) *$S_k(t)$ is strongly continuous for $t \ge 0$;*
(ii) *$S_k(t)[D_A] \subset D_A$, $S_k(t)Az = AS_k(t)z$ for all $z \in D_A$, $t \ge 0$;*
(iii) *$S_k(t)z_k$ is a solution of the Cauchy problem*

$$D^l z(0) = 0, \quad l \in \{0, 1, \dots, m-1\} \setminus \{k\}, \quad D^k z(0) = z_k \tag{7}$$

to Equation (6) *for any $z_k \in D_A$.*

Remark 1. *Thus, a k-resolving family $\{S_k(t) \in \mathcal{L}(\mathcal{Z}) : t \ge 0\}$ consists of operators, such that $S_k(t)$ for $t \ge 0$ maps arbitrary $z_k \in D_A$ into the value $z(t) = S_k(t)z_k$ at the point t of a solution of Cauchy problem* (6) *and* (7)*. Therefore, the families $\{S_k(t) \in \mathcal{L}(\mathcal{Z}) : t \ge 0\}$, $k = 0, 1, \dots, m-1$, entirely describe the solution of the complete Cauchy problem* (5) *and* (6)*.*

A resolving family of operators is called *analytic* if it has an analytic continuation to a sector Σ_{ψ_0} for some $\psi_0 \in (0, \pi/2]$. An analytic resolving family of operators $\{S(t) \in \mathcal{L}(\mathcal{Z}) : t \ge 0\}$ has a type (ψ_0, a_0) for some $\psi_0 \in (0, \pi/2]$, $a_0 \in \mathbb{R}$, if for any $\psi \in (0, \psi_0)$, $a > a_0$ there exists $C(\psi, a) > 0$, such that for every $t \in \Sigma_\psi$ the inequality $\|S(t)\|_{\mathcal{L}(\mathcal{Z})} \le C(\psi, a)e^{a\mathrm{Re}\, t}$ holds.

Remark 2. *Similar notions of analytic resolving families of operators are used in the study of integral evolution equations [31] and fractional differential equations [32]. They generalize the notion of an analytic resolving semigroup of operators for the first order equation $D_t^1 z(t) = Az(t)$ (see [33–35]).*

Denote $\rho(A) := \{\lambda \in \mathbb{C} : (\lambda I - A)^{-1} \in \mathcal{L}(\mathcal{Z})\}$ for an operator $A \in \mathcal{C}l(\mathcal{Z})$, i.e., $\rho(A)$ is the resolvent set of A. Define a class $\mathcal{A}_W(\theta_0, a_0)$ (see [29]) of all operators $A \in \mathcal{C}l(\mathcal{Z})$, such that:

(i) there exist $\theta_0 \in (\pi/2, \pi]$, $a_0 \ge 0$, such that $W(\lambda) \in \rho(A)$ for every $\lambda \in S_{\theta_0, a_0}$;
(ii) for every $\theta \in (\pi/2, \theta_0)$, $a > a_0$ there exists $K(\theta, a) > 0$, such that for all $\lambda \in S_{\theta,a}$

$$\|(W(\lambda)I - A)^{-1}\|_{\mathcal{L}(\mathcal{Z})} \le \frac{|\lambda|K(\theta, a)}{|W(\lambda)||\lambda - a|}.$$

Remark 3. *The classes $\mathcal{A}_W(\theta_0, a_0)$ in works [26–28] are partial cases of this class with the same denotation $\mathcal{A}_W(\theta_0, a_0)$ due to the more general construction of the distributed derivative in the present work. If μ is a constant, excluding a unique jump in the point $\alpha = c$, class $\mathcal{A}_W(\theta_0, a_0)$ coincides with the class $\mathcal{A}_c(\theta_0, a_0)$, defined in [32]. For $c = 1$, this class contains generators of analytic operator semigroups [33–35].*

Remark 4. *If $A \in \mathcal{L}(\mathcal{Z})$, then $A \in \mathcal{A}_W(\theta_0, a_0)$ for some $\theta_0 \in (\pi/2, \pi)$, $a_0 \ge 0$ (see [29]).*

For an operator $A \in \mathcal{A}_W(\theta_0, a_0)$, the operators

$$Z_k(t) := \frac{1}{2\pi i} \int_\Gamma \frac{W_k(\lambda)}{\lambda^{k+1}} (W(\lambda)I - A)^{-1} e^{\lambda t} d\lambda, \quad k = 0, 1, \dots, m-1,$$

are defined for $t > 0$, where $\Gamma = \Gamma_+ \cup \Gamma_- \cup \Gamma_0$, $\Gamma_\pm = \{\mu \in \mathbb{C} : \mu = a + re^{\pm i\theta}, \ r \in (\delta, \infty)\}$, $\Gamma_0 = \{\mu \in \mathbb{C} : \mu = a + \delta e^{i\varphi}, \ \varphi \in (-\theta, \theta)\}$ for some $\delta > 0$, $a > a_0$, $\theta \in (\pi/2, \theta_0)$.

Theorem 1 ([29]). *Let $b, c \in \mathbb{R}$, $b < c$, $m-1 < c \leq m \in \mathbb{N}$, $\mu \in BV((b,c];\mathbb{C})$, c be a variation point of the measure $d\mu(\alpha)$. Then, there exists an analytic 0-resolving family of operators of the type $(\theta_0 - \pi/2, a_0)$ for Equation (6), if and only if $A \in \mathcal{A}_W(\theta_0, a_0)$. In this case, there exists a unique k-resolving family of operators for every $k = 0, 1, \dots, m-1$, and it has the form $\{Z_k(t) \in \mathcal{L}(\mathcal{Z}) : t \geq 0\}$.*

Remark 5. *The theorem shows that the condition $A \in \mathcal{A}_W(\theta_0, a_0)$ is not only sufficient, but also necessary for the analytic resolving families existence, in other words, for the unique solvability of problem* (5) *and* (6) *in the considered sense.*

Theorem 2 ([29]). *Let $b, c \in \mathbb{R}$, $b < c$, $2 < c$, $\mu \in BV((b,c];\mathbb{C})$, c be a variation point of the measure $d\mu(\alpha)$, $\mu(\alpha) \in \mathbb{R}$ for all α from some left neighborhood of c, $A \in \mathcal{A}_W(\theta_0, a_0)$ for some $\theta_0 \in (\pi/2, \pi)$, $a_0 \geq 0$. Then, $A \in \mathcal{L}(\mathcal{Z})$.*

Denote for $t > 0$

$$Z(t) := \frac{1}{2\pi i} \int_\Gamma e^{\lambda t} (W(\lambda)I - A)^{-1} d\lambda.$$

Recall that $C^\gamma([0,T];\mathcal{Z})$ with $\gamma \in (0,1]$ is the class of functions $f : [0,T] \to \mathcal{Z}$, such that for all $t, s \in [0,T]$ the Hölder condition $\|f(t) - f(s)\|_\mathcal{Z} \leq C|t-s|^\gamma$ is satisfied with some $C > 0$.

Theorem 3 ([29]). *Let $b, c \in \mathbb{R}$, $b < c$, $m-1 < c \leq m \in \mathbb{N}$, $\mu \in BV((b,c];\mathbb{C})$, c be a variation point of the measure $d\mu(\alpha)$, $\theta_0 \in (\pi/2, \pi]$, $a_0 \geq 0$, $A \in \mathcal{A}_W(\theta_0, a_0)$, $g \in C([0,T];D_A) \cup C^\gamma([0,T];\mathcal{Z})$, $\gamma \in (0,1]$, $z_k \in D_A$, $k = 0, 1, \dots, m-1$. Then, the function*

$$z(t) = \sum_{k=0}^{m-1} Z_k(t) z_k + \int_0^t Z(t-s) g(s) ds$$

is a unique solution of Cauchy problem (5) *for the equation*

$$\int_b^c D^\alpha z(t) d\mu(\alpha) = Az(t) + f(t).$$

3. Some Properties of Distributed Derivatives

For $t_0, T, \beta \in \mathbb{R}$, $t_0 < T$, denote the space $C^{m-1,\beta}([t_0,T];\mathcal{Z}) := \{z \in C^{m-1}([t_0,T];\mathcal{Z}) : D^\beta z \in C([t_0,T];\mathcal{Z})\}$ with the norm

$$\|z\|_{C^{m-1,\beta}([t_0,T];\mathcal{Z})} := \|z\|_{C^{m-1}([t_0,T];\mathcal{Z})} + \|D^\beta z\|_{C([t_0,T];\mathcal{Z})}.$$

It is evident, that $C^{m-1,\beta}([t_0,T];\mathcal{Z}) = C^{m-1}([t_0,T];\mathcal{Z})$, if and only if $\beta \leq m-1$. It can be proved directly that even for $\beta > m-1$ the space $C^{m-1,\beta}([t_0,T];\mathcal{Z})$ is complete.

Lemma 2. *Let $m-1<\beta\le m\in\mathbb{N}$, $z\in C^{m-1,\beta}([t_0,T];\mathcal{Z})$. Then, for every $\alpha\in[0,\beta]$ $D^\alpha z\in C([t_0,T];\mathcal{Z})$, moreover, there exists $C>0$, such that for all $\alpha\in[0,\beta]$, $z\in C^{m-1,\beta}([t_0,T];\mathcal{Z})$*

$$\|D^\alpha z\|_{C([t_0,T];\mathcal{Z})}\le C\|z\|_{C^{m-1,\beta}([t_0,T];\mathcal{Z})}.$$

Proof. If $m-1<\alpha<\beta<m\in\mathbb{N}$, we have for $y\in C^m([t_0,T];\mathcal{Z})$

$$D^m J^{\beta-\alpha}y(t)=D^m\int_0^{t-t_0}\frac{s^{\beta-\alpha-1}}{\Gamma(\beta-\alpha)}y(t-s)ds=$$

$$=D^{m-1}\left(\frac{(t-t_0)^{\beta-\alpha-1}}{\Gamma(\beta-\alpha)}y(t_0)+\int_0^{t-t_0}\frac{s^{\beta-\alpha-1}}{\Gamma(\beta-\alpha)}D^1y(t-s)ds\right)=$$

$$=D^{m-2}\left(\frac{(t-t_0)^{\beta-\alpha-2}}{\Gamma(\beta-\alpha-1)}y(t_0)+\frac{(t-t_0)^{\beta-\alpha-1}}{\Gamma(\beta-\alpha)}D^1y(t_0)+\int_0^{t-t_0}\frac{s^{\beta-\alpha-1}}{\Gamma(\beta-\alpha)}D^2y(t-s)ds\right)=$$

$$=\dots=\sum_{l=0}^{m-1}\frac{(t-t_0)^{\beta-\alpha-m+l}}{\Gamma(\beta-\alpha-m+l+1)}D^ly(t_0)+\int_{t_0}^{t}\frac{(t-s)^{\beta-\alpha-1}}{\Gamma(\beta-\alpha)}D^my(s)ds=$$

$$=\sum_{l=0}^{m-1}\frac{(t-t_0)^{\beta-\alpha-m+l}}{\Gamma(\beta-\alpha-m+l+1)}D^ly(t_0)+J^{\beta-\alpha}D^my(t). \tag{8}$$

Therefore, for $z\in C^{m-1,\beta}([t_0,T];\mathcal{Z})$

$$D^\alpha z(t)=D^mJ^{\beta-\alpha}J^{m-\beta}\left(z(t)-\sum_{k=0}^{m-1}D^kz(t_0)\frac{(t-t_0)^k}{k!}\right)=$$

$$=\sum_{l=0}^{m-1}\frac{(t-t_0)^{\beta-\alpha-m+l}}{\Gamma(\beta-\alpha-m+l+1)}D^ly(t_0)+J^{\beta-\alpha}D^my(t)=J^{\beta-\alpha}D^\beta z(t),$$

where

$$y(t)=J^{m-\beta}\left(z(t)-\sum_{k=0}^{m-1}D^kz(t_0)\frac{(t-t_0)^k}{k!}\right)\in C^m([t_0,T];\mathcal{Z}),$$

$$D^ly(t_0)=0,\ l=0,1,\dots,m-1.$$

Hence, $D^\alpha z\in C([t_0,T];\mathcal{Z})$ and

$$\|D^\alpha z\|_{C([t_0,T];\mathcal{Z})}\le\frac{(T-t_0)^{\beta-\alpha}}{\Gamma(\beta-\alpha+1)}\|D^\beta z\|_{C([t_0,T];\mathcal{Z})}\le\max_{s\in[0,\beta]}\frac{(T-t_0)^s}{\Gamma(s+1)}\|D^\beta z\|_{C([t_0,T];\mathcal{Z})}.$$

Let $m-1<\alpha<m=\beta$, then $D^\alpha z(t)=J^{m-\alpha}D^mz(t)$, and we have the same result.

In the case $n-1<\alpha<n\le m-1<\beta<m$, for $z\in C^{m-1,\beta}([t_0,T];\mathcal{Z})$ we can obtain similarly

$$D^\alpha z(t)=D^nJ^{n-\alpha}\left(z(t)-\sum_{k=0}^{n-1}D^kz(t_0)\frac{(t-t_0)^k}{k!}\right)=$$

$$=D^mJ^{m-n}J^{n-\alpha}\left(z(t)-\sum_{k=0}^{n-1}D^kz(t_0)\frac{(t-t_0)^k}{k!}\right)=$$

$$=D^mJ^{\beta-\alpha}J^{m-\beta}\left(z(t)-\sum_{k=0}^{n-1}D^kz(t_0)\frac{(t-t_0)^k}{k!}\right)=J^{\beta-\alpha}D^\beta z(t)$$

due to (8), since for $k = 0, 1, \ldots, m-1$

$$\|D^k J^{m-\beta} z(t)\|_{\mathcal{Z}} \leq \frac{(t-t_0)^{\beta-m}}{\Gamma(\beta-m+1)} \|z\|_{C^{m-1}([t_0,t];\mathcal{Z})} \to 0, \quad t \to t_0.$$

Consequently,

$$\|D^\alpha z\|_{C([t_0,T];\mathcal{Z})} \leq \max_{s\in[0,\beta]} \frac{(T-t_0)^s}{\Gamma(s+1)} \|D^\beta z\|_{C([t_0,T];\mathcal{Z})}.$$

If $\beta = m$ here, then

$$D^\alpha z(t) = D^m J^{m-\alpha}\left(z(t) - \sum_{k=0}^{n-1} D^k z(t_0)\frac{(t-t_0)^k}{k!}\right) =$$

$$= \sum_{l=n}^{m-1} \frac{(t-t_0)^{l-\alpha}}{\Gamma(l-\alpha+1)} D^l z(t_0) + J^{m-\alpha} D^m y(t)$$

and

$$\|D^\alpha z\|_{C([t_0,T];\mathcal{Z})} \leq m \max_{s\in[0,\beta]} \frac{(T-t_0)^s}{\Gamma(s+1)} \|z\|_{C^{m-1,\beta}([t_0,T];\mathcal{Z})}.$$

Finally, in the case $\alpha \in \{0, 1, \ldots, m-1\}$, we have the estimate $\|D^\alpha z\|_{C([t_0,T];\mathcal{Z})} \leq \|z\|_{C^{m-1}([t_0,T];\mathcal{Z})}$. □

Corollary 1. *Let $z \in C^{m-1}([t_0, T]; \mathcal{Z})$. Then, $D^\alpha z \in C([t_0, T]; \mathcal{Z})$ for all $\alpha \in [0, m-1]$, besides, there exists $C > 0$, such that for all $\alpha \in [0, m-1]$, $z \in C^{m-1}([t_0, T]; \mathcal{Z})$*

$$\|D^\alpha z\|_{C([t_0,T];\mathcal{Z})} \leq C\|z\|_{C^{m-1}([t_0,T];\mathcal{Z})}.$$

Proof. Take $\beta = m-1$ in the proof of Lemma 2. □

Remark 6. *If $z \in C([t_0, T]; \mathcal{Z})$ and $\alpha < 0$, then it is obvious that $D^\alpha z \in C([t_0, T]; \mathcal{Z})$.*

Corollary 2. *Let $m-1 < c \leq m \in \mathbb{N}$, $b < c$, $\mu \in BV((b, c]; \mathbb{C})$, c be a variation point of the measure $d\mu(\alpha)$, $z \in C^{m-1,c}([t_0, T]; \mathcal{Z})$. Then, $\int_b^c D^\alpha z(t) d\mu(\alpha) \in C([t_0, T]; \mathcal{Z})$, besides, there exists $C_1 > 0$, such that for all $z \in C^{m-1,c}([t_0, T]; \mathcal{Z})$*

$$\left\| \int_b^c D^\alpha z(t) d\mu(\alpha) \right\|_{C([t_0,T];\mathcal{Z})} \leq C_1 \|z\|_{C^{m-1,c}([t_0,T];\mathcal{Z})}.$$

Proof. Indeed, due to Lemma 2

$$\left\| \int_b^c D^\alpha z(t) d\mu(\alpha) \right\|_{C([t_0,T];\mathcal{Z})} \leq C V_b^c(\mu) \|z\|_{C^{m-1,c}([t_0,T];\mathcal{Z})},$$

where $V_b^c(\mu)$ is the variation of μ on $(b, c]$. □

Lemma 3. *Let $\beta \in (0, 1)$, $z, D^\beta z \in C([t_0, T]; \mathcal{Z})$. Then, $z \in C^\beta([t_0, T]\mathcal{Z})$, moreover, there exists $C > 0$, such that for all $t, \tau \in [t_0, T]$*

$$\|h(t) - h(\tau)\| \leq \frac{\|D^\beta z\|_{C([t_0,T];\mathcal{Z})}}{\Gamma(\beta+1)} |t-\tau|^\beta.$$

Proof. If $t_0 \le \tau < t \le T$, then

$$\|h(t) - h(\tau)\|_{\mathcal{Z}} = \|J^\beta D^\beta h(t) - J^\beta D^\beta h(\tau)\|_{\mathcal{Z}} \le$$

$$\le \frac{(t-t_0)^\beta - (\tau - t_0)^\beta}{\Gamma(\beta+1)} \|D^\beta h\|_{C([t_0,T];\mathcal{Z})} \le \frac{(t-\tau)^\beta}{\Gamma(\beta+1)} \|D^\beta h\|_{C([t_0,T];\mathcal{Z})},$$

since the function

$$\frac{(t-t_0)^\beta - (\tau - t_0)^\beta}{(t-\tau)^\beta}$$

decreases with respect to $\tau \in [t_0, t)$ at $\beta \in (0,1)$. □

Corollary 3. *Let $m-1 < c \le m \in \mathbb{N}$, $b < c < \beta$, $\mu \in BV((b,c];\mathbb{C})$, c be a variation point of the measure $d\mu(\alpha)$. Then, for every $z \in C^{m-1,\beta}([t_0,T];\mathcal{Z})$, $\varepsilon \in (0, \beta - c)$ we have $\int_b^c D^\alpha z(t) d\mu(\alpha) \in C^{\beta-c-\varepsilon}([t_0,T];\mathcal{Z})$. Additionally, there exists $C > 0$, such that for all $z \in C^{m-1,c}([t_0,T];\mathcal{Z})$, $s,t \in [t_0,T]$*

$$\left\| \int_b^c D^\alpha z(t) d\mu(\alpha) - \int_b^c D^\alpha z(s) d\mu(\alpha) \right\|_{\mathcal{Z}} \le C\|z\|_{C^{m-1,\beta}([t_0,T];\mathcal{Z})} |t-s|^{\beta-c-\varepsilon}.$$

Proof. Indeed, due to Lemmas 2 and 3 for every s, t, such that $t_0 \le s < t \le T$, we have

$$\left\| \int_b^c D^\alpha z(t) d\mu(\alpha) - \int_b^c D^\alpha z(s) d\mu(\alpha) \right\|_{\mathcal{Z}} \le$$

$$\le \frac{|t-s|^{\beta-c-\varepsilon}}{\Gamma(\beta-c-\varepsilon+1)} \left\| D^{\beta-c-\varepsilon} \int_b^c D^\alpha z(t) d\mu(\alpha) \right\|_{C([t_0,T];\mathcal{Z})} \le C\|z\|_{C^{m-1,\beta}([t_0,T];\mathcal{Z})} |t-s|^{\beta-c-\varepsilon}.$$

□

4. Nonlocal Unique Solvability of Quasilinear Equation

A solution on a segment $[t_0, T]$ of the Cauchy problem

$$D^k z(t_0) = z_k, \quad k = 0,1,\dots,m-1, \tag{9}$$

for the equation

$$\int_b^c D^\alpha z(t) d\mu(\alpha) = Az(t) + B\left(t, \int_{b_1}^{c_1} D^\alpha z(t) d\mu_1(\alpha), \dots, \int_{b_n}^{c_n} D^\alpha z(t) d\mu_n(\alpha)\right), \tag{10}$$

where $b < c$, $m-1 < c \le m \in \mathbb{N}$, $b_l < c_l$, $m_l - 1 < c_l \le m_l \in \mathbb{Z}$, $c_1 \le c_2 \le \dots \le c_n < c$, $\mu \in BV((b,c];\mathbb{C})$, $\mu_l \in BV((b_l,c_l];\mathbb{C})$, $l = 1,2,\dots,n$, $T > t_0$, $g \in C([t_0,T];\mathcal{Z})$, is a function $z \in C^{m-1}([t_0,T];\mathcal{Z}) \cap C((t_0,T];D_A)$, such that $\int_b^c D^\alpha z(t) d\mu(\alpha) \in C((t_0,T];\mathcal{Z})$, $\int_{b_l}^{c_l} D^\alpha z(t) d\mu_l(\alpha) \in C([t_0,T];\mathcal{Z})$, $l = 1,2,\dots,n$, and equalities (9) and (10) for $t \in (t_0,T]$ are fulfilled.

Lemma 4. *Let $m-1 < c \le m \in \mathbb{N}$, $b < c$, $\mu \in BV((b,c];\mathbb{C})$, c be a variation point of the measure $d\mu(\alpha)$, $n \in \mathbb{N}$, $c_1 \le c_2 \le \dots \le c_n < c$, $\mu_l \in BV((b_l,c_l];\mathbb{C})$, c_l be a variation point of the measure $d\mu_l(\alpha)$, $l = 1,2,\dots,n$, $A \in \mathcal{A}_W(\theta_0, a_0)$ for some $\theta_0 \in (\pi/2, \pi)$, $a_0 \ge 0$, $z_k \in D_A$,*

$k = 0, 1, \ldots, m-1$, $B \in C([t_0, T] \times \mathcal{Z}^n; D_A)$. *Then a function* z *is a solution of problem* (9) *and* (10) *on the segment* $[t_0, T]$, *if and only if* $z \in C^{m-1,c_n}([t_0, T]; \mathcal{Z})$ *and for all* $t \in [t_0, T]$ *the equality*

$$z(t) = \sum_{k=0}^{m-1} Z_k(t - t_0)z_k + \int_{t_0}^{t} Z(t - s)B^z(s)ds \tag{11}$$

holds, where

$$B^z(s) := B\left(s, \int_{b_1}^{c_1} D^{\alpha}z(s)d\mu_1(\alpha), \int_{b_2}^{c_2} D^{\alpha}z(s)d\mu_2(\alpha), \ldots, \int_{b_n}^{c_n} D^{\alpha}z(s)d\mu_n(\alpha)\right).$$

Proof. If z is a solution of problem (9) and (10), then there exists $D^{c_n}z \in C([t_0, T]; \mathcal{Z})$, since c_n is a variation point of the measure $d\mu_n(\alpha)$. Therefore, $z \in C^{m-1,c_n}([t_0, T]; \mathcal{Z})$ and due to Corollary 2 the mapping

$$t \to B\left(t, \int_{b_1}^{c_1} D^{\alpha}z(t)d\mu(\alpha), \int_{b_2}^{c_2} D^{\alpha}z(t)d\mu(\alpha), \ldots, \int_{b_n}^{c_n} D^{\alpha}z(t)d\mu(\alpha)\right) \tag{12}$$

acts continuously from $[t_0, T]$ into D_A, since $B \in C([t_0, T] \times \mathcal{Z}^n; D_A)$. Consequently, by Theorem 3, equality (11) is valid.

Let $z \in C^{m-1,c_n}([t_0, T]; \mathcal{Z})$ and for all $t \in [t_0, T]$ equality (11) holds. Then, by Corollary 2, mapping (12) belongs to the class $C([t_0, T]; D_A)$ in the case $B \in C([t_0, T] \times \mathcal{Z}^n; D_A)$. By Theorem 3, z is a solution of problem (9) and (10). □

A mapping $B : [t_0, T] \times \mathcal{Z}^n \to \mathcal{Z}$ is called Lipschitz continuous, if there exists $C_L > 0$, such that for all $t \in [t_0, T]$, $x_1, x_2, \ldots, x_n, y_1, y_2, \ldots, y_n \in \mathcal{Z}$

$$\|B(t, x_1, x_2, \ldots, x_n) - B(t, y_1, y_2, \ldots, y_n)\|_{\mathcal{Z}} \leq C_L \sum_{l=1}^{n} \|x_l - y_l\|_{\mathcal{Z}}.$$

Theorem 4. *Let* $m - 1 < c \leq m \in \mathbb{N}$, $b < c$, $\mu \in BV((b, c]; \mathbb{C})$, c *be a variation point of the measure* $d\mu(\alpha)$, $n \in \mathbb{N}$, $c_1 \leq c_2 \leq \cdots \leq c_n < c$, $b_l < c_l$, $\mu_l \in BV((b_l, c_l]; \mathbb{C})$, c_l *be a variation point of the measure* $d\mu_l(\alpha)$, $l = 1, 2, \ldots, n$, $A \in \mathcal{A}_W(\theta_0, a_0)$ *for some* $\theta_0 \in (\pi/2, \pi)$, $a_0 \geq 0$, $z_k \in D_A$, $k = 0, 1, \ldots, m-1$, *a mapping* $B \in C([t_0, T] \times \mathcal{Z}^n; D_A)$ *be Lipschitz continuous. Then, problem* (9) *and* (10) *have a unique solution on the segment* $[t_0, T]$.

Proof. Due to Lemma 4, it is sufficient to prove that the integro-differential Equation (11) has a unique solution in the Banach space $C^{m-1,c_n}([t_0, T]; \mathcal{Z})$.

For $z \in C^{m-1,c_n}([t_0, T]; \mathcal{Z})$ define the operator

$$G(z)(t) := \sum_{k=0}^{m-1} Z_k(t - t_0)z_k + \int_{t_0}^{t} Z(t - s)B^z(s)\,ds, \ \ t \in [t_0, T].$$

Since mapping (12) belongs to $C([t_0, T]; D_A)$, due to Theorem 3, we find that $G(z) \in C^{m-1}([t_0, T]; \mathcal{Z})$, $D^kG(z)(t_0) = z_k$ for $k = 0, 1, \ldots, m-1$.

If $c_n < k$, then the form of Z_k implies that by (4)

$$\text{Lap}[D^{c_n}Z_k(t)z_k](\lambda) = \lambda^{c_n-1-k}W_k(\lambda)R_{W(\lambda)}(A)z_k,$$

$$\left\|\lambda^{c_n-1-k}W_k(\lambda)R_{W(\lambda)}(A)z_k\right\|_{\mathcal{Z}} \leq \frac{C\|z_k\|_{\mathcal{Z}}}{|\lambda|^{k+1-\varepsilon-c_n}}$$

for some $\varepsilon \in (0, k - c_n)$ due to Lemma 1. Hence, $D^{c_n} Z_k(0) z_k = 0$ and $D^{c_n} Z_k(t - t_0) z_k \in C([t_0, T]; \mathcal{Z})$. It is known that $D^k Z_k(t - t_0) z_k \in C([t_0, T]; \mathcal{Z})$. In the case $c_n > k$, we have due to equality (4)

$$\mathrm{Lap}[D^{c_n} Z_k(t) z_k](\lambda) = \lambda^{c_n - 1 - k} W_k(\lambda) R_{W(\lambda)}(A) z_k - \lambda^{c_n - 1 - k} z_k =$$

$$= \lambda^{c_n - 1 - k} [W_k(\lambda) - W(\lambda)] R_{W(\lambda)}(A) z_k + \lambda^{c_n - 1 - k} R_{W(\lambda)}(A) A z_k,$$

$$\left\| \lambda^{c_n - 1 - k} [W_k(\lambda) - W(\lambda)] R_{W(\lambda)}(A) z_k \right\|_{\mathcal{Z}} \leq \frac{C \|z_k\|_{\mathcal{Z}}}{|\lambda|^{c - \varepsilon + 1 - c_n}},$$

$$\left\| \lambda^{c_n - 1 - k} R_{W(\lambda)}(A) A z_k \right\|_{\mathcal{Z}} \leq \frac{C \|z_k\|_{D_A}}{|\lambda|^{c - \varepsilon + 1 - c_n + k}},$$

for some $\varepsilon \in (0, c - c_n)$ by Lemma 1. Therefore, $D^{c_n} Z_k(0) z_k = 0$ and $D^{c_n} Z_k(t - t_0) z_k \in C([t_0, T]; \mathcal{Z})$.

Due to [29] Lemma 4 $D^k Z(0) = 0$, $k = 0, 1, \ldots, m - 2$, $\|D^{m-1} Z(t)\|_{\mathcal{L}(\mathcal{Z})} = O(t^{c - \varepsilon - m})$ as $t \to 0+$. Therefore,

$$\left\| \int_{t_0}^{t} Z(t - s) B^z(s) ds \right\|_{\mathcal{Z}} = O((t - t_0)^{c - \varepsilon}), \ t \to t_0+,$$

$$D^k|_{t = t_0} \int_{t_0}^{t} Z(t - s) B^z(s) ds = 0, \quad k = 0, 1, \ldots, m - 1,$$

since B^z is continuous on $[t_0, T]$ for $z \in C^{m-1, c_n}([t_0, T]; \mathcal{Z})$ by Corollary 2. We have

$$\|\mathrm{Lap}[D^{m_n} J^{m_n - c_n} Z(t)](\lambda)\|_{\mathcal{L}(\mathcal{Z})} = \|\lambda^{c_n} R_{W(\lambda)}(A)\|_{\mathcal{L}(\mathcal{Z})} \leq \frac{C}{|\lambda|^{c - \varepsilon - c_n}}$$

with $\varepsilon \in (0, c - c_n)$, consequently, $\|D^{m_n} J^{m_n - c_n} Z(t)\|_{\mathcal{L}(\mathcal{Z})} = O(t^{c - \varepsilon - c_n - 1})$ as $t \to 0+$. Since,

$$D^{c_n} \int_{t_0}^{t} Z(t - s) B^z(s) ds = D^{m_n} J^{m_n - c_n} \int_{t_0}^{t} Z(t - s) B^z(s) ds = \int_{t_0}^{t} D^{m_n} J^{m_n - c_n} Z(t - s) B^z(s) ds,$$

we have

$$\left\| \int_{t_0}^{t} D^{m_n} J^{m_n - c_n} Z(t - s) B^z(s) ds \right\|_{\mathcal{L}(\mathcal{Z})} = O(t^{c - \varepsilon - c_n}).$$

Thus, $G(z) \in C^{m-1, c_n}([t_0, T]; \mathcal{Z})$.

Let G^j be the j-th degree of the operator G, $j \in \mathbb{N}$. For the sake of certainty, we consider that $T - t_0 \geq 1$. In the case $T - t_0 < 1$, further reasoning will remain valid after the replacement $T - t_0$ by 1.

Arguing as before, we can find that for $k = 0, 1, \ldots, m - 1$ and for small $\varepsilon > 0$ the inequality $\|D^k Z(t)\|_{\mathcal{L}(\mathcal{Z})} \leq C t^{c - \varepsilon - 1 - k}$ is valid. Consequently, for $x, y \in C^{m-1, c_n}([t_0, T]; \mathcal{Z})$, we have in the case $c_n > m - 1$

$$\|D^k G(x)(t) - D^k G(y)(t)\|_{\mathcal{Z}} \leq C \int_{t_0}^{t} (t - s)^{c - \varepsilon - 1 - k} \|B^x(s) - B^y(s)\|_{\mathcal{Z}} ds \leq$$

$$\leq C_1 \|x - y\|_{C^{m-1, c_n}([t_0, t]; \mathcal{Z})} (t - t_0)^{c - \varepsilon - k} \leq$$

$$\leq C_2 \|x - y\|_{C^{m-1, c_n}([t_0, t]; \mathcal{Z})} [(t - t_0)^c + (t - t_0)^{c - \varepsilon - c_n}],$$

$$\|D^{c_n}G(x)(t) - D^{c_n}G(y)(t)\|_{\mathcal{Z}} \le C\int_{t_0}^{t}(t-s)^{c-\varepsilon-1-c_n}\|B^x(s) - B^y(s)\|_{\mathcal{Z}}ds \le$$

$$\le C_2\|x-y\|_{C^{m-1,c_n}([t_0,t];\mathcal{Z})}[(t-t_0)^c + (t-t_0)^{c-\varepsilon-c_n}].$$

Therefore,

$$\|G(x)-G(y)\|_{C^{m-1,c_n}([t_0,t];\mathcal{Z})} \le C_2(m+1)\|x-y\|_{C^{m-1,c_n}([t_0,t];\mathcal{Z})}[(t-t_0)^c + (t-t_0)^{c-\varepsilon-c_n}].$$

Then, for $k = 0, 1, \dots, m-1$

$$\|D^kG^2(x)(t) - D^kG^2(y)(t)\|_{\mathcal{Z}} \le C\int_{t_0}^{t}(t-s)^{c-\varepsilon-1-k}\|B^{G(x)}(s) - B^{G(y)}(s)\|_{\mathcal{Z}}ds \le$$

$$\le C_2(T-t_0)^c\int_{t_0}^{t}\|G(x)-G(y)\|_{C^{m-1,c_n}([t_0,s];\mathcal{Z})}ds \le$$

$$\le C_2^2(m+1)(T-t_0)^c\|x-y\|_{C^{m-1,c_n}([t_0,t];\mathcal{Z})}[(t-t_0)^{c+1} + (t-t_0)^{c-\varepsilon-c_n+1}],$$

$$\|D^{c_n}G^2(x)(t) - D^{c_n}G^2(y)(t)\|_{\mathcal{Z}} \le C\int_{t_0}^{t}(t-s)^{c-\varepsilon-1-c_n}\|B^{G(x)}(s) - B^{G(y)}(s)\|_{\mathcal{Z}}ds \le$$

$$\le C_2^2(m+1)(T-t_0)^c\|x-y\|_{C^{m-1,c_n}([t_0,t];\mathcal{Z})}[(t-t_0)^{c+1} + (t-t_0)^{c-\varepsilon-c_n+1}],$$

$$\|G^2(x)-G^2(y)\|_{C^{m-1,c_n}([t_0,t];\mathcal{Z})} \le$$

$$\le C_2^2(m+1)^2(T-t_0)^c\|x-y\|_{C^{m-1,c_n}([t_0,t];\mathcal{Z})}[(t-t_0)^{c+1} + (t-t_0)^{c-\varepsilon-c_n+1}].$$

By the same way, we obtain

$$\|G^3(x)-G^3(y)\|_{C^{m-1,c_n}([t_0,t];\mathcal{Z})} \le$$

$$\le C_2^3(m+1)^3(T-t_0)^{2c}\|x-y\|_{C^{m-1,c_n}([t_0,t];\mathcal{Z})}\frac{(t-t_0)^{c+2} + (t-t_0)^{c-\varepsilon-c_n+2}}{2}.$$

Similarly, we obtain for $t \in [t_0, T]$, $j \in \mathbb{N}$, $x, y \in C^{m-1,c_n}([t_0,T];\mathcal{Z})$ that

$$\|G^j(x)-G^j(y)\|_{C^{m-1,c_n}([t_0,t];\mathcal{Z})} \le$$

$$\le \frac{C_0^j[(t-t_0)^{c+j-1} + (t-t_0)^{c-\varepsilon-c_n+j-1}]}{(j-1)!}\|x-y\|_{C^{m-1,c_n}([t_0,t];\mathcal{Z})}$$

with $C_0 = C_2(m+1)(T-t_0)^c$. Consequently,

$$\|G^j(x)-G^j(y)\|_{C^{m-1,c_n}([t_0,T];\mathcal{Z})} \le \frac{2C_0^j(T-t_0)^{c+j-1}}{(j-1)!}\|x-y\|_{C^{m-1,c_n}([t_0,T];\mathcal{Z})}.$$

Hence, for a large enough j, the mapping G^j is a contraction in the space $C^{m-1,c_n}([t_0,T];\mathcal{Z})$ and it has a unique fixed point in this space, which is known to be the unique fixed point in $C^{m-1,c_n}([t_0,T];\mathcal{Z})$ of the mapping G. Due to Lemma 4, z is the fixed point of G, if and only if it is a unique solution of problem (9) and (10).

If $c_n \le m-1$, then we will omit the estimates for the derivatives of the order c_n. □

Lemma 5. *Let $m-1 < c \le m \in \mathbb{N}$, $b < c$, $\mu \in BV((b,c];\mathbb{C})$, c be a variation point of the measure $d\mu(\alpha)$, $n \in \mathbb{N}$, $c_1 \le c_2 \le \dots \le c_n < \beta < c$, $\mu_l \in BV((b_l,c_l];\mathbb{C})$, c_l be a variation point of the measure $d\mu_l(\alpha)$, $l = 1,2,\dots,n$, $A \in \mathcal{A}_W(\theta_0, a_0)$ for some $\theta_0 \in (\pi/2, \pi)$, $a_0 \ge 0$,*

$z_k \in D_A$, $k = 0, 1, \ldots, m-1$, $B \in C([t_0, T] \times \mathcal{Z}^n; \mathcal{Z})$ be Lipschitz continuous. Then, a function $z \in C^{m-1,\beta}([t_0, T]; \mathcal{Z})$ is a solution of problem (9) *and* (10) *on the segment $[t_0, T]$, if and only if for all $t \in [t_0, T]$ it satisfies equality* (11).

Proof. If $z \in C^{m-1,\beta}([t_0, T]; \mathcal{Z})$ is a solution of problem (9) and (10), then due to Lipschitz continuity of B and by Corollary 3 the function B^z satisfies the Hölder condition. Due to Theorem 3, equality (11) is valid.

Let $z \in C^{m-1,\beta}([t_0, T]; \mathcal{Z})$ and for all $t \in [t_0, T]$ equality (11) is valid. Then, by Corollary 3, the function B^z is Hölderian. By Theorem 3, the function z is a solution of problem (9) and (10). □

Theorem 5. *Let $m-1 < c \leq m \in \mathbb{N}$, $b < c$, $\mu \in BV((b, c]; \mathbb{C})$, c be a variation point of the measure $d\mu(\alpha)$, $n \in \mathbb{N}$, $c_1 \leq c_2 \leq \cdots \leq c_n < c$, $b_l < c_l$, $\mu_l \in BV((b_l, c_l]; \mathbb{C})$, c_l be a variation point of the measure $d\mu_l(\alpha)$, $l = 1, 2, \ldots, n$, $A \in \mathcal{A}_W(\theta_0, a_0)$ for some $\theta_0 \in (\pi/2, \pi)$, $a_0 \geq 0$, $z_k \in D_A$, $k = 0, 1, \ldots, m-1$, a mapping $B \in C([t_0, T] \times \mathcal{Z}^n; \mathcal{Z})$ be Lipschitz continuous. Then, problem* (9) *and* (10) *have a unique solution on the segment $[t_0, T]$.*

Proof. Choose $\beta \in (c_n, c)$ and for $z \in C^{m-1,\beta}([t_0, T]; \mathcal{Z})$ consider the operator

$$G(z)(t) := \sum_{k=0}^{m-1} Z_k(t - t_0) z_k + \int_{t_0}^{t} Z(t-s) B^z(s)\, ds, \quad t \in [t_0, T].$$

Since B is Lipschitz continuous and by Corollary 3 all the arguments of B satisfy the Hölder condition, hence, B^z is Hölderian also. Consequently, by Theorem 3, we have $G(z) \in C^{m-1}([t_0, T]; \mathcal{Z})$, $D^k G(z)(t_0) = z_k$ for $k = 0, 1, \ldots, m-1$.

If $c_n \geq m-1$, then $\beta > m-1$ and, as in the proof of the previous theorem, it can be shown that $G(z) \in C^{m-1,\beta}([t_0, T]; \mathcal{Z})$, for sufficiently large j, the mapping G^j is a contraction in $C^{m-1,\beta}([t_0, T]; \mathcal{Z})$ and G has a unique fixed point in $C^{m-1,\beta}([t_0, T]; \mathcal{Z})$. Due to Lemma 5, the unique fixed point is a unique solution of problem (9) and (10).

If $c_n < m-1$, we can take $\beta = m-1$ and the proof will be simpler. □

5. Local Unique Solvbability of Quasilinear Equation

Now, the nonlinear operator B is defined on some open subset U of $\mathbb{R} \times \mathcal{Z}^n$. A solution on some segment $[t_0, t_1]$, $t_1 > t_0$, of Cauchy problem (9) for Equation (10) is a function $z \in C^{m-1}([t_0, t_1]; \mathcal{Z}) \cap C((t_0, t_1]; D_A)$, such that $\int_b^c D^\alpha z(t) d\mu(\alpha) \in C((t_0, t_1]; \mathcal{Z})$, $\int_{b_l}^{c_l} D^\alpha z(t) d\mu_l(\alpha) \in C([t_0, t_1]; \mathcal{Z})$, $l = 1, 2, \ldots, n$, equalities (9), inclusion

$$\left(t, \int_{b_1}^{c_1} D^\alpha z(t) d\mu_1(\alpha), \int_{b_2}^{c_2} D^\alpha z(t) d\mu_2(\alpha), \ldots, \int_{b_n}^{c_n} D^\alpha z(t) d\mu_n(\alpha) \right) \in U$$

for $t \in [t_0, t_1]$ and equality (10) for $t \in (t_0, t_1]$ are satisfied.

As before, here $b < c$, $m-1 < c \leq m \in \mathbb{N}$, $\mu \in BV((b, c]; \mathbb{C})$, $b_l < c_l$, $m_l - 1 < c_l \leq m_l \in \mathbb{Z}$, $c_1 \leq c_2 \leq \cdots \leq c_n < c$, $\mu_l \in BV((b_l, c_l]; \mathbb{C})$, $l = 1, 2, \ldots, n$.

A mapping $B : U \to \mathcal{Z}$ is called locally Lipschitz continuous, if for every point $(t, x_1, x_2, \ldots, x_n) \in U$ there exists its vicinity $V \subset U$ and a constant $C > 0$, such that for all $(s, y_1, y_2, \ldots, y_n), (s, v_1, v_2, \ldots, v_n) \in V$

$$\|B(s, y_1, y_2, \ldots, y_n) - B(s, v_1, v_2, \ldots, v_n)\|_{\mathcal{Z}} \leq C \sum_{l=1}^{n} \|y_l - v_l\|_{\mathcal{Z}}. \tag{13}$$

Denote for $z_k \in D_A$, $k = 0, 1, \dots, m-1$, from initial conditions (9)

$$\tilde{z}(t) = z_0 + z_1(t - t_0) + \cdots + z_{m-1}\frac{(t - t_0)^{m-1}}{(m-1)!}, \ \tilde{z}_l = \int_{b_l}^{c_l} D^\alpha \tilde{z}(t_0) d\mu_l(\alpha), \ l = 1, 2, \dots, n.$$

Theorem 6. *Let $m - 1 < c \le m \in \mathbb{N}$, $b < c$, $\mu \in BV((b,c];\mathbb{C})$, c be a variation point of the measure $d\mu(\alpha)$, $n \in \mathbb{N}$, $c_1 \le c_2 \le \cdots \le c_n < c$, $b_l < c_l$, $\mu_l \in BV((b_l,c_l];\mathbb{C})$, c_l be a variation point of the measure $d\mu_l(\alpha)$, $l = 1, 2, \dots, n$, $A \in \mathcal{A}_W(\theta_0, a_0)$ for some $\theta_0 \in (\pi/2, \pi)$, $a_0 \ge 0$, $z_k \in D_A$, $k = 0, 1, \dots, m-1$, $(t_0, \tilde{z}_1, \tilde{z}_2, \dots, \tilde{z}_n) \in U$, a mapping $B \in C(U; D_A)$ be locally Lipschitz continuous. Then, there exists $t_1 > t_0$, such that problem* (9) *and* (10) *have a unique solution on the segment $[t_0, t_1]$.*

Proof. Take a sufficiently small $\delta > 0$, such that in the neighborhood

$$V := \{(t, x_1, x_2, \dots, x_n) \in \mathbb{R} \times \mathcal{Z}^n : |t - t_0| \le \delta, \ \|x_l - \tilde{z}_l\|_{\mathcal{Z}} \le \delta\}$$

the inequality (13) holds with some $C > 0$. Define

$$\mathfrak{M}_{t_1} := \left\{ z \in C^{m-1,c_n}([t_0,t_1];\mathcal{Z}) : \left\| \int_{b_l}^{c_l} D^\alpha z(t) d\mu_l(\alpha) - \tilde{z}_l \right\|_{\mathcal{Z}} \le \delta, \right.$$

$$\left. t \in [t_0, t_1], \ l = 1, 2, \dots, n \right\}.$$

Due to Corollary 2 $\mathfrak{M}_{t_1}$ is a complete metric space with the metric

$$d(x,y) = \|x - y\|_{C^{m-1,c_n}([t_0,t_1];\mathcal{Z})}.$$

For $z \in \mathfrak{M}_{t_1}$, define the operator

$$G(z)(t) := \sum_{k=0}^{m-1} Z_k(t - t_0) z_k + \int_{t_0}^{t} Z(t-s) B^z(s)\, ds, \ \ t \in [t_0, t_1].$$

Since B^z belongs to $C([t_0,t_1];D_A)$, we have $G(z) \in C^{m-1}([t_0,t_1];\mathcal{Z})$, $D^k G(z)(t_0) = z_k$ for $k = 0, 1, \dots, m-1$. As in the proof of Theorem 4, we have $G(z) \in C^{m-1,c_n}([t_0,t_1];\mathcal{Z})$, therefore, $G(z) \in \mathfrak{M}_{t_1}$. If necessary, we can reduce t_1 here. Due to Corollary 2

$$\int_{b_l}^{c_l} D^\alpha G(z)(t) d\mu_l(\alpha) \in C([t_0,t_1];\mathcal{Z}), \quad l = 1, 2, \dots, n.$$

Consequently, for small enough $t_1 - t_0$ $G(z) \in \mathfrak{M}_{t_1}$.

Arguing as in the proof of Theorem 4, we have for $k = 0, 1, \dots, m-1$ and small $\varepsilon > 0$ $\|D^k Z(t)\|_{\mathcal{L}(\mathcal{Z})} \le C t^{c-\varepsilon-1-k}$. Therefore, for $x, y \in \mathfrak{M}_{t_1}$

$$\|D^k G(x)(t) - D^k G(y)(t)\|_{\mathcal{Z}} \le C \int_{t_0}^{t} (t-s)^{c-\varepsilon-1-k} \|B^x(s) - B^y(s)\|_{\mathcal{Z}} ds \le$$

$$\le C_1 \|x - y\|_{C^{m-1,c_n}([t_0,t_1];\mathcal{Z})} (t_1 - t_0)^{c-\varepsilon-k} \le \frac{\|x - y\|_{C^{m-1,c_n}([t_0,t_1];\mathcal{Z})}}{2(m+1)}, \quad k = 0, 1, \dots, m-1,$$

$$\|D^{c_n} G(x)(t) - D^{c_n} G(y)(t)\|_{\mathcal{Z}} \le C_1 \|x - y\|_{C^{m-1,c_n}([t_0,t_1];\mathcal{Z})} (t_1 - t_0)^{c-\varepsilon-c_n} \le$$

$$\leq \frac{\|x-y\|_{C^{m-1,c_n}([t_0,t_1];\mathcal{Z})}}{2(m+1)},$$

for sufficiently small $t_1 - t_0$, hence,

$$\|G(x)-G(y)\|_{C^{m-1,c_n}([t_0,t_1];\mathcal{Z})} \leq \frac{1}{2}\|x-y\|_{C^{m-1,c_n}([t_0,t_1];\mathcal{Z})}.$$

Thus, the mapping G is a contraction in the metric space $\mathfrak{M}_{t_1}$. By the Banach theorem on a fixed point, G has a unique fixed point z in this space. Due to Lemma 4, the fixed point z is a unique solution of problem (9) and (10) on $[t_0, t_1]$. □

Theorem 7. *Let $m-1 < c \leq m \in \mathbb{N}$, $b < c$, $\mu \in BV((b,c];\mathbb{C})$, c be a variation point of the measure $d\mu(\alpha)$, $n \in \mathbb{N}$, $c_1 \leq c_2 \leq \cdots \leq c_n < c$, $b_l < c_l$, $\mu_l \in BV((b_l,c_l];\mathbb{C})$, c_l be a variation point of the measure $d\mu_l(\alpha)$, $l = 1,2,\dots,n$, $A \in \mathcal{A}_W(\theta_0, a_0)$ for some $\theta_0 \in (\pi/2,\pi)$, $a_0 \geq 0$, $z_k \in D_A$, $k = 0,1,\dots,m-1$, $(t_0, \tilde{z}_1, \tilde{z}_2, \dots, \tilde{z}_n) \in U$, a mapping $B \in C([t_0,T]\times\mathcal{Z}^n;\mathcal{Z})$ be locally Lipschitz continuous. Then, there exists $t_1 > t_0$, such that problem* (9) *and* (10) *have a unique solution on the segment $[t_0, t_1]$.*

Proof. For a fixed $\beta \in (c_n, c)$ take a small enough $\delta > 0$, such that in

$$V := \{(t, x_1, x_2, \dots, x_n) \in \mathbb{R}\times\mathcal{Z}^n : |t-t_0| \leq \delta,\ \|x_l - \tilde{z}_l\|_{\mathcal{Z}} \leq \delta\}.$$

the inequality (13) is satisfied with a constant $C > 0$. Define

$$\mathfrak{M}_{t_1} := \left\{ z \in C^{m-1,\beta}([t_0,t_1];\mathcal{Z}) : \left\| \int_{b_l}^{c_l} D^\alpha z(t)\,d\mu_l(\alpha) - \tilde{z}_l \right\|_{\mathcal{Z}} \leq \delta, \right.$$

$$\left. t \in [t_0,t_1],\ l = 1,2,\dots,n \right\}.$$

For $z \in \mathfrak{M}_{t_1}$, define the operator

$$G(z)(t) := \sum_{k=0}^{m-1} Z_k(t-t_0)z_k + \int_{t_0}^{t} Z(t-s)B^z(s)\,ds,\ \ t \in [t_0,t_1].$$

Due to the Lipschitz continuity of B by Corollary 3, B^z satisfies the Hölder condition. Due to Theorem 3, $G(z) \in C^{m-1}([t_0,t_1];\mathcal{Z})$, $D^kG(z)(t_0) = z_k$, $k = 0,1,\dots,m-1$.

If $c_n \geq m-1$, then $\beta > m-1$. Reasoning by the same way as in the proof of Theorem 6, we can obtain that $G(z) \in C^{m-1,\beta}([t_0,T];\mathcal{Z})$ and the mapping G is a contraction in $\mathfrak{M}_{t_1}$ and has a unique fixed point in the metric space $\mathfrak{M}_{t_1}$. By Lemma 5, the fixed point is a unique solution of problem (9) and (10) on the segment $[t_0, t_1]$.

If $c_n < m-1$, we take $\beta = m-1$. □

6. Application to a Nonlinear Initial-Boundary Value Problem

In the framework of the Cauchy problem for a quasilinear equation in Banach space, we can investigate initial-boundary value problems for partial differential equations with time-distributed derivatives. For this aim, we need to choose an appropriate space $\mathcal{Z}$ and an operator A. Now, we will demonstrate this with the example of the following problem.

Consider a bounded region $\Omega \subset \mathbb{R}^d$ with a smooth boundary $\partial\Omega$, $\beta, \gamma, \nu \in \mathbb{R}$, $c \in (1,2)$, $b < c$, $\alpha_1 < \alpha_2 < \cdots < \alpha_n \leq c$, $\omega_k \in \mathbb{R}\setminus\{0\}$, $k = 1,2,\dots,n$, $\omega \in C([b,c];\mathbb{R})$; if $\alpha_n < c$, then $\omega(c) \neq 0$ in a some left vicinity of c; $\beta_l < c$, $b_l < c_l < c$, $\mu_l \in BV((b_l,c_l];\mathbb{R})$, $l = 1,2$. Consider the initial-boundary value problem

$$u(s,0) = u_0(s), \quad v(s,0) = v_0(s), \quad s \in \Omega, \tag{14}$$

$$\frac{\partial u}{\partial t}(s,0) = u_1(s), \quad \frac{\partial v}{\partial t}(s,0) = v_1(s), \quad s \in \Omega, \tag{15}$$

$$u(s,t) = v(s,t) = 0, \quad (s,t) \in \partial\Omega \times (0,T], \tag{16}$$

for the nonlinear system of equations in $\Omega \times (0,T]$

$$\sum_{k=1}^{n} \omega_k D_t^{\alpha_k} u(s,t) + \int_b^c \omega(\alpha) D_t^{\alpha} u(s,t) d\alpha = \Delta u(s,t) - \Delta v(s,t) + \\ + F_1\left(s, D^{\beta_1} u(s,t), \int_{b_1}^{c_1} D^{\alpha} v(s,t) d\mu_1(\alpha)\right), \tag{17}$$

$$\sum_{k=1}^{n} \omega_k D_t^{\alpha_k} v(s,t) + \int_b^c \omega(\alpha) D_t^{\alpha} v(s,t) d\alpha = \nu\Delta v(s,t) + \beta u(s,t) + \gamma v(s,t) + \\ + F_2\left(s, D^{\beta_2} v(s,t), \int_{b_2}^{c_2} D^{\alpha} u(s,t) d\mu_2(\alpha)\right). \tag{18}$$

Remark 7. *If* $\omega_2 = \omega_3 = \cdots = \omega_n = 0$, $\alpha_1 = 1$, $\omega(\alpha) \equiv 0$ *for all* $\alpha \in (b,c)$, $F_1 \equiv 0$, $\beta_2 = 0$, $\mu_2 \equiv 0$, *after linear replacement of unknown functions* $u(s,t) = \widetilde{u}(s,t) + \frac{\kappa}{2}\widetilde{v}(s,t)$, $v(s,t) = \frac{\kappa}{2}\widetilde{v}(s,t)$, $\kappa \in \mathbb{R}$, *systems* (17) *and* (18) *are the linearization of the phase field system of equations [36,37].*

Define $\Lambda_1 z = \Delta z$, $D_{\Lambda_1} = H_0^{j+2}(\Omega) \subset H^j(\Omega)$. By $\{\varphi_k : k \in \mathbb{N}\}$, denote an orthonormal in the inner product $\langle \cdot, \cdot \rangle$ in $L_2(\Omega)$ eigenfunctions of Λ_1, which are enumerated in the non-increasing order of the eigenvalues $\{\lambda_k : k \in \mathbb{N}\}$ taking into account their multiplicities.

Take the Sobolev space $\mathcal{Z} = (H^j(\Omega))^2$ for some $j \in \mathbb{N} \cup \{0\}$, such that $j > d/2$,

$$A = \begin{pmatrix} \Lambda_1 & -\Lambda_1 \\ \beta I & \gamma I + \nu\Lambda_1 \end{pmatrix}, \quad D_A = (H_0^{j+2}(\Omega))^2, \tag{19}$$

where $H_0^{j+2}(\Omega) := \{z \in H^{j+2}(\Omega) : z(s) = 0,\, s \in \partial\Omega\}$. Consequently, $A \in \mathcal{C}l(\mathcal{Z})$.

Theorem 8 ([29])**.** *Let* $c \in (1,2)$, $\nu > 0$, $\beta, \gamma \in \mathbb{R}$, *then there exist* $\theta_0 \in (\pi/2, \pi)$, $a_0 \geq 0$, *such that* $A \in \mathcal{A}_W(\theta_0, a_0)$.

Theorem 9. *Let* $c \in (1,2)$, $b < c$, $\nu > 0$, $\beta, \gamma \in \mathbb{R}$, $\alpha_1 < \alpha_2 < \cdots < \alpha_n \leq c$, $\omega_k \in \mathbb{R} \setminus \{0\}$, $k = 1,2,\ldots,n$, $\omega \in C([b,c];\mathbb{R})$; *if* $\alpha_n < c$, *then* $\omega(c) \neq 0$ *in the left vicinity of* c; $\beta_l < c$, $b_l < c_l < c$, $\mu_l \in BV((b_l,c_l];\mathbb{R})$, c_l *be a variation point of the measure* $d\mu_l(\alpha)$, $l = 1,2$, $u_0, u_1, v_0, v_1 \in H_0^{j+2}(\Omega)$, $F_1, F_2 \in C^\infty(\mathbb{R}^n;\mathbb{R})$. *Then, there exists a unique solution of problem* (14)–(18) *on a segment* $[0,t_1]$ *with some* $t_1 > 0$. *If the first order partial derivatives of functions* F_1, F_2 *with respect to the second and the third variables are bounded, then there exists a unique solution of problem* (14)–(18) *on a segment* $[0,T]$ *with every* $T > 0$.

Proof. We can consider problem (14)–(18) as Cauchy problem (9) and (10) in the space $\mathcal{Z} = (H^j(\Omega))^2$ with the operator A, which is defined by (19). Note that the left sides of Equations (17) and (18) are the same distributed derivative. By Theorem 8 $A \in \mathcal{A}_W(\theta_0, a_0)$ for some $\theta_0 \in (\pi/2, \pi)$, $a_0 \geq 0$ and it remains to show that the nonlinear operator $B(x,y)(\cdot) = (F_1(\cdot, x(\cdot), y(\cdot)), F_2(\cdot, x(\cdot), y(\cdot)))$ satisfies the conditions of Theorem 7. Due to [38] (Proposition 1 in Appendix B) for $x, y \in H^j(\Omega)$, we have $F_l(x,y) \in H^j(\Omega)$, $l = 1,2$, since $j > d/2$. Moreover, by [38] (Proposition 1 in Appendix B), $B \in C^\infty((H^j(\Omega))^2; H^j(\Omega))$. Hence, B is locally Lipschitz continuous and in the case of boundedness of the first order partial derivatives of functions F_1, F_2 with respect to the second and the third variables B is Lipschitz continuous. It remains to apply Theorem 7 or Theorem 5, respectively. □

7. Conclusions

Using the form of the unique solution for the Cauchy problem to the linear inhomogeneous equation in a Banach space with a distributed Gerasimov–Caputo fractional derivative and with a linear closed operator A, which generates an analytic resolving family, we reduce the Cauchy problem for an analogous quasilinear equation to an equation of the form $z = G(z)$, where the mapping $G(z)$ uses the forms of k-resolving families of operators of the initial linear equation. It allows us to prove the fulfillment of the conditions of the Banach theorem on a fixed point in a specially constructed spaces of functions. Thus, in this paper, it is shown how the linear theory of resolving families of operators made it possible to make the transition from the study of linear equations with a distributed derivative to the study of the corresponding quasilinear equations. The obtained results will allow us to study the unique solvability issues for new initial-boundary value problems for equations and systems of equations with distributed Gerasimov–Caputo partial derivatives.

Using the approach developed in this paper, we plan to investigate the initial problems for quasilinear equations with distributed Riemann–Liouville, Hilfer, φ-Hilfer fractional derivatives [39], as well as other integrodifferential operators.

Author Contributions: Conceptualization, V.E.F.; methodology, V.E.F. and N.V.F.; software, N.V.F.; validation, N.V.F.; formal analysis, N.V.F.; investigation, V.E.F. and N.V.F.; resources, N.V.F.; data curation, N.V.F.; writing—original draft preparation, V.E.F.; writing—review and editing, V.E.F.; visualization, N.V.F.; supervision, V.E.F.; project administration, V.E.F.; funding acquisition, V.E.F. All authors have read and agreed to the published version of the manuscript.

Funding: The work was performed as part of research conducted in the Ural Mathematical Center with the financial support of the Ministry of Science and Higher Education of the Russian Federation (Agreement number 075-02-2023-913).

Data Availability Statement: No new data were created.

Conflicts of Interest: The authors declare no conflict of interest. The funders had no role in the design of the study; in the collection, analysis, or interpretation of data; in the writing of the manuscript; or in the decision to publish the results.

References

1. Gerasimov, A.N. Generalization of linear laws of deformation and their application to problems of internal friction. *Prikl. Mat. Mekhanika* **1948**, *12*, 251–260. (In Russian)
2. Caputo, M. Linear model of dissipation whose Q is almost frequancy independent. II. *Geophys. J. R. Soc.* **1967**, *13*, 529–539. [CrossRef]
3. Nakhushev, A.M. *Fractional Calculus ant Its Applications*; Fizmatlit: Moscow, Russia, 2003. (In Russian)
4. Uchaykin, V.V. *Fractional Derivatives for Physicists and Engineers*; Higher Education Press: Beijing, China, 2012.
5. Samko, S.G.; Kilbas, A.A.; Marichev, O.I. *Fractional Integrals and Derivatives. Theory and Applications*; Gordon and Breach Science: Philadelphia, PA, USA, 1993.
6. Podlubny, I. *Fractional Differential Equations*; Academic: Boston, MA, USA, 1999.
7. Pskhu, A.V. *Partial Differential Equations of Fractional Order*; Nauka: Moscow, Russia, 2005. (In Russian)
8. Kilbas, A.A.; Srivastava, H.M.; Trujillo, J.J. *Theory and Applications of Fractional Differential Equations*; Elsevier Science Publishing: Amsterdam, The Netherlands; Boston, MA, USA; Heidelberg, Germany, 2006.
9. Caputo, M. Mean fractional order derivatives. Differential equations and filters. *Ann. dell'Universita di Ferrara. Sez. VII Sci. Mat.* **1995**, *41*, 73–84. [CrossRef]
10. Caputo, M. Distributed order differential equations modeling dielectric induction and diffusion. *Fract. Calc. Appl. Anal.* **2001**, *4*, 421–442.
11. Kochubei, A.N. Distributed order calculus and equations of ultraslow diffusion. *J. Math. Anal. Appl.* **2008**, *340*, 252–280. [CrossRef]
12. Sokolov, I.M.; Chechkin, A.V.; Klafter, J. Distributed-order fractional kinetics. *Acta Phys. Pol. B* **2004**, *35*, 1323–1341.
13. Nakhushev, A.M. Positiveness of the operators of continual and discrete differentiation and integration, which are quite important in the fractional calculus and in the theory of mixed-type equations. *Differ. Equ.* **1998**, *34*, 103–112.
14. Bagley, R.L.; Torvik, P.J. On the existence of the order domain and the solution of distributed order equations. Part 1. *Int. J. Appl. Math.* **2000**, *2*, 865–882.

15. Lorenzo, C.F.; Hartley, T.T. Variable order and distributed order fractional operators. *Nonlinear Dyn.* **2002**, *29*, 57–98. [CrossRef]
16. Jiao, Z.; Chen, Y.; Podlubny, I. *Distributed-Order Dynamic System. Stability, Simulations, Applications and Perspectives*; Springer: London, UK, 2012.
17. Pskhu, A.V. On the theory of the continual and integro-differentiation operator. *Differ. Equ.* **2004**, *40*, 128–136. [CrossRef]
18. Pskhu, A.V. Fractional diffusion equation with a discretely distributed differentiation operator. *Sib. Elektron. Math. Rep.* **2016**, *13*, 1078–1098.
19. Umarov, S.; Gorenflo, R. Cauchy and nonlocal multi-point problems for distributed order pseudo-differential equations. *Z. Anal. Anwend.* **2005**, *24*, 449–466.
20. Atanacković, T.M.; Oparnica, L.; Pilipović, S. On a nonlinear distributed order fractional differential equation. *J. Math. Anal. Appl.* **2007**, *328*, 590–608. [CrossRef]
21. Efendiev, B.I. Steklov problem for a second-order ordinary differential equation with a continual derivative. *Differ. Equ.* **2013**, *49*, 450–456. [CrossRef]
22. Efendiev, B.I. Lagrange formula for ordinary continual second-order differential equations. *Differ. Equ.* **2017**, *53*, 736–744. [CrossRef]
23. Streletskaya, E.M.; Fedorov, V.E.; Debbouche, A. The Cauchy problem for distributed order equations in Banach spaces. *Math. Notes NEFU* **2018**, *25*, 63–72. (In Russian)
24. Fedorov, V.E.; Streletskaya, E.M. Initial-value problems for linear distributed-order differential equations in Banach spaces. *Electron. J. Differ. Equ.* **2018**, *2018*, 176.
25. Fedorov, V.E.; Phuong, T.D.; Kien, B.T.; Boyko, K.V.; Izhberdeeva, E.M. A class of semilinear distributed order equations in Banach spaces. *Chelyabinsk Phys. Math. J.* **2020**, *5*, 343–351.
26. Fedorov, V.E. On generation of an analytic in a sector resolving operators family for a distributed order equation. *Zap. POMI* **2020**, *489*, 113–129. [CrossRef]
27. Fedorov, V.E. Generators of analytic resolving families for distributed order equations and perturbations. *Mathematics* **2020**, *8*, 1306. [CrossRef]
28. Fedorov, V.E.; Filin, N.V. Linear equations with discretely distributed fractional derivative in Banach spaces. *Tr. Instituta Mat. i Mekhaniki UrO RAN* **2021**, *27*, 264–280.
29. Sitnik, S.M.; Fedorov, V.E.; Filin, N.V.; Polunin, V.A. On the solvability of equations with a distributed derivative given by the Stieltjes integral. *Mathematics* **2022**, *10*, 2979. [CrossRef]
30. Novozhenova, O.G. Life and science of Alexey Gerasimov, one of the pioneers of fractional calculus in Soviet Union. *Fract. Calc. Appl. Anal.* **2017**, *20*, 790–809. [CrossRef]
31. Prüss, J. *Evolutionary Integral Equations and Applications*; Springer: Basel, Switzerland, 1993.
32. Bajlekova, E.G. Fractional Evolution Equations in Banach Spaces. Ph.D. Thesis, Eindhoven University of Technology, Eindhoven, The Netherlands, 2001.
33. Hille, E.; Phillips, R.S. *Functional Analysis and Semi-Groups*; American Mathematical Society: Providence, RI, USA, 1957.
34. Yosida, K. *Functional Analysis*; Springer: Berlin/ Heidelberg, Germany, 1965.
35. Kato, K. *Perturbation Theory for Linear Operators*; Springer: Berlin/Heidelberg, Germany, 1966.
36. Caginalp, G. An analysis of a phase field model of a free boundary. *Arch. Ration. Mech. Anal.* **1986**, *92*, 205–245. [CrossRef]
37. Caginalp, G. Stefan and Hele–Shaw type models as asymptotic limits of the phase-field equations. *Phys. Rev. A* **1989**, *39*, 5887–5896. [CrossRef] [PubMed]
38. Hassard, B.D.; Kazarinoff, N.D.; Wan, Y.-H. *Theory and Applications of Hopf Bifurcation*; Cambridge University Press: Cambridge, UK, 1981.
39. Guechi, S.; Dhayal, R.; Debbouche, A.; Malik, M. Analysis and optimal control of φ-Hilfer fractional semilinear equations involving nonlocal impulsive conditions. *Symmetry* **2021**, *13*, 2084. [CrossRef]

mathematics

MDPI

Article

Trace Formulae for Second-Order Differential Pencils with a Frozen Argument

Yi-Teng Hu [1,*] and Murat Şat [2]

[1] School of Mathematics and Statistics, Xidian University, Xi'an 710126, China
[2] Department of Mathematics, Faculty of Science and Art, Erzincan Binali Yildirim University, Erzincan 24100, Turkey; msat@erzincan.edu.tr
* Correspondence: ythu@xidian.edu.cn

Abstract: This paper deals with second-order differential pencils with a fixed frozen argument on a finite interval. We obtain the trace formulae under four boundary conditions: Dirichlet–Dirichlet, Neumann–Neumann, Dirichlet–Neumann, Neumann–Dirichlet. Although the boundary conditions and the corresponding asymptotic behaviour of the eigenvalues are different, the trace formulae have the same form which reveals the impact of the frozen argument.

Keywords: differential pencils; regularized trace formulae; frozen argument

MSC: 34A55; 34K29; 47E05

Citation: Hu, Y.-T.; Şat, M. Trace Formulae for Second-Order Differential Pencils with a Frozen Argument. *Mathematics* **2023**, *11*, 3996. https://doi.org/10.3390/math11183996

Academic Editors: Luigi Rodino and Simeon Reich

Received: 30 June 2023
Revised: 4 September 2023
Accepted: 18 September 2023
Published: 20 September 2023

1. Introduction

In this paper, we consider boundary value problem generated by

$$-y''(x) + [q_0(x) + \rho q_1(x)]y(a) = \rho^2 y(x), \quad 0 < x < 1, \tag{1}$$

and boundary conditions

$$y^{(\alpha)}(0) = y^{(\beta)}(1) = 0, \tag{2}$$

where $a \in (0,1)$, $q_j(x), j = 0,1$ are complex-valued functions in Sobolev space $W_2^j[0,1]$, ρ is the spectral parameter and $\alpha, \beta \in \{0,1\}$. We note that $(\alpha,\beta) = (0,0),(1,1),(0,1)$ or $(1,0)$ represent Dirichlet–Dirichlet, Neumann–Neumann, Dirichlet–Neumann, Neumann–Dirichlet boundary conditions correspondingly. We denote the corresponding operator by $L_{\alpha,\beta} = L(q_0,q_1,\alpha,\beta,a)$. We call $L_{\alpha,\beta}$ the second-order differential pencils with frozen argument. Specifically, we deduce the trace formulae for $L_{\alpha,\beta}$.

Trace is an important conserved quantity in matrix theory. In finite dimensional space, the sum of principal diagonal elements of a matrix equals to the sum of eigenvalues which we call the trace. While considering the differential operators in the Hilbert space, however, a sum of infinitely many eigenvalues leads to a divergence series. In 1953, for the first time, Gelfand and Levitan [1] introduced an interesting formula for the Sturm–Liouville operator:

$$\sum_{n=0}^{\infty}\left[\lambda_n - n^2 - \frac{1}{\pi}\int_0^{\pi} q(t)dt\right] = \frac{1}{4}[q(0)+q(\pi)] - \frac{1}{2\pi}\int_0^{\pi} q(t)dt, \tag{3}$$

where the operator was generated by the Neumann–Neumann-type boundary problem

$$-y''(x) + q(x)y(x) = \lambda y(x), \quad y'(0) = y'(\pi) = 0,$$

and $q(x) \in C^1[0,\pi]$, λ_n are the corresponding eigenvalues. After that, many scholars put attention to this quantity, which has many applications in integrable system theory and the inverse problem [2–6]. Also, it turned out that regularized trace formulae had physical meaning, as discussed by Sadovnichii and Podol'skii [7]: *"The meaning of the regularized trace as a measure of the defect of the total energy of the system when it is perturbed in the case when the total energy itself of the system (more precisely, of the model under consideration) is infinite $\cdots$ "*.

In physics, the interactions between colliding particles is of fundamental significance. For example, Jaulent and Jean [8] describe this phenomenon by s-wave Schrödinger equation with a radial static potential $V(x)$:

$$-y''(x) + V(E,x)y(x) = Ey(x),$$

where $V(E,x)$ is the following form for the energy dependence:

$$V(E,x) = Q(x) + 2\sqrt{E}P(x).$$

With an additional condition $Q(x) = -P^2(x)$, the above Schrödinger equation reduces to the Klein–Gordon equation for a particle of zero mass and energy E, which could serve as part of Lax pair in a two-component Camassa–Holm Equation [9]. Due to the nonlinear dependence on the spectral parameter, the corresponding inverse spectral problem and inverse scattering theory are difficult; we refer to papers [10–16]. With the method of asymptotic estimation on the family of contours [17], Cao and Zhuang [18] studied regularized traces of the Schrödinger equation with energy-dependent potential, in which the final quantity only contains the term with $P(x)$. Further, Yang [19–21] obtained some new formulae related to both $P(x)$ and $Q(x)$.

Recently, the Sturm–Liouville equation with the frozen argument of the form

$$-y''(x) + q(x)y(a) = \lambda y(x), \quad x \in (0,1) \tag{4}$$

has attracted much attention. This equation can be classified as a special case of a functional differential equation with a deviating argument. Especially Equation (4) belongs to the class of loaded equations [22] which arise in mathematical physics, such as groundwater dynamics [23,24], heat conduction [25,26], system with energy feedback [27].

For the inverse spectral problem of differential operators with frozen argument, the classical approaches like the method of spectral mappings and the Gelfand–Levian–Marchenko method do not work. Albeverio et al. [28] and Nizhnik [29,30] studied some special cases where the nonlocal boundary condition guarantees the self-adjointness of the corresponding operator. Bondarenko et al. [31] studied Equation (4) with Boundary conditions (2) where $1/a \in \mathbb{N}$ and $\alpha, \beta \in \{0,1\}$. They classified two cases: degenerate and non-degenerate, depending on the values of α, β and on the parity of $k = 1/a$. Moreover, Bondarenko et al. established the unique solvability of the inverse problem. For the study of different aspects of this operator, such as arbitrary $a \in (0,1)$, non-separated boundary conditions, etc., we refer to [32–40]. Namely, Kuznetsova [41,42] proved the well-posedness of the inverse spectral problem generated by (4) and (2) by a new approach, which is effective in both the rational and irrational cases. Bondarenko [43] explained the relation between the Sturm–Liouville operators with frozen argument and the Laplace operator with integral matching conditions on a star-shaped graph. Also, as pointed out by Buterin [44], the frozen argument term appeared naturally in the study of a Sturm–Liouville operator with constant delay.

However, there are few works on differential pencils with frozen argument. Equation (1) appears, for example, after applying the Fourier method of separation of variables to the following loaded hyperbolic equation:

$$\frac{\partial^2}{\partial t^2}u(x,t) = \frac{\partial^2}{\partial x^2}u(x,t) - (\lambda r(x) + q(x))u(a,t),\ 0 < x < 1,\ t > 0,$$

where $a \in (0,1)$, λ is a spectral parameter, $r(x)$ is called the loss function and $q(x)$ the impedance function. This model arises in the study of inverse scattering in lossy layered media [45]; moreover, we assume that the model is affected by a magnetic field exerting a force-per-unit mass represented by $-(\lambda r(x)+q(x))u(a,t)$, i.e., depending on the lateral displacement $u(a,t)$ at point a at time t. The author of [46] studied the inverse spectral problem for $L_{\alpha,\beta}$ by using the approach suggested in [31]. Namely, $L_{\alpha,\beta}$ share the same degenerate and non-degenerate conditions with boundary value problem (4) and (2).

Motivated by the works of Cao and Zhuang [18], we focus on regularized traces of $L_{\alpha,\beta}$. For the next section, we recall some basic facts from [46], i.e., the integral equation for the characteristic functions of $L_{\alpha,\beta}$ and asymptotic behaviour of the corresponding eigenvalues; then, we provide the main results. Finally, we offer a conclusion.

2. Preliminaries and Main Results

We let $C(x,\rho),\ S(x,\rho)$ be the solutions of Equation (1) under the initial conditions

$$C(a,\rho)=S'(a,\rho)=1,\quad S(a,\rho)=C'(a,\rho)=0.$$

It is easy to verify that

$$C(x,\rho)=\cos\rho(x-a)+\int_a^x q_1(t)\sin\rho(x-t)dt+\int_a^x q_0(t)\frac{\sin\rho(x-t)}{\rho}dt, \tag{5}$$

$$S(x,\rho)=\frac{\sin\rho(x-a)}{\rho}. \tag{6}$$

Integrating by parts the second term of (5), we obtain

$$\begin{aligned} C(x,\rho)=&\cos\rho(x-a)+\frac{1}{\rho}(q_1(x)-q_1(a)\cos\rho(x-a))\\ &-\frac{1}{\rho}\int_a^x q_1'(t)\cos\rho(x-t)dt+\frac{1}{\rho}\int_a^x q_0(t)\sin\rho(x-t)dt. \end{aligned} \tag{7}$$

We define

$$\Delta_{\alpha,\beta}(\rho)=\begin{vmatrix} C^{(\alpha)}(0,\rho) & S^{(\alpha)}(0,\rho)\\ C^{(\beta)}(1,\rho) & S^{(\beta)}(1,\rho)\end{vmatrix}; \tag{8}$$

then, it is easy to verify that the eigenvalues of $L_{\alpha,\beta}$ coincide with the zeros of $\Delta_{\alpha,\beta}(\rho)$.

We note that the spectrum of the operator $L(q_0(1-x),q_1(1-x),1-a,\alpha,\beta)$ coincides with the one of $L(q_0(x),q_1(x),a,\alpha,\beta)$; without loss of generality, we assume $0<a\leq 1/2$ for definiteness.

Theorem 1 ([46]). *The characteristic functions $\Delta_{\alpha,\beta}(\rho)$ of the problem $L_{\alpha,\beta}$ have the form of*

$$\Delta_{\alpha,\alpha}(\rho)=\rho^{2\alpha-2}\left(\rho\sin\rho-W_{\alpha,\alpha}(a,\rho)+\int_0^1(U_{\alpha,\alpha}(t)\cos\rho t+V_{\alpha,\alpha}(t)\sin\rho t)\,dt\right), \tag{9}$$

if $\alpha=\beta$, and

$$\Delta_{\alpha,\beta}(\rho)=\rho^{-1}\left((-1)^{\alpha}(\rho\cos\rho-W_{\alpha,\beta}(a,\rho))+\int_0^1(U_{\alpha,\beta}(t)\sin\rho t+V_{\alpha,\beta}(t)\cos\rho t)\,dt\right), \tag{10}$$

if $\alpha \neq \beta$, *where*

$$W_{\alpha,\beta}(a,\rho) = \begin{cases} q_1(a)\sin\rho - q_1(0)\sin\rho(1-a) - q_1(1)\sin\rho a, & (\alpha,\beta) = (0,0), \\ q_1(a)\sin\rho, & (\alpha,\beta) = (1,1), \\ q_1(a)\cos\rho - q_1(0)\cos\rho(1-a), & (\alpha,\beta) = (0,1), \\ q_1(a)\cos\rho - q_1(1)\cos\rho a, & (\alpha,\beta) = (1,0). \end{cases}$$

Moreover, the functions $U_{\alpha,\beta}(t)$ *and* $V_{\alpha,\beta}(t)$ *have the following form:*

$$U_{\alpha,\beta}(t) = \frac{(-1)^{\alpha\beta}}{2} \begin{cases} q_0(1-a+t) + dq_0(1-a-t), & t \in (0,a), \\ cq_0(1+a-t) + dq_0(1-a-t), & t \in (a,1-a), \\ c(q_0(1+a-t) + q_0(t-1+a)), & t \in (1-a,1) \end{cases} \tag{11}$$

and

$$V_{\alpha,\beta}(t) = \frac{(-1)^{\gamma}}{2} \begin{cases} -q_1'(1-a+t) + dq_1'(1-a-t), & t \in (0,a), \\ cq_1'(1+a-t) + dq_1'(1-a-t), & t \in (a,1-a), \\ c(q_1'(1+a-t) - q_1'(t-1+a)), & t \in (1-a,1), \end{cases} \tag{12}$$

where $c = (-1)^{1+\beta}$, $d = (-1)^{\alpha+\beta}$ *and* $\gamma = \max\{\alpha,\beta\}$.

We let $\mathbb{Z}_0 := \mathbb{Z} \setminus \{0\}$, $\mathbb{Z}_1 := \{\pm 0, \pm 1, \pm 2, \cdots\}$ and $\mathbb{Z}_2 := \mathbb{Z}$. From this, we stipulate that if n denotes an index for eigenvalues, then $n \in \mathbb{Z}_0$ for $(\alpha,\beta) = (0,0)$, $n \in \mathbb{Z}_1$ for $(\alpha,\beta) = (1,1)$ and $n \in \mathbb{Z}_2$ for $(\alpha,\beta) = (0,1)$ or $(\alpha,\beta) = (1,0)$.

Theorem 2 ([46]). *The eigenvalues of* $L_{\alpha,\beta}$ *can be numbered as* $\{\rho_{n,\alpha,\beta}\}$, *counting with their multiplicities, such that the following asymptotics hold:*

(i) For $(\alpha,\beta) = (0,0)$,

$$\rho_{n,0,0} = n\pi + \frac{q_1(0) + (-1)^{n+1}q_1(1)}{n\pi}\sin n\pi a + \frac{\kappa_{0,0,n}}{n}, \{\kappa_{n,0,0}\} \in l_2; \tag{13}$$

(ii) For $(\alpha,\beta) = (1,1)$,

$$\rho_{n,1,1} = n\pi + \frac{\kappa_{1,1,n}}{n}, \ \{\kappa_{n,1,1}\} \in l_2; \tag{14}$$

(iii) For $(\alpha,\beta) = (0,1)$,

$$\rho_{n,0,1} = \left(n - \frac{1}{2}\right)\pi + \frac{q_1(0)}{n\pi}\sin\left(n - \frac{1}{2}\right)\pi a + \frac{\kappa_{0,1,n}}{n}, \ \{\kappa_{n,0,1}\} \in l_2; \tag{15}$$

(iv) For $(\alpha,\beta) = (1,0)$,

$$\rho_{n,1,0} = \left(n - \frac{1}{2}\right)\pi + \frac{(-1)^{n+1}q_1(1)}{n\pi}\cos\left(n - \frac{1}{2}\right)\pi a + \frac{\kappa_{1,0,n}}{n}, \ \{\kappa_{n,1,0}\} \in l_2. \tag{16}$$

In order to obtain the trace formulae of $L_{\alpha,\beta}$, we need the following lemma.

Lemma 1 ([17]). *Let* $\omega(z)$ *and* $\omega_0(z)$ *be two entire functions on a* $z-$*plane and have no zeros on some closed contour* Γ. *Suppose that* $\omega(z)\backslash\omega_0(z) = 1 + \theta(z)$, *where* $|\theta(z)| \leq \delta$ *on* Γ, $0 < \delta < 1$; *then,*

$$\sum_{\Gamma}(\lambda_n^\sigma - \mu_n^\sigma) = -\frac{1}{2\pi i}\oint_\Gamma \sigma z^{\sigma-1}\ln\frac{\omega(z)}{\omega_0(z)}dz, \tag{17}$$

where λ_n *and* μ_n *are zeros of* $\omega(z)$ *and* $\omega_0(z)$ *inside* Γ *correspondingly, and* σ *is a positive integer.*

We let $\{\tau_{n,\alpha,\beta}\}$ be the spectrum of $L(0,0,\alpha,\beta,a)$, $\alpha,\beta\in\{0,1\}$.

Theorem 3. *The following formulae hold:*

$$\sum_{n\in\mathbb{Z}_j}(\rho_{n,\alpha,\beta}-\tau_{n,\alpha,\beta})=q_1(a), \tag{18}$$

where $j=0,1,2$.

Proof. We let Γ_N, $N=1,2,\cdots$ be the counterclockwise square contours $A_NB_NC_ND_N$ with

$$A_N=\left(N+\frac{3}{4}\right)(1-i),\ B_N=\left(N+\frac{3}{4}\right)(1+i),$$
$$C_N=\left(N+\frac{3}{4}\right)(-1+i),\ D_N=\left(N+\frac{3}{4}\right)(-1-i).$$

Formulae (13)–(16) imply that, for sufficiently large N, the eigenvalues $\rho_{n,\alpha,\beta}$, $|n|\le N$ are inside Γ_N, and the eigenvalues $\rho_{n,\alpha,\beta}$ with $|n|>N$ are outside Γ_N. Also, since $\{\tau_{n,\alpha,\beta}\}$ is the the spectrum of $L(0,0,\alpha,\beta,a)$, we have $\{\tau_{n,\alpha,\beta}\}\cap\Gamma_N=\varnothing$.

Now we prove the theorem for the case $(\alpha,\beta)=(0,0)$; the other cases are similar. We let $\Delta^{\circ}_{0,0}(\rho)=\sin\rho/\rho$ be the characteristic function of $L(0,0,0,0,a)$. By using (9), (11) and (12), we estimate the fraction $\Delta_{0,0}(\rho)/\Delta^{\circ}_{0,0}(\rho)$ on the contour Γ_N for sufficiently large N:

$$\frac{\Delta_{0,0}(\rho)}{\Delta^{\circ}_{0,0}(\rho)}=1+\frac{q_1(0)\sin\rho(1-a)+q_1(1)\sin\rho a}{\rho\sin\rho}-\frac{q_1(a)}{\rho}+o\left(\frac{1}{\rho}\right),\rho\in\Gamma_N.$$

Using the Taylor series expansion, we have

$$\ln\frac{\Delta_{0,0}(\rho)}{\Delta^{\circ}_{0,0}(\rho)}=\frac{q_1(0)\sin\rho(1-a)+q_1(1)\sin\rho a}{\rho\sin\rho}-\frac{q_1(a)}{\rho}+o\left(\frac{1}{\rho}\right),\rho\in\Gamma_N.$$

Calculating the contour integral by (17) and using residue calculation, we obtain that for sufficiently large N,

$$\begin{aligned}\sum_{n=-N}^{N}(\rho_{0,0,n}-\mu_{0,0,n})&=-\frac{1}{2\pi i}\oint_{\Gamma_N}\ln\frac{\Delta_{0,0}(\rho)}{\Delta^{\circ}_{0,0}(\rho)}d\rho\\&=q_1(a)+q_1(0)\left(2\sum_{n=1}^{N}\theta_n-(1-a)\right)+q_1(1)\left(2\sum_{n=1}^{N}\zeta_n-a\right)+o(1),\end{aligned}$$

where

$$\theta_n=(-1)^{n+1}\frac{\sin n\pi(1-a)}{n\pi},\qquad \zeta_n=(-1)^{n+1}\frac{\sin n\pi a}{n\pi}.$$

Together with the Fourier series

$$x=2\sum_{n=1}^{\infty}(-1)^{n+1}\frac{\sin n\pi x}{n\pi},\quad x\in(-1,1),$$

we arrive at (18) for $(\alpha,\beta)=(0,0)$ by taking $N\to\infty$. Note that for the cases $(\alpha,\beta)=(0,1)$ and $(\alpha,\beta)=(1,0)$, we need the Fourier series expansion

$$\frac{1}{2}=2\sum_{n=1}^{\infty}(-1)^{n+1}\frac{\cos(n-\frac{1}{2})\pi x}{(n-\frac{1}{2})\pi},x\in(-1,1).$$

□

3. Conclusions

In this paper, we deduce the trace formulae of second-order differential pencils with frozen argument. By applying the methods in complex analysis, we calculate the regularized sum of infinite eigenvalues of $L_{\alpha,\beta}$ in the Gelfand–Levitan sense. Let us mention some advantages of our approach:

1. Operator $L_{\alpha,\beta}$ is non-selfadjoint which may have complex eigenvalues with multiplicity; however, the method we use allows us dealing with the regularized sum of eigenvalues in the whole meaning.
2. The regularized trace of $L_{\alpha,\beta}$ depends only on the value of $q_1(x)$ at the frozen point a, regardless of the boundary conditions and the potential $q_0(x)$.
3. In the study of inverse spectral problem of $L_{\alpha,\beta}$, the rationality of frozen argument a is important. Whether a is rational leads to different approachs of inverse spectral problem. However, we do not need this distinction while calculating the trace formulae.

Author Contributions: Methodology, Y.-T.H.; Writing—original draft, Y.-T.H.; Writing—review & editing, M.Ş. All authors have read and agreed to the published version of the manuscript.

Funding: The author Hu was supported by the Fundamental Research Funds for the Central Universities (XJS220703, ZYTS23049).

Acknowledgments: The authors would like to thank the anonymous referees and Natalia P. Bondarenko, Department of Applied Mathematics and Physics, Samara National Research University, Samara, Russia, for valuable suggestions, which helped to improve the readability and quality of the paper.

Conflicts of Interest: The authors declare no conflict of interest.

References

1. Gelfand, I.M.; Levitan, B.M. On a formula for eigenvalues of a differential operator of second order. *Dokl. Akad. Nauk SSSR* **1953**, *88*, 593–596. (In Russian)
2. Gesztesy, F.; Holden, H. On trace formulas for Schrödinger-type operators. In *Multiparticle Quantum Scattering with Applications to Nuclear, Atomic and Molecular Physics*; The IMA Volumes in Mathematics and Its Applications Series; Springer: New York, NY, USA, 1977; Volume 89, pp. 121–145.
3. Trubowitz, E. The inverse problem for periodic potentials. *Comm. Pure Appl. Math.* **1977**, *30*, 321–337. [CrossRef]
4. Lax, P.D. Trace formulas for the Schrödinger operator. *Comm. Pure Appl. Math.* **1994**, *47*, 503–512. [CrossRef]
5. Gesztesy, F.; Ratnaseelan, R.; Teschl, G. The KdV hierarchy and associated trace formulas. In Proceedings of the International Conference on Applications of Operator Theory, Winnipeg, MB, Canada, 2–6 October 1994 ; Gohberg, I., Lancaster, P., Shivakumar, P.N., Eds.; Operator Theory: Advances and Applications Series; Birkhäuser: Basel, Switzerland, 1996; Volume 87, pp. 125–163.
6. Gesztesy, F.; Simon, B. The xi function. *Acta Math.* **1996**, *176*, 49–71. [CrossRef]
7. Sadovnichii, V.A.; Podol'skii, V.E. Traces of operators. *Russ. Math. Surv.* **2006**, *61*, 885–953. [CrossRef]
8. Jaulent, M.; Jean, C. 1972 The inverse s-wave scattering problem for a class of potentials depending on energy. *Commun. Math. Phys.* **1972**, *28*, 177–220. [CrossRef]
9. Chen, M.; Liu, S.Q.; Zhang, Y. A two-component generalization of the Camassa-Holm equation and its solutions. *Lett. Math. Phys.* **2006**, *75*, 1–15. [CrossRef]
10. Gasymov, M.G.; Guseinov, G.S. Determination of diffusion operator from spectral data. *Akad. Nauk Azerb. SSR Dokl.* **1981**, *37*, 19–23. (In Russian)
11. Guseinov, I.M.; Nabiev, I.M. A class of inverse problems for a quadratic pencil of Sturm-Liouville operators. *Diff. Equ.* **2000**, *36*, 471–473. [CrossRef]
12. Guseinov, I.M.; Nabiev, I.M. The inverse spectral problem for pencils of differential operators. *Sb. Math.* **2007**, *198*, 1579–1598. [CrossRef]
13. Buterin, S.A.; Yurko, V.A. Inverse problems for second-order differential pencils with Dirichlet boundary conditions. *J. Inverse Ill-Posed Probl.* **2012**, *20*, 855–881. [CrossRef]
14. Hryniv, R.O.; Pronska, N. Inverse spectral problems for energy-dependent Sturm-Liouville equations. *Inverse Probl.* **2012**, *28*, 085008. [CrossRef]
15. Pronska, N. Reconstruction of energy-dependent Sturm-Liouville quations from two spectra. *Integral Equ. Oper. Theory* **2013**, *76*, 403–419. [CrossRef]
16. Hryniv, R.O.; Manko, S.S. Inverse scattering on the half-line for energy-dependent Schrödinger equations. *Inverse Probl.* **2020**, *36*, 095002. [CrossRef]

17. Cao, C. Asymptotic traces of non-self-adjoint Sturm-Liouville operators. *Acta Math. Sci.* **1981**, *241*, 84–94. (In Chinese)
18. Cao, C.; Zhuang, D. Some trace formulas for the Schrödinger equation with energy-dependent potential. *Acta Math. Sci.* **1985**, *5*, 131–140. [CrossRef]
19. Yang, C.F. New trace formulae for a quadratic pencil of the Schrödinger operator. *J. Math. Phys.* **2010**, *51*, 033506. [CrossRef]
20. Yang, C.F.; Huang, Z.Y.; Wang, Y.P. Trace formulae for the Schrödinger equation with energy-dependent potential. *J. Phys. A Math. Theor.* **2010**, *43*, 415207. [CrossRef]
21. Yang, C.F. Identities for eigenvalues of the Schrödinger equation with energy-dependent potential. *Z. Naturforsch. A* **2011**, *66*, 699–704. [CrossRef]
22. Nakhushev, A.M. *Loaded Equations and Their Applications*; Nauka: Moscow, Russia, 2012.
23. Nakhushev, A.M.; Borisov, V.N. Boundary value problems for loaded parabolic equations and their applications to the prediction of ground water level. *Differ. Equ.* **1977**, *13*, 105–110.
24. Nakhushev, A.M. An approximate method for solving boundary value problems for differential equations and its application to the dynamics of ground moisture and ground water. *Differ. Equ.* **1982**, *18*, 72–81.
25. Iskenderov, A.D. The first boundary-value problem for a loaded system of quasilinear parabolic equations. *Differ. Equ.* **1971**, *7*, 1911–1913.
26. Dikinov, K.; Kerefov, A.A.; Nakhushev, A.M. A certain boundary value problem for a loaded heat equation. *Differ. Equ.* **1976**, *12*, 177–179.
27. Krall, A.M. The development of general differential and general differential-boundary systems. *Rocky Mt. J. Math.* **1975**, *5*, 493–542. [CrossRef]
28. Albeverio, S.; Hryniv, R.O.; Nizhnik, L.P. Inverse spectral problems for non-local Sturm-Liouville operators. *Inverse Probl.* **2007**, *23*, 523–535. [CrossRef]
29. Nizhnik, L.P. Inverse eigenvalue problems for nonlocal Sturm-Liouville operators. *Methods Funct. Anal. Topol.* **2009**, 15, 41–47.
30. Nizhnik, L.P. Inverse nonlocal Sturm-Liouville problem. *Inverse Probl.* **2010**, *26*, 125006. [CrossRef]
31. Bondarenko, N.P.; Buterin, S.V.; Vasiliev, S.V. An inverse spectral problem for Sturm-Liouville operators with frozen argument. *J. Math. Anal. Appl.* **2019**, *472*, 1028–1041. [CrossRef]
32. Buterin, S.A.; Vasiliev, S.V. On recovering a Sturm-Liouville-type operator with the frozen argument rationally proportioned to the interval length. *J. Inverse Ill-Posed Probl.* **2019**, 27, 429–438. [CrossRef]
33. Buterin, S.A.; Kuznetsova, M. On the inverse problem for Sturm-Liouville-type operators with frozen argument: Rational case. *Comput. Appl. Math.* **2020**, 39, 15. [CrossRef]
34. Hu, Y.T.; Bondarenko, N.P.; Yang, C.F. Traces and inverse nodal problem for Sturm-Liouville operators with frozen argument. *Appl. Math. Lett.* **2020**, 102, 106096. [CrossRef]
35. Hu, Y.T.; Huang, Z.Y.; Yang, C.F. Traces for Sturm-Liouville operators with frozen argument on star graphs. *Results Math.* **2020**, 75, 9. [CrossRef]
36. Buterin, S.A.; Hu, Y.T. Inverse spectral problems for Hill-type operators with frozen argument. *Anal. Math. Phys.* **2021**, 11, 22. [CrossRef]
37. Wang, Y.P.; Zhang, M.; Zhao, W.; Wei, X. Reconstruction for Sturm-Liouville operators with frozen argument for irrational cases. *Appl. Math. Lett.* **2021**, *111*, 106590. [CrossRef]
38. Bondarenko, N.P. Finite-difference approximation of the inverse Sturm-Liouville problem with frozen argument. *Appl. Math. Comput.* **2022**, *413*, 126653. [CrossRef]
39. Dobosevych, O.; Hryniv, R.O. Reconstruction of differential operators with frozen argument. *Axioms* **2022**, *11*, 24. [CrossRef]
40. Tsai, T.M.; Liu, H.F.; Buterin, S.A.; Chen, L.H.; Shieh, C.T. Sturm-Liouville-type operators with frozen argument and Chebyshev polynomials. *Math. Methods Appl. Sci.* **2022**, *45*, 9635–9652. [CrossRef]
41. Kuznetsova, M. Necessary and sufficient conditions for the spectra of the Sturm-Liouville operators with frozen argument. *Appl. Math. Lett.* **2022**, *131*, 108035. [CrossRef]
42. Kuznetsova, M. Uniform stability of recovering the Sturm-Liouville operators with frozen argument. *Results Math.* **2023**, *78*, 169. [CrossRef]
43. Bondarenko, N.P. Inverse problem for a differential operator on a star-shaped graph with nonlocal matching condition. *Boletn de la Sociedad Matemática Mex.* **2023**, *29*, 27. [CrossRef]
44. Buterin, S.; Vasilev, S. An inverse Sturm-Liouville-type problem with constant delay and non-zero initial function. *arXiv* **2023**, arXiv:2304.05487.
45. Borcea, L.; Vladimir, D.; Jörn, Z. A reduced order model approach to inverse scattering in lossy layered media. *J. Sci. Comput.* **2021**, *89*, 36. [CrossRef]
46. Hu, Y.T.; Sat, M. Inverse spectral problem for differential pencils with a frozen argument. *arXiv* **2023**, arXiv:2305.02529.

mathematics

MDPI

Article

Spectral, Scattering and Dynamics: Gelfand–Levitan–Marchenko–Krein Equations

Sergey Kabanikhin [1,2,3], **Maxim Shishlenin** [1,2,3,*], **Nikita Novikov** [1,2,3,*] **and Nikita Prokhoshin** [3]

[1] Institute of Computational Mathematics and Mathematical Geophysics SB RAS, Prospect Akademika Lavrentjeva, 6, 630090 Novosibirsk, Russia; kabanikhin@sscc.ru
[2] Sobolev Institute of Mathematics SB RAS, Akad. Koptyug Avenue, 4, 630090 Novosibirsk, Russia
[3] Department of Mathematics and Mechanics, Novosibirsk State University, Pirogova St., 2, 630090 Novosibirsk, Russia; nikita.prokhoshin@gmail.com
[*] Correspondence: mshishlenin@ngs.ru (M.S.); novikov-1989@yandex.ru (N.N.)

Abstract: In this paper, we consider the Gelfand–Levitan–Marchenko–Krein approach. It is used for solving a variety of inverse problems, like inverse scattering or inverse problems for wave-type equations in both spectral and dynamic formulations. The approach is based on a reduction of the problem to the set of integral equations. While it is used in a wide range of applications, one of the most famous parts of the approach is given via the inverse scattering method, which utilizes solving the inverse problem for integrating the nonlinear Schrodinger equation. In this work, we present a short historical review that reflects the development of the approach, provide the variations of the method for 1D and 2D problems and consider some aspects of numerical solutions of the corresponding integral equations.

Keywords: Gelfand–Levitan–Krein–Marchenko equation; inverse coefficient problem; inverse scattering problem

MSC: 35R30; 65M32; 65R32; 65N21

Citation: Kabanikhin, S.; Shishlenin, M.; Novikov, N.; Prokhoshin, N. Spectral, Scattering and Dynamics: Gelfand–Levitan–Marchenko–Krein Equations. *Mathematics* **2023**, *11*, 4458. https://doi.org/10.3390/math11214458

Academic Editor: Natalia Bondarenko

Received: 25 June 2023
Revised: 15 October 2023
Accepted: 16 October 2023
Published: 27 October 2023

1. Introduction

The origin of Gelfand–Levitan–Krein–Marchenko (GLKM) approach is related to the inverse problem of recovering the differential operator via the spectral data. The first work in this field was conducted in 1929 by V.M. Ambartsumyan [1] and this was later followed by results of G. Borg [2] and N. Levinson [3,4]. The foundation of the GLKM method is connected with the paper [5], where I.M. Gelfand and B.M. Levitan presented a method to reconstruct a Sturm–Liouville operator by the spectral function and provided conditions sufficient for a given monotonic function to be the spectral function of the operator. A few years later, in 1954, a paper [6] by M.G. Krein was published, proposing algorithms for solving the inverse problem for a wave equation. These two works can be used to denote two directions of the method's development in the second half of the XX century. On the one hand, the continued study of inverse scattering and its applications was used by C.S. Gardner et al. in [7], where the authors used the theory of the inverse scattering problem to integrate the Korteweg de Vries equation and, later, the non-linear Schrodinger equation. This (and the fact the Schrodinger equation is widely used in photonics) determined the importance of developing the methods and algorithms of inverse scattering. On the other hand, results of M.G. Krein led to the dynamic version of the approach, which was applied to a variety of inverse problems for hyperbolic equations, and has strong connections with geophysical problems. The feature of the approach that is important for geophysical applications is its direct nature, with respect to methods, which reduces the inverse problem to optimization.

Since both "branches" of the approach have different variations and applications, the motivation behind this paper is to provide a structured and generalized review of the approach that considers the history of the method and its main variations (and multidimensional analogs as well) from both the theoretical and numerical points of view.

The paper is organized as follows. In the current section we, after a brief introduction, provide a review of the large amount of works related to the GLKM approach, divided into the two groups—papers related to the spectral and scattering inverse problems, and papers related to the inverse problems for hyperbolic equations.

In Section 2, we provide a description of the result that was previously obtained in a one-dimensional case. We provide the formulation of the I.M. Gelfand–B.M. Levitan equation in both spectral and dynamic formulations in Section 2.1. In Section 2.2, we consider the inverse scattering problem, which is reduced to the V.A. Marchenko equation and also describe the way to use this equation to obtain the solution of the Cauchy problem for the KdV equation (the inverse scattering method). We provide more details in Section 2.2, because the general scheme behind the results of Section 2.1 is covered by their two-dimensional analogs. Section 2.3 considers the M.G. Krein equation that arises in the 1D acoustic inverse problem. We also use this opportunity to consider the boundary-control method (which is often referred to as the BC method) for that inverse problem. We formulate the essence of the method and show that both the Krein method and the BC method have the same discrete form in one-dimensional cases. Then, in Section 2.5 we formulate how to use the already mentioned one-dimensional results in order to solve the inverse problem for the 1D seismic inverse problem.

Section 3 is devoted to the multi-dimensional analog of the Gelfand–Levitan equation (Section 3.1) and the Krein equation (Section 3.2), correspondingly. In Section 4, we present a review of the numerical methods that are applicable to solve the GLKM equation in both 1D and 2D. We also provide some numerical results to illustrate the two-dimensional variation of the GLKM method in Section 5. And finally, in Section 6 we provide some new results regarding using the approach to recover the speed of sound and the density from the acoustic equation in cases in which both functions depend on both space variables.

1.1. Inverse Spectral Problems—Inverse Scattering Problems and Method

Let us consider the short history of the development of the theory of inverse spectral and inverse scattering problems. In this section, we will also mention the inverse scattering method because it is very difficult to separate the scientific results obtained.

1929—V.M. Ambartsumyan [1] established that "a homogeneous string is uniquely determined by the set of eigenfrequencies". Specifically, if eigenvalues λ_n of the boundary value problem

$$-y'' + q(x)y = \lambda y, \quad x \in [0, \pi];$$
$$y'(0) = y'(\pi) = 0$$

are $\lambda_n = n^2$ and $q(x)$ is a real continuous function, then $q(x) \equiv 0$.

1943—W. Heisenberg [8,9] considered the inverse scattering problem and proved that in order to solve the inverse scattering problem, it is sufficient to know the asymptotic behavior of the wave function.

1946—G. Borg [2] investigated the inverse problem for a Sturm–Liouville operator. He shown that the result, provided by V.M. Ambartsumyan was quite rare, in the sense of recovering the potential, using only one spectrum. More specifically, he proved the following:

Theorem 1. *Let $\lambda_0 < \lambda_1 < \lambda_2 < \ldots$ be the eigenvalues of the operator*

$$-y'' + q(x)y = \lambda y, \quad 0 \le x \le \pi,$$
$$y'(0) - hy(0) = 0, \quad y'(\pi) + Hy(\pi) = 0, \quad h, H \in \mathbb{R},$$

Let $\mu_0 < \mu_1 < \mu_2 < \ldots$ be the eigenvalues of the operator

$$-y'' + q(x)y = \mu y, \quad 0 \le x \le \pi,$$
$$y'(0) - hy(0) = 0, \quad y'(\pi) + H_1 y(\pi) = 0, \quad H_1 \ne H \in \mathbb{R},$$

Then, the sequences $\{\lambda_m\}_{m=0}^{\infty}$ *and* $\{\mu_n\}_{n=0}^{\infty}$ *uniquely determine the function* $q(x)$ *and the numbers* h, H *and* H_1.

Over the next several years, the subject was actively studied and then advanced in 1949 by a series of interconnected works of N. Levinson and V. Bargmann. N. Levinson [3] provided simpler proofs for some of the results obtained by G. Borg [2]. Then, he [4] tackled an inverse problem of the quantum theory of scattering. He proved that in the absence of negative eigenvalues a scattering phase given for all positive energies and any fixed angular momentum uniquely determines the potential. Meanwhile V. Bargmann [10,11] proved that in the general case the spherically symmetric potential (at any fixed angular momentum) is not uniquely determined by the scattering phase.

1949—A.N. Tikhonov [12] continued (in a way) the works of G. Borg and N. Levinson in studying the properties of data that can guarantee the uniqueness of the solution. He proved the theorem of uniqueness of solving the inverse Sturm–Liouville problem on a semi-axis via a given Weyl function. These works were also the first where the inverse problem on a semi-axis was considered.

1950—V.A. Marchenko [13] investigated several questions for differential operators. This work can be considered the first in a series of fundamental results that formed the essence of the GLKM approach. He proved that the spectral function of a Sturm–Liouville operator (given in a half-line or finite interval) uniquely determines the operator. One of the main features of his work was to use transformation operators to investigate inverse problems.

1951—I.M. Gelfand and B.M. Levitan [5] developed a method to reconstruct a Sturm–Liouville operator via the spectral data. This work was the first that introduced the integral equations, named after authors. They also formulated conditions sufficient for a given monotonic function to be the spectral function of the operator (in a half-line or finite interval). It follows that for two sequences of real numbers $\{\lambda_n\}_0^{\infty}$, $\{\alpha_n\}_0^{\infty}$, $\alpha_n > 0$, to be the spectrum and normalization numbers of the Sturm–Liouville operator, it is sufficient that for some constants a_0, a_1, b_0, b_1

$$\sqrt{\lambda_n} = n + \frac{a_0}{n} + \frac{a_1}{n^3} + O\left(\frac{1}{n^4}\right), \qquad \alpha_n = \frac{\pi}{2} + \frac{b_0}{n^2} + \frac{b_1}{n^4} + O\left(\frac{1}{n^4}\right).$$

The results of I.M. Gelfand and B.M. Levitan were given in an intuitive account by N. Levinson in [14].

1952—V.A. Marchenko extended his results [15], obtained earlier, and provided a more systematic approach for usage of transmutation operators for studying the Sturm–Liouville inverse problems. The results of V.A. Marchenko generalize the results of G. Borg and N. Levinson and also explain the results of V. Bagrmann.

We should also mention a series of papers published in the same years by M.G. Krein [16–20] that are closely connected with works of Gelfand, Levitan and Marchenko. He developed an efficient method to construct the Sturm–Liouville operator via the spectral function and two spectra. Later, he used these results to solve the inverse problem for the string equations that started the usage of the approach for hyperbolic equations.
Over the next ten years, the method that originated in the mentioned works was extensively studied in a number of works [21–28].

The continued study of inverse scattering eventually led to a method that uses the solving of inverse problem for integrating several nonlinear equations. As a first step of the method's development, one can consider the work of E. Fermi et al. [29] in 1954, where the authors detected computationally an abnormally slow stochastization of a dynamic system in the form of a chain of nonlinear oscillators.

1962—R. Newton, using the results of I.M. Gelfand and B.M. Levitan, studied the construction of scattering potentials from the phase shifts at fixed energy [30]. A few years later, in 1965, M.D. Kruskal and N.J. Zabussky [31], using the work of E. Fermi et al., discovered using numerical simulation that collisions of solitons in the Korteweg–de Vries (KdV) equation are elastic. As a result, an endless series of conservation laws was discovered soon after.

1967—C.S. Gardner et al. [7] proposed the method of inverse scattering and integrated the KdV equation:

$$u_t - 6uu_x + u_{xxx} = 0, \qquad x \in \mathbb{R}, t > 0$$

with the initial condition

$$u(x,0) = f(x)$$

by going from the potential of the one-dimensional Schrodinger equation

$$-\frac{d^2\psi}{dx^2} + q(x)\psi = k^2\psi$$

to the reflection coefficient $r(k)$ of this potential. One of the crucial steps of the method is to solve the inverse problem and restore the potential of the Schrodinger equation, which was provided by results of Gelfand, Levitan and Marchenko.

Over the next ten years, such fundamental result led to a large number of papers where the inverse scattering method was studied and generalized.

1968–1969—P.D. Lax [32] discovered an algebraic mechanism forming a basis of the method of inverse scattering.

1970—B.B. Kadomtsev and V.I. Petviashvili obtained the generalization of the KdV equation when studying the stability of solitary waves in weakly dispersive media [33]. The equation was named after them.

1971—C.S. Gardner [34] introduced a theory of the KdV equation as a Hamiltonian system.

1971—V.E. Zakharov and A.B. Shabat [35] applied the method of inverse scattering to the nonlinear Schrodinger equation.

1971—V.E. Zakharov and L.D. Faddeev [36], independently of C.S. Gardner, constructed a theory of the KdV equation as a Hamiltonian system.

1973—A.B. Shabat [37] constructed a class of quasi-linear equations that can be reduced to linear equations.

1974—S.P. Novikov [38], P.D. Lax [39] and V.A. Marchenko [40] studied a periodic Cauchy problem for the KdV equation.

The new result for the periodic problem obtained from the method introduced by C.S. Gardner et al. [7] was the Faddeev–Zakharov theorem: the eigenvalues of the operator L are commutating integrals of the KdV equation as a Hamiltonian system.

1974—V.E. Zakharov and A.B. Shabat [41] proposed a general scheme of the inverse scattering method to integrate the nonlinear differential equations.

1974—L.D. Faddeev [42] published a paper that contains the first multi-dimensional analogs of the Gelfand–Levitan method. Another example of solving scattering problems for more than one dimension can be found in [43].

1974—V.E. Zakharov and S.V. Manakov [44] showed that the nonlinear Shrodinger equation considered as a Hamiltonian system is fully integrable. This can be done using a scattering matrix of the one-dimensional Dirac operator.

1976—S.V. Manakov [45] generalized the Lax pair for two-dimensional time-dependent equations.

1976—V.E. Zakharov and S.V. Manakov [46] showed that each one-dimensional differential operator whose coefficient depends on an arbitrary set of parameters can be associated with a series of multidimensional nonlinear partial differential equations integrable via the inverse scattering method.

1976—P.D. Lax [47] considered the almost periodic behavior (in time) of the periodic solutions to the KdV equation. He presented a new proof based on Lennart's recursion.

1979—V.E. Zakharov and A.B. Shabat [48] extended the method developed in 1976 to spectral problems as rational functions of the spectral parameter. They obtained a description of new classes of equations integrable via the method of inverse scattering and an algorithm to construct their exact solutions.

In the early 1980s, new nonlinear equations integrable via the method of inverse scattering (in particular, the nonlinear Shrodinger equation, the sin-Gordon equation, etc.) were found.

1980—V.E. Zakharov et al. [49] provided a systematic description of the method of inverse scattering.

1982—L.P. Nizhnik and M.D. Pochinayko [50] investigated the nonlinear two-dimensional (in space) Shrodinger equation and used the inverse scattering method for its integration.

1984—A.P. Veselov and S.P. Novikov [51] considered a two-dimensional generalization of the KdV equation (the Veselov–Novikov equation) with the help of the two-dimensional potential Shrodinger operator.

1984—R.G. Novikov and G.M. Henkin [52] applied (and adapted) the inverse scattering method to obtain weakly localized solutions to a KdV equation in which the transmission coefficient of the scattering matrix may be zero for a finite set of pulses.

1984—A.P. Veselov and S.P. Novikov obtained a two-dimensional integrable extension of the KdV equation [51].

1985–1986—P. Grinevich et al. [53–56] gave the first results on the inverse scattering method for the Veselov–Novikov equation with decaying at infinity potential at fixed energy E.

The method of inverse scattering was investigated by Boiti et al. [57], Tsai [58], Nachman [59], Bogdanov et al. [60], Lassas et al. [61], Lassas et al. [62,63], Music [64] and Perry [65].

1986—R.G. Novikov [53] used it for physical inverse scattering, i.e., for inverse scattering in a physical sense, for the first time.

1992—J.-P. Francoise and R.G. Novikov [66] investigated the hierarchy of the Calogero–Moser system for the Kadomtsev–Petviashvili and Veselov–Novikov equations.

1999—R.G. Novikov [67] proposed some inverse scattering algorithm, which requires the solution of the linear integral equation with a specific kernel. That equation can be considered as a two-dimensional analog of the the Gelfand–Levitan–Marchenko–Krein equations.

2011—A.V. Kazeykina and R.G. Novikov [68] studied the asymptotics of solutions to a Cauchy problem for the Veselov–Novikov equation with positive energy.

2013— M. Music et al. [69] studiet nonlinear scattering transform for the two-dimensional Schrodinger equation at zero energy with a radial potential.

2020 and 2021—N. Bondarenko obtained the spectral data characterization for the matrix Sturm–Liouville operator with the general self-adjoint boundary conditions [70]. This result implies the characterization of the spectral data for the Sturm–Liouville operators on geometrical graphs of arbitrary structure with rationally dependent edge lengths. It is worth mentioning that the spectral data characterization for the Sturm–Liouville operator on a star-shaped graph was previously obtained in [71].

2021—S.A. Avdonin et al. [72] investigated the inverse problem of recovering the matrix potential from the dynamical Neumann-to-Dirichlet operator for a dynamical system with boundary control for the vector Schrodinger equation on the interval with a non-self-adjoint matrix potential.

2023—X.-C. Xu and N. Bondarenko [73] proved the local solvability and stability of the inverse Robin–Regge problem in the general case, taking eigenvalue multiplicities into account. The new approach was developed based on the reduction of this inverse problem to the recovery of the Sturm–Liouville potential from the Cauchy data.

1.2. Inverse Problems for Hyperbolic Equations

1954—M.G. Krein [6] was the first to use the GLK method for inverse problems for hyperbolic equations. He considered the so-called string problem and formulated theorems on the solvability of an inverse problem. The nonlinear inverse problem for the string equation was reduced to an integral equation (the Krein equation).

1968—B.S. Pariiskii [74] studied a one-dimensional inverse problem for the wave equation with a perturbation at some depth and derived the Krein equation.

1970–1971—B. Gopinath and M. Sondhi [75,76], independently of each other, proposed an integral equation (also in a time domain) for reconstructing human speech from acoustic measurements.

1971—A.S. Blagoveshchenskii [77] obtained another proof of Krein results. He showed that the dependence of the sought-for coefficient on the additional information is local. In contrast, earlier such problems were studied by using Fourier (or Laplace) transforms in time. Later, that differential equation coefficient was actually reconstructed from the properties of the eigenfunctions of the corresponding spectral problem.

1975—A.S. Alekseev and V.I. Dobrinskii [78] used a discrete analog of the Gelfand–Levitan method to study numerical algorithms for solving the one-dimensional inverse dynamic problem of seismology. Later, ideas from the I.M. Gelfand and B.M. Levitan approach were used to deal with monochromatic seismic problems [79].

1977—B.S. Pariiskii [80] published a detailed review of the numerical methods for solving Gelfand–Levitan equation.

1979—W. Symes [81] applied nonlinear integral equations for an inverse problem in time domain.

1980—R. Burridge [82] attempted to apply the Gelfand–Levitan–Marchenko equations for elasticity theory in a time domain and found a relation between them and the Gopinath–Sondhi equation.

1982—F. Santosa [83] developed an exact method for solving an inverse problem of plane wave propagation via the Gelfand–Levitan method, tested a numerical scheme for solving the integral equation, investigated the stability and analyzed the numerical errors and approximations.

1988—S.I. Kabanikhin [84] proposed a new algorithm for solving the Gelfand–Levitan equation using a sufficient condition for solvability of the inverse problem.

1991—V.G. Romanov and S.I. Kabanikhin [85] applied a dynamic version of the Gelfand–Levitan method to the one-dimensional inverse problem of geoelectrics for a quasi-stationary approximation of the system of Maxwell equations.

1998—A.S. Alekseev and V.S. Belonosov [86] used the spectral method to reconstruct the acoustic impedance in a one-dimensional wave equation.

1987—M.I. Belishev [87] developed the first multi-dimensional analogs of the Gelfand–Levitan–Krein equations for hyperbolic inverse problems.

1988—S.I. Kabanikhin [84,88] proposed another multi-dimensional analog of the Gelfand–Levitan–Krein equations.

1992—M.I. Belishev and A.S. Blagoveshchenskii [89] proposed a multidimensional analog of the Gelfand–Levitan equation based on the boundary-control method.

2004—S.I. Kabanikhin and M.A. Shishlenin [90] showed that the discrete analogs of the Krein and boundary-control methods are the same for the one-dimensional coefficient inverse problem of acoustics.

2005—S.I. Kabanikhin et al. [91] published a book on numerical methods for solving two-dimensional analogs of the Gelfand–Levitan–Krein equation for coefficient inverse problems for the wave and acoustics equations.

2011—S.I. Kabanikhin and M.A. Shishlenin [92] developed a numerical method for solving the Krein equation for the Nth approximation of the two-dimensional inverse acoustic problem. The Krein equation for the Nth approximation was obtained in matrix form, for which a numerical method was constructed based on the singular value decomposition method.

2011—M.A. Shishlenin and N.S. Novikov [93] conducted a comparative analysis of two numerical methods for solving the Gelfand–Levitan equation and developed the Monte-Carlo method for solving the Gelfand–Levitan equation.

2016—V. Druskin et al. [94] proposed the method of finding the velocity propagation speed in the acoustic wave equation based on the discrete form of the GLKM method.

2018—L. Borcea et al. [95] proposed new linear-algebraic algorithm that uses a reduced-order model to compare data with data corresponding to the Born model with single scattering.

2021—V. Druskin et al. [96] combined data-driven reduced-order models with the Lippmann–Schwinger integral equation to obtain a direct nonlinear inversion method. Numerically, it has been shown that the proposed inversion is much better than the Born inversion.

2021—V.G. Romanov [97] justified the scheme related to the construction of the infinite system of integral equations in the case when the potential is analytic in x.

The Gelfand–Levitan–Krein method was applied for solving acoustic [98–100], elasticity [101] and seismic [102,103] coefficient inverse problems. We also should mention series of works of A.V. Baev [104–107], where he recently obtained some new results, related to solving inverse problems for hyperbolic equations by using variations of GLK approach.

An advantage of the Gelfand–Levitan–Krein approach for solving coefficient inverse problems for hyperbolic equations is that the direct problems need not be solved many times. We note the boundary control (BC) method created by M.I. Belishev [108–111] and a globally convergent method of [112–118] by M.V. Klibanov. In the next section, we will briefly consider the BC method and its connection to the Krein approach in the one-dimensional case.

2. One-Dimensional Problems

2.1. I.M. Gelfand–B.M. Levitan Equation

First, we consider the direct Sturm–Liouville problem:

$$l_q y(x) := -y'' + q(x)y, \tag{1}$$

defined on a set of functions $y \in W_2^2(0,\pi)$ and satisfying the relations

$$\begin{aligned} l_q y(x) = \lambda y, \quad x \in (0,\pi), \\ y'(0) - hy(0) = 0, \quad y'(\pi) + Hy(\pi) = 0. \end{aligned} \tag{2}$$

Here, h, $H \in \mathbb{R}$, $q(x) \in L_2(0,\pi)$. Let λ_n be an eigenvalue and $\varphi(x,\lambda_n)$, an eigenfunction of the operator l_q and $\varphi(0,\lambda) = 1$, $\varphi'(0,\lambda) = h$. Let

$$\alpha_n = \int_0^\pi \varphi^2(x,\lambda_n)dx, \tag{3}$$

The set $\{\lambda_n, \alpha_n\}_{n\geq 0}$ is called the spectral data of the operator l_q, with the following asymptotic properties [119]:

$$\sqrt{\lambda_n} = n + \frac{\omega}{\pi n} + \frac{\beta_n}{n}, \quad \alpha_n = \frac{\pi}{2} + \frac{\beta_{1n}}{n}, \quad \{\beta_n\}, \{\beta_{1n}\} \in l_2,$$

$$\alpha_n > 0, \qquad \lambda_n \neq \lambda_m, \quad (n \neq m).$$

The inverse Sturm–Liouville problem consists in reconstructing the potential $q(x)$ and coefficients h, H from the spectral data. Using a function $F(x,t)$:

$$F(x,t) = \sum_{n=0}^{\infty} \left(\frac{\cos\sqrt{\lambda_n}x \cos\sqrt{\lambda_n}t}{\alpha_n} - \frac{\cos nx \cos nt}{\alpha_n^0} \right), \tag{4}$$

where

$$\alpha_n^0 = \begin{cases} \frac{\pi}{2}, & n > 0, \\ \pi, & n = 0, \end{cases}$$

one can reduce the inverse Sturm–Liouville problem to the Gelfand–Levitan equation:

$$G(x,t) + F(x,t) + \int_0^x G(x,s)F(s,t)ds = 0, \quad 0 < t < x. \tag{5}$$

The solution to the Gelfand–Levitan equation makes it possible to determine the solution to the inverse problem via the formula

$$q(x) = 2\frac{d}{dx}G(x,x), \quad h = G(+0,+0), \quad H = \omega - h - \frac{1}{2}\int_0^\pi q(t)\mathrm{d}t. \tag{6}$$

One can also consider the following coefficient dynamic inverse problem: find an even $q(x)$ that satisfies the following:

$$u_{tt} = u_{xx} - q(x)u, \qquad x \in \mathbb{R}, \quad t > 0; \tag{7}$$

$$u|_{t=0} = 0, \quad u_t|_{t=0} = \delta(x);$$

$$u|_{x=0} = f(t). \tag{8}$$

The inverse problem can be reduced to the Gelfand–Levitan equation [120]:

$$\tilde{w}(x,t) + \int_0^x [f'(t-\tau) + f'(t+\tau)]\tilde{w}(x,\tau)d\tau = -\frac{1}{2}[f'(t-x) + f'(t+x)], \quad t \in [0,x). \tag{9}$$

Here, $f(t)$ is an odd continuation of the inverse problem's data to the negative t and the derivative of $f(t)$ is taken at the points of continuity only. The solution to the inverse problems (7) and (8) can be found via the formula

$$q(x) = 4\frac{\mathrm{d}}{\mathrm{d}x}\tilde{w}(x,x-0), \quad x > 0. \tag{10}$$

We should also mention the paper of R.G. Novikov [121], where he proposed an explicit formula to solve the inverse scattering problem for the Sturm–Liouville operator (in dimension 1) up to smooth functions.

2.2. V.A. Marchenko Equation—The Inverse Scattering Method

Let us consider the direct scattering problem: given $q(x)$ such that $q(x) \to 0$ for $x \to \pm\infty$, find eigenfunctions and eigenvalues of the problem ([122]):

$$-y'' + q(x)y = k^2 y. \tag{11}$$

Equation (11) for $q(x) < 0$ and $k^2 > 0$ has a continuous spectrum of eigenvalues. If $k^2 < 0$ then the spectrum is discrete. Let us suppose that

$$\int_{-\infty}^{\infty} (1 + |x|)|q(x)|\mathrm{d}x < \infty.$$

Equation (11) has a fundamental system of the solutions:

$$\psi_1(x,k) \cong \mathrm{e}^{-ikx} + o(1), \qquad \psi_2(x,k) \cong \mathrm{e}^{ikx} + o(1), \qquad x \to +\infty, \tag{12}$$

$$\varphi_1(x,k) \cong \mathrm{e}^{-ikx} + o(1), \qquad \varphi_2(x,k) \cong \mathrm{e}^{ikx} + o(1), \qquad x \to -\infty. \tag{13}$$

Note that

$$\psi_1(x,k) = \psi_2^*(x,k) = \psi_2(x,-k), \tag{14}$$

$$\varphi_1(x,k) = \varphi_2^*(x,k) = \varphi_2(x,-k). \tag{15}$$

Here, * is the complex conjugation. Equation (11) has two linear independent solutions; therefore, each solution can be represented as a linear combination

$$\varphi(x,k) = a(k)\psi(x,k) + b(k)\psi^*(x,k), \tag{16}$$

$$\varphi^*(x,k) = c(k)\psi(x,k) + d(k)\psi^*(x,k). \tag{17}$$

Therefore, we have that

$$d(k) = a^*(k), \qquad c(k) = b^*(k).$$

It follows from (16) that

$$\frac{\varphi(x,k)}{a(k)} = \psi(x,k) + \frac{b(k)}{a(k)}\psi^*(x,k), \tag{18}$$

and using asymptotic for $x \to +\infty$ we obtain that

$$\frac{\varphi(x,k)}{a(k)} = \mathrm{e}^{-ikx} + r(k)\mathrm{e}^{ikx} + o(1). \tag{19}$$

Here, $r(k) = b(k)/a(k)$ is the reflection coefficient. For $x \to -\infty$, we have

$$\varphi(x,k) \cong \mathrm{e}^{-ikx}. \tag{20}$$

Then, the last wave has the form

$$\frac{\varphi(x,k)}{a(k)} \cong t(k)\mathrm{e}^{-ikx}.$$

Here, $t(k) = 1/a(k)$ is the completion rate. The discrete spectrum of the Schrodinger operator is $k_n^2 = -\chi_n^2$ ($\chi_n > 0$). If we consider the asymptotic for $x \to -\infty$ in the form

$$\varphi^{(n)}(x) = \mathrm{e}^{\chi_n x} + o(\mathrm{e}^{\chi_n x}), \tag{21}$$

then for $x \to +\infty$ we have the wave function in the form

$$\varphi^{(n)}(x) = b_n \mathrm{e}^{-\chi_n x} + o(\mathrm{e}^{-\chi_n x})$$

Eigenfunctions corresponding to the discrete spectrum and the eigenvalues are real-valued. Let us enumerate eigenvalues

$$\chi_1^2 > \chi_2^2 > \ldots > \chi_n^2 > 0$$

and suppose that the $\varphi^{(1)}(x)$ wave function corresponding to χ_1^2 has no zeros for $x \in (-\infty, \infty)$. Then, $\varphi^{(n)}(x)$ has $(n-1)$ zeros using oscillatory theorem [123] and we have that $b_n = (-1)^{n-1}|b_n|$.

The set

$$S = \{r(k),\ \chi_n,\ |b_n|,\ n = \overline{1,N}\}$$

is called scattering data. The direct scattering problem is to find scattering data via the given $q(x)$.

Let us consider the following functions

$$\chi_-(x,k) = \varphi(x,k)\mathrm{e}^{\mathrm{i}kx}, \tag{22}$$

$$\chi_+(x,k) = \psi(x,k)\mathrm{e}^{\mathrm{i}kx}, \tag{23}$$

and assume the following boundary conditions hold

$$\lim_{x\to-\infty} \chi_-(x,k) = 1, \tag{24}$$

$$\lim_{x\to+\infty} \chi_+(x,k) = 1. \tag{25}$$

Using the connection between Equation (11) and the Volterra integral equations, one can obtain that the functions φ, ψ are the solutions of the following integral equations [15,42]:

$$\begin{aligned}\varphi(x,k) &= \mathrm{e}^{-\mathrm{i}kx} + \int_{-\infty}^{x} \frac{\sin k(x-\xi)}{k} q(\xi)\varphi(\xi,k)\mathrm{d}\xi,\\ \psi(x,k) &= \mathrm{e}^{-\mathrm{i}kx} - \int_{x}^{+\infty} \frac{\sin k(x-\xi)}{k} q(\xi)\psi(\xi,k)\mathrm{d}\xi.\end{aligned} \tag{26}$$

Functions $\chi_-(x,k)$ and $\chi_+(x,k)$ solve the following equation

$$-\chi_{\pm xx}(x,k) + 2\mathrm{i}k\chi_{\pm x}(x,k) + q(x)\chi_\pm(x,k) = 0, \tag{27}$$

and instead of (26) we obtain

$$\chi_-(x,k) = 1 + \int_{-\infty}^{x} \frac{\mathrm{e}^{2\mathrm{i}k(x-\xi)} - 1}{2\mathrm{i}k} q(\xi)\chi_-(\xi,k)\mathrm{d}\xi.$$

$$\chi_+(x,k) = 1 - \int_{x}^{\infty} \frac{\mathrm{e}^{2\mathrm{i}k(x-\xi)} - 1}{2\mathrm{i}k} q(\xi)\chi_+(\xi,k)\mathrm{d}\xi.$$

For $k \to \infty$, we have

$$\chi_+(x,k) = 1 + \int_{x}^{\infty} \frac{q(\xi)}{2\mathrm{i}k}\mathrm{d}\xi + o\left(\frac{1}{k^2}\right). \tag{28}$$

It follows from (16) that

$$\frac{\varphi(x,k)\mathrm{e}^{\mathrm{i}ky}}{a(k)} = \psi(x,k)\mathrm{e}^{\mathrm{i}ky} + r(k)\psi^*(x,k)\mathrm{e}^{\mathrm{i}ky}. \tag{29}$$

Let us integrate (29) with respect to k:

$$\int_{-\infty}^{\infty} \frac{\varphi(x,k)\mathrm{e}^{\mathrm{i}ky}}{a(k)}\mathrm{d}k = \int_{-\infty}^{\infty} \psi(x,k)\mathrm{e}^{\mathrm{i}ky}\mathrm{d}k + \int_{-\infty}^{\infty} r(k)\psi^*(x,k)\mathrm{e}^{\mathrm{i}ky}\mathrm{d}k. \tag{30}$$

In the left-hand side of (30)

$$\int_{-\infty}^{\infty} \frac{\varphi(x,k)\mathrm{e}^{\mathrm{i}ky}}{a(k)}\mathrm{d}k = 2\pi\mathrm{i}\sum_{n=1}^{N} \frac{\varphi(x,\mathrm{i}\chi_n)}{a_n(\mathrm{i}\chi_n)}\mathrm{e}^{-\chi_n y}. \tag{31}$$

We have

$$\varphi(x, \mathrm{i}\chi_n) = b_n\psi^*(x, \mathrm{i}\chi_n) = b_n\psi(x, -\mathrm{i}\chi_n),$$

then (31) can be rewritten in the form

$$\int_{-\infty}^{\infty} \frac{\varphi(x,k)\mathrm{e}^{\mathrm{i}ky}}{a(k)}\mathrm{d}k = 2\pi\mathrm{i}\sum_{n=1}^{N} \frac{b_n\psi(x,-\mathrm{i}\chi_n)}{a_n(\mathrm{i}\chi_n)}\mathrm{e}^{-\chi_n y}. \tag{32}$$

Let us introduce a new function $K(x,y)$, such that

$$\psi(x,k) = \mathrm{e}^{-\mathrm{i}kx} + \int_{x}^{\infty} K(x,y)\mathrm{e}^{-\mathrm{i}ky}\mathrm{d}y, \tag{33}$$

then (32) is rewritten as follows

$$\int_{-\infty}^{\infty} \frac{\varphi(x,k)\mathrm{e}^{\mathrm{i}ky}}{a(k)}\mathrm{d}k = 2\pi\mathrm{i}\sum_{n=1}^{N} \frac{b_n\mathrm{e}^{-\chi_n y}}{a_n(\mathrm{i}\chi_n)}\left[\mathrm{e}^{-\chi_n x} + \int_{x}^{\infty} K(x,z)\mathrm{e}^{-\chi_n z}\mathrm{d}z\right]. \tag{34}$$

It follows from (30) that

$$\int_{-\infty}^{\infty} \frac{\varphi(x,k)\mathrm{e}^{\mathrm{i}ky}}{a(k)}\mathrm{d}k = 2\pi\int_{x}^{\infty} K(x,z)\delta(y-z)\mathrm{d}z +$$

$$+\int_{-\infty}^{\infty} r(k)\mathrm{e}^{\mathrm{i}k(y+x)}\mathrm{d}k + \int_{x}^{\infty} K(x,z)\left[\int_{-\infty}^{\infty} r(k)\mathrm{e}^{\mathrm{i}k(z+y)}\mathrm{d}k\right]\mathrm{d}z. \tag{35}$$

Therefore, we obtain

$$2\pi\mathrm{i}\sum_{n=1}^{N}\frac{b_n}{a_k(\mathrm{i}\chi_n)}\mathrm{e}^{-\chi_n(x+y)} + 2\pi\mathrm{i}\int_{x}^{\infty} K(x,z)\sum_{n=1}^{\infty}\frac{b_n}{a_n(\mathrm{i}\chi_n)}\mathrm{e}^{-\chi_n(x+y)}\mathrm{d}z =$$

$$= 2\pi K(x,y) + \int_{-\infty}^{\infty} r(k)\mathrm{e}^{\mathrm{i}k(y+x)}\mathrm{d}k + \int_{-\infty}^{\infty} K(x,z)\left[\int_{-\infty}^{\infty} r(k)\mathrm{e}^{\mathrm{i}k(z+y)}\mathrm{d}k\right]\mathrm{d}z. \tag{36}$$

Define the function $F(x)$ which consists of scattering data

$$F(x) = \sum_{n=1}^{N}\frac{b_n\mathrm{e}^{-\chi_n x}}{\mathrm{i}a_n(\mathrm{i}\chi_n)} + \frac{1}{2\pi}\int_{-\infty}^{\infty} r(k)\mathrm{e}^{\mathrm{i}kx}\mathrm{d}k, \tag{37}$$

then Equation (36) can be rewritten in the form of the V.A. Marchenko integral equation

$$K(x,y) + F(x+y) + \int_{x}^{\infty} K(x,s)F(s+y)\mathrm{d}s = 0. \tag{38}$$

Using (33), we obtain that

$$\chi_+(x,k) = 1 + \int_{x}^{\infty} K(x,y)\mathrm{e}^{\mathrm{i}k(x-y)}\mathrm{d}y = 1 - \frac{1}{\mathrm{i}k}K(x,y)\mathrm{e}^{\mathrm{i}k(x-y)}\Big|_{y=x}^{y=\infty} + o\left(\frac{1}{k}\right). \tag{39}$$

We have

$$\frac{1}{\mathrm{i}k}K(x,y)\mathrm{e}^{\mathrm{i}k(x-y)} \to 0, \quad y \to \infty$$

then

$$\chi_+(x,k) = 1 + \frac{1}{\mathrm{i}k} K(x,x). \tag{40}$$

It follows from (28) that

$$K(x,x) = \frac{1}{2} \int_x^\infty q(\xi) \mathrm{d}\xi. \tag{41}$$

Then, the function $q(x)$ in the equation is reconstructed via the formula

$$q(x) = -2 \frac{\mathrm{d}}{\mathrm{d}x} K(x,x).$$

Now, before moving further to the inverse scattering method, we would like to illustrate the connection between the considered inverse problems for 1D via Figure 1.

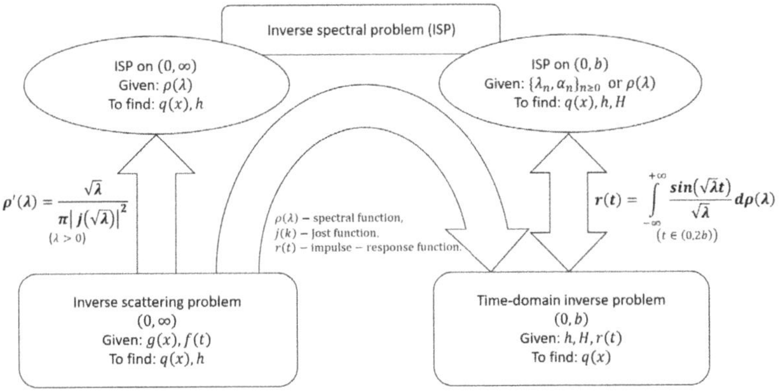

Figure 1. The connection between the inverse spectral problem, the inverse scattering problem and the inverse problem in time domain.

Let us consider the inverse scattering method, which appears while studying some nonlinear equations of mathematical physics. The method, proposed by C.S. Gardner, J.M. Green, M.D. Kruskal and R.M. Miura in 1967 [7] represent the nonlinear equation under study as a compatibility condition for a system of linear equations. An initial version of the method based on the theory of scattering for differential operators (hence, the name of the method) was applied to the Korteweg–de Vries equation

$$u_t - 6uu_x + u_{xxx} = 0, \tag{42}$$

and can be represented in the form of the solution of linear equations

$$\begin{aligned} \psi_{xx} + (\lambda - u)\psi = 0, \\ \psi_t + \psi_{xxx} - 3(\lambda + u)\psi_x = C(t)\psi. \end{aligned} \tag{43}$$

It is a compatibility condition of the overspecified linear system of equations

$$(L - \lambda)\psi = 0, \tag{44}$$

$$\psi_t + A\psi = 0,$$

where

$$L = -\frac{\mathrm{d}^2}{\mathrm{d}x^2} + u(x,t), \quad A = \frac{\mathrm{d}^3}{\mathrm{d}x^3} - 3\left[u\frac{\mathrm{d}}{\mathrm{d}x} + \frac{\mathrm{d}}{\mathrm{d}x}u\right],$$

and is equivalent to the operator relation (Lax representation)

$$\frac{\partial L}{\partial t} = [L, A]. \tag{45}$$

Let us consider the initial condition for the Korteweg–de Vries Equation (42)

$$u(x,0) = f(x). \tag{46}$$

We suppose that

$$\int_{-\infty}^{\infty} (1+|x|)|f(x)|\mathrm{d}x < \infty.$$

Let it be that for the known $f(x)$ we find from the scattering data

$$S_n(0) = \{\lambda_n(0); r(k,0); b_n(0); n = \overline{1,N}\}. \tag{47}$$

The wave function in Equation (43) depends on the time variable t:

$$\varphi(x,k,t) = a(k,t)\psi(x,k,t) + b(k,t)\psi^*(x,k,t), \tag{48}$$

and we have the following asymptotics when $x \to +\infty$

$$\varphi(x,k,t) = a(k,t)\mathrm{e}^{-\mathrm{i}kx} + b(k,t)\mathrm{e}^{\mathrm{i}kx} + o(1). \tag{49}$$

Substituting (49) into (43) we obtain

$$\begin{aligned} &\dot{a} + 4\mathrm{i}k^3 a - ca = 0, \\ &\dot{b} - 4\mathrm{i}k^3 b - cb = 0. \end{aligned} \tag{50}$$

Therefore, solving (50) we obtain

$$r(k,t) = \frac{b(k,t)}{a(k,t)} = r(k,0)\mathrm{e}^{8\mathrm{i}k^3 t}. \tag{51}$$

$$b_n(t) = b_n(0)\mathrm{e}^{8\chi_n^3 t}, \qquad n = \overline{1,N}. \tag{52}$$

Therefore, if by the given initial data $f(x)$ we find $S(t=0)$, then $S(t)$ has the following form

$$S(t) = \left\{r(k,0)\mathrm{e}^{8\mathrm{i}k^3 t};\ \chi_n(0);\ b_n(0)\mathrm{e}^{8\chi_n^3 t}, \qquad n = \overline{1,N}\right\}. \tag{53}$$

Let us denote

$$F(x,t) = \sum_{n=1}^{N} \frac{b_n(0)\mathrm{e}^{-\chi_n x + 8\chi_n^3 t}}{\mathrm{i}a_n(\mathrm{i}\chi_n)} + \frac{1}{2\pi}\int_{-\infty}^{\infty} r(k,0)\mathrm{e}^{\mathrm{i}kx + 8\mathrm{i}k^3 t}\mathrm{d}k. \tag{54}$$

Then, we solve the Marchenko integral equation for the function $K(x,y,t)$ to solve the inverse problem of scattering:

$$K(x,y,t) + F(x+y,t) + \int_{x}^{\infty} K(x,s,t)F(s+y,t)\mathrm{d}s = 0. \tag{55}$$

Then, we find the solution of the KdV equation via the formula

$$u(x,t) = -2\frac{\partial}{\partial x}K(x,x,t). \tag{56}$$

Therefore, in the first step we find the scattering data $S(t=0)$ for the known $f(x)$. In the second step, we find the scattering data $S(t)$. Then, the function $F(x,t)$ is found via (54). In the fourth step, we solve the Marchenko Equation (55). In the last step, we find the solution of the KdV equation via Equation (56). The scheme of the method is illustrated in Figure 2:

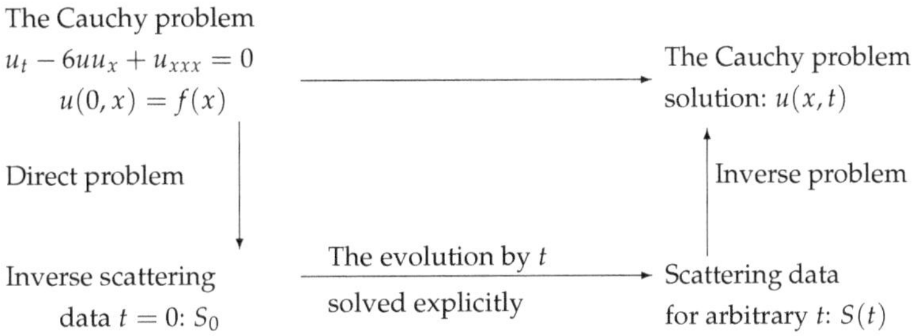

Figure 2. Scheme of solving the KdV equation by solving the inverse scattering problem.

To solve these problems efficiently, numerical calculations are required. An advantage of the inverse scattering method is that it allows advancing in time as far as needed without loss of accuracy.

2.3. *Krein Equation*

We consider the following inverse problem:

$$u_{tt} = u_{xx} - \frac{\sigma'(x)}{\sigma(x)} u_x, \qquad x > 0, \quad t > 0; \tag{57}$$

$$u\big|_{t<0} \equiv 0; \tag{58}$$

$$u_x\big|_{x=+0} = \delta(t); \tag{59}$$

$$u(+0,t) = f(t). \tag{60}$$

In [77], the inverse problem of acoustics is reduced to the Krein equation

$$-2f(+0)\Phi(x,t) - \int_{-x}^{x} f'(t-s)\Phi(x,s)\mathrm{d}s = \frac{1}{\sigma(+0)}, \qquad |t| < x, \tag{61}$$

in which the function $r(t)$ is oddly extended to the negative t.

The solution to the inverse problems (57)–(60) can be found via the formula

$$\sigma(x) = \frac{1}{4\sigma(+0)\Phi^2(x,x-0)}. \tag{62}$$

2.4. *Boundary-Control Method in One-Dimensional Case*

Let us consider the inverse problem with arbitrary source:

$$\frac{\partial^2 u^g}{\partial t^2} = \frac{\partial^2 u^g}{\partial x^2} - \frac{\sigma'(x)}{\sigma(x)}\frac{\partial u^g}{\partial x}, \qquad x > 0, \quad t > 0;$$
$$u^g\big|_{t<0} \equiv 0;$$
$$\frac{\partial u^g}{\partial x}\bigg|_{x=+0} = g(t);$$
$$u^g(0,t) = f^g(t).$$

$f^g(t)$ is data measured for $g(t)$. The inverse problem consists in finding $\sigma(x)$ by the given functions $g(t)$ and $f^g(t)$. Let us define

$$w^{gh}(s,t) = \int_0^L \frac{u^g(x,s)\,u^h(x,t)}{\sigma(x)}dx := \Big(u^g(\cdot,s), u^h(\cdot,t)\Big)_{H'} \tag{63}$$

which for $s = t = \ell$ can be expressed as:

$$w^{gh}(\ell,\ell) = \frac{1}{2\sigma(+0)} \int_0^\ell \int_\xi^{2\ell-\xi} \Big[h(\tau)f^g(\xi) - g(\xi)f^h(\tau)\Big]d\tau d\xi. \tag{64}$$

Note, we can find the response to the arbitrary source $g(t)$ by knowing response $f(t)$ (57)–(60):

$$u^g(0,t) := f^g(t) = \int_0^t f(t-s)\,g(s)ds. \tag{65}$$

Let us consider dense system of functions $\{\psi_k(t)\} \in L_2(0,\ell)$, $k = \overline{1,M}$ and

$$u^g(x,\ell) = \theta(\ell - x). \tag{66}$$

Here, $\theta(\cdot)$ is a Heaviside theta-function. Therefore, we find the approximate source

$$g(t) \approx g_M(t) = \sum_{k=1}^M \alpha_k \psi_k(t). \tag{67}$$

The coefficients $\{\alpha_k\}$, $k = \overline{1,M}$ solve the following system

$$\Gamma\alpha = b, \tag{68}$$

where

$$\Gamma_{jk} = \Big(u_j(\cdot,T), u_k(\cdot,T)\Big)_{H'}, \quad b_j = \Big(a(\cdot), u_j(\cdot,T)\Big)_{H'}, \quad j,k = \overline{1,M}. \tag{69}$$

Using (65), (69) and taking into account that $\{\psi_k(t)\}$ are given on $[0,\ell]$, one can show that coefficients of matrix Γ and components of vector b are defined from $\{\psi_k(t)\}$, $k = \overline{1,M}$:

$$\Gamma_{jk} = -\frac{f(+0)}{\sigma(+0)} \int_0^\ell \int_0^{\ell-\tau} \psi_j(\xi)d\xi \int_0^{\ell-\tau} \psi_k(\eta)d\eta d\tau -$$
$$-\frac{1}{2\sigma(+0)} \int_0^\ell \int_0^{\ell-\eta} \psi_k(\eta')d\eta' \int_0^\ell [r'(\tau+\eta) + r'(|\tau-\eta|)] \times \int_0^{\ell-\tau} \psi_j(\xi)d\xi d\tau d\eta; \tag{70}$$

$$b_j = -\frac{1}{\sigma(+0)} \int_0^\ell (\ell - t)\psi_j(t)dt. \tag{71}$$

Here, $j,k = \overline{1,M}$. One can obtain:

$$\|a\|_H^2 \approx \|a_M\|_H^2 = \sum_{k=1}^M \alpha_k b_k. \tag{72}$$

Therefore

$$\int_0^\ell \frac{dx}{\sigma(x)} = \|a\|_H^2 \approx \|a_M\|_H^2 = \sum_{k=1}^M \alpha_k b_k. \tag{73}$$

It follows from (70) and (71) that all components of system (68) depend on parameter ℓ. Therefore, differentiating by ℓ we derive from (73):

$$\frac{d}{d\ell}\|a\|_H^2(\ell) = \frac{d}{d\ell}\int_0^\ell \frac{dx}{\sigma(x)} = \frac{1}{\sigma(\ell)} \approx \frac{d}{d\ell}\left[\sum_{k=1}^M \alpha_k b_k\right]. \tag{74}$$

Using (73) and (74), an approximate solution of the inverse problem is given by the formula:

$$\sigma(\ell) \approx \left\{ \frac{d}{d\ell} \left[\sum_{k=1}^{M} \alpha_k b_k \right] \right\}^{-1}. \tag{75}$$

Now, we will consider the connection between the BC method and the Krein equation for the 1D acoustic problem. In order to do that, we consider the discrete version of Equations (61) and (68).

Let $h = \ell/N$, $\psi_k^n = \psi_k(nh)$, $k = \overline{1,N}$, $n = \overline{0,N-1}$. We consider the following basis functions:

$$\psi_k^n = \frac{1}{h^2}\left[\delta_{N-k+1,n} - \delta_{N-k,n}\right], \quad n = \overline{0,N-1}, \qquad k = \overline{1,N},$$

where

$$\delta_{n,k} = \begin{cases} 1, & n = k \\ 0, & n \neq k \end{cases}.$$

We use the following formula for numerical integration:

$$\int_0^\ell f(t)dt = h \sum_{n=0}^{N-1} f^n + o(h).$$

Let us find the discrete analogs of components of vector b and matrix Γ.

Lemma 1. *The components of the vector can be represented in the form*

$$b_j = \frac{1}{\sigma_0} + o(h), \quad j = \overline{1,N}. \tag{76}$$

Lemma 2. *The following equality holds true*

$$\sum_{n=0}^{N-m-1} \psi_k^n = -\frac{1}{h^2}\delta_{m+1,k}. \tag{77}$$

Lemma 3. *The coefficients Γ_{jk} can be represented in the form*

$$\Gamma_{jk} = -\frac{f^0}{h\sigma_0}\delta_{k,j} - \frac{1}{2\sigma_0}\left[f'^{j+k-2} + f'^{|j-k|}\right] + o(h). \tag{78}$$

Using (78), we rewrite the initial Equation (68) in discrete form:

$$2\frac{f^0}{\sigma_0}\alpha_j + \frac{h}{\sigma_0}\sum_{k=1}^{N}\left[f'^{j+k-2} + f'^{|j-k|}\right]\alpha_k = -2\frac{h}{\sigma_0}, \qquad j = \overline{1,N}. \tag{79}$$

Denote $\alpha_k^N = \alpha_k(Nh)$. Then, we obtain that

$$\sum_{k=1}^{N} \alpha_k^N b_k^N = \frac{1}{\sigma_0}\sum_{k=1}^{N}\alpha_k^N. \tag{80}$$

Let us find the finite-difference derivative of (80):

$$\left(\sum_{k=1}^{N}\alpha_k^N b_k^N\right)_{\bar{\ell}} = \frac{1}{h\sigma_0}\left[\sum_{k=1}^{N}\alpha_k^N - \sum_{k=1}^{N-1}\alpha_k^{N-1}\right] = \frac{1}{\sigma_0}\sum_{k=1}^{N-1}\frac{\alpha_k^N - \alpha_k^{N-1}}{h} + \frac{\alpha_N^N}{h}. \tag{81}$$

Let us denote

$$\beta_k^N = \frac{1}{2h\sigma_0}\alpha_k^N.$$

Then, Equation (79) has the form

$$2f^0\beta_j^N + h\sum_{k=1}^{N}\left[f'^{j+k-2} + f'^{|j-k|}\right]\beta_k^N = -\frac{1}{\sigma_0}, \qquad j = \overline{1,N}. \tag{82}$$

Equation (82) is solved for fixed N. Then, the solution to the inverse problem can be found in discrete form

$$\sigma_N = \left[\left(\sum_{k=1}^{N}\alpha_k^N b_k^N\right)_{\bar{\ell}}\right]^{-1} = \frac{1}{2}\left[\sum_{k=1}^{N-1}(\beta_k^N - \beta_k^{N-1}) + \beta_N^N\right]^{-1}.$$

$$\sigma_N = \frac{1}{4\sigma_0\,[\Psi_N^N]^2}$$

Note that the system of Equation (82) coincides with the Krein Equation (61). Thus, we have shown that the basic relations of the boundary-control method and the Krein method coincide in the one-dimensional case in a discrete form. Both methods allow us to find a solution to the inverse problem at a specific point x_0 of depth without any special calculations of unknown coefficients in the interval $(0, x_0)$.

2.5. One-Dimensional Inverse Seismic Problem

A.S. Alekseev used the one-dimensional Gelfand–Levitan approach for solving the theory of elasticity's inverse problem [22]. The problem is to determine elastic properties of the medium from the following system:

$$\begin{gathered}
\rho\frac{\partial^2\mathbf{U}}{\partial t^2} = (\lambda+\mu)\text{grad div}\mathbf{U} + \mu\Delta\mathbf{U} + \text{grad}\lambda\text{div}\mathbf{U} + \sum_{i=1}^{3}\text{grad}\mu\left(\frac{\partial\mathbf{U}}{\partial x_i} + \text{grad}U_{x_i}\right)\mathbf{e}_i; \\
\mathbf{U}(x,y,z,t)|_{t<0} \equiv 0; \\
\sigma_z|_{z=0} = \lambda_0\left[\frac{\partial U_x}{\partial x} + \frac{\partial U_y}{\partial y} + \frac{\partial U_z}{\partial z}\right] + 2\mu_0\frac{\partial U_z}{\partial z} = g_1(x,y,t); \\
\tau_{xz}|_{z=0} = \mu_0\left(\frac{\partial U_x}{\partial z} + \frac{\partial U_z}{\partial x}\right) = g_2(x,y,t); \\
\tau_{yz}|_{z=0} = \mu_0\left(\frac{\partial U_y}{\partial z} + \frac{\partial U_z}{\partial y}\right) = g_3(x,y,t); \\
U_x(x,y,0,t) = f_1(x,y,t); \quad U_y(x,y,0,t) = f_2(x,y,t); \quad U_z(x,y,0,t) = f_3(x,y,t).
\end{gathered} \tag{83}$$

The first Equation in (83) describes the propagation of elastic waves through the medium. Functions $g_i, i = 1,2,3$ set the seismic load, which causes the propagation of the seismic waves through the medium. The problem is to determine Lame parameters λ, μ, ρ by using the measurements $f_j, j = 1,2,3$, given by the last ratios in (83) that correspond to measuring the components of the displacement vector $\mathbf{U}$ by using receivers located on the surface $z = 0$.

If the desirable elastic parameters λ, μ, ρ depend only on the depth z, then we can solve the inverse problem (83) by reducing it to several families of integral equations of Gelfand–Levitan type. As shown in [22], if we apply it to the medium surficial moment of force with the intensity $\delta(t)$, then the propagated wave is the SH wave. The inverse problem can be reduced then, using the Hankel transform and the symmetry of the problem, to the following family of problems:

$$\begin{aligned} \frac{1}{v_s^2}\frac{\partial^2 U}{\partial t^2} &= \frac{\partial^2 U}{\partial z^2} + \frac{\partial \ln(\mu)}{\partial z}\frac{\partial U}{\partial z} - k^2 U; \\ U(z,t;k)|_{t<0} &\equiv 0; \\ \frac{\partial U}{\partial z}|_{z=0} &= \frac{1}{4\pi\mu_0}\delta(t); \\ U(z,t;k)|_{z=0} &= f_k(t). \end{aligned} \tag{84}$$

Here, $v_s = \sqrt{\frac{\mu}{\rho}}$ is the shear wave's velocity, k is the integer parameter, μ_0 is the known value of the function $\mu(z)$ at the surface $z = 0$ and $f_k(t)$ are known functions. We can use travel-time coordinates to rewrite the first equation in (84) as follows:

$$\frac{\partial^2 V}{\partial t^2} = \frac{\partial^2 V}{\partial x^2} - q(x;k)V; \tag{85}$$

Here, $x = \int_0^z \frac{d\xi}{v_s(\xi)}$, $q(x;k) = k^2 v_s^2 - \frac{1}{2}\frac{\sigma''}{\sigma} + \frac{3}{4}(\frac{\sigma'}{\sigma})^2$ and $\sigma(x) = \frac{1}{\sqrt{\mu\rho}}$ is the acoustic impedance of the medium. A.S. Alekseev proposed solving the inverse problem for Equation (85) by reducing it to the inverse Sturm–Liouville problem and using the Gelfand–Levitan approach (9) in the spectral domain. This allows us to calculate shear wave velocity $v_s(x)$ and the density of the medium $\rho(x)$. After that, as shown in [22], we can reconstruct p-wave velocity $v_p(z) = \sqrt{\frac{\lambda+2\mu}{\rho}}(z)$ in the same manner by using the point force type source with intensity $\delta(t)$.

3. Two-Dimensional Analogs of the Approach

3.1. A Two-Dimensional Analog of Gelfand–Levitan Equation

Consider the following sequence of direct problems ($k = 0, \pm 1, \pm 2, \ldots$):

$$u_{tt}^{(k)} = u_{xx}^{(k)} + u_{yy}^{(k)} - q(x,y)u^{(k)}, \qquad x \in \mathrm{R}, \quad y \in \mathrm{R}, \quad t > 0; \tag{86}$$

$$u^{(k)}|_{t=0} = 0, \quad u_t^{(k)}|_{t=0} = \delta(x)\mathrm{e}^{\mathrm{i}ky}. \tag{87}$$

We suppose that $q(x,y)$ is a 2π-periodic function with respect to y. Consider an inverse problem: determine the even function $q(x,y)$ from the additional information

$$u^{(k)}|_{x=0} = f^{(k)}(y,t), \qquad k = 0, \pm 1, \pm 2, \ldots \tag{88}$$

The uniqueness of the solution to the inverse problems (86)–(88) can be proved using a technique proposed in [124,125], based on properties of the Dirichlet-to-Neumann map and the finite dependance of the solution on the boundary conditions. Now, we consider the following auxiliary sequence of direct problems ($m = 0, \pm 1, \pm 2, \ldots$) [91,126]:

$$w_{tt}^{(m)} = w_{xx}^{(m)} + w_{yy}^{(m)} - q(x,y)w^{(m)}, \qquad x > 0, \quad y \in \mathrm{R}, \quad t \in \mathrm{R}; \tag{89}$$

$$w^{(m)}|_{x=0} = \mathrm{e}^{\mathrm{i}my}\delta(t), \qquad w_x^{(m)}|_{x=0} = 0. \tag{90}$$

Now, using the d'Alembert formula for problems (89) and (90), we can obtain [91,126]:

$$w^{(m)}(x,y,t) = \frac{1}{2}\mathrm{e}^{\mathrm{i}my}[\delta(t-x)+\delta(t+x)]+ \\ +\frac{1}{2}\int_0^x \int_{t-x+\xi}^{t+x-\xi} \big[-w_{yy}^{(m)} + q(x,y)w^{(m)}\big](\xi,y,\tau)\mathrm{d}\xi\mathrm{d}\tau. \quad (91)$$

The following condition takes place: $w^{(m)}(x,y,t) \equiv 0, 0 < |x| < t, y \in \mathbb{R}$.
We denote

$$\tilde{w}^{(m)}(x,y,t) = w^{(m)}(x,y,t) - \frac{1}{2}\mathrm{e}^{\mathrm{i}my}\big[\delta(t-x)+\delta(t+x)\big]. \quad (92)$$

Using (92) in (91), we can obtain:

$$\tilde{w}^{(m)}(x,y,t) = \frac{1}{4}\mathrm{e}^{\mathrm{i}my}\theta(x-|t|)\Big[xm^2 + Q(x,y,t)\Big]+ \\ +\frac{1}{2}\int_0^x \int_{t-x+\xi}^{t+x-\xi} \big[-\tilde{w}_{yy}^{(m)} + q(x,y)\tilde{w}^{(m)}\big](\xi,y,\tau)\mathrm{d}\xi\mathrm{d}\tau. \quad (93)$$

Here,

$$Q(x,y,t) = \int_0^{\frac{x+t}{2}} q(\xi,y)\mathrm{d}\xi + \int_0^{\frac{x-t}{2}} q(\xi,y)\mathrm{d}\xi. \quad (94)$$

Then,

$$\tilde{w}^{(m)}(x,y,x-0) = \frac{1}{4}\mathrm{e}^{\mathrm{i}my}\Big[xm^2 + \int_0^x q(\xi,y)\mathrm{d}\xi\Big]. \quad (95)$$

Inverse problems (86)–(88) can be reduced formally to a system of integral equations ($k = 0, \pm 1, \pm 2, \ldots$) of the first kind

$$\int_{-x}^{x} \sum_m f_m^{(k)}(t-s)\tilde{w}^{(m)}(x,y,s)\mathrm{d}s = -\frac{1}{2}\Big[f^{(k)}(y,t-x) + f^{(k)}(y,t+x)\Big], \quad (96)$$

or the second kind

$$\tilde{w}^{(k)}(x,y,t) + \int_{-x}^{x} \sum_m f_m^{(k)\prime}(t-s)\tilde{w}^{(m)}(x,y,s)\mathrm{d}s = -\frac{1}{2}\Big[f_t^{(k)}(y,t-x) + f_t^{(k)}(y,t+x)\Big]. \quad (97)$$

Here, $|t| < x$, $y \in \mathrm{R}$. The systems of Equations (96) and (97) are two-dimensional analogs of the Gelfand–Levitan equation. Note that, according to Equation (95), $q(x,y)$ can be calculated, for instance, via the formula

$$q(x,y) = 4\frac{\mathrm{d}}{\mathrm{d}x}\tilde{w}^{(0)}(x,y,x-0). \quad (98)$$

3.2. *A Two-Dimensional Analog of Krein Equation*

Consider the following sequence of the direct problems ($k = 0, \pm 1, \pm 2, \ldots$) [92]:

$$u_{tt}^{(k)} = u_{xx}^{(k)} + u_{yy}^{(k)} - \nabla \ln \rho(x,y)\, \nabla u^{(k)}, \qquad x > 0, \quad y \in \mathrm{R}, \quad t > 0; \quad (99)$$

$$u^{(k)}|_{t<0} \equiv 0, \qquad u_x^{(k)}(+0,y,t) = \mathrm{e}^{\mathrm{i}ky}\,\delta(t). \quad (100)$$

An inverse problem is to determine the function $\rho(x,y)$ from the additional information

$$u^{(k)}(+0,y,t) = f^{(k)}(y,t), \qquad k = 0, \pm 1, \pm 2, \ldots \quad (101)$$

We suppose that $\ln \rho(x,y)$ is a 2π-periodic function. The necessary condition of the inverse problem's solvability can be obtained [84,88,92]:

$$f^{(k)}(y,+0) = -\mathrm{e}^{iky}, \qquad y \in (-\pi,\pi), \quad k \in \mathbb{Z}. \tag{102}$$

We also consider the following sequence of the auxiliary problem:

$$w_{tt}^{(m)} = w_{xx}^{(m)} + w_{yy}^{(m)} - \nabla \ln \rho(x,y)\, \nabla w^{(m)}, \qquad x>0, \quad y \in (-\pi,\pi), \quad t \in \mathbb{R}, \quad m \in \mathbb{Z}; \tag{103}$$

$$w^{(m)}|_{x=0} = \mathrm{e}^{imy}\delta(t), \qquad w_x^{(m)}|_{x=0} = 0. \tag{104}$$

Then, we have the following form of the solution of problems (103) and (104):

$$w^{(m)}(x,y,t) = \frac{1}{2}\mathrm{e}^{imy}\sqrt{\frac{\rho(x,y)}{\rho(0,y)}}[\delta(x+t)+\delta(x-t)] + \widetilde{w}^{(m)}(x,y,t), \tag{105}$$

Here, $\widetilde{w}^{(m)}(x,y,t)$ is the piecewise-smooth function. Solutions of sequences (99), (100) and (103), (104) are connected:

$$u^{(k)}(x,y,t) = \int_R \sum_m f_m^{(k)}(t-s) w^{(m)}(x,y,s)ds, \quad x>0,\ y \in (-\pi,\pi),\ t \in \mathbb{R}; \tag{106}$$

Here,

$$f^{(k)}(y,t) = \sum_m f_m^{(k)}(t)\mathrm{e}^{imy}. \tag{107}$$

Then, we extend functions $f^{(k)}$ and $u^{(k)}$ for $t<0$ as an odd continuation:

$$f^{(k)}(y,t) = -f^{(k)}(y,-t), \quad t<0; \tag{108}$$

$$u^{(k)}(x,y,t) = -u^{(k)}(x,y,-t), \quad t<0; \tag{109}$$

Now, we apply the operator

$$\int_0^x \frac{(.)}{\rho(\xi,y)} d\xi \tag{110}$$

to equality (106). Let us denote

$$V^{(m)}(x,y,t) = \int_0^x \frac{w^{(m)}(\xi,y,t)}{\rho(\xi,y)} d\xi, \tag{111}$$

Then, we can obtain:

$$\frac{\partial}{\partial t}\int_0^x \frac{u^{(k)}(\xi,y,t)}{\rho(\xi,y)} d\xi = \frac{\partial}{\partial t}\int_R \sum_m V^{(m)}(x,y,s) f_m^{(k)}(t-s)ds = \\ = -2V^{(k)}(x,y,t) + \int_{-x}^{x} \sum_m V^{(m)}(x,y,s) f_m^{(k)\prime}(t-s)ds. \tag{112}$$

It was shown [84,92] that the left part of Equation (112) does not depend on x, t, and satisfies the following ratio:

$$\frac{\partial}{\partial t}\int_{-\pi}^{\pi}\int_0^x \frac{u^{(k)}(\xi,y,t)}{\rho(\xi,y)} d\xi dy = -\int_{-\pi}^{\pi} \frac{e^{iky}}{\rho(0,y)} \mathrm{d}y. \tag{113}$$

Let us denote

$$\Phi^{(m)}(x,t) = \int_{-\pi}^{\pi} V^{(m)}(x,y,t)\mathrm{d}y. \tag{114}$$

We can obtain from (112) and (113):

$$2\Phi^{(k)}(x,t) - \sum_m \int_{-x}^{x} (f_m^k)'(t-s)\Phi^{(m)}(x,s)ds = -\int_{-\pi}^{\pi} \frac{e^{iky}}{\rho(0,y)} dy, \quad k \in \mathbb{Z}. \tag{115}$$

For every fixed value of x, Equation (115) is a linear Fredholm integral equation of the second kind. The set of Equation (115) is the multi-dimensional analog of the M.G. Krein equation [84,91]. It was proved [91,126] that

$$V^{(m)}(x,y,x-0) = \frac{e^{imy}}{2\sqrt{\rho(x,y)\rho(0,y)}}. \tag{116}$$

Therefore,

$$\Phi^{(m)}(x,x-0) = \int_{-\pi}^{\pi} \frac{e^{imy}}{2\sqrt{\rho(x,y)\rho(0,y)}} dy. \tag{117}$$

The solution to the inverse problem $\rho(x,y)$ can be found via the formula

$$\rho(x,y) = \frac{\pi^2}{\rho(0,y)} \Big[\sum_m \Phi^m(x,x-0)e^{-imy} \Big]^{-2}. \tag{118}$$

To find the solution to the inverse problem $\rho(x,y)$ at a point $x_0 > 0$, we solve the system (115) setting $x = x_0$ and calculate $\rho(x_0,y)$ via Equation (118).

To numerically solve the two-dimensional analog of the Krein equation (see Figures 3–6), we use the Nth approximation [127,128] of the Krein equation [91]. That is, in the system (115) we set $\Phi^k(x,t) \equiv 0$ for all $N < |k|$ [92]. Discrete analogs of the Gelfand–Levitan equation are investigated in [91,129–132].

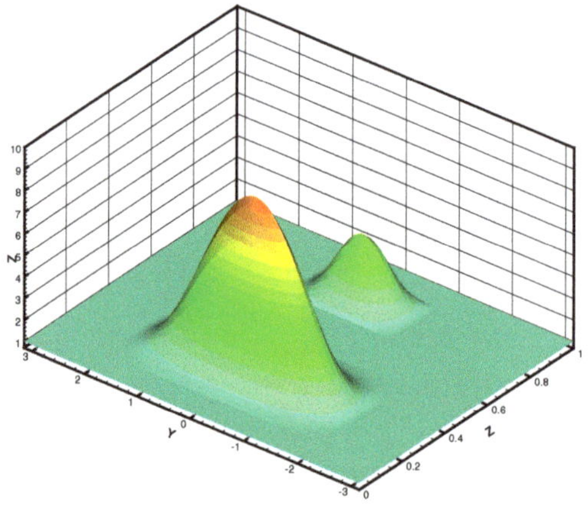

Figure 3. Exact solution of the inverse problem.

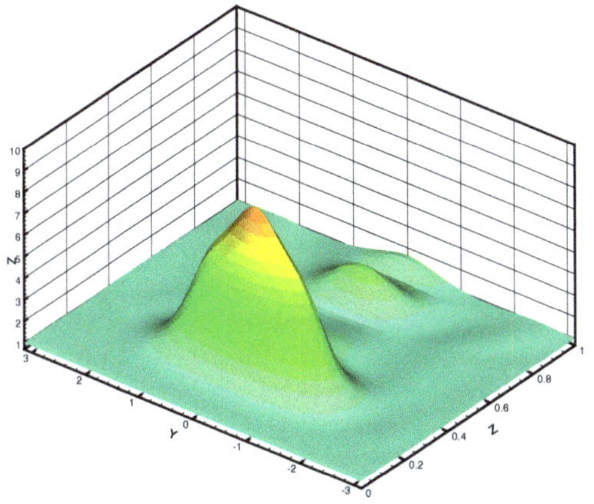

Figure 4. Approximate solution of the inverse problem, $N = 5$, $\varepsilon = 0$.

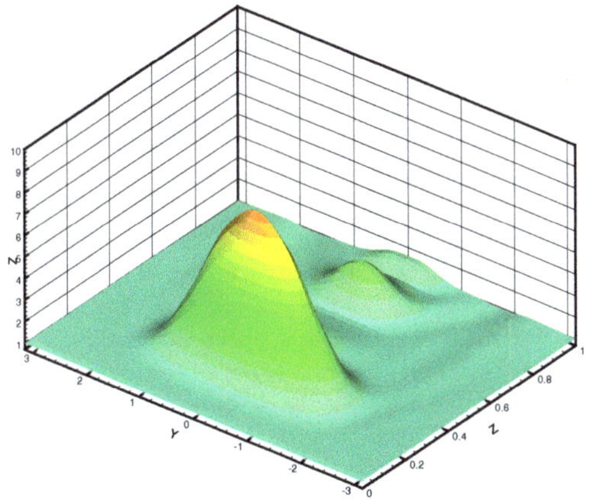

Figure 5. Approximate solution of the inverse problem, $N = 10$, $\varepsilon = 0$.

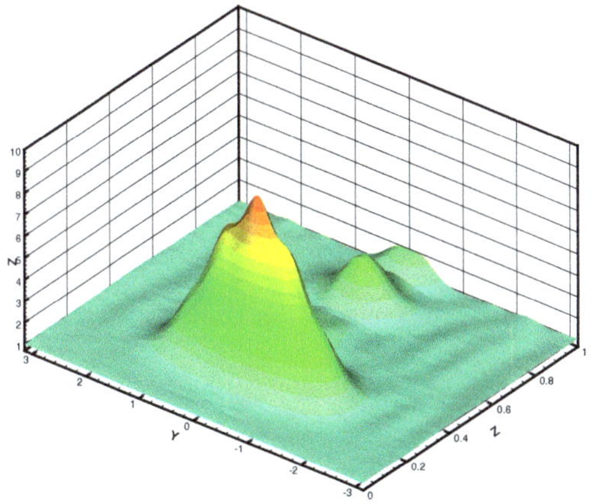

Figure 6. Approximate solution of the inverse problem, $N = 10$, $\varepsilon = 0.05$.

4. Numerical Methods for Solving Gelfand–Levitan and Krein Equations

In this section, we present a short review of numerical methods that can be used for solving arising equations. In order to find the solution of one of the considered inverse problems, we have to solve the family of corresponding integral equations for every value of the parameter x, which is correlated to the current depth. On the one hand, Gelfand–Levitan ((9) and (91)) and Krein ((61) and (115)) equations, despite their differences, have similar basic structure, which is provided by the form of the Fredholm linear integral equation with a convolution-type kernel and the fact that we need only one component of the solution to solve the inverse problem. On the other hand, this structure makes several numerical techniques suitable for solving these equations. The efficiency of one or another algorithm results from its ability to utilize the structure of the equation.

In [133,134], the authors proposed the Monte-Carlo method for solving the 2D analog of the Gelfand–Levitan Equation (91). The idea is to represent the solution of the equation by the sum of Neumann series. This sum is calculated as mean value of some random variable. Such a scheme allows one to estimate specific components of the solution, which is important due to the fact that we only need one component of the solution of the Gelfand–Levitan-type equation to restore the solution of the inverse problem. Another stochastic approach was proposed in [135], where Equation (91) was solved by a randomized version of the Kaczmarz algorithm. Randomization of the well-known Kaczmarz projection method allows one to achieve (under some assumptions) second-order complexity of the method. It is also possible to modify the algorithm to use more information about the structure of the discrete version of Equation (91), like block structure of the matrix. Moreover, unlike most Monte-Carlo methods, this algorithm does not require the convergence of Neumann series.

We should also mention the approaches based on the convolution-type kernels of the Gelfand–Levitan and Krein equations. Such kernels allow one to reduce the equations to linear systems with Toeplitz (or block-Toeplitz) matrices. In [136], Equation (115) was solved via the block version of the Levinson–Durbin recursion method. It was shown that such an approach allows one to use the connection between Equation (115) for different values of depth parameter x. The proposed recursion method allows one to calculate the solutions of the whole family of Krein equations during the solution of only one linear system. Both methods allow one to obtain the solution of the inverse problem in every point of the mesh for $O(N^2)$ operations, where N is the number of mesh points.

5. Numerical Calculations

In this section, we present an example of the numerical reconstruction of the inverse problem's solution, based on the GLK approach. For the numerical solution of inverse problems (99)–(101), we used a regularization technique, based on a projection of the problem on N-dimensional subspace, produced by the basis $\{\mathrm{e}^{\mathrm{i}ky}\}_{k=0,\pm1,\dots,\pm N}$ [91,92,128]. This approach reduces the two-dimensional problem to a finite system of one-dimensional inverse problems [91]. We suppose that the solution of problems (99) and (100) can be represented as a series:

$$u^{(k)}(x,y,t) = \sum_m u_m^{(k)}(x,t)\mathrm{e}^{\mathrm{i}my}; \tag{119}$$

We also suppose that the function $\rho(x,y)$ also has the same representation:

$$\ln\rho(x,y) = \sum_m a_m(x)\mathrm{e}^{\mathrm{i}my}; \tag{120}$$

In this case, problems (99)–(101) can be rewritten as follows

$$\frac{\partial^2 u_n^{(k)}}{\partial t^2} = \frac{\partial^2 u_n^{(k)}}{\partial x^2} - n^2 u_n^{(k)}(x,t) - \sum_{m\in\mathbb{Z}} \frac{\partial a_m}{\partial x}(x)\frac{\partial u_{n-m}^{(k)}}{\partial x}(x,t) +$$
$$+ \sum_{m\in\mathbb{Z}} m(m-n)a_m(x)u_{n-m}^{(k)}(x,t), \quad x \in \mathbb{R}, \quad t > 0, \quad k,n \in \mathbb{Z};$$

$$u_n^{(k)}|_{t=0} = 0, \qquad \frac{\partial u_n^{(k)}}{\partial t}|_{t=0} = \delta_{kn}\delta(x);$$
$$u_n^{(k)}|_{x=0} = f_n^{(k)}(t).$$

Here, δ_{kn} is the Kronecker symbol:

$$\delta_{kn} = \begin{cases} 1, & k = n \\ 0, & \text{else} \end{cases}.$$

Now, we suppose that all the Fourier coefficients with number greater than N vanish and consider then the following problem:

$$\frac{\partial^2 \overline{V}_N^{(k)}}{\partial t^2} = \frac{\partial^2 \overline{V}_N^{(k)}}{\partial x^2} - K\overline{V}_N^{(k)} + A(x)\overline{V}_N^{(k)} - B(x)\frac{\partial \overline{V}_N^{(k)}}{\partial x}, \qquad x > 0, \quad t > 0; \tag{121}$$

$$\overline{V}_N^{(k)}|_{t<0} \equiv 0, \qquad \frac{\partial \overline{V}_N^{(k)}}{\partial x}|_{x=0} = I_N^{(k)}\delta(t); \tag{122}$$

$$\overline{V}_N^{(k)}|_{x=0} = \overline{F}_N^{(k)}. \tag{123}$$

Problems (121)–(123) are called an N-approximation of inverse problems (99) and (100). Here, A, K, B are square matrices of size $2N+1$ with elements:

$$K_{km} = m^2\delta_{km}; \tag{124}$$

$$A_{km}(x) = m(k-m)a_{k-m}(x), \qquad k,m = -N\ldots N; \tag{125}$$

$$B_{km}(x) = \frac{\partial a_{k-m}}{\partial x}, \qquad k,m = -N\ldots N. \tag{126}$$

Using the technique proposed in [137], one can obtain that, as $N \to \infty$, the N-approximation converges to the solution of systems (99) and (100). The N-approximation of the Krein Equation (115) can be also obtained:

$$\Phi^k(x,t) = \frac{1}{2}\int_{-x}^{x} \sum_{|m|<N} (f_m^k)'(t-s)\Phi^m(x,s)ds + G^k, \quad k = -N,\ldots,N; \tag{127}$$

Numerical calculations (see Figures 3–6) are used to find an approximate solution to the inverse problem. The two-dimensional inverse problem is approximated via a finite system of one-dimensional inverse problems [91,92,128]. The problem is solved in the domain $x \in (0,1)$, $y \in (-\pi,\pi)$, $t \in (0,2)$. Equation (127) is approximated via an SLAE, the size of which depends on both the number of grid points and the number of harmonics considered. The spatial dimension of the grid is 100×100. The number of Fourier harmonics can be associated with the number of sources and receivers that one has on the surface $x = 0$. While the dependence of the solution on the number of harmonics required was considered in several works related to the numerical solution of 2D problems (that we mentioned in previous section), in this paper we consider some values of that number only to illustrate the behaviour of the numerical reconstruction. The number of Fourier harmonics is $N = 5$ in Figure 4 and $N = 10$ in Figures 5 and 6.

In order to illustrate the impact of the noise on the reconstruction, we consider the noisy data in the following form:

$$f^{\varepsilon}(y,t) = f(y,t) + \varepsilon\alpha(y,t)(f_{\max} - f_{\min}). \tag{128}$$

Here, ε is a noise level in the data, $\alpha(y,t)$ is a uniformly distributed random number on the interval $[-1,1]$ for a fixed y and t and $f_{\max}$ and $f_{\min}$ are the maximum and minimum of the exact data, respectively. Below, we provide the results of the inverse problem's solution for 5% error in the data. We chose this number in an arbitrary way to illustrate the stability of the method. On the other hand, such a level of errors, introduced in the data without any pre-processing, could be considered as a way to simulate the real-case scenario.

One can see that the accuracy of the numerical reconstruction was acceptable. However, the second peak, which is located further from the daylight surface (and, thus, receivers), was reconstructed worse than the first one. On the one hand, this result fits well into the physics beside the inverse problem—the large obstacle deters part of the information about the object to get to the receivers, located on the daylight surface. On the other hand, from a mathematical point of view the convergence of the numerical solution to the exact one can be provided by increasing the numbers of Fourier coefficients and grid points, but only in the case of smooth parameters. The stability of the method is also acceptable and can be improved further by using some processing of the noised data (because the kernel of the equation depends on the first derivative of the data).

6. Reconstruction of the Velocity $c(x,y)$ and the Density $\rho(x,y)$

Let us consider the following inverse problem: find the velocity $c(x,y)$ and the density $\rho(x,y)$ from the sequence of relations ($k = 0, \pm1, \pm2, \ldots$):

$$c^{-2}(x,y)u^{(k)}_{tt} = \Delta u^{(k)} - \nabla \ln \rho(x,y) \nabla u^{(k)}, \qquad x \in \mathrm{R}, \quad y \in \mathrm{R}, \quad t > 0;$$

$$u^{(k)}|_{t=0} = 0, \qquad u^{(k)}_t|_{t=0} = \mathrm{e}^{\mathrm{i}ky}\,\delta(x).$$

$$u^{(k)}(0,y,t) = f^{(k)}(y,t), \qquad u^{(k)}_x(+0,y,t) = 0.$$

Let $\tau(x,y)$ be a solution of the Cauchy problem for the the eikonal equation

$$\begin{gathered} \tau_x^2 + \tau_y^2 = c^{-2}(x,y), \qquad x > 0, \quad y \in \mathrm{R}; \\ \tau|_{x=0} = 0, \qquad \tau_x|_{x=0} = c^{-1}(0,y), \qquad y \in \mathrm{R}. \end{gathered} \tag{129}$$

Let us introduce new variables $z = \tau(x,y), y = y$ and new functions

$$v^{(k)}(z,y,t) = u^{(k)}(x,y,t), \qquad b(z,y) = c(x,y). \tag{130}$$

Since the velocity is supposed to be strictly positive, this change of variables is not degenerate at least in some interval $x \in (0,h)$.

Let us consider the sequence of the auxiliary problems ($m = 0, \pm1, \pm2, \ldots$) [91,126]:

$$\begin{gathered} w^{(m)}_{tt} = w^{(m)}_{zz} + b^2 w^{(m)}_{yy} + q w^{(m)}_{yz} + p w^{(m)}_{z} + r w^{(m)}_{y}, \qquad z > 0, \quad y \in \mathrm{R}, \quad t \in \mathrm{R}; \\ w^{(m)}(0,y,t) = \mathrm{e}^{\mathrm{i}my}\,\delta(t), \qquad w^{(m)}_z(0,y,t) = 0. \end{gathered} \tag{131}$$

Here,

$$q(z,y) = 2b^2\tau_y, \tag{132}$$

$$p(z,y) = b^2(z,y)(\tau_{xx} + \tau_{zz}) - (\ln\rho)_z - \frac{1}{2}q(z,y)(\ln\rho)_y, \tag{133}$$

$$r(z,y) = -b^2(z,y)(\ln\rho)_y - \frac{1}{2}q(z,y)(\ln\rho)_z. \tag{134}$$

We suppose that $c(0,y) = b(0,y)$ is known and for simplicity $b(0,y) \equiv 1$ for $y \in R$.

In the neighborhood of the plane $t = z$, the solution of the direct problem (131) has the form [91,126]:

$$w^{(m)}(z,y,t) = S^{(m)}(z,y)\delta(z-t) + Q^{(m)}(z,y)\theta(z-t) + \tilde{w}^{(m)}(z,y,t). \tag{135}$$

Here, $\tilde{w}^{(m)}$ is continuous function and functions $S^{(m)}$ and $Q^{(m)}$ solve the following problems:

$$\begin{gathered} 2(S^{(m)})_z + q(z,y)(S^{(m)})_y + Sp(z,y) = 0, \qquad t > 0, \quad y \in \mathbb{R}; \\ S^{(m)}|_{t=0} = \frac{1}{2}\mathrm{e}^{imy}. \end{gathered} \tag{136}$$

$$\begin{aligned} &2(Q^{(m)})_z + q(z,y)(Q^{(m)})_y + Qh(z,y) = \\ &= -\left[(S^{(m)})_{zz} + b(z,y)(S^{(m)})_{yy} + q(S^{(m)})_{yz} + p(S^{(m)})_z + q(S^{(m)})_y\right], t > 0, \quad y \in R; \end{aligned} \tag{137}$$

$$Q^{(m)}|_{t=0} = 0. \tag{138}$$

The 2D analogy of the M.G. Krein equation follows from (135) ($m = 0, \pm1, \pm2, \ldots$):

$$\sum_m S^{(m)}(z,y) f_m^{(k)'}(t-z) + \tilde{w}^{(k)}(z,y,t) + \sum_m \int_{-z}^{z} f_m^{(k)'}(t-s)\tilde{w}^{(m)}(z,y,s)\mathrm{d}s = 0, \quad |t| < z. \tag{139}$$

So, to solve the inverse problem we can solve the system (136)–(139) using the projection method and find functions $c(x,y)$ and $\rho(x,y)$ using the following iterative algorithm.

First, we introduce the N-approximation of the system (136)–(139), e.g., let $\tilde{w}^{(m)}$, $S^{(m)}$ and $Q^{(m)}$ be equal to 0 for all $|m| > N$. Let us suppose that $c_n(x,y)$ is known. Then, we calculate $\tau_n(x,y)$ from (129), $b_n(z,y)$ from (130) and $q_n(z,y)$ and $p_n(z,y)$ from (134). Function $S_n^{(m)}(t,y)$ is calculated from (136). Then, solving the 2D analogy of M.G. Krein Equation (139), we find $\tilde{w}_n^{(m)}(z,y,t)$ for $|m| \leq N$. It follows from (135) that $Q_n^{(m)}(t,y) = \tilde{w}_n^{(m)}(t+0,y,t)$. Then, from Equations (136) and (137), we find function $b_{n+1}(z,y)$ and after that the new value $c_{n+1}(x,y) = b_{n+1}(z,y)$ is calculated.

7. Conclusions

In this paper, we reviewed existing works related to the Gelfand–Levitan–Marchenko–Krein approach that allows one to solve some inverse problems by reducing them to sets of integral equations. We discussed spectral and dynamic variations of the method, as well as the connection between the GLKM approach and the BC method and the inverse scattering method that utilizes the connection of some nonlinear equations (we considered the KdV equation in the manuscript) and the inverse scattering problem to formulate the algorithm to integrate the equations. Also, we mentioned different approaches to the numerical solution of the GLKM equations.

When considering the further development of the approach, one should mention that, while the one-dimensional version of the method is well-developed, there are still a lot of aspects that have to be considered in the multi-dimensional case, both in theory and numerics. Also, the fact that the method belongs to the class of the direct ones provides the natural usage of the approach as the first step of the data processing and parameter estimations that can be later improved via other techniques of solving the inverse problems.

Author Contributions: Conceptualization, S.K.; methodology, S.K., M.S. and N.N.; software, M.S. and N.N.; formal analysis, S.K. and N.P.; writing—original draft preparation, M.S., N.N. and N.P.; writing—review and editing, S.K. and N.N.; supervision, S.K.; project administration, M.S.; funding acquisition, S.K., M.S., N.N. All authors have read and agreed to the published version of the manuscript.

Funding: The work is supported by the Mathematical Center in Akademgorodok under the agreement No. 075-15-2022-281 with the Ministry of Science and Higher Education of the Russian Federation.

Data Availability Statement: Data sharing is not applicable to this article.

Conflicts of Interest: The authors declare no conflicts of interest.

References

1. Ambarzumijan, V.A. Uber eine Frage der Eigenwerttheorie. *Z. Phys.* **1929**, *53*, 690–695. [CrossRef]
2. Borg, G. Eine Umkehrung der Sturm Liouvilleschen Eigenwertanfgabe. Bestimmung der Differentialgleichung durch die Eigenwarte. *Acta Math.* **1946**, *78*, 1–96. [CrossRef]
3. Levinson, N. The inverse Sturm–Lionville problem. *Math. Tidsskr. B* **1949**, *36*, 25–30.
4. Levinson, N. On the uniqueness of the potential in a Schrodinger equation for a given asymptotic phase. *Danske Vid. Selsk. Mat. Fys. Medd.* **1949**, *25*, 9.
5. Gelfand, I.M.; Levitan, B.M. On the determination of a differential equation from its spectral function. *Izv. Akad. Nauk SSSR* **1951**, *15*, 309–336.
6. Krein, M.G. On a method of effective solution of an inverse boundary value problem. *Dokl. Akad. Nauk SSSR.* **1954**, *94*, 6.
7. Gardner, C.S.; Greene, J.M.; Kruskal, M.D.; Miura, R.M. Method for Solving the Korteweg–de Vries Equation. *Phys. Rev. Lett.* **1967**, *19*, 1095–1097. [CrossRef]
8. Heisenberg, W. Die beobachtbaren Grossen in der Theorie der Elementarteilchen—I. *Z. Phys.* **1943**, *120*, 513–538. [CrossRef]
9. Heisenberg, W. Die beobachtbaren Grossen in der Theorie der Elementarteilchen—II. *Z. Phys.* **1943**, *120*, 673–702. [CrossRef]
10. Bargmann, V. On the connection between phase shifts and scattering potential. *Revue Mod. Phys.* **1949**, *21*, 488–493. [CrossRef]
11. Bargmann, V. Remarks on the determination of a central field of force from the elastic scattering phase shifts. *Phys. Rev.* **1949**, *75*, 301–302. [CrossRef]
12. Tikhonov, A.N. On the uniqueness of the solution to the problem of electrical prospecting. *Dokl. Akad. Nauk SSSR* **1949**, *69*, 6.
13. Marchenko, V.A. Some questions in the theory of differential operator of the second order. *Dokl. Akad. Nauk SSSR* **1950**, *72*, 3.
14. Levinson, N. Certain relations between phase shifts and scattering potential. *Phys. Rev.* **1953**, *89*, 755. [CrossRef]
15. Marchenko, V.A. *Some Questions in the Theory of One-Dimensional Linear Differential Operators of the Second Order—I*; Moscow Mathematical Society: Moscow, Russia, 1952; Volume I.
16. Krein, M.G. Solution of the inverse Sturm–Liouville problem. *Dokl. Akad. Nauk SSSR* **1951**, *76*, 1.
17. Krein, M.G. Determination of the density of an inhomogeneous symmetric string from its frequency spectrum. *Dokl. Akad. Nauk SSSR* **1951**, *76*, 3.
18. Krein, M.G. On inverse problems for an inhomogeneous string. *Dokl. Akad. Nauk SSSR* **1952**, *82*, 5.
19. Krein, M.G. On the transition function of the one-dimensional boundary value problem of the second order. *Dokl. Akad. Nauk SSSR* **1953**, *88*, 3.
20. Krein, M.G. On some cases of effective determination of the density of a non-homogeneous string by its spectral function. *Dokl. Akad. Nauk SSSR* **1953**, *93*, 4.
21. Agranovich, Z.S.; Marchenko, V.A. *The Inverse Problem of Scattering Theory*; Kharkiv National University: Kharkov, Ukraine, 1960.
22. Alekseev, A.S. Some inverse problems in wave propagation theory. *Dokl. Akad. Nauk SSSR* **1962**, *11*, 405–408.
23. Blokh, A.S. On the Determination of a Differential Equation from its Spectral Function–Matrix. *Dokl. Akad. Nauk SSSR* **1953**, *92*, 2.
24. Kay, I.; Moses, H.E. The determination of the scattering potential from the spectral measure function. *Nuovo C.* **1956**, *3*, 276–304. [CrossRef]
25. Levin, B.Y. *Distribution of the Roots of Integer Functions*; Gostekhizdat: Moscow, Russia, 1956; p. 632.
26. Levitan, B.M.; Gasymov, M.G. Determination of a differential equation by two spectra. *Uspekhi Matem. Nauk.* **1964**, *19*, 2. [CrossRef]
27. Marchenko, V.A. Reconstruction of the potential energy from the phases of scattered waves. *Dokl. Akad. Nauk SSSR* **1955**, *104*, 5.
28. Regge, T. Introduction to complex orbital momenta. *Nuovo C.* **1959**, *14*, 951–976. [CrossRef]
29. Fermi, E.; Pasta, J.; Ulam, S. *Studies of Nonlinear Problems—I*; Report; Los Alamos Scintific Laboratory of the University of California: Los Almos, NM, USA, 1954.
30. Newton, R.G. Construction of potentials from the phase shifts at fixed energy. *J. Math. Phys.* **1962**, *3*, 75–82. [CrossRef]
31. Kruskal, N.J.; Zabussky, M.D. Interaction of "Solitons" in a Collisionless Plasma and the Recurrence of Initial States. *Phys. Rev. Lett.* **1965**, *15*, 240.

32. Lax, P.D. Integrals of Nonlinear Equations of Evolution and Solitary Waves. *Commun. Pure Appl. Math.* **1968**, *21*, 467–490; *Matematika* **1969**, *13*, 128–150. [CrossRef]
33. Kadomtsev, B.B.; Petviashvili, V.I. On the stability of solitary waves in weakly dispersive media. *Sov. Phys. Dokl.* **1970**, *15*, 539–541.
34. Gardner, C.S. The Korteweg–de Vries Equation and Generalizations—IV. The Korteweg–de Vries Equation as a Hamiltonian System. *J. Math. Phys.* **1971**, *12*, 1548–1551. [CrossRef]
35. Zakharov, V.E.; Shabat, A.B. Exact theory of two-dimensional self-focusing and onedimensional self-modulation of waves in nonlinear media. *Zh. Eksperim. Teoret. Fiz.* **1971**, *61*, 118.
36. Zakharov, V.E.; Faddeev, L.D. The Korteweg–de Vries equation: A completely integrable Hamiltonian system. *Funkts. Anal. Ego Prilozheniya* **1971**, *5*, 18–27.
37. Shabat, A.B. On the Korteweg–de Vries equation. *Dokl. Akad. Nauk SSSR* **1973**, *211*, 6.
38. Novikov, S.P. A periodic problem for the KdV equation. *Funkts. Anal. Ego Prilozheniya* **1974**, *8*, 236–246. [CrossRef]
39. Lax, P.D. Periodic solutions of the KdV equation. *Lect. Appl. Math.* **1974**, *15*, 51. [CrossRef]
40. Marchenko, V.A. The periodic KdV problem. *Dokl. Akad. Nauk SSSR* **1974**, *217*, 1052–1056.
41. Zakharov, V.E.; Shabat, A.B. A scheme for integrating the nonlinear equations of mathematical physics by the method of the inverse scattering problem—I. *Funkts. Anal. Ego Prilozheniya* **1974**, *8*, 43–53.
42. Faddeev, L.D. The inverse problem in the quantum theory of scattering—II. In *Current Problems in Mathematics*; Loose Errata; Akad. Nauk SSSR Vsesojuz. Inst. Nauch. i Tehn. Informacii: Moscow, Russia, 1974; Volume 3, pp. 93–180. (In Russian)
43. Newton, R.G. Inverse Schrodinger scattering in three dimensions. In *Text and Monographs in Physics*; Springer: Berlin, Germany, 1989.
44. Zakharov, V.E.; Manakov, S.V. On the complete integrability of a nonlinear Schrödinger equation. *Teoret. Matem. Fiz.* **1974**, *19*, 332–343.
45. Manakov, S.V. The inverse scattering method and two-dimensional evolution equations. *Uspekhi Mat. Nauk.* **1976**, *31*, 245–246.
46. Zakharov, V.E.; Manakov, S.V. Generalization of the inverse scattering problem method. *Teor. Matem. Fizika* **1976**, *27*, 283–287. [CrossRef]
47. Lax, P.D. Almost periodic solutions of the KdV equation. *SIAM Rev.* **1976**, *18*, 351–375. [CrossRef]
48. Zakharov, V.E.; Shabat, A.B. Integration of nonlinear equations of mathematical physics by the method of the inverse scattering problem—II. *Funkts. Anal. Ego Prilozheniya* **1979**, *13*, 13–22.
49. Zakharov, V.E.; Manakov, S.V.; Novikov, S.P.; Pitaevskii, L.P. *The Theory of Solitons: The Inverse Problem Method*; Nauka: Novosibirsk, Russia, 1980.
50. Nizhnik, L.P.; Pochinayko, M.D. Integration of the nonlinear Schrodinger equation in two spatial dimensions by the inverse scattering method. Funkts. *Anal. Ego Prilozheniya* **1982**, *16*, 80–82.
51. Veselov, A.P.; Novikov, S.P. Finite-zone two-dimensional Schrodinger operators. Potential operators. *Dokl. Akad. Nauk SSSR* **1984**, *279*, 4.
52. Novikov, R.G.; Khenkin, G.M. Oscillating weakly localized solutions of the Korteweg–de Vries equation. *Theor. Math. Phys.* **1984**, *61*, 1089–1099. [CrossRef]
53. Novikov, R.G. Construction of a two-dimensional Schrödinger operator with a given scattering amplitude at fixed energy. *Theoret. Math. Phys.* **1986**, *66*, 154–158. [CrossRef]
54. Grinevich, P.G.; Novikov, R.G. Analogues of multisoliton potentials for the two-dimensional Schrodinger operator. *Funct. Anal. Appl.* **1985**, *19*, 276–285. [CrossRef]
55. Grinevich, P.G.; Novikov, R.G. Analogues of multisoliton potentials for the two-dimensional Schrödinger equations and a nonlocal Riemann problem. *Soviet Math. Dokl.* **1986**, *33*, 9–12.
56. Grinevich, P.G.; Manakov, S.V. Inverse scattering problem for the two-dimensional Schrödinger operator, the $\bar{\partial}$-method and nonlinear equations. *Funct. Anal. Appl.* **1986**, *20*, 94–103. [CrossRef]
57. Boiti, M.; Leon, J.P.; Manna, M.; Pempinelli, F. On a spectral transform of a KdV-like equation related to the Schrödinger operator in the plane. *Inverse Probl.* **1987**, *3*, 25–36. [CrossRef]
58. Tsai, T.-Y. The Schrödinger operator in the plane. *Inverse Probl.* **1993**, *9*, 763–787. [CrossRef]
59. Nachman, A.I. Global uniqueness for a two-dimensional inverse boundary value problem. *Ann. Math.* **1996**, *143*, 71–96. [CrossRef]
60. Bogdanov, L.V.; Konopelchenko, B.G.; Moro, A. Symmetric reductions of a real dispersionless Veselov–Novikov equation. *Fund. Prikl. Matem.* **2004**, *10*, 5–15.
61. Lassas, M.; Mueller, J.L.; Siltanen, S. Mapping properties of the nonlinear Fourier transform in dimension two. *Comm. Partial Differ. Equ.* **2007**, *32*, 591–610. [CrossRef]
62. Lassas, M.; Mueller, J.L.; Siltanen, S.; Stahel, A. The Novikov-Veselov equation and the inverse scattering method: II. Computation. *Nonlinearity* **2012**, *25*, 1799–1818. [CrossRef]
63. Lassas, M.; Mueller, J.L.; Siltanen, S.; Stahel, A. The Novikov-Veselov equation and the inverse scattering method, Part I: Analysis. *Phys. D* **2012**, *241*, 1322–1335. [CrossRef]
64. Music, M. The nonlinear Fourier transform for two-dimensional subcritical potentials. *Inverse Probl. Imaging* **2014**, *8*, 1151–1167. [CrossRef]
65. Perry, P. Miura maps and inverse scattering for the Novikov-Veselov equation. *Anal. Partial. Differ. Equ.* **2014**, *7*, 311–343. [CrossRef]

66. Francoise, J.-P.; Novikov, R.G. Solutions rationnelles des equations de type Korteweg–de Vries en dimension 2+ 1 et problemes a m corps sur la droite. Comptes rendus de l'Academie des sciences. Serie 1. *Mathematique* **1992**, *314*, 109–113.
67. Novikov, R.G. Approximate solution of the inverse problem of quantum scattering theory with fixed energy in dimension 2. *Tr. Mat. Inst. Steklova* **1999**, *225*, 301–318.
68. Kazeykina, A.V.; Novikov, R.G. A large time asymptotics for transparent potentials for the Novikov–Veselov equation at positive energy. *J. Nonlinear Math. Phys.* **2011**, *18*, 377–400. [CrossRef]
69. Music, M.; Perry, P.; Siltanen, S. Exceptional Circles of Radial Potentials. *Inverse Probl.* **2013**, *29*, 045004. [CrossRef]
70. Bondarenko, N.P. Inverse problem solution and spectral data characterization for the matrix Sturm–Liouville operator with singular potential, Anal. *Math. Phys.* **2021**, *11*, 145. [CrossRef]
71. Bondarenko, N.P. Spectral data characterization for the Sturm–Liouville operator on the star-shaped graph. *Anal. Math. Phys.* **2020**, *10*, 83. [CrossRef]
72. Avdonin, S.A.; Mikhaylov, A.S.; Mikhaylov, V.S.; Park, J.C. Inverse problem for the Schrodinger equation with non-self-adjoint matrix potential. *Inverse Probl.* **2021**, *37*, 035002. [CrossRef]
73. Xu, X.-C.; Bondarenko, N. Local solvability and stability of the generalized inverse Robin–Regge problem with complex coefficients. *J. Inverse Ill-Posed Probl.* **2023**, *31*, 711–721. [CrossRef]
74. Pariiskii, B.S. *The Inverse Problem for a Wave Equation with a Depth Effect. Some Direct and Inverse Problems of Seismology*; Nauka: Moscow, Russia, 1968; pp. 25–40.
75. Gopinath, B.; Sondhi, M. Determination of the shape of the human vocal tract from acoustical measurements. *Bell System Tech. J.* **1970**, *49*, 1195–1214. [CrossRef]
76. Gopinath, B.; Sondh, M. Inversion of telegraph equation and synthesis of nonuniform lines. *Proc. IEEE* **1971**, *59*, 383–392. [CrossRef]
77. Blagoveshchenskii, A.S. *The Local Method of Solution of the Nonstationary Inverse Problem for an Inhomogeneous String*; Trudy Matem Inst. Im. Steklova Akad. Nauk SSSR: Moscow, Russia, 1971; Volume 115, pp. 28–38.
78. Alekseev, A.S.; Dobrinskii, V.I. Some questions of practical application of dynamic inverse problems of seismics. In *Mathematical Problems of Geophysics*; Computing Center of the Siberian Branch of USSR Academic Science: Novosibirsk, Russia, 1975; pp. 7–53.
79. Yu, L.; Brodov, V.V.; Loctsik, V.M.; Markushevich, N.N.; Novikova, V.E.; Fedorov, S.B. *Sinjunkhina, Monochromatic Sounding of the Upper Part of a Velocity Profile by a Horizontal Vibrator, Selected Papers from Volumes 24 and 25 of Vychislitel'naya Seysmologiya*; American Geophysical Union: Washington, DC, USA, 2013; pp. 150–155.
80. Pariiskii, B.S. *Economical Methods for the Numerical Solutions of Convolution Equations and of Systems of Algebraic Equations with Töplitz Matrices*; Computing Center of the Siberian Branch of USSR Academic Science: Novosibirsk, Russia, 1977.
81. Symes, W.W. Inverse boundary value problems and a theorem of Gel'fand and Levitan. *J. Math. Anal. Appl.* **1979**, *71*, 378–402. [CrossRef]
82. Burridge, R. The Gelfand–Levitan, the Marchenko and the Gopinath-Sondhi integral equation of inverse scattering theory, regarded in the context of inverse impulse-response problems. *Wave Motion.* **1980**, *2*, 305–323. [CrossRef]
83. Santosa, F. Numerical scheme for the inversion of acoustical impedance profile based on the Gelfand–Levitan method. *Geophys. J. Roy. Astr. Soc.* **1982**, *70*, 229–244. [CrossRef]
84. Kabanikhin, S.I. *Projection-Difference Methods of Determination of the Coefficients of Hyperbolic Equations*; Nauka: Novosibirsk, Russia, 1988.
85. Romanov, V.G.; Kabanikhin, S.I. *Inverse Problems of Geoelectrics*; Nauka: Moscow, Russia, 1991.
86. Alekseev, A.S.; Belonosov, V.S. *Spectral Methods in One-Dimensional Problems of Wave Propagation Theory*; Institute of Computational Mathematics and Mathematical Geophysics: Novosibirsk, Russia, 1998.
87. Belishev, M.I. On an approach to multidimensional inverse problems for the wave equation. *Dokl. Akad. Nauk SSSR* **1987**, *297*, 524–527.
88. Kabanikhin, S.I. *Linear Regularization of Multidimensional Inverse Problems for Hyperbolic Equations*; Preprint No. 27; Institute of Mathematics of the Siberian Branch of the Russian Academic Science: Novosibirsk, Russia, 1988.
89. Belishev, M.I.; Blagoveshchenskii, A.S. Multidimensional analogs of equations of the Gelfand–Levitan–Krein type in the inverse problem for the wave equation. In *Ill-Posed Problems of of Mathematical Physics and Analysis*; American Mathematical Society: Ann Arbor, MI, USA, 1992; pp. 50–63.
90. Kabanikhin, S.I.; Shishlenin, M.A. Boundary control and Gelfand–Levitan–Krein methods in inverse acoustic problem. *J. Inv. Ill-Posed Probl.* **2004**, *12*, 125–144. [CrossRef]
91. Kabanikhin, S.I.; Satybaev, A.D.; Shishlenin, M.A. *Direct Methods of Solving Inverse Hyperbolic Problems*; VSP: Amsterdam, The Netherlands, 2005.
92. Kabanikhin, S.I.; Shishlenin, M.A. Numerical algorithm for two-dimensional inverse acoustic problem based on Gel'fand-Levitan–Krein equation. *J. Inverse Ill-Posed Probl.* **2011**, *18*, 979–995. [CrossRef]
93. Shishlenin, M.A.; Novikov, N.S. Comparative analysis of two numerical methods for solving the Gelfand–Levitan–Krein equation. *Sib. Electron. Math. Rep.* **2011**, *8*, 379–393.
94. Druskin, V.; Mamonov, A.V.; Thaler, A.E.; Zaslavsky, M. Direct, Nonlinear Inversion Algorithm for Hyperbolic Problems via Projection-Based Model Reduction. *Siam J. Imaging Sci.* **2016**, *9*, 684–747. [CrossRef]

95. Borcea, L.; Druskin, V.; Mamonov, A.; Zaslavsky, M. Untangling the nonlinearity in inverse scattering with data-driven reduced order models. *Inverse Probl.* **2018**, *34*, 065008. [CrossRef]
96. Druskin, V.; Moskow, S.; Zaslavsky, M. Lippmann–Schwinger–Lanczos algorithm for inverse scattering problems. *Inverse Probl.* **2021**, *37*, 075003. [CrossRef]
97. Romanov, V.G. Justification of the Gelfand–Levitan–Krein Method for a Two-Dimensional Inverse Problem. *Sib. Math. J.* **2021**, *62*, 908–924. [CrossRef]
98. Kabanikhin, S.; Shishlenin, M.; Novikov, N. Multidimensional analogs of Gelfand–Levitan–Krein equations. In Proceedings of the 6th International Conference on Control and Optimization with Industrial Applications, Khalilov, Baku, 11–13 July 2018; Institute of Applied Mathematics: Khalilov, Baku, 2018; pp. 31–33.
99. Shishlenin, M.A.; Izzatulah, M.; Novikov, N.S. Comparative Study of Acoustic Parameter Reconstruction by using Optimal Control Method and Inverse Scattering Approach. *J. Phys. Conf. Ser.* **2021**, *2092*, 012004. [CrossRef]
100. Novikov, N.; Shishlenin, M. Direct Method for Identification of Two Coefficients of Acoustic Equation. *Mathematics* **2023**, *11*, 3029. [CrossRef]
101. Kabanikhin, S.I.; Novikov, N.S.; Shishlenin, M.A. Gelfand–Levitan–Krein method in one-dimensional elasticity inverse problem. *J. Phys. Conf. Ser.* **2021**, *2092*, 012022. [CrossRef]
102. Kabanikhin, S.; Novikov, N. Shishlenin, M. Linear seismic data processing of area observing systems. In Proceedings of the 6th International Conference on Control and Optimization with Industrial Applications, Khalilov, Baku, 11–13 July 2018; Institute of Applied Mathematics: Khalilov, Baku, 2018; pp. 219–221.
103. Kabanikhin, S.I.; Shishlenin, M.A. Digital field. Georesursy. *Georesources* **2018**, *20*, 139–141. [CrossRef]
104. Baev, A.V. On t-local solvability of inverse scattering problems in two-dimensional layered media. *Comput. Math. Math. Phys.* **2015**, *55*, 1033–1050. [CrossRef]
105. Baev, A.V. Solution of an inverse scattering problem for the acoustic wave equation in three-dimensional media. *Comput. Math. Math. Phys.* **2016**, *56*, 2043–2055. [CrossRef]
106. Baev, A.V. Imaging of layered media in inverse scattering problems for an acoustic wave equation. *Math. Model. Comput. Simulations* **2016**, *8*, 689–702. [CrossRef]
107. Baev, A.V.; Gavrilov, S.V. The Inverse Scattering Problem in a Nonstationary Medium. *Comput. Math. Model.* **2019**, *30*, 218–229. [CrossRef]
108. Belishev, M.I. Boundary control in reconstruction of manifolds and metrics (the BC method). *Inverse Probl.* **1997**, *13*, R1–R45. [CrossRef]
109. Belishev, M.I.; Blagoveshchenskii, A.S. *Dynamic Inverse Problems of Wave Theory*; Publication House of Saint-Petersburg State University: Saint-Petersburg, Russia, 1999.
110. Belishev, M.I. Recent progress in the boundary-control method. *Inverse Probl.* **2007**, *23*, R1–R67. [CrossRef]
111. Belishev, M.I. Boundary control and inverse problems: The one-dimensional variant of the BC method. In *Mathematical Problems in the Theory of Wave Propagation*; Zap. Nauchn. Sem. POMI: Saint-Petersburg, Russia, 2008; Volume 37.
112. Beilina, L.; Klibanov, M.V.; Kokurin, M.Y. Adaptivity with relaxation for ill-posed problems and global convergence for a coefficient inverse problem. *J. Math. Sci.* **2010**, *167*, 279–325. [CrossRef]
113. Klibanov, M.V.; Timonov, A. A comparative study of two globally convergent numerical methods for acoustic tomography. *Commun. Anal. Comput.* **2023**, *1*, 12–31. [CrossRef]
114. Klibanov, M.V.; Jingzhi, L. *Inverse Problems and Carleman Estimates: Global Uniqueness, Global Convergence and Experimental Data*; De Gruyter: Berlin, Germany, 2021. [CrossRef]
115. Klibanov, M. Carleman estimates for the regularization of ill-posed Cauchy problems. *Appl. Numer. Math.* **2015**, *94*, 46–74. [CrossRef]
116. Klibanov Michael, V. Global convexity in a three-dimensional inverse acoustic problem. *Siam J. Math. Anal.* **1997**, *28*, 1371–1388. [CrossRef]
117. Kuzhuget, A.V.; Beilina, L.; Klibanov, M.V.; Sullivan, A.; Nguyen, L. Blind backscattering experimental data collected in the field and an approximately globally convergent inverse algorithm. *Inverse Probl.* **2012**, *28*, 095007. [CrossRef]
118. Klibanov Michael, V. Carleman estimates for global uniqueness, stability and numerical methods for coefficient inverse problems. *J. Inverse Ill-Posed Probl.* **2013**, *21*, 477–560. [CrossRef]
119. Yurko, V.A. *Introduction to the Theory of Inverse Spectral Problems*; Fizmatlit: Moscow, Russia, 2007.
120. Romanov, V.G. *Inverse Problems of Mathematical Physics*; Nauka: Moscow, Russia, 1984.
121. Novikov, R.G. Inverse scattering for the Schrodinger equation in dimension 1 up to smooth functions. *Bull. Des Sci. Math.* **1996**, *120*, 473–491.
122. Kudryashov, N.A. *Analytical Theory of Nonlinear Differential Equations*; Ijevsk: Moscow, Russia, 2004; p. 361. (In Russian)
123. Landau, L.D.; Lifshitz, E.M. *Quantum Mechanics*; MSU Quantum Technology Centre: Moscow, Russia, 1963; p. 704. (In Russian)
124. Rakesh. An inverse problem for the wave equation in the half plane. *Inverse Probl.* **1993**, *9*, 433–441. [CrossRef]
125. Sylvester, J.; Uhlmann, G. A global uniqueness theorem for an inverse boundary value problem. *Ann. Math.* **1987**, *125*, 153–169. [CrossRef]
126. Kabanikhin, S.I. On linear regularization of multidimensional inverse problems for hyperbolic equations. *Sov. Math. Dokl.* **1990**, *40*, 579–583.

127. Kabanikhin, S.I. Regularization of multidimensional inverse problems for hyperbolic equations based on a projection method. *Doklady Akademii Nauk.* **1987**, *292*, 534–537.
128. Kabanikhin, S.I.; Shishlenin, M.A. Quasi-solution in inverse coefficient problems. *J. Inverse Ill-Posed Probl.* **2008**, *16*, 705–713. [CrossRef]
129. Gladwell, G.M.L.; Willms, N.B. A discrete Gelfand–Levitan method for band-matrix inverse eigenvalue problems. *Inverse Probl.* **1989**, *5*, 165–179. [CrossRef]
130. Kabanikhin, S.I.; Bakanov, G.B. Discrete Analogy of Gelfand–Levitan Method. *Doklady Akademii Nauk* **1997**, *356*, 157–160.
131. Kabanikhin, S.I.; Bakanov, G.B. A discrete analog of the Gelfand–Levitan method in a two-dimensional inverse problem for a hyperbolic equation. *Sib. Math. J.* **1999**, *40*, 262–280. [CrossRef]
132. Natterer, F. *A Discrete Gelfand–Levitan Theory*; Technical Report; Institut fuer Numerische und Instrumentelle Mathematik Universitaet Munster: Muenster, Germany, 1994.
133. Kabanikhin, S.; Sabelfeld, K.; Novikov, N.; Shishlenin, M. Numerical solution of the multidimensional Gelfand–Levitan equation. *J. Inverse Ill-Posed Probl.* **2015**, *23*, 439–450. [CrossRef]
134. Novikov, N.S. Comparative analysis of numerical methods for solving two-dimensional Gelfand–Levitan equation. *Sib. Electron. Math. Rep.* **2014**, *23*, 132–144.
135. Kabanikhin, S.; Sabelfeld, K.; Novikov, N.; Shishlenin, M. Numerical solution of an inverse problem of coefficient recovering for a wave equation by a stochastic projection methods. *Monte Carlo Methods Appl.* **2015**, *21*, 189–203. [CrossRef]
136. Kabanikhin, S.I.; Novikov, N.S.; Oseledets, I.V.; Shishlenin, M.A. Fast Toeplitz linear system inversion for solving two-dimensional acoustic inverse problem. *J. Inverse Ill-Posed Probl.* **2015**, *23*, 687–700. [CrossRef]
137. Kabanikhin, S.I.; Scherzer, O.; Shishlenin, M.A. Iteration methods for solving a two dimensional inverse problem for a hyperbolic equation. *J. Inverse Ill-Posed Probl.* **2003**, *11*, 87–109. [CrossRef]

Σ mathematics MDPI

Article

To the Question of the Solvability of the Ionkin Problem for Partial Differential Equations

Aleksandr I. Kozhanov

Sobolev Institute of Mathematics, Acad. Koptyug, 4, Novosibirsk 630090, Russia; kozhanov@math.nsc.ru

Abstract: We study the solvability of the Ionkin problem for some differential equations with one space variable. These equations include parabolic and quasiparabolic, hyperbolic and quasihyperbolic, pseudoparabolic and pseudohyperbolic, elliptic and quasielliptic equations and equations of many other types. For the above equations, the following theorems are proved with the use of the splitting method: the existence of regular solutions—solutions that all have weak derivatives in the sense of S. L. Sobolev and occur in the corresponding equation.

Keywords: spatial nonlocal problems; Ionkin condition; splitting method; regular solutions; existence; uniqueness

MSC: 35G40

1. Introduction

The Ionkin problem (N. I. Ionkin [1]) has been studied by many authors and for many classes of partial differential equations, and at the same time, the original method, proposed by N. I. Ionkin himself, has almost always been used. This is the method of decomposing the solution in some special biorthogonal systems of functions. In 2006, in A. M. Nakhushev's book [2] (see also [3–5]) and in the recent works [6,7] of the author of this paper, new approaches were applied to studying the Ionkin problem and close nonlocal problems—in A. M. Nakhushev's work, for second-order parabolic equations, and in the works of the author of this paper, for quasiparabolic equations, parabolic equations with an arbitrary evolution direction and elliptic equations.

In the present article, A. M. Nakhushev's approach is further developed: we show that this approach is applicable to a wide class of differential equations and that with its help, one can obtain a number of substantially new results on the solvability of the Ionkin problem and some other nonlocal problems close to it.

In 1986 in [8], N. I. Yurchuk proposed his approach to studying the solvability of the Ionkin problem for second-order parabolic equations. This approach was based on a priori estimates but it gave the existence of solutions belonging to some weighted Sobolev space. Let us clarify that in contrast to N. I. Yurchuk's approach, the approach of [6,7] gives the existence of solutions belonging to classical Sobolev spaces. The splitting method proposed below also gives the existence of regular solutions.

Citation: Kozhanov, A.I. To the Question of the Solvability of the Ionkin Problem for Partial Differential Equations. *Mathematics* **2024**, *12*, 487. https://doi.org/10.3390/math12030487

Academic Editor: Natalia Bondarenko

Received: 1 December 2023
Revised: 15 January 2024
Accepted: 26 January 2024
Published: 2 February 2024

2. Statement of the Problems

Let Ω be the interval $(0,1)$ of the Ox axis, Q be the rectangle $\Omega\times(0,T)$ of the variables x and t be finite height T. Denote by D_x^k and D_t^k the derivatives $\frac{\partial^k}{\partial x^k}$ and $\frac{\partial^k}{\partial t^k}$, respectively. Furthermore, let

$$P_k(t,D_t)=\sum_{j=1}^{p_k}\alpha_{kj}(t)D_k^j,\quad k=1,\dots,m,$$

be operators with real coefficients and L be the differential operator

$$L = \sum_{k=0}^{m} P_k(t, D_t) D_x^{2k} \quad .$$

For the operator L, define the conditions

$$U_j(x, 0) = 0, \quad j = 1, \dots m_1, \quad x \in \Omega, \tag{1}$$

$$U_j(x, T) = 0, \quad j = m_1 + 1, \dots, p_0 = \max(p_1, \dots, p_m), \quad x \in \Omega. \tag{2}$$

Nonlocal Problem I: *Find a function $u(x,t)$ that is a solution in Q to the equation*

$$Lu = f(x, t)$$

and satisfies conditions (1) *and* (2) *and also the conditions*

$$D_x^{2k} u(x, t)\Big|_{x=0} = 0, \quad k = 0, \dots, m-1, \quad t \in (0, T), \tag{3}$$

$$D_x^{2k+1} u(x, t)\Big|_{x=0} - D_x^{2k+1} u(x, t)\Big|_{x=1} = 0, \quad k = 0, \dots, m-1, \quad t \in (0, T). \tag{4}$$

Nonlocal Problem II: *Find a function $u(x,t)$ that is a solution in Q to the equation*

$$Lu = f(x, t)$$

and satisfies conditions (1) *and* (2) *and also the conditions*

$$D_x^{2k+1} u(x, t)\Big|_{x=0} = 0, \quad k = 0, \dots, m-1, \quad t \in (0, T), \tag{5}$$

$$D_x^{2k} u(x, t)\Big|_{x=0} - D_x^{2k} u(x, t)\Big|_{x=1} = 0, \quad k = 0, \dots, m-1, \quad t \in (0, T). \tag{6}$$

Nonlocal Problem III: *Find a function $u(x,t)$ that is a solution in Q to the equation*

$$Lu = f(x, t)$$

and satisfies conditions (1), (2), *and* (3) *and also the condition*

$$D_x^{2k+1} u(x, t)\Big|_{x=0} + D_x^{2k+1} u(x, t)\Big|_{x=1} = 0, \quad k = 0, \dots, m-1, \quad t \in (0, T). \tag{7}$$

Nonlocal Problem IV: *Find a function $u(x,t)$ that is a solution in Q to the equation*

$$Lu = f(x, t)$$

and satisfies conditions (1), (2) *and* (5), *and also the condition*

$$D_x^{2k} u(x, t)\Big|_{x=0} + D_x^{2k} u(x, t)\Big|_{x=1} = 0, \quad k = 0, \dots, m-1, \quad t \in (0, T). \tag{8}$$

For $m = 1$, $P_0 = D_t$, $P_1 = -I$, Nonlocal Problem I is the Ionkin problem [1] (see also [9]). Nonlocal Problem II for the same operators P_0 and P_1 can be called the problem conjugate to the Ionkin problem. If in Problems I and II, the operators P_0 and P_1 are not the same as in the Ionkin problem, then these problems can be called a generalization of the Ionkin problem and the conjugate Ionkin problem.

Nonlocal Problems III and IV for the equation $Lu = f$ have not been studied previously.

Define the linear space H:

$$H = \{v(x, t) : v(x, t) \in L_2(Q),\ D_t^{p_k} D_x^{2k} v(x, t) \in L_2(Q), \quad k = 0, \dots, m\}$$

(here the derivatives are understood as weak derivatives in the sense of S. L. Sobolev).

Obviously, H is a Banach space with respect to the norm

$$\|v\|_H = \left(\int\limits_Q \left[v^2 + \sum_{k=0}^{m} \left(D_t^{p_k} D_x^{2k} v \right)^2 \right] dx\, dt \right)^{1/2}.$$

The aim of this article is to prove the existence of solutions to Nonlocal Problems I–IV belonging to H.

3. Main Results

We put $F(x,t) = f(x,t) + f(1-x,t)$.
Consider two auxiliary problems.

Problem 1. *Find a function $v(x,t)$ that is a solution in Q to the equation*

$$Lv = F(x,t)$$

and satisfies conditions (1) *and* (2) *and also the condition*

$$D_x^{2k+1} v(x,t)\Big|_{x=0} = D_x^{2k+1} v(x,t)\Big|_{x=1} = 0, \quad k = 0,\ldots,m-1, \quad t \in (0,T). \tag{9}$$

Problem 2. *Find a function $w(x,t)$ that is a solution in Q to the equation*

$$Lw = f(x,t)$$

and satisfies the conditions

$$D_x^{2k} w(x,t)\Big|_{x=0} = 0, \quad k = 0,\ldots,m-1, \quad t \in (0,T),$$

$$D_x^{2k} w(x,t)\Big|_{x=1} = D_x^{2k} v(x,t)\Big|_{x=0}, \quad k = 0,\ldots,m-1, \quad t \in (0,T) \tag{10}$$

($v(x,t)$ is a solution to Problem 1).

The Main Condition: *The operators P_k, $k = 0,\ldots,m$, the function $f(x,t)$, and conditions* (1) *and* (2) *are such that boundary value Problems A and B are uniquely solvable in H.*

Theorem 1. *Suppose the fulfillment of the Main Condition. Then the solution $w(x,t)$ to Problem 2 is a solution to Nonlocal Problem I in H.*

Proof. Alongside $v(x,t)$, the function $v(1-x,t)$ is also a solution to Problem 1. Since a solution to Problem 1 is unique, for $(x,t) \in Q$ we have

$$v(x,t) = v(1-x,t). \tag{11}$$

Next, using the solution $w(x,t)$ to Problem 2, define the function $V(x,t)$:

$$V(x,t) = w(x,t) + w(1-x,t).$$

This function satisfies the equalities

$$D_x^{2k} V(x,t)\Big|_{x=0} = D_x^{2k} w(x,t)\Big|_{x=1} = D_x^{2k} v(x,t)\Big|_{x=0}, \tag{12}$$

$$D_x^{2k} V(x,t)\Big|_{x=1} = D_x^{2k} w(x,t)\Big|_{x=1} = D_x^{2k} v(x,t)\Big|_{x=0} = D_x^{2k} v(x,t)\Big|_{x=1} \tag{13}$$

(the last equality follows from (11)). □

These equalities imply that the functions $v(x,t)$ and $V(x,t)$ satisfy identical boundary conditions, and they are both solutions to the same equation. Due to the uniqueness of solutions, we have

$$V(x,t) = v(x,t).$$

However, then

$$D_x^{2k+1}w(x,t)\Big|_{x=0} - D_x^{2k+1}w(x,t)\Big|_{x=1}$$
$$= D_x^{2k+1}v(x,t)\Big|_{x=0} + D_x^{2k+1}v(x,t)\Big|_{x=1} = 0, \quad t \in (0,T).$$

In other words, the function $w(x,t)$ satisfies the desired boundary conditions of Nonlocal Problem I.

The fulfillment of conditions (1) and (2) for $w(x,t)$, the validity of the equation $Lw = f$ and the membership $w(x,t) \in H$ are obvious.

Therefore, $w(x,t)$ is a desired solution to Nonlocal Problem I.

The theorem is proved.

We put $f_1(x,t) = \int\limits_0^x f(y,t)\,dy$.

Theorem 2. *Suppose that the function $f_1(x,t)$ satisfies the Main Condition. Then, Nonlocal Problem II has a solution belonging to H.*

Proof. Let $\underline{u}(x,t)$ be a solution to Nonlocal Problem I for $f_1(x,t)$. We put $u(x,t) = \underline{u}_x(x,t)$. The function $u(x,t)$ will be the desired solution to Nonlocal Problem II.

The theorem is proved. □

For proving the solvability of Nonlocal Problems III and IV, we need a modified Main Condition.

We put $F_1(x,t) = f(x,t) - f(1-x,t)$.

Consider two auxiliary problems:

Problem 3. *Find a function $v(x,t)$ that is a solution in Q to the equation*

$$Lv = F_1(x,t)$$

and satisfies conditions (1) *and* (2) *and also condition* (9).

Problem 4. *Find a function $w(x,t)$ that is a solution in Q to the equation*

$$Lw = f(x,t)$$

and satisfies the conditions

$$D_x^{2k}w(x,t)\Big|_{x=0} = 0, \quad D_x^{2k}w(x,t)\Big|_{x=1} = -D_x^{2k}v(x,t)\Big|_{x=0},$$
$$k = 0,\ldots,m-1, \quad t \in (0,T)$$

($v(x,t)$ is a solution to Problem 3).

The Modified Main Condition: *The operators P_k, $k = 0,\ldots,m$, the function $f(x,t)$, and conditions* (1) *and* (2) *are such that boundary value Problems A_1 and B_1 are uniquely solvable in H.*

Theorem 3. *Suppose the fulfillment of the Modified Main Condition. Then the solution $w(x,t)$ to Problem 4 is a solution to Nonlocal Problem III from H.*

Proof. Like $v(x,t)$, the function $-v(1-x,t)$ is a solution to Problem 3. Since the solution to Problem 3 is unique, then for $(x,t) \in Q$, we have $v(x,t) = -v(1-x,t)$. We put

$V_1(x,t) = w(x,t) - w(1-x,t)$. Obviously, $w(x,t)$ satisfies all conditions to Nonlocal Problem III. The membership of $w(x,t)$ in H follows from the Modified Main Condition.

The theorem is proved. □

Theorem 4. *Suppose that $f_1(x,t)$ satisfies the Modified Main Condition. Then, Nonlocal Problem IV has a solution belonging to H.*

The proof of this theorem is obvious.

4. Examples

Theorems 1–4 imply that for proving the solvability of Nonlocal Problems I–IV (and, in particular, the solvability of the Ionkin problem), it suffices to check the fulfillment of the Main Condition or the Modified Main Condition. Let us give several examples when these conditions either hold or are easy to be seen to hold.

Example 1. *Quasiparabolic Equations of Arbitrary Order.*

Let P_0 and P_m be the operators

$$P_0 = (-1)^{p+1} D_t^{2p+1}, \quad P_m = (-1)^{m+1} I.$$

In the rectangle Q, consider the equation

$$P_0 u + P_m D_x^{2m} u = f(x,t). \tag{14}$$

For $p = 0$, $m = 1$, this equation is the heat equation; the Ionkin problem (Nonlocal Problem I) was studied in this case (by expanding the solution in the series in special biorthogonal function systems) in [1,9]. In the more general case of second-order parabolic equations with arbitrary coefficients, the solvability of Nonlocal Problems I and II was established in [6,8]. Next, the solvability of Nonlocal Problem I in the special case of $p = 0$, $m = 2$, was studied in [10] (also with the use of expanding the solution in special biorthogonal systems).

In the general case of $p \geq 0$, $m \geq 1$, Nonlocal Problems I–IV have not been studied before.

As was shown in Section 2, for proving the solvability of Nonlocal Problems I–IV in H, it suffices to prove that they satisfy the Main Condition or the Modified Main Condition.

As with conditions (1) and (2), for Equation (14), choose either the conditions

$$D_t^k u(x,t)\Big|_{t=0} = 0, \quad k = 0,\dots,p, \quad x \in \Omega,$$

$$D_t^k u(x,t)\Big|_{t=T} = 0, \quad k = 0,\dots,p-1, \quad x \in \Omega \tag{15}$$

or the conditions

$$D_t^k u(x,t)\Big|_{t=0} = 0, \quad k = 0,\dots,p, \quad x \in \Omega,$$

$$D_t^k u(x,t)\Big|_{t=T} = 0, \quad k = p+1,\dots,2p-1, \quad x \in \Omega. \tag{16}$$

The solvability of boundary value Problem 1 in H for Equation (14) (with conditions (15) or (16)) is not hard to prove by the classical Fourier method. Obviously, this solution is unique, and $f(x,t) \in L_2(Q)$ is a sufficient condition for the solvability (existence and uniqueness) of Problem 1.

We show that under some additional assumptions on $f(x,t)$, Problem 2 is also uniquely solvable in H.

Transform Problem 2: turn it into a problem with zero boundary conditions by setting

$$w(x,t) = W(x,t) - \varphi(x,t),$$

$$\varphi(x,t) = a_{2(m-1)}(x)D_x^{2(m-1)}v(0,t) + a_{2(m-2)}(x)D_x^{2(m-2)}v(0,t) + \ldots + a_0(x)v(0,t).$$

Here, the coefficients $a_k(x)$ are polynomials of degree of at most $2m-1$, and they are chosen so that the conditions

$$D_x^{2k}\varphi(x,t)\Big|_{x=0} = 0, \quad D_x^{2k}\varphi(x,t)\Big|_{x=1} = D_x^{2k}v(x,t)\Big|_{x=0}$$

$$k = 0,\ldots,m-1, \quad t \in (0,T),$$

hold. Obviously, the function $W(x,t)$ must satisfy the equation

$$LW = f(x,t) - P_0\varphi(x,t) = \widetilde{f}(x,t)$$

in Q.

The boundary value problem for this equation with zero boundary data for Problem 2 has a solution belonging to H if $\widetilde{f}(x,t) \in L_2(Q)$. Since $f(x,t) \in L_2(Q)$, for the validity of the desired inclusion for the function $\widetilde{f}(x,t)$, it suffices that the equations

$$D_t^{2p+1}D_x^{2k}v(0,t) \in L_2([0,T]), \quad k = 0,\ldots,m-1, \tag{17}$$

hold. We show that under some additional conditions for $f(x,t)$, the solution $v(x,t)$ to boundary value Problem 1 (with conditions (15) and (16)) is such that the equations in (17) hold.

Proposition 1. *Suppose that the functions $D_x^k f(x,t)$, $k = 0,\ldots,2m-1$, belong to $L_2(Q)$ and for $m \geq 2$ we have*

$$D_x^{2k-1}f(x,t)\Big|_{x=0} - D_x^{2k-1}f(x,t)\Big|_{x=1} = 0, \quad k = 1,\ldots,m-1, \quad t \in (0,T). \tag{18}$$

Then, boundary value Problem 1 with conditions (15) *or* (16) *for Equation* (14) *has a solution $v(x,t)$ such that $v(x,t) \in H$, $D_t^{2p+1}D_x^{2m-1}v(x,t) \in L_2(Q)$, $D_x^{3m-1}v(x,t) \in L_2(Q)$.*

Proof. Consider the auxiliary problem: Find a function $v(x,t)$ that is a solution to Q to the equation

$$Lv + \varepsilon(-1)^p D_t^{2p+1}D_x^{4m-2}v = F(x,t) \tag{19}$$

($\varepsilon > 0$) and satisfies conditions (15) and (9) and also the conditions

$$D_x^{2k+1}v(x,t)\Big|_{x=0} = D_x^{2k+1}v(x,t)\Big|_{x=1}, \quad k = m,\ldots,2(m-1), \quad t \in (0,T). \tag{20}$$

Define the space H_1:

$$H_1 = \{v(x,t) : v(x,t) \in H,\ D_t^{2p+1}D_x^{4m-2}v(x,t) \in L_2(Q)\}.$$

Boundary Value Problem (9), (15), (19), (20) has a solution $v(x,t)$ belonging to H_1; this is not hard to prove by using by the classical Fourier method or by the Galerkin method with the choice of a special basis (see, for example, [11]).

Multiply (19) by the function $(-1)^{p+1}(T_0-t)D_t^{2p+1}D_x^{4m-2}v(x,t)$, $T_0>0$, and integrate it by using Q. After easy calculations, we infer that solutions $v(x,t)$ to the boundary value problem (9), (15), (19), (20) satisfy the a priori estimate

$$\begin{gathered}\int\limits_Q\left\{\left[D_t^{2p+1}D_x^{2m-1}v(x,t)\right]^2+\left[D_t^pD_x^{3m-1}v(x,t)\right]^2\right\}dx\,dt\\ +\varepsilon\int\limits_Q\left[D_t^{2p+1}D_x^{4m-2}v(x,t)\right]^2dx\,dt\le C\sum_{k=0}^{2m-1}\int\limits_Q\left[D_x^kf(x,t)\right]^2dx\,dt,\end{gathered}\tag{21}$$

in which the number C is defined only by T.

Estimate (21) and the reflexivity of a Hilbert space (see [12,13]) implies that there exist sequences $\{\varepsilon_l\}_{l=1}^{\infty}$ of positive numbers and $\{v_l(x,t)\}_{l=1}^{\infty}$ of solutions to problem (9), (15), (19), (20) with $\varepsilon=\varepsilon_l$ such that as $l\to\infty$, the convergences

$$\varepsilon_l\to 0,$$

$$v_l(x,t)\to v(x,t)\quad \textit{weakly in } H,$$

$$\varepsilon_lD_t^{2p+1}D_x^{4m-2}v_l(x,t)\to 0\quad \textit{weakly in } L_2(Q),$$

hold. Obviously, the limit function $v(x,t)$ is a desired solution to Problem 1 under condition (15).

For condition (16), all the arguments are analogous to those given above.

The proposition is proved. □

The proposition implies that, for solutions to Problem 1, under the above conditions on $f(x,t)$, equations (17) hold. As we said above, Nonlocal Problem I satisfies the Main Condition under conditions (15) or (16).

Summing up what was said above, we obtain the following theorem:

Theorem 5. *For any function $f(x,t)$ such that $D_x^kf(x,t)$, $k=0,\dots,2m-1$, belong to $L_2(Q)$ and satisfying* (18) *for $m\ge 2$, Nonlocal Problem I with conditions* (15) *and* (16) *has a solution $u(x,t)$ belonging to H.*

The solvability of Nonlocal Problem II with conditions (15) or (16) with respect to t is not hard to prove with the use of Theorem 2.

The solvability of nonlocal Problem III with conditions (15) or (16) is not hard to prove with the use of Theorem 3. We only specify that here we must also use the assertion about the presence of the additional equations in (17) for solutions $v(x,t)$ to Problem 3 and that condition (18) must be replaced by the condition *for $m\ge 2$,*

$$D_x^{2k-1}f(x,t)\Big|_{x=0}+D_x^{2k-1}f(x,t)\Big|_{x=1}=0,\quad k=0,\dots,2(m-1),\quad t\in(0,T).\tag{22}$$

The solvability of Nonlocal Problem IV is not hard to prove with the use of Theorem 4.

We do not give the exact statements of Nonlocal Problems II–IV due to their obviousness.

Example 2. *Hyperbolic and Quasihyperbolic Equations.*

We confine ourselves to the case $m=1$.

In the rectangle Q, consider the equation

$$Lv\equiv(-1)^{p+1}D_t^{2p}u-u_{xx}=f(x,t)\tag{23}$$

For $p=1$, this equation is the usual wave equation; the nonlocal Ionkin problem for this equation was studied in [14]. For $p>1$, Equation (23) is not hyperbolic (and, in particular, the classical initial boundary value problem for it is ill-posed); Nonlocal Problems I–IV have not been studied for it before.

As with conditions (1) and (2), we use either the conditions

$$D_t^k u(x,t)\Big|_{t=0} = 0, \quad k = 0,\dots,p, \quad x \in \Omega,$$

$$D_t^k u(x,t)\Big|_{t=T} = 0, \quad k = 1,\dots,p-1, \quad x \in \Omega, \tag{24}$$

or the conditions

$$D_t^k u(x,t)\Big|_{t=0} = 0, \quad k = 0,\dots,p, \quad x \in \Omega,$$

$$D_t^k u(x,t)\Big|_{t=T} = 0, \quad k = p+1,\dots,2p-1, \quad x \in \Omega. \tag{25}$$

The solvability of Problem 1 for Equation (23) with conditions (24) with respect to t in H was established in [15,16], and the solvability of Problem 1 for Equation (23) with conditions (25) with respect to t in H was shown in [17]; in both cases, the solution $v(x,t)$ is unique, and in both cases, the following memberships for $f(x,t)$ are required: $f(x,t) \in L_2(Q)$, $f_t(x,t) \in L_2(Q)$.

We show that the solution $v(x,t)$ to Problem 1 with conditions (24) or (25) under some additional constraints on $f(x,t)$ is such that $D_t^{2p} D_x v(x,t) \in L_2(Q)$.

Proposition 2. *Suppose that the functions $f(x,t)$, $f_t(x,t)$, $f_{xt}(x,t)$ belong to $L_2(Q)$. Then Boundary Problem 1 with conditions* (24) *or* (25) *for Equation* (23) *has a solution $v(x,t)$ such that $D_t^{2p} D_x v(x,t) \in L_2(Q)$, $D_x^3 v(x,t) \in L_2(Q)$.*

Proof. Consider the auxiliary problem: *Find a function $v(x,t)$ that is a solution in Q to the equation*

$$Lv + \varepsilon(-1)^{p+1} D_t^{2p} D_x^4 v = F(x,t) \tag{26}$$

($\varepsilon > 0$) and satisfies (24) and the condition

$$D_x^{2k+1} v(x,t)\Big|_{x=0} = D_x^{2k+1} v(x,t)\Big|_{x=1} = 0, \quad k = 0,1, \quad t \in (0,T). \tag{27}$$

Using the Fourier method or the Galerkin method with the choice of a special basis, it is not hard to see that problem (24), (26), (27) has a solution $v(x,t)$ such that $v(x,t) \in H$, $D_t^{2p} D_x^4 v(x,t) \in L_2(Q)$. Demonstrate that $v(x,t)$ satisfies a priori estimates uniform in ε.

Multiply Equation (26) by the function $(T_0 - t) D_t D_x^4 v(x,t)$, $T_0 > T$, and integrate the result over Q. After easy calculations, we infer that solutions $v(x,t)$ to problem (24), (26), (27) satisfy the estimate

$$\begin{aligned} &\int_Q \left\{ \left[D_t^p D_x^2 v(x,t)\right]^2 + \left[D_x^3 v(x,t)\right]^2 \right\} dx\,dt \\ &+\varepsilon \int_Q \left[D_t^p D_x^4 v(x,t)\right]^2 dx\,dt + \int_\Omega \left[D_x^3 v(x,T)\right]^2 \\ &\le C_1 \int_Q \left[f^2(x,t) + f_t^2(x,t) + f_{xt}^2(x,t)\right] dx\,dt, \end{aligned} \tag{28}$$

where the constant C_1 is determined only by T.

At the next step, multiply (26) by $D_t^{2p} D_x^2 v(x,t)$ and integrate the result over Q. Using Young's inequality and estimate (28), we conclude that solutions $v(x,t)$ to problem (24), (26), (27) admit the estimate

$$\begin{aligned} &\int_Q \left[D_t^{2p} D_x v(x,t)\right]^2 dx\,dt + \varepsilon \int_Q \left[D_t^{2p} D_x^3 v(x,t)\right]^2 dx\,dt \\ &\leq C_2 \int_Q \left[f^2(x,t) + f_t^2(x,t) + f_x^2(x,t) + f_{xt}^2(x,t)\right] dx\,dt, \end{aligned} \tag{29}$$

where the constant C_2 is determined only by T.

Estimates (28) and (29) are quite enough for passing to the limit in problem (24), (26), (27). Using (28) and (29) and the reflexivity of a Hilbert space, passing to the limit in the corresponding subsequence, we conclude that Problem 1 with condition (24) for Equation (23) has a desired solution $v(x,t)$.

If in Problem 1 condition (25) is given for Equation (23), then completely analogous arguments again yield the existence of a desired solution $v(x,t)$.

The proposition is proved. □

Proposition 2 means that Nonlocal Problem I satisfies the Main Condition. Therefore, we have the following theorem:

Theorem 6. *For any function $f(x,t)$ such that $f(x,t) \in W_2^1(Q)$, $f_{xt}(x,t) \in L_2(Q)$, Nonlocal Problem I with conditions* (24) *or* (25) *has a solution $u(x,t) \in H$.*

It is not hard to prove the solvability of Nonlocal Problems II–IV with conditions (24) or (25) with respect to t using the algorithm of Section 2 and the technique of obtaining a priori estimates presented in the proof of Proposition 2.

Example 3. *Elliptic and Quasielliptic Equations.*

We again confine ourselves to the case $m = 1$.

In the rectangle Q, consider the equation

$$(-1)^{p+1} D_t^{2p} u + u_{xx} = f(x,t) \tag{30}$$

The Ionkin problem (in its generalized statement) for Equation (30) in the case $p = 1$ (i.e, for an elliptic equation) was studied in [7], whereas for $p > 1$, Nonlocal Problems I–IV for (30) have not been studied before.

Let us consider two versions of conditions (1) and (2) again: the condition

$$D_t^k u(x,t)\Big|_{t=0} = D_t^k u(x,t)\Big|_{t=T} = 0, \quad k = 0,\ldots,p-1, \quad x \in \Omega, \tag{31}$$

or the condition

$$D_t^{2k} u(x,t)\Big|_{t=0} = D_t^{2k} u(x,t)\Big|_{t=T} = 0, \quad k = 0,\ldots,p-1, \quad x \in \Omega. \tag{32}$$

The Main Condition will be fulfilled for Nonlocal Problem I if the solution $v(x,t)$ to Problem 1 for Equation (30) with conditions (31) or (32) satisfies the membership $D_t^{2p} D_x v(x,t) \in L_2(Q)$; the proof of what is required is similar to the proofs of Propositions 1 and 2 (i.e., involves regularizations and a priori estimates).

Theorem 7. *For any function $f(x,t)$ such that $f(x,t) \in L_2(Q)$, $f_x(x,t) \in L_2(Q)$, Nonlocal Problem I with conditions* (31) *and* (32) *has a solution $u(x,t)$ belonging to $L_2(Q)$.*

The proof of this theorem is obvious.

The solvability of Nonlocal Problems II–IV for Equation (30) is also easy to prove by using Theorems 2–4.

Example 4. *Equations of Sobolev Type.*

Nonlocal Problems I–IV for equations of Sobolev type have rarely been studied—we can only mention the works [6,18], in which the solvability of Problems I and II was investigated for equations called pseudohyperbolic [19,20] and pseudoparabolic [21,22] in the literature. Let us show that the technique presented in Section 2 makes it possible to obtain existence theorems for solutions to Nonlocal Problems I–IV and for some other classes of Sobolev-type equations.

In the rectangle Q, consider the equation

$$u_{tt} - \alpha u_{xx} - u_{xxtt} = f(x,t). \tag{33}$$

The equation arises in the mathematical modeling of processes of plasma physics, in describing the dynamics of long waves on water, in electrodynamics and in elasticity theory (see [20–26]).

As with conditions (1), (2), we use either the Cauchy conditions

$$u(x,0) = u_t(x,0) = 0, \quad x \in \Omega, \tag{34}$$

or the Dirichlet conditions

$$u(x,0) = u(x,T) = 0, \quad x \in \Omega, \tag{35}$$

The solvability of Problem 1 for Equation (33) with conditions (34) or (35) in H for $f(x,t) \in L_2(Q)$ is obvious. Moreover, a solution $v(x,t)$ to Problem 1 for Equation (33) with conditions (34) for $f(x,t) \in L_2(Q)$ and arbitrary α, and with conditions (35) for $f(x,t) \in L_2(Q)$, $\alpha \leq 0$, satisfies the membership $v_{xtt}(x,t) \in L_2(Q)$. Consequently, Problem 2 for Equation (33) is also solvable in H. Thus, the Main Condition for Nonlocal Problem I is fulfilled both for condition (34) and for condition (35). This means that the following theorem holds:

Theorem 8. *For any function $f(x,t) \in L_2(Q)$, Nonlocal Problem I with condition* (34) *is solvable in H for any α. If condition* (35) *is defined in Nonlocal Problem I, then a solution from H exists provided that $f(x,t) \in L_2(Q)$, $\alpha \leq 0$.*

The solvability of Nonlocal Problems II–IV is easily proved with the use of Theorems 2–4.

Example 5. *Degenerating Equations.*

In all the above examples, the equations under consideration were equations with constant coefficients. At the same time, all equations could have coefficients depending on t. Moreover, the corresponding equations could degenerate, i.e., some of the coefficients defining the type of the equation could vanish.

Now, consider the degenerating elliptic equation

$$u_{tt} + h(t)u_{xx} + \mu(t) = f(x,t), \tag{36}$$

in Q, in which $h(t)$ is a nonnegative function on $[0,T]$.

The Ionkin problem for Equation (36) was studied in [27,28] for $h(t) = t^m$ by the method based on representing the solution as a series in special biorthogonal function systems, where the function $\mu(t)$ also had a model form (subordinate to $h(t)$). Let us demonstrate that both for the Ionkin problem (i.e., Nonlocal Problem I) and for Nonlocal Problems II–IV, it is not hard to also obtain results on solvability in H for more general equations.

As with conditions (1), (2), we use a Dirichlet condition; namely, the condition

$$u(x,0) = u(x,T) = 0, \quad x \in \Omega, \tag{37}$$

Proposition 3. *Suppose the fulfillment of the conditions*

$$h(t) \in C([0,T]), \quad h(t) \geq 0 \quad \textit{for } t \in [0,T]; \tag{38}$$

$$\mu(t) \in C([0,T]), \quad \mu(t) \leq 0 \quad \textit{for } t \in [0,T]. \tag{39}$$

Then, for any function $f(x,t)$ for which one of the conditions

$$f(x,t) \in L_2(Q), \quad f_x(x,t) \in L_2(Q), \quad f_{xx}(x,t) \in L_2(Q),$$

$$f_x(0,t) - f_x(1,t) = 0, \quad t \in (0,T), \tag{40}$$

or

$$f(x,t) \in L_2(Q), \quad f_x(x,t) \in L_2(Q), \quad h^{-\frac{1}{2}}(t) f_x(x,t) \in L_2(Q), \tag{41}$$

holds, Problem 1 for Equation (36) *with condition* (37) *has a solution $v(x,t)$ such that $v(x,t) \in H$, $v_{xtt}(x,t) \in L_2(Q)$, $h^{\frac{1}{2}}(t) v_{xxx}(x,t) \in L_2(Q)$.*

Proof. Consider the auxiliary problem: Find a function $v(x,t)$ that is a solution in Q to the equation

$$v_{tt} + h(t) v_{xx} - \varepsilon v_{xxtt} + \mu v = f(x,t) \tag{42}$$

($\varepsilon > 0$) and satisfies (37) and also the condition

$$v_x(0,t) = v_{xxx}(0,t) = v_x(1,t) = v_{xxx}(0,t) = 0, \quad t \in (0,T). \tag{43}$$

The existence of a regular solution (of a solution having all square-integrable derivatives occurring in the equation) to this problem under the conditions of the theorem is obvious. Multiplying (42) first by $-v_{xxxx}(x,t)$, then by $-v_{xxtt}(x,t)$, integrating over Q and using the hypotheses of the theorem, it is not hard to see that solutions $v(x,t)$ to the boundary value problem (36), (42), (43) satisfy the estimate

$$\int_Q \left[v_{xtt}^2 + h(t) v_{xxx}^2 \right] dx\,dt + \varepsilon \int_Q \left[v_{xxtt}^2 + v_{xxxx}^2 \right] dx\,dt \leq C_0(f), \tag{44}$$

where the constant $C_0(f)$ does not depend on ε. This estimate and the reflexivity of a Hilbert space imply the possibility of choosing a sequence converging to a desired solution to Problem 1 for Equation (36) with condition (37).

The proposition is proved. □

Proposition 3 means that the Main Condition is fulfilled for Nonlocal Problem I. Therefore, the following Theorem holds:

Theorem 9. *Suppose the fulfillment of conditions* (38) *and* (39) *and also of one of conditions* (40) *or* (41). *Then, Nonlocal Problem I has a solution $u(x,t) \in H$.*

Using Theorems 2–4, it is not hard to obtain theorems on the solvability of Equation (36) with condition (37) to Nonlocal Problems II–IV.

5. Comments and Supplements

5.1. The splitting method proposed in this article makes it possible to study further properties of solutions to Nonlocal Problems I–IV without investigating the properties of the corresponding function series. For example, knowing the properties of solutions to Problems A and B or to Problems A_1 and B_1, it is not hard to obtain theorems on increasing the smoothness, the boundedness of the solutions, the behavior of the solutions, etc.

5.2. In the examples presented in Section 3, some specific conditions (1) and (2) are used. Obviously, other conditions can be used—for example, for quasihyperbolic equations (23), one can use the conditions from [17]; for quasielliptic equations (30), along with conditions (31) or (32), we can use mixed conditions, etc.

5.3. The examples of Section 3 do not exhaust all classes of equations for which the splitting method is applicable. Observe first of all that in Examples 1–4, we consider equations with constant coefficients but in fact all equations can have variable coefficients (with the unconditional type preserved). In Examples 2–5, instead of the case $m = 1$, it is quite possible to consider the case $m > 1$ (in this case, conditions on the function $f(x,t)$ of the type of conditions (18) or (22) can appear). Equations of Sobolev type are certainly not limited to the simplest pseudoparabolic and pseudohyperbolic equations discussed in Example 4; for example, Nonlocal Problems I–IV can also be effectively studied for the general pseudohyperbolic equations of [25,29], etc.

5.4. It seems that the splitting method can be effectively used for studying the solvability of Nonlocal Problems I–IV with fractional derivatives.

Funding: This work was supported by the Russian Science Foundation, project 23–21–00269.

Data Availability Statement: Data supporting reported results can be requested from the corresponding author.

Conflicts of Interest: The author declares no conflict of interest.

References

1. Ionkin, N.I. Solution of a Boundary-Value Problem in Heat Conduction with a Nonclassical Boundary Condition. *Differ. Equ.* **1977**, *13*, 204–211.
2. Nakhushev, A.M. *Problems with Shift for Partial Differential Equation*; Nauka: Moscow, Russia, 2006. (In Russian)
3. Sadybekov, M.A. Initial-Boundary Value Problem for a Heat Equation with not Strongly Regular Boundary Conditions. In *Functional Analysis in Interdisciplinary Applications*; Springer Proceedings in Mathematics & Statistics; Springer: Cham, Switzerland, 2017; pp. 330–348.
4. Orazov, I.; Sadybekov, M.A. On a Class of Problems of Determining the Temperature and Density of Heat Sources Given Initial and Final Temperature. *Sib. Math. J.* **2012**, *53*, 146–151. [CrossRef]
5. Nakhusheva, Z.A. The Samarskii Problem for the Fractal Diffusion Equation. *Math. Notes* **2014**, *95*, 815–819. [CrossRef]
6. Kozhanov, A.I. Nonlocal Problems with Generalized Samarskii–Ionkin Condition for some Classes for Nonstationary Differential Equations. *Dokl. Math.* **2023**, *107*, 40–43. [CrossRef]
7. Kozhanov, A.I.; Dyuzheva, A.V. Well–Posedness of the Generalized Samarskii–Ionkin Problem for Elliptic Equations in a Cylindrical Domain. *Differ. Equ.* **2023**, *59*, 230–242. [CrossRef]
8. Yurchuk, N.I. Mixed Problem with an Integral Condition for Certain Parabolic Equation. *Differ. Equ.* **1986**, *22*, 1457–1463.
9. Ionkin, N.I. The Stability of a Problem in the Theory of Heat Equations with Nonclassical Boundary Conditions. *Diff. Uravn.* **1979**, *15*, 1279–1283.
10. Berdyshev, A.S.; Cabada, A.; Kadirkulov, B.J. The Samarskii–Ionkin Type Problem for the Fourth Order Parabolic Equation with Fractional Differential Operator. *Comput. Math. Appl.* **2011**, *62*, 3884–3893. [CrossRef]
11. Lions, J.-L. *Quelques Methods de Resolution des Problems aux Limites Nonlineaires*; Dunod Gauthier: Villars, Paris, France, 1969.
12. Edwards, R.E. *Functional Analysis*; Holt, Rinehart and Winston: New York, NY, USA, 1965.
13. Trenogin, V.A. *Functional Analysis*; Nauka: Moscow, Russia, 1980. (In Russian)
14. Beilin, S.A. Existence of Solutions for One–Dimentional Wave Equations with Nonlocal Conditions. *Election J. Differ. Equ.* **2001**, *76*, 1–8.
15. Vragov, V.N. On the Theory of Boundary Value Problems for Equations of Mixed Type. *Diff. Uravn.* **1977**, *13*, 1098–1105. (In Russian)
16. Egorov, I.E.; Fedorov, V.E. *Higher–Order Nonclassical Equations of Mathematical Physics*; Computer Center Press: Novosibirsk, Russia, 1995. (In Russian)
17. Kozhanov, A.I.; Pinigina, N.R. Boundary-value problems for some higher-order nonclassical differential equations. *Math. Notes* **2017**, *101*, 467–474. [CrossRef]
18. Kozhanov, A.I.; Tarasova, G.I. The Samarsky–Ionkin Problem with Integral Perturbation for a Pseudoparabolic Equation. The Bulletin of Irkutsk State University. *Ser. Math.* **2022**, *42*, 59–74. (In Russian)

19. Larkin, N.A. Existence Theorems for Quasilinear Pseudohyperbolic Equations. *Rep. Ussr Acad. Sci.* **1982**, *6*, 1316–1319. (In Russian)
20. Khudaverdiev, K.I.; Veliyev, A.A. *Study of a One-Dimensional Mixed Problem for One Class of Third-Order Pseudohyperbolic Equations with a Nonlinear Operator Right-Hand Side*; Chashyoglu: Baku, Azerbaijan, 2010; p. 168.
21. Showalter, R.E.; Ting, T.W. Pseudoparabolic Partial Differential Equations. *SIAM J. Math. Anal.* **1970**, *1*, 1–26. [CrossRef]
22. Sveshnikov, A.G.; Alshin, A.B.; Korpusov, M.O.; Pletner, Y.D. *Linear and Nonlinear Equations of Sobolev Type*; Fizmatlit: Moscow, Russian, 2007.
23. Whitham, G.B. *Linear and Nonlinear Waves*; John Wiley and Sons: New York, NY, USA, 1974.
24. Lonngren, K.; Scott, A. (Eds.) *Solitons in Action*; Academic: New York, NY, USA, 1978.
25. Demidenko, G.V.; Uspenskii, S.V. *Partial Differential Equations and Systems not Solvable with Respect to Highest Order Derivatives*; Marcel Dekker: New York, NY, USA; Basel, Switzerland, 2003.
26. Zhegalov, V.I.; Mironov, A.N.; Utkina, E.A. *Equations with Dominant Partial Derivative*; Kazan Federal University: Kazan, Russia, 2014.
27. Moiseev, E.I. Solvability of a Nonlocal Boundary Value Problem. *Differ. Equ.* **2001**, *37*, 1643–1646. [CrossRef]

28. Moiseev, E.I. On the Solution of a Nonlocal Boundary Value Problem by the Spectral Method. *Differ. Equ.* **1999**, *35*, 1105–1112.
29. Bondar, L.N.; Demidenko, G.V. On the Cauchy Problem for Pseudohyperbolic Equations with Lower Order Terms. *Mathematics* **2023**, *11*, 3943. [CrossRef]

mathematics

Article

An Approach to Solving Direct and Inverse Scattering Problems for Non-Selfadjoint Schrödinger Operators on a Half-Line

Vladislav V. Kravchenko and Lady Estefania Murcia-Lozano *

Department of Mathematics, Cinvestav, Campus Querétaro, Libramiento Norponiente #2000, Fracc. Real de Juriquilla, Querétaro 76230, Mexico; vkravchenko@math.cinvestav.edu.mx
* Correspondence: emurcia@math.cinvestav.mx

Abstract: In this paper, an approach to solving direct and inverse scattering problems on the half-line for a one-dimensional Schrödinger equation with a complex-valued potential that is exponentially decreasing at infinity is developed. It is based on a power series representation of the Jost solution in a unit disk of a complex variable related to the spectral parameter by a Möbius transformation. This representation leads to an efficient method of solving the corresponding direct scattering problem for a given potential, while the solution to the inverse problem is reduced to the computation of the first coefficient of the power series from a system of linear algebraic equations. The approach to solving these direct and inverse scattering problems is illustrated by several explicit examples and numerical testing.

Keywords: non-selfadjoint Schrödinger operator; Jost solution; direct scattering problem; inverse scattering problem

MSC: 34A55; 34L05; 34L16; 34L25; 34L40; 65L09; 65L15

Citation: Kravchenko, V.V.; Murcia-Lozano, L.E. An Approach to Solving Direct and Inverse Scattering Problems for Non-Selfadjoint Schrödinger Operators on a Half-Line. *Mathematics* **2023**, *11*, 3544. https://doi.org/10.3390/math11163544

Academic Editor: Natalia Bondarenko

Received: 19 July 2023
Revised: 6 August 2023
Accepted: 7 August 2023
Published: 16 August 2023

1. Introduction

Consider the one-dimensional Schrödinger equation

$$l[y] := -y'' + q(x)y = \lambda y, \ \ x \in (0, \infty), \tag{1}$$

with $\lambda \in \mathbb{C}$ and a complex-valued potential $q(x)$ satisfying the condition

$$\int_0^\infty e^{\varepsilon x}|q(x)|dx < \infty \tag{2}$$

for some $\varepsilon > 0$. By ρ, we denote the square root of λ such that $\rho \in \overline{\mathbb{C}^+} := \{w \in \mathbb{C} : \operatorname{Im}(w) \geq 0\}$. In the present work, an approach to solving direct and inverse scattering problems for (1) under Condition (2) is developed.

Complex-valued potentials arise when studying parity time (PT)-symmetric potentials [1] (Chapter 1), [2], quasi-exactly solvable (QES) potentials [3,4], hydrodynamics, and magnetohydrodynamics [5]; see also [6–8].

Studying a Zakharov–Shabat system, even with a real-valued potential, naturally leads to a couple of equations of the form (1) with complex-valued potentials; see [9]. Indeed, consider the Zakharov–Shabat system

$$\vec{v}_x(x) = \begin{pmatrix} v_1(x) \\ v_2(x) \end{pmatrix}_x = \begin{pmatrix} -i\rho & u(x) \\ -u(x) & i\rho \end{pmatrix} \vec{v}(x), \ 0 < x < \infty \tag{3}$$

where ρ is a complex spectral parameter and $u(x)$ is a real-valued potential.

The further transformation of $\overrightarrow{v}(x)$ is as follows:

$$y_1(x) = v_2(x) - iv_1(x),$$
$$y_2(x) = v_2(x) + iv_1(x)$$

This leads to a pair of Schrödinger equations with complex-valued potentials

$$-y_1''(x) + (-iu'(x) - u^2(x))y_1(x) = \rho^2 y_1(x), \quad (4)$$
$$-y_2''(x) + (iu'(x) - u^2(x))y_2(x) = \rho^2 y_2(x). \quad (5)$$

Thus, the results of the present work are applicable to direct and inverse scattering problems for a Zakharov–Shabat system.

A direct scattering problem for (1) with a complex-valued potential was studied in a number of publications ([10–13]). Equation (1) under Condition (2) was considered in [12] (p. 292), [14–20] (p. 353), and [21,22].

It is well-known (see, e.g., [12] (p. 443), [18]) that (1) admits a unique solution, which we denote by $e(\rho, x)$, satisfying the asymptotic equality

$$e(\rho, x) = e^{i\rho x}(1 + o(1)), \ x \to \infty.$$

This solution is called the Jost solution of (1). It admits the Levin integral representation [12] (see also [18,23,24])

$$e(\rho, x) = e^{i\rho x} + \int_x^\infty A(x,t)e^{i\rho t}dt, \ \operatorname{Im}\rho > -\frac{\varepsilon}{2}, \ x \geq 0 \quad (6)$$

where for every fixed x, the kernel $A(x,t)$ belongs to $\mathcal{L}_2(x,\infty)$. In [25] (see also [26]) a Fourier–Laguerre series representation for $A(x,t)$ was proposed in the form

$$A(x,t) = \sum_{n=0}^{\infty} a_n(x)L_n(t-x)e^{\frac{x-t}{2}}, \quad (7)$$

where $L_n(\tau)$ stands for the Laguerre polynomial of order n. A recurrent integration procedure was developed in [27] to calculate the coefficients $a_n(x)$. The substitution of (7) into (6) was found to lead to a series representation for the Jost solution [25,26]

$$e(\rho, x) = e^{i\rho x}\left(1 + (z+1)\sum_{n=0}^{\infty}(-1)^n z^n a_n(x)\right), \ x \geq 0, \ \rho \in \overline{\mathbb{C}^+} \quad (8)$$

where

$$z = z(\rho) = \frac{\left(\frac{1}{2} + i\rho\right)}{\left(\frac{1}{2} - i\rho\right)}. \quad (9)$$

In the present work, we consider the direct and inverse scattering problems for (1) subject to the homogeneous Dirichlet condition

$$y(0) = 0, \quad (10)$$

however, the approach developed here is also applicable in the case of other boundary conditions, such as

$$y'(0) - hy(0) = 0$$

with $h \in \mathbb{C}$.

The problem (1) and (10) under Condition (2) possesses a continuous spectrum coinciding with the positive semi-axis $\lambda > 0$, and may have a point spectrum that coincides with the squares of the non-real roots of the Jost function

$$e(\rho) := e(\rho, 0),$$

if such roots exist. Let us denote them as $\rho_1, \ldots, \rho_\alpha$. Their multiplicity may be greater than one. In this case, instead of norming constants associated to the eigenvalues, the corresponding normalization polynomials $X_k(x)$ naturally arise (see Section 3.3 below).

As a component of the scattering data for (1), the scattering function

$$s(\rho) := \frac{e(-\rho)}{e(\rho)}$$

is considered in the strip $|\mathrm{Im}(\rho)| < \varepsilon_0$ where ε_0 is sufficiently small (see Section 3.2 below).

The direct scattering problem for (1) and (10) consists of obtaining the set of the scattering data

$$\left\{ \{\rho_k, m_k, X_k(x)\}_{k=1}^{\alpha}, s(\rho) \right\}. \tag{11}$$

The overall approach developed in the present work to solve this problem is based on the representation (8). Indeed, the calculation of $\{\rho_k\}_{k=1}^{\alpha}$ is easily realizable with the aid of the argument principle theorem applied to find zeros of (8) in the unit disc. To the best of our knowledge, there has been no practical way of calculating the normalization polynomials. We propose a simple procedure for computing their coefficients by solving a finite system of linear algebraic equations. For this, an auxiliary result for the derivatives $\frac{\partial^m}{\partial z^m} e(\rho(z), x)$ is obtained.

The calculation of the scattering function $s(p)$ requires an analytic extension of the Jost function $e(\rho)$ obtained from (8), onto the strip $-\varepsilon_0 < \mathrm{Im}(\rho) < 0$. We explore different possibilities for such an extension, including the Padé approximants (see [28,29]) and the power series analytic continuation [30] (p. 150), [31]. This results in an efficient numerical method for solving the direct scattering problem.

The inverse scattering problem consists of recovering the potential $q(x)$ from the set of the scattering data. A general theory of this inverse problem can be found in [12,13,20] (p. 353), [24,32–35]. Here, we use the representation (7) for the numerical solution of the problem, thus extending the approach developed in [25,26,36–38] to the non-selfadjoint situation. The inverse Sturm–Liouville problem is reduced to an infinite system of linear algebraic equations. The potential $q(x)$ is recovered from the first component of the solution vector, which coincides with $a_0(x)$ in (7).

The reduction to the infinite system of linear algebraic equations is based on the substitution of the series representation (7) for the kernel $A(x,t)$ into the Gelfand–Levitan equation (see [39]),

$$A(x,t) = \int_x^\infty A(x,u) f(u+t) du + f(x+t), \ 0 \le x \le t < \infty, \tag{12}$$

where the function f can be computed from the set of scattering data (11):

$$f(x) := \frac{1}{2\pi} \int_{-\infty+i\eta}^{\infty+i\eta} (s(\rho) - 1) e^{ix\rho} d\rho - \sum_{k=1}^{\alpha} X_k(x) e^{i\rho_k x}, \ 0 < \eta < \varepsilon_0.$$

To approximate the complex-valued function $a_0(x)$, we consider the truncated system of linear algebraic equations, for which the existence, uniqueness and stability of the solution is proved.

Finally, we illustrate the proposed approach by numerical calculations performed in Matlab2021a.

We discuss the details of the numerical implementation of the method: its convergence, stability and accuracy. In a couple of examples, we show the "in-out" performance of the approach, i.e., we solve the direct problem numerically and use the results of our computation as the input data to solve the inverse problem.

The approach based on the representations (7) and (8) leads to efficient numerical methods for solving both direct and inverse scattering problems.

In Section 2, we recall the series representations for the kernel $A(x,t)$ and for the Jost solution, then prove additional results related to these representations. In Section 3, we recall the set of scattering data and put forward an algorithm for solving the direct scattering problem. Additionally, we present analytical examples. In Section 4, the approach for solving the inverse scattering problem is developed. Analytical examples from Section 2 are considered in order to illustrate the approach. In Section 5, we discuss the numerical implementation of the algorithms proposed for solving the direct and inverse scattering problems. Section 6 contains some concluding remarks.

2. Series Representations for the Transmutation Operator Kernel and Jost Solution

Consider the one-dimensional Schrödinger equation on the half-line (1) where $\lambda = \rho^2 \in \mathbb{C}$ is the spectral parameter. The potential $q(x)$ is a complex-valued function satisfying Condition (2) for some $\varepsilon > 0$.

Equation (1) is considered on the class of functions $\mathcal{D}(l) = \{y \in \mathcal{W}^{2,2}(0,\infty) : l[y] \in \mathcal{L}_2(0,\infty)\}$.

Series Representation for Solutions of the One-Dimensional Schrödinger Equation

Equation (1) possesses the unique so-called Jost solution $e(\rho,x)$ (see, e.g., [12] (p. 443), [18]), which for all $x \geq 0$ is a holomorphic function of ρ in the half-plane $\operatorname{Im}\rho \geq 0$ and satisfies the asymptotic relation

$$e(\rho,x) = e^{i\rho x}(1+o(1)) \text{ when } x \to \infty \text{ and } \operatorname{Im}\rho \geq 0. \tag{13}$$

The function $e(\rho) := e(\rho,0)$ is called the Jost function.

Remark 1. *Under the assumption $q(x) \in \mathcal{L}(0,\infty)$ instead of (2), for every $x \geq 0$ the solution $e(\rho,x)$ is continuous with respect to ρ for $\rho \in \overline{\mathbb{C}^+} \setminus \{0\}$ and holomorphic with respect to ρ for $\rho \in \mathbb{C}^+$. If in addition $(1+x)q(x) \in \mathcal{L}(0,\infty)$, the functions $e^{(\nu)}(\rho,x)$, $\nu = 0,1$ are continuous for $\rho \in \overline{\mathbb{C}^+}$, $x \geq 0$ (see [24] (p. 105)).*

Remark 2. *Under Condition (2), the Jost solution satisfies the asymptotic relations*

$$\frac{\partial^j}{\partial x^j} e(\rho,x) = (ix)^{(j)} e^{i\rho x} + o\left(e^{-\frac{\varepsilon}{2}x}\right), \quad j = 0,1,\ldots \text{ when } x \to \infty,$$

provided the existence of these derivatives; see [21].

The solution $e(\rho,x)$ admits the Levin integral representation [12]

$$e(\rho,x) = e^{ix\rho} + \int_x^\infty A(x,t) e^{i\rho t} dt, \ \operatorname{Im}\rho \geq 0, \ x \geq 0, \tag{14}$$

where $A(x,t)$ is a complex-valued continuous function for $0 \leq x \leq t < \infty$. Denote $Q(x) := \|q\|_{\mathcal{L}(x,\infty)}$. The kernel $A(x,t)$ admits the bound [24] (p. 108)

$$|A(x,t)| \leq \frac{1}{2} Q\left(\frac{x+t}{2}\right) \exp\left(\|Q\|_{\mathcal{L}(x,\infty)} - \|Q\|_{\mathcal{L}(\frac{x+t}{2},\infty)}\right). \tag{15}$$

Under Condition (2), the Jost solution is extensible onto the half-plane $\operatorname{Im}\rho > -\frac{\varepsilon}{2}$ through the Levin representation (14). The extension satisfies (13) for $\operatorname{Im}\rho > -\frac{\varepsilon}{2}$.

Proposition 1. *Under Condition (2), the kernel $A(x,t)$ admits the bound*

$$|A(x,t)| \leq \frac{1}{2}e^{-\varepsilon\left(\frac{x+t}{2}\right)}\left(\int_{\frac{x+t}{2}}^{\infty} e^{\varepsilon\tau}|q(\tau)|d\tau\right)\exp\left(\frac{C_\varepsilon}{\varepsilon}\left(e^{-\varepsilon x} - e^{-\varepsilon\frac{x+t}{2}}\right)\right), 0 \leq x \leq t < \infty, \quad (16)$$

where $C_\varepsilon = \int_0^\infty e^{\varepsilon t}|q(t)|dt$.

Proof. Under Condition (2), the potential $q(x)$ satisfies the inequality

$$Q(x) \leq e^{-\varepsilon x}\int_x^\infty e^{\varepsilon t}|q(t)|dt = C_\varepsilon e^{-\varepsilon x}. \quad (17)$$

Moreover, for any fixed $x \in [0,\infty)$ we have [12] (p. 317)

$$\|Q\|_{\mathcal{L}(x,\infty)} \leq \frac{C_\varepsilon}{\varepsilon}e^{-\varepsilon x}. \quad (18)$$

Thus, substitution of (17) and (18) into (15) gives us (16). □

Additionally, the kernel $A(x,t)$ has first continuous derivatives that satisfy the inequalities [12] (p. 305)

$$|A_x(x,t)|, |A_t(x,t)| \leq \frac{1}{4}\left|q\left(\frac{x+t}{2}\right)\right| + C_\varepsilon \exp\left(-\varepsilon\left(\frac{3}{2}x + t\right)\right), \quad (19)$$

and the equality [12] (p. 328)

$$A(x,x) = \frac{1}{2}\int_x^\infty q(t)dt. \quad (20)$$

As was pointed out in [25], since $A(x,\cdot) \in \mathcal{L}_2(x,\infty)$, the function

$$a(x,t) := e^{\frac{t}{2}}A(x,t+x) \quad (21)$$

belongs to $L_2([0,\infty);e^{-t})$ and hence admits the series representation

$$a(x,t) = \sum_{n=0}^\infty a_n(x)L_n(t), \quad (22)$$

where $L_n(t)$ stands for the Laguerre polynomial of order n and $a_n(x)$ are complex-valued functions such that $\{a_n(x)\}_{n=0}^\infty \in l_2$ for any $x \geq 0$. For all $x \geq 0$, the series (22) converges in the norm of $\mathcal{L}_2([0,\infty);e^{-t})$. Thus,

$$A(x,t) = \sum_{n=0}^\infty a_n(x)L_n(t-x)e^{\frac{x-t}{2}} \quad (23)$$

and

$$\sum_{n=0}^\infty a_n(x) = A(x,x) = \frac{1}{2}\int_x^\infty q(t)dt. \quad (24)$$

This series representation was obtained in [25] for real-valued $q(x)$. However, (23) remains true in the non-selfadjoint case as well.

Proposition 2. *For any fixed $x \geq 0$, the series*

$$a(x,t) = \sum_{n=0}^\infty a_n(x)L_n(t),\ t \in [0,\infty) \quad (25)$$

converges pointwise.

Proof. We use [40] (Theorem 6.5), and thus need to verify that the following assertions are true.

1. $a(x,\cdot)$ is of class $\mathcal{L}([0,\infty);e^{-t})$.
2. $a(x,\cdot)$ is γ-Hölder continuous, i.e., there exists $0<\gamma\le 1$, such that

$$|a(x,t_0)-a(x,t)|\le M|t_0-t|^{\gamma},$$

for some constant $M>0$ and arbitrary $t,t_0\in[0,\infty)$.

3. The integrals

$$\int_0^1 t^{-3/4}|a(x,t)|dt,\ \int_1^{\infty} e^{-t/2}|a(x,t)|dt \tag{26}$$

exist.

To prove the first assertion, it is enough to consider estimate (15). Indeed,

$$\begin{aligned}\int_0^{\infty} e^{-t}|a(x,t)|dt &\le \int_0^{\infty}|A(x,x+t)|dt\\ &\le \frac{1}{2}\int_0^{\infty} Q\left(\frac{2x+t}{2}\right)\exp\left(\|Q\|_{\mathcal{L}(x,\infty)}-\|Q\|_{\mathcal{L}(\frac{2x+t}{2},\infty)}\right)dt\\ &\le \frac{\exp\left(\|Q\|_{\mathcal{L}(x,\infty)}\right)}{2}\int_0^{\infty} Q\left(\frac{2x+t}{2}\right)\exp\left(-\|Q\|_{\mathcal{L}(\frac{2x+t}{2},\infty)}\right)dt\\ &=\exp\left(\|Q\|_{\mathcal{L}(x,\infty)}\right)\int_x^{\infty} Q(\tau)\exp\left(-\|Q\|_{\mathcal{L}(\tau,\infty)}\right)d\tau.\end{aligned}$$

Note that $\frac{d}{d\tau}\|Q\|_{\mathcal{L}(\tau,\infty)}=Q(\tau)$ and therefore

$$\int_x^{\infty} Q(\tau)\exp\left(-\|Q\|_{\mathcal{L}(\tau,\infty)}\right)d\tau=1-\exp\left(-\|Q\|_{\mathcal{L}(x,\infty)}\right). \tag{27}$$

Thus,

$$\int_0^{\infty} e^{-t}|a(x,t)|dt\le \exp\left(\|Q\|_{\mathcal{L}(x,\infty)}\right)-1<\infty.$$

The second assertion follows from the inclusion $A(x,\cdot)\in C^1(x,\infty)$.

The existence of the first integral in (26) follows from the continuity of $a(x,\cdot)$. Finally, for the second integral we have

$$\int_1^{\infty} e^{-t/2}|a(x,t)|dt=\int_1^{\infty}|A(x,x+t)|dt\le\int_0^{\infty}|A(x,x+t)|dt,$$

and thus, from the proof of the first assertion, we obtain $\int_1^{\infty} e^{-t/2}|a(x,t)|dt<\infty$.

Now, the application of Theorem 6.5 from [40] completes the proof. □

Following [25] (see also [26] (p. 63)), the substitution of (23) into (14) and termwise integration lead to the series representation (8) for the Jost solution.

The series (8) is convergent in the open unit disk of the complex z-plane, $\mathbb{D}:=\{z\in\mathbb{C}:|z|<1\}$, and for every x, the function $e(\rho,x)e^{-i\rho x}$ belongs to the Hardy space $H^2(\mathbb{D})$ as a function of z [26].

Proposition 3. *Let $q(x)(1+x)\in\mathcal{L}(0,\infty)$. Then, the kernel $A(x,t)$ admits the representation (23), where for any x fixed the series converges in the norm of $\mathcal{L}_2(x,\infty)$, and the complex-valued coefficients $a_n(x)$ satisfy the system of equations*

$$-l[a_0]-a_0'=q, \tag{28}$$

$$-l[a_n]-a_n'=-l[a_{n-1}]+a_{n-1}',\ n=1,2,\ldots, \tag{29}$$

as well as the inequality

$$|a_n(x)| \leq \exp\left(\|Q\|_{\mathcal{L}(x,\infty)}\right) - 1,\ n = 0,1,2,\dots. \tag{30}$$

Proof. The proof of (28) and (29) from [26] (Theorem 10.1, p. 66) given for the case of a real-valued q remains valid in this more general situation as well.

Note that

$$a_n(x) = \int_0^\infty a(x,t)L_n(t)e^{-t}dt.$$

From estimate (15) and inequality ([41] (p. 164)) $|L_n(t)| \leq e^{t/2}$, $t \geq 0$, we have

$$\begin{aligned} |a_n(x)| &\leq \int_0^\infty |A(x,x+t)L_n(t)|e^{-t/2}dt \leq \tfrac{1}{2}\int_0^\infty Q\left(\tfrac{2x+t}{2}\right)\exp\left(\|Q\|_{\mathcal{L}(x,\infty)} - \|Q\|_{\mathcal{L}(\frac{2x+t}{2},\infty)}\right)dt \\ &= \exp\left(\|Q\|_{\mathcal{L}(x,\infty)}\right) - 1 \end{aligned} \tag{31}$$

(see (27)). □

Corollary 1. *Under Condition (2), the coefficients a_n satisfy the inequality*

$$|a_n(x)| \leq \exp\left(\frac{C_\varepsilon}{\varepsilon}e^{-\varepsilon x}\right) - 1. \tag{32}$$

Proof. Substitution of (17) and (18) into (31) yields (32). □

Remark 3. *Under the assumption that functions $a^{(\nu)}(x,t)$ are absolutely continuous with respect to t in $[0,\infty)$ for $\nu = 0,1,2$, the convergence of the power series in (8) for $z \in \overline{\mathbb{D}}$ can be proved with the aid of a result from [42], which states that*

$$|a_n(x)| \leq \frac{V}{\sqrt{n(n-1)(n-2)}},$$

provided that

$$\lim_{t\to+\infty} e^{-t/2}t^{1+j}a^{(j)}(x,\cdot) = 0 \quad j = 0,1,2 \tag{33}$$

and

$$V = \sqrt{\int_0^\infty t^3e^{-t}\left[a^{(3)}(x,\cdot)\right]^2dt} < \infty. \tag{34}$$

Moreover,

$$\left\|a(x,t) - \sum_{n=0}^{N} a_n(x)L_n(t)\right\|_{\mathcal{L}_2(0,\infty;e^{-t})} \leq \frac{V\sqrt{N}}{\sqrt{(N-1)(N-2)(N-3)}}.$$

To ensure Condition (33) for $j = 0$, notice that from (16) we have

$$|a(x,t)| = e^{\frac{t}{2}}|A(x,t+x)| \leq Ce^{\frac{t}{2}}e^{-\varepsilon\left(\frac{x+t}{2}\right)}. \tag{35}$$

For $j = 1$, Condition (33) holds due to (19). However, the fulfillment of (33) for $j = 2$ as well as that of (34) requires the additional regularity of $q(x)$, ensuring the possibility of the differentiation of the integral equation for the kernel

$$\begin{aligned} A(x,t) = {} & \frac{1}{2}\int_{(x+t)/2}^\infty q(\xi)d\xi + \frac{1}{2}\int_x^{(x+t)/2} q(\xi)\left(\int_{t+x-\xi}^{t+\xi-x} A(\xi,\eta)d\eta\right)d\xi \\ & + \frac{1}{2}\int_{(t+x)/2}^\infty q(\xi)\left(\int_\xi^{t+\xi-x} A(\xi,\eta)d\eta\right)d\xi, \quad 0 \leq x \leq t < \infty \end{aligned}$$

at least three times [12] (p. 296).

Remark 4. *Denote*

$$e_N(\rho,x)=e^{i\rho x}\left(1+(z+1)\sum_{n=0}^{N}(-1)^n z^n a_n(x)\right),\ \rho\in\mathbb{C}^+. \tag{36}$$

In [27], the following statements were proved in the case of a real-valued potential.
1. If $\operatorname{Im}\rho>0$*, then*

$$|e(\rho,x)-e_N(\rho,x)|\le\varepsilon_N(x)\frac{e^{-\operatorname{Im}\rho\, x}}{\sqrt{2\operatorname{Im}\rho}}$$

where

$$\varepsilon_N(x):=\left(\sum_{n=N+1}^{\infty}\left|a_n(x)^2\right|\right)^{1/2}=\left(\int_0^{\infty}e^{-t}|a(x,t)-a_N(x,t)|^2dt\right)^{1/2}. \tag{37}$$

2. If $\rho\in\mathbb{R}$*, then*

$$\|e(\cdot,x)-e_N(\cdot,x)\|_{\mathcal{L}_2(-\infty,\infty)}=\sqrt{2\pi}\varepsilon_N(x).$$

These results remain valid in the case of a complex-valued potential. Moreover, under the assumptions of Remark 3, we obtain the inequality

$$\varepsilon_N(x)\le\frac{V\sqrt{N}}{\sqrt{(N-1)(N-2)(N-3)}}.$$

Remark 5. *The substitution of* $\rho=\frac{i}{2}$ *into (8) leads to the equality* $a_0(x)=e\left(\frac{i}{2},x\right)e^{x/2}-1$.
Moreover, note that we have

$$q(x)=\frac{a_0''(x)-a_0'(x)}{a_0(x)+1}. \tag{38}$$

By $\omega(\rho,x)$, we denote the solution of (1), satisfying the initial conditions

$$\omega(\rho,0)=0,\ \frac{d}{dx}\omega(\rho,0)=1. \tag{39}$$

We also need the solution

$$\Omega(\rho,x)=\frac{2i\rho\omega(\rho,x)}{e(\rho)}. \tag{40}$$

3. Direct Problem

3.1. Spectrum of (1) and (10)

Consider the problem (1) and (10) under Condition (2). Let us recall some definitions and facts from [12] (p. 452) (see also [18]). The continuous spectrum fills the entire semi-axis $\lambda>0$.

Definition 1. *We call the roots of* $e(\rho)$ *that lie in* $\overline{\mathbb{C}^+}\setminus\{0\}$ *the singular numbers of the problem (1) and (10).*

If they exist, their number is finite. Let us denote the non-real singular numbers by $\rho_1,\dots,\rho_\alpha$. *The numbers* $\lambda_k=\rho_k^2$ *constitute the point spectrum of the problem, and the multiplicities of the zeros* ρ_k $(k=1,\dots,\alpha)$ *are called the multiplicities of the singular numbers and denoted by* m_k*, respectively.*

Thus, we are interested in the zeros z_k of the Jost function

$$e(\rho) = 1 + (z+1)\sum_{n=0}^{\infty}(-1)^n z^n a_n(0) \tag{41}$$

to obtain the eigenvalues from $\lambda_k = -\left(\frac{z_k-1}{2(z_k+1)}\right)^2$.

For an estimate of the number of the eigenvalues, we refer to [43].

3.2. Scattering Function $s(\rho)$

Let us introduce ε_1 as the distance from the real axis to the non-real roots of the function $e(\rho)$. Let $\varepsilon_0 = \min(\varepsilon_1, \frac{\varepsilon}{2})$ when $\varepsilon_1 \neq 0$ ($\varepsilon_1 = 0$ means that there are no non-real roots), or $\varepsilon_0 = \frac{\varepsilon}{2}$ otherwise.

The scattering function $s(\rho)$ is defined by

$$s(\rho) := \frac{e(-\rho)}{e(\rho)}, \ |\mathrm{Im}(\rho)| < \varepsilon_0. \tag{42}$$

Let us recall some properties of the Jost function $e(\rho)$ and scattering function $s(\rho)$ (see, e.g., [39]).

1. $e(\rho)$ is holomorphic for $\mathrm{Im}\,\rho > -\varepsilon_0$, and for every $0 < \eta < \varepsilon_0$ it satisfies the asymptotic relation
$$e(\rho) = 1 + O\left(\frac{1}{\rho}\right), \text{ as } |\rho| \to \infty \tag{43}$$
uniformly in the strip $|\mathrm{Im}\,\rho| \leq \eta$.
2. $s(\rho)$ is meromorphic in the strip $|\mathrm{Im}\,\rho| < \varepsilon_0$, and for every $0 < \eta < \varepsilon_0$:
$$s(\rho) = 1 + O\left(\frac{1}{\rho}\right), \text{ as } |\rho| \to \infty \tag{44}$$
uniformly in the strip $|\mathrm{Im}\,\rho| \leq \eta$.
3. $s(\rho)$ has no non-real poles in the strip $|\mathrm{Im}\,\rho| < \varepsilon_0$.
4. $s(\rho)s(-\rho) = 1$.
5. $s(0) = \pm 1$.

A function satisfying properties 2–5 is said to be of S-type in the strip $|\mathrm{Im}\,\rho| < \varepsilon_0$. The following examples illustrate some of the above definitions.

Example 1 ([44,45])**.** *Consider the potential*

$$q_1(x) := 10ie^{-x}, \ x \geq 0$$

with $0 < \varepsilon < 1$ *in (2). With the aid of Wolfram Mathematica v.12 the Jost solution can be obtained in a closed form,*

$$e_1(\rho, x) = \left(\sqrt{-10i}\right)^{2i\rho} J_{-2i\rho}((2-2i)\sqrt{5}e^{-x/2})\Gamma(1-2i\rho), \ x \geq 0, \ \mathrm{Im}(\rho) > -1/2, \tag{45}$$

where $J_\nu(z)$ *stands for the Bessel function of the first kind of order* ν*.*

Hence,

$$e_1(\rho) = \left(\sqrt{-10i}\right)^{2i\rho} J_{-2i\rho}((2-2i)\sqrt{5})\Gamma(1-2i\rho),$$

and the eigenvalues are the squares of the values $\rho \in \mathbb{C}^+$ *such that*

$$J_{-2i\rho}((2-2i)\sqrt{5}) = 0.$$

From here, we obtain the only singular number

$$\rho_0 \approx 1.7840658475274275761348792\,32 + 0.60878868120671863102402200\,34i. \tag{46}$$

The scattering function has the form

$$s_1(\rho) = \frac{(1-i)^{-4i\rho}5^{-2i\rho}J_{2i\rho}((2-2i)\sqrt{5})\Gamma(1+2i\rho)}{J_{-2i\rho}((2-2i)\sqrt{5})\Gamma(1-2i\rho)}. \tag{47}$$

It is well-defined in the domain

$$D(s_1) = \left\{\rho \in \mathbb{C} : J_{-2i\rho}\left((2-2i)\sqrt{5}\right) \neq 0 \wedge \left(\mathrm{Im}(\rho) > -\frac{1}{2} \vee -2i\rho \notin \mathbb{Z}\right) \wedge \left(\mathrm{Im}(\rho) < \frac{1}{2} \vee 2i\rho \notin \mathbb{Z}\right)\right\}$$

and is an S-type function in the strip $|\mathrm{Im}(\rho)| < \frac{1}{2}$.

Example 2. *Consider the potential*

$$q_2(x) := -4i\,\mathrm{sech}(2x)\tanh(2x) - 4\,\mathrm{sech}^2(2x),\ x \geq 0,$$

which satisfies Condition (2) for $0 < \varepsilon < 2$. *The Schödinger equation with this potential comes from a Zakharov–Shabat system (3) with the potential* $u(x) = 2\,\mathrm{sech}(2x)$ *and its reduction to Equation (5).*

The corresponding Jost solution $e_2(\rho, x)$ *is obtained from the Jost solution of a Zakharov–Shabat system (see [46]) with the potential* $u(x)$,

$$e_2(\rho, x) = \frac{\rho - \tanh(2x)\,\mathrm{sech}(2x)}{\rho + i}e^{i\rho x}, x \geq 0,\ \mathrm{Im}(\rho) > -\frac{1}{2}.$$

Thus, the Jost function is

$$e_2(\rho) = \frac{\rho + 1}{\rho + i},\ \mathrm{Im}(\rho) > -\frac{1}{2}.$$

It has one root, $\rho_* = -1$, *which corresponds to the spectral singularity* $\lambda_* = \rho_*^2 = 1$. *The scattering function is given by*

$$s_2(\rho) = \left(\frac{1-\rho}{i-\rho}\right)\left(\frac{\rho+i}{\rho+1}\right), \tag{48}$$

which is an S-type function in the strip $|\mathrm{Im}(\rho)| < \frac{1}{2}$.

Example 3 ([21]). *Consider the potentials of the form*

$$q(x) = -2a^2\,\mathrm{sech}^2(ax+b),\ x \geq 0,\ b \in \mathbb{C},\ a > 0 \tag{49}$$

satisfying Condition (2) for $0 < \varepsilon < 2a$. *The Jost solution has the form*

$$e(\rho, x) = \frac{\rho + ia\tanh(b + ax)}{\rho + ia}e^{i\rho x},\ x \geq 0,\ \mathrm{Im}(\rho) > -a,$$

from which the Jost function is obtained

$$e(\rho) = \frac{\rho + ia\tanh(b)}{\rho + ia},\quad \mathrm{Im}(\rho) > -a$$

with the single root $\rho = -ia\tanh(b)$.

The square of this ρ represents the discrete spectrum of the problem. The potential (49) is complex-valued when b is not purely imaginary. The scattering function has the form

$$s(\rho) = \frac{(\rho - ia\tanh(b))(\rho + ia)}{(\rho + ia\tanh(b))(\rho - ia)},$$

which is an S-type function in $|\mathrm{Im}(\rho)| < \min\{a, \mathrm{Im}(-ia\tanh(b))\}$ in the case of a complex-valued potential. In the case of a real-valued potential, $s(\rho)$ is an S-type function in $|\mathrm{Im}(\rho)| < a$.

To present an explicit example, we fix $a = 1$ and $b = -1 - i$ in (49). Then,

$$q_3(x) = -2\,\mathrm{sech}^2(x - 1 - i),\ x \geq 0$$

with $0 < \varepsilon < 2$ in Condition (2) and the Jost solution is

$$e_3(\rho, x) = \frac{\rho + i\tanh(x - 1 - i)}{\rho + i} e^{i\rho x},\ x \geq 0,\ \mathrm{Im}(\rho) > -1.$$

Thus, the Jost function has the form

$$e_3(\rho) = \frac{\rho - i\tanh(1 + i)}{\rho + i},\quad \mathrm{Im}(\rho) > -1,$$

and one eigenvalue exists: $\lambda = -\tanh^2(1 + i)$.

The scattering function

$$s_3(\rho) = \frac{(\rho + i\tanh(1 + i))(\rho + i)}{(\rho - i\tanh(1 + i))(\rho - i)} \tag{50}$$

is an S-type function in the strip $|\mathrm{Im}(\rho)| < 1$.

3.3. Normalization Polynomials

The normalization polynomial $X_k(x)$ of degree $m_k - 1$, associated with the eigenvalue ρ_k^2 (m_k is the algebraic multiplicity of ρ_k as zero of $e(\rho)$), defined by the equation [18]

$$i\,\mathrm{Res}(\Omega(\rho, x); \rho_k) = e^{i\rho_k x} X_k(x) + \int_x^\infty A(x, t) X_k(t) e^{i\rho_k t} dt, \tag{51}$$

where $\Omega(\rho, x)$ is defined by (40). Using the series representation (23) of the kernel $A(x, t)$, we can obtain a method to compute the coefficients of $X_k(x)$.

Remark 6. *Note that the series (8) can be written as*

$$e(\rho, x) = e^{i\rho x}\left(1 + (z + 1)\sum_{n=0}^{\infty}(-1)^n a_n(x) P_n^{(-n,0)}(1 + 2z)\right),\ \rho \in \mathbb{C}^+ (z \in \mathbb{D}) \tag{52}$$

in terms of the Jacobi polynomials $P_n^{(\alpha,\beta)}(\tau)$.

Let us write Equation (51) in terms of the Jost solution and Jacobi polynomials, as follows.

Proposition 4. *Let $\lambda_k = \rho_k^2$, $k = 1, \ldots, \alpha$ be an eigenvalue of problem (1) and (10) and m_k be its multiplicity. For the normalization polynomial $X_k(x)$, the equality holds*

$$i\,\mathrm{Res}(\Omega(\rho, x); \rho_k) = X_k(x) e(\rho_k, x) + e^{i\rho_k x}\sum_{n=0}^{\infty}(-1)^n a_n(x)\sum_{j=1}^{m_k - 1}\frac{d^j X_k(x)}{dx^j}(z_k + 1)^{j+1} P_n^{(j-n,0)}(1 + 2z_k),\ x \geq 0. \tag{53}$$

Proof. The substitution of (23) into (51) yields

$$i\,\mathrm{Res}(\Omega(\rho,x);\rho_k) = e^{i\rho_k x}\left(X_k(x) + \int_0^\infty \sum_{n=0}^\infty a_n(x)L_n(s)e^{-(\frac{1}{2}-i\rho_k)s}X_k(x+s)ds\right).$$

Here, we change the order of summation and integration due to Parseval's identity [47] (p. 16) and additionally use the equality

$$X_k(x+s) = \sum_{j=0}^{m_k-1}\frac{s^j}{j!}\frac{d^jX_k(x)}{dx^j}.$$

Thus,

$$i\,\mathrm{Res}(\Omega(\rho,x);\rho_k) = e^{i\rho_k x}\left(X_k(x) + \sum_{n=0}^\infty a_n(x)\sum_{j=0}^{m_k-1}\frac{1}{j!}\frac{d^jX_k(x)}{dx^j}\int_0^\infty L_n(s)e^{-(\frac{1}{2}-i\rho_k)s}s^jds\right).$$

The last integral can be explicitly evaluated [48] (Formula 7.414 (7))

$$\int_0^\infty \left(L_n(s)e^{-(\frac{1}{2}-i\rho_k)s}\right)s^jds = j!(z_k+1)^{j+1}F(-n,j+1,1;z_k+1) = j!(z_k+1)^{j+1}(-1)^nP_n^{(j-n,0)}(1+2z_k),$$

where $F(a,b,c;z)$ stands for the hypergeometric function [49] (p. 56). Thus, we have the equation

$$i\,\mathrm{Res}(\Omega(\rho,x);\rho_k) = e^{i\rho_k x}\left[X_k(x) + \sum_{n=0}^\infty(-1)^na_n(x)\sum_{j=0}^{m_k-1}(z_k+1)^{j+1}\frac{d^jX_k(x)}{dx^j}P_n^{(j-n,0)}(1+2z_k)\right],$$

and due to Remark 6, we obtain (53). □

Hereinafter $C_k^n = \begin{pmatrix} n \\ k \end{pmatrix}$ denotes the binomial coefficient.

Lemma 1. *The m-th derivative of the Jost solution $e(\rho,x)$ with respect to the variable z admits the representation*

$$\frac{\partial^m}{\partial z^m}(e(\rho,x)) = e(\rho,x)\sum_{j=0}^{m-1}(-1)^jC_j^{m-1}\frac{m!}{(m-j)!}x^{m-j}(z+1)^{j-2m}$$
$$+ e^{i\rho x}\sum_{n=0}^\infty(-1)^na_n(x)\sum_{j=2}^{m+1}P_n^{(j-n-1,0)}(1+2z)\sum_{s=j-1}^{m}(-1)^{s+j+1}C_{s-j+1}^{m-1}\frac{m!}{(m-s)!}x^{m-s}(z+1)^{s+1-2m}, \quad (54)$$

where $\rho \in \mathbb{C}^+(z \in \mathbb{D})$ and $x \geq 0$.

Proof. We use the identity [50] (p. 3)

$$F_z(-n,j,1,z+1)(z+1) = jF(-n,j+1,1,z+1) - jF(-n,j,1,z+1), \quad (55)$$

where F_z means the derivative with respect to z, and j, n are integers.

Let us prove the lemma by induction. For $m = 1$, from (52), we have

$$\frac{\partial}{\partial z}(e(\rho,x)) = \left(\frac{e^{i\rho x}x}{(z+1)^2}\right)\left(1 + (z+1)\sum_{n=0}^\infty(-1)^na_n(x)P_n^{(-n,0)}(1+2z)\right)$$
$$+ e^{i\rho x}\left(\sum_{n=0}^\infty a_n(x)F(-n,1,1,z+1) + (z+1)\sum_{n=0}^\infty a_n(x)F_z(-n,1,1,z+1)\right).$$

The application of (55) gives

$$\frac{\partial}{\partial z}(e(\rho,x)) = \frac{xe(\rho,x)}{(z+1)^2} + e^{i\rho x}\sum_{n=0}^{\infty}(-1)^n a_n(x) P_n^{(1-n,0)}(1+2z_k). \tag{56}$$

Consider Formula (54) as the induction hypothesis for $m = k$. The idea is to prove the equation

$$\begin{aligned}
&\frac{\partial}{\partial z}\left(e(\rho,x)\sum_{j=0}^{k-1}(-1)^j C_j^{k-1}\frac{k!}{(k-j)!}x^{k-j}(z+1)^{j-2k}\right)\\
&-e^{i\rho x}\sum_{n=0}^{\infty}a_n(x)F(-n,2,1,z+1)\left(\sum_{j=0}^{k-1}(-1)^j C_j^{k-1}\frac{k!}{(k-j)!}x^{k-j}(z+1)^{j-2k}\right)\\
=&e(\rho,x)\sum_{j=0}^{k}(-1)^j C_j^{k}\frac{(k+1)!}{(k+1-j)!}x^{k+1-j}(z+1)^{j-2k-2}
\end{aligned} \tag{57}$$

and the equality

$$\begin{aligned}
&\frac{\partial}{\partial z}\left(e^{i\rho x}\sum_{n=0}^{\infty}a_n(x)\sum_{j=2}^{k+1}F(-n,j,1,z+1)\right.\\
&\left.\left(\sum_{m=j-1}^{k}(-1)^{j-m+1}C_{m-j+1}^{k-1}\frac{k!}{(k-m)!}x^{k-m}(z+1)^{m-2k+1}\right)\right)\\
&+e^{i\rho x}\sum_{n=0}^{\infty}a_n(x)F(-n,2,1,z+1)\left(\sum_{j=0}^{k-1}(-1)^j C_j^{k-1}\frac{k!}{(k-j)!}x^{k-j}(z+1)^{j-2k}\right)\\
&=e^{i\rho x}\sum_{n=0}^{\infty}a_n(x)\left(\sum_{j=2}^{k+2}F(-n,j,1,z+1)\right.\\
&\left.\left(\sum_{m=j-1}^{k+1}(-1)^{m+j+1}C_{m-j+1}^{k}\frac{(k+1)!}{(k-m+1)!}x^{k-m+1}(z+1)^{m-2k-1}\right)\right).
\end{aligned} \tag{58}$$

Then, noting that the second terms on the left-hand side of (57) and (58) coincide up to the sign, the desired result is obtained by summing up both equations.

The proof of Equations (57) and (58) is presented in Appendix A, which completes the proof of the Lemma. □

As long as there is no possible misunderstanding, we consider a fixed $\rho = \rho_k$ with a multiplicity $m = m_k$ and the corresponding normalization polynomial $X(x) = X_k(x)$. Thus, the index k is omitted along the following two statements.

Lemma 2. *The coefficients b_j of a normalization polynomial $X(x)$ of degree $m-1$*

$$X(x) = \sum_{j=0}^{m-1} b_j x^j \tag{59}$$

satisfy the equation

$$i\,\mathrm{Res}(\Omega(\rho,x);\rho_k) = b_0 e(\rho_k,x) + \sum_{n=1}^{m-1}\sum_{r=0}^{n-1} b_n\frac{n!}{(r+1)!}C_r^{n-1}\left(\frac{\partial^{r+1}}{\partial z^{r+1}}(e(\rho,x))\right)\Bigg|_{z=z_k}(z_k+1)^{n+r+1}. \tag{60}$$

Proof. Comparing (53) with (60) we see that, in fact, we need to prove the equality

$$X(x)e(\rho,x)+e^{i\rho x}\sum_{n=0}^{\infty}(-1)^{n}a_{n}(x)\sum_{j=1}^{m-1}\frac{d^{j}X(x)}{dx^{j}}(z+1)^{j+1}P_{n}^{(j-n,0)}(1+2z)$$
$$=b_{0}e(\rho,x)+\sum_{n=1}^{m-1}\sum_{r=0}^{n-1}b_{n}\frac{n!}{(r+1)!}C_{r}^{n-1}\frac{\partial^{r+1}}{\partial z^{r+1}}(e(\rho,x))(z+1)^{n+r+1}. \tag{61}$$

Note that

$$X(x)e(\rho,x)+e^{i\rho x}\sum_{n=0}^{\infty}(-1)^{n}a_{n}(x)\sum_{j=1}^{m-1}\frac{d^{j}X(x)}{dx^{j}}(z+1)^{j+1}P_{n}^{(j-n,0)}(1+2z)$$
$$=\sum_{s=0}^{m-1}b_{s}x^{s}e(\rho,x)+e^{i\rho x}\sum_{n=0}^{\infty}(-1)^{n}a_{n}(x)\sum_{j=1}^{m-1}\frac{d^{j}}{dx^{j}}\left(\sum_{s=0}^{m-1}b_{s}x^{s}\right)(z+1)^{j+1}P_{n}^{(j-n,0)}(1+2z)$$
$$=\sum_{s=0}^{m-1}b_{s}x^{s}e(\rho,x)+e^{i\rho x}\sum_{n=0}^{\infty}(-1)^{n}a_{n}(x)\sum_{s=1}^{m-1}\left(\sum_{j=1}^{s}\frac{s!}{(s-j)!}b_{s}x^{s-j}\right)(z+1)^{j+1}P_{n}^{(j-n,0)}(1+2z)$$
$$=b_{0}e(\rho,x)+\sum_{s=1}^{m-1}b_{s}\left(e(\rho,x)x^{s}+e^{i\rho x}\sum_{n=0}^{\infty}(-1)^{n}a_{n}(x)\sum_{j=1}^{s}\frac{s!}{(s-j)!}x^{s-j}(z+1)^{j+1}P_{n}^{(j-n,0)}(1+2z)\right). \tag{62}$$

Then, upon comparison of (61) with (62), it can be observed that proving (61) is equivalent to proving the equality

$$\sum_{r=0}^{s-1}\frac{s!}{(r+1)!}C_{r}^{s-1}\frac{\partial^{r+1}}{\partial z^{r+1}}(e(\rho,x))(z+1)^{s+r+1} \tag{63}$$
$$=e(\rho,x)x^{s}+e^{i\rho x}\sum_{n=0}^{\infty}\sum_{j=1}^{s}(-1)^{n}a_{n}(x)\frac{s!}{(s-j)!}x^{s-j}(z+1)^{j+1}P_{n}^{(j-n,0)}(1+2z) \tag{64}$$

for some natural number $s\leq m-1$. Thus, we are going to prove (64). The substitution of the term with the derivative in (63) by Formula (54) for $m=r+1$ is enough to obtain (64) as follows

$$\sum_{r=0}^{s-1}\frac{s!}{(r+1)!}C_{r}^{s-1}\frac{\partial^{r+1}}{\partial z^{r+1}}(e(\rho,x))(z+1)^{s+r+1}$$
$$=\sum_{r=0}^{s-1}\frac{s!}{(r+1)!}C_{r}^{s-1}e(\rho,x)\left(\sum_{j=0}^{r}(-1)^{j}C_{j}^{r}\frac{(r+1)!}{(r+1-j)!}x^{r-j+1}(z+1)^{j-2r-2}\right.$$
$$\left.+e^{i\rho x}\sum_{n=0}^{\infty}a_{n}(x)\sum_{j=2}^{r+2}F(-n,j,1,z+1)\sum_{s=j-1}^{r+1}(-1)^{s+j+1}C_{s-j+1}^{r}\frac{(r+1)!}{(r-s+1)!}x^{r-s+1}(z+1)^{s-2r-1}\right)$$

$$=\sum_{r=0}^{s-1}\frac{s!}{(r+1)!}C_{r}^{s-1}e(\rho,x)\left(\sum_{j=0}^{r}(-1)^{j}C_{j}^{r}\frac{(r+1)!}{(r+1-j)!}x^{r-j+1}(z+1)^{j-2r-2}\right.$$
$$\left.+e^{i\rho x}\sum_{n=0}^{\infty}a_{n}(x)\sum_{j=1}^{r+1}F(-n,j+1,1,z+1)\sum_{s=j}^{r+2}(-1)^{s+j}C_{s-j}^{r}\frac{(r+1)!}{(r-s+1)!}x^{r-s+1}(z+1)^{s-2r-1}\right)$$
$$=e(\rho,x)x^{s}+e^{i\rho x}\sum_{n=0}^{\infty}\sum_{j=1}^{s}a_{n}(x)\frac{s!}{(s-j)!}x^{s-j}(z+1)^{j+1}F(-n,j+1,1;z+1).$$

This completes the proof of the Lemma. □

Equation (60) provides us with a simple method for computing the coefficients b_j in (59), and consequently for calculating the normalization polynomials.

Theorem 1. *The coefficients b_j of a normalization polynomial $X_k(x) = \sum_{j=0}^{m_k-1} b_j x^j$ corresponding to a complex singular number ρ_k satisfy the system of linear algebraic equations*

$$A \cdot B = D, \tag{65}$$

where A is an $m \times m_k$ matrix with entries defined by

$$A_{jn} = \begin{cases} e(\rho_k, x_j), & n = 1, \\ \sum_{r=0}^{n-2} \frac{n!}{(r+1)!} C_r^{n-1} \left(\frac{\partial^{r+1}}{\partial z^{r+1}} e(\rho_k, x_j) \right) \Big|_{z=z_k} (z_k+1)^{n+r+1}, & 1 < n \leq m_k - 1. \end{cases} \tag{66}$$

Here, $x_j \geq 0$ are distinct points, $j = 1, \ldots, m$ ($m \geq m_k$). B is an m_k vector with its entries being the normalization polynomial coefficients $B_n = b_{n-1}, n = 1, \ldots, m_k$, and D is an m vector defined by

$$D_j = i \operatorname{Res}(\Omega(\rho, x_j); \rho_k). \tag{67}$$

Proof. The proof consists of observing that each row in (65) is just Formula (60) corresponding to a point x_j. The number of rows must be at least m_k; otherwise, the system (65) is underdetermined. □

Thus, the coefficients of the normalization polynomial are obtained from the system (65).

Definition 2. *A set*

$$J = \left\{ \{\rho_k, m_k, X_k(x)\}_{k=1,\ldots,\alpha}, s(\rho) \right\} \tag{68}$$

is called the scattering data set of problem (1) and (10).

Here, ρ_k are the non-real singular numbers, m_k their multiplicities, $X_k(x)$ the corresponding normalization polynomials, and $s(\rho)$ is the scattering function (S-type function in the strip $|\operatorname{Im} \rho| < \varepsilon_0$).

In order to recall a result on the characterization of the scattering data, we need the following definition [39].

Definition 3. *Let $s(\rho)$ be an S-type function in the strip $|\operatorname{Im} \rho| < \varepsilon_0$ and let $\mathcal{L}$ be a curve lying in the strip and running from $-\infty$ to $+\infty$, such that all roots (poles) of $s(\rho)$ are situated above (below) $\mathcal{L}$. The increment divided by 2π of a continuous branch of* $\operatorname{Arg} s(\rho)$, *when ρ runs along $\mathcal{L}$ from $-\infty$ to $+\infty$, is called the* index *of $s(\rho)$ and denoted by* $\operatorname{Ind} s$.

Let us assume that a set J as in Definition 2 is given. A necessary and sufficient condition (obtained in [18]) to ensure that this set represents the scattering data for a problem (1) and (10) with Condition (2) is the following relation

$$\operatorname{Ind} s + 2m + \varkappa = 0 \tag{69}$$

where

$$m = m_1 + \ldots + m_\alpha, \; \varkappa = \frac{1}{2}[1 - s(0)] = \begin{cases} 0 & \text{for } e(0) \neq 0, \\ 1 & \text{for } e(0) = 0. \end{cases}$$

In the case when $m_k = 1$, the notion of the Birkhoff solution is useful for computing the corresponding norming constants.

Remark 7. *Let $E(\rho,x)$ denote the Birkhoff solution of Equation (1) (see [24] (p. 113)), i.e., a solution satisfying the asymptotic relation*

$$E(\rho,x) = (-i\rho)^{(\nu)} e^{-i\rho x}(1+o(1)),\ x\to\infty,\ \nu = 0,1$$

uniformly for $|\rho|\geq\delta$, for each $\delta>0$. For $\operatorname{Im}\rho>0$, this solution is not unique. Indeed, if $E_0(\rho,x)$ is a Birkhoff solution, then $E(\rho,x)=E_0(\rho,x)+ce(\rho,x)$ is also a Birkhoff solution of (1) for any constant $c\in\mathbb{C}$. Note that for $\rho=\rho_k$ (a singular number of the problem), the values of all Birkhoff solutions at the origin coincide. We have $E(\rho_k):=E(\rho_k,0)=E_0(\rho_k,0)$, because $e(\rho_k)=0$. Moreover,

$$E(\rho_k) = \frac{2i\rho_k}{e'(\rho_k,0)}, \tag{70}$$

which can be observed by considering the Wronskian $W[e(\rho,x),E(\rho,x)]=-2i\rho$.

Remark 8. *The solution $\omega(\rho,x)$ satisfying Conditions (39) has the form*

$$\omega(\rho,x) = \frac{E(\rho)e(\rho,x)-e(\rho)E(\rho,x)}{2i\rho},\ \rho\in\overline{\mathbb{C}^+}. \tag{71}$$

Note that ρ_k is a pole of $\Omega(\rho,x)$ in the upper half-plane of the complex variable ρ if and only if it is a root of the Jost function $e(\rho)$ (see (40)). Thus, in case of a simple pole ρ_k in Equation (67), the residue can be computed as follows

$$\begin{aligned}\operatorname{Res}(\Omega(\rho,x);\rho_k) &= \operatorname{Res}\left(\frac{2i\rho}{e(\rho)}\omega(\rho,x);\rho_k\right) = \operatorname{Res}\left(\frac{E(\rho)e(\rho,x)-e(\rho)E(\rho,x)}{e(\rho)};\rho_k\right)\\ &= \frac{E(\rho_k)e(\rho_k,x)-e(\rho_k)E(\rho_k,x)}{\dot{e}(\rho_k)} = \frac{E(\rho_k)e(\rho_k,x)}{\dot{e}(\rho_k)},\end{aligned}$$

where $\dot{e}(\rho):=\frac{d}{d\rho}e(\rho)$, and the corresponding normalization polynomial (in fact normalization constant) is given by

$$c_k = \frac{\operatorname{Res}(\Omega(\rho,x);\rho_k)}{e(\rho_k,x)}i = \frac{E(\rho_k)}{\dot{e}(\rho_k)}i. \tag{72}$$

Moreover, due to (70), we have

$$c_k = -\frac{2\rho_k}{\dot{e}(\rho_k)e'(\rho_k,0)}. \tag{73}$$

Similarly to the case of a real-valued potential [51] (p. 95), one can see that

$$c_k = \frac{1}{\int_0^\infty e^2(\rho_k,x)dx} = \frac{1}{(e'(\rho_k,0))^2\int_0^\infty \omega^2(\rho_k,x)dx}.$$

If $|q(x)|\leq c_1\exp(-c_2|x|^\gamma)$, $\gamma>1$ (for some constants $c_1,c_2>0$), then $e(\rho)$ is an entire function of ρ (see [51] (p. 95)). In this case, as a Birkhoff solution $E(\rho,x)$, one can consider the Jost solution $e(-\rho,x)$, $\operatorname{Im}(\rho)>0$, and hence from (72) we obtain

$$c_k = \frac{e(-\rho_k)}{\dot{e}(\rho_k)}i. \tag{74}$$

Example 4. *According to Remark 8, the normalization constant associated with the unique eigenvalue of the operator from Example 3 is*

$$c_1 = \frac{e(ia\tanh(b))}{\dot{e}(-ia\tanh(b))}i = \left(\frac{2a\tanh(b)}{\tanh(b)+1}\right)(\tanh(b)-1),$$

and, in particular, for $q_3(x)$, we have

$$c_1 = \frac{2(1+\tanh(1+i))\tanh(1+i)}{1-\tanh(1+i)}.$$

Example 5. *With the aid of Remark 8, an approximate value of the normalization constant for the eigenvalue λ_0 from Example 1 is obtained*

$$c_0 \approx 16.339391035537 + 40.670169841396i.$$

3.4. Numerical Algorithm

The approximate solution of the direct problem can be performed with the following steps.

1. Compute the Jost function using (41) and the recurrent integration procedure from [27], for $\mathrm{Im}(\rho) \geq 0$.
2. Extend the Jost function $e(\rho)$ to $-\varepsilon_0 < \mathrm{Im}(\rho) < 0$ using any convenient technique, such as the classic analytic continuation, Padé approximants [28] or some other approach [52,53].
3. Obtain the scattering function $s(\rho)$ for $0 \leq |\mathrm{Im}(\rho)| < \varepsilon_0$ by Formula (42).
4. To locate the eigenvalues, find the non-real poles of the function $\Omega(\rho, x)$, which is equivalent to finding zeroes of the function $e(\rho)$ in the unit disk in terms of z. This can be achieved with the aid of the argument principle theorem. In particular, in the present work, we compute the change in the argument along rectangular contours γ. If the change in the argument along γ is zero, consider another contour. Otherwise, subdivide the region within the contour until the desired accuracy is attained. Note that for a sufficiently large N, zeros of $e_N(\rho)$, approximate the square roots of the eigenvalues of the problem arbitrarily closely. The proof is analogous to that in [54] and is based on the Rouché theorem from complex analysis.
5. Obtain the normalization polynomials.
   - 5.1 For simple poles, use Remark 8 to obtain the normalization constants.
   - 5.2 Otherwise, for higher multiplicities, solve the linear system of Equation (65) for the coefficients b_{n_k}, $n_k = 0, 1, \ldots, m_k - 1$ computing $A_{j,n}$ and D_j defined in Equations (66) and (67) for several values of x_j.

4. Inverse Problem

In order to reconstruct the potential in (1) from the scattering data, it is convenient to introduce the function [39]

$$\phi_s(x) = \frac{1}{2\pi}\int_{-\infty+i\eta}^{\infty+i\eta} (s(\rho)-1)e^{ix\rho}d\rho, \tag{75}$$

where η is a number satisfying the inequalities $0 < \eta < \varepsilon_0$ (ε_0s, defined in Section 3.2), and the function

$$f(x) = \phi_s(x) - \sum_{k=1}^{\alpha} X_k(x)e^{i\rho_k x}, \; x \geq 0. \tag{76}$$

Remark 9. *Hereinafter, we use the notation*

$$\int_{\mathcal{L}_\eta} = \int_{-\infty+i\eta}^{\infty+i\eta} \tag{77}$$

for $0 < \eta < \varepsilon_0$ where $\mathcal{L}_\eta$ represents a line parallel to the real axis crossing $i\eta$.

The kernel $A(x,t)$ and the function $f(x)$ satisfy the following Gel'fand–Levitan (G-L) equation [39] (Theorem 10.1)

$$A(x,t) = \int_x^\infty A(x,u)f(u+t)du + f(x+t),\ 0 \le x \le t < \infty. \tag{78}$$

4.1. Infinite Linear Algebraic System for Coefficients $a_n(x)$

Following [38], from the G-L Equation (78), we deduce the following system of linear algebraic equations for the coefficients $a_n(x)$ from the series representation (23).

Theorem 2. *The complex-valued functions $a_n(x)$ satisfy the equations*

$$a_m(x) - \sum_{n=0}^\infty a_n(x)A_{mn}(x) = f_m(2x),\ m = 0,1,\ldots \tag{79}$$

where

$$f_m(x) := \int_0^\infty f(s+x)L_m(s)e^{-\frac{s}{2}}ds, \tag{80}$$
$$A_{mn}(x) := \int_0^\infty f_n(2x+s)L_m(s)e^{-\frac{s}{2}}ds.$$

Proof. Substitution of the series representation (23) into (78) leads to the equalities

$$\begin{aligned} f(x+t) &= \sum_{n=0}^\infty a_n(x)L_n(t-x)e^{\frac{x-t}{2}} - \sum_{n=0}^\infty a_n(x)\int_x^\infty L_n(u-x)e^{\frac{x-u}{2}}f(u+t)du \\ &= \sum_{n=0}^\infty a_n(x)L_n(t-x)e^{\frac{x-t}{2}} - \sum_{n=0}^\infty a_n(x)\int_0^\infty L_n(y)e^{-\frac{y}{2}}f(x+y+t)dy, \end{aligned} \tag{81}$$

where the change in the order of summation and integration is justified by the general Parseval identity [47] (p. 16).

We have

$$\begin{aligned} \int_x^\infty A(x,u)f(u+t)du &= \left\langle A(x,x+u), \overline{f(u+x+t)} \right\rangle_{\mathcal{L}_2(0,\infty)} \\ &= \sum_{n=0}^\infty \left\langle A(x,x+u), e^{-u/2}L_n(u) \right\rangle_{\mathcal{L}_2(0,\infty)} \left\langle e^{-u/2}L_n(u), \overline{f(u+x+t)} \right\rangle_{\mathcal{L}_2(0,\infty)} \\ &= \sum_{n=0}^\infty a_n(x)\int_0^\infty e^{-\frac{u}{2}}L_n(u)f(u+x+t)du. \end{aligned}$$

Denote $s = t - x$. Equation (81) is equivalent to

$$f(s+2x) = \sum_{n=0}^\infty a_n(x)L_n(s)e^{-\frac{s}{2}} - \sum_{n=0}^\infty a_n(x)\int_0^\infty L_n(y)e^{-\frac{y}{2}}f(s+2x+y)dy. \tag{82}$$

Multiplying the last equation by $L_m(s)e^{-\frac{s}{2}}$ and integrating this, we obtain

$$\begin{aligned} \int_0^\infty f(s+2x)L_m(s)e^{-\frac{s}{2}}ds &= \sum_{n=0}^\infty a_n(x)\int_0^\infty L_n(s)L_m(s)e^{-s}ds \\ &\quad - \sum_{n=0}^\infty a_n(x)\left(\int_0^\infty L_m(s)e^{-\frac{s}{2}}\left(\int_0^\infty f(s+2x+y)L_n(y)e^{-\frac{y}{2}}dy\right)ds\right). \end{aligned} \tag{83}$$

Note that

$$\int_0^\infty L_n(s)L_m(s)e^{-s}ds = \delta_{mn},$$

and

$$\int_0^\infty f(s+2x+y)L_n(y)e^{-\frac{y}{2}}dy = f_n(2x+s).$$

Thus, from (83) we obtain (79). □

4.2. Expressions for $f_m(x)$ and $A_{mn}(x)$

It is convenient to regard the functions $f_m(x)$ and $A_{mn}(x)$ as a sum of the components corresponding to the continuous $f_{m,c}(x)$, $A_{mn,c}(x)$ and discrete spectra $f_{m,d}(x)$, $A_{mn,d}(x)$, and simplify these expressions with the aid of the formula ([48], Formula 7.414 (6))

$$\int_0^\infty L_m(s)e^{s(i\rho-\frac{1}{2})}ds = \frac{(-1)^m\left(\frac{1}{2}+i\rho\right)^m}{\left(\frac{1}{2}-i\rho\right)^{m+1}}. \tag{84}$$

The continuous and discrete components for the function $f_m(x)$ have the form

$$\begin{aligned} f_{m,c}(x) &:= \int_0^\infty \phi_s(s+x)L_m(s)e^{-\frac{s}{2}}ds = \frac{1}{2\pi}\int_{\mathcal{L}_\eta}(s(\rho)-1)e^{i\rho x}\int_0^\infty L_m(s)e^{i\rho s-\frac{s}{2}}dsd\rho \\ &= \frac{(-1)^m}{2\pi}\int_{\mathcal{L}_\eta}(s(\rho)-1)\frac{\left(\frac{1}{2}+i\rho\right)^m}{\left(\frac{1}{2}-i\rho\right)^{m+1}}e^{i\rho x}d\rho, \end{aligned} \tag{85}$$

and

$$\begin{aligned} f_{m,d}(x) &:= -\sum_{k=1}^{\alpha}\int_0^\infty X_k(s+x)e^{i(s+x)\rho_k}L_m(s)e^{-\frac{s}{2}}ds, \\ &= -\sum_{k=1}^{\alpha}e^{i\rho_k x}\int_0^\infty\left(\sum_{j=0}^{m_k-1}\left(\frac{1}{j!}\right)\frac{d^jX_k(x)}{dx^j}s^j\right)e^{i\rho_k s}L_m(s)e^{-\frac{s}{2}}ds, \\ &= -\sum_{k=1}^{\alpha}e^{i\rho_k x}\sum_{j=0}^{m_k-1}\left(\frac{1}{j!}\right)\frac{d^jX_k(x)}{dx^j}\int_0^\infty s^jL_m(s)e^{-(\frac{1}{2}-i\rho_k)s}ds, \\ &= (-1)^{m+1}\sum_{k=1}^{\alpha}\sum_{j=0}^{m_k-1}e^{i\rho_k x}\frac{d^jX_k(x)}{dx^j}(z_k+1)^{j+1}P_m^{(j-m,0)}(1+2z_k). \end{aligned} \tag{86}$$

For the function $A_{mn,c}(x)$, we have

$$\begin{aligned} A_{mn,c}(x) &:= \int_0^\infty L_m(s)f_{n,c}(2x+s)e^{-\frac{s}{2}}ds \\ &= \frac{(-1)^n}{2\pi}\int_{\mathcal{L}_\eta}(s(\rho)-1)\frac{\left(\frac{1}{2}+i\rho\right)^n}{\left(\frac{1}{2}-i\rho\right)^{n+1}}\left(\int_0^\infty L_m(s)e^{-(\frac{1}{2}-i\rho)s}ds\right)e^{2i\rho x}d\rho \\ &= \frac{(-1)^{n+m}}{2\pi}\int_{\mathcal{L}_\eta}(s(\rho)-1)\frac{\left(\frac{1}{2}+i\rho\right)^{n+m}}{\left(\frac{1}{2}-i\rho\right)^{n+m+2}}e^{2i\rho x}d\rho, \end{aligned} \tag{87}$$

and for $A_{mn,d}(x)$, we use (84) to obtain

$$
\begin{aligned}
A_{mn,d}(x) &= \int_0^\infty L_m(s) f_{n,d}(2x+s) e^{-\frac{s}{2}} ds \\
&= -\int_0^\infty L_m(s) \sum_{k=1}^{\alpha} \sum_{j=0}^{m_k-1} \frac{d^j}{dx^j}(X_k(2x+s))(z_k+1)^{j+1} F(-n, j+1, 1; z_k+1) e^{2i\rho_k x} e^{-(\frac{1}{2}-i\rho_k)s} ds \\
&= -\sum_{k=1}^{\alpha} \sum_{j=0}^{m_k-1} (z_k+1)^{j+1} F(-n, j+1, 1; z_k+1) e^{2i\rho_k x} \\
&\int_0^\infty \frac{d^j}{dx^j} \left(\sum_{p=0}^{m_k-1} \left(\frac{s^p}{2^p p!} \frac{d^p}{dx^p}(X_k(2x)) \right) \right) L_m(s) e^{-(\frac{1}{2}-i\rho_k)s} ds \\
&= -\sum_{k=1}^{\alpha} \sum_{j=0}^{m_k-1} \sum_{p=0}^{m_k-1} \frac{1}{2^p p!} \frac{d^{p+j}}{dx^{p+j}}(X_k(2x))(z_k+1)^{j+1} F(-n, j+1, 1; z_k+1) e^{2i\rho_k x} \\
&\int_0^\infty s^p L_m(s) e^{-(\frac{1}{2}-i\rho_k)s} ds \\
&= (-1)^{m+n+1} \sum_{k=1}^{\alpha} \sum_{j=0}^{m_k-1} \sum_{p=0}^{m_k-1-j} \frac{1}{2^p} \frac{d^{p+j}}{dx^{p+j}}(X_k(2x))(z_k+1)^{p+j+2} P_n^{(j-n,0)}(1+2z_k) \\
&P_m^{(p-m,0)}(1+2z_k) e^{2i\rho_k x}.
\end{aligned} \tag{88}
$$

Remark 10. *When an eigenvalue ρ_k^2 is simple and the corresponding normalization polynomial $X_k(x)$ is just a normalization constant c_k, expressions (86) and (88) can be written in the form*

$$
f_{m,d}(x) = -\sum_{k=1}^{\alpha} e^{ix\rho_k}(-z_k)^m (z_k+1) c_k, \tag{89}
$$

$$
A_{mn,d}(x) = -\sum_{k=1}^{\alpha} e^{2ix\rho_k}(-z_k)^{m+n} (z_k+1)^2 c_k. \tag{90}
$$

We illustrate the calculation of the functions (85)–(88) with some examples.

Example 6. *Consider the scattering function obtained in Example 2:*

$$
s_2(\rho) = \left(\frac{1-\rho}{i-\rho}\right)\left(\frac{\rho+i}{\rho+1}\right) \tag{91}
$$

in the strip $0 \le \mathrm{Im}(\rho) < 1$, with no discrete spectrum and thus no normalization polynomials. Let us compute the function $\phi_s(x)$ defined by (75), where the line $\mathcal{L}_\eta$ lies in the strip $0 < \mathrm{Im}(\rho) < 1$. Since the function $s_2(\rho)$ is analytic in the strip $0 < \mathrm{Im}(\rho) < 1$, the value of the integral is independent of the choice of $0 < \eta < 1$. Using Jordan's lemma to calculate the integral in (75), we obtain

$$
f(x) = \phi_s(x) = -2ie^{-x}.
$$

Now, computing the functions $f_m(x)$ and $A_{mn}(x)$ from Formula (85) and (87) and using the residue theorem, we obtain

$$
f_m(x) = -4i \cdot 3^{-(m+1)} e^{-x}, \; A_{nm}(x) = -8i \cdot 3^{-(n+m+2)} e^{-2x}.
$$

Thus, in the case of the potential $q_2(x)$, the system of Equation (79) can be written explicitly.

Example 7. *Consider the scattering function $s_3(\rho)$ from Example 3. It has two poles in the upper half-plane: at i and $i\tanh(1+i)$. Hence, using the residue theorem, we find that*

$$\phi_s(x) = \frac{2(1+\tanh(1+i))e^{-x(1+\tanh(1+i))}\left(e^{x\tanh(1+i)} - e^x\tanh(1+i)\right)}{\tanh(1+i)-1},$$
$$f(x) = -2e^{(2-x+2i)},\ f_m(x) = -4\cdot 3^{-(m+1)}e^{(2-x+2i)},\ A_{mn}(x) = -8e^{2-2x+2i}3^{-(m+n+2)}.$$

Again, the corresponding system of Equation (79) can be written explicitly.

Example 8. *Consider $s_1(\rho)$ from Example 1. To compute $\phi_s(x)$, we consider the singularities of $s_1(\rho)$ in the upper half-plane. From the set $D(s_1)$ (see Example 1), we have that $s_1(\rho)$ has an infinite number of isolated singularities at the points $\rho_k = \frac{ik}{2}$ with $k \in \mathbb{N}\setminus\{0\}$ and a singular number ρ_0; see (46). Using properties of the gamma function, we obtain*

$$\underset{\substack{\rho=\rho_k\\k=1,2,\dots}}{\mathrm{Res}}\,(s_1(\rho)-1) = \left(\frac{(1-i)^{-4i\rho_k}5^{-2i\rho_k}J_{2i\rho_k}((2-2i)\sqrt{5})}{J_{-2i\rho_k}((2-2i)\sqrt{5})\Gamma(1-2i\rho_k)}\right)\underset{\substack{\rho=\rho_k\\k=1,2,\dots}}{\mathrm{Res}}\,\Gamma(1+2i\rho)$$
$$= \left(\frac{(1-i)^{2k}5^kJ_{-k}((2-2i)\sqrt{5})}{J_k((2-2i)\sqrt{5})\Gamma(1+k)}\right)\frac{(-1)^k i}{2(k-1)!} = \frac{(5(1-i)^2)^k}{2k!(k-1)!}i,$$

and

$$\underset{\rho=\rho_0}{\mathrm{Res}}\,(s_1(\rho)-1) = \frac{e(-\rho_0)}{e'(\rho_0)} = -ic_0,$$

where c_0 is the normalization constant obtained in Example 5. Therefore, for $x>0$, we have

$$\phi_s(x) = ie^{ix\rho_0}\underset{\rho=\rho_0}{\mathrm{Res}}\,(s_1(\rho)-1) + i\sum_{k=1}^{\infty}e^{-\frac{xk}{2}}\underset{\rho=\rho_k}{\mathrm{Res}}\,(s_1(\rho)-1)$$
$$= c_0e^{ix\rho_0} - \frac{1}{2}\sum_{k=1}^{\infty}\frac{\left(-e^{-\frac{x}{2}}\right)^k}{(k-1)!} = c_0e^{ix\rho_0} + \frac{e^{-e^{-x/2}-x/2}}{2}.$$

Hence, the function $f(x)$ has the form

$$f(x) = c_0e^{ix\rho_0} + \frac{e^{-e^{-x/2}-x/2}}{2} - c_0e^{ix\rho_0} = \frac{e^{-e^{-x/2}-x/2}}{2},$$

and we obtain the functions $f_{m,c}(x)$ and $f_{m,d}(x)$ in terms of $z_0 = \frac{\frac{1}{2}+i\rho_0}{\frac{1}{2}-i\rho_0}$ (see (9)) as follows

$$f_{n,c}(x) = c_0e^{ix\rho_0}(-1)^n z_0^n(z_0+1) + (-1)^n\sum_{k=1}^{\infty}\frac{\left(-e^{-\frac{x}{2}}\right)^k}{(k-1)!}\frac{(1-k)^n}{(1+k)^{n+1}}, \tag{92}$$

and

$$f_{n,d}(x) = (-1)^{n+1}c_0e^{ix\rho_0}(z_0+1)z_0^n. \tag{93}$$

Thus, from Equations (92) and (93) we obtain

$$f_n(x) = (-1)^n\sum_{k=1}^{\infty}\frac{\left(-e^{-\frac{x}{2}}\right)^k}{(k-1)!}\frac{(1-k)^n}{(1+k)^{n+1}}.$$

Likewise, applying the residue theorem, we have

$$A_{mn}(x) = 2(-1)^{n+m} \sum_{k=1}^{\infty} \frac{(-e^{-x})^k}{(k-1)!} \frac{(1-k)^{n+m}}{(1+k)^{n+m+2}}.$$

Thus, as in the previous two examples, the system of Equation (79) can be written explicitly.

The cancellation of terms when summing up (92) with (93) is not incidental and is generalized below in Remark 12.

To calculate the integrals in functions f_m and A_{mn} in the case when the scattering function is given explicitly, we implement Jordan's lemma and the residue theorem considering the asymptotics (44). However, often the function $s(\rho)$ is not given in a closed form but as a table of data—then, the following techniques can be useful to compute the integrals. First, we recall a widely used technique for the quadrature of highly oscillatory integrals through approximations of the Fourier sine and cosine transform. This is illustrated below in Example 16. A second option is a transformation of integrals in f_m and A_{mn} into integrals over a finite interval providing a certain advantage for its numerical implementation. This is illustrated below in Example 21.

Remark 11. *We mainly discuss the calculation of the functions f_m. The calculation of A_{mn} is analogous.*

1. *Suppose $s(\rho)$ is given in a closed form. By* Pol *we denote the set of its poles in the open upper half-plane. Since $s(\rho)$ satisfies the asymptotics (44), the integral in (85) can be computed with the aid of Jordan's lemma and the residue theorem as follows*

$$f_{m,c}(x) = (-1)^m i \sum_{\rho_j \in \mathrm{Pol}} \operatorname*{Res}_{\rho=\rho_j} (z+1)z^m (s(\rho)-1)e^{i\rho x}, \tag{94}$$

provided the series on the right-hand side is convergent; see [55] (p. 459).
If Pol *contains only simple poles, we obtain*

$$f_{m,c}(x) = (-1)^m i \sum_{\rho_j \in \mathrm{Pol}} (z_j+1)z_j^m e^{i\rho_j x} \operatorname*{Res}_{\rho=\rho_j} (s(\rho)-1). \tag{95}$$

2. *Consider the integral in (85)*

$$f_{m,c}(x) = \frac{(-1)^m e^{-\eta x}}{2\pi} \int_{-\infty}^{\infty} (s(\sigma+i\eta)-1) \frac{\left(\frac{1}{2}+i(\sigma+i\eta)\right)^m}{\left(\frac{1}{2}-i(\sigma+i\eta)\right)^{m+1}} e^{i\sigma x} d\sigma \tag{96}$$

for some $0 < \eta < \epsilon_0$. Following the approach from [56] (p. 236), denote

$$g(\sigma) := (s(\sigma+i\eta)-1) \frac{\left(\frac{1}{2}+i(\sigma+i\eta)\right)^m}{\left(\frac{1}{2}-i(\sigma+i\eta)\right)^{m+1}},$$

and set $\psi(\sigma) = g(\sigma) + g(-\sigma)$, $\phi(\sigma) = g(\sigma) - g(-\sigma)$. Then

$$\int_{-\infty}^{\infty} (s(\sigma+i\eta)-1) \frac{\left(\frac{1}{2}+i(\sigma+i\eta)\right)^m}{\left(\frac{1}{2}-i(\sigma+i\eta)\right)^{m+1}} e^{i\sigma x} d\sigma = \int_0^{\infty} \psi(\sigma)\cos(\sigma x) d\sigma + i \int_0^{\infty} \phi(\sigma)\sin(\sigma x) d\sigma. \tag{97}$$

The integrals on the right hand side (the Fourier cosine and sine transforms) are approximated by the corresponding sums

$$h\sum_{k=0}^{N}\psi\left(\left(k+\frac{1}{2}\right)h\right)\cos\left(x\left(k+\frac{1}{2}\right)\right) \text{ and } h\sum_{k=0}^{N}\phi(kh)\sin(xkh), \tag{98}$$

where h and N are chosen to be sufficiently small and large, respectively.

3. *Transform the line $\mathcal{L}_\eta$ into a circle centered at $\frac{-2\eta}{1+2\eta}$ of radius $\frac{1}{2+\eta}$ with the aid of the formulas*

$$\rho=\frac{i(1+4\eta-\exp(i\theta))}{2(1+\exp(i\theta))},\ d\rho=\exp(i\theta)\frac{(1+2\eta)}{(1+\exp(i\theta))^2}d\theta. \tag{99}$$

This enables us to consider the integral in (85) in the form

$$f_{m,c}(x)=\frac{(-1)^m}{2\pi}\int_0^{2\pi}\left(s\left(\frac{i(1+4\eta-\exp(i\theta))}{2(1+\exp(i\theta))}\right)-1\right)\exp\left(-x\left(\frac{1+4\eta-\exp(i\theta)}{2(1+\exp(i\theta))}\right)\right)\cdot$$
$$\frac{\left(\frac{1}{2}-\left(\frac{1+4\eta-\exp(i\theta)}{2(1+\exp(i\theta))}\right)\right)^m}{\left(\frac{1}{2}+\left(\frac{1+4\eta-\exp(i\theta)}{2(1+\exp(i\theta))}\right)\right)^{m+1}}\exp(i\theta)\frac{(1+2\eta)}{(1+\exp(i\theta))^2}d\theta,$$

reducing the integration to a finite interval.

Remark 12. *Suppose that the eigenvalues are simple, and Formula (74) is applicable. Denote the set $K=\{\rho_1,\ldots,\rho_\alpha\}$ of non-real singular values. From (89), (90) and (95), we have that the functions $f_m(x)$ and $A_{mn}(x)$ can be computed as*

$$f_m(x)=(-1)^m i\sum_{\rho_j\in \mathrm{Pol}\setminus K}\frac{1}{\dot{e}(\rho_j)}(z_j)^m(z_j+1)e^{i\rho_j x}\mathrm{Res}\big(e(-\rho);\rho=\rho_j\big),$$
$$A_{mn}(x)=(-1)^{n+m}i\sum_{\rho_j\in \mathrm{Pol}\setminus K}\frac{1}{\dot{e}(\rho_j)}(z_j)^{n+m}(z_j+1)^2e^{2i\rho_j x}\mathrm{Res}\big(e(-\rho);\rho=\rho_j\big).$$

4.3. Stability of the System and Its Solution

Consider the truncated system (79):

$$a_m(x)-\sum_{n=0}^{M}a_n(x)A_{mn}(x)=f_m(2x),\ m=0,\ldots,M. \tag{100}$$

Denote its solution as $U_M=\{a_m^M\}_{m=0}^M$. In the following two theorems, we prove the unique solvability of (100), the convergence of its solution to the exact one as well as its stability.

Theorem 3. *Let $x\geq 0$ be fixed. Consider the system (100) truncated to $M+1$ equations. Then, for a sufficiently large M, the truncated system is uniquely solvable, and*

$$a_m^M(x)\to a_m(x),\ \ M\to\infty, m=0,1\ldots. \tag{101}$$

Proof. Since $\{f_m(2x)\}_{m=0}^\infty\in\ell_2$ and $\{A_{m,n}(x)\}_{m,n=0}^\infty\in\ell_2\otimes\ell_2$ and we look for $\{a_m(x)\}_{m=0}^\infty\in\ell_2$, the assertion of the theorem for the truncated system follows directly from the general theory presented in [57] (Chapter 14, §3). □

Theorem 4. *The approximate solution $\{a_m^M(x)\}_{m=0}^M$ of the system is stable.*

Proof. Note that the truncated system (100) coincides with that obtained by applying the Bubnov–Galerkin procedure to the G-L Equation (78) with the orthonormal system of Laguerre polynomials in $\mathcal{L}^2([0,\infty);e^{-x})$; see [58] (§14). Let I_M denote the $(M+1)\times(M+1)$ identity matrix, $L_M=\{A_{m,n}(x)\}_{m,n=0}^{M}$ be the coefficient matrix of the truncated system and $R_M=\{f_m(2x)\}_{m=0}^{M}$ the right hand side of (100). Following [58] (§9), consider a system called inexact

$$(I_M+L_M+\Gamma_M)V_M=R_M+\delta_M,$$

where Γ_M is an $(M+1)\times(M+1)$ matrix representing errors in the coefficients $A_{m,n}$, and δ_M is the column vector representing errors in the coefficients f_m. Let V_M be a solution of the non-exact system. The solution of the Bubnov–Galerkin procedure is said to be stable if there exist constants $c_1,c_2>0$, such that for $\|\Gamma_M\|\le r$ and arbitrary δ_M the non-exact system is solvable, and the following inequality holds

$$\|U_M-V_M\|\le c_1\|\Gamma_M\|+c_2\|\delta_M\|. \tag{102}$$

Now, since in the case under consideration, the inequality (102) is true (see [58] (Theorems 14.1 and 14.2)), the approximate solution is stable. □

4.4. Algorithm to Recover the Potential

Given a scattering data set J as in Definition 2, the algorithm to recover $q(x)$ consists of the following steps.

1. Compute the functions $f_m(x)$ and $A_{mn}(x)$ with the aid of (85)–(88).
2. Solve the truncated system of linear algebraic Equation (100) to obtain the coefficient $a_0(x)$.
3. Recover the potential $q(x)$ from (38).

5. Numerical Examples

We implemented the algorithms proposed in Sections 3.4 and 4.4 to solve the direct and inverse problems, respectively, with machine precision and with the aid of Matlab2021. Several examples are discussed, some of which have been introduced in previous sections.

5.1. Direct Problem

In this subsection, we discuss the computation of the scattering data, based on the series representation of the Jost solution (8). We deal with the approximate solution obtained by truncating the series (36).

The computation of the coefficients $a_n(x)$ is performed with the aid of the recurrent integration procedure from [27].

First of all, we discuss the choice of the number N in (36). Below, we show that a satisfactory accuracy is attained for a relatively small N (from several units to several dozens), and a reliable indicator

$$\varepsilon_N=\max\left|\sum_{n=0}^{N}a_n(x)-\frac{1}{2}\int_x^{\infty}q(t)dt\right| \tag{103}$$

can be used to choose an appropriate N.

In the case of simple singular numbers ρ_k, the norming constants can be computed with the aid of (73):

$$c_k\approx-\frac{2\rho_k}{\dot{e}_N(\rho_k)e_N'(\rho_k,0)}. \tag{104}$$

Another possibility consists of using (74) in the form

$$c_k\approx\frac{[m,n]_{e_N(\rho)}(-\rho_k)}{\dot{e}_N(\rho_k)}, \tag{105}$$

where $[m,n]_{e_N(\rho)}(-\rho_k)$ stands for the Padé approximant of $e_N(\rho)$ at $\rho = -\rho_k$. This can be achieved when the accuracy of this rational approximation in the upper half-plane is satisfactory, i.e., when one has a suitable small value of $\max\left|[m,n]_{e_N(\rho)}(\rho) - e_N(\rho)\right|$ in a sufficiently large region in the upper half-plane of the complex variable ρ.

A reliable algorithm to compute derivatives of (36) in (104) is proposed in [27].

To obtain the scattering function (42) in the strip $|\mathrm{Im}(\rho)| < \varepsilon_0$ we consider two options depending on how the computation of the Jost function is performed for ρ in the lower half-plane. The first one uses

$$s(\rho) \approx \frac{[m,n]_{e_N(\rho)}(-\rho)}{e_N(\rho)},$$

provided $[m,n]_{e_N(\rho)}(\rho)$ extends $e_N(\rho)$ analytically onto a certain strip in the lower half ρ-plane. A second option for the computation of $s(\rho)$ is

$$s(\rho) \approx \frac{e_N(-\rho)}{e_N(\rho)}, \tag{106}$$

where the expression (36) is calculated at points ρ of a parallel line sufficiently close to the real axis and contained in the lower half ρ-plane.

Remark 13. *The notation for the approximate Jost solution (Jost function) may contain two indices, k and N: $e_{k,N}(\rho, x)$ ($e_{k,N}(\rho)$), where k denotes the solution associated with the Schrödinger equation with the potential $q_k(x)$ and N is the parameter from* (36).

Example 9. *Consider the potential $q_2(x)$ from Example 2. We present the indicator ε_N in Table 1 for different values of N in (103).*

Table 1. Example 9: indicator ε_N for different values of N.

N	ε_N **for** $x \in [0, 12]$	N	ε_N **for** $x \in [0, 12]$
2	1.57×10^{-1}	30	6.7×10^{-12}
3	5.24×10^{-2}	35	6.02×10^{-12}
5	5.82×10^{-3}	40	5.21×10^{-12}
10	2.39×10^{-5}	45	4.98×10^{-12}
15	9.86×10^{-8}	55	4.34×10^{-12}
20	4.11×10^{-10}	180	1.98×10^{-12}

Figure 1 shows the real and imaginary parts of the approximate and exact Jost solution computed from (36) *at a sample point $\rho = 1 + i/3$ with $N = 30$, i.e., $e_{2,30}(1 + i/3, x)$. The maximum absolute error of the computed Jost solution for x in the interval $[0, 12]$ is 2.14×10^{-13}.*

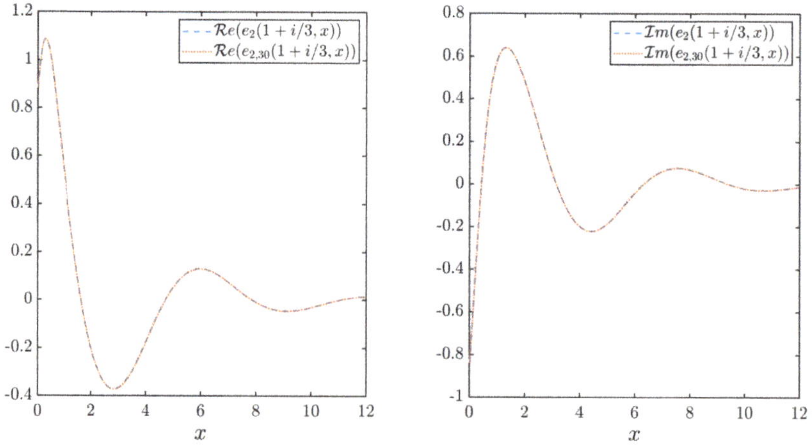

Figure 1. Real (**left**) and imaginary (**right**) parts of $e_2(1+i/3,x)$ and approximate Jost solution $e_{2,30}(1+i/3,x)$.

Table 2 presents the maximum absolute and relative errors of the approximate Jost function $e_{2,N}(\rho(z))$ for $z \in \overline{\mathbb{D}}$ for different values of N.

Table 2. Maximum absolute and relative errors of the approximate Jost function $e_{2,N}(\rho(z))$ in $\overline{\mathbb{D}}$.

N	2	5	20	25	30	40	180
Abs. Error	1.57×10^{-1}	5.82×10^{-3}	4.06×10^{-10}	1.68×10^{-12}	2.48×10^{-14}	1.98×10^{-14}	1.96×10^{-14}
Rel. Error	1.78×10	6.58×10^{-1}	4.58×10^{-8}	1.89×10^{-10}	9.66×10^{-13}	3.17×10^{-13}	1.57×10^{-13}

Figure 2 shows the function $|e_{2,30}(\rho(z))|$ for $z \in \overline{\mathbb{D}}$. Here, we illustrate the existence of a unique singular number. Indeed, this singular number $\rho = -1$ corresponds to $z = -0.6 - 0.8i$ and the value $e_{2,30}(-1)$ is $-5.55 \times 10^{-15} + 2.66 \times 10^{-15}i$.

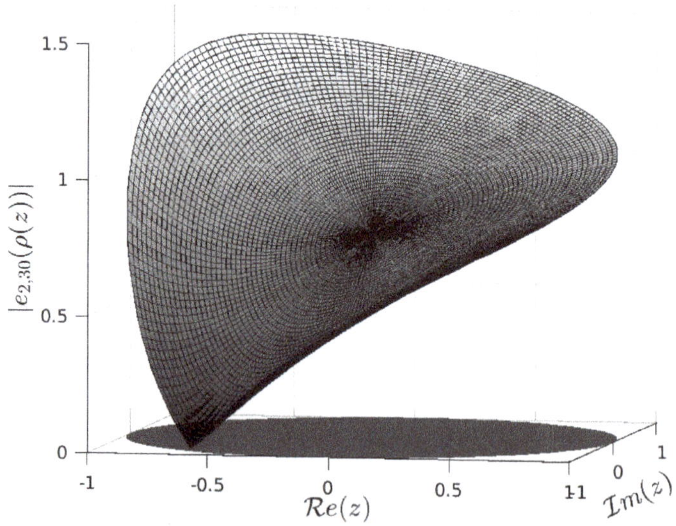

Figure 2. Function $|e_{2,30}(\rho(z))|$ for $z \in \overline{\mathbb{D}}$.

The distribution of the absolute and relative errors of the approximate Jost function is presented in Figure 3 and Figure 4 (respectively), where the maximum absolute error is 1.98×10^{-14} *and the maximum relative error is* 3.17×10^{-13}.

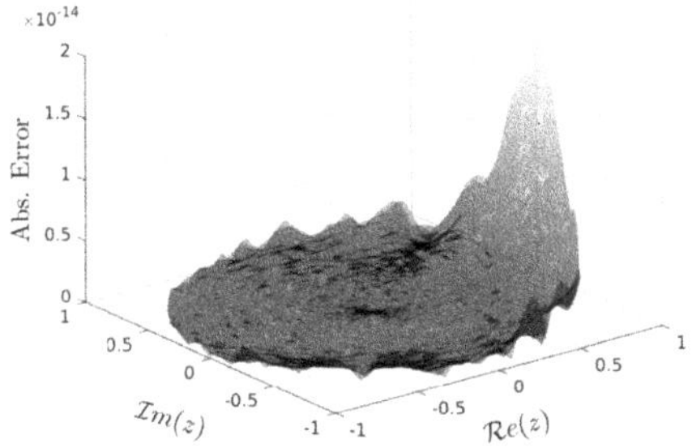

Figure 3. Absolute error of approximate Jost function $e_{2,30}(\rho(z))$ for $z \in \overline{\mathbb{D}}$.

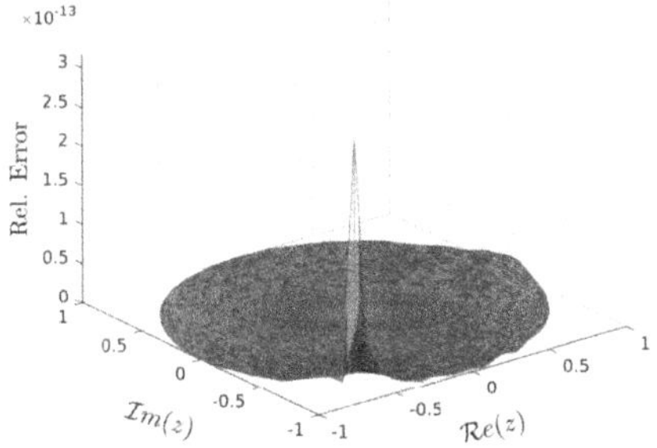

Figure 4. Relative error of approximate Jost function $e_{2,30}(\rho(z))$ for $z \in \overline{\mathbb{D}}$.

Furthermore, a good approximation of the derivative of the Jost function becomes essential for the argument principle algorithm performance. This is necessary to obtain the eigenvalues as the squares of non-real zeros of the approximate Jost function. In Figure 5, we illustrate $\frac{de_{2,30}(\rho(z))}{dz}$*, and Figures 6 and 7 depict the distribution of the absolute and relative errors, respectively. The maximum absolute error is* 9.6×10^{-13} *and the maximum relative error is* 7.21×10^{-13}.

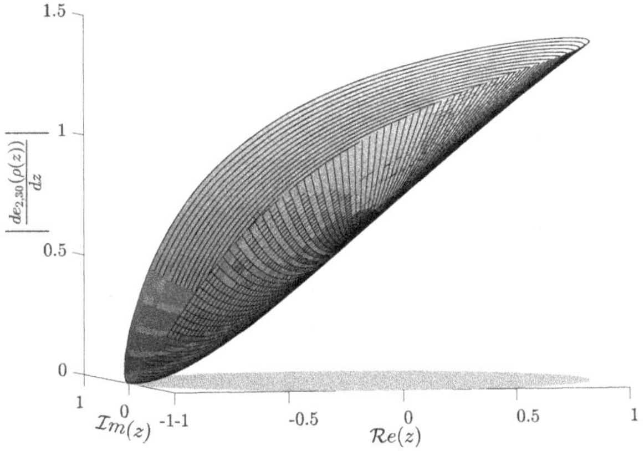

Figure 5. Function $\left|\frac{de_{2,30}(\rho(z))}{dz}\right|$ for $z \in \overline{\mathbb{D}}$.

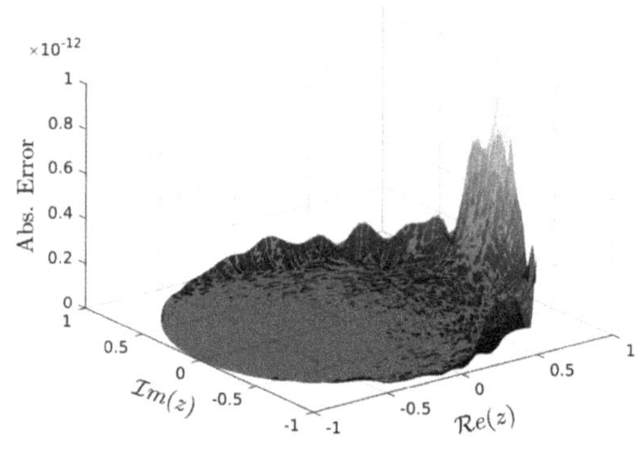

Figure 6. Absolute error of $\frac{de_{2,30}(\rho(z))}{dz}$, $z \in \overline{\mathbb{D}}$.

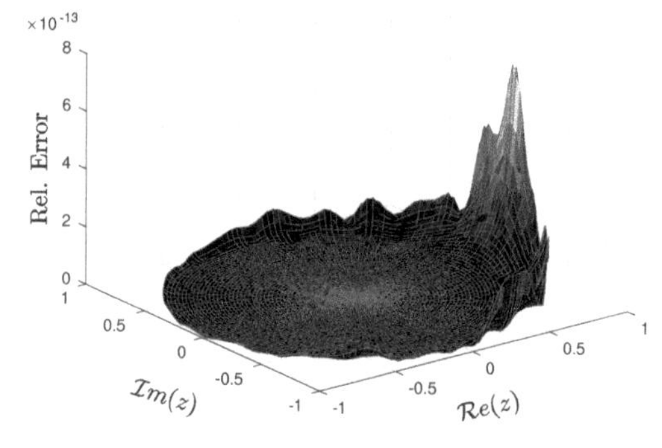

Figure 7. Relative error of $\frac{de_{2,30}(\rho(z))}{dz}$, $z \in \overline{\mathbb{D}}$.

To find the singular numbers, we consider the circle $\{z \in \mathbb{C} : |z| = 1\}$ *(real axis in* ρ*) and a cubic spline interpolation of the approximate Jost function (*$N = 30$*). For the spline interpolation, we use the Matlab routine* csapi. *To locate the zeros of the spline, we use* slmsolve *from the Shape Language Modeling (SLM) toolbox, version 1.14 by John D'Errico [59], available for Matlab2021a. The value* $\rho_1 = -1.000000000000003$ *was obtained with an absolute error of* 3.11×10^{-15}. *Additionally, the argument principle algorithm applied to* $e_{2,30}(\rho(z))$ *in* $\mathbb{D}$ *discarded any eigenvalue of the problem (non-real zero* ρ*).*

The second step of the algorithm from Section 3.4 requires computing the Jost function in the strip $-\varepsilon_0 < \text{Im}(\rho) < 0$ *(*$\varepsilon_0 = \frac{\varepsilon}{2} = 1$*). In this example, we extend* $e_{2,30}(\rho(z))$ *analytically via Padé's approximation. The Padé approximant* $[m,n]_{e_{2,30}(\rho)}$ *was computed in Matlab2021a using the routine* pade.

In Table 3, we computed the maximum absolute and relative errors of the Padé approximant $[1,1]_{e_{2,N}(\rho)}$ *of* $e_{2,N}(\rho)$ *with* $N = 3, 5, 20, 30, 40, 50$ *and* 180 *for* $0 \leq \text{Im}(\rho) < 1$. *These values indicate the possibility of dealing with this Padé approximant when computing the set of scattering data. Additionally, from Table 4, we confirm that this approximant satisfactorily extends the Jost function to a desirable strip in the lower half-plane (the strip is related to the one needed for the calculation of the scattering function* $s_2(\rho)$*).*

Table 3. Maximum absolute and relative errors of the Padé approximant $[1,1]_{e_{2,N}(\rho)}$ with respect to $e_{2,N}(\rho)$ in a strip in the upper half ρ-plane.

N	Max. Abs. Error of $[1,1]_{e_{2,N}(\rho)}(\rho)$ with $0 \leq \mathrm{Im}(\rho) < 1$	N	Max. Rel. Error of $[1,1]_{e_{2,N}(\rho)}(\rho)$ with $0 \leq \mathrm{Im}(\rho) < 1$
3	7.18×10^{-1}	3	2.97
5	5.45×10^{-1}	5	2.14
20	1.26×10^{-6}	20	4.3×10^{-6}
30	4.6×10^{-11}	30	1.58×10^{-10}
40	8.71×10^{-12}	40	3.10×10^{-11}
50	1.93×10^{-11}	50	6.56×10^{-11}
180	1.05×10^{-10}	180	3.56×10^{-10}

Table 4. Maximum absolute and relative errors of the Padé approximant $[1,1]_{e_{2,N}(\rho)}$ with respect to the exact Jost function $e_2(\rho)$ in a strip in the lower half ρ-plane.

N	Max. Abs. Error of $[1,1]_{e_{2,N}(\rho)}(\rho)$ in $-1 < \mathrm{Im}(\rho) \leq 0$	N	Max. Rel. Error of $[1,1]_{e_{2,N}(\rho)}(\rho)$ in $-1 < \mathrm{Im}(\rho) \leq 0$
3	3.07	3	2.99
5	2.66	5	2.16
20	7.99×10^{-6}	20	4.34×10^{-6}
30	2.89×10^{-10}	30	1.59×10^{-10}
40	5.65×10^{-11}	40	3.12×10^{-11}
50	1.23×10^{-10}	50	6.61×10^{-11}
180	6.54×10^{-10}	180	3.59×10^{-10}

To obtain $s_2(\rho)$ *numerically on the strip* $0 < \mathrm{Im}(\rho) < \varepsilon_0 = 1$, *we use the truncated series* $e_{2,30}(\rho)$ *and the Padé approximant* $[1,1]_{e_{2,30}(\rho)}$:

$$s_2(\rho) \approx \frac{[1,1]_{e_{2,30}(\rho)}(-\rho)}{e_{2,30}(\rho)}.$$

The maximum absolute error inside the region $\mathcal{R} = [-30,30] \times [10^{-2}i, (\varepsilon_0 - 10^{-2})i]$ *is* 9.64×10^{-10}.

Remark 14. *The order of the Padé approximant used for the Jost function is not arbitrary. Although the maximum absolute errors inside the region* $\mathcal{R}$ *of other approximations of* $s_2(\rho)$ *using* $[2,2]_{e_{2,30}(\rho)}(9.35 \times 10^{-11})$, $[3,3]_{e_{2,30}(\rho)}$ (1.34×10^{-11}), $[4,4]_{e_{2,30}(\rho)}$ (1.5×10^{-11}) *and* $[7,7]_{e_{2,30}(\rho)}$ (6.37×10^{-12}) *are better in comparison with* $[1,1]_{e_{2,30}(\rho)}$, *we choose* $[1,1]_{e_{2,30}(\rho)}$ *as the most suitable option to avoid the appearance of Froissart doublets. Indeed, the use of the Padé approximants when there is no available information about the smoothness of the function to be approximated is challenging. Some publications propose modified algorithms [60], even using the Toeplitz matrix theory with many numerical implementations in Maple, Wolfram Mathematica (see [61]) or Matlab (see [62]). For the purposes of this paper, it is sufficient to use only the information obtained from the truncated series* $e_N(\rho)$ *and the argument principle algorithm to construct the approximant. Consider the number K of zeros counting multiplicities of the approximate Jost function* $e_N(\rho)$ *(singular numbers being calculated using the argument principle algorithm) located inside* $\mathbb{D}$ *as the degree of the polynomial in the numerator in the Padé approximant. Recalling that, in most cases, an accurate Padé's approximation is obtained on the diagonal approximant types for analytical functions, it is reasonable to choose the Padé approximant as* $[K,K]_{e_N(\rho)}$.

Example 10. *Consider the potential $q_3(x)$ from Example 3. The approximate Jost function $e_{3,N}(\rho)$ is computed in the strip $0 \le \operatorname{Im}(\rho) < \frac{\varepsilon}{2} = 1$ for several values of N. In Table 5, the maximum absolute error of the approximate Jost function is presented.*

Table 5. Maximum absolute error of the Jost function $e_{3,N}(\rho)$ for $N = 2, 5, 20, 30, 50$ and 180.

N	2	5	20	30	50	180
Abs. Error	2.34×10^{-1}	8.65×10^{-3}	6.03×10^{-10}	5.57×10^{-14}	4.72×10^{-14}	4.67×10^{-14}

Similarly to the previous example, a search for real singular numbers was performed; however, none were detected. Subsequently, the argument principle algorithm located a non-real singular number in $\mathbb{D}$, with the value $z_1 \approx -0.386709149322063 - 0.105221869864471i$ ($\rho_1 \approx -0.271752585319512 + 1.083923327338694i$). Its absolute error is 8×10^{-15}. The contour refinement is not a concern, since the performed algorithm from [54] is based on the argument principle algorithm followed by several Newton iterations.

Additionally, the Jost function was extended to the strip $|\operatorname{Im}(\rho)| < \varepsilon_0 = 1$ through the Padé approximant

$$[1,1]_{e_{3,30}(\rho)} = \frac{\rho(4.69 \times 10^{15} + 4.94 \times 10^{13}i) + 1.94 \times 10^{15} - 4.92 \times 10^{15}i}{\rho(4.69 \times 10^{15} + 6.12 \times 10^{14}i) - 6.11 \times 10^{14} + 4.69 \times 10^{15}i}.$$

The corresponding maximum absolute error of $[1,1]_{e_{3,30}(\rho)}(\rho)$ inside the rectangle $\mathcal{R}_1 := [-20,20] \times [0i, (\varepsilon_0 - 10^{-2})i]$ in the complex ρ-plane is 8.43×10^{-11}. Inside $\mathcal{R}_2 := [-20,20] \times [(-\varepsilon_0 + 10^{-2})i, (\varepsilon_0 - 10^{-2})i]$, the maximum absolute error is 8.42×10^{-11}.

Next, an approximate value of the normalization constant corresponding to ρ_1 was computed

$$c_1 \approx \frac{[1,1]_{e_{3,30}(\rho)}(-\rho_1)}{\dot{e}_{3,30}(\rho_1)} i \approx -10.317711295453737 + 12.894194226972697i$$

with an absolute error of 2.8×10^{-9}.

Finally, we calculate the scattering function by

$$s_3(\rho) \approx \frac{[1,1]_{e_{3,30}(\rho)}(-\rho)}{e_{3,30}(\rho)}. \tag{107}$$

The maximum absolute error of the approximation of $s_3(\rho)$ in $\mathcal{R}_1$ is 1×10^{-9} (see Figure 8).

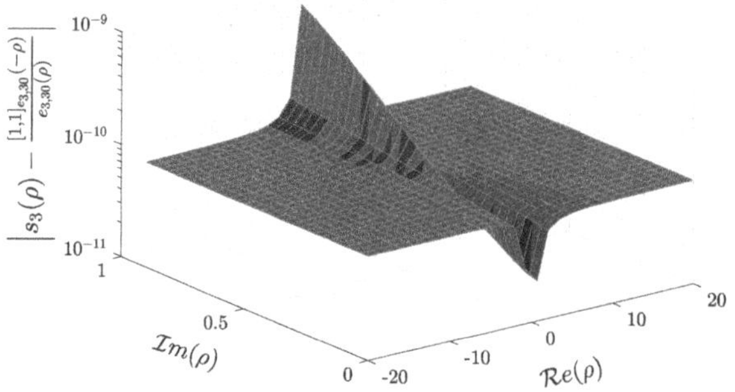

Figure 8. Absolute error of the approximate scattering function $s_3(\rho) \approx \frac{[1,1]_{e_{3,30}(\rho)}(-\rho)}{e_{3,30}(\rho)}$.

Example 11. *Consider the potential $q_1(x)$ from Example 1. Table 6 shows the parameter ε_N for some values of N.*

Table 6. Example 11: parameter ε_N for different values of N.

N	ε_N **for** $x \in [0, 12]$	N	ε_N **for** $x \in [0, 12]$
2	4.43	38	1.12×10^{-5}
3	2.91	48	6.97×10^{-7}
5	1.13	58	5.17×10^{-8}
10	1.32×10^{-1}	88	4.46×10^{-11}
15	1.88×10^{-2}	98	5.12×10^{-12}
20	3.17×10^{-3}	108	9.48×10^{-13}
28	2.27×10^{-4}	178	1.49×10^{-12}

Note that the approximation of the Jost function in this example requires more terms in the series representation than in previous examples. To control the accuracy of the approximation, in addition to the parameter ε_N, one can use the asymptotic relation for the Jost function from [24] (p. 105),

$$e(\rho) = 1 + \frac{\omega(0)}{i\rho} - \frac{q(0)}{(2i\rho)^2} + \frac{\omega^2(0)}{(2i\rho)^2} + o\left(\frac{1}{\rho^2}\right), \ |\rho| \to \infty, \rho \in \mathbb{C}^+ \setminus \{0\}, \tag{108}$$

where $\omega(x) = -\frac{1}{2}\int_x^\infty q(s)ds$. This relation is valid for $q(x)$ with first and second summable derivatives.

Figure 9 depicts the Jost function computed with $N = 98$ and the singular number $\rho_0 \approx 1.784065846059995 + 0.608788673578742i$. Figure 10 shows the fulfillment of the asymptotic relation (108), namely the graph of $\left|e_{1,98}(\rho) - \frac{\omega(0)}{i\rho} + \frac{q(0)}{(2i\rho)^2} - \frac{\omega^2(0)}{(2i\rho)^2}\right|$, which tends to 1 when $|\rho| \to \infty$.

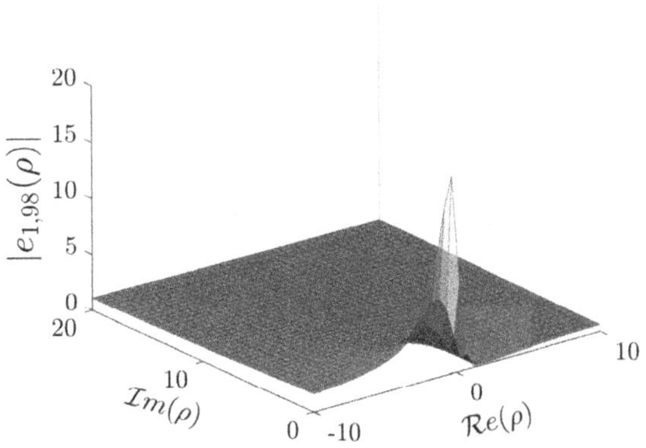

Figure 9. Absolute value of $e_{1,98}(\rho)$ in the upper half ρ-plane. The marked point is the singular number ρ_0.

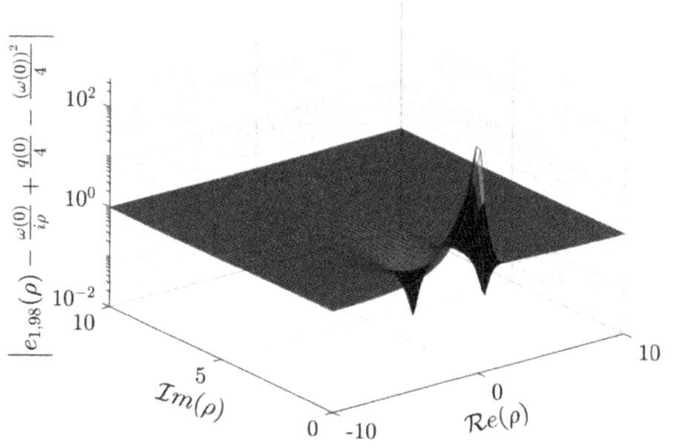

Figure 10. Graph of $\left|e_{1,98}(\rho)-\frac{\omega(0)}{i\rho}+\frac{q(0)}{(2i\rho)^2}-\frac{\omega^2(0)}{(2i\rho)^2}\right|$ tending to 1 when $|\rho|\to\infty$.

The eigenvalue is computed numerically as a zero of the exact Jost function with the aid of Wolfram Mathematica v.12 (Wolfram Research, Inc., Champaign, IL, USA) $\lambda_0 \approx \lambda_0^* := 2.8122672899483 + 2.1722381890043i$. *This "exact" eigenvalue is compared with the approximation* $2.812267289948449 + 2.172238189004328i$ *obtained as the square of the approximate* ρ_0. *The absolute error is* 1.52×10^{-13}.

For the numerical calculation of the analytic extension of $e_{1,98}(\rho)$ *onto the strip* $-\frac{1}{2} < \mathrm{Im}(\rho) < 0$, *it is not possible to consider the Padé approximant* $[1,1]_{e_{1,98}(\rho)}$. *This does not approximate* $e_{1,98}(\rho)$ *accurately even in the upper half-plane of the complex variable* ρ. *Using the Padé approximant* $[7,7]_{e_{1,98}(\rho)}$ *the absolute error was* 0.17.

Instead of using Formula (105) to compute the normalization constant c_0, *Formula (104) is applied to obtain the approximation* $c_0 \approx 16.339391965970112 + 40.670169715260290i$ *with absolute error* 3.05×10^{-12}.

To compute the scattering function $s_1(\rho)$ *on a line parallel to the real axis contained in the strip* $|\mathrm{Im}(\rho)| < \varepsilon_0 = |\mathrm{Im}(\rho_0)| \approx 0.608788673578742i$, *Formula (106) was used. The function* $e_1(\rho)$ *is represented by* (36) *for* ρ *on a line in the lower half* ρ*-plane parallel and sufficiently close to the real axis. Having calculated these series representations for the functions involved in* $s_1(\rho)$, *we compute*

$$s_1(\rho) \approx \frac{e_{1,98}(-\rho)}{e_{1,98}(\rho)}$$

with a maximum absolute error 1.45×10^{-7} *along the line* $\mathcal{L}_{\eta=0.1}$ *(see (77)).*

In this example, we obtain a satisfactory accuracy in the calculations of the scatterin data set using the expression (36) *alone and the derivatives required by (104).*

Example 12 ([44]). *Consider the potential*

$$q_4(x) := Ri\sin(x)e^{-x},$$

with R being a constant (Reynolds number). When $R > 0$ *is sufficiently large, the eigenvalues may exist. For example, for* $R = 10$, *there is one eigenvalue in the box* $\mathcal{B} := 1.6043912_{44}^{58} + 1.7978849i_{67}^{81}$ *[44] (see also [45]).*

The Jost solution is not available in a closed form. In order to check the validity of the numerical calculation of the coefficients $a_n(x)$ *for* $e_{4,N}(\rho.x)$, *we consider the indicator* ε_N *(Table 7).*

Table 7. Example 12: indicator ε_N for different values of N (potential $q_4(x)$ with $R = 10$).

N	ε_N **for** $x \in [0, 12]$	N	ε_N **for** $x \in [0, 12]$
2	4.43	55	3.92×10^{-7}
3	2.91	105	9.77×10^{-11}
5	5.97×10^{-1}	136	1.06×10^{-12}
10	7.85×10^{-2}	137	9.19×10^{-13}
15	1.17×10^{-2}	155	6.11×10^{-13}
20	1.70×10^{-3}	175	7.77×10^{-13}
45	6.95×10^{-6}	200	9.72×10^{-13}

Figure 11 depicts the Jost function computed with $N = 137$ *and the approximation of the singular number* $\rho_1 \approx 1.416695330664399 + 0.634534798062634i$, *with its square belonging to the box* $\mathcal{B}$. *Additionally, Figure 12 shows the fulfillment of the asymptotic relation (108).*

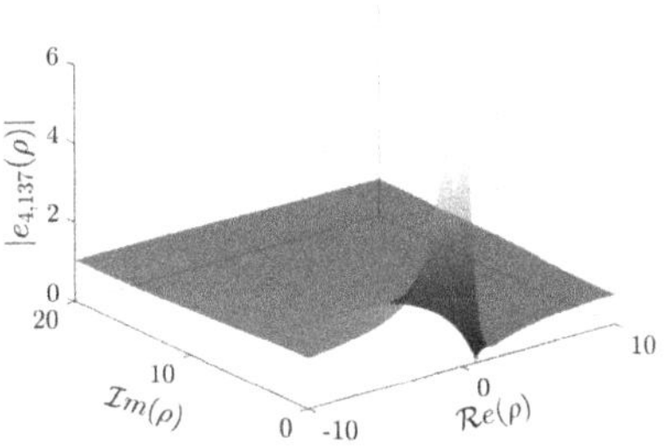

Figure 11. Absolute value of $e_{4,137}(\rho)$ ($R = 10$) in the upper half ρ-plane and the marked point is the approximate singular number ρ_1.

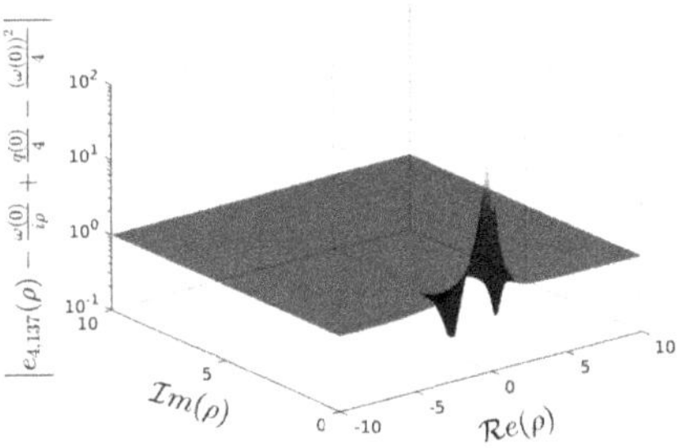

Figure 12. Graph of $\left| e_{4,137}(\rho) - \frac{\omega(0)}{i\rho} + \frac{q(0)}{(2i\rho)^2} - \frac{\omega^2(0)}{(2i\rho)^2} \right|$ tending to 1 when $|\rho| \to \infty$ ($R = 10$).

The normalization constant c_1 *is calculated using (104),*

$$c_1 \approx 0.423317609673475 + 10.608764849282464i.$$

Finally, the scattering function is approximated by $\dfrac{e_{4,137}(-\rho)}{e_{4,137}(\rho)}$.

Now, take $R = 30$ *in the potential* $q_4(x)$. *In this case, two boxes localizing the only two eigenvalues* λ_1 *and* λ_2 *were obtained in [44],*

$$\mathcal{B}_1 := 2.555641614^{35}_{19} + 7.688187018^{19}_{03}i, \quad \mathcal{B}_2 := 6.37465^{91}_{12} + 2.4699^{55}_{46}i.$$

Table 8 provides the values of the indicator ε_N for several values of N.

Table 8. Example 12: indicator ε_N for different values of N ($R = 30$).

N	ε_N **for** $x \in [0, 12]$	N	ε_N **for** $x \in [0, 12]$
5	1.09×10^{1}	105	1.27×10^{-7}
10	2.86	155	2.30×10^{-10}
15	7.71×10^{-1}	175	2.78×10^{-11}
20	1.17×10^{-1}	200	5.70×10^{-12}
45	1.66×10^{-3}	220	6.46×10^{-12}
55	1.65×10^{-4}	230	7.04×10^{-12}

Approximate eigenvalues computed from $e_{4,N}(\rho)$ for different values of N, are presented in Tables 9 and 10.

Table 9. Approximate eigenvalue $\tilde{\lambda}_1$ computed using different values of N in $e_{4,N}(\rho)$ ($R = 30$).

N	$\tilde{\lambda}_1$
35	$2.555647790300414 + 7.688132784089897i$
65	$2.555641614022092 + 7.688187017680658i$
105	$2.555641614273991 + 7.688187018110548i$
200	$2.555641614273991 + 7.688187018110548i$
230	$2.555641614273991 + 7.688187018110548i$

Table 10. Approximate eigenvalue $\tilde{\lambda}_2$ computed using different values of N in $e_{4,N}(\rho)$ ($R = 30$).

N	$\tilde{\lambda}_2$
35	$6.368733224187178 + 2.460948309337657i$
65	$6.374657558248066 + 2.460950123973226i$
105	$6.374654410969357 + 2.460950093296220i$
200	$6.374654410861196 + 2.460950093077938i$
230	$6.374654410861196 + 2.460950093077938i$

Note that $\tilde{\lambda}_1 \in \mathcal{B}_1$ and $\tilde{\lambda}_2 \in \mathcal{B}_2$ for $N = 200$. Finally, the normalization constants are calculated using $e_{4,200}(\rho)$ in (104),

$$c_1 \approx 1.669128547357084 \times 10^2 - 1.694940279771396 \times 10^2 i$$
$$c_2 \approx -54.578951306154920 + 45.276710620944780i.$$

Although, in this example, more powers for the series representation of the Jost function were used, the method proved to be applicable to obtaining the scattering data set without any additional informatio. The good accuracy achieved is confirmed by the ability t use the scattering data obtained as input data to solve the inverse scattering problem to recover the potential $q_4(x)$ with $R = 30$ below in Example 22.

5.2. Inverse Problem

In the present section, we discuss the accuracy, convergence and stability of the proposed method for solving the inverse scattering problem.

Remark 15. *By $q_{k,M}(x)$, we denote the approximation of the potential $q_k(x)$ ($k = 1,2,3,4,5$) obtained by solving the truncated system (100) with the sum up to M, i.e., with $M+1$ equations.*

5.2.1. Convergence and Accuracy

Example 13. *Consider the scattering data calculated in Example 3:*

$$J = \left\{ \left\{ \rho_1 = i\tanh(1+i), m_1 = 1, c_1 = \frac{2(i + i\tanh(1+i))\tanh(1+i)}{i - i\tanh(1+i)} \right\}, \right.$$

$$\left. s_3(\rho) = \frac{(\rho + i\tanh(1+i))(\rho+i)}{(\rho - i\tanh(1+i))(\rho-i)} \right\}$$

where $s_3(\rho)$ is an S-type function in the strip $0 \le |\mathrm{Im}(\rho)| < \varepsilon_0 = 1$.

We shall recover the potential $q_3(x) = -2\,\mathrm{sech}^2(x-1-i)$. The system (100) of linear algebraic equations for this example is obtained in a closed form (see Example 7). For a different number of equations in the truncated system, we obtain a solution symbolically by using the Matlab routine solve. *The potential $q_3(x)$ is recovered from (38). Figure 13 presents the recovered potential in each case.*

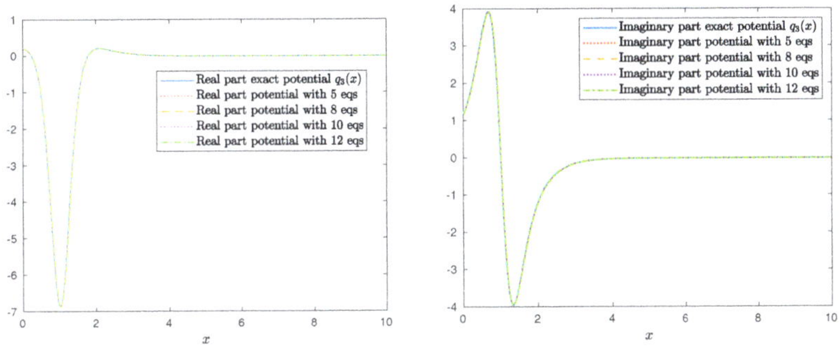

Figure 13. Real (**left**) and imaginary (**right**) part of the recovered potential $q_{3,M}(x)$ for $M = 4,7,9$ and 11.

The corresponding absolute and relative errors are presented in Figure 14 and Figure 15, respectively. Note that a high accuracy is attained even in the case of a very reduced number of equations in the truncated system. Moreover, a very fast convergence of the method can be appreciated.

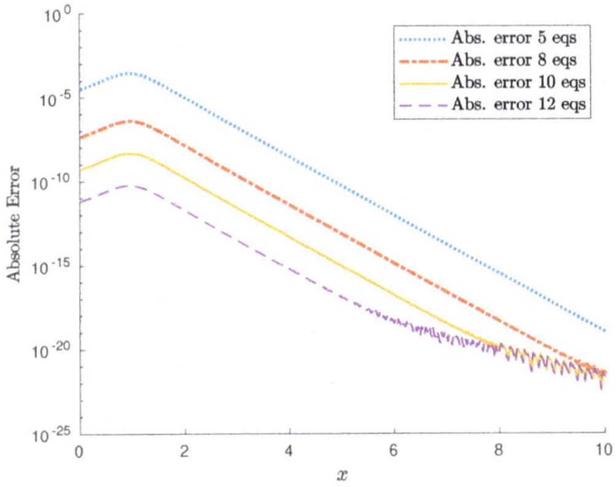

Figure 14. Absolute error of the recovered potential $q_{3,M}(x)$ with $M = 4,7,9$ and 11.

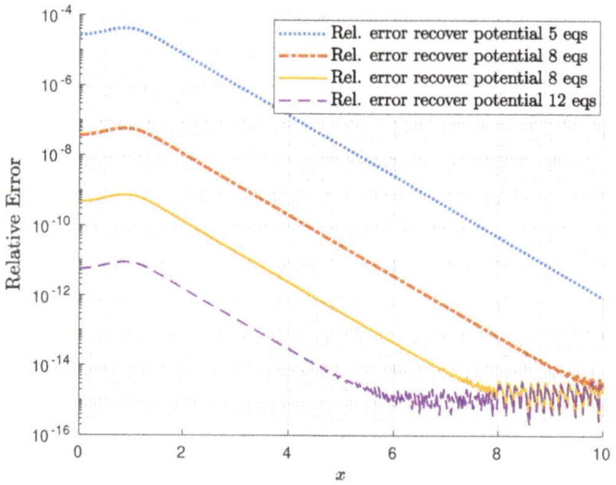

Figure 15. Relative error of the recovered potential $q_{3,M}(x)$ with $M = 4, 7, 9$ and 11.

Example 14. *Consider the scattering data $J = \{s_2(\rho)\}$ from Example 2. As was shown above (Example 6), the system (100) for this example can be written explicitly. Again, when solving the corresponding truncated system for different values of M we observe a fast convergence and remarkable accuracy even for small values of M (see Table 11 and Figure 16).*

Table 11. Maximum absolute error of the approximate potential $q_{2,M}(x)$ for some values of M.

M **in (100)**	**0**	**1**	**2**	**3**	**6**	**8**
$\max_{x\in[0,6]} \|q_2(x) - q_{2,M}(x)\|$	1.05	1.13×10^{-1}	1.25×10^{-2}	1.4×10^{-3}	1.91×10^{-6}	2.35×10^{-8}

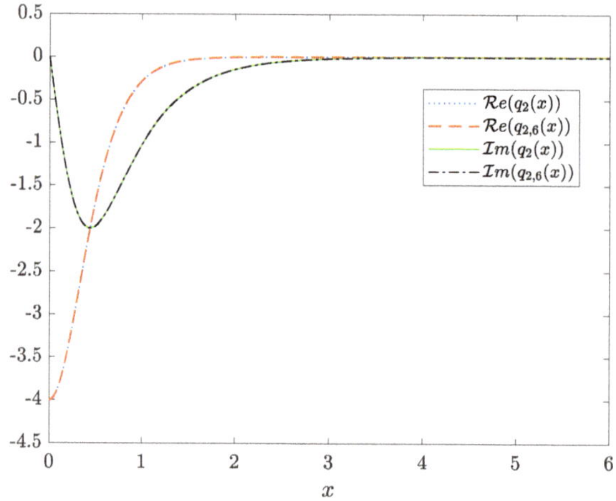

Figure 16. Exact and computed potential $q_{2,6}(x)$.

Example 15. *Consider the closed form of the scattering function $s_1(\rho)$ from Example 1. We compute functions $f_{m,c}(x)$ and $A_{mn,c}(x)$ using the first option from Remark 11. Some poles and residues are given in Table 12 (computed with the aid of the package* Numerical Calculus *of Mathematica v.12).*

Table 12. Poles and residues in the upper half-plane of the function $s_1(\rho) - 1$.

Poles	Residues
$0.5i$	5
$1.784065846059995 + 0.608788673578742i$	$40.670169841396 - 16.339391035537i$
$1i$	$-25i$
$1.5i$	$-\frac{125}{3}$
$2i$	$\frac{625}{18}i$
$2.5i$	$\frac{625}{36}$
$3i$	$-\frac{625}{108}i$
$3.5i$	$-\frac{1743}{1265}$
$4i$	$\frac{249}{1012}i$
$6i$	$\frac{7}{267{,}683}i$
$7i$	$-\frac{1}{10{,}857{,}221}i$

Note that the absolute value of the residues decreases considerably as the poles move away from the origin on the imaginary axis. This allows us to use a small number of poles for the calculation of the functions $f_{m,c}(x)$ and $A_{mn,c}(x)$.

The convergence of the method in this case results to be slower; see Figure 17, although a satisfactory accuracy is attained for $M = 9$.

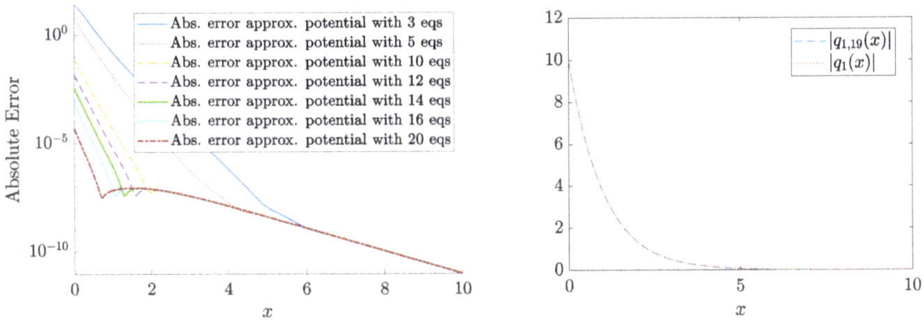

Figure 17. Absolute error for different values of M in the truncated system (100) (**left**) and the absolute value of the recovered potential $q_{1,19}(x)$ computed with 20 equations (**right**) for $x \in [0, 10]$.

5.2.2. Stability of the System

Since the stability of the method was proved in Theorem 4, we are able to work efficiently with noisy scattering data. First, we consider the natural noise arising from the numerical implementations of the last two procedures in Remark 11, i.e., calculation of the approximate matrix in (100) from the scattering function $s(\rho)$ given in a closed form. Another situation considered in this subsection is the recovery of the potential from a uniformly noisy scattering function.

Remark 16. *Henceforth, denote by $\tilde{f}_m(x)$, $\tilde{f}_{m,c}(x)$, $\tilde{A}_{mn}(x)$ and $\tilde{A}_{mn,c}(x)$ the numerical approximation of $f_m(x)$, $f_{m,c}(x)$, $A_{mn}(x)$ and $A_{mn,c}(x)$.*

Remark 17. *In the last step of the algorithm from Section 4.4, for recovering q with the aid of (38), the coefficient a_0 needs to be differentiated twice. This was performed by interpolating $a_0(x)$ with a quintic spline through the Matlab routine* spapi *and a posterior differentiation with the Matlab command* fnder.

Example 16. *Let us consider the scattering data from Example 2. The recovery of the potential $q_2(x)$ from the exact scattering function $s_2(\rho)$, obtained by using approximate functions $\tilde{f}_m(x)$ and $\tilde{A}_{mn}(x)$ in the truncated system (100), is presented. The computation of functions $f_m(x)$ and $A_{mn}(x)$ requires numerical integration along the line $\mathcal{L}_{\eta=0.5}$ (see (77)). For this purpose, the last two procedures in Remark 11 were applied.*

Method 1. The second option in Remark 11 is implemented. With the scattering function (91) at points $\rho = \sigma + 0.5i$ and $\sigma = -(k+1/2)h$ for $k = 0,1,\ldots,N(x)$, where $N(x) = 55000/x$ and $h = 0.145454545$, the calculation of the Fourier transforms in (98) is carried out. In Table 13, the maximum absolute error of $\tilde{f}_m(x)$ is presented for 4 values of the parameter m.

Table 13. Example 16: maximum absolute error of $\tilde{f}_{m,c}(x)$ calculated with the second procedure in Remark 11.

m	0	1	2	3
$\max_{x\in(0,20)} \lvert f_m(x) - \tilde{f}_m(x)\rvert$	1.102×10^{-9}	7.995×10^{-10}	7.988×10^{-10}	8.757×10^{-10}

Now, we compute $\tilde{A}_{mn,c}(x)$ using the same numerical integration method with parameters $N(x) = 5500/x$ and $h = 0.127272727$.

Table 14 shows the maximum absolute error of $\tilde{A}_{mn}(x)$ for parameters $m, n = 0,1,2,3$.

Table 14. Example 16: maximum absolute error of $\tilde{A}_{mn}(x)$ calculated with the second procedure in Remark 11.

m	$\max_{x\in(0,20)} \lvert A_{mn}(x) - \tilde{A}_{mn}(x)\rvert$			
	$n=0$	$n=1$	$n=2$	$n=3$
0	4.721×10^{-10}	4.712×10^{-10}	4.711×10^{-10}	4.714×10^{-10}
1	4.712×10^{-10}	4.711×10^{-10}	4.714×10^{-10}	4.712×10^{-10}
2	4.711×10^{-10}	4.714×10^{-10}	4.712×10^{-10}	4.699×10^{-10}
3	4.714×10^{-10}	4.712×10^{-10}	4.699×10^{-10}	4.678×10^{-10}

The system (100) constructed with $\tilde{f}_m(x)$ and $\tilde{A}_{mn}(x)$ is solved numerically in Matlab for several values of M. Maximum absolute and relative errors of the approximation of the potential $q_{2,M}$ are shown in Table 15.

Table 15. Example 16: maximum absolute and relative errors of the approximation of the potential $q_{2,M}(x)$ by the recovered potential for some values of M in (100).

M in (100)	0	1	2	3	5	7
Abs. Error of $q_{2,M}(x)$	9.3×10^{-1}	1×10^{-1}	1.1×10^{-2}	1.21×10^{-3}	1.06×10^{-4}	1.14×10^{-4}
Rel. Error of $q_{2,M}(x)$	1.6	1.74×10^{-2}	1.93×10^{-3}	2.14×10^{-4}	4.15×10^{-6}	2.04×10^{-6}

Figure 18 presents the absolute value of the recovered q_2 potential from 4 equations in (100).

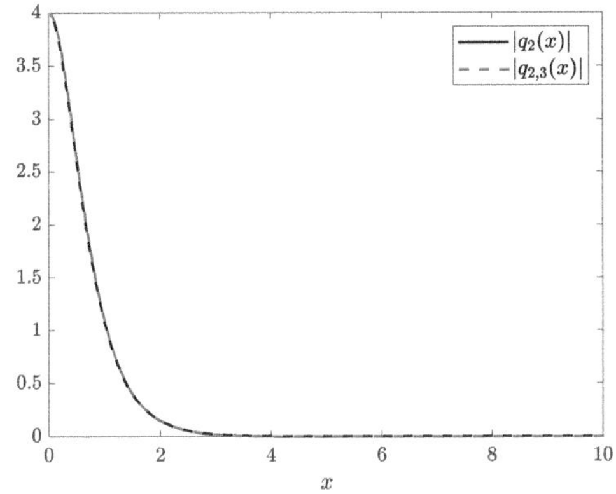

Figure 18. Recovered potential $q_{2,3}(x)$ by Method 1 in Example 16.

Method 2. Now, we compute the approximate functions $\tilde{f}_m(x)$ (see Table 16) and $\tilde{A}_{mn}(x)$ (see Table 17) following the third procedure in Remark 11.

Table 16. Example 16: maximum absolute error of $\tilde{f}_m(x)$ after applying the third procedure in Remark 11.

m	0	1	2	3
$\max_{x\in(0,20)} \left\|f_m(x) - \tilde{f}_m(x)\right\|$	6.05×10^{-5}	2.02×10^{-5}	6.73×10^{-6}	2.24×10^{-6}

Table 17. Example 16: maximum absolute error of $\tilde{A}_{mn}(x)$ after applying the third procedure in Remark 11.

m	$\max_{x\in(0,20)} \left\|A_{mn}(x) - \tilde{A}_{mn}(x)\right\|$			
	$n = 0$	$n = 1$	$n = 2$	$n = 3$
0	1.27×10^{-13}	4.26×10^{-14}	1.42×10^{-14}	4.74×10^{-15}
1	1.22×10^{-9}	1.42×10^{-14}	4.74×10^{-15}	1.58×10^{-15}
2	4.07×10^{-10}	1.36×10^{-10}	1.58×10^{-15}	5.29×10^{-16}
3	1.36×10^{-10}	4.52×10^{-11}	1.5×10^{-11}	1.68×10^{-12}

In Table 18, the absolute error of the recovered potential for some values of M in (100) is presented.

Table 18. Example 16: maximum absolute error of recovered potential $q_{2,M}(x)$ using the third option in Remark 11.

M in (100)	0	1	2	3	5	7
$\max_{x\in(0,20)} \left\|q_2(x) - q_{2,M}(x)\right\|$	8.25×10^{-1}	2.34×10^{-1}	5.25×10^{-2}	9.71×10^{-3}	5.71×10^{-3}	2.23×10^{-3}
Rel. Error of $q_{2,M}(x)$	1.24	1.21×10^{-1}	5.2×10^{-1}	8×10^{-2}	6.19×10^{-2}	7.25×10^{-3}

Both methods (procedures 2 and 3 from Remark 11) illustrated in the above example have proven to be suitable for calculating the functions f_m and A_{mn} from a table of values for the $s_2(\rho)$. Nevertheless, it is worth mentioning that although the first method (procedure 2) produced slightly more accurate results, this approach might be sensitive to the choice of the $N(x)$ and h parameters, whereas the second method (procedure 3) only requires the implementation of *trapz*, the Matlab integration routine on a dense set of points defined in the interval $(0, 2\pi)$. Hence, for the purposes of this paper, it is sufficient to consider procedure 3 from Remark 11 in the following examples, so as to obtain satisfactory approximations of f_m and A_{mn}.

As expected from the results of Example 14 for this potential, the numerical method for recovering the potential $q_2(x)$ converges very fast. Indeed, an acceptable approximation of $q_2(x)$ is achieved with only four equations in this case, where an inexact matrix in the linear system (100) is considered. In fact, the difference between the approximate and the exact potential presented in Figure 18 is indistinguishable.

In the following examples, a noisy scattering function with a uniformly distributed noise $\varepsilon(\rho)$ added to the *rand* routine of Matlab is considered.

Example 17. *Consider the scattering function $s_2(\rho)$ and denote the noisy scattering function by $\hat{s}_2(\rho) := s_2(\rho) + \varepsilon(\rho)$. Here, $\varepsilon(\rho)$ is $\pm 5\%$ uniformly distributed complex-valued noise (the percentage of the noise is applied pointwise to the modulus and argument of the value of $s_2(\rho)$). The maximum absolute error of $\hat{s}_2$ on the line $\mathcal{L}_{\eta=0.5}$ is 2.46×10^{-1}. The potential was recovered using five equations with a maximum absolute error of 5.2×10^{-1}. The real and imaginary parts of the potential and the absolute error of its recovery are shown in Figure 19.*

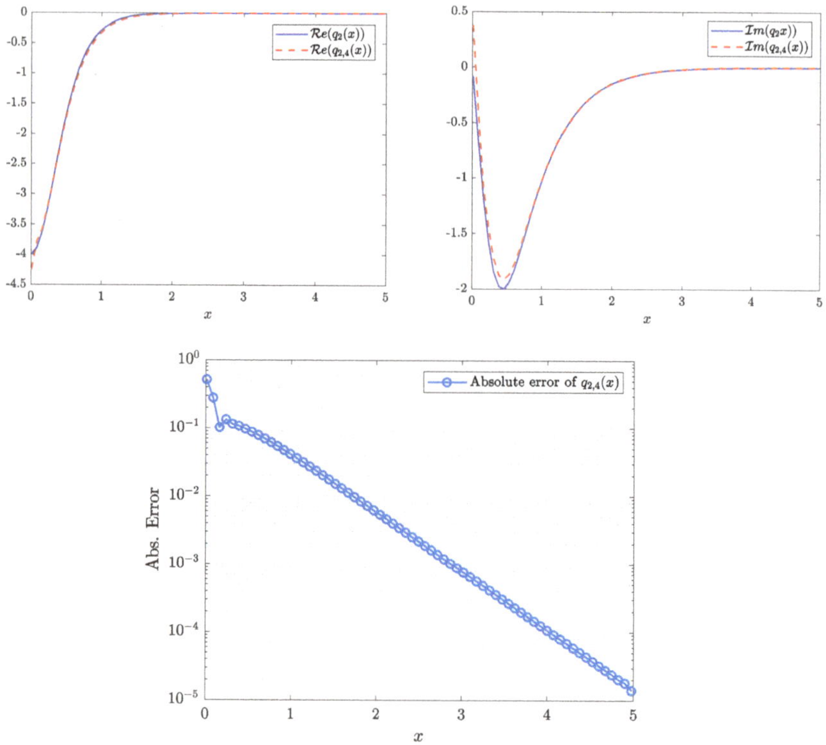

Figure 19. Example 17: figures on the top part show the recovered potential $q_{2,4}(x)$, and the bottom figure shows the absolute error of the recovered potential.

Despite the noise that $\hat{s}_2(\rho)$ produces in the matrix of the system (100), the method recovers the shape of the potential q_2 with reasonable accuracy.

Example 18. *Consider the scattering function $s_3(\rho)$ and define $\hat{s}_3(\rho) := s_3(\rho) + \varepsilon(\rho)$ where $\varepsilon(\rho)$ is a $\pm 10\%$ uniformly distributed complex-valued noise (considered as in the previous example).*

The maximum absolute error of $\hat{s}_3$ on the line $\mathcal{L}_{\eta=0.5}$ is 1.75. The potential was recovered using eight equations with a maximum absolute error of 8.6×10^{-1}. The real and imaginary parts of the potential as well as the absolute error of its recovery are shown in Figure 20.

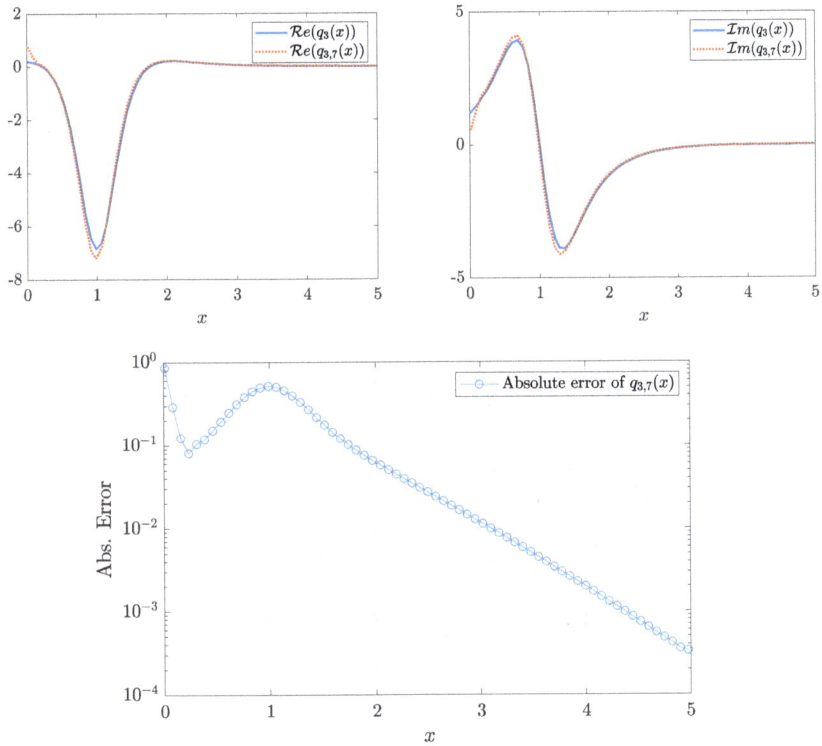

Figure 20. Example 18: figures on the top part show the potential $q_{3,7}(x)$ recovered with eight equations, and the bottom figure presents the absolute error of the recovered potential.

Although, in this case, the absolute error of $\hat{s}_3(\rho)$ is larger, the shape of the recovered potential is still quite close to that of the exact one.

5.2.3. In-Out

In this subsection, we consider the results obtained in Section 5.1 as input data for the inverse problem.

Example 19. *We use the approximate scattering function $s_3(\rho)$ from Example 10 calculated by (107). Particularly, the form in which it is given allows for us to approximate functions $\tilde{f}_{m,c}(x)$ and $\tilde{A}_{mn,c}(x)$ with the aid of the numerical calculus of residues, i.e., the first procedure in Remark 11 (see Tables 19 and 20).*

Table 19. Example 19: maximum absolute error of the approximation of the function $f_{m,c}(x)$ using calculus of residues.

m	0	1	2	3
$\max_{x\in(0,15)} \lvert f_{m,c}(x) - \tilde{f}_{m,c}(x) \rvert$	4.795×10^{-10}	1.911×10^{-10}	7.494×10^{-11}	2.861×10^{-11}

Table 20. Example 19: maximum absolute error of the approximation of the function $A_{mn,c}(x)$ using calculus of residues.

m	$\max_{x\in(0,15)} \left\|A_{mn,c}(x) - \tilde{A}_{mn,c}(x)\right\|$			
	$n = 0$	$n = 1$	$n = 2$	$n = 3$
0	2.995×10^{-10}	1.206×10^{-10}	4.835×10^{-11}	1.895×10^{-11}
1	1.206×10^{-10}	4.835×10^{-11}	1.895×10^{-11}	7.238×10^{-12}
2	4.835×10^{-11}	1.895×10^{-11}	7.238×10^{-12}	2.73×10^{-12}
3	1.895×10^{-11}	7.238×10^{-12}	2.73×10^{-12}	1.057×10^{-12}

The potential $q_3(x)$ was recovered with an absolute error of 1.8×10^{-5} *in the interval* $(0, 15)$ *using* 8 *equations.*

Example 20. *Consider the approximate scattering function $s_1(\rho)$ from Example 11. The coefficient $a_0(x)$ was recovered using 14 equations with a maximum absolute error of* 4.29×10^{-3}, *from which the potential was recovered with a maximum absolute error of* 0.23, *Figure 21.*

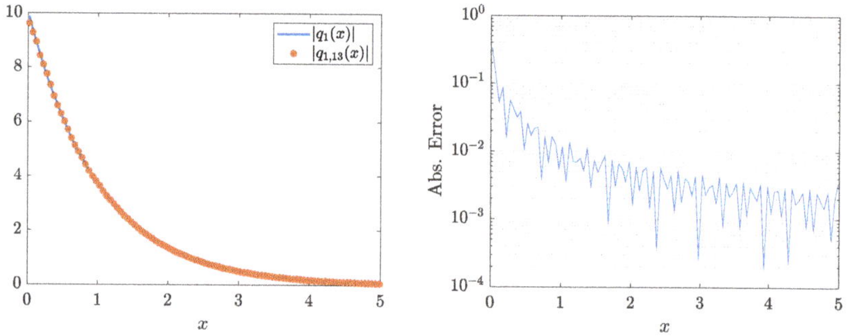

Figure 21. The (**left**) figure shows the recovered potential $q_{1,13}(x)$, and the (**right**) figure presents the absolute error of the recovered potential.

Example 21. *Consider the approximate scattering data obtained in Example 9. The approximate functions $\tilde{f}_m(x)$ (see Table 21) and $\tilde{A}_{mn}(x)$ (see Table 22) were obtained accurately enough to recover the potential (see Table 23).*

Table 21. Example 21: maximum absolute error of the approximation of the function $f_m(x)$.

m	0	1	2	3
$\max_{x\in(0,100)} \left\|f_m(x) - \tilde{f}_m(x)\right\|$	8.92×10^{-3}	2.97×10^{-3}	9.91×10^{-4}	3.3×10^{-4}

Table 22. Example 21: maximum absolute error of the approximation of the function $A_{mn}(x)$.

m	$\max_{x\in(0,15)} \left\|A_{mn}(x) - \tilde{A}_{mn}(x)\right\|$			
	$n = 0$	$n = 1$	$n = 2$	$n = 3$
0	1.5×10^{-7}	4.8×10^{-8}	1.6×10^{-8}	5.4×10^{-9}
1	2.7×10^{-6}	1.6×10^{-8}	5.4×10^{-9}	1.8×10^{-9}
2	9×10^{-6}	2.9×10^{-6}	1.8×10^{-9}	6×10^{-10}
3	3×10^{-6}	9.9×10^{-7}	3.3×10^{-7}	2×10^{-10}

Table 23. Maximum absolute error of the approximation of the potential by the recovered potential $q_{2,M}(x)$.

M in (100)	$\max_{x\in(0,20)} \lvert q_2(x) - q_{2_M}(x)\rvert$
0	1.09
1	1.63×10^{-1}
2	6.44×10^{-2}
3	5.41×10^{-2}
5	5.3×10^{-2}
7	5.3×10^{-2}

Figure 22 illustrates the stability and convergence of the method with the absolute error stabilized at 5.3×10^{-2}.

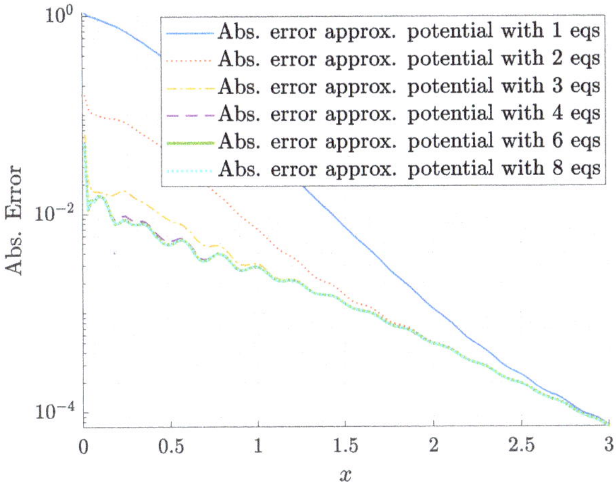

Figure 22. Absolute error of the approximation of the potential by the recovered potential $q_{2,M}(x)$ using 1, 2, 3, 4, 6 and 8 equations.

Example 22. *Consider the potential* $q_4(x) = 30i\sin(x)\exp(-x)$ *introduced in Example 12. Using the results of the solution of the direct scattering problem from Example 12, we recover* $q_4(x)$ *using 20 equations with a maximum absolute error of* 8.67×10^{-1} *(Figure 23).*

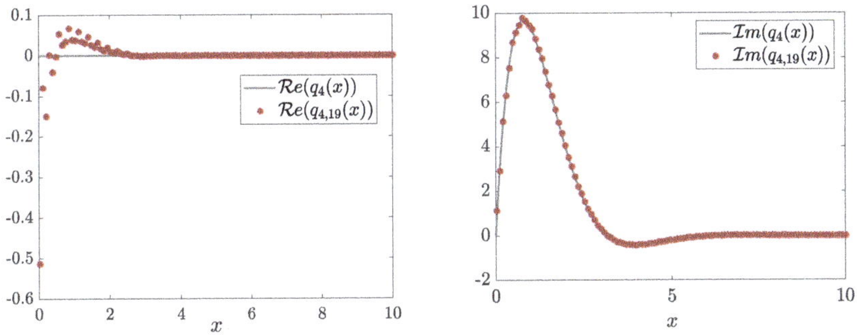

Figure 23. Real (**left**) and imaginary (**right**) parts of the exact potential and the recovered $q_{4,19}(x)$ $(R = 30)$.

It is worth mentioning that the coefficient $a_0(x)$ is recovered with an absolute error of 2.28×10^{-2} *(Figure 24). The error is calculated and compared with the solution of the Cauchy problem*

$$a_0''(x) - a_0'(x) = q_4(x)(a_0(x) + 1), \qquad (109)$$
$$a_0(b) = 0,\ a_0'(b) = -\frac{1}{2},$$

for a sufficiently large value of $b > 0$*, obtained using* ode45 *routine of Matlab2021a.*

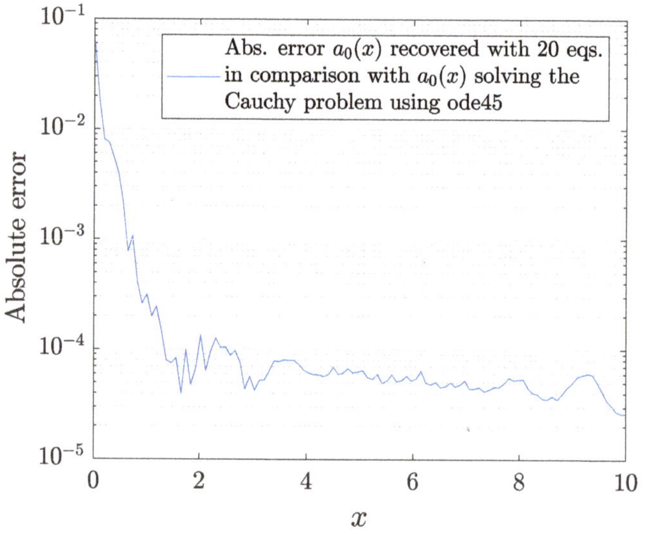

Figure 24. Example 22: absolute error of the recovered coefficient $a_0(x)$ with 20 equations.

This is a case where closed formulas for the scattering data set are unavailable. Therefore, the In–Out procedure confirms a satisfactory accuracy in the solution of both the direct and inverse scattering problems.

Example 23. *Consider the singular potential*

$$q_5(x) = \frac{\exp(-2.5x)}{\left(x - \frac{\pi}{2}\right)^{1/3}}.$$

In Table 24, we present the parameter ε_N *for different values of N in (103).*

Table 24. Example 23: indicator ε_N for different values of N.

N	ε_N **for** $x \in [0, 4]$	N	ε_N **for** $x \in [0, 4]$
2	1.57×10^{-1}	35	1.10×10^{-3}
3	5.24×10^{-2}	40	1.03×10^{-3}
5	3.97×10^{-3}	45	9.57×10^{-4}
10	2.25×10^{-3}	150	4.84×10^{-4}
15	1.77×10^{-3}	250	3.60×10^{-4}
25	1.33×10^{-3}	450	2.56×10^{-4}

Using data from Table 24, we computed the scattering data with $N = 45$*. No eigenvalue was detected, so the scattering data set consists of the scattering function approximated by the expression*

$$s_5(\rho) \approx \frac{e_{5,45}(-\rho)}{e_{5,45}(\rho)}, \rho \in \mathbb{R}.$$

Using this scattering data set to solve the inverse problem, we obtained the coefficient $a_0(x)$ *as shown in Figure 25. The maximum absolute error resulted in* 1.9×10^{-4}*.*

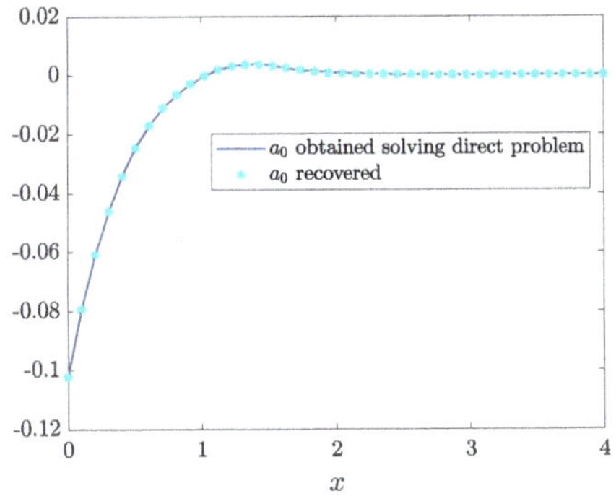

Figure 25. Example 23: coefficient $a_0(x)$ with four equations.

The potential is recovered as shown in Figure 26. The corresponding absolute error is presented in Figure 27. Indeed, the maximum absolute error is 9.82×10^{-2}*.*

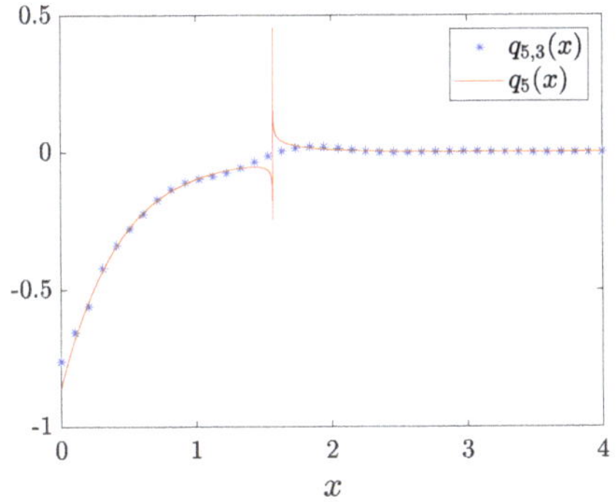

Figure 26. Recovered potential $q_{5,3}(x)$.

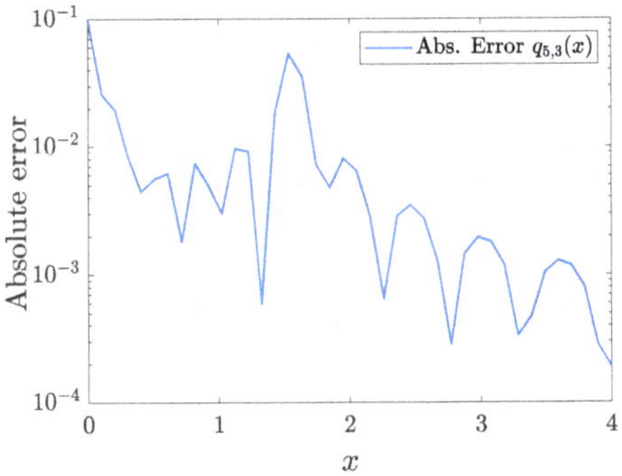

Figure 27. Absolute error of $q_{5,3}(x)$.

This example shows the applicability of the proposed algorithms to both the solution of the direct and inverse scattering problems in the case of non-smooth potentials.

6. Conclusions

An approach to solving the direct and inverse scattering problems on the half-line for the one-dimensional Schrödinger equation with an exponentially decreasing complex-valued potential is developed. It is based on a series representation of the Jost solution from [25], which is shown in the present work to remain valid in a non-selfadjoining case.

When solving the direct problem, this representation is used to calculate the scattering data set through a simple and efficient procedure, which includes a proposed algorithm for computing normalization polynomials (which are part of the scattering data set) by solving a finite system of linear algebraic equations for its coefficients.

When solving the inverse problem, the use of the series representation combined with the Gel'fand–Levitan equation reduces the problem to a system of linear algebraic equations for the series coefficients, and the knowledge of the first coefficient is sufficient to recover the potential.

The numerical results illustrate the remarkable accuracy of the proposed algorithms in solving both the direct and inverse scattering problems.

Author Contributions: Conceptualization, formal analysis, methodology, funding acquisition, investigation, project administration, software, supervision, writing—review and editing, V.V.K.; formal analysis, investigation, software, writing—original draft, writing—review and editing, L.E.M.-L. All authors have read and agreed to the published version of the manuscript.

Funding: Research was supported by CONAHCYT, Mexico via the project 284470.

Data Availability Statement: The data that support the findings of this study are available upon reasonable request.

Conflicts of Interest: This work does not have any conflict of interest.

Appendix A. Proofs of Auxiliary Equalities for Lemma 1

Let us prove Equation (57). Consider

$$\begin{aligned}
&\frac{\partial}{\partial z}\left(e(\rho,x)\sum_{j=0}^{k-1}(-1)^j C_j^{k-1}\frac{k!}{(k-j)!}x^{k-j}(z+1)^{j-2k}\right)\\
&-e^{i\rho x}\sum_{n=0}^{\infty}a_n(x)P_n^{(0,1-n)}\sum_{j=0}^{k-1}(-1)^j C_j^{k-1}\frac{k!}{(k-j)!}x^{k-j}(z+1)^{j-2k}\\
&=e_z(\rho,x)\sum_{j=0}^{k-1}(-1)^j C_j^{k-1}\frac{k!}{(k-j)!}x^{k-j}(z+1)^{j-2k}\\
&+e(\rho,x)\sum_{j=0}^{k-1}(-1)^j C_j^{k-1}\frac{k!}{(k-j)!}x^{k-j}(-2k+j)(z+1)^{j-2k-1}\\
&-e^{i\rho x}\sum_{n=0}^{\infty}a_n(x)P_n^{(0,1-n)}\sum_{j=0}^{k-1}(-1)^j C_j^{k-1}\frac{k!}{(k-j)!}x^{k-j}(z+1)^{j-2k}.
\end{aligned}$$

Using the representation (52) for the Jost solution, the last expression can be written as follows

$$\begin{aligned}
&\frac{xe(\rho,x)}{(z+1)^2}\sum_{j=0}^{k-1}(-1)^j C_j^{k-1}\frac{k!}{(k-j)!}x^{k-j}(z+1)^{j-2k}\\
&+e(\rho,x)\sum_{j=0}^{k-1}(-1)^j C_j^{k-1}\frac{k!}{(k-j)!}x^{k-j}(j-2k)(z+1)^{j-2k-1}
\end{aligned}$$

$$=e(\rho,x)\sum_{j=0}^{k-1}(-1)^j C_j^{k-1}\frac{k!}{(k-j)!}x^{k-j+1}(z+1)^{j-2k-2}\Big(1+(j-2k)\Big(x^{-1}(z+1)\Big)\Big)$$
$$=e(\rho,x)\sum_{j=0}^{k-1}(-1)^j C_j^{k}\frac{(k+1)!}{(k+1-j)!}x^{k-j+1}(z+1)^{j-2k-2}$$
$$+e(\rho,x)(-1)^k C_k^k\frac{(k+1)!}{(k+1-k)!}x(z+1)^{-k-2}$$
$$=e(\rho,x)\sum_{j=0}^{k}(-1)^j C_j^{k}\frac{(k+1)!}{(k+1-j)!}x^{k-j+1}(z+1)^{j-2k-2}.$$

Now, let us prove equality (58). Consider

$$\frac{\partial}{\partial z}\Bigg(e^{i\rho x}\sum_{n=0}^{\infty}a_n(x)\sum_{j=2}^{k+1}F(-n,j,1,z+1)$$
$$\sum_{m=j-1}^{k}(-1)^{1-m+j}C_{m-j+1}^{k-1}\frac{k!}{(k-m)!}x^{k-m}(z+1)^{m-2k+1}\Bigg)$$
$$+e^{i\rho x}\sum_{n=0}^{\infty}a_n(x)F(-n,2,1,z+1)\sum_{j=0}^{k-1}(-1)^j C_j^{k-1}\frac{k!}{(k-j)!}x^{k-j}(z+1)^{j-2k}$$

$$=\frac{xe^{i\rho x}}{(z+1)^2}\sum_{n=0}^{\infty}a_n(x)\sum_{j=2}^{k+1}F(-n,j,1,z+1)\sum_{m=j-1}^{k}(-1)^{j-m+1}C_{m-j+1}^{k-1}\frac{k!}{(k-m)!}x^{k-m}(z+1)^{m-2k+1}$$
$$+e^{i\rho x}\sum_{n=0}^{\infty}a_n(x)\Bigg(\sum_{j=2}^{k+1}F_z(-n,j,1,z+1)\sum_{m=j-1}^{k}(-1)^{j-m+1}C_{m-j+1}^{k-1}\frac{k!}{(k-m)!}x^{k-m}(z+1)^{m-2k+1}$$
$$+\sum_{j=2}^{k+1}F(-n,j,1,z+1)\sum_{m=j-1}^{k}(-1)^{j-m+1}C_{m-j+1}^{k-1}\frac{k!}{(k-m)!}x^{k-m}(m-2k+1)(z+1)^{m-2k}\Bigg)$$
$$+e^{i\rho x}\sum_{n=0}^{\infty}a_n(x)F(-n,2,1,z+1)\sum_{j=0}^{k-1}(-1)^j C_j^{k-1}\frac{k!}{(k-j)!}x^{k-j}(z+1)^{j-2k}.$$

Note that the last expression can be written as $e^{i\rho x}\sum_{n=0}^{\infty}a_n(x)\mathcal{F}_n(z)$ where

$$\mathcal{F}_n(z)=\frac{x}{(z+1)^2}\sum_{j=2}^{k+1}F(-n,j,1,z+1)\sum_{m=j-1}^{k}(-1)^{j-m+1}C_{m-j+1}^{k-1}\frac{k!}{(k-m)!}x^{k-m}(z+1)^{m-2k+1}$$
$$+\sum_{j=2}^{k+1}F_z(-n,j,1,z+1)\sum_{m=j-1}^{k}(-1)^{j-m+1}C_{m-j+1}^{k-1}\frac{k!}{(k-m)!}x^{k-m}(z+1)^{m-2k+1}$$
$$+\sum_{j=2}^{k+1}F(-n,j,1,z+1)\sum_{m=j-1}^{k}(-1)^{j-m+1}C_{m-j+1}^{k-1}\frac{k!}{(k-m)!}x^{k-m}(m-2k+1)(z+1)^{m-2k}$$
$$+F(-n,2,1,z+1)\sum_{j=0}^{k-1}(-1)^j C_j^{k-1}\frac{k!}{(k-j)!}x^{k-j}(z+1)^{j-2k}.$$

Associating terms in the expression for $\mathcal{F}_n(z)$, we obtain

$$\mathcal{F}_n(z) = \sum_{j=2}^{k+1} F(-n,j,1,z+1) \sum_{m=j-1}^{k} (-1)^{j-m+1} C_{m-j+1}^{k-1} \frac{k!}{(k-m)!} x^{k-m} (z+1)^{m-2k+1}$$
$$\left(\frac{x}{(z+1)^2} - \frac{2k-m-1}{z+1}\right)$$
$$+ \sum_{j=2}^{k+1} F_z(-n,j,1,z+1) \sum_{m=j-1}^{k} (-1)^{j-m+1} C_{m-j+1}^{k-1} \frac{k!}{(k-m)!} x^{k-m} (z+1)^{m-2k+1}$$
$$+ F(-n,2,1,z+1) \sum_{j=0}^{k-1} (-1)^j C_j^{k-1} \frac{k!}{(k-j)!} x^{k-j} (z+1)^{j-2k}.$$

Simplification of the last expression results in

$$\mathcal{F}_n(z) = \sum_{j=2}^{k+1} F(-n,j,1,z+1) \sum_{m=j-1}^{k} (-1)^{j-m+1} C_{m-j+1}^{k-1} \frac{k!}{(k-m)!} x^{k-m} (z+1)^{m-2k+1}$$
$$\left(\frac{x}{(z+1)^2} - \frac{2k-1-m}{z+1}\right)$$
$$+ \sum_{j=2}^{k+1} (jF(-n,j+1,1,z+1) - jF(-n,j,1,z+1)) \sum_{m=j-1}^{k} (-1)^{j-m+1} C_{m-j+1}^{k-1} \frac{k!}{(k-m)!} x^{k-m} (z+1)^{m-2k}$$
$$+ F(-n,2,1,z+1) \sum_{j=0}^{k-1} (-1)^j C_j^{k-1} \frac{k!}{(k-j)!} x^{k-j} (z+1)^{j-2k},$$

where we applied Formula (55). Thus,

$$\mathcal{F}_n(z) = \sum_{j=2}^{k+1} F(-n,j,1,z+1) \sum_{m=j-1}^{k} (-1)^{j-m+1} C_{m-j+1}^{k-1} \frac{k!}{(k-m)!} x^{k-m} (z+1)^{m-2k+1}$$
$$\left(\frac{x}{(z+1)^2} - \frac{2k-1-m}{z+1} - \frac{j}{z+1}\right)$$
$$+ \sum_{j=2}^{k+1} jF(-n,j+1,1,z+1) \sum_{m=j-1}^{k} (-1)^{j-m+1} C_{m-j+1}^{k-1} \frac{k!}{(k-m)!} x^{k-m} (z+1)^{m-2k}$$
$$+ F(-n,2,1,z+1) \sum_{j=0}^{k-1} (-1)^j C_j^{k-1} \frac{k!}{(k-j)!} x^{k-j} (z+1)^{j-2k}$$

$$= \sum_{j=2}^{k+1} F(-n,j,1,z+1) \sum_{m=j-1}^{k} (-1)^{j-m+1} C_{m-j+1}^{k-1} \frac{k!}{(k-m)!} x^{k-m} (z+1)^{m-2k+1}$$
$$\left(\frac{x}{(z+1)^2} - \frac{2k-1-m}{z+1} - \frac{j}{z+1}\right)$$
$$+ \sum_{j=2}^{k} jF(-n,j+1,1,z+1) \sum_{m=j-1}^{k} (-1)^{j-m+1} C_{m-j+1}^{k-1} \frac{k!}{(k-m)!} x^{k-m} (z+1)^{m-2k}$$
$$+ (k+1)F(-n,k+2,1,z+1) \sum_{m=k}^{k} (-1)^{k-m+2} \binom{k-1}{m-k} \frac{k!}{(k-m)!} x^{k-m} (z+1)^{m-2k}$$
$$+ F(-n,2,1,z+1) \sum_{j=0}^{k-1} (-1)^j C_j^{k-1} \frac{k!}{(k-j)!} x^{k-j} (z+1)^{j-2k}$$

$$
\begin{aligned}
&=\sum_{j=2}^{k+1} F(-n,j,1,z+1) \sum_{m=j-1}^{k} (-1)^{j-m+1} C_{m-j+1}^{k-1} \frac{k!}{(k-m)!} x^{k-m} (z+1)^{m-2k+1} \\
&\left(\frac{x}{(z+1)^2} - \frac{2k-1-m}{z+1} - \frac{j}{z+1}\right) \\
&+\sum_{j=2}^{k} jF(-n,j+1,1,z+1) \sum_{m=j-1}^{k} (-1)^{j-m+1} C_{m-j+1}^{k-1} \frac{k!}{(k-m)!} x^{k-m} (z+1)^{m-2k} \\
&++\frac{k+1}{(z+1)^k} F(-n,k+2,1,z+1)k! + F(-n,2,1,z+1) \sum_{j=0}^{k-1} (-1)^j C_j^{k-1} \frac{k!}{(k-j)!} x^{k-j} (z+1)^{j-2k} \\
&=\sum_{j=2}^{k+1} F(-n,j,1,z+1) \sum_{m=j-1}^{k} (-1)^{j-m+1} C_{m-j+1}^{k-1} \frac{k!}{(k-m)!} x^{k-m} (z+1)^{m-2k+1} \\
&\left(\frac{x}{(z+1)^2} - \frac{2k-1-m}{z+1} - \frac{j}{z+1}\right) \\
&+\sum_{j=2}^{k} jF(-n,j+1,1,z+1) \sum_{m=j-1}^{k} (-1)^{j-m+1} C_{m-j+1}^{k-1} \frac{k!}{(k-m)!} x^{k-m} (z+1)^{m-2k} \\
&+F(-n,2,1,z+1) \sum_{j=0}^{k-1} (-1)^j C_j^{k-1} \frac{k!}{(k-j)!} x^{k-j} (z+1)^{j-2k} \\
&+F(-n,k+2,1,z+1)(k+1)!(z+1)^{(-k)}
\end{aligned}
$$

$$
\begin{aligned}
&=\sum_{j=2}^{k+1} F(-n,j,1,z+1) \sum_{m=j-1}^{k+1} (-1)^{(m+j+1)} C_{m-j+1}^{k} \frac{(k+1)!}{(k-m+1)!} x^{k-m+1} (z+1)^{m-2k-1} \\
&+F(-n,k+2,1,z+1)(k+1)!(z+1)^{-k} \\
&=\sum_{j=2}^{k+1} F(-n,j,1,z+1) \sum_{m=j-1}^{k+1} (-1)^{(m+j+1)} C_{m-j+1}^{k} \frac{(k+1)!}{(k-m+1)!} x^{(k-m+1)} (z+1)^{(m-2k-1)} \\
&+F(-n,k+2,1,z+1)(-1)^{(2k+4)} C_0^k \frac{(k+1)!}{(0)!} x^0 (z+1)^{-k} \\
&=\sum_{j=2}^{k+2} F(-n,j,1,z+1) \sum_{m=j-1}^{k+1} (-1)^{m+j+1} C_{m-j+1}^{k} \frac{(k+1)!}{(k-m+1)!} x^{k-m+1} (z+1)^{m-2k-1}
\end{aligned}
$$

which is the desired result.

References

1. Bagarello, F.; Gazeau, J.P.; Szafraniec, F.H.; Znojil, M. *Non-Selfadjoint Operators in Quantum Physics*; John Wiley & Sons: Hoboken, NJ, USA, 2015.
2. Bender, C.M.; Boettcher, S. Real spectra in non-Hermitian Hamiltonians having P T symmetry. *Phys. Rev. Lett.* **1998**, *80*, 5243. [CrossRef]
3. Bender, C.M.; Boettcher, S. Quasi-exactly solvable quartic potential. *J. Phys. A Math. Gen.* **1998**, *31*, L273. [CrossRef]
4. Ushveridze, A.G. *Quasi-Exactly Solvable Models in Quantum Mechanics*; CRC Press: New York, NY, USA, 2017.
5. Chandrasekhar, S. On characteristic value problems in high order differential equations which arise in studies on hydrodynamic and hydromagnetic stability. *Am. Math.* **1954**, *61*, 32–45. [CrossRef]
6. Dolph, C.L. Recent developments in some non-self-adjoint problems of mathematical physics. *Bull. Am. Math.* **1961**, *67*, 1–69. [CrossRef]
7. Moiseyev, N. *Non-Hermitian Quantum Mechanics*; Cambridge University Press: Cambridge, UK, 2011.
8. Lombard, R.J.; Mezhoud, R.; Yekken, R.; Ezzouar, U.B. Complex potentials with real eigenvalues and the inverse problem. *Rom. J. Phys.* **2018**, *63*, 101.

9. Hryniv, O.R.; Manko, S.S. Inverse scattering on the half-line for ZS-AKNS systems with integrable potentials. *Integr. Oper. Theory* **2016**, *84*, 323–355. [CrossRef]
10. Bairamov, E.; Seyyidoglu, M.S. Non-Self-Adjoint singular Sturm-Liouville problems with boundary conditions dependent on the eigenparameter. In *Abstract and Applied Analysis*; Hindawi: London, UK, 2010; Volume 2010.
11. Gasymov, M.G. On the decomposition in a series of eigenfunctions for a nonselfconjugate boundary value problem of the solution of a differential equation with a singularity at a zero point. *Dokl. Akad. Nauk Russ. Acad. Sci.* **1965**, *165*, 261–264.
12. Naimark, M.A. Linear *Differential Operators. Part II: Linear Differential Operators in Hilbert Space*; Dawson, E.R., Everitt, W.N., Translators; Ungar Publishing: New York, NY, USA, 1968.
13. Yurko, V.A. *Introduction to the Theory of Inverse Spectral Problems*; Fizmatlit: Moscow, Russia, 2007. (In Russian)
14. Barrera-Figueroa, V. Analysis of the spectral singularities of Schrödinger operator with complex potential by means of the SPPS method. *J. Phys. Conf. Ser.* **2016**, *698*, 012029. [CrossRef]
15. Kir, E. Spectrum and principal functions of the non-self-adjoint Sturm–Liouville operators with a singular potential. *Appl. Math.* **2005**, *18*, 1247–1255. [CrossRef]
16. Naimark, M.A. Investigation of the spectrum and the expansion in eigenfunctions of a nonselfadjoint operator of the second order on a semi-axis. *Tr. Mosk. Mat. Obs.* **1954**, *3*, 181–270. (In Russian)
17. Pavlov, B.S. On the non-selfadjoint operator $-y'' + q(x)y$ on a semiaxis. *Dokl. Akad. Nauk SSSR* **1961**, *141*, 807–810. (In Russian)
18. Pavlov, B.S. *On a Non-Selfadjoint Schrödinger operator, Probl. Math. Phys., No. I, Spectral Theory and Wave Processes*; Izdat. Leningrad. Univ.: Leningrad, Russia, 1966; pp. 102–132. (In Russian)
19. Pavlov, B.S. *On a Non-Selfadjoint Schrödinger Operator II, Problems of Mathematical Physics, No. 2, Spectral Theory, Diffraction Problems*; Izdat. Leningrad. Univ: Leningrad, Russia, 1967; pp. 133–157. (In Russian)
20. Sabatier, P.C. *Applied Inverse Problems*; Lecture Notes in Physics; Springer: Heidelberg, Germany, 1978; Volume 85.
21. Samsonov, B.F. Spectral singularities of non-Hermitian Hamiltonians and SUSY transformations. *J. Phys. A Math. Gen.* **2005**, *38*, L571. [CrossRef]
22. Sitenko, A.G. *Lectures in Scattering Theory*; Pergamon Press: Oxford, UK, 1971.
23. Agranovich, Z.S.; Marchenko, V.A. *The Inverse Problem of Scattering Theory*; Courier Dover Publications: New York, NY, USA, 2020.
24. Freiling, G.; Yurko, V. *Inverse Sturm–Liouville Problems and Their Applications*; NOVA Science Publishers: New York, NY, USA, 2001.
25. Kravchenko, V.V. On a method for solving the inverse scattering problem on the line. *Math. Methods Appl. Sci.* **2019**, *42*, 1321–1327. [CrossRef]
26. Kravchenko, V.V. *Direct and Inverse Sturm-Liouville Problems: A Method of Solution, Frontiers in Mathematics*; Birkhäuser: Cham, Switzerland, 2020.
27. Delgado, B.B.; Khmelnytskaya, K.V.; Kravchenko, V.V. A representation for Jost solutions and an efficient method for solving the spectral problem on the half line. *Math. Methods Appl. Sci.* **2020**, *43*, 9304–9319. [CrossRef]
28. Baker, G.A., Jr.; Graves-Morris, P.R. *Padé Approximants, Part I: Basic Theory*; Cambridge University Press: Cambridge, UK, 1981.
29. Baker, G.A., Jr.; Graves-Morris, P.R. *Padé Approximants, Part 2: Extensions and Applications*; Cambridge University Press: Cambridge, UK, 1981.
30. Henrici, P. *Applied and Computational Complex Analysis*; Power Series, Integration, Conformal Mapping, Location of Zeros; John Wiley & Sons: New York, NY, USA, 1974; Volume 1.
31. Suetin, S.P. Padé approximants and efficient analytic continuation of a power series. *Russ. Math. Surv.* **2002**, *57*, 43. [CrossRef]
32. Bondarenko, E.I.; Rofe-Beketov, F.S. Inverse scattering problem on the semiaxis for the system with the triangle matrix potential. *Zhurnal Mat. Fiz. Anal. Geom. [J. Math. Phys. Anal. Geom.]* **2003**, *10*, 412–424.
33. Levitan, B.M. *Inverse Sturm-Liouville Problems*; VSP: Zeist, The Netherlands, 1987.
34. Lyantse, V.É. The inverse problem for a nonselfadjoint operator. *Dokl. Akad. Nauk. Russ. Acad. Sci.* **1966**, *166*, 30–33.
35. Xu, X.C.; Bondarenko, N.P. Stability of the inverse scattering problem for the self-adjoint matrix Schrödinger operator on the half line. *Stud. Appl. Math.* **2022**, *149*, 815–838. [CrossRef]
36. Delgado, B.B.; Khmelnytskaya, K.V.; Kravchenko, V.V. The transmutation operator method for efficient solution of the inverse Sturm-Liouville problem on a half-line. *Math. Methods Appl. Sci.* **2019**, *42*, 7359–7366. [CrossRef]
37. Grudsky, S.M.; Kravchenko, V.V.; Torba, S.M. Realization of the inverse scattering transform method for the Korteweg-de Vries equation. *Math. Methods Appl. Sci.* **2023**, *46*, 9217–9251. [CrossRef]
38. Kravchenko, V.V.; Shishkina, E.L.; Torba, S.M. A transmutation operator method for solving the inverse quantum scattering problem. *Inverse Probl.* **2020**, *36*, 125007. [CrossRef]
39. Lyantse, V.É. An analog of the inverse problem of scattering theory for a nonselfadjoint operator. *Mat. Sb.* **1967**, *1*, 485. (In Russian)
40. Suetin, P.K. *Classical Orthogonal Polynomials*, 3rd ed.; Fizmatlit: Moscow, Russia, 2005. (In Russian)
41. Szego, G. *Orthogonal Polynomials, American Mathematical Society Colloquium Publications, 23*; American Mathematical Society: New York, NY, USA, 1939.
42. Xiang, S. Asymptotics on Laguerre or Hermite polynomial expansions and their applications in Gauss quadrature. *J. Math. Anal. Appl.* **2012**, *393*, 434–444. [CrossRef]
43. Frank, R.L.; Laptev, A.; Safronov, O. On the number of eigenvalues of Schrödinger operators with complex potentials. *J. Lond. Math. Soc.* **2016**, *94*, 377–390. [CrossRef]

44. Brown, B.M.; Langer, M.; Marletta, M.; Tretter, C.; Wagenhofer, M. Eigenvalue bounds for the singular Sturm–Liouville problem with a complex potential. *J. Phys. A Math. Gen.* **2003**, *36*, 3773. [CrossRef]
45. Chanane, B. Computing the spectrum of non-self-adjoint Sturm–Liouville problems with parameter-dependent boundary conditions. *J. Comput. Appl. Math.* **2007**, *206*, 229–237. [CrossRef]
46. Satsuma, J.; Yajima, N.B. Initial value problems of one-dimensional self-modulation of nonlinear waves in dispersive media. *Prog. Theor. Phys. Suppl.* **1974**, *55*, 284–306. [CrossRef]
47. Akhiezer, N.I.; Glazman, I.M. *Theory of Linear Operators in Hilbert Space*; Dover: New York, NY, USA, 1993.
48. Gradshteyn, I.S.; Ryzhik, I.M. *Table of Integrals, Series, and Products*; Academic Press: New York, NY, USA, 2007.
49. Erdélyi, A.; Magnus, W.; Oberhettinger, F.; Tricomi, F.G. *Higher Transcendental Functions*; McGraw-Hill Book Co.: New York, NY, USA, 1953; Volume 1.
50. Ancarani, L.U.; Gasaneo, G. Derivatives of any order of the Gaussian hypergeometric function ${}_2F_1(a,b,c;z)$ with respect to the parameters a, b and c. *J. Phys. A Math. Theor.* **2009**, *42*, 395208. [CrossRef]
51. Ramm, A.G. *Inverse Problems: Mathematical and Analytical Techniques with Applications to Engineering*; Springer: New York, NY, USA, 2005.
52. Trefethen, L.N. *Approximation Theory and Approximation Practice*; SIAM: Philadelphia, PA, USA, 2013.
53. Trefethen, L.N. Quantifying the ill-conditioning of analytic continuation. *BIT Numer. Math.* **2020**, *60*, 901–915. [CrossRef]
54. Kravchenko, V.V.; Torba, S.M.; Velasco-García, U. Spectral parameter power series for polynomial pencils of Sturm-Liouville operators and Zakharov-Shabat systems. *J. Math. Phys.* **2015**, *56*, 073508. [CrossRef]
55. Roussos, I.M. *Improper Riemann Integrals*; Taylor & Francis Group: Boca Raton, FL, USA, 2013.
56. Davis, P.J.; Rabinowitz, P. *Methods of Numerical Integration*, 2nd ed.; Dover Publishers: New York, NY, USA, 2007.
57. Kantorovich, L.V.; Akilov, G.P. *Functional Analysis*, 2nd ed.; Silcock, H.L., Translator; Pergamon Press: Oxford-Elmsford, NY, USA, 1982.
58. Mikhlin, S.G. *The Numerical Performance of Variational Methods*; Wolters-Noordhoff Publishing: Groningen, The Netherlands, 1971.
59. D'Errico, J. SLM-Shape Language Modeling. 2009. Available online: http://www.mathworks.commatlabcentral/fileexchange/24443-slm-shape-language-modeling:Mathworks (accessed on 10 March 2022).
60. Sarnari, A.J.; Živanović, R. Robust Padé approximation for the holomorphic embedding load flow. In Proceedings of the 2016 Australasian Universities Power Engineering Conference (AUPEC), Brisbane, Australia, 25–28 September 2016; pp. 1–6.
61. Ibryaeva, O.L.; Aduko, V.M. An algorithm for computing a Padé approximant with minimal degree denominator. *J. Comput. Appl. Math.* **2013**, *237*, 529–541. [CrossRef]
62. Ibryaeva, O.L. A new algorithm for computing Padé approximants and its implementation in Matlab. *Bull. South Ural. State Univ. Ser. Math. Model. Program.* **2011**, *10*, 99–107. (In Russian)

mathematics

MDPI

Article

Schatten Index of the Sectorial Operator via the Real Component of Its Inverse

Maksim V. Kukushkin [1,2]

[1] National Research University Higher School of Economics, 101000 Moscow, Russia; kukushkinmv@rambler.ru
[2] Institute of Applied Mathematics and Automation, Kabardino-Balkarian Scientific Center, Russian Academy of Sciences, 360000 Nalchik, Russia

Abstract: In this paper, we study spectral properties of **non-self-adjoint** operators with the discrete spectrum. The main challenge is to represent a complete description of belonging to the Schatten class through the properties of the Hermitian real component. The method of estimating the singular values is elaborated by virtue of the established asymptotic formulas. The latter fundamental result is advantageous since, of many theoretical statements based upon it, one of them is a concept on the root vectors series expansion, which leads to a wide spectrum of applications in the theory of evolution equations. In this regard, the evolution equations of fractional order with the sectorial operator in the term not containing the time variable are involved. The concrete well-known operators are considered and the advantage of the represented method is convexly shown.

Keywords: strictly accretive operator; Abel–Lidskii basis property; Schatten–von Neumann class; convergence exponent; counting function

MSC: 47B28; 47A10; 47B12; 47B10; 34K30; 58D25

1. Introduction

Erhard Schmidt, whose advisor had been David Hilbert, studied the integral equations with nonsymmetric kernels and introduced singular values (s-numbers), which afterwards were interpreted by the brilliant Allakhverdiyev theorem as a measure of deviation between a compact operator and finite-dimensional ones. From that time on, singular values have become a most popular tool for studying spectral properties of non-self-adjoint operators. However, although the history could have developed in a different way, the fact is that the eigenvalues of the operator real component are no less suitable for this study. The last idea fully reflects the plot of this paper.

The idea to write this paper originates from the concept of decomposition of an element of the abstract Hilbert space on the root vectors series. The latter concept lies in the framework of abstract functional analysis, and its appearance arises from elaboration of methods of solving evolution equations investigated in the recent century by Lidskii V.B. [1], Markus A.S., Matsaev V.I. [2], Agaranovich M.S. [3], and others. In its simple reduced form, applicably to self-adjoint operators, the concept admits the interpretation through the well-known fact that the eigenvectors of the compact self-adjoint operator form a basis in the closure of its range. The question of what happens in the case when the operator is non-self-adjoint is rather complicated and deserves to be considered as a separate part of the spectral theory.

We should make a brief digression and explain that relevance appears just in the case when a senior term of a considered operator is not self-adjoint, for there is a number of papers [2,4–8] devoted to the perturbed self-adjoint operators. The fact is that most of them deal with a decomposition of the operator on a sum, where the senior term must be either a self-adjoint or normal operator. In other cases, the methods of the papers [9,10] become relevant and allow us to study spectral properties of operators

Citation: Kukushkin, M.V. Schatten Index of the Sectorial Operator via the Real Component of Its Inverse. *Mathematics* **2024**, *12*, 540. https://doi.org/10.3390/math12040540

Academic Editor: Natalia Bondarenko

Received: 24 November 2023
Revised: 28 January 2024
Accepted: 31 January 2024
Published: 8 February 2024

whether we have the abovementioned representation or not; moreover, they have a natural mathematical origin that appears brightly while we are considering abstract constructions expressed in terms of the semigroup theory [10].

Generally, the aims of the mentioned part of the spectral theory are propositions on the convergence of the root vectors series in one or another sense to an element belonging to the closure of the operator range; by this, we mean Bari, Riesz, and Abel–Lidskii senses of the series convergence [11]. The main condition in terms of which the propositions are mostly described is the asyptotics of the operator singular numbers; here, we should note that it is originally formulated in terms of the operator belonging to the Schatten class. However, Agaranovich M.S. made an attempt to express the sufficient conditions of the root vector series basis property, in the abovementioned generalized sense, through the asymptotics of the eigenvalues of the real component [3]. The paper by Markus A.S. and Matsaev V.I. [2] can be also considered within the scope since it establishes the relationship between the asymptotics of the operator eigenvalues absolute value and eigenvalues of the real component.

Thus, the interest in how to express root vectors series decomposition theorems through the asymptotics of the real component eigenvalues arose previously, and the obvious technical advantage in finding the asymptotics creates a prerequisite to investigate the issue properly. We should point out that under the desired relationship between asymptotics, we are able to reformulate theorems on the root vectors series expansion in terms of the assumptions related to the real component of the operator. The latter idea is relevant, since in many cases, the calculation of the real component eigenvalues asymptotics is simpler than direct calculation of the singular numbers' asymptotics.

If we make a comparison analysis between the methods of root vectors decomposition by Lidskii V.B. [1] and Agaranovich M.S. [3], we will see that the first one formulated the conditions in terms of the singular values but the second one did so in terms of the real component eigenvalues. In this regard, we will show that the real component eigenvalue asymptotics are stronger than that of the singular numbers; however, Agaranovich M.S. [3] imposed the additional condition—the spectrum belongs to the domain of the parabolic type. From this point of view, the results by Lidskii V.B. [1] are more advantageous since the convergence in the Abel–Lidskii sense was established for an operator class wider than the class of sectorial operators. Apparently, a reasonable question that may appear is about minimal conditions that guarantee the desired result, which, in particular, is considered in this paper.

Here, we can obviously extend the results devoted to operators with the discrete spectrum to operators with the compact resolvent, for they can be easily reformulated from one realm to another. In this regard, we should give warning that the latter fact does not hold for real components since the real component of the inverse operator does not coincide with the inverse of the operator real component. However, such a complication was diminished due to the results of [9], where the asymptotic equivalence between the eigenvalues of the mentioned operators was established.

The following are a couple of words on the applied relevance of the issue. The abstract approach to the Cauchy problem for the fractional evolution equation is a classic one [12,13]. In its framework, the application of results connected with the basis property covers many problems in the theory of evolution equations [1,10,14–16]. In its general statement, the problem appeals to many applied ones, and we can produce a number of papers dealing with differential equations which can be studied by the abstract methods [17–22]. Apparently, the main advantage of this paper is a method that enables the implementation of the existence and uniqueness theorem abstract condition verification for concrete evolution equations. The latter concept may be interesting for the reader, for it allows broadening of the condition under which the Abel–Lidskii method works, which, in turn, gives a wide spectrum of applications in the theory of differential equations. Thus, we can claim that the offered approach is undoubtedly novel from the abstract theory point of view, and is relevant from the applied one.

2. Preliminaries

Let $C, C_i,\ i \in \mathbb{N}_0$ be real constants. We assume that a value of C is positive and can be different in various formulas, but values of C_i are certain. Denote by $\operatorname{int} M,\ \operatorname{Fr} M$ the interior and the set of boundary points of the set M, respectively. Everywhere further, if the contrary is not stated, we consider linear densely defined operators acting on a separable complex Hilbert space $\mathfrak{H}$. Denote by $\mathcal{B}(\mathfrak{H})$ the set of linear bounded operators on $\mathfrak{H}$. Denote by $\tilde{L}$ the closure of an operator L. We establish the following agreement on using symbols $\tilde{L}^i := (\tilde{L})^i$, where i is an arbitrary symbol. Denote by $\mathrm{D}(L),\ \mathrm{R}(L),\ \mathrm{N}(L)$ the *domain of definition* , the *range*, and the *kernel*, or *null space*, of an operator L, respectively. The deficiency (codimension) of $\mathrm{R}(L)$, dimension of $\mathrm{N}(L)$ are denoted by $\operatorname{def} L,\ \operatorname{nul} L$, respectively. In some places, if it is necessary from the stylistic point of view, we use the following notation: $L^{-1} := I/L$. Assume that L is a closed operator acting on $\mathfrak{H}$, $\mathrm{N}(L) = 0,$; let us define a Hilbert space $\mathfrak{H}_L := \big\{f, g \in \mathrm{D}(L),\ (f,g)_{\mathfrak{H}_L} = (Lf, Lg)_{\mathfrak{H}}\big\}$. Considering a pair of complex Hilbert spaces $\mathfrak{H}, \mathfrak{H}_+$, the notation $\mathfrak{H}_+ \subset\subset \mathfrak{H}$ means that $\mathfrak{H}_+$ is dense in $\mathfrak{H}$ as a set of elements and we have a bounded embedding provided by the inequality

$$\|f\|_{\mathfrak{H}} \leq C_0 \|f\|_{\mathfrak{H}_+},\ C_0 > 0,\ f \in \mathfrak{H}_+;$$

moreover, any bounded set with respect to the norm $\mathfrak{H}_+$ is compact with respect to the norm $\mathfrak{H}$. Let L be a closed operator for any closable operator S such that $\tilde{S} = L$, its domain $\mathrm{D}(S)$ will be called a core of L. Denote by $\mathrm{D}_0(L)$ a core of a closeable operator L. Let $\mathrm{P}(L)$ be the resolvent set of an operator L and $R_L(\zeta),\ \zeta \in \mathrm{P}(L),\ [R_L := R_L(0)]$ denotes the resolvent of an operator L. Denote by $\lambda_i(L),\ i \in \mathbb{N}$ the eigenvalues of an operator L, we numerate them in order of increasing (decreasing) of their absolute values. Suppose L is a compact operator and $N := (L^*L)^{1/2},\ r(N) := \dim \mathrm{R}(N)$; then the eigenvalues of the operator N are called the *singular values* (*s-numbers*) of the operator L and are denoted by $s_i(L),\ i = 1, 2, \ldots, r(N)$. If $r(N) < \infty$, then we use by definition $s_i = 0,\ i = r(N) + 1, 2, \ldots$. Let $\mathfrak{S}_p(\mathfrak{H}),\ 0 < p < \infty$ be the Schatten–von Neumann class (Schatten class) and $\mathfrak{S}_\infty(\mathfrak{H})$ be the set of compact operators, by definition use

$$\mathfrak{S}_p(\mathfrak{H}) := \left\{ L : \mathfrak{H} \to \mathfrak{H},\ \sum_{n=1}^{\infty} s_n^p(L) < \infty,\ 0 < p < \infty \right\}.$$

According to the terminology of the monograph [11], the dimension of the root vectors subspace corresponding to a certain eigenvalue λ_k is called the *algebraic multiplicity* of the eigenvalue λ_k. Let $\nu(L)$ denote the sum of all algebraic multiplicities of an operator L. Denote by $n(r)$ a function equal to a number of the elements of the sequence $\{a_n\}_1^\infty,\ |a_n| \uparrow \infty$ within the circle $|z| < r$. Let A be a compact operator, denoted by $n_A(r)$ *counting function* a function $n(r)$ corresponding to the sequence $\{s_i^{-1}(A)\}_1^\infty$. Let $\mathfrak{S}_p(\mathfrak{H}),\ 0 < p < \infty$ be a Schatten–von Neumann class and $\mathfrak{S}_\infty(\mathfrak{H})$ be the set of compact operators. Suppose L is an operator with a compact resolvent and $s_n(R_L) \leq C n^{-\mu},\ n \in \mathbb{N},\ 0 \leq \mu < \infty$; then we denote by $\mu(L)$ order of the operator L in accordance with the definition given in the paper [8]. Denote by $\mathfrak{Re} L := (L + L^*)/2,\ \mathfrak{Im} L := (L - L^*)/2i$ the real and imaginary components of an operator L, respectively. In accordance with the terminology of the monograph [23], the set $\Theta(L) := \{z \in \mathbb{C} : z = (Lf, f)_{\mathfrak{H}},\ f \in \mathrm{D}(L),\ \|f\|_{\mathfrak{H}} = 1\}$ is called the *numerical range* of an operator L. An operator L is called *sectorial* if its numerical range belongs to a closed sector $\mathfrak{L}_\iota(\theta) := \{\zeta : |\arg(\zeta - \iota)| \leq \theta < \pi/2\}$, where ι is the vertex and θ is the semiangle of the sector $\mathfrak{L}_\iota(\theta)$. If we want to stress the correspondence between ι and θ, then we will write θ_ι. An operator L is called *bounded from below* if the following relation holds: $\mathrm{Re}(Lf, f)_{\mathfrak{H}} \geq \gamma_L \|f\|^2_{\mathfrak{H}},\ f \in \mathrm{D}(L),\ \gamma_L \in \mathbb{R}$, where γ_L is called a lower bound of L. An operator L is called *accretive* if $\gamma_L = 0$. An operator L is called *strictly accretive* if $\gamma_L > 0$. An operator L is called *m-accretive* if the following relation holds: $(A + \zeta)^{-1} \in \mathcal{B}(\mathfrak{H}),\ \|(A + \zeta)^{-1}\| \leq (\mathrm{Re}\zeta)^{-1},\ \mathrm{Re}\zeta > 0$. An operator L is called *m-sectorial* if L is sectorial and $L + \beta$ is m-accretive for some constant β. An operator L is called *symmetric* if one is densely defined and the following equality holds: $(Lf,g)_{\mathfrak{H}} = (f,Lg)_{\mathfrak{H}}$, $f,g \in \mathrm{D}(L)$. Let B be a bounded operator acting in $\mathfrak{H}$, and assume that $\{\varphi_n\}_1^\infty$, $\{\psi_n\}_1^\infty$ are a pair of orthonormal bases in $\mathfrak{H}$. Define the *absolute operator norm* as follows:

$$\|B\|_2 := \left(\sum_{n,k=1}^{\infty} |(B\varphi_n,\psi_k)_{\mathfrak{H}}|^2\right)^{1/2} < \infty.$$

Everywhere further, unless otherwise stated, we use notations of the papers [11,23–26].

2.1. Sectorial Sesquilinear Forms and the Hermitian Components

Consider the Hermitian components of an operator (not necessarily bounded):

$$\mathfrak{Re}L := \frac{L+L^*}{2}, \quad \mathfrak{Im}L := \frac{L-L^*}{2i},$$

where it is clear that in the case when the operator L is unbounded but densely defined we need agreement between the domain of definition of the operator and its adjoint, since in other cases, the real component may be not densely defined. However, the latter claim requires concrete examples; in this regard, we can refer to Remark 4 [10].

Consider a sesquilinear form $t[\cdot,\cdot]$ (see [23]) defined on a linear manifold of the Hilbert space $\mathfrak{H}$. Denote by $t[\cdot]$ the quadratic form corresponding to the sesquilinear form $t[\cdot,\cdot]$. Let

$$\mathfrak{h} = (t+t^*)/2, \, \mathfrak{k} = (t-t^*)/2i$$

be a real and imaginary component of the form t, respectively, where $t^*[u,v] = \overline{t[v,u]}$, $\mathrm{D}(t^*) = \mathrm{D}(t)$. In accordance with the definitions, we have $\mathfrak{h}[\cdot] = \mathrm{Re}\, t[\cdot]$, $\mathfrak{k}[\cdot] = \mathrm{Im}\, t[\cdot]$. Denote by $\tilde{t}$ the closure of a form t. The range of a quadratic form $t[f]$, $f \in \mathrm{D}(t)$, $\|f\|_{\mathfrak{H}} = 1$ is called *range* of the sesquilinear form t and is denoted by $\Theta(t)$. A form t is called *sectorial* if its range belongs to a sector having a vertex ι situated at the real axis and a semiangle $0 \leq \theta_\iota < \pi/2$. Suppose t is a closed sectorial form; then a linear manifold $\mathrm{D}_0(t) \subset \mathrm{D}(t)$ is called the *core* of t, if the restriction of t to $\mathrm{D}_0(t)$ has the closure t (see [23], p. 166).

Suppose L is a sectorial densely defined operator and $t[u,v] := (Lu,v)_{\mathfrak{H}}$, $\mathrm{D}(t) = \mathrm{D}(L)$; then due to Theorem 1.27 ([23], p. 318), the corresponding form t is closable, and due to Theorem 2.7 ([23], p. 323), there exists a unique m-sectorial operator $T_{\tilde{t}}$ associated with the form $\tilde{t}$. In accordance with the definition ([23], p. 325), the operator $T_{\tilde{t}}$ is called a *Friedrichs extension* of the operator L.

Due to Theorem 2.7 ([23], p. 323), there exist unique m-sectorial operators $T_t, T_{\mathfrak{h}}$ associated with the closed sectorial forms $t, \mathfrak{h}$, respectively. The operator $T_{\mathfrak{h}}$ is called a *real part* of the operator T_t and is denoted in accordance with the original definition [23] by $\mathrm{Re}\, T_t$.

Here, we should stress that the construction of the real part in some cases is obviously coincident with that of the real component; however, the latter does not require the agreement between the domain of definitions mentioned above. The condition represented below reflects the nature of uniformly elliptic operators being the direct generalization of the one considered in the context of the theory of Sobolev spaces.

H1: *There exists a Hilbert space $\mathfrak{H}_+ \subset\subset \mathfrak{H}$ and a linear manifold $\mathfrak{M}$ that is dense in $\mathfrak{H}_+$. The closed operator W is defined on $\mathfrak{M}$ and the latter set is its core.*

H2: $|(Wf,g)_{\mathfrak{H}}| \leq C_1\|f\|_{\mathfrak{H}_+}\|g\|_{\mathfrak{H}_+}, \ \mathrm{Re}(Wf,f)_{\mathfrak{H}} \geq C_2\|f\|^2_{\mathfrak{H}_+}, \ f,g \in \mathfrak{M}, \ C_1, C_2 > 0.$

Consider a condition $\mathfrak{M} \subset \mathrm{D}(W^*)$; in this case, the real Hermitian component $\mathcal{H} := \mathfrak{Re}\, W$ of the operator is defined on $\mathfrak{M}$, and the fact is that $\tilde{\mathcal{H}}$ is self-adjoint, bounded from below (see Lemma 3 [9]). Hence, a corresponding sesquilinear form (denote this form by h) is symmetric and bounded from below also (see Theorem 2.6 [23], p. 323). The conditions H1, H2 allow us to claim that the form t corresponding to the operator W is a closed sectorial form; consider the corresponding form $\mathfrak{h}$. It can be easily shown that $h \subset \mathfrak{h}$, but, using this fact, we cannot claim in general that $\tilde{\mathcal{H}} \subset H$, where $H := \mathrm{Re}W$ (see [23], p. 330). We just have an inclusion $\tilde{\mathcal{H}}^{1/2} \subset H^{1/2}$ (see [23], p. 332). Note that the fact $\tilde{\mathcal{H}} \subset H$ follows from a condition $\mathrm{D}_0(\mathfrak{h}) \subset \mathrm{D}(h)$ (see Corollary 2.4 [23], p. 323). However, it is proved (see proof of Theorem 4 [9]) that relation H2 guarantees that $\tilde{\mathcal{H}} = H$. Note that the last relation is very useful in applications, since in most concrete cases we can find a concrete form of the operator $\mathcal{H}$.

2.2. Previously Obtained Results

Here, we represent previously obtained results that will undergo thorough study since our principal challenge is to obtain an accurate description of the Schatten–von Neumann class index of a non-self-adjoint operator.

Further, we consider Theorem 1 [10] statements separately under assumptions H1, H2. Note that in terms of Theorem 1 [10] the operator W is a closure of the restriction of the operator L on the set $\mathfrak{M}$. Without loss of generality, we can assume that W is closed since the conditions H1, H2 guarantee that it is closeable. Thus, the given above version of the conditions H1, H2 allows us to avoid redundant notations, more detailed information in this regard is given in the paper [10].

We have the following classification in terms of the operator order μ, where it is defined as follows $\lambda_n(R_H) = O(n^{-\mu}),\ n \to \infty$.

(A) The following Schatten classification holds:

$$R_W \in \mathfrak{S}_p,\ \inf p \leq 2/\mu,\ \mu \leq 1,\ R_W \in \mathfrak{S}_1,\ \mu > 1.$$

Moreover, under assumptions $\lambda_n(R_H) \geq C n^{-\mu},\ 0 \leq \mu < \infty$, the following implication holds: $R_W \in \mathfrak{S}_p,\ p \in [1,\infty), \Rightarrow \mu > 1/p$.

Observe that the above-given classification is far from the exact description of the Schatten–von Neumann class index p. However, having analyzed the above implications, we can see that it makes a prerequisite to establish a hypotheses $R_W \in \mathfrak{S}_p,\ \inf p = 1/\mu$. The following narrative is devoted to its verification.

Let us thoroughly analyze the technical tools involved in the proof of the statement in order to absorb and contemplate the scheme of reasonings. Consider the statement, if $\mu \leq 1$, then $R_W \in \mathfrak{S}_p,\ \inf p \leq 2/\mu$. The main result, on which it is based, is the asymptotic equivalence between the inverse of the real component and the real component of the resolvent. Indeed, due to application of some technicalities, we have a relation

$$(|R_W|^2 f, f)_{\mathfrak{H}} = \|R_W f\|^2_{\mathfrak{H}} \leq C \cdot \mathrm{Re}(R_W f, f)_{\mathfrak{H}} = C \cdot (\mathfrak{Re} R_W f, f)_{\mathfrak{H}};$$

using the minimax principle, we obtain the s-numbers asymptotics through the asymptotics of the real component eigenvalues.

Consider the statement that if $\lambda_n(R_H) \geq C n^{-\mu},\ 0 \leq \mu < \infty$, then the following implication holds: $R_W \in \mathfrak{S}_p,\ p \in [1,\infty), \Rightarrow \mu > 1/p$. The main results that guarantee the fulfilment of the latter relation are inequality (7.9) ([11], p. 123), Theorem 3.5 [10], in accordance with which we obtain

$$\sum_{i=1}^{\infty} |s_i(R_W)|^p \geq \sum_{i=1}^{\infty} |(R_W \varphi_i, \varphi_i)_{\mathfrak{H}}|^p \geq \sum_{i=1}^{\infty} |\mathrm{Re}(R_W \varphi_i, \varphi_i)_{\mathfrak{H}}|^p =$$

$$= \sum_{i=1}^{\infty} |(\mathfrak{Re} R_W \varphi_i, \varphi_i)_{\mathfrak{H}}|^p = \sum_{i=1}^{\infty} |\lambda_i(\mathfrak{Re} R_W)|^p \geq C \sum_{i=1}^{\infty} i^{-\mu p},\ p \geq 1.$$

Thus, we see that estimation of the series is involved; in this regard, we will make a more detailed remark further.

Below, we represent the second statement of Theorem 1 [10], where the peculiar result related to the asymptotics of the eigenvalue absolute value is given.
(B) In the case $\nu(R_W) = \infty$, $\mu \neq 0$, the following relation holds:

$$|\lambda_n(R_W)| = o(n^{-\tau}),\ n \to \infty,\ 0 < \tau < \mu.$$

It is based on the Theorem 6.1 ([11], p. 81), in accordance with which we have

$$\sum_{m=1}^{k} |\mathrm{Im}\,\lambda_m(B)|^p \leq \sum_{m=1}^{k} |\lambda_m(\mathfrak{Im}B)|^p,\ (k = 1, 2, \ldots, \nu_{\mathfrak{I}}(B)),\ 1 \leq p < \infty, \tag{1}$$

where $\nu_{\mathfrak{I}}(B) \leq \infty$ is the sum of all algebraic multiplicities corresponding to the not-real eigenvalues of the bounded operator B, $\mathfrak{Im}B \in \mathfrak{S}_\infty$ (see [11], p. 79).

Note that the statement (B) allows us to arrange brackets in the series that converges in the Abel–Lidskii sense (see [1,14]), which would be an advantageous achievement in the theory constructed further. However, it has a harmonious correspondence with the case where we do not have the exact index of the Schatten class, for in this case, due to the convergence test, we obtain a relation

$$R_W \in \mathfrak{S}_p, \Rightarrow s_n = o(n^{-1/p}),$$

which gives us a relation $|\lambda_n(R_W)| = o\left(n^{-1/p}\right)$ in accordance with the connection of the asymptotics (see Chapter II, §3 [11]). Note that the latter relation does not contradict (B) if we assume $p > 1/\mu$. Thus, along the abovementioned implication $R_W \in \mathfrak{S}_p,\ p \in [1, \infty), \Rightarrow p > 1/\mu$, it makes the prerequisite to observe the hypotheses $\inf p = 1/\mu$.

Apparently, the used technicalities appeal to the so-called nondirect estimates for singular values realized due to estimates of the series. As we will see further, the main advantage of the series estimation is the absence of the conditions imposed on the type of the asymptotics; it may be not one of the power type. However, we will show that under the restriction imposed on the type of the asymptotics, assuming that one is of the power type, we can obtain direct estimates for singular values. In the reminder, let us note that the classes of differential operators have the asymptotics of the power type, which make the issue rather relevant.

3. Main Results

The Main Refinement of the Result A

The reasonings produced below appeal to a compact operator B, which represents a most general case in the framework of the decomposition on the root vectors theory; however, to obtain more peculiar results, we are compelled to deploy some restricting conditions. In this regard, we involve hypotheses H1, H2 if it is necessary. The result represented below gives us the upper estimate for the singular values; it is based on the result by Ky Fan [27], which can be found as a corollary of the well-known Allakhverdiyev theorem (see Corollary 2.2 [11]).

Lemma 1. *Assume that B is a compact sectorial operator with the vertex situated at the point zero, then*

$$s_{2m-1}(B) \leq \sqrt{2}\,\sec\theta \cdot \lambda_m(\mathfrak{Re}B),\ \ s_{2m}(B) \leq \sqrt{2}\,\sec\theta \cdot \lambda_m(\mathfrak{Re}B),\ \ m = 1, 2, \ldots.$$

Proof. Consider the Hermitian components

$$\mathfrak{Re}B := \frac{B + B^*}{2},\ \ \mathfrak{Im}B := \frac{B - B^*}{2i},$$

where it is clear that they are compact self-adjoint operators, since B is compact and due to the technicalities of the given algebraic constructions. Note that the following relation can be established by direct calculation:

$$\mathfrak{Re}^2 B + \mathfrak{Im}^2 B = \frac{B^* B + BB^*}{2},$$

from what follows the inequality

$$\frac{1}{2} \cdot B^* B \leq \mathfrak{Re}^2 B + \mathfrak{Im}^2 B. \tag{2}$$

Having analyzed the latter formula, we see that it is rather reasonable to think over the opportunity of applying the corollary of the minimax principle, pursuing the aim to estimate the singular values of the operator B. For this purpose, consider the following relation: $\mathfrak{Re}^2 B\, f_n = \lambda_n^2 f_n$, where f_n, λ_n are the eigenvectors and the eigenvalues of the operator $\mathfrak{Re} B$, respectively. Since the operator $\mathfrak{Re} B$ is self-adjoint and compact, then its set of eigenvalues form a basis in $\overline{\mathrm{R}(\mathfrak{Re} B)}$. Assume that there exists a nonzero eigenvalue of the operator $\mathfrak{Re}^2 B$ that is different from $\{\lambda_n^2\}_1^\infty$, then, in accordance with the well-known fact of the operator theory, the corresponding eigenvector is orthogonal to the eigenvectors of the operator $\mathfrak{Re} B$. Taking into account the fact that the latter form a basis in $\overline{\mathrm{R}(\mathfrak{Re} B)}$, we come to the conclusion that the eigenvector does not belong to $\overline{\mathrm{R}(\mathfrak{Re} B)}$. Thus, the obtained contradiction proves the fact $\lambda_n(\mathfrak{Re}^2 B) = \lambda_n^2(\mathfrak{Re} B)$. Implementing the same reasonings, we obtain $\lambda_n(\mathfrak{Im}^2 B) = \lambda_n^2(\mathfrak{Im} B)$.

Further, we need a result by Ky Fan [27] (see Corollary 2.2) [11] (Chapter II, § 2.3), in accordance with which we have

$$s_{m+n-1}(\mathfrak{Re}^2 B + \mathfrak{Im}^2 B) \leq \lambda_m(\mathfrak{Re}^2 B) + \lambda_n(\mathfrak{Im}^2 B), \quad m, n = 1, 2, \ldots .$$

Choosing $n = m$ and $n = m + 1$, we obtain, respectively,

$$s_{2m-1}(\mathfrak{Re}^2 B + \mathfrak{Im}^2 B) \leq \lambda_m(\mathfrak{Re}^2 B) + \lambda_m(\mathfrak{Im}^2 B),$$

$$s_{2m}(\mathfrak{Re}^2 B + \mathfrak{Im}^2 B) \leq \lambda_m(\mathfrak{Re}^2 B) + \lambda_{m+1}(\mathfrak{Im}^2 B) \quad m = 1, 2, \ldots .$$

At this stage of reasoning we need involve the sectorial property $\Theta(B) \subset \mathfrak{L}_0(\theta)$, which gives us $|\mathrm{Im}(Bf, f)| \leq \tan\theta\, \mathrm{Re}(Bf, f)$. Applying the corollary of the minimax principle to the latter relation, we obtain $|\lambda_n(\mathfrak{Im} B)| \leq \tan\theta\, \lambda_n(\mathfrak{Re} B)$. Therefore,

$$s_{2m-1}(\mathfrak{Re}^2 B + \mathfrak{Im}^2 B) \leq \lambda_m(\mathfrak{Re}^2 B) + \lambda_m(\mathfrak{Im}^2 B) \leq \sec^2\theta \cdot \lambda_m^2(\mathfrak{Re} B),$$

$$s_{2m}(\mathfrak{Re}^2 B + \mathfrak{Im}^2 B) \leq \sec^2\theta \cdot \lambda_m^2(\mathfrak{Re} B) \quad m = 1, 2, \ldots .$$

Applying the minimax principle to formula (2), we obtain

$$s_{2m-1}(B) \leq \sqrt{2} \sec\theta \cdot \lambda_m(\mathfrak{Re} B), \quad s_{2m}(B) \leq \sqrt{2} \sec\theta \cdot \lambda_m(\mathfrak{Re} B), \quad m = 1, 2, \ldots .$$

This gives us the upper estimate for the singular values of the operator B. □

However, to obtain the lower estimate, we need involve Lemma 3.1 ([23], p. 336), Theorem 3.2 ([23], p. 337). Consider an unbounded operator T, $\Theta(T) \subset \mathfrak{L}_0(\theta)$; in accordance with the first representation theorem ([23], p. 322), we can consider its Friedrichs extension—the m-sectorial operator W, in turn, due to the results ([23], p. 337), it has a real part H which coincides with the Hermitian real component if we deal with a bounded operator. Note that by virtue of the sectorial property, the operator H is non-negative. Further, we consider the case $\mathrm{N}(H) = 0$; it follows that $\mathrm{N}(H^{\frac{1}{2}}) = 0$. To prove this fact we should note that $\mathrm{def} H = 0$; considering inner product with the element belonging to $\mathrm{N}(H^{\frac{1}{2}})$, we easily obtain the fact that it must equal zero. Having analyzed the proof of

Theorem 3.2 ([23], p. 337) we see that its statement remains true in the modified form even in the case where we lift the m-accretive condition; thus, under the sectorial condition imposed upon the closed densely defined operator T, we obtain the following inclusion:

$$T \subset H^{1/2}(I + iG)H^{1/2},$$

where the symbol G denotes a bounded self-adjoint operator in $\mathfrak{H}$. However, to obtain the asymptotic formula established in Theorem 5 [9], we cannot be satisfied by the made assumptions but require the existence of the resolvent at the point zero and its compactness. In spite of the fact that we can proceed our narrative under the weakened conditions regarding the operator W in comparison with H1, H2, we can claim that the statement of Theorem 5 [9] remains true under the assumptions made above, and we prefer to deploy H1, H2, which guarantees the conditions we need and at the same time provides a description of the issue under the natural point of view.

Lemma 2. *Assume that the conditions H1, H2 hold for the operator W, moreover,*

$$\|\mathfrak{Im} W/\mathfrak{Re} W\|_2 < 1,$$

then

$$\lambda_{2n}^{-1}(\mathfrak{Re} W) \leq C s_n(R_W),\; n \in \mathbb{N}.$$

Proof. Firstly, let us show that $\mathrm{D}(W^2)$ is a dense set in $\mathfrak{H}_+$. Since the operator W is closed and strictly accretive, then in accordance with Theorem 3.2 ([23], p. 268), we have $\mathrm{R}(W) = \mathfrak{H}$; hence, there exists the preimage of the set $\mathfrak{M}$—let us denote it by $\mathfrak{M}'$. Consider an arbitrary set of elements $\{x_n\}_0^\infty \subset \mathfrak{H}$ and denote their preimages by x_n'. Using the strictly accretive property of the operator, we have

$$\|x_0 - x_n\|_{\mathfrak{H}} = \|W(x_0' - x_n')\|_{\mathfrak{H}} \geq C\|x_0' - x_n'\|_{\mathfrak{H}_+}.$$

Choosing a sequence

$$\{x_n\}_1^\infty \subset \mathfrak{M},\; x_n \xrightarrow{\mathfrak{H}} x_0,$$

we obtain the fact that the set $\mathfrak{M}'$ is dense in $\mathrm{D}(W)$ in the sense of the norm $\mathfrak{H}_+$; hence, it is dense in $\mathfrak{H}_+$ since $\mathfrak{M} \subset \mathrm{D}(W)$ is dense in $\mathfrak{H}_+$ in accordance with condition H1. Therefore, the set $\mathrm{D}(W^2)$ is dense in $\mathfrak{H}_+$ since $\mathfrak{M}' \subset \mathrm{D}(W^2)$. Thus, we have proved the fulfilment of condition H1 for the operator W^2 with respect to the same pair of Hilbert spaces.

Note that under the assumptions H1, H2, using the reasonings of Theorem 3.2 ([23], p. 337), we have the following representation

$$W = H^{1/2}(I + iG)H^{1/2},\; W^* = H^{1/2}(I - iG)H^{1/2}.$$

It follows easily from this formula that the Hermitian components of the operator W are defined, and we have $\mathfrak{Re} W = H,\;\; \mathfrak{Im} W = H^{1/2}GH^{1/2}$. Using the decomposition $W = \mathfrak{Re} W + i\mathfrak{Im} W,\; W^* = \mathfrak{Re} W - i\mathfrak{Im} W$, we easily obtain

$$\left(\frac{W^2 + W^{*2}}{2} f, f\right)_{\mathfrak{H}} = \|\mathfrak{Re} W f\|_{\mathfrak{H}}^2 - \|\mathfrak{Im} W f\|_{\mathfrak{H}}^2;$$

$$\left(\frac{W^2 - W^{*2}}{2i} f, f\right)_{\mathfrak{H}} = (\mathfrak{Im} W\, \mathfrak{Re} W f, f)_{\mathfrak{H}} + (\mathfrak{Re} W\, \mathfrak{Im} W f, f)_{\mathfrak{H}},\; f \in \mathrm{D}(W^2).$$

Using simple reasonings, we can rewrite the above formulas in terms of Theorem 3.2 ([23], p. 337); we have

$$\mathrm{Re}(W^2 f, f)_{\mathfrak{H}} = \|Hf\|_{\mathfrak{H}}^2 - \|H^{1/2}GH^{1/2}f\|_{\mathfrak{H}}^2,\; \mathrm{Im}(W^2 f, f)_{\mathfrak{H}} = \mathrm{Re}(H^{1/2}GH^{1/2}f, Hf)_{\mathfrak{H}},$$

$$f \in \mathrm{D}(W^2). \tag{3}$$

Consider a set of eigenvalues $\{\lambda_n\}_1^\infty$ and a complete system of orthonormal vectors $\{e_n\}_1^\infty$ of the operator H, the conditions H1, H2 guarantee the existence of the system $\{e_n\}_1^\infty$ since R_H is compact (see Theorem 3 [10]); using the matrix form of the operator G, we have

$$\|Hf\|_{\mathfrak{H}}^2 = \sum_{n=1}^{\infty} |\lambda_n|^2 |f_n|^2, \ \|H^{1/2}GH^{1/2}f\|_{\mathfrak{H}}^2 = \sum_{n=1}^{\infty} \lambda_n \left| \sum_{k=1}^{\infty} b_{nk} \sqrt{\lambda_k} f_k \right|^2,$$

$$\mathrm{Re}(H^{1/2}GH^{1/2}f, Hf)_{\mathfrak{H}} = \mathrm{Re}\left(\sum_{n=1}^{\infty} \lambda_n^{3/2} f_n \sum_{k=1}^{\infty} b_{nk} \sqrt{\lambda_k} \bar{f}_k \right),$$

where b_{nk} are the matrix coefficients of the operator G. Applying the Cauchy–Schwartz inequality, we obtain

$$\|H^{1/2}GH^{1/2}f\|_{\mathfrak{H}}^2 \leq \sum_{n=1}^{\infty} \lambda_n \left| \sum_{k=1}^{\infty} |\lambda_k f_k|^2 \sum_{k=1}^{\infty} |b_{nk}|^2/\lambda_k \right| \leq \|Hf\|_{\mathfrak{H}}^2 \sum_{n,k=1}^{\infty} |b_{nk}|^2 \lambda_n/\lambda_k;$$

$$|\mathrm{Re}(H^{1/2}GH^{1/2}f, Hf)_{\mathfrak{H}}| \leq \|Hf\|_{\mathfrak{H}} \left(\sum_{n=1}^{\infty} \left| \sum_{k=1}^{\infty} \bar{b}_{nk} \sqrt{\lambda_n \lambda_k} f_k \right|^2 \right)^{1/2} \leq \|Hf\|_{\mathfrak{H}}^2 \left(\sum_{n,k=1}^{\infty} |b_{nk}|^2 \lambda_n/\lambda_k \right)^{1/2}.$$

In accordance with the definition of the sectorial property, we require

$$|\mathrm{Im}(W^2 f, f)_{\mathfrak{H}}| \leq \tan\theta \cdot \mathrm{Re}(W^2 f, f)_{\mathfrak{H}}, \ 0 < \theta < \pi/2.$$

Therefore, the sufficient conditions of the sectorial property can be expressed as follows:

$$\|Hf\|_{\mathfrak{H}}^2 \left(\sum_{n,k=1}^{\infty} |b_{nk}|^2/\lambda_k \right)^{1/2} \leq \|Hf\|_{\mathfrak{H}}^2 \left(1 - \sum_{n,k=1}^{\infty} |b_{nk}|^2 \lambda_n/\lambda_k \right) \tan\theta;$$

$$\sum_{n,k=1}^{\infty} |b_{nk}|^2 \lambda_n/\lambda_k + \cot\theta \left(\sum_{n,k=1}^{\infty} |b_{nk}|^2 \lambda_n/\lambda_k \right)^{1/2} \leq 1,$$

where θ is the semiangle of the supposed sector. Solving the corresponding quadratic equation, we obtain the desired estimate:

$$\left(\sum_{n,k=1}^{\infty} |b_{nk}|^2 \lambda_n/\lambda_k \right)^{1/2} < \frac{1}{2}\left\{ \sqrt{\cot^2\theta + 4} - \cot\theta \right\}. \tag{4}$$

Having noticed the fact that the right-hand side of (4) tends to one from below when θ tends to $\pi/2$, we obtain the condition of the sectorial property expressed in terms of the absolute norm:

$$\|H^{1/2}GH^{-1/2}\|_2 := \left(\sum_{n,k=1}^{\infty} |b_{nk}|^2 \lambda_n/\lambda_k \right)^{1/2} < 1, \tag{5}$$

in this case, we can choose the semiangle of the sector using the following relation:

$$\tan\theta = \frac{N}{1 - N^2} + \varepsilon, \ N := \|H^{1/2}GH^{-1/2}\|_2,$$

where ε is an arbitrary small positive number. Thus, we can assume that if the value of the absolute norm is less than one, then the operator W^2 is sectorial and the value of the absolute norm defines the semiangle. Note that coefficients $b_{nk}\sqrt{\lambda_n/\lambda_k}$, $\overline{b_{kn}}\sqrt{\lambda_n/\lambda_k}$ correspond to the matrices of the operators, respectively,

$$H^{1/2}GH^{-1/2}f=\sum_{n=1}^{\infty}\lambda_n^{1/2}e_n\sum_{k=1}^{\infty}b_{nk}\lambda_k^{-1/2}f_k,\; H^{-1/2}GH^{1/2}f=\sum_{n=1}^{\infty}\lambda_n^{-1/2}e_n\sum_{k=1}^{\infty}b_{nk}\lambda_k^{1/2}f_k.$$

Thus, if the absolute operator norm exists, i.e.,

$$\|H^{1/2}GH^{-1/2}\|_2<\infty,$$

then both of them belong to the so-called Hilbert–Schmidt class; however, it is clear without involving the absolute norm since the above operators are adjoint. It is remarkable that we can formally write the obtained estimate in terms of the Hermitian components of the operator, i.e.,

$$\|\mathfrak{Im}W/\mathfrak{Re}W\|_2<1.$$

Below, for a convenient form of writing, we will use a short-hand notation $A:=R_W$, where it is necessary. The next step is to establish the asymptotic formula

$$\lambda_n\left(\frac{A^2+A^{2*}}{2}\right)\asymp\lambda_n^{-1}\left(\mathfrak{Re}W^2\right),\; n\to\infty. \tag{6}$$

However, we cannot directly apply Theorem 5 [9] to the operator W^2; thus, we are compelled to modify the proof having taken into account weaker conditions and the additional condition (5).

Let us observe that the compactness of the operator $R_W(\lambda)$, $\lambda\in\mathrm{P}(W)$ gives us the compactness of the operator W^{-2}. Since the latter is sectorial, it follows easily that $R_{W^2}(\lambda)$, $\lambda\in\mathrm{P}(W^2)$ is compact, since the outside of the sector belongs to the resolvent set and the resolvent compact, at least at one point, is compact everywhere on the resolvent set. Note that due to the reasonings given above, the following relation holds:

$$\mathrm{Re}(W^2f,f)_{\mathfrak{H}}\geq C\|Hf\|^2_{\mathfrak{H}}\geq C\|f\|^2_{\mathfrak{H}_+},\; f\in\mathrm{D}(W^2), \tag{7}$$

where the latter inequality can be obtained easily (see (28) [9]). Thus, we obtain the fact that the operator W^2 is a sectorial, strictly accretive operator; hence, it falls in the scope of the first representation theorem (see Theorem 2.1 [23], p. 322) in accordance with which there exists one-to-one correspondence between the closed densely defined sectorial forms and m-sectorial operators. Using this fact, we can claim that the real part $H_1:=\mathrm{Re}W^2$ is defined and the following relations hold in accordance with the second representation theorem, i.e., Theorem 3.2 ([23], p. 337).

$$W^2=H_1^{1/2}(I+iG_1)H_1^{1/2},\; W^{2*}=H_1^{1/2}(I+iG_2)H_1^{1/2},$$

where G_1,G_2 are self-adjoint bounded operators. Now, by direct calculation, we can verify that $H_1=\mathfrak{Re}W^2$, and we should also note that $\mathrm{D}(W^2)$ is a core of the corresponding closed densely defined sectorial form $\mathfrak{h}$ placed in correspondence to the operator H_1 by virtue of the first representation theorem, i.e., $\mathrm{D}_0(\mathfrak{h})=\mathrm{D}(W^2)$. Let us show that $G_1=-G_2$. We have

$$H_1f=\frac{1}{2}\left[H_1^{\frac{1}{2}}(I+iG_1)+H_1^{\frac{1}{2}}(I+iG_2)\right]H_1^{\frac{1}{2}}=$$

$$=H_1f+\frac{i}{2}H_1^{\frac{1}{2}}(G_1+G_2)H_1^{\frac{1}{2}}f,\; f\in\mathfrak{M}'.$$

By virtue of inequality (7), we see that the operator H_1 is strictly accretive, therefore $\mathrm{N}(H_1) = 0;\ (G_1 + G_2)H_1^{1/2} = 0$. Since

$$\mathfrak{H} = \overline{\mathrm{R}(H_1^{1/2})} \oplus \mathrm{N}(H_1^{1/2}),$$

then $G_1 = G_2 =: G'$. Applying the reasonings represented in Theorem 5 [9], we obtain the fact that $H_1^{-1/2}$ is a bounded operator defined on $\mathfrak{H}$. Using the properties of the operator G', we obtain $\|(I + iG')f\|_{\mathfrak{H}} \cdot \|f\|_{\mathfrak{H}} \geq \mathrm{Re}([I + iG']f, f)_{\mathfrak{H}} = \|f\|^2_{\mathfrak{H}},\ f \in \mathfrak{H}$. Hence, $\|(I + iG')f\|_{\mathfrak{H}} \geq \|f\|_{\mathfrak{H}},\ f \in \mathfrak{H}$. It implies that the operator $I + iG'$ is invertible. The reasonings corresponding to the operator $I - iG'$ are absolutely analogous. Therefore,

$$A^2 = H_1^{-\frac{1}{2}}(I + iG')^{-1}H_1^{-\frac{1}{2}},\ A^{2*} = H_1^{-\frac{1}{2}}(I - iG')^{-1}H_1^{-\frac{1}{2}}. \tag{8}$$

Using simple calculation based upon the operator properties established above, we obtain

$$\mathfrak{Re}A^2 = \frac{1}{2}H_1^{-\frac{1}{2}}(I + G'^2)^{-1}H_1^{-\frac{1}{2}}. \tag{9}$$

Therefore,

$$\left(\mathfrak{Re}A^2 f, f\right)_{\mathfrak{H}} = \left(H_1^{-\frac{1}{2}}(I + G'^2)^{-1}H_1^{-\frac{1}{2}}f, f\right)_{\mathfrak{H}} \leq \|(I + G'^2)^{-1}\| \cdot (R_{H_1}f, f)_{\mathfrak{H}},\ f \in \mathfrak{H}.$$

On the other hand, it is easy to see that $((I + G'^2)^{-1}f, f)_{\mathfrak{H}} \geq \|(I + G'^2)^{-1}f\|^2_{\mathfrak{H}}$. At the same time, it is obvious that the operator $I + G'^2$ is bounded and we have $\|(I + G'^2)^{-1}f\|_{\mathfrak{H}} \geq \|I + G'^2\|^{-1}\|f\|_{\mathfrak{H}}$. Applying these estimates, we obtain

$$\left(\mathfrak{Re}A^2 f, f\right)_{\mathfrak{H}} = \left((I + G'^2)^{-1}H_1^{-\frac{1}{2}}f, H_1^{-\frac{1}{2}}f\right)_{\mathfrak{H}} \geq \|(I + G'^2)^{-1}H_1^{-\frac{1}{2}}f\|^2_{\mathfrak{H}} \geq$$

$$\geq \|I + G'^2\|^{-2} \cdot (R_{H_1}f, f)_{\mathfrak{H}},\ f \in \mathfrak{H}.$$

Using relation (7), we obtain the fact that the resolvent R_{H_1} is compact, and the fact that $\mathfrak{Re}A^2$ is compact is obvious. Thus, analogously to the reasonings of Theorem 5 [9], applying the minimax principle, we obtain the desired asymptotic formula (6). Further, we will use the following formula obtained due to the positiveness of the squared Hermitian imaginary component of the operator A, and we have

$$\frac{A^2 + A^{2*}}{2} = \frac{A^2 + A^{*2}}{2} \leq A^*A + AA^*.$$

Applying the corollary of the well-known Allakhverdiyev theorem (Ky Fan [27]), see Corollary 2.2 [11] (Chapter II, § 2.3), we have

$$\lambda_{2n}(A^*A + AA^*) \leq \lambda_n(A^*A) + \lambda_n(AA^*),\ n \in \mathbb{N}.$$

Taking into account the fact $s_n(A) = s_n(A^*)$, using the minimax principle, we obtain the estimate

$$s_n^2(A) \geq C\lambda_{2n}\left(\frac{A + A^{2*}}{2}\right),\ n \in \mathbb{N},$$

and applying (6), we obtain

$$s_n^2(A) \geq C\lambda_{2n}^{-1}\left(\mathfrak{Re}W^2\right),\ n \in \mathbb{N}.$$

Here, it is rather reasonable to apply formula (3), which gives us

$$\|f\|^2_{\mathfrak{H}} \leq \|f\|^2_{\mathfrak{H}_+} \leq \left(\mathfrak{Re}W^2 f, f\right)_{\mathfrak{H}} \leq (Hf, Hf)_{\mathfrak{H}},\ f \in \mathrm{D}(W^2),$$

which, in turn, collaboratively with the minimax principle, leads us to the theorem statement. Here, we should remark that in order to apply the minimax principle, we need a compact embedding of the energetic space, which is provided by the estimate from below. □

Remark 1. *It is remarkable that the central point of the proof is the representation theorems; in accordance with the first one, we have a plain construction of the operator real part equaling the Hermitian real component. These allow us to implement the simplified scheme of reasonings represented in* [9].

Consider a rather wide operator class including the operators having the asymptotics of the resolvent singular values or one of the real component eigenvalues of the power type, i.e.,

$$C_1 n^{\mu} \leq \lambda_n \leq C_2 n^{\mu},\ \mu < 0.$$

In order to apply the obtained theoretical results to the class, we can reformulate them in the following stylistically convenient form.

Theorem 1. *Assume that the hypotheses H1, H2 hold for the operator W, moreover,*

$$\|\mathfrak{Im} W / \mathfrak{Re} W\|_2 < 1,$$

then

$$s_n(R_W) \asymp \lambda_n^{-1}(\mathfrak{Re} W).$$

Proof. Since conditions H1, H2 hold, then the resolvent R_W is a compact sectorial operator with the vertex situated at the point zero (see Theorem 3 [10]). The estimates from the above and below for the singular values follow from the application of Lemmas 1 and 2, respectively; here, we should take into account the fact that $(Cn)^{\gamma} \asymp n^{\gamma},\ \gamma \in \mathbb{R}$ and the fact that $\lambda_n(\mathfrak{Re} R_W) \asymp \lambda_n^{-1}(\mathfrak{Re} W)$, which is the claim of Theorem 5 [9]. □

4. Mathematical Applications

4.1. The Low Bound for the Schatten Index of the Perturbed Differential Operator

1. Trying to show an application of Lemma 1, we produce an example of a non-self-adjoint operator that is not completely subordinated in the sense of forms (see [8,9]). The pointed-out fact means that we cannot deal with the operator applying methods [8] for they do not work.

   Consider a differential operator acting in the complex Sobolev space:

$$\mathcal{L}f := (c_k f^{(k)})^{(k)} + (c_{k-1} f^{(k-1)})^{(k-1)} + \ldots + c_0 f,$$

$$\mathrm{D}(\mathcal{L}) = H^{2k}(I) \cap H_0^k(I),\ k \in \mathbb{N},$$

where $I := (a,b) \subset \mathbb{R}$, and the complex-valued coefficients $c_j(x) \in C^{(j)}(\bar{I})$ satisfy the condition $\mathrm{sign}(\mathrm{Re} c_j) = (-1)^j,\ j = 1,2,\ldots,k$. Consider a linear combination of the Riemann–Liouville fractional differential operators (see [26], p .44) with the constant real-valued coefficients:

$$\mathcal{D}f := p_n D_{a+}^{\alpha_n} + q_n D_{b-}^{\beta_n} + p_{n-1} D_{a+}^{\alpha_{n-1}} + q_{n-1} D_{b-}^{\beta_{n-1}} + \ldots + p_0 D_{a+}^{\alpha_0} + q_0 D_{b-}^{\beta_0},$$

$$\mathrm{D}(\mathcal{D}) = H^{2k}(I) \cap H_0^k(I),\ n \in \mathbb{N},$$

where $\alpha_j, \beta_j \geq 0$, $0 \leq [\alpha_j], [\beta_j] < k$, $j = 0, 1, \ldots, n.$,

$$q_j \geq 0, \ \text{sign}\, p_j = \begin{cases} (-1)^{\frac{[\alpha_j]+1}{2}}, \ [\alpha_j] = 2m-1, \ m \in \mathbb{N}, \\ (-1)^{\frac{[\alpha_j]}{2}}, \ [\alpha_j] = 2m, \ m \in \mathbb{N}_0. \end{cases}$$

The following result is represented in the paper [9]; consider the operator

$$G = \mathcal{L} + \mathcal{D},$$

$$\mathrm{D}(G) = H^{2k}(I) \cap H_0^k(I).$$

It is clear that it is an operator with a compact resolvent; however, for the accuracy we will prove this fact. Moreover, we will produce a pair of Hilbert spaces so that conditions H1, H2 hold. It follows that the resolvent is compact; thus, we are able to observe the problem related to calculating the Schatten index. Apparently, it may happen that the direct calculation of the singular values or their estimation is rather complicated since we have the following relation:

$$GG^* \supset (\mathcal{L} + \mathcal{D})(\mathcal{L}^* + \mathcal{D}^*) \supset \mathcal{L}\mathcal{L}^* + \mathcal{D}\mathcal{L}^* + \mathcal{L}\mathcal{D}^* + \mathcal{D}\mathcal{D}^*,$$

where inclusions must satisfy some conditions connected with the core of the operator form, for in other cases, we have the risk of losing some singular values. In spite of the fact that the shown difficulties, in many cases, can be eliminated, the offered method of singular values estimation becomes apparently relevant.

Let us prove the fulfilment of the conditions H1, H2 under the assumptions $\mathfrak{H} := L_2(I)$, $\mathfrak{H}^+ := H_0^k(I)$, $\mathfrak{M} := C_0^\infty(I)$. The fulfillment of the condition H1 is obvious; let us show the fulfilment of the condition H2. It is easy to see that

$$\mathrm{Re}(\mathcal{L}f, f)_{L_2(I)} \geq \sum_{j=0}^{k} |\mathrm{Re} c_j| \, \|f^{(j)}\|^2_{L_2(I)} \geq C \|f^{(j)}\|^2_{H_0^k(I)}, \ f \in \mathrm{D}(\mathcal{L}).$$

On the other hand,

$$|(\mathcal{L}f, f)_{L_2(I)}| = \left| \sum_{j=0}^{k} (-1)^j (c_j f^{(j)}, g^{(j)})_{L_2(I)} \right| \leq \sum_{j=0}^{k} \left| (c_j f^{(j)}, g^{(j)})_{L_2(I)} \right| \leq$$

$$\leq C \sum_{j=0}^{k} \|f^{(j)}\|_{L_2(I)} \|g^{(j)}\|_{L_2(I)} \leq \|f\|_{H_0^k(I)} \|g\|_{H_0^k(I)}, \ f \in \mathrm{D}(\mathcal{L}).$$

Consider fractional differential Riemann–Liouville operators of arbitrary non-negative order α (see [26], p. 44) defined by the expressions

$$D_{a+}^{\alpha} f = \left(\frac{d}{dx} \right)^{[\alpha]+1} I_{a+}^{1-\{\alpha\}} f; \ D_{b-}^{\alpha} f = \left(-\frac{d}{dx} \right)^{[\alpha]+1} I_{b-}^{1-\{\alpha\}} f,$$

where the fractional integrals of arbitrary positive order α, defined by

$$(I_{a+}^{\alpha} f)(x) = \frac{1}{\Gamma(\alpha)} \int_a^x \frac{f(t)}{(x-t)^{1-\alpha}} dt, \ (I_{b-}^{\alpha} f)(x) = \frac{1}{\Gamma(\alpha)} \int_x^b \frac{f(t)}{(t-x)^{1-\alpha}} dt, f \in L_1(I).$$

Suppose $0 < \alpha < 1$, $f \in AC^{l+1}(\bar{I})$, $f^{(j)}(a) = f^{(j)}(b) = 0$, $j = 0, 1, \ldots, l$; then the next formula follows from Theorem 2.2 ([26], p. 46):

$$D_{a+}^{\alpha+l} f = I_{a+}^{1-\alpha} f^{(l+1)}, \ D_{b-}^{\alpha+l} f = (-1)^{l+1} I_{b-}^{1-\alpha} f^{(l+1)}. \quad (10)$$

Further, we need the following inequalities (see [28]):

$$Re(D_{a+}^{\alpha}f,f)_{L_2(I)} \geq C\|f\|_{L_2(I)}^2, \; f \in I_{a+}^{\alpha}(L_2),$$

$$Re(D_{b-}^{\alpha}f,f)_{L_2(I)} \geq C\|f\|_{L_2(I)}^2, \; f \in I_{b-}^{\alpha}(L_2), \tag{11}$$

where $I_{a+}^{\alpha}(L_2), I_{b-}^{\alpha}(L_2)$ are the classes of the functions representable by the fractional integrals (see [26]). Consider the following operator with the constant real-valued coefficients:

$$\mathcal{D}f := p_n D_{a+}^{\alpha_n} + q_n D_{b-}^{\beta_n} + p_{n-1} D_{a+}^{\alpha_{n-1}} + q_{n-1} D_{b-}^{\beta_{n-1}} + \ldots + p_0 D_{a+}^{\alpha_0} + q_0 D_{b-}^{\beta_0},$$

$$\mathrm{D}(\mathcal{D}) = H^{2k}(I) \cap H_0^k(I), \; n \in \mathbb{N},$$

where $\alpha_j, \beta_j \geq 0, \; 0 \leq [\alpha_j], [\beta_j] < k, \; j = 0, 1, \ldots, n.$,

$$q_j \geq 0, \; \mathrm{sign}\, p_j = \begin{cases} (-1)^{\frac{[\alpha_j]+1}{2}}, \; [\alpha_j] = 2m-1, \; m \in \mathbb{N}, \\ (-1)^{\frac{[\alpha_j]}{2}}, \; [\alpha_j] = 2m, \; m \in \mathbb{N}_0. \end{cases}$$

Using (10) and (11), we obtain

$$(p_j D_{a+}^{\alpha_j} f, f)_{L_2(I)} = p_j \left(\left(\frac{d}{dx} \right)^m D_{a+}^{m-1+\{\alpha_j\}} f, f \right)_{L_2(I)} = (-1)^m p_j \left(I_{a+}^{1-\{\alpha_j\}} f^{(m)}, f^{(m)} \right)_{L_2(I)} \geq$$

$$\geq C \left\| I_{a+}^{1-\{\alpha_j\}} f^{(m)} \right\|_{L_2(I)}^2 = C \left\| D_{a+}^{\{\alpha_j\}} f^{(m-1)} \right\|_{L_2(I)}^2 \geq C \left\| f^{(m-1)} \right\|_{L_2(I)}^2,$$

where $f \in \mathrm{D}(\mathcal{D})$ is a real-valued function and $[\alpha_j] = 2m-1, \; m \in \mathbb{N}$. Similarly, we obtain for orders $[\alpha_j] = 2m, \; m \in \mathbb{N}_0$

$$(p_j D_{a+}^{\alpha_j} f, f)_{L_2(I)} = p_j \left(D_{a+}^{2m+\{\alpha_j\}} f, f \right)_{L_2(I)} = (-1)^m p_j \left(D_{a+}^{m+\{\alpha_j\}} f, f^{(m)} \right)_{L_2(I)} =$$

$$= (-1)^m p_j \left(D_{a+}^{\{\alpha_j\}} f^{(m)}, f^{(m)} \right)_{L_2(I)} \geq C \left\| f^{(m)} \right\|_{L_2(I)}^2.$$

Thus in both cases, we have

$$(p_j D_{a+}^{\alpha_j} f, f)_{L_2(I)} \geq C \left\| f^{(s)} \right\|_{L_2(I)}^2, \; s = [[\alpha_j]/2].$$

In the same way, we obtain the inequality

$$(q_j D_{b-}^{\alpha_j} f, f)_{L_2(I)} \geq C \left\| f^{(s)} \right\|_{L_2(I)}^2, \; s = [[\alpha_j]/2].$$

Hence, in the complex case, we have

$$\mathrm{Re}(\mathcal{D}f, f)_{L_2(I)} \geq C\|f\|_{L_2(I)}^2, \; f \in \mathrm{D}(\mathcal{D}).$$

Combining Theorem 2.6 ([26], p. 53) with (10), we obtain

$$\left\| p_j D_{a+}^{\alpha_j} f \right\|_{L_2(I)} = \left\| I_{a+}^{1-\{\alpha_j\}} f^{([\alpha_j]+1)} \right\|_{L_2(I)} \leq C \left\| f^{([\alpha_j]+1)} \right\|_{L_2(I)} \leq C\|f\|_{H_0^k(I)};$$

$$\left\| q_j D_{b-}^{\alpha_j} f \right\|_{L_2(I)} \leq C\|f\|_{H_0^k(I)}, \; f \in \mathrm{D}(\mathcal{D}).$$

Hence, we obtain

$$\|\mathcal{D}f\|_{L_2(I)} \leq C\|f\|_{H_0^k(I)},\ f \in \mathrm{D}(\mathcal{D}).$$

Taking into account the relation

$$\|f\|_{L_2(I)} \leq C\|f\|_{H_0^k(I)},\ f \in H_0^k(I),$$

combining the above estimates, we obtain

$$\mathrm{Re}(Gf,f)_{L_2(I)} \geq C\|f\|^2_{H_0^k(I)},\ |(Gf,g)_{L_2(I)}| \leq \|f\|_{H_0^k(I)}\|g\|_{H_0^k(I)},\ f,g \in C_0^\infty(I).$$

Thus, we have obtained the desired result.

To deploy the minimax principle for eigenvalues estimating, we come to the following relation:

$$C_1\|f\|^2_{H_0^k(I)} \leq (\mathfrak{Re}Gf,f)_{L_2(I)} \leq C_2\|f\|^2_{H_0^k(I)},$$

from which follows easily, due to the asymptotic formula for the eigenvalues of a self-adjoint operator (see [29]), the fact that

$$\lambda_n(\mathfrak{Re}G) \asymp n^{2k},\ n \in \mathbb{N};$$

therefore, applying Lemma 1 collaboratively with the asymptotic equivalence formula (see Theorem 5 [9])

$$\lambda_n^{-1}(\mathfrak{Re}G) \asymp \lambda_n(\mathfrak{Re}R_G),\ n \in \mathbb{N},$$

we obtain the fact that

$$R_G \in \mathfrak{S}_p,\ \inf p \leq 1/2k.$$

Thus, it gives us an opportunity to establish the range of the Schatten index.

2. Let us show the application of Lemma 2; firstly, consider the following reasonings:

$$\|\mathfrak{Im}WH^{-1}\|_2 = \|H^{-1}\mathfrak{Im}W\|_2 = \sum_{n,k=1}^{\infty}\left|(\mathfrak{Im}We_n, H^{-1}e_k)_{\mathfrak{H}}\right|^2 = \sum_{n,k=1}^{\infty}\lambda_n^{-2}(H)|(e_n,\mathfrak{Im}We_k)_{\mathfrak{H}}|^2 =$$

$$= \sum_{n=1}^{\infty}\lambda_n^{-2}(H)||\mathfrak{Im}We_n||^2_{\mathfrak{H}},$$

where $\{e_n\}_1^\infty$ is the orthonormal set of the eigenvectors of the operator H. Thus, we obtain the following condition:

$$\sum_{n=1}^{\infty}\lambda_n^{-2}(H)||\mathfrak{Im}We_n||^2_{\mathfrak{H}} < 1, \tag{12}$$

which guarantees the fulfilment of the conditions expressed in terms of absolute norm in Lemma 2. It is remarkable that this form of the condition is quite convenient if we consider perturbations of differential operators. Below, we observe a simplified case of the operator considered in the previous paragraph. Consider

$$Lf := -f'' + \xi D^\alpha_{0+}f,\ \mathrm{D}(L) = H^2(I) \cap H_0^1(I),\ I = (0,\pi),\ \alpha \in (0,1/2),\ \xi \in \mathbb{R},$$

then

$$C_0(L_1f,f)_{L_2(I)} \leq (\mathfrak{Re}Lf,f)_{L_2(I)} \leq C_1(L_1f,f)_{L_2(I)},\ L_1f := -f'',\ \mathrm{D}(L_1) = \mathrm{D}(L).$$

It is a well-known fact that

$$\lambda_n(L_1) = n^2,\ e_n = \sin nx.$$

It is also clear that

$$\mathfrak{Im}L \supset \xi(D^{\alpha}_{0+} - D^{\alpha}_{\pi-})/2i.$$

In accordance with the first representation theorem (see Theorem 2.1 [23], p. 322), we have that $H^2(I) \cap H^1_0(I)$ is a core of the form corresponding to the operator L^*; hence,

$$\mathfrak{Im}L = \xi(D^{\alpha}_{0+} - D^{\alpha}_{\pi-})/2i.$$

Note that

$$(D^{\alpha}_{0+}e_n)(x) = \frac{n}{\Gamma(1-\alpha)}\int\limits_0^x (x-t)^{-\alpha}\cos nt\,dt.$$

Applying the generalized Minkowski inequality, we obtain

$$\left(\int\limits_0^{\pi}|(D^{\alpha}_{a+}e_n)(x)|^2dx\right)^{1/2} = \frac{n}{\Gamma(1-\alpha)}\left(\int\limits_0^{\pi}\left|\int\limits_0^x (x-t)^{-\alpha}\cos nt\,dt\right|^2\right)^{1/2} \leq$$

$$\leq \frac{n}{\Gamma(1-\alpha)}\int\limits_0^{\pi}\cos nt\,dt\left(\int\limits_t^{\pi}(x-t)^{-2\alpha}dx\right)^{1/2} = \frac{n}{\sqrt{(1-2\alpha)}\Gamma(1-\alpha)}\int\limits_0^{\pi}(\pi-t)^{1/2-\alpha}\cos nt\,dt \leq$$

$$\leq \frac{n\pi^{1/2-\alpha}}{\sqrt{(1-2\alpha)}\Gamma(1-\alpha)}.$$

Analogously, we obtain

$$\left(\int\limits_0^{\pi}|(D^{\alpha}_{\pi-}e_n)(x)|^2dx\right)^{1/2} \leq \frac{n\pi^{1/2-\alpha}}{\sqrt{(1-2\alpha)}\Gamma(1-\alpha)}.$$

Hence,

$$\|\mathfrak{Im}Le_n\| \leq \frac{n\xi\pi^{1/2-\alpha}}{\sqrt{(1-2\alpha)}\Gamma(1-\alpha)}.$$

Therefore,

$$\sum_{n=1}^{\infty}\lambda_n^{-2}(\mathfrak{Re}L)||\mathfrak{Im}Le_n||^2 < \frac{\xi^2\pi^{1-2\alpha}}{(1-2\alpha)\Gamma^2(1-\alpha)}\sum_{n=1}^{\infty}\frac{1}{n^2} = \frac{\xi^2\pi^{3-2\alpha}}{6(1-2\alpha)\Gamma^2(1-\alpha)}.$$

Using this relation, we can obviously impose a condition on ξ that guarantees the fulfilment of relation (12), i.e.,

$$\xi < \frac{\sqrt{6(1-2\alpha)}\Gamma(1-\alpha)}{\pi^{3/2-\alpha}}.$$

In accordance with Theorem 1, the last condition follows that

$$s_n^{-1}(R_L) \asymp n^2,\; R_L \in \mathfrak{S}_p,\; \inf p = 1/2.$$

4.2. Existence and Uniqueness Theorems for Evolution Equations via Obtained Results

In this paragraph, we consider applications to differential equations in concrete Hilbert spaces and involve such operators as Riemann–Liouville operator, Kipriyanov operator, and Riesz potential, difference operator. Moreover, we produce the artificially constructed normal operator for which the clarification of the Lidskii results relevantly works.

Further, we consider a Hilbert space $\mathfrak{H}$ which consists of element-functions $u : \mathbb{R}_+ \to \mathfrak{H}$, $u := u(t)$, $t \geq 0$ and we assume that if u belongs to $\mathfrak{H}$ then the fact holds for all values of the variable t. Notice that under such an assumption all standard topological properties, such as completeness, compactness, etc., remain correctly defined. We understand such operations as differentiation and integration in the generalized sense that is caused by the topology of the Hilbert space $\mathfrak{H}$; more detailed information can be found in Chapter 4 Krasnoselskii M.A. [30]. Consider an arbitrary compact operator B; we can form the operators corresponding to the groups of its eigenvalues, i.e.,

$$\mathcal{P}_\nu(B,\alpha,t) \Leftrightarrow \lambda_{N_\nu+1}, \lambda_{N_\nu+2}, \ldots, \lambda_{N_{\nu+1}},$$

where $\{N_\nu\}_0^\infty$ is a sequence of natural numbers,

$$\mathcal{P}_\nu(B,\alpha,t) = \frac{1}{2\pi i} \int\limits_{\vartheta_\nu(B)} e^{-\lambda^\alpha t} B(I - \lambda B)^{-1} d\lambda, \; \alpha > 0,$$

$\vartheta_\nu(B)$ is a contour on the complex plain containing the eigenvalues $\lambda_{N_\nu+1}, \lambda_{N_\nu+2}, \ldots, \lambda_{N_{\nu+1}}$ only and no more eigenvalues.

The root vectors of the operator B are called by the Abel–Lidskii basis if

$$\sum_{\nu=0}^\infty \mathcal{P}_\nu(B,\alpha,t) \to I, \; t \to 0,$$

where convergence is understood as the operator pointwise convergence in the Hilbert space.

The correspondence between the series and the element, given due to the formula, is known as a convergence in the Abel–Lidskii sense. We can compare this definition with the main principle of the spectral theorem—the unit decomposition. We place the following contour in correspondence to the operator:

$$\vartheta(B) := \{\lambda : |\lambda| = r > 0, |\arg\lambda| \leq \theta + \varepsilon\} \cup \{\lambda : |\lambda| > r, |\arg\lambda| = \theta + \varepsilon\}.$$

Consider the following hypotheses:

S1: *Under the assumptions $B \in \mathfrak{S}_p$, $\inf p \leq \alpha$, $\Theta(B) \subset \mathfrak{L}_0(\theta)$, a sequence of natural numbers $\{N_\nu\}_0^\infty$ can be chosen so that*

$$\frac{1}{2\pi i} \int\limits_{\vartheta(B)} e^{-\lambda^\alpha t} B(I - \lambda B)^{-1} f d\lambda = \sum_{\nu=0}^\infty \mathcal{P}_\nu(B,\alpha,t)f, \; f \in \mathfrak{H},$$

the latter series is absolutely convergent in the sense of the norm.

Combining the generalized integrodifferential operations, we can consider a fractional differential operator in the Riemann–Liouville sense, i.e., in the formal form, we have

$$\mathfrak{D}_-^{1/\alpha} f(t) := -\frac{1}{\Gamma(1 - 1/\alpha)} \frac{d}{dt} \int\limits_0^\infty f(t+x) x^{-1/\alpha} dx, \; \alpha > 1.$$

Let us study a Cauchy problem:

$$\mathfrak{D}_-^{1/\alpha} u = Wu, \; u(0) = h \in D(W). \tag{13}$$

Note that it is possible to apply the Abel–Lidskii concept using the methods [1,10,14–16] in the case $R_W \in \mathfrak{S}_p$, $\inf p \leq \alpha$. We can assume that the central result of the above-listed papers is to find conditions under which the hypotheses S1 holds. We can generalize the results related to the existence and uniqueness theorem (see Theorem 4 [31], Theorem 1 [16], Theorem 6 [15]), as follows:

Theorem 2. *Assume that S1 holds, then there exists a solution of Cauchy problem* (13) *in the form*

$$u(t)=\sum_{\nu=0}^{\infty}\mathcal{P}_{\nu}(B,\alpha,t)h.$$

Apparently, under this point of view, the results of the paper become relevant since, applying Theorem 1, we can find the exact value of the Schatten index p. Therefore, we can decrease the value of α, satisfying the condition $\inf p\leq\alpha$ in accordance with S1.

To demonstrate the claimed result, we produce an example dealing with well-known operators. Consider a rectangular domain in the space $\mathbb{R}^n$, defined as follows: $\Omega:=\{x_j\in[0,\pi],\,j=1,2,\ldots,n\}$; and consider the Kipriyanov fractional differential operator defined in the paper [25] by the formal expression

$$\mathfrak{D}^{\beta}f(Q)=\frac{\beta}{\Gamma(1-\beta)}\int\limits_{0}^{r}\frac{[f(Q)-f(T)]}{(r-t)^{\beta+1}}\left(\frac{t}{r}\right)^{n-1}dt+(n-1)!f(Q)r^{-\beta}/\Gamma(n-\beta),$$

$$\beta\in(0,1),\,P\in\partial\Omega,$$

where $Q:=P+\mathbf{e}r,\ P:=P+\mathbf{e}t,\ \mathbf{e}$ is a unit vector having a direction from the fixed point of the boundary P to an arbitrary point Q belonging to Ω. Consider the perturbation of the Laplace operator by the Kipriyanov operator:

$$L:=D^{2k}+\xi\mathfrak{D}^{\beta},\,\mathrm{D}(L)=H_0^k(\Omega)\cap H^{2k}(\Omega),$$

where $\xi>0$,

$$D^{2k}f=(-1)^k\sum_{j=1}^{n}\mathcal{D}_j^{2k}f.$$

It was proved in the paper [10] that

$$C_0(D^{2k}f,f)_{L_2(\Omega)}\leq(\mathfrak{Re}Lf,f)_{L_2(\Omega)}\leq C_1(D^{2k}f,f)_{L_2(\Omega)},\ f\in\mathrm{D}(L).$$

Therefore,

$$\lambda_n(\mathfrak{Re}L)\asymp n^{2k/n}.$$

On the other hand, we have the following eigenfunctions of D^{2k} in the rectangular domain:

$$e_{\bar{l}}=\prod_{j=1}^{n}\sin l_jx_j,\ \bar{l}:=\{l_1,l_2,\ldots,l_n\},\,l_s\in\mathbb{N},\,s=1,2,\ldots,n.$$

It is clear that

$$D^{2k}e_{\bar{l}}=\lambda_{\bar{l}}e_{\bar{l}},\ \lambda_{\bar{l}}=\sum_{j=1}^{n}l_j^{2k}.$$

Since the search for the below-given information in the literature (however, it is a well-known fact) can bring some difficulties, we would like to represent it. Let us prove that the system $\{e_{\bar{l}}\}$ is complete in the Hilbert space $L_2(\Omega)$. We will show it if we prove that the element that is orthogonal to every element of the system is a zero. Assume that

$$\int\limits_0^{\pi}\sin l_1x_1dx_1\int\limits_0^{\pi}\sin l_2x_2dx_2\ldots\int\limits_0^{\pi}\sin l_nx_nf(x_1,x_2,\ldots,x_n)dx_n=(e_{\bar{l}},f)_{L_2(\Omega)}=0.$$

In accordance with the fact that the system $\{\sin mx\}_1^\infty$ is a complete system in $L_2(0,\pi)$, we conclude that

$$\int_0^\pi \sin l_2 x_2 dx_2 \ldots \int_0^\pi \sin l_n x_n f(x_1, x_2, \ldots, x_n) dx_n = 0.$$

Having repeated the same reasonings step by step, we obtain the desired result. Taking into account the following inequality (see [10]) and the embedding theorems, we obtain

$$\|\mathfrak{D}^\beta f\|_{L_2(\Omega)} \leq C_\beta \|f\|_{H_0^1(\Omega)} \leq C_{\beta,k,n} \|f\|_{H_0^k(\Omega)}, \tag{14}$$

where the constant C_β is defined through the infinitesimal generator J of the corresponding semigroup of contraction (shift semigroup in the direction) (9) [10]. Now it is clear that the conditions H1, H2 are satisfied, where $\mathfrak{H} := L_2(\Omega)$, $\mathfrak{H}_+ := H_0^k(\Omega)$, $\mathfrak{M} := C_0^\infty(\Omega)$. Using the intermediate inequality (14), by direct calculation, we obtain

$$\sum_{l_1, l_2, \ldots l_n = 1}^\infty \lambda_{\bar{l}}^{-2}(\mathfrak{Re} L)_{L_2(\Omega)} \, \|\mathfrak{Im} L e_{\bar{l}}\|^2_{L_2(\Omega)} \leq (\xi C_\beta)^2 \sum_{l_1, l_2, \ldots l_n = 1}^\infty \frac{\lambda_{\bar{l}}(D^2)}{\lambda^2_{\bar{l}}(D^{2k})}.$$

Therefore, if the following condition holds,

$$\sum_{l_1, l_2, \ldots l_n = 1}^\infty \frac{l_1^2 + l_2^2 + \ldots + l_n^2}{(l_1^{2k} + l_2^{2k} + \ldots + l_n^{2k})^2} < (\xi C_\beta)^{-2}, \tag{15}$$

then the conditions of Lemma 2 are satisfied. Applying Lemma 2, we can consider the values of the parameters k, n such that the last series is convergent, and at the same time, $R_L \in \mathfrak{S}_p$, $\inf p = n/2k > 1$. The latter fact gives us the argument showing the relevance of Lemma 2 since we can find the range of α appropriate for the Abel–Lidskii method applicability. Below, we produce the corresponding reasonings.

Assume that the following condition holds:

$$\frac{n}{2} + 1 < 2k < n.$$

Consider the vector function

$$\psi(\bar{l}) = \frac{(l_1^{2k} + l_2^{2k} + \ldots + l_n^{2k})^2}{l_1^2 + l_2^2 + \ldots + l_n^2},$$

then $\psi(\bar{t}) = nt^{2(2k-1)}$, $\bar{t} = \{t, t, \ldots t\}$. It is clear that the number s of values $\psi(\bar{l})$, $l_i \leq t$ equals t^n, i.e., $s = t^n$. Therefore,

$$\psi(\bar{t}) = ns^{\frac{2(2k-1)}{n}}, \; \psi(\overline{t-1}) = n(s^{1/n} - 1)^{2(2k-1)};$$

$$n(s^{1/n} - 1)^{2(2k-1)} \leq \psi(\bar{l}) \leq ns^{\frac{2(2k-1)}{n}}, \; t - 1 \leq l_i \leq t, \; i = 1, 2, \ldots, n.$$

Having arranged the values in the order corresponding to their absolute value increasing, we obtain

$$n(s^{1/n} - 1)^{2(2k-1)} \leq \psi_j \leq ns^{\frac{2(2k-1)}{n}}, \; (s^{1/n} - 1)^n < j < s.$$

Therefore,

$$\frac{(s^{1/n} - 1)^{2(2k-1)}}{s^{\frac{2(2k-1)}{n}}} < \frac{\psi_j}{nj^{\frac{2(2k-1)}{n}}} < \frac{s^{\frac{2(2k-1)}{n}}}{(s^{1/n} - 1)^{2(2k-1)}},$$

from which follows the convergence of the following series, since if we take into account the condition $n/2 + 1 < 2k$, we obtain

$$\sum_{j=1}^{\infty} \psi_j^{-1} < \infty.$$

In other words, we have proved that series (15) is convergent. Thus, we have considered the case showing the relevance of Lemma 2. We can claim that the Abel–Lidskii method in its classical form is not applicable to the fractional evolution equation for the values of α less than $n/2k$. This rather ridiculous result, from one point of view, gives us a better comprehension of methodology and allows us to avoid disturbing calculation and difficulties of any kind connected with the verification of opportunity to apply the method.

5. Conclusions

In this paper, we represent an efficient tool for finding the asymptotics of operator singular values. However, it may be interesting itself since it appeals to the spectral properties of the operator real component, which are undoubtedly relevant in the framework of the abstract spectral theory. Some difficulties in the application of the Abel–Lidskii method were considered under the point of view of the created concept, where the the mathematical applications cover integrodifferential operators of the real order.

Funding: This research received no external funding.

Data Availability Statement: Data are contained within the article.

Acknowledgments: The author is sincerely grateful to Natalia P. Bondarenko for exhaustive remarks, witty comments, and careful proofreading.

Conflicts of Interest: The authors declare no conflict of interest.

References

1. Lidskii, V.B. Summability of series in terms of the principal vectors of non-selfadjoint operators. *Tr. Mosk. Mat. Obs.* **1962**, *11*, 3–35.
2. Markus, A.S.; Matsaev, V.I. Operators generated by sesquilinear forms and their spectral asymptotics. *Mat. Issled* **1981**, *61*, 86–103.
3. Agranovich, M.S. On series with respect to root vectors of operators associated with forms having symmetric principal part. *Funct. Anal. Its Appl.* **1994**, *28*, 151–167. [CrossRef]
4. Katsnelson, V.E. Conditions under which systems of eigenvectors of some classes of operators form a basis. *Funct. Anal. Appl.* **1967**, *1*, 122–132. [CrossRef]
5. Krein, M.G. Criteria for completeness of the system of root vectors of a dissipative operator. *Amer. Math. Soc. Transl. Ser. Amer. Math. Soc.* **1963**, *26*, 221–229.
6. Markus, A.S. Expansion in root vectors of a slightly perturbed selfadjoint operator. *Soviet Math. Dokl.* **1962**, *3*, 104–108.
7. Motovilov, A.K.; Shkalikov, A.A. Preserving of the unconditional basis property under non-self-adjoint perturbations of self-adjoint operators. *Funktsional. Anal. i Prilozhen.* **2019**, *53*, 45–60. [CrossRef]
8. Shkalikov, A.A. Perturbations of selfadjoint and normal operators with a discrete spectrum. *Russ. Math. Surv.* **2016**, *71*, 113–174. [CrossRef]
9. Kukushkin, M.V. On one method of studying spectral properties of non-selfadjoint operators. *Abstr. Appl. Anal.* **2020**, *2020*, 1461647. [CrossRef]
10. Kukushkin, M.V. Abstract fractional calculus for m-accretive operators. *Int. J. Appl. Math.* **2021**, *34*. [CrossRef]
11. Gohberg, I.C.; Krein, M.G. *Introduction to the Theory of Linear Non-Selfadjoint Operators in a Hilbert Space*; Fizmatlit: Moscow, Russia, 1965.
12. Bazhlekova, E. The abstract Cauchy problem for the fractional evolution equation. *Fract. Calc. Appl. Anal.* **1998**, *1*, 255–270.
13. Clément, P.; Gripenberg, G.; Londen, S.-O. Hölder regularity for a linear fractional evolution equation. In *Topics in Nonlinear Analysis, The Herbert Amann Anniversary Volume*; Birkhäuser: Basel, Switzerland, 1998.
14. Kukushkin, M.V. Natural lacunae method and Schatten-von Neumann classes of the convergence exponent. *Mathematics* **2022**, *10*, 2237. [CrossRef]
15. Kukushkin, M.V. Evolution Equations in Hilbert Spaces via the Lacunae Method. *Fractal Fract.* **2022**, *6*, 229. [CrossRef]
16. Kukushkin, M.V. Abstract Evolution Equations with an Operator Function in the Second Term. *Axioms* **2022**, *11*, 434. [CrossRef]
17. Mainardi, F. The fundamental solutions for the fractional diffusion-wave equation. *Appl. Math. Lett.* **1966**, *9*, 23–28. [CrossRef]

18. Mamchuev, M.O. Solutions of the main boundary value problems for the time-fractional telegraph equation by the Green function method. *Fract. Calc. Appl. Anal.* **2017**, *20*, 190–211. [CrossRef]
19. Mamchuev, M.O. Boundary value problem for the time-fractional telegraph equation with Caputo derivatives. *Math. Model. Nat. Phenom.* **2017**, *12*, 82–94. [CrossRef]
20. Moroz, L.; Maslovskaya, A.G. Hybrid stochastic fractal-based approach to modeling the switching kinetics of ferroelectrics in the injection mode. *Math. Model. Comput. Simulations* **2020**, *12*, 348–356. [CrossRef]
21. Pskhu, A.V. The fundamental solution of a diffusion-wave equation of fractional order. *Izv. Math.* **2009**, *73*, 351–392. [CrossRef]
22. Wyss, W. The fractional diffusion equation. *J. Math. Phys.* **1986**, *27*, 2782–2785. [CrossRef]
23. Kato, T. *Perturbation Theory for Linear Operators*; Springer: Berlin/Heidelberg, Germany; New York, NY, USA, 1980.
24. Kipriyanov, I.A. On spaces of fractionally differentiable functions. *Izv. Akad. Nauk SSSR Ser. Mat.* **1960**, *24*, 865–882.
25. Kipriyanov, I.A. The operator of fractional differentiation and powers of the elliptic operators. *Proc. Acad. Sci. USSR* **1960**, *131*, 238–241.
26. Samko, S.G.; Kilbas, A.A.; Marichev, O.I. *Fractional Integrals and Derivatives: Theory and Applications*; Gordon and Breach Science Publishers: Philadelphia, PA, USA, 1993.
27. Fan, K. Maximum properties and inequalities for the eigenvalues of completely continuous operators. *Proc. Nat. Acad. Sci. USA* **1951**, *37*, 760–766. [CrossRef]
28. Kukushkin, M.V. Spectral properties of fractional differentiation operators. *Electron. J. Differ. Equations* **2018**, *2018*, 1–24.
29. Rozenblyum, G.V.; Solomyak, M.Z.; Shubin, M.A. Spectral theory of differential operators. *Results Sci. Technol. Ser. Mod. Probl. Math. Fundam. Dir.* **1989**, *64*, 5–242.
30. Krasnoselskii, M.A.; Zabreiko, P.P.; Pustylnik, E.I.; Sobolevskii, P.E. *Integral Operators in the Spaces of Summable Functions*; Fizmatlit: Moscow, Russia, 1966.
31. Kukushkin, M.V. Cauchy Problem for an Abstract Evolution Equation of Fractional Order. *Fractal Fract.* **2023**, *7*, 111. [CrossRef]

mathematics MDPI

Article

Keller–Osserman Phenomena for Kardar–Parisi–Zhang-Type Inequalities

Andrey B. Muravnik

Nikol'skii Mathematical Institute, Peoples Friendship University of Russia, Miklukho–Maklaya ul. 6, 117198 Moscow, Russia; amuravnik@mail.ru

Abstract: For coercive quasilinear partial differential inequalities containing nonlinearities of the Kardar–Parisi–Zhang type, we find conditions guaranteeing the absence of global positive solutions. These conditions extend both the classical result of Keller and Osserman and its recent Kon'kov–Shishkov generalization. Additionally, they complement the results for the noncoercive case, which had been previously established by the same author.

Keywords: coercive quasilinear inequalities; KPZ nonlinearities; blow-up

MSC: 35R45; 35J62

Citation: Muravnik, A.B. Keller–Osserman Phenomena for Kardar–Parisi–Zhang-Type Inequalities. *Mathematics* **2023**, *11*, 3787. https://doi.org/10.3390/math11173787

Academic Editor: Natalia Bondarenko

Received: 13 August 2023
Revised: 1 September 2023
Accepted: 2 September 2023
Published: 4 September 2023

1. Introduction

According to the Pokhozhaev paradigm, blow-up phenomena are equivalent to the absence of global solutions. This approach is based on the following reasoning. It is quite frequent for real models of mathematical physics that an equation (or inequality) is resolvable "in small" (i.e., in a neighborhood of the ground state). In this case, if we can prove that no global solutions exist, then there exists a point where the solution is destroyed, which means the blow-up phenomenon. For a thorough explanation of this approach, readers are addressed to the famous monograph [1] providing the foundation of the global nonexistence theory and containing a lot of blow-up results for various semilinear and quasilinear equations, inequalities, and boundary-value problems.

In this paper, the said phenomena are investigated for coercive inequalities with nonlinearities of the Kardar–Parisi–Zhang-type (KPZ-type nonlinearities), i.e., coercive inequalities containing the second power of the first derivative of the desired function (note that this kind of nonlinearity is not covered by the authors of [1]). The motivation to study KPZ-type nonlinearities is well known; for instance, a comprehensive list of recent publications illustrating their applications not covered by other kinds of nonlinearities (for example, interface dynamics and directed polymer models) is provided in [2], which is a review of various results about equations and inequalities with KPZ-type nonlinearities. Their theoretical value is caused by the following circumstance: the second power is the greatest one such that a priori L_∞ estimates of first-order derivatives of the solution via the L_∞-norm of the solution itself hold (see, for example, [3–5]).

For noncoercive KPZ-type nonlinearities, i.e., for the case where the highest-order linear part is dominated by the (low-order) nonlinear one, the above phenomenon is investigated earlier (see [2] and references therein). The coercive case was an open problem up to now, though the result for semilinear coercive inequalities (studied in regards to the problem of the equilibrium of a charged gas in a container) has been known longer than six decades: in [6,7], a sufficient condition of the absence of global solutions is proved for inequalities of the kind

$$\Delta\nu \geq \mu(\nu), \tag{1}$$

where Δ denotes the Laplacian: $\Delta = \dfrac{\partial^2}{\partial x_1^2} + \cdots + \dfrac{\partial^2}{\partial x_n^2}$.

2. Regular Case

In $\mathbb{R}^n$, consider the inequality

$$\Delta u + \sum_{j=1}^{n} g_j(x,u)\left(\frac{\partial u}{\partial x_j}\right)^2 \geq \omega(u), \tag{2}$$

assuming that there exists a function g continuous and locally summable over the positive semiaxis and such that $g_j(x,s) \leq g(s)$ in $\mathbb{R}^n \times (0,\infty), j = 1,2,\dots,n$.

Introduce the function

$$f(s) := \int_0^s e^{\int_0^x g(\tau)d\tau} dx. \tag{3}$$

Then, $f'(s) = e^{\int_0^s g(\tau)d\tau} > 0$, i.e., f is monotone. Hence, the function f^{-1} is well defined on the range of the function f and is monotone as well. Denote f^{-1} by ψ.

The main result of this section is preceded by the following classical theorem (see [6,7]).

Theorem 1 (Keller–Osserman). *If*

$$\int_1^\infty \frac{d\tau}{\sqrt{\int_1^\tau \mu(s)ds}} < \infty, \tag{4}$$

then inequality (1) *has no positive global solutions.*

The following assertion is valid.

Theorem 2. *If*

$$\int_1^\infty \frac{d\tau}{\sqrt{\int_1^\tau \frac{\omega[\psi(s)]}{\psi'(s)} ds}} < \infty,$$

then inequality (2) *has no positive global solutions.*

Proof. Suppose, to the contrary, that the assumptions of the theorem are satisfied, but there exists a positive function $u(x)$ satisfying inequality (2) in $\mathbb{R}^n$. Then, the function $u(x)$ satisfies the inequality

$$\Delta u + g(u)|\nabla u|^2 \geq \omega(u) \tag{5}$$

in $\mathbb{R}^n$ as well.

Introduce $v(x) := f[u(x)]$, where the function f is defined by relation (3). Then,

$$\frac{\partial v}{\partial x_j} = f'(u)\frac{\partial u}{\partial x_j} \text{ and } \frac{\partial^2 v}{\partial x_j^2} = f''(u)\left(\frac{\partial u}{\partial x_j}\right)^2 + f'(u)\frac{\partial^2 u}{\partial x_j^2}, j = \overline{1,n}.$$

Further, $f''(s) = g(s)e^{\int_0^s g(\tau)d\tau}$, i.e., $g(s) = \frac{f''(s)}{f'(s)}$ and, therefore,

$$\Delta v = f'(u)\left[\Delta u + g(u)|\nabla u|^2\right], \text{ i.e., } \psi'(v)\Delta v = \left[\Delta u + g(u)|\nabla u|^2\right].$$

Now, taking into account that $v(x)$ is positive provided that $u(x)$ is positive, we conclude that $v(x)$ is a positive solution of inequality (1) with $\mu(s) = \dfrac{\omega[\psi(s)]}{\psi'(s)}$. However, due to Theorem 1, inequality (1) has no global positive solutions under Condition (4).

The obtained contradiction completes the proof. □

3. Singular Case

The case where $g(s) = \dfrac{\text{const}}{s}$ is not covered by the previous section because the local integrability condition is violated. In that case, another ansatz is used. More exactly, the following assertion is valid.

Theorem 3. *If there exists* α *from* $(-1, \infty)$ *such that* $g_j(x,s) \le \dfrac{\alpha}{s}$ *in* $\mathbb{R}^n \times (0,\infty), j = 1, 2, \ldots, n,$ *and*

$$\int_1^\infty \frac{d\tau}{\sqrt{\int_1^{\tau^{\frac{1}{\alpha+1}}} s^{2\alpha}\omega(s)ds}} < \infty, \tag{6}$$

then inequality (2) *has no positive global solutions.*

Proof. Suppose, to the contrary, that the assumptions of the theorem are satisfied, but there exists a positive function $u(x)$ satisfying inequality (2) in $\mathbb{R}^n$. Then, the function $u(x)$ satisfies inequality

$$\Delta u + \frac{\alpha}{u}|\nabla u|^2 \ge \omega(u) \tag{7}$$

in $\mathbb{R}^n$ as well.

Denoting $u^{\alpha+1}(x)$ by $v(x)$, we see that

$$\frac{\partial v}{\partial x_j} = (\alpha+1)u^\alpha \frac{\partial u}{\partial x_j}, \frac{\partial^2 v}{\partial x_j^2} = \alpha(\alpha+1)u^{\alpha-1}\left(\frac{\partial u}{\partial x_j}\right)^2 + (\alpha+1)u^\alpha \frac{\partial^2 u}{\partial x_j^2} j = \overline{1,n},\ \Delta v = (\alpha+1)u^\alpha\left[\Delta u + \frac{\alpha}{u}|\nabla u|^2\right],$$

and, therefore,

$$\frac{\Delta v}{(\alpha+1)u^\alpha} \ge \omega(u).$$

Now, taking into account that $u(x) = v^{\frac{1}{\alpha+1}}(x)$, and $u(x)$ is positive everywhere, we conclude that $v(x)$ is a positive solution of the inequality

$$\Delta v \ge (\alpha+1)v^{\frac{\alpha}{\alpha+1}}\omega\left(v^{\frac{1}{\alpha+1}}\right). \tag{8}$$

Due to [6,7], the last inequality has no global positive solutions provided that

$$\int_1^\infty \frac{d\tau}{\sqrt{\int_1^\tau \rho^{\frac{\alpha}{\alpha+1}}\omega\left(\rho^{\frac{1}{\alpha+1}}\right)d\rho}} < \infty. \tag{9}$$

Now, consider inequality (6) and use the substitution $s = \rho^{\frac{1}{\alpha+1}}$ in its internal integral. We see that Condition (6) implies the validity of inequality (9). Hence, inequality (8) has no global positive solutions.

The obtained contradiction completes the proof. □

4. Critical Case

If $\alpha = -1$ in inequality (7), then the substitution from the previous section cannot be used. However, the following (weaker) result is still valid for this critical case.

Theorem 4. *Let* $g_j(x,s) \le -\frac{1}{s}$ *in* $\mathbb{R}^n \times (0,\infty), j = 1,2,\dots,n$, *and there exists a positive constant* β *such that*

$$\int_1^\infty \frac{d\tau}{\sqrt{\int_{\beta e}^{\beta e^\tau} \frac{\omega(s)}{s^2} ds}} < \infty. \tag{10}$$

Then, inequality (2) *has no global solutions exceeding* β *everywhere.*

Proof. Suppose, to the contrary, that the assumptions of the theorem are satisfied, but there exists a function $u(x)$ satisfying inequality (2) in $\mathbb{R}^n$ such that $u(x) > \beta$ in $\mathbb{R}^n$. Then, the function $u(x)$ satisfies the inequality

$$\Delta u - \frac{|\nabla u|^2}{u} \ge \omega(u) \tag{11}$$

in $\mathbb{R}^n$ as well.

Denoting $v(x) = \ln \frac{u(x)}{\beta}$ by $v(x)$, we see that $v(x)$ is positive everywhere,

$$\frac{\partial v}{\partial x_j} = \frac{1}{u(x)} \frac{\partial u}{\partial x_j} = \frac{\beta}{u(x)} \frac{1}{\beta} \frac{\partial u}{\partial x_j}, \quad \text{and} \quad \frac{\partial^2 v}{\partial x_j^2} = \frac{\frac{\partial^2 u}{\partial x_j^2} u - \left(\frac{\partial u}{\partial x_j}\right)^2}{u^2(x)} = \frac{1}{u(x)} \left[\frac{\partial^2 u}{\partial x_j^2} - \frac{1}{u(x)} \left(\frac{\partial u}{\partial x_j}\right)^2\right], j = \overline{1,n}.$$

Therefore,

$$\Delta v = \frac{1}{u}\left(\Delta u - \frac{|\nabla u|^2}{u}\right),$$

i.e., the left-hand side of inequality (11) is equal to $u \Delta v$.

Now, taking into account that $e^v = \frac{u}{\beta}$, we conclude that $u(x) = \beta e^v$. Thus, $\beta e^v \Delta v \ge \omega\left(\beta e^v\right)$, i.e., $v(x)$ is a positive solution of the inequality

$$\Delta v \ge \frac{e^{-v}}{\beta} \omega\left(\beta e^v\right). \tag{12}$$

Due to [6,7], this inequality has no global positive solutions provided that

$$\int_1^\infty \frac{d\tau}{\sqrt{\int_1^\tau \frac{\omega\left(\beta e^\rho\right)}{e^\rho} d\rho}} < \infty. \tag{13}$$

Now, using the substitution $s = \beta e^\rho$, we conclude that Condition (13) is equivalent to Condition (10). Hence, inequality (12) has no global positive solutions.

The obtained contradiction completes the proof. □

5. Examples

5.1. Inequalities with Constant Coefficients at Principal Nonlinear Terms

Since 1957, the classical result of Keller and Osserman was substantially strengthened. In particular, it is extended for the case of variable principal coefficients, i.e., for the case where the left-hand side of inequality (1) is changed for

$$\sum_{i,j=1}^{n} \frac{\partial^2}{\partial x_i \partial x_j} a_{i,j}(x,u), \tag{14}$$

where each principal coefficient satisfies the following restriction for the growth with respect to the second independent variable: $a_{i,j}(x,s) \leq \text{const}|s|$ (see [8]). Since only sufficient blow-up conditions are provided in [8], the question whether the above growth restriction is essential remained open up to now. However, using the above results for KPZ-type inequalities, one can show that the said coefficients are allowed to grow much faster. To do that, it suffices to consider inequality (5), assigning the coefficient $g(s)$ to be equal to a positive constant (denote it by α). That inequality can be represented in the form

$$\sum_{i,j=1}^{n} \frac{\partial^2 u}{\partial x_i \partial x_j} \left(\delta_i^j \frac{1}{\alpha} e^{\alpha u} \right) \geq e^{\alpha u} \omega(u).$$

Its left-hand side is represented in form (14), but the growth restriction is not satisfied: the coefficients are allowed to grow exponentially. However, Theorem 2 provides the following sufficient blow-up condition:

$$\int_1^\infty \frac{d\tau}{\sqrt{\int_1^\tau (\alpha s+1)^2 \omega\left[\frac{\ln(\alpha s+1)}{\alpha}\right] ds}} < \infty.$$

Indeed, in the considered case, we have the relations $f(s) = \int_0^s e^{\alpha x} dx = \frac{e^{\alpha s}-1}{\alpha}$, i.e., $\psi(s) = \frac{\ln(\alpha s+1)}{\alpha}$, $\psi'(s) = \frac{1}{\alpha s+1}$, and, therefore, the function $\mu(s)$ from Condition (4) takes the form

$$e^{\ln(\alpha s+1)} \omega\left[\frac{\ln(\alpha s+1)}{\alpha}\right](\alpha s+1) = (\alpha s+1)^2 \omega\left[\frac{\ln(\alpha s+1)}{\alpha}\right].$$

5.2. Case of Emden–Fowler Nonlinearities at Right-Hand Sides

Consider the following singular case (see Section 3 above), where the right-hand side of the equation is a power function (the so-called nonlinearity of the Emden–Fowler kind):

$$\Delta u + \frac{\alpha}{u} |\nabla u|^2 \geq u^p. \tag{15}$$

According to Theorem 3, this inequality has no global positive solutions provided that

$$\int_1^\infty \frac{d\tau}{\sqrt{\int_1^{\tau^{\frac{1}{\alpha+1}}} s^{2\alpha+p} ds}} < \infty.$$

The internal integral is equal to

$$\frac{s^{2\alpha+p+1}}{2\alpha+p+1}\Bigg|_1^{\tau^{\frac{1}{\alpha+1}}} = \frac{\tau^{\frac{2\alpha+p+1}{\alpha+1}}-1}{2\alpha+p+1} = \frac{\tau^{1+\frac{\alpha+p}{\alpha+1}}-1}{2\alpha+p+1} = \frac{\tau^{q+1}-1}{2\alpha+p+1},$$

where $q = \dfrac{\alpha+p}{\alpha+1}$.

Then, the left-hand side of the last inequality is equal to $\int\limits_1^\infty \dfrac{d\tau}{\sqrt{\dfrac{\tau^{q+1}-1}{2\alpha+p+1}}}$. Its convergence is equivalent to the convergence of the integral $\int\limits_1^\infty \dfrac{d\tau}{\sqrt{\tau^{q+1}-1}}$. Apply the substitution $z := \tau^{q+1}-1$. Then, $\tau^{q+1} = z+1$, $\tau = (z+1)^{\frac{1}{q+1}}$, and, therefore, $d\tau = \dfrac{(z+1)^{\frac{1}{q+1}-1}}{q+1}dz = \dfrac{dz}{(q+1)(z+1)^{\frac{q}{q+1}}}$. Thus, the last integral is equal to $\frac{1}{(q+1)}\int\limits_0^\infty \dfrac{dz}{\sqrt{z}(z+1)^{\frac{q}{q+1}}}$. The singularity of the integrand function at the origin is integrable. Its singularity at infinity is integrable under the assumption that $\dfrac{1}{2}+\dfrac{q}{q+1} > 1$, i.e., $q > 1$, which is equivalent to the inequality $p > 1$.

We see that inequality (15) has no global solutions provided that $p > 1$.

To compare this coercive example with the noncoercive case, consider the noncoercive inequality

$$\Delta u + \frac{\alpha}{u}|\nabla u|^2 + u^p \leq 0, \tag{16}$$

for positive values of α (assuming that $n \geq 3$).

As in Section 3, assume that there exist its global positive solution $u(x)$ and introduce the function $v(x) := u^{\alpha+1}(x)$. Then $\Delta u + \frac{\alpha}{u}|\nabla u|^2 = \dfrac{\Delta v}{(\alpha+1)u^\alpha}$ (see Section 3). Now, taking into account that $u(x) = v(x)^{\frac{1}{\alpha+1}}$, we conclude that

$$\frac{\Delta v}{(\alpha+1)v^{\frac{\alpha}{\alpha+1}}} + v^{\frac{p}{\alpha+1}} \leq 0, \text{ i.e., } -\Delta v \geq (\alpha+1)v^{\frac{\alpha+p}{\alpha+1}}.$$

Since $\alpha+1 > 0$, it follows that $v(x)$ is a global positive solution of the inequality

$$-\Delta v \geq v^{\frac{\alpha+p}{\alpha+1}}.$$

According to [9], the last inequality has no global positive solutions provided that $1 < \dfrac{\alpha+p}{\alpha+1} < \dfrac{n}{n-2}$. This condition is equivalent to the condition $1 < p < \dfrac{n+2\alpha}{n-2}$ (cf. the condition $p > 1$ obtained for the *coercive* case above).

6. Conclusions

In this paper, we investigate quasilinear partial differential inequalities of kind (2), where the coefficients $g_i(x,s)$ at the principal nonlinear terms are majorized either by locally summable (with respect to s) functions or by functions with singularities of the kind $\dfrac{\text{const}}{s}$. For both cases, we provide sufficient conditions of the absence of global positive solutions (or, which is the same, necessary conditions of their existence). The obtained results generalize both the classical Keller–Osserman result in [6,7] and its recent Kon'kov–Shishkov extension (see [8]).

Funding: This work is supported by the Ministry of Science and Higher Education of the Russian Federation (project number FSSF-2023-0016).

Acknowledgments: The author expresses his profound gratitude to A. L. Skubachevskii for his valuable considerations and his permanent attention to this work.

Conflicts of Interest: The author declares no conflict of interest.

References

1. Mitidieri, È.; Pokhozhaev, S.I. A priori estimates and the absence of solutions of nonlinear partial differential equations and inequalities. *Proc. Steklov Inst. Math.* **2001**, *234*, 1–362.
2. Muravnik, A.B. Qualitative properties of solutions of equations and inequalities with KPZ-type nonlinearities. *Mathematics* **2023**, *11*, 990. [CrossRef]
3. Amann, H.; Crandall, M.G. On some existence theorems for semi-linear elliptic equations. *Ind. Univ. Math. J.* **1978**, *27*, 779–790. [CrossRef]
4. Kazdan, I.L.; Kramer, R.I. Invariant criteria for existence of solutions to second-order quasilinear elliptic equations. *Comm. Pure Appl. Math.* **1978**, *31*, 619–645. [CrossRef]
5. Pohožaev, S. Equations of the type $\Delta u = f(x, u, Du)$. *Mat. Sb.* **1980**, *113*, 324–338.
6. Keller, J.B. On solutions of $\Delta u = f(u)$. *Comm. Pure Appl. Math.* **1957**, *10*, 503–510. [CrossRef]
7. Osserman, R. On the inequality $\Delta u \geq f(u)$. *Pac. J. Math.* **1957**, *7*, 1641–1647. [CrossRef]
8. Kon'kov, A.A.; Shishkov, A.E. Generalization of the Keller–Osserman theorem for higher order differential inequalities. *Nonlinearity* **2019**, *32*, 3012–3022. [CrossRef]
9. Gidas, B.; Spruck, J. Global and local behavior of positive solutions of nonlinear elliptic equations. *Comm. Pure Appl. Math.* **1981**, *34*, 525–598. [CrossRef]

mathematics MDPI

Article

Direct Method for Identification of Two Coefficients of Acoustic Equation

Nikita Novikov [1,2,3,*] and Maxim Shishlenin [1,2,3]

1 Sobolev Institute of Mathematics, 630090 Novosibirsk, Russia; mshishlenin@ngs.ru
2 Institute of Computational Mathematics and Mathematical Geophysics, 630090 Novosibirsk, Russia
3 Department of Mathematics and Mechanics, Novosibirsk State University, 630090 Novosibirsk, Russia
* Correspondence: novikov-1989@yandex.ru

Abstract: We consider the coefficient inverse problem for the 2D acoustic equation. The problem is recovering the speed of sound in the medium (which depends only on the depth) and the density (function of both variables). We describe the method, based on the Gelfand–Levitan–Krein approach, which allows us to obtain both functions by solving two sets of integral equations. The main advantage of the proposed approach is that the method does not use the multiple solution of direct problems, and thus has quite low CPU time requirements. We also consider the variation of the method for the 1D case, where the variation of the wave equation is considered. We illustrate the results with numerical experiments in the 1D and 2D case and study the efficiency and stability of the approach.

Keywords: acoustic equation; inverse problems; direct methods; integral equations

MSC: 35R30; 45B05

Citation: Novikov, N.; Shishlenin, M. Direct Method for Identification of Two Coefficients of Acoustic Equation. *Mathematics* **2023**, *11*, 3029. https://doi.org/10.3390/math11133029

Academic Editors: Natalia Bondarenko and Hongyu LIU

Received: 4 May 2023
Revised: 20 June 2023
Accepted: 3 July 2023
Published: 7 July 2023

1. Introduction

In this paper we consider the coefficient inverse problems for the second-order hyperbolic acoustic equation. Such problems (that can be interpreted as the recovering of the structure of the subsurface by using measurements) obtained on the daylight surface are one of the basic problems in seismology. The applied nature of the coefficient inverse problems for hyperbolic equations explains the wide range of methods, developed for its solution, from kinematic methods to reverse-time migration. While we did not plan to consider the detailed review of existing methods (one can find such survey in [1], for example), we should mention that approaches that utilize the dynamic characteristics of the wave field have tended to become more popular recently. For example, one of the rapidly developing methods is full waveform inversion [2], which is based on the reduction of the problem for the optimization of some misfit functional.

However, methods based on the optimization approach require the solution of the corresponding direct problem on each iteration of the scheme. Despite the recent progress of computational algorithms and hardware, the amount of computations required is still concerning, especially in multi-dimensional cases. The other feature of optimization approaches is their reliance on the basic structure of the medium, which, on the mathematical level, leads to the usage of prior information. In case of poor knowledge of prior information, the efficiency of the methods decreases due to the fact that the corresponding misfit functional usually have several local minima.

In this paper we use the alternate approach, developed originally in works of I.M. Gelfand, B.M. Levitan, M.G. Krein, and V.A. Marchenko [3–5]. During the second half of the XX century, the ideas of I.M. Gelfand, B.M. Levitan, and M.G. Krein were developed by A.S. Alekseev [6], A.S. Blagoveschenckiy [7], W. Symes [8], R. Burridge [9], F. Santosa [10], and many other researchers, and these applied for several problems of acoustics, seismics, and geoelectrics [11–14]. The multi-dimensional variation of the Gelfand–Levitan–Krein

(G–L–K) approach was considered in [14,15]. Different numerical algorithms for solving G–L–K equations were considered in [16–18]. The G–L–K method was applied for solving coefficient inverse problems of acoustics [19], elasticity [20], and seismics [21].

G–L–K methods allow us to reduce the nonlinear inverse problem to a family of linear integral equations. Such methods that provide the inversion of the forward problem are called direct ones. Another direct method, suitable for solving such problems, is the boundary control method [22,23]. The other important feature of this approach is that one does not have to use the prior information when solving G–L–K equations. There are a few other approaches that do not rely on the usage of a priori information. Aside from boundary control, one can also mention the global convergence method [24–28].

In this work, we use the G–L–K method to recover both the density and the speed of sound from the acoustic equation, where density depends on two variables, while the speed of sound is the function of the depth only. The recovering of the speed of sound as the 2D function by the G–L–K approach was considered in [29], where it was reduced to quite complicated system of equations. In this work, we consider a more simple case of the structure of the speed of sound in order to recover two coefficients of the equation. One should also mention the paper [30], where the wave speed in the acoustic wave equation was estimated from boundary measurements by constructing a reduced-order model matching discrete time–domain data. The success of the proposed algorithm hinges on the data-driven Gram–Schmidt orthogonalization of the snapshots that suppresses multiple reflections and can be viewed as a discrete form of the Marchenko–Gelfand–Levitan–Krein algorithm.

There are some other papers that consider the identification of several parameters, but most of them are dealing with Maxwell's equations and based on the Carleman estimates [31–34] and/or the optimization scheme [35]. Similar approaches were used for the system of elasticity in [36] for the problem of recovering two coefficients from the acoustic equation using interior data in [37], and for the system of the first-order acoustic equations in [38].

A method of finding the complex permittivity and permeability of a sample of isotropic material partially filling the rectangular waveguide cross section was presented in [39].

It was proved that the electromagnetic material parameters are uniquely determined by boundary measurements for the time-harmonic Maxwell equations in certain anisotropic settings [40].

In [41], the reconstruction of a complex-valued anisotropic tensor from the knowledge of several internal magnetic fields, which satisfies the anisotropic Maxwell system on a bounded domain with prescribed boundary conditions, was considered.

The recovering of conductivity, permittivity, and the electrokinetic mobility parameter in Maxwell's equations were analyzed with internal measurements, while allowing the magnetic permeability to be a variable function [42]. It was shown that the knowledge of two internal data sets associated with well-chosen boundary electrical sources uniquely determines these parameters.

In [43], an inverse boundary value problem for the time-harmonic Maxwell equations was considered. The authors showed that the electromagnetic material parameters were determined by boundary measurements where part of the boundary data is measured on a possibly very small set.

It was justified [44] that some elements of the scheme related to the construction of the infinite system of integral equations in the case when the potential is analytic in x. In particular, the convergence of the series in these equations was proved and the conditions for the N-approximation of the system was found.

This paper has the following structure. In the introduction we have considered a brief review of different approaches related to the solution of inverse problems for hyperbolic equations. In Section 2 we formulate the inverse problem for the acoustic equation in 2D and use the G–L–K approach to recover the acoustic impedance. In Section 3 we obtain the additional family of integral equations to recover the auxiliary function, that, with the acoustic impedance, gives us the necessary information to obtain both the density of the medium (which is considered to be a 2D function) and the speed of sound (that depends only on the depth). In Section 4 we consider the 1D problem for the acoustic-type equation, which is connected to the seismic inverse problem. In order to find both the speed of sound and the velocity, we use the property of the solution of the obtained integral equation to recover the unknown functions. In Section 5 we present numerical examples of the solution of the inverse problem. In the discussion we summarize the results of numerical experiments and consider ways for further develop the approach.

2. Two-Dimensional Acoustic Inverse Problem

We consider the following equation that describes the propagation of acoustic waves in the medium:

$$\frac{1}{c^2(z,y)}u_{tt} = \Delta u - \nabla \ln \rho(z,y)\, \nabla u.$$

Here $c(z,y) > 0$ is the speed of the propagation of the acoustic waves, $\rho(z,y) > 0$ is the density of the medium, and $u(z,y,t)$ is the exceeded pressure. As we mentioned in the introduction, we consider that the speed of sound is dependent only on the depth, and the density is dependent on both variables (we should also mention that the method proposed can be naturally extended to the 3D case with respect to the density).

Let us consider the sequence of the direct problems ($y^{(0)} \in \mathbb{R}$—the set of real numbers):

$$\frac{1}{c^2(z)}u_{tt}(z,y,t) = \Delta_{z,y}u - \nabla_{z,y} \ln \rho(z,y) \nabla_{z,y}u, \qquad z > 0, \quad t > 0, \quad y \in \mathbb{R}, \tag{1}$$

$$u|_{t<0} \equiv 0, \qquad z > 0, \quad y \in \mathbb{R}, \tag{2}$$

$$u_z|_{z=0} = \delta(t)\delta(y - y^{(0)}), \qquad t > 0, \quad y \in \mathbb{R}. \tag{3}$$

The boundary condition (3) describes the source of the acoustic waves, which is located on the surface $z = 0$ in point y_0 and has the form of the Dirac delta function in the time domain.

The inverse problem consists in finding the functions $c(z)$, $\rho(z,y)$ using the additional information on the surface $z = 0$:

$$u(+0,y,t) = f(y,t;y^{(0)}), \qquad t > 0, \quad y \in \mathbb{R}. \tag{4}$$

Now, in order to rewrite Equation (1), we consider the time travel coordinates:

$$x = \int_0^z \frac{d\xi}{c(\xi)}$$

Then, the inverse problem (1)–(4) can be rewritten as follows:

$$u_{tt} = \sigma(x,y)\left[\frac{\partial}{\partial x}\left(\frac{u_x}{\sigma}\right) + c^2(x)\frac{\partial}{\partial y}\left(\frac{u_y}{\sigma}\right)\right]; \tag{5}$$

$$u|_{t<0} \equiv 0, \qquad x > 0, \quad y \in \mathbb{R} \tag{6}$$

$$u_x|_{x=0} = c_0\delta(t)\delta(y - y^{(0)}), \qquad t > 0, \quad y \in \mathbb{R}. \tag{7}$$

$$u(+0,y,t) = f(y,t;y^{(0)}), t > 0, y \in \mathbb{R} \tag{8}$$

Here function $\sigma(x,y) = c(x)\rho(x,y)$ describes the acoustic impedance of the medium. We will also assume that the value $c_0 = c(+0)$ is known, and the function $\rho(0,y)$ is also known and sufficiently smooth.

First, we tend to recover the function $\sigma(x,y)$. In order to do that, we apply the Fourier transform with respect to $y^{(0)}$ to the problem (5)–(8). For simplicity, we consider the case when all the functions considered are 2π periodical with respect to y. Then, we obtain the following sequence of ratios (here $k \in \mathbb{Z}$):

$$u_{tt}^{(k)} = \sigma(x,y)\left[\frac{\partial}{\partial x}\left(\frac{u_x^{(k)}}{\sigma}\right) + c^2(x)\frac{\partial}{\partial y}\left(\frac{u_y^{(k)}}{\sigma}\right)\right]; \tag{9}$$

$$u^{(k)}|_{t<0} \equiv 0, x > 0, y \in (-\pi, \pi); \tag{10}$$

$$u^{(k)}|_{y=-\pi} = u^{(k)}|_{y=\pi}; \tag{11}$$

$$u_x^{(k)}|_{x=0} = c_0 e^{iky}\delta(t), t > 0, y \in \mathbb{R}. \tag{12}$$

$$u^{(k)}(+0, y, t) = f^{(k)}(y, t), t > 0, y \in \mathbb{R} \tag{13}$$

The problem is to recover $\sigma(x,y)$ by the given $f^{(k)}(y,t), k \in \mathbb{Z}$ set of integers. The problem in (9)–(13) was considered in [14,16], in case of $c(x) = 1$. Using the same approach, based on a combination of the projection method and a 2D analogue of the M.G. Krein approach, we reduce the problem in (9)–(13) to the following set of systems of linear integral equations:

$$\Phi^{(k)}(x,t) - \frac{1}{2c_0}\sum_{m\in\mathbb{Z}}\int_{-x}^{x} f_m^{(k)\prime}(t-s)\Phi^{(m)}(x,s)ds = \frac{1}{2c_0}\int_{-\pi}^{\pi}\frac{e^{iky}}{\rho(0,y)}dy. \tag{14}$$

Here $x < 0$, $t \in (-x, x)$, $k \in \mathbb{Z}$, and function $f_m^{(k)}$ is the Fourier coefficient of the inverse problem's data with respect to y:

$$f^{(k)}(y,t) = \sum_{m\in\mathbb{Z}} f_m^{(k)}(t)e^{imy}, t > 0; f^{(k)}(y,t) = -f^{(k)}(y,-t), t < 0.$$

For every given x, the Equation (14) is a system of linear integral equations of the second kind with respect to the unknown functions $\Phi^{(k)}(x,t)$. These functions are connected with $\sigma(x,y)$ by the values on characteristic lines $t = x$:

$$\Phi^{(m)}(x, x-0) = \int_{-\pi}^{\pi}\frac{e^{imy}}{2\sqrt{\sigma(x,y)\rho(0,y)}}dy.$$

Therefore, when Equation (14) is solved, the acoustic impedance $\sigma(x,y)$ can be calculated as follows:

$$\sigma(x,y) = \frac{\pi^2}{\rho(0,y)}\left[\sum_{m\in\mathbb{Z}}\Phi^{(m)}(x, x-0)e^{-imy}\right]^{-2} \tag{15}$$

3. Obtaining the Density and the Speed of Sound

Now we suppose that the function $\sigma(x,y)$ is known. However, the given acoustic impedance is not enough to calculate the density or the speed of sound instantly. Therefore, we consider another substitution. Let us consider the function $v(x,y,t)$, which is connected with $u(x,y,t)$ as follows:

$$u(x,y,t) = c_0\sqrt{\frac{\sigma(x,y)}{\sigma(0,y)}}v(x,y,t).$$

Let us also suppose, that the function $\sigma(x,y)$ satisfies the condition $\sigma_x|_{x=0} = 0$. Then, the problem (9)–(12) can be rewritten as follows:

$$v_{tt} = \Delta_{x,y}v(x,y,t) + q(x,y)v; \tag{16}$$

$$x > 0, \quad y \in (-\pi, \pi), \quad t > 0,$$

$$v^{(k)}|_{t<0} \equiv 0, \tag{17}$$

$$v_x^{(k)}|_{x=0} = \delta(t)e^{iky}. \tag{18}$$

The function $q(x,y)$ has the following structure:

$$q(x,y) = \left[\frac{1}{2}\left(\frac{\sigma_x}{\sigma}\right)_x - \frac{1}{4}\left(\frac{\sigma_x}{\sigma}\right)^2\right] + c^2(x)\left[\frac{1}{2}\left(\frac{\sigma_y}{\sigma}\right)_y - \frac{1}{4}\left(\frac{\sigma_y}{\sigma}\right)^2\right]. \quad (19)$$

The condition (13) provides the data of the inverse problem:

$$v^{(k)}(+0,y,t) = \frac{1}{c_0}f^{(k)}(y,t). \quad (20)$$

The inverse problem (16)–(20) was extensively studied by S.I. Kabanikhin. We reduce it to the 2D analogue of the I.M. Gelfand–B.M. Levitan equation [15,17,18]:

$$c_0 w^{(k)}(x,y,t) + \int_{-x}^{x} \sum_{m\in\mathbb{Z}} f_m^{(k)\prime}(t-s)w^{(m)}(x,y,s)ds =$$
$$= -\frac{1}{2}\left[f^{(k)\prime}(y,t-x) + f^{(k)\prime}(y,t+x)\right], \qquad k\in\mathbb{Z}. \quad (21)$$

The connection between the function $q(x,y)$ and the solution of Equation (21) has the following form:

$$q(x,y) = -4\frac{d}{dx}w^{(0)}(x,y,x-0). \quad (22)$$

Thus, the problem of recovering functions $c(x)$ and $\rho(x,y)$ for every $x > 0$ is reduced to the following procedure:

1. Solve Equation (14) and use (15) to obtain $\sigma(x,y)$;
2. Solve Equation (21) and use (22) to obtain $q(x,y)$;
3. Use the representation (19) to calculate $c(x)$, when $q(x,y), \sigma(x,y)$ are known;
4. Recover the density $\rho(x,y) = \frac{\sigma(x,y)}{c(x)}$.

When $c(x)$ is calculated, one can reverse the travel-time transform and return to the original coordinates $z = \int_0^x c(\xi)d\xi$.

However, we should mention that the proposed scheme requires that function

$$q_2(x,y) = \left(\frac{\sigma_y}{\sigma}\right)_y - \frac{1}{2}\left(\frac{\sigma_y}{\sigma}\right)^2$$

takes a non-zero value in at least one point y. Thus, the horizontal inhomogeneity of the density is necessary. The case when both the coefficients depend only on the depth is considered in the following section, when we have to consider a different form of equation and propose a different approach of the parameter's reconstruction.

4. One-Dimensional Acoustic Inverse Problem

The 1D formulation of the Equation (1) has the form:

$$\frac{1}{c^2(z)}\frac{\partial^2 U}{\partial t^2} = \frac{\partial^2 U}{\partial z^2} + \frac{\partial ln(\rho)}{\partial z}\frac{\partial U}{\partial z}.$$

When using the travel-time transform one can obtain:

$$\frac{\partial^2 U}{\partial t^2} = \frac{\partial^2 U}{\partial x^2} - \frac{\sigma'(x)}{\sigma(x)}\frac{\partial U}{\partial x}$$

Where, as in the previous section, $\sigma(x) = c(x)\rho(x)$. Thus, as we mentioned previously, in this case one can only recover the acoustic impedance and not the speed and the density independently. Therefore, we alter the formulation of the inverse problem and consider it as follows:

$$\frac{1}{c^2(z)}\frac{\partial^2 U}{\partial t^2} = \frac{\partial^2 U}{\partial z^2} - \frac{\partial ln\rho}{\partial z}\frac{\partial U}{\partial z} - k^2 U; \tag{23}$$

$$U(z,t;k)|_{t<0} \equiv 0; \tag{24}$$

$$\frac{\partial U}{\partial z}|_{z=0} = \delta(t); \tag{25}$$

$$U(z,t;k)|_{z=0} = f_k(t). \tag{26}$$

Here $k \neq 0$ is some given parameter. The inverse problem (23)–(26) is to recover functions $\rho(z)$, $c(z)$ by the given function $f(t)$. Such formulation is connected with the seismic inverse problem in the case of the horizontally layered structure and was considered by A.S. Alekseev [6] in the spectral domain. The connection between the two mentioned problems is based on the special form of the sounding wave and the Hankel transform. Once again, in the travel-time coordinates, we consider the following formulation (for simplicity we also assume that $c(0) = 1$):

$$\frac{\partial^2 U}{\partial t^2} = \frac{\partial^2 U}{\partial x^2} - \frac{\sigma'(x)}{\sigma(x)}\frac{\partial U}{\partial x} - k^2 c^2 U; \tag{27}$$

$$U(x,t)|_{t<0} \equiv 0; \tag{28}$$

$$\frac{\partial U}{\partial x}|_{x=0} = \delta(t); \tag{29}$$

$$U(x,t)|_{x=0} = f(t). \tag{30}$$

The inverse problem (27)–(30) can be considered in the same manner as in the previous section. Yet, we use the 1D case to illustrate the different scheme of recovering two parameters. While the first one is based on the solution of both M.G. Krein and I.M. Gelfand–B.M. Levitan Equations (14) and (21), in this section, we show the possibility of substituting the solution of (21) with some finite-difference ratios.

Once again, we start with the recovering of the acoustic impedance. The 1D analogue of (14) gives us the M.G. Krein equation:

$$-2f_0(+0)V(x,t) - \int_{-x}^{x} V(x,s)f_0'(t-s)ds = 1, t \in (-x,x) \tag{31}$$

The solution $V(x,t)$ of Equation (31) is connected with the function $\sigma(x)$ as follows:

$$\sigma(x) = \frac{V(0,0)}{2V^2(x,x)}. \tag{32}$$

Thus, in order to recover $\sigma(x)$, one has to use only the one component of the solution $V(x,t)$ of Equation (31). While this property can be used to increase the efficiency of numerical algorithms [17], now we will assume that the solution of Equation (31) gives us the full set of values of $V(x,t), x > 0, t \in (-x,x)$.

Using the framework of [7,14], one can obtain that the function $V(x,t)$ solves the following system:

$$\frac{\partial^2 V}{\partial t^2} = \frac{\partial^2 V}{\partial x^2} + \frac{\sigma'(z)}{\sigma(z)}\frac{\partial V}{\partial x} - k^2\int_0^x c^2(\xi)\sigma(\xi)V_x(\xi,t)d\xi, x > 0, t \in \mathbb{R}; \tag{33}$$

$$V|_{x=0} = 0; \tag{34}$$

$$V_x|_{x=0} = \delta(t). \tag{35}$$

Using the initial conditions (34) and (35), one can obtain that $V(x,t) = 0$ for $|t| > x$, $V(x,-t) = V(x,t), t > 0$, and $V(x, x-0) = \frac{1}{2\sqrt{\sigma(x)\sigma(0)}}$. Therefore, we replace (33)–(35) with the problem:

$$\frac{\partial^2 V}{\partial t^2} = \frac{\partial^2 V}{\partial x^2} + \frac{\sigma'(z)}{\sigma(z)}\frac{\partial V}{\partial x} - k^2 \int_0^x c^2(\xi)\sigma(\xi)V_x(\xi,t)d\xi, x > 0, t \in (-x,x);$$

$$V|_{t=x} = V|_{t=-x} = \frac{1}{2\sqrt{\sigma(x)\sigma(0)}}.$$

Let us introduce the uniform grid $t_k = kh, x_i = ih$ and consider the finite-difference approximation of (33):

$$\frac{V_i^{k-1} - 2V_i^k + V_i^{k+1}}{h^2} = \frac{V_{i-1}^k - 2V_i^k + V_{i+1}^k}{h^2} + \frac{V_{i+1}^k - V_{i-1}^k}{2h}\frac{\sigma_{i+1} - \sigma_i}{h\sigma_i} + hk^2 \sum_{s=1}^{i} c_s^2 P_s^k; \quad (36)$$

Here $\sigma_i = \sigma(x_i)$, $c_s^2 = c^2(x_s)$, and

$$P_s^k = \sigma(x_s)V_x(x_s, t_k) \approx \sigma_s \frac{V_s^k - V_{s-1}^k}{h}$$

Using (36) for $t = 0$, one can obtain:

$$hk^2 c_i^2 P_i^0 = \frac{1}{h^2}\left(V_i^{-1} + V_i^1 - V_{i-1}^0(1 - \frac{\sigma_{i+1} - \sigma_i}{2\sigma_i}) - V_{i-1}^0(1 + \frac{\sigma_{i+1} - \sigma_i}{2\sigma_i})\right) - hk^2 \sum_{s=1}^{i-1} c_s^2 P_s^0. \quad (37)$$

The ratio (37) provides the recurrent procedure, which allows us to reconstruct the value of $c^2(x)$ for each depth by using values σ_i and V_i^k, obtained from (31) and (32).

5. Numerical Results

In this section, we consider the results of the numerical experiments of the reconstruction of velocity and density of the medium in the 1D and 2D cases. We consider the synthetic data obtained by solving the direct problem. For the 1D case, we consider the layered structure of the medium corresponded to the Yurubcheno-Tokhomskoe gas field [45]. The values of the parameters are presented in Table 1.

Table 1. The 1D model—parameters of the layers.

Depth, km	**0.17**	**0.47**	**0.87**	**1.070**	**1.320**	**1.600**	**2.100**	**2.200**	**2.300**
$v_s(z)$, km/s	0.9	1.7	3.1	3.5	2.7	3.2	2.85	3.4	2.8
$\rho(z)$, 10^3 kg/m^3	2.1	2.4	2.65	2.75	2.5	2.7	2.6	2.75	2.6

In order to solve the inverse problem, first we consider the solution of the 1D Krein Equation (31) and calculate the function $V(x,t)$. We solve the integral Equation (31) by discretizing it and reducing it to the SLAE. If $x_M = Mh$, then the system has the following form:

$$(-2f_0 I - h\tilde{A}_{[M]})\overline{V}_{[M]} = \overline{1}_{[M]}.$$

where I is an identity matrix, and

$$\tilde{A}_{[M]} = \begin{bmatrix} f_0 & f_1 & \cdots & f_{2M-1} \\ f_{-1} & f_0 & \cdots & f_{2M-2} \\ \ddots & \ddots & \ddots & \ddots \\ f_{-2M+2} & \cdots & f_0 & f_1 \\ f_{-2M+1} & \cdots & f_{-1} & f_0 \end{bmatrix}.$$

Thus, the structure of Equation (31) allows to reduce it to the system with the Toeplitz matrix. We solve the system by using an adaptation of the Levinson–Durbin algorithm. A full description can be found in [16]. When the Krein equation is solved, we use (37) to recover the speed of sound.

The results of the reconstruction are presented in Figure 1. We used the uniform mesh with $N_t = N_x = 200$ during the reconstruction. As one can see, the layered structure is fully recovered and the accuracy is acceptable.

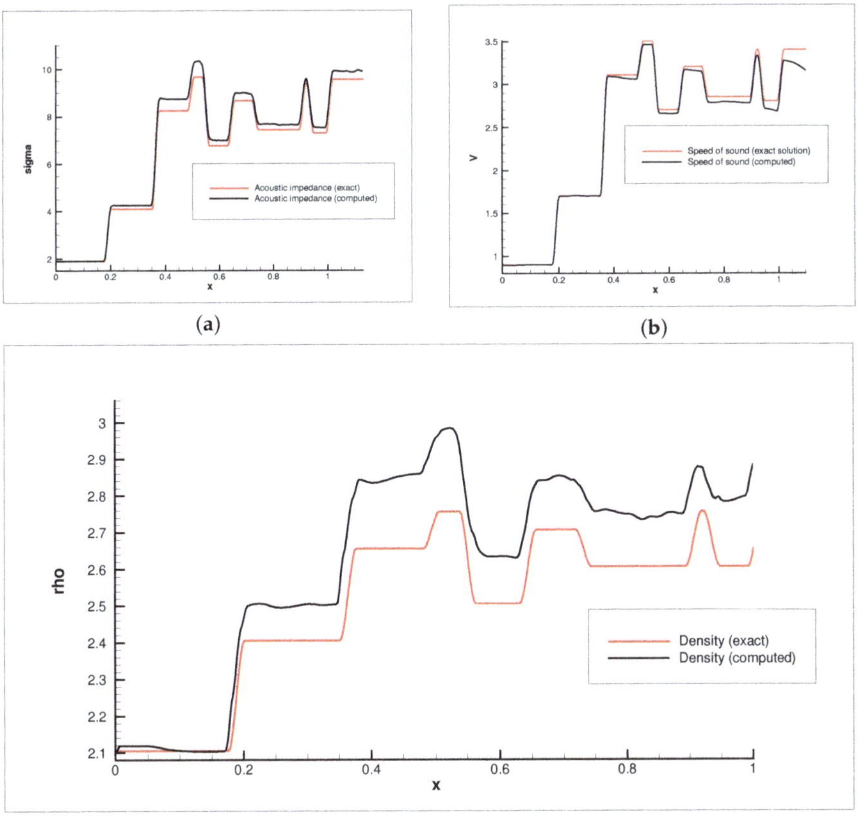

(c)

Figure 1. The 1D case—the result of the reconstruction in travel-time coordinates: (**a**) the reconstruction of the acoustic impedance; (**b**) the reconstruction of the speed of sound; (**c**) the reconstruction of the density of the medium.

We also reverse the travel-time transform and return to the original depth. The results of the reconstruction of the speed of sound is presented in Figure 2.

We are now moving to the 2D case. We used synthetic data during the numerical experiments, obtained from the solution of the direct problem (9)–(13) by using a combination of the projection method and the finite-difference scheme. We should mention that we used different grid parameters during the solution of the direct and inverse problem (the grid parameters were chosen as $N_{dir} = 250$ and $N_{inv} = 100$ for the direct and inverse problems, correspondingly). We also introduced random errors into the data, as discussed further.

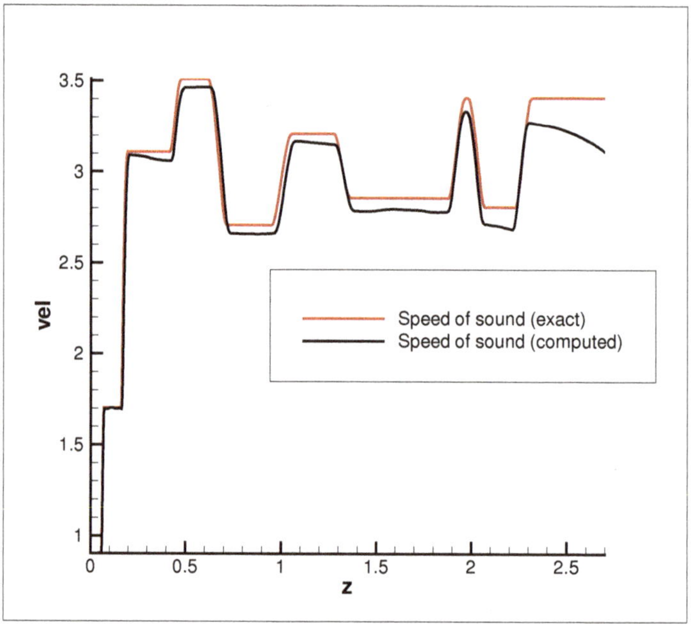

Figure 2. The 1D case—the result of the reconstruction of the speed of sound in the original coordinates.

First we consider the case of smooth parameters. The speed of sound was chosen as a function, closed to linear (the red line in Figure 3), while the density varies with respect to both variables and is presented in Figure 4.

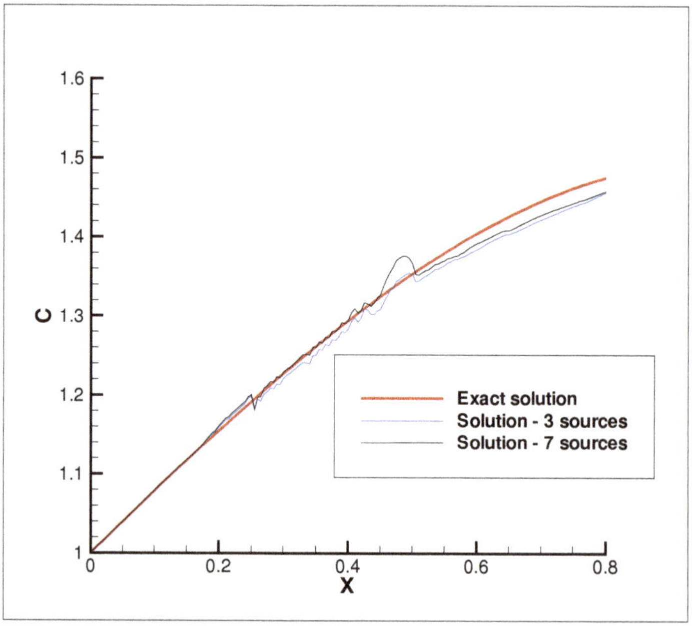

Figure 3. The 2D case—the speed of sound reconstruction.

The important parameter of the 2D reconstruction is the number of sources and receivers that describe the quantity of data available during the inverse problems solution. The Figures 3 and 5 illustrate the dependence of the results on the amount of given sources/receivers.

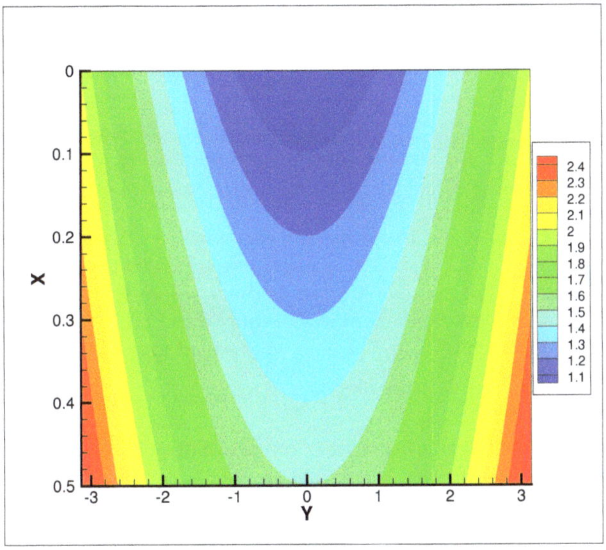

Figure 4. The 2D smooth case—density of the medium (exact values).

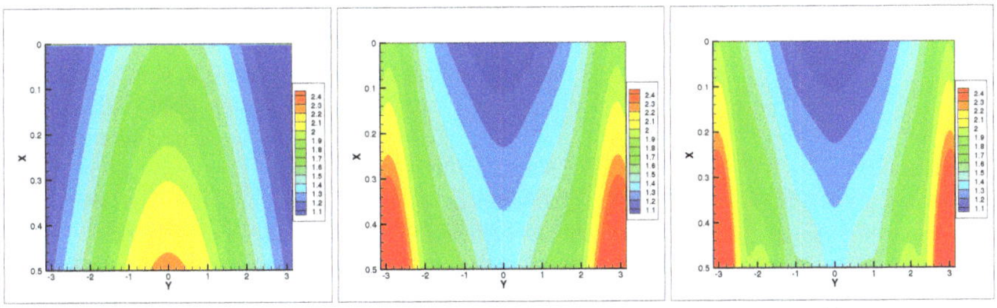

(a) (b) (c)

Figure 5. The 2D smooth case—density reconstruction: (**a**) 3 sources/receivers; (**b**) 7 sources/receivers; (**c**) 11 sources/receivers.

We also present the result of computations for non-smooth parameters, illustrated by Figure 6. However, due to the complex connection between impedance, speed of sound, and the solution of Equation (21) provided by the formula (19), one should consider additional regularizing procedures during the simultaneous solution of Equations (14) and (21). We plan to study the restoration of parameters in the non-smooth case in future work. For now we have focused on the problem of recovering the density of the medium, while we suppose that the speed of sound is known.

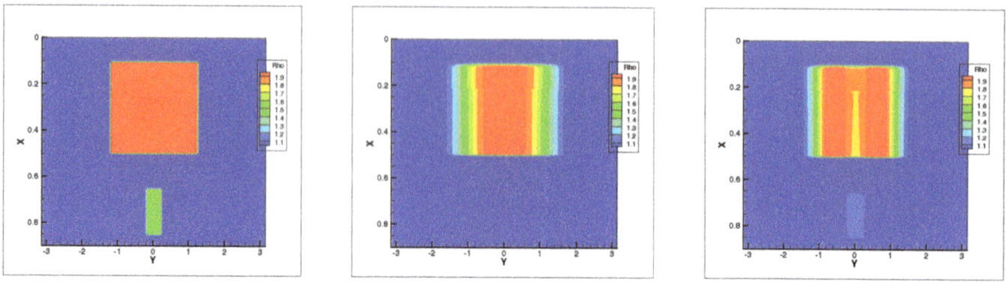

(a) (b) (c)

Figure 6. The 2D case—density reconstruction: (**a**) exact solution; (**b**) computed solution (five sources/receivers); (**c**) computed solution (nine sources/receivers).

During the experiment we considered the model with two inclusions into the homogeneous media (the smaller one is located beneath the larger one). As illustrated by Figure 6, it is possible to locate both inclusions when using more data obtained.

The next experiment was to study the effects of the noise in the data on the solution of the inverse problem. The noise was added to the data according to the formula:

$$f^{(k)}_{err}(y,t) = f^{(k)}(y,t) + \varepsilon\alpha(y,t)f^{(k)}(y,t),$$

where $f^{(k)}(y,t)$ is the value of computed data, α is the random variable, which is uniformly distributed on $(-1,1)$, and ε is the level of noise. The results of computations with noised data are presented in Figures 7 and 8.

The structure of the density of the model, used during experiments with noised data, is presented in Figure 7a. It consists of three non-homogeneous layers and two elliptic inclusions located between the layers. The next picture illustrates the result of the reconstruction for the noiseless data. We should mention that due to the relative complexity of the model, one has to use more data to obtain a relatively accurate solution. Depicted in Figure 7c is the result of adding the noise to the data. One can see the distortion caused by the noise, which becomes more observable with depth. One of the possible ways to decrease the influence of noise is, as presented in the last part of Figure 7, to reduce the number of sources and receivers used during the solution of the equations. In this case, the number of sources can be considered as a regularization parameter.

In the last series of tests we tried to compare the proposed method, based on the G–L–K approach, with the optimization approach. The latter method was broadly described in [46,47] and based on the optimization of the misfit functional by gradient-based methods. The direct and adjoint problem is solved on each iteration by using the finite volume scheme. Since the mentioned method was considered for dealing with problems of acoustic tomography, we chose another model for the last test. The model consists of a round object with several inclusions. Figure 9 provides the comparative analysis of the density reconstruction, while Table 2 describes the elapsed CPU time for both methods (the computations were carried out on a laptop). We should also mention that a comparative analysis of the G–L–K method and the optimal control method was carried out in [19], and the comparison with boundary control was considere in [48,49].

Table 2. G–L–K method and optimization scheme—time–cost comparison.

Time (s)	G–L–K Method				Optimization
	2 sources	5 sources	8 sources	11 sources	8 sources
$N_x = 100$	1.37	5.22	11.87	23.016	300
$N_x = 200$	5.17	20.81	52.17	89.25	1700

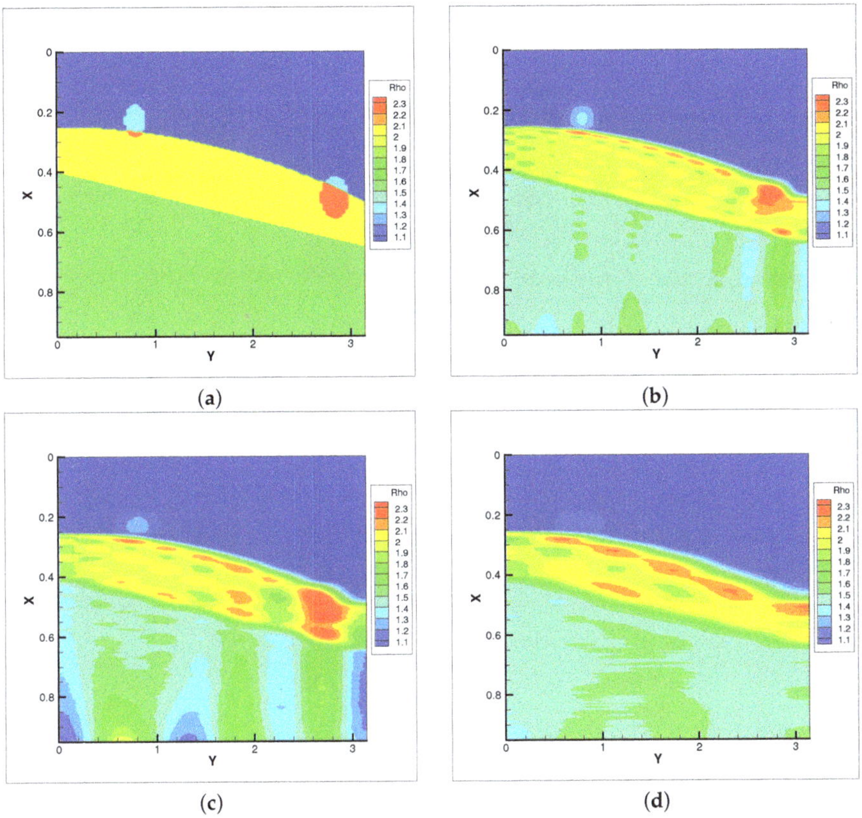

Figure 7. The 2D case—density reconstruction (noised data): (**a**) exact solution; (**b**) computed solution (15 sources/receivers, noiseless data); (**c**) computed solution (15 sources/receivers, 5% noise in the data); (**d**) computed solution (7 sources/receivers, 5% noise in the data).

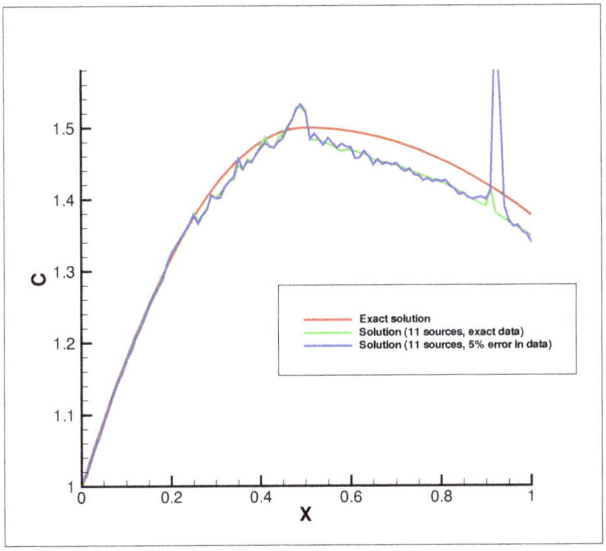

Figure 8. The 2D case—the speed of sound reconstruction (noised data).

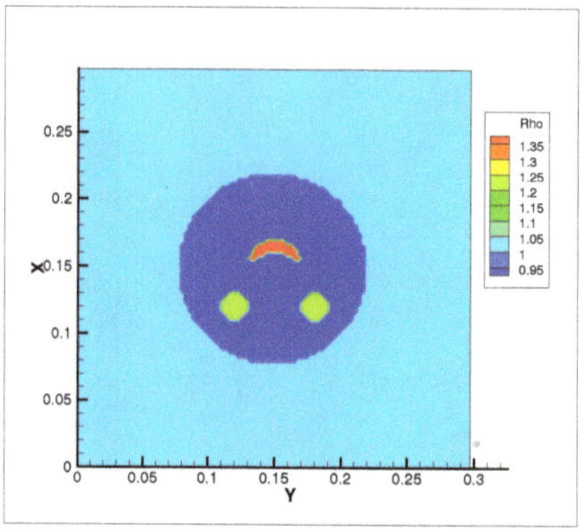

(a)

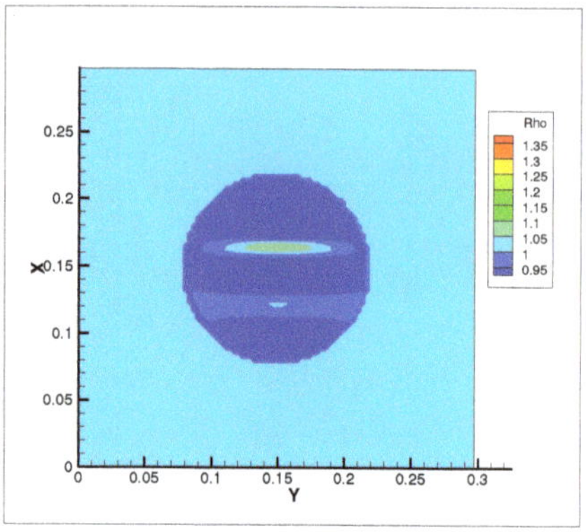

(b)

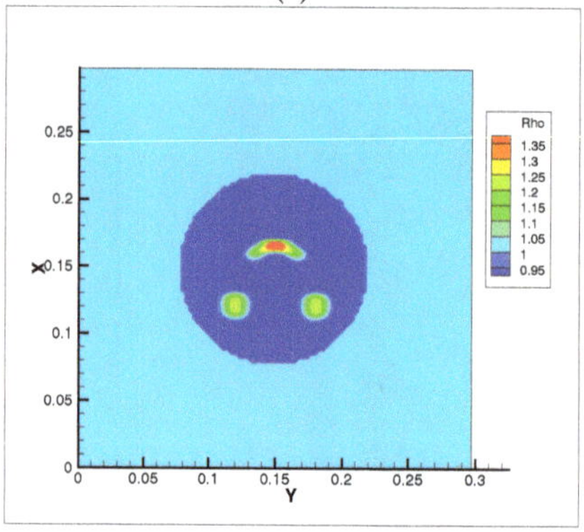

(c)

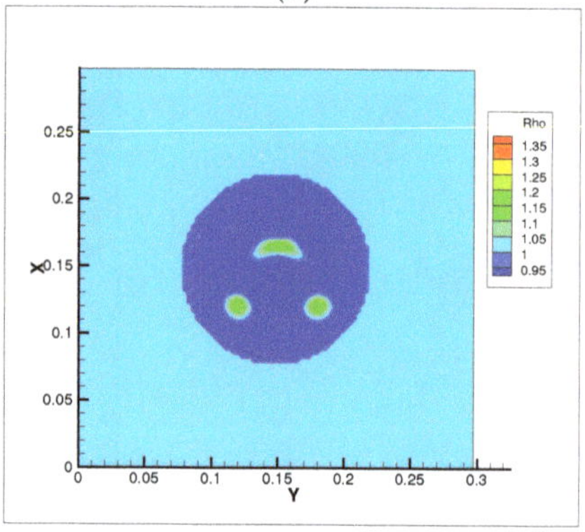

(d)

Figure 9. The 2D case—the result of the density reconstruction: (**a**) exact solution; (**b**,**c**) solution obtained by G–L–K method; ((**b**) 5 sources/receivers; (**c**) 11 sources/receivers); (**d**) solution obtained by optimization approach (8 sources/receivers).

6. Discussion

In this paper, we considered the problem of recovering two parameters from the acoustic equation when the speed of sound in the medium depends only on the depth but the density is the function of two variables. As for the 1D case, we considered the variation of the wave equation, which allows for the reconstruction of two parameters. We used the direct method, based on the approach of G–L–K, to solve the mentioned inverse problems and provided several numerical experiments to illustrate the results.

The proposed algorithm demonstrated acceptable accuracy during the synthetic tests. The time cost of the method is also low, because the convolution type of the kernels of the obtained equations allows to use a specific numerical algorithm, based on the Toeplitz matrix inversion. The stability of the method is yet to be improved. One should mention that the issue of stability becomes more complicated due to the fact that we use the derivative of the data for the computation, and because the effects, provided by the noise, tend to accumulate with increasing depth (which is caused by the structure of Equations (14) and (21)). On the other hand, due to the over-determination of the considered inverse problem, one can decrease the impact of the noise by using the quantity of data as the regularization parameter.

We also have to mention that the comparison of the proposed approach and the approach based on the optimization scheme was carried out only on a basic level. We plan to study it in detail in future work. The efficiency of both methods depends on several factors, such as the method of optimization of the functional, the method used for solving the integral Equations (14) and (21), the complexity of the governing equation, etc. However, the main difference of the methods is based on their core structure—the direct nature of the G–L–K approach requires less computation time, while the optimization methods are more versatile and allow us to control the residual on each iteration.

We should mention, that the proposed ideas could be used for solving the inverse problems, in cases when the speed of sound can be considered as a sum of the main trend function, which depends only on the depth, and a small enough horizontal non-homogeneous part. Indeed, let us consider the inverse problem:

$$\frac{1}{c^2(z,y)}u_{tt}(z,y,t) = \rho(z,y)\mathrm{div}\left(\frac{\mathrm{grad}\ u}{\rho}\right), z > 0, t > 0, y \in \mathbb{R} \tag{38}$$

$$u|_{t<0} \equiv 0, x > 0, y \in \mathbb{R} \tag{39}$$

$$u_z|_{z=0} = \delta(t)\delta(y - y^{(0)}), t > 0, y \in \mathbb{R}. \tag{40}$$

$$u(+0,y,t) = f(y,t;y^{(0)}), t > 0, y \in \mathbb{R} \tag{41}$$

Let us assume that

$$c(z,y) = c_0(z) + c_1(z,y).$$

In this case, one could linearize the direct problem's operator and solve (38)–(41) in two steps:

1. Use the G–L–K approach to recover $c_0(z)$ by solving (1)–(4);
2. Recover $c_1(z,y)$ by solving the linearized problem

$$\frac{1}{c_0^2(z)}u_{tt}^{(1)}(z,y,t) = \rho(z,y)\mathrm{div}\left(\frac{\mathrm{grad,}\ u^{(1)}}{\rho}\right) + 2c_1(z,y)Q(x,y,t), z > 0, t > 0, y \in \mathbb{R}, \tag{42}$$

$$u|_{t<0} \equiv 0, x > 0, y \in \mathbb{R} \tag{43}$$

$$u_z|_{z=0} = 0, t > 0, y \in \mathbb{R}. \tag{44}$$

$$u(+0,y,t) = \left[f - f^{(0)}\right](y,t;y^{(0)}), t > 0, y \in \mathbb{R} \tag{45}$$

Here the functions $Q(x,y,t) = \frac{1}{c(z)}\mathrm{div}\left(\frac{\mathrm{grad}\ u^{(0)}}{\rho}\right)$ and $f^{(0)}(y,t;y^{(0)}) = u^{(0)}(+0,y,t)$ depend only on the solution $u^{(0)}(x,y,t)$ of the direct problem (1)–(3) and can be calculated, since the coefficients of the direct problem were found in the previous step.

Since the last formulated problem is linear, it can be solved by several methods, direct or iterational. In that case that the approach of G–L–K is used in the first step and provides the initial approximation of the parameters, this is improved in the next step of data processing.

Author Contributions: Methodology, N.N. and M.S.; software, N.N.; formal analysis, N.N. and M.S.; writing—original draft, N.N.; writing—review & editing, N.N. and M.S. All authors have read and agreed to the published version of the manuscript.

Funding: The work has been supported by the Russian Science Foundation under grant 20-71-00128 "Development of new algorithms for parameters identification of geophysics based on the direct methods of data processing".

Data Availability Statement: Data sharing is not applicable to this article.

Conflicts of Interest: The authors declare no conflict of interest.

References

1. Anikiev, D.V.; Kazei, V.V.; Kashtan, B.M.; Ponomarenko, A.V.; Troyan, V.N.; Shigapov, R.A. Methods of Seismic Waveform Inversion. *Russ. J. Geophys. Technol.* **2014**, *1*, 38–58. (In Russian)
2. Agudo, O.; Silva, N.; Warner, M.; Morgan, J. Acoustic full-waveform inversion in an elastic world. *Geophysics* **2018**, *83*, R257–R271. [CrossRef]

3. Gel'fand, I.M.; Levitan, B.M. On the determination of a differential equation from its spectral function. *Izv. Akad. Nauk SSSR Ser. Mat.* **1951**, *15*, 309–360.
4. Krein, M.G. On a method of effective solution of an inverse boundary problem. *Dokl. Akad. Nauk SSSR* **1954**, *94*, 987–990.
5. Marchenko, V.A. Restoration of the potential energy by the phase of the dissipated waves. *Dokl. Akad. Nauk SSSR* **1955**, *104*, 695–698.
6. Alekseev, A.S. Inverse dynamical problems of seismology. In *Some Methods and Algorithms for Interpretation of Geophysical Data*; Nauka: Russia, Moscow, 1967; pp. 9–84.
7. Blagoveschenskii, A.S. The local method of solution of the non-stationary inverse problem for an inhomogeneous string. *Proc. Math. Steklov Inst.* **1971**, *115*, 28–38.
8. Symes, W. Inverse boundary value problems and a theorem of Gel'fand and Levitan. *Math. Anal. Appl.* **1979**, *71*, 379–402. [CrossRef]
9. Burridge, R. The Gelfand–Levitan, the Marchenko and the Gopinath-Sondhi integral equation of inverse scattering theory, regarded in the context of inverse impulse-response problems. *Wave Motion* **1980**, 2, 305–323. [CrossRef]
10. Santosa, F. Numerical scheme for the inversion of acoustical impedance profile based on the Gelfand–Levitan method. *Geophys. J. R. Astr. Soc.* **1982**, *70*, 229–243. [CrossRef]
11. Kunetz, G. Essai d'analyse de traces sismiques. *Geophys. Prospect.* **1961**, *8*, 317–341. [CrossRef]
12. Alekseev, A.S.; Belonosov, V.S. Spectral methods in one-dimensional problems of wave propagation theory. *Mat. Model. Geofiz. Proc. ICMMG RAS* **1998**, *11*, 7–39.
13. Baev, A.V. Solution of an inverse scattering problem for the acoustic wave equation in three-dimensional media. *Comput. Math. Math. Phys.* **2016**, *56*, 2043–2055 [CrossRef]
14. Kabanikhin, S.I.; Shishlenin, M.A. Numerical algorithm for two-dimensional inverse acoustic problem based on Gel'fand-Levitan-Krein equation. *J. Inverse-Ill-Posed Probl.* **2011**, *18*, 979–995. [CrossRef]
15. Kabanikhin, S.I.; Shishlenin, M.A. Two-dimensional analogs of the equations of Gelfand, Levitan, Krein, and Marchenko. *Eurasian J. Math. Comput. Appl.* **2015**, *3*, 70–99.
16. Kabanikhin, S.I.; Novikov, N.S.; Oseledets, I.V.; Shishlenin, M.A. Fast Toeplitz linear system inversion for solving two-dimensional acoustic inverse problem. *J. Inverse-Ill-Posed Probl.* **2015**, *23*, 687–700. [CrossRef]
17. Kabanikhin, S.I.; Sabelfeld, K.K.; Novikov, N.S.; Shishlenin, M.A. Numerical solution of the multidimensional Gelfand-Levitan equation. *J. Inverse-Ill-Posed Probl.* **2015**, *23*, 439–450. [CrossRef]
18. Kabanikhin, S.I.; Sabelfeld, K.K.; Novikov, N.S.; Shishlenin, M.A. Numerical solution of an inverse problem of coefficient recovering for a wave equation by a stochastic projection methods. *Monte Carlo Methods Appl.* **2015**, *21*, 189–203. [CrossRef]
19. Shishlenin, M.A.; Izzatulah, M.; Novikov, N.S. Comparative Study of Acoustic Parameter Reconstruction by using Optimal Control Method and Inverse Scattering Approach. *J. Phys. Conf. Ser.* **2021**, *2092*, 012004. [CrossRef]
20. Kabanikhin, S.I.; Novikov, N.S.; Shishlenin, M.A. Gelfand-Levitan-Krein method in one-dimensional elasticity inverse problem. *J. Phys. Conf. Ser.* **2021**, *2092*, 012022. [CrossRef]
21. Kabanikhin, S.I.; Shishlenin, M.A. Digital field. *Georesursy Georesources* **2018**, *20*, 139–141. [CrossRef]
22. Belishev, M. Local Boundary Controllability in Classes of Differentiable Functions for the Wave Equation. *J. Math. Sci.* **2019**, *238*, 591–601. [CrossRef]
23. Belishev, M.; Blagoveshchensky, A.; Karazeeva, N. Simplest Test for the Three-Dimensional Dynamical Inverse Problem (The BC-Method). *J. Math. Sci.* **2021**, *252*, 576–591. [CrossRef]
24. Klibanov, M.V.; Li, J.; Zhang, W. Convexification for the inversion of a time dependent wave front in a heterogeneous medium. *SIAM J. Appl. Math.* **2019**, *79*, 1722–1747. [CrossRef]
25. Klibanov, M.V. Travel time tomography with formally determined incomplete data in 3D. *Inverse Probl. Imaging* **2019**, *13*, 1367–1393. [CrossRef]
26. Xin, J.; Beilina, L.; Klibanov, M. Globally convergent numerical methods for some coefficient inverse problems. *Comput. Sci. Eng.* **2010**, *12*, 64–76.
27. Beilina, L.; Klibanov, M.V. Globally strongly convex cost functional for a coefficient inverse problem. *Nonlinear Anal. Real World Appl.* **2015**, *22*, 272–288. [CrossRef]
28. Klibanov, M.V. On the travel time tomography problem in 3D. *J. Inverse-Ill-Posed Probl.* **2019**, *27*, 591–607. [CrossRef]
29. Kabanikhin, S.I.; Satybaev, A.D.; Shishlenin, M.A. *Direct Methods of Solving Multidimensional Inverse Hyperbolic Problems*; VSP Utrecht: Boston, MA, USA, 2004.
30. Druskin, V.; Mamonov, A.V.; Thaler, A.E.; Zaslavsky, M. Direct, Nonlinear Inversion Algorithm for Hyperbolic Problems via Projection-Based Model Reduction. *SIAM J. Imaging Sci.* **2016**, *9*, 684–747. [CrossRef]
31. Bellassoued, M.; Jellal, D.; Yamamoto, M. Lipschitz stability in in an inverse problem for a hyperbolic equation with a finite set of boundary data. *Appl. Anal.* **2008**, *87*, 1105–1119. [CrossRef]
32. Klibanov, M.V. Carleman estimates for global uniqueness, stability and numerical methods for coefficient inverse problems. *J. Inverse-Ill-Posed Probl.* **2013**, *21*, 477–560. [CrossRef]
33. Li, S.; Yamamoto, M. An inverse problem for Maxwell's equations in anisotropic media in two dimensions. *Chin. Ann. Math. Ser. B* **2007**, *28*, 35–54. [CrossRef]

34. Bellassoued, M.; Cristofol, M.; Soccorsi, E. Inverse boundary value problem for the dynamical heterogeneous Maxwell's system. *Inverse Probl.* **2012**, *28*, 095009 [CrossRef]
35. Beilina, L.; Cristofol, M.; Niinimaki, K. Optimization approach for the simultaneous reconstruction of the dielectric permittivity and magnetic permeability functions from limited observations. *Inverse Probl. Imaging* **2015**, *9*, 1–25. [CrossRef]
36. Imanuvilov, O.Y.; Isakov, V.; Yamamoto, M. An inverse problem for the dynamical Lame system with two sets of boundary data. *Commun. Pure Appl. Math.* **2003**, *56*, 1366–1382. [CrossRef]
37. Beilina, L.; Cristofol, M.; Li, S.; Yamamoto, M. Lipschitz stability for an inverse hyperbolic problem of determining two coefficients by a finite number of observations. *Inverse Probl.* **2017**, *34*, 015001. [CrossRef]
38. Klyuchinskiy, D.V.; Novikov, N.S.; Shishlenin, M.A. CPU-time and RAM memory optimization for solving dynamic inverse problems using gradient-based approach. *J. Comput. Phys.* **2021**, *439*, 110374. [CrossRef]
39. Pitarch, J.; Contelles-Cervera, M.; Peñaranda-Foix, F.L.; Catalá-Civera, J.M. Determination of the permittivity and permeability for waveguides partially loaded with isotropic samples. *Meas. Sci. Technol.* **2005**, *17*, 145–152. [CrossRef]
40. Kenig, C.E.; Salo, M.; Uhlmann, G. Inverse problems for the anisotropic maxwell equations. *Duke Math. J.* **2011**, *157*, 369–419. [CrossRef]
41. Guo, C.; Bal, G. Reconstruction of complex-valued tensors in the Maxwell system from knowledge of internal magnetic fields. *Inverse Probl. Imaging* **2014**, *8*, 1033–1051. [CrossRef]
42. Chen, J.; De Hoop, M. The inverse problem for electroseismic conversion: Stable recovery of the conductivity and the electrokinetic mobility parameter. *Inverse Probl. Imaging* **2016**, *10*, 641–658.
43. Chung, F.J.; Ola, P.; Salo, M.; Tzou, L. Partial data inverse problems for Maxwell equations via Carleman estimates. *Ann. L'Institut Henri Poincare (C) Anal. Non Lineaire* **2018**, *35*, 605–624.
44. Romanov, V.G. Justification of the Gelfand-Levitan-Krein Method for a Two-Dimensional Inverse Problem. *Sib. Math. J.* **2021**, *62*, 908–924. [CrossRef]
45. Gorshkalev, S.B.; Karsten, V.V.; Afonina, E.V.; Vishnevskiy, D.M.; Khogoeva, E.E. Polarization analysis of reflected PS-waves in subsurface with varing cracks orientation. *Technol. Seism. Explor.* **2016**, *7*, 52–60.
46. Klyuchinskiy, D.; Novikov, N.; Shishlenin, M.A. Modification of gradient descent method for solving coefficient inverse problem for acoustics equations. *Computation* **2020**, *8*, 73. [CrossRef]
47. Klyuchinskiy, D.; Novikov, N.; Shishlenin, M. Recovering density and speed of sound coefficients in the 2d hyperbolic system of acoustic equations of the first order by a finite number of observations. *Mathematics* **2021**, *9*, 199. [CrossRef]
48. Kabanikhin, S.I.; Shishlenin, M.A. Comparative Analysis of boundary control and Gel'fand-Levitan methods of solving inverse acoustic problem. In *Inverse Problems in Engineering Mechanics IV, Proceedings of the International Symposium on Inverse Problems in Engineering Mechanics (ISIP 2003), Nagano, Japan, 18-21 February 2003*; Elsevier: Amsterdam, The Netherlands, 2003; pp. 503–512.
49. Kabanikhin, S.I.; Shishlenin, M.A. Boundary control and Gelfand-Levitan-Krein methods in inverse acoustic problem. *J. Inverse-Ill-Posed Probl.* **2004**, *12*, 125–144. [CrossRef]

mathematics MDPI

Article

Multi-Dimensional Integral Transform with Fox Function in Kernel in Lebesgue-Type Spaces

Sergey Sitnik [1,*] **and Oksana Skoromnik** [2]

[1] Department of Applied Mathematics and Computer Modeling, Belgorod State National Research University (BelGU), Pobedy St. 85, 308015 Belgorod, Russia
[2] Faculty of Computer Science and Electronics, Euphrosyne Polotskaya State University of Polotsk, Blokhin St. 29, 211440 Novopolotsk, Belarus; skoromnik@gmail.com
* Correspondence: sitnik@bsu.edu.ru

Abstract: This paper is devoted to the study of the multi-dimensional integral transform with the Fox H-function in the kernel in weighted spaces with integrable functions in the domain $\mathbb{R}^n_+$ with positive coordinates. Due to the generality of the Fox H-function, many special integral transforms have the form studied in this paper, including operators with such kernels as generalized hypergeometric functions, classical hypergeometric functions, Bessel and modified Bessel functions and so on. Moreover, most important fractional integral operators, such as the Riemann–Liouville type, are covered by the class under consideration. The mapping properties in Lebesgue-weighted spaces, such as the boundedness, the range and the representations of the considered transformation, are established. In special cases, it is applied to the specific integral transforms mentioned above. We use a modern technique based on the extensive use of the Mellin transform and its properties. Moreover, we generalize our own previous results from the one-dimensional case to the multi-dimensional one. The multi-dimensional case is more complex and needs more delicate techniques.

Keywords: multi-dimensional integral transform; Fox H-function; Melling transform; weighted space; fractional integrals and derivatives

MSC: 44A30; 33C60; 35A22

Citation: Sitnik, S.; Skoromnik, O. Multi-Dimensional Integral Transform with Fox Function in Kernel in Lebesgue-Type Spaces. *Mathematics* **2024**, *12*, 1829. https://doi.org/10.3390/math12121829

Academic Editor: Farhad Jafari

Received: 7 April 2024
Revised: 25 May 2024
Accepted: 12 June 2024
Published: 12 June 2024

1. Introduction

We consider the multi-dimensional H-integral transform ([1], Formula (43)):

$$(\mathrm{H}f)(\mathbf{x}) = \int_0^\infty \mathrm{H}^{\mathbf{m},\mathbf{n}}_{\mathbf{p},\mathbf{q}}\left[\mathbf{xt}\,\middle|\,\begin{matrix}(\mathbf{a}_i,\overline{\alpha}_i)_{1,p}\\(\mathbf{b}_j,\overline{\beta}_j)_{1,q}\end{matrix}\right]f(\mathbf{t})\mathbf{dt},\qquad \mathbf{x}>0; \tag{1}$$

where (see [1,2], ch. 28; [3], ch. 1) $\mathbf{x} = (x_1, x_2, \ldots, x_n) \in \mathbb{R}^n$; $\mathbf{t} = (t_1, t_2, \ldots, t_n) \in \mathbb{R}^n$, $\mathbb{R}^n$ is the n-dimensional Euclidean space; $\mathbf{x}\cdot\mathbf{t} = \sum_{n=1}^{n} x_n t_n$ denotes their scalar product; in particular, $\mathbf{x}\cdot\mathbf{1} = \sum_{n=1}^{n} x_n$ for $\mathbf{1}$ = (1, 1, …, 1). The inequality $\mathbf{x} > \mathbf{t}$ means that $x_1 > t_1, \ldots, x_n > t_n$, and inequalities $\geq, <, \leq$ have similar meanings; $\int_0^\infty = \int_0^\infty\int_0^\infty\cdots\int_0^\infty$; by $\mathbb{N} = \{1, 2, \ldots\}$, we denote the set of natural numbers, $\mathbb{N}_0 = \mathbb{N}\bigcup\{0\}$, $\mathbb{N}_0^n = \mathbb{N}_0 \times \cdots \times \mathbb{N}_0$; $k = (k_1, k_2, \ldots, k_n) \in \mathbb{N}_0^n$ ($k_i \in \mathbb{N}_0$, $i = 1, 2, \ldots, n$) is a multi-index with $k! = k_1! \cdots k_n!$ and $|k| = k_1 + \cdots + k_n$; $\mathbb{R}^n_+ = \{\mathbf{x} \in \mathbb{R}^n, \mathbf{x} > 0\}$; for $\kappa = (\kappa_1, \kappa_2, \ldots, \kappa_n) \in \mathbb{R}^n_+$ $\mathbf{D}^\kappa = \frac{\partial^{|\kappa|}}{(\partial x_1)^{\kappa_1}\cdots(\partial x_n)^{\kappa_n}}$; $\mathbf{dt} = dt_1 \cdots dt_n$; $\mathbf{t}^\kappa = t^{\kappa_1}t^{\kappa_2}\cdots t^{\kappa_n}$; $f(\mathbf{t}) = f(t_1, t_2, \ldots, t_n)$; $\mathbb{C}^n$ ($n \in \mathbb{N}$) is the n-dimensional space of n complex numbers $z = (z_1, z_2, \cdots, z_n)$ ($z_j \in \mathbb{C}, j = 1, 2, \cdots, n$); $\overline{\lambda} = (\lambda_1, \lambda_2, \ldots, \lambda_n) \in \mathbb{C}^n$; $\overline{h} = (h_1, h_2, \ldots, h_n) \in \mathbb{R}^n_+$; $\frac{\mathrm{d}}{\mathrm{dx}} = \frac{d}{dx_1\cdot dx_2\cdots dx_n}$;

$\mathbf{m} = (m_1, m_2, \ldots, m_n) \in \mathbb{N}_0^n$ and $m_1 = m_2 = \cdots = m_n$; $\mathbf{n} = (\overline{n}_1, \overline{n}_2, \ldots, \overline{n}_n) \in \mathbb{N}_0^n$ and $\overline{n}_1 = \overline{n}_2 = \cdots = \overline{n}_n$; $\mathbf{p} = (p_1, p_2, \ldots, p_n) \in \mathbb{N}_0^n$ and $p_1 = p_2 = \cdots = p_n$; $\mathbf{q} = (q_1, q_2, \ldots, q_n) \in \mathbb{N}_0^n$ and $q_1 = q_2 = \cdots = q_n$ $(0 \leq \mathbf{m} \leq \mathbf{q}, 0 \leq \mathbf{n} \leq \mathbf{p})$;

$\mathbf{a}_i = (a_{i_1}, a_{i_2}, \ldots, a_{i_n})$, $1 \leq i \leq \mathbf{p}$, $a_{i_1}, a_{i_2}, \ldots, a_{i_n} \in \mathbb{C}$ $(i_1 = 1, 2, \ldots, p_1; \ldots; i_n = 1, 2, \ldots, p_n)$;

$\mathbf{b}_j = (b_{j_1}, b_{j_2}, \ldots, b_{j_n})$, $1 \leq j \leq \mathbf{q}$, $b_{j_1}, b_{j_2}, \ldots, b_{j_n} \in \mathbb{C}$ $(j_1 = 1, 2, \ldots, q_1; \ldots; j_n = 1, 2, \ldots, q_n)$;

$\overline{\alpha}_i = (\alpha_{i_1}, \alpha_{i_2}, \ldots, \alpha_{i_n})$, $1 \leq i \leq \mathbf{p}$, $\alpha_{i_1}, \alpha_{i_2}, \ldots, \alpha_{i_n} \in \mathbb{R}_1^+$ $(i_1 = 1, 2, \ldots, p_1; \ldots; i_n = 1, 2, \ldots, p_n)$;

$\overline{\beta}_j = (\beta_{j_1}, \beta_{j_2}, \ldots, \beta_{j_n})$, $1 \leq j \leq \mathbf{q}$, $\beta_{j_1}, \beta_{j_2}, \ldots, \beta_{j_n} \in \mathbb{R}_1^+$ $(j_1 = 1, 2, \ldots, q_1; \ldots; j_n = 1, 2, \ldots, q_n)$.

The function in the kernel of (1)

$$\mathrm{H}_{\mathbf{p},\mathbf{q}}^{\mathbf{m},\mathbf{n}}\left[\mathbf{xt}\,\middle|\, \begin{matrix}(\mathbf{a}_i, \overline{\alpha}_i)_{1,p} \\ (\mathbf{b}_j, \overline{\beta}_j)_{1,q}\end{matrix}\right] = \prod_{k=1}^{n} H_{p_k,q_k}^{m_k,\overline{n}_k}\left[x_k t_k \,\middle|\, \begin{matrix}(a_{i_k}, \alpha_{i_k})_{1,p_k} \\ (b_{j_k}, \beta_{j_k})_{1,q_k}\end{matrix}\right] \tag{2}$$

is the product of H-functions $H_{p,q}^{m,n}[z]$:

$$H_{p,q}^{m,n}[z] \equiv H_{p,q}^{m,n}\left[z \,\middle|\, \begin{matrix}(a_i, \alpha_i)_{1,p} \\ (b_j, \beta_j)_{1,q}\end{matrix}\right] = \frac{1}{2\pi i}\int_L \mathcal{H}_{p,q}^{m,n}(s) z^{-s} ds, \; z \neq 0, \tag{3}$$

where

$$\mathcal{H}_{p,q}^{m,n}(s) \equiv \mathcal{H}_{p,q}^{m,n}\left[\begin{matrix}(a_i, \alpha_i)_{1,p} \\ (b_j, \beta_j)_{1,q}\end{matrix} \middle| s\right] = \frac{\prod\limits_{j=1}^{m} \Gamma(b_j + \beta_j s) \prod\limits_{i=1}^{n} \Gamma(1 - a_i - \alpha_i s)}{\prod\limits_{i=n+1}^{p} \Gamma(a_i + \alpha_i s) \prod\limits_{j=m+1}^{q} \Gamma(1 - b_j - \beta_j s)}. \tag{4}$$

In the representation (3), L is a specifically chosen infinite contour, and the empty products, if any, are taken to be one.

The H-function (3) is the most general of the known special functions and includes, as special cases, elementary functions and special functions of the hypergeometric and Bessel type, as well as the Meyer G-function. One may find its properties, for example, in the books by Mathai and Saxena ([4], Ch. 2); Srivastava, Gupta and Goyal ([5], ch. 1); Prudnikov, Brychkov and Marichev ([6], Section 8.3); Kiryakova [7]; and Kilbas and Saigo ([8], Ch.1–Ch.4).

Due to the generality of the Fox H-function, many special integral transforms have the form studied in this paper, including operators with such kernels as generalized hypergeometric functions, classical hypergeometric functions, Bessel and modified Bessel functions and so on. Moreover, most important fractional integral operators, such as the Riemann–Liouville type, are covered by the class under consideration. The mapping properties in Lebesgue-weighted spaces, such as the boundedness, the range and the representations of the considered transformation, are established. In special cases, it is applied to the specific integral transforms mentioned above. We use a modern technique based on the extensive use of the Mellin transform and its properties.

Our paper is devoted to the study of the H-transform (1) in Lebesgue-type weighted spaces $\mathfrak{L}_{\overline{\nu},\overline{2}}$ of functions $f(\mathbf{x}) = f(x_1, x_2, \ldots, x_n)$ on $\mathbb{R}_+^{\mathbf{n}}$, such that

$$\|f\|_{\overline{\nu},\overline{2}} = \{\int_{\mathbb{R}_+^1} x_n^{2\cdot\nu_n - 1}\{\cdots\{\int_{\mathbb{R}_+^1} x_2^{2\cdot\nu_2\cdot - 1} \times$$

$$[\int_{\mathbb{R}_+^1} x_1^{2\cdot\nu_1 - 1}|f(x_1, \ldots, x_n)|^2 dx_1] dx_2\} \cdots\} dx_n\}^{1/2} < \infty,$$

$\overline{\nu} = (\nu_1, \nu_2, \ldots, \nu_n) \in \mathbb{R}^n$, $\nu_1 = \nu_2 = \cdots = \nu_n$, and $\overline{2} = (2, 2, \ldots, 2)$.

In this paper, we apply some our previous results to obtain mapping properties such as the boundedness, the range and the representations for the H-transform (1).

The research results for transformation (1) generalize those obtained earlier for the corresponding one-dimensional transformation (see [8], Ch. 3):

$$(\mathrm{H}f)(x) = \int_0^\infty \mathrm{H}^{m,n}_{p,q}\left[xt\middle| \begin{array}{c}(a_i,\alpha_i)_{1,p}\\(b_j,\beta_j)_{1,q}\end{array}\right] f(t)dt, \qquad x > 0; \tag{5}$$

in the space $\mathfrak{L}_{\nu,2}$ of Lebesgue measurable functions f on $\mathbb{R}^1_+ = (0,\infty)$, such that

$$\int_0^\infty |t^\nu f(t)|^2 \frac{dt}{t} < \infty \quad (\nu \in \mathbb{R}).$$

The H-transform (5) generalizes many integral transforms: transforms with the Meijer G-function, Laplace and Hankel transforms, transforms with Gauss hypergeometric functions and transforms with other hypergeometric and Bessel functions in the kernels. One may find a survey of results and a bibliography in this field for the one-dimensional case in a monograph ([8], Sections 6–8). Note that a very important class of transforms under consideration is the class of Buschman–Erdélyi operators; they have many important properties and applications. The topic of this paper is also strongly connected with transmutation theory, cf. [9].

Note that, in transmutation theory applied to differential equations, its solutions are represented as integral transforms; in this way, solutions of perturbed differential equations are represented via more simple solutions of unperturbed equations. Through the results of this paper and similar ones, such a representation may also be accompanied by norm estimates in classical functional spaces. It helps to estimate the norms of perturbed equations and analyze their smoothness or singularity conditions, cf. [9].

2. Preliminaries

The properties of the H-function $H^{m,n}_{p,q}[z]$ (3) depend on the following numbers ([8], Formulas 1.1.7–1.1.15):

$$a^* = \sum_{i=1}^{n}\alpha_i - \sum_{i=n+1}^{p}\alpha_i + \sum_{j=1}^{m}\beta_j - \sum_{j=m+1}^{q}\beta_j; \; \Delta = \sum_{j=1}^{q}\beta_j - \sum_{i=1}^{p}\alpha_i; \tag{6}$$

$$\delta = \prod_{i=1}^{p}\alpha_i^{-\alpha_i}\prod_{j=1}^{q}\beta_j^{\beta_j}; \tag{7}$$

$$\mu = \sum_{j=1}^{q} b_j - \sum_{i=1}^{p} a_i + \frac{p-q}{2}; \tag{8}$$

$$a_1^* = \sum_{j=1}^{m}\beta_j - \sum_{i=n+1}^{p}\alpha_i; \; a_2^* = \sum_{i=1}^{n}\alpha_i - \sum_{j=m+1}^{q}\beta_j; \; a_1^* + a_2^* = a^*, \; a_1^* - a_2^* = \Delta; \tag{9}$$

$$\xi = \sum_{j=1}^{m} b_j - \sum_{j=m+1}^{q} b_j + \sum_{i=1}^{n} a_i - \sum_{i=n+1}^{p} a_i; \tag{10}$$

$$c^* = m + n - \frac{p+q}{2}. \tag{11}$$

The empty sum in (6), (8), (9), (10) and the empty product in (7), if they occur, are taken to be zero and one, respectively.

The following assertions hold.

Lemma 1 ([8], Lemma 1.2). *For σ, $t \in \mathbb{R}$, the following estimate holds*

$$|\mathcal{H}_{p,q}^{m,n}(\sigma + it)| \sim C|t|^{\Delta\sigma + \text{Re}(\mu)} \exp^{-\pi[|t|a^* + \text{Im}(\xi)\text{sign}(t)]/2} \ (|t| \to \infty) \tag{12}$$

uniformly in σ on any bounded interval in $\mathbb{R}$, where

$$C = (2\pi)^{c^*} \exp^{-c^* - \Delta\sigma - \text{Re}(\mu)} \delta^{\sigma} \prod_{i=1}^{p} \alpha_i^{1/2 - \text{Re}(a_i)} \prod_{j=1}^{q} \beta_j^{\text{Re}(b_j) - 1/2} \tag{13}$$

and ξ and c^* are defined in (10) *and* (11).

Theorem 1 ([8],Theorem 3.4). *Let $\alpha < \zeta < \beta$ and either of the conditions $a^* > 0$ or $a^* = 0$ and $\Delta\zeta + \text{Re}(\mu) < -1$ hold. Then, for $x > 0$, except for $x = \delta$ when $a^* = 0$ and $\Delta = 0$, the relation*

$$H_{p,q}^{m,n}\left[x \middle| \begin{matrix} (a_p, \alpha_p) \\ (b_p, \beta_p) \end{matrix}\right] = \frac{1}{2\pi i} \int_{\gamma - i\infty}^{\gamma + i\infty} \mathcal{H}_{p,q}^{m,n}\left[\begin{matrix} (a_p, \alpha_p) \\ (b_p, \beta_p) \end{matrix} \middle| t\right] x^{-t} dt \tag{14}$$

holds and the estimate

$$|H_{p,q}^{m,n}\left[x \middle| \begin{matrix} (a_p, \alpha_p) \\ (b_p, \beta_p) \end{matrix}\right]| \leq A_{\zeta} x^{-\zeta} \tag{15}$$

is valid, where A_{ζ} is a positive constant depending only on ζ.

A set of bounded linear operators acting from a Banach space X into a Banach space Y is denoted by $[X, Y]$.

The multi-dimensional Mellin integral transform $(\mathfrak{M}f)(\mathbf{x})$ of function $f(\mathbf{x}) = f(x_1, x_2, \ldots, x_n)$, $\mathbf{x} = (x_1, x_2, \ldots, x_n) \in \mathbb{R}_+^n$, is determined by the formula

$$(\mathfrak{M}f)(\mathbf{s}) = \int_0^{\infty} f(\mathbf{t})\mathbf{t}^{\mathbf{s}-1}\mathbf{dt}, \ \text{Re}(\mathbf{s}) = \overline{\nu}, \tag{16}$$

$\mathbf{s} = (s_1, s_2, \ldots, s_n) \in \mathbb{C}^n$. The inverse multi-dimensional Mellin transform has the form

$$(\mathfrak{M}^{-1}g)(\mathbf{x}) = \frac{1}{(2\pi i)^n} \int_{\gamma_1 - i\infty}^{\gamma_1 + i\infty} \cdots \int_{\gamma_n - i\infty}^{\gamma_n + i\infty} \mathbf{x}^{-\mathbf{s}} g(\mathbf{s})\mathbf{ds}, \tag{17}$$

$\mathbf{x} \in \mathbb{R}_+^n$, $\gamma_j = \text{Re}(s_j)$ $(j = 1, \cdots, n)$. The theory of multi-dimensional integral transformations (16) and (17) can be recognized, for example, in books ([3], Ch. 1; [10,11]).

We will need the following spaces. As usual, by $\text{L}_{\overline{p}}(\mathbb{R}^n)$, we understand the space of functions $f(\mathbf{x}) = f(x_1, x_2, \ldots, x_n)$, for which

$$\|f\|_{\overline{p}} = \left\{\int_{\mathbb{R}^n} |f(\mathbf{x})|^{\overline{p}} \mathbf{dx}\right\}^{1/\overline{p}} < \infty, \ \overline{p} = (p_1, p_2, \ldots, p_n), \ 1 \leq \overline{p} < \infty.$$

If $\overline{p} = \infty$, then the space $\text{L}_{\infty}(\mathbb{R}^n)$ is defined as the collection of all measurable functions with a finite norm

$$\|f\|_{\text{L}_{\infty}(\mathbb{R}^n)} = esssup|f(\mathbf{x})|,$$

where $esssup\,|f(\mathbf{x})|$ is the essential supremum of the function $|f(\mathbf{x})|$ [12].

We need the following properties of the Mellin transform (16).

Lemma 2 ([1], Lemma 1). *Let $\overline{\nu} = (\nu_1, \nu_2, \ldots, \nu_n) \in \mathbb{R}^n$, $\nu_1 = \nu_2 = \cdots = \nu_n$. The following properties of the Mellin transform* (16) *are valid.*

(a) Transformation (16) *is a unitary mapping of the space $\mathfrak{L}_{\overline{\nu}, \overline{2}}$ onto the space $\text{L}_{\overline{2}}(\mathbb{R}^n)$.*

(b) For $f \in \mathfrak{L}_{\overline{\nu},\overline{2}}$, the following holds

$$f(\mathbf{x}) = \frac{1}{(2\pi i)^n} \lim_{R\to\infty} \int_{\nu_1-iR}^{\nu_1+iR} \int_{\nu_2-iR}^{\nu_2+iR} \cdots \int_{\nu_n-iR}^{\nu_n+iR} (\mathfrak{M}f)(\mathbf{s})\mathbf{x}^{-\mathbf{s}}\mathrm{d}\mathbf{s}, \tag{18}$$

where the limit is taken in the topology of the space $\mathfrak{L}_{\overline{\nu},\overline{2}}$ and where

if $F(\overline{\nu} + i\mathbf{t}) = \prod_{i=1}^{n} F_j(\nu_j + it_j)$, $F_j(\nu_j + it_j) \in L_1(-R, R)$, $j = 1, 2, \ldots, n$, then

$$\int_{\nu_1-iR}^{\nu_1+iR} \int_{\nu_2-iR}^{\nu_2+iR} \cdots \int_{\nu_n-iR}^{\nu_n+iR} F(\mathbf{s})\mathrm{d}\mathbf{s} = i^n \int_{-R}^{R} \int_{-R}^{R} \cdots \int_{-R}^{R} F(\overline{\nu} + i\mathbf{t})\mathrm{d}\mathbf{t}.$$

(c) For functions $f \in \mathfrak{L}_{\overline{\nu},\overline{2}}$ and $g \in \mathfrak{L}_{1-\overline{\nu},\overline{2}}$, the following equality holds

$$\int_0^\infty f(\mathbf{x})g(\mathbf{x})\mathrm{d}\mathbf{x} = \frac{1}{(2\pi i)^n} \int_{\overline{\nu}-i\infty}^{\overline{\nu}+i\infty} (\mathfrak{M}f)(\mathbf{s})(\mathfrak{M}g)(1-\mathbf{s})\mathbf{x}^{-\mathbf{s}}\mathrm{d}\mathbf{s}. \tag{19}$$

In [1], we consider the general multi-dimensional integral transform ([1], Formula (1))

$$(\mathrm{K}f)(\mathbf{x}) = \overline{h}\mathbf{x}^{1-(\overline{\lambda}+1)/\overline{h}} \frac{\mathrm{d}}{\mathrm{d}\mathbf{x}} \mathbf{x}^{(\overline{\lambda}+1)/\overline{h}} \int_0^\infty \mathrm{k}[\mathbf{xt}]f(\mathbf{t})\mathrm{d}\mathbf{t} \ \ (\mathbf{x} > 0), \tag{20}$$

where the function $\mathrm{k}[\mathbf{xt}]$ in the kernel of (20) is the product of one type of special function:

$$\mathrm{k}[\mathbf{xt}] = \mathrm{k}[x_1t_1] \cdot \mathrm{k}[x_2t_2] \cdots \mathrm{k}[x_nt_n].$$

Transformation (20) satisfies the following theorem.

Theorem 2 ([1], Theorem 1). *Let $\overline{\nu} = (\nu_1, \nu_2, \ldots, \nu_n) \in \mathbb{R}^n$ $(\nu_1 = \nu_2 = \cdots = \nu_n)$, $\overline{h} = (h_1, h_2, \ldots, h_n) \in \mathbb{R}^n_+$, and $\overline{\lambda} = (\lambda_1, \lambda_2, \ldots, \lambda_n) \in \mathbb{C}^n$.*

(a) If the transformation operator (20) *satisfies the condition $\mathrm{K} \in [\mathfrak{L}_{\overline{\nu},\overline{2}}, \mathfrak{L}_{1-\overline{\nu},\overline{2}}]$, then the kernel on the right side of* (20) *$\mathrm{k} \in \mathfrak{L}_{1-\overline{\nu},\overline{2}}$. If we set, for $\nu_j \neq 1 - (\mathrm{Re}(\lambda_j) + 1)/h_j$, $j = 1, 2, \ldots, n$,*

$$(\mathfrak{M}\mathrm{k})(1 - \overline{\nu} + i\mathbf{t}) = \frac{\theta(\mathbf{t})}{\overline{\lambda} + 1 - (1 - \overline{\nu} + i\mathbf{t})\overline{h}}$$

$$= \prod_{j=1}^{n} \frac{\theta(t_j)}{\lambda_j + 1 - (1 - \nu_j + it_j)h_j} \tag{21}$$

almost everywhere, then function $\theta \in L_\infty(\mathbb{R}^n)$, and, for $f \in \mathfrak{L}_{\overline{\nu},\overline{2}}$, the relation

$$(\mathfrak{M}\mathrm{K}f)(1 - \overline{\nu} + i\mathbf{t}) = \theta(\mathbf{t})(\mathfrak{M}f)(\overline{\nu} - i\mathbf{t}) \tag{22}$$

holds almost everywhere.

(b) Conversely, for a given function $\theta \in L_\infty(\mathbb{R}^n)$, there is a transform $\mathrm{K} \in [\mathfrak{L}_{\overline{\nu},\overline{2}}, \mathfrak{L}_{1-\overline{\nu},\overline{2}}]$ so that the equality (22) *holds for $f \in \mathfrak{L}_{\overline{\nu},\overline{2}}$. Moreover, if $\nu_j \neq 1 - (\mathrm{Re}(\lambda_j) + 1)/h_j$, $j = 1, 2, \ldots, n$, then transformation $\mathrm{K}f$* (20) *is representable in the form* (20) *with the kernel* k *defined by* (21).

(c) Based on statement (a) or (b) with $\theta \neq 0$, K is a one-to-one transformation from the space $\mathfrak{L}_{\overline{\nu},\overline{2}}$ into the space $\mathfrak{L}_{1-\overline{\nu},\overline{2}}$, and if, in addition, $1/\theta \in L_\infty(\mathbb{R}^n)$, then K maps $\mathfrak{L}_{\overline{\nu},\overline{2}}$ onto $\mathfrak{L}_{1-\overline{\nu},\overline{2}}$, and, for functions f, $g \in \mathfrak{L}_{\overline{\nu},\overline{2}}$, the relation

$$\int_0^\infty f(\mathbf{x})(\mathrm{K}g)(\mathbf{x})\mathrm{d}\mathbf{x} = \int_0^\infty (\mathrm{K}f)(\mathbf{x})g(\mathbf{x})\mathrm{d}\mathbf{x} \tag{23}$$

is valid.

3. $\mathfrak{L}_{\overline{\nu},2}$-Theory for the Multi-Dimensional H-Transform

To formulate the results for the transform Hf (1), we need the following constants ([1]), which are analogous for the one-dimensional case defined via the parameters of the H-function (3) ([8], (3.4.1), (3.4.2), (1.1.7), (1.1.8), (1.1.10)).

Let $\widetilde{\alpha} = (\widetilde{\alpha}_1, \widetilde{\alpha}_2, \ldots, \widetilde{\alpha}_n)$ and $\widetilde{\beta} = (\widetilde{\beta}_1, \widetilde{\beta}_2, \ldots, \widetilde{\beta}_n)$, where

$$\widetilde{\alpha}_1 = \begin{cases} -\min\limits_{1 \le j_1 \le m_1} \left[\frac{\mathrm{Re}(b_{j_1})}{\beta_{j_1}} \right], & m_1 > 0, \\ -\infty, & m_1 = 0, \end{cases} \qquad \widetilde{\beta}_1 = \begin{cases} \min\limits_{1 \le i_1 \le \overline{n}_1} \left[\frac{1 - \mathrm{Re}(a_{i_1})}{\alpha_{i_1}} \right], & \overline{n}_1 > 0, \\ \infty, & \overline{n}_1 = 0, \end{cases}$$

$$\widetilde{\alpha}_2 = \begin{cases} -\min\limits_{1 \le j_2 \le m_2} \left[\frac{\mathrm{Re}(b_{j_2})}{\beta_{j_2}} \right], & m_2 > 0, \\ -\infty, & m_2 = 0, \end{cases} \qquad \widetilde{\beta}_2 = \begin{cases} \min\limits_{1 \le i_2 \le \overline{n}_2} \left[\frac{1 - \mathrm{Re}(a_{i_2})}{\alpha_{i_2}} \right], & \overline{n}_2 > 0, \\ \infty, & \overline{n}_2 = 0, \end{cases}$$

and

$$\widetilde{\alpha}_n = \begin{cases} -\min\limits_{1 \le j_n \le m_n} \left[\frac{\mathrm{Re}(b_{j_n})}{\beta_{j_n}} \right], & m_n > 0, \\ -\infty, & m_2 = 0, \end{cases} \qquad \widetilde{\beta}_n = \begin{cases} \min\limits_{1 \le i_n \le \overline{n}_n} \left[\frac{1 - \mathrm{Re}(a_{i_n})}{\alpha_{i_n}} \right], & \overline{n}_n > 0, \\ \infty, & \overline{n}_n = 0; \end{cases} \tag{24}$$

and let $a^* = (a_1^*, a_2^*, \ldots, a_n^*)$, $\Delta = (\Delta_1, \Delta_2, \ldots, \Delta_n)$ and

$$a_1^* = \sum_{i=1}^{\overline{n}_1} \alpha_{i_1} - \sum_{i=\overline{n}_1+1}^{p_1} \alpha_{i_1} + \sum_{j=1}^{m_1} \beta_{j_1} - \sum_{j=m_1+1}^{q_1} \beta_{j_1}, \; \Delta_1 = \sum_{j=1}^{q_1} \beta_{j_1} - \sum_{i=1}^{p_1} \alpha_{i_1},$$

$$a_2^* = \sum_{i=1}^{\overline{n}_2} \alpha_{i_2} - \sum_{i=\overline{n}_2+1}^{p_2} \alpha_{i_2} + \sum_{j=1}^{m_2} \beta_{j_2} - \sum_{j=m_1+1}^{q_2} \beta_{j_2}, \; \Delta_2 = \sum_{j=1}^{q_2} \beta_{j_2} - \sum_{i=1}^{p_2} \alpha_{i_2},$$

and

$$a_n^* = \sum_{i=1}^{\overline{n}_n} \alpha_{i_n} - \sum_{i=\overline{n}_n+1}^{p_n} \alpha_{i_n} + \sum_{j=1}^{m_n} \beta_{j_n} - \sum_{j=m_n+1}^{q_n} \beta_{j_n}; \Delta_n = \sum_{j=1}^{q_n} \beta_{j_n} - \sum_{i=1}^{p_n} \alpha_{i_n}; \tag{25}$$

and let $\mu = (\mu_1, \mu_2, \ldots, \mu_n)$ and

$$\mu_1 = \sum_{j=1}^{q_1} b_{j_1} - \sum_{i=1}^{p_1} a_{i_1} + \frac{p_1 - q_1}{2}, \mu_2 = \sum_{j=1}^{q_2} b_{j_2} - \sum_{i=1}^{p_2} a_{i_2} + \frac{p_2 - q_2}{2}, \ldots,$$

$$\mu_n = \sum_{j=1}^{q_n} b_{j_n} - \sum_{i=1}^{p_n} a_{i_n} + \frac{p_n - q_n}{2}; \tag{26}$$

The exceptional set $\mathcal{E}_{\overline{\mathcal{H}}}$ of a function $\overline{\mathcal{H}}^{\mathbf{m},\mathbf{n}}_{\mathbf{p},\mathbf{q}}(\mathbf{s})$

$$\overline{\mathcal{H}}^{\mathbf{m},\mathbf{n}}_{\mathbf{p},\mathbf{q}}(\mathbf{s}) \equiv \overline{\mathcal{H}}^{\mathbf{m},\mathbf{n}}_{\mathbf{p},\mathbf{q}} \left[\begin{matrix} (\mathbf{a}_i, \overline{\alpha}_i)_{1,\mathbf{p}} \\ (\mathbf{b}_j, \overline{\beta}_j)_{1,\mathbf{q}} \end{matrix} \middle| \mathbf{s} \right] = \prod_{k=1}^{n} \mathcal{H}^{m_k, \overline{n}_k}_{p_k, q_k} \left[\begin{matrix} (a_{i_k}, \alpha_{i_k})_{1,p_k} \\ (b_{j_k}, \beta_{j_k})_{1,q_k} \end{matrix} \middle| s \right], \tag{27}$$

is called a set of vectors $\overline{\nu} = (\nu_1, \nu_2, \ldots, \nu_n) \in \mathbb{R}^n$ ($\nu_1 = \nu_2 = \cdots = \nu_n$), such that $\widetilde{\alpha}_k < 1 - \nu_k < \widetilde{\beta}_k$, $k = 1, 2, \ldots n$, where the parameters $\widetilde{\alpha}_k$, $\widetilde{\beta}_k (k = 1, 2, \ldots, n)$ are defined by Formula (24), and functions $\mathcal{H}^{m_k, \overline{n}_k}_{p_k, q_k}(s_k)$ $(k = 1, 2, \ldots, n)$ of the view (4) have zeros on lines $\mathrm{Re}(s_k) < 1 - \nu_k$ $(k = 1, 2, \ldots, n)$, respectively.

Applying the multi-dimensional Mellin transformation (16) to (1), formally, we obtain

$$(\mathfrak{M}\mathrm{H}f)(\mathbf{s}) = \overline{\mathcal{H}}^{\mathbf{m},\mathbf{n}}_{\mathbf{p},\mathbf{q}} \left[\begin{matrix} (\mathbf{a}_i, \alpha_i)_{1,\mathbf{p}} \\ (\mathbf{b}_j, \beta_j)_{1,\mathbf{q}} \end{matrix} \middle| \mathbf{s} \right] (\mathfrak{M}f)(1 - \mathbf{s}). \tag{28}$$

Theorem 3. *Suppose that*

$$\widetilde{\alpha}_k < 1 - \nu_k < \widetilde{\beta}_k;\ \nu_k = \nu_l,\ k \neq l\ (k, l = 1, 2, \ldots, n); \tag{29}$$

and that either of the conditions

$$a_k^* > 0\ (k = 1, 2, \ldots, n); \tag{30}$$

or

$$a_k^* = 0, \Delta_k[1 - \nu_k] + \mathrm{Re}(\mu_k) \leq 0\ (k = 1, 2, \ldots, n) \tag{31}$$

holds. Then, we have the following results.

(a) There exists a one-to-one transform $\mathrm{H} \in [\mathfrak{L}_{\overline{\nu},\overline{2}}, \mathfrak{L}_{1-\overline{\nu},\overline{2}}]$ *so that the relation (28) holds for* $\mathrm{Re}(\mathbf{s}) = 1 - \overline{\nu}$ *and* $f \in \mathfrak{L}_{\overline{\nu},\overline{2}}$.

If $a_k^* = 0, \Delta_k[1 - \nu_k] + \mathrm{Re}(\mu_k) = 0\ (k = 1, 2, \ldots, n)$*, and* $\overline{\nu}$ *does not belong to an exceptional set* $\mathcal{E}_{\overline{\mathcal{H}}}$*, then the operator* H *maps* $\mathfrak{L}_{\overline{\nu},\overline{2}}$ *onto* $\mathfrak{L}_{1-\overline{\nu},\overline{2}}$.

(b) If $f \in \mathfrak{L}_{\overline{\nu},\overline{2}}$ *and* $g \in \mathfrak{L}_{\overline{\nu},\overline{2}}$*, then, for* H*, we have the relation* (23)

$$\int_0^\infty f(\mathbf{x})(\mathrm{H}g)(\mathbf{x})\mathbf{dx} = \int_0^\infty (\mathrm{H}f)(\mathbf{x})g(\mathbf{x})d\mathbf{x}. \tag{32}$$

(c) Let $f \in \mathfrak{L}_{\overline{\nu},\overline{2}}$, $\overline{\lambda} = (\lambda_1, \lambda_2, \ldots, \lambda_n) \in \mathbb{C}^n$, $\overline{h} = (h_1, h_2, \ldots, h_n) \in \mathbb{R}_+^n$. *If* $\mathrm{Re}(\overline{\lambda}) > (1 - \overline{\nu})\overline{h} - 1$*, then* $\mathrm{H}f$ *is given by the formula*

$$(\mathrm{H}f)(\mathbf{x}) = \overline{h}\mathbf{x}^{1-(\overline{\lambda}+1)/\overline{h}}$$

$$\times \frac{\mathbf{d}}{\mathbf{dx}}\mathbf{x}^{(\overline{\lambda}+1)/\overline{h}} \int_0^\infty \mathrm{H}^{\mathbf{m},\mathbf{n}+1}_{\mathbf{p}+1,\mathbf{q}+1}\left[\mathbf{xt}\left|\begin{matrix}(-\overline{\lambda}, \overline{h}), (\mathbf{a}_i, \alpha_i)_{1,\mathbf{p}} \\ (\mathbf{b}_j, \beta_j)_{1,\mathbf{q}}, (-\overline{\lambda} - 1, \overline{h})\end{matrix}\right.\right] f(\mathbf{t})\mathbf{dt}. \tag{33}$$

When $\mathrm{Re}(\overline{\lambda}) < (1 - \overline{\nu})\overline{h} - 1$, $\mathrm{H}f$ *is given by*

$$(\mathrm{H}f)(\mathbf{x}) = -\overline{h}\mathbf{x}^{1-(\overline{\lambda}+1)/\overline{h}}$$

$$\times \frac{\mathbf{d}}{\mathbf{dx}}\mathbf{x}^{(\overline{\lambda}+1)/\overline{h}} \int_0^\infty \mathrm{H}^{\mathbf{m}+1,\mathbf{n}}_{\mathbf{p}+1,\mathbf{q}+1}\left[\mathbf{xt}\left|\begin{matrix}(\mathbf{a}_i, \alpha_i)_{1,\mathbf{p}}, (-\overline{\lambda}, \overline{h}) \\ (-\overline{\lambda} - 1, \overline{h}), (\mathbf{b}_j, \beta_j)_{1,\mathbf{q}}\end{matrix}\right.\right] f(\mathbf{t})\mathbf{dt}. \tag{34}$$

(d) The transform H *is independent of* $\overline{\nu}$ *in the sense that, for* $\overline{\nu}$ *and* $\widetilde{\overline{\nu}}$ *satisfying the assumptions (29), and either (30) or (31), and for the respective transforms* H *on* $\mathfrak{L}_{\overline{\nu},\overline{2}}$ *and* $\widetilde{\mathrm{H}}$ *on* $\mathfrak{L}_{\widetilde{\overline{\nu}},\overline{2}}$ *given in (28), then* $\mathrm{H}f = \widetilde{\mathrm{H}}f$ *for* $f \in \mathfrak{L}_{\overline{\nu},\overline{2}} \cap \mathfrak{L}_{\widetilde{\overline{\nu}},\overline{2}}$.

Proof. Let $\overline{\omega}(\mathbf{t}) = \overline{\mathcal{H}}(1 - \overline{\nu} + i\mathbf{t}) = \prod_{k=1}^{n} \mathcal{H}(1 - \nu_k + it_k)$. By virtue of (4), (24), and the conditions (29), the functions $\mathcal{H}^{m_1,\overline{n}_1}_{p_1,q_1}(s_1), \mathcal{H}^{m_2,\overline{n}_2}_{p_2,q_2}(s_2), \ldots, \mathcal{H}^{m_n,\overline{n}_n}_{p_n,q_n}(s_n)$ are analytic in the strips $\widetilde{\alpha}_1 < 1 - \nu_1 < \widetilde{\beta}_1, \ldots, \widetilde{\alpha}_n < 1 - \nu_n < \widetilde{\beta}_n, \nu_1 = \nu_2 = \cdots = \nu_n$, respectively. In accordance with (12) and conditions (30) or (31), $\overline{\omega}(\mathbf{t}) = O(1)$ as $|\mathbf{t}| \to \infty$. Therefore, $\overline{\omega} \in \mathrm{L}_\infty(\mathbb{R}^n)$, and hence we obtain from Theorem 2 (b) that there exists a transform $\mathrm{H} \in [\mathfrak{L}_{\overline{\nu},\overline{2}}, \mathfrak{L}_{1-\overline{\nu},\overline{2}}]$ such that

$$(\mathfrak{M}\mathrm{H}f)(\mathbf{s})(1 - \overline{\nu} + i\mathbf{t}) = \overline{\mathcal{H}}(1 - \overline{\nu} + i\mathbf{t})(\mathfrak{M}f)(\overline{\nu} - i\mathbf{t})$$

for $f \in \mathfrak{L}_{\overline{\nu},\overline{2}}$. This means that the equality (28) holds when condition $\mathrm{Re}(\mathbf{s}) = 1 - \overline{\nu}$ is met. Since the functions $\mathcal{H}^{m_1,\overline{n}_1}_{p_1,q_1}(s_1), \mathcal{H}^{m_2,\overline{n}_2}_{p_2,q_2}(s_2), \ldots, \mathcal{H}^{m_n,\overline{n}_n}_{p_n,q_n}(s_n)$ are analytic in the strips $\widetilde{\alpha}_1 < 1 - \nu_1 < \widetilde{\beta}_1, \ldots, \widetilde{\alpha}_n < 1 - \nu_n < \widetilde{\beta}_n, \nu_1 = \nu_2 = \cdots = \nu_n$, respectively, and have isolated zeros, then $\overline{\omega}(\mathbf{t}) \neq 0$ almost everywhere. Thus, it follows from Theorem 2 (c) that $\mathrm{H} \in [\mathfrak{L}_{\overline{\nu},\overline{2}}, \mathfrak{L}_{1-\overline{\nu},\overline{2}}]$ is a one-to-one transform. If $a_k^* = 0$, $\Delta_k(1 - \nu_k) + \mathrm{Re}(\mu_k) = 0$

$(k = 1, 2, \ldots n)$ and $\overline{\nu}$ is not in the exceptional set $\mathcal{E}_{\overline{\mathcal{H}}}$ of $\overline{\mathcal{H}}$, then $1/\overline{\omega} \in \mathrm{L}_\infty(\mathbb{R}^n)$, and, from Theorem 2 (c), we have that H transforms the space $\mathfrak{L}_{\overline{\nu},\overline{2}}$ onto $\mathfrak{L}_{1-\overline{\nu},\overline{2}}$. This completes the proof of the statement (a) of the theorem.

According to the statement of the Theorem 2 (c), if $f \in \mathfrak{L}_{\overline{\nu},\overline{2}}$ and $g \in \mathfrak{L}_{\overline{\nu},\overline{2}}$, then the relation (32) is valid. Thus, the assertion (b) is true.

Let us prove the validity of the representation (33). Suppose that $f \in \mathfrak{L}_{\overline{\nu},\overline{2}}$ and $\mathrm{Re}(\overline{\lambda}) > (1-\overline{\nu})\overline{h} - 1$. To show the relation (33), it is sufficient to calculate the kernel k in the transform (20) for such $\overline{\lambda}$. From (21), we obtain the equality

$$(\mathfrak{M}\mathrm{k})(1-\overline{\nu}+i\mathbf{t}) = \overline{\mathcal{H}}(1-\overline{\nu}+i\mathbf{t})\frac{1}{\overline{\lambda}+1-(1-\overline{\nu}+i\mathbf{t})\overline{h}}$$

$$= \prod_{k=1}^{n} \mathcal{H}(1-\nu_k+it_k)\frac{1}{\lambda_k+1-(1-\nu_k+it_k)h_k}$$

or, for $\mathrm{Re}(\mathbf{s}) = 1-\overline{\nu}$,

$$(\mathfrak{M}\mathrm{k})(\mathbf{s}) = \overline{\mathcal{H}}(\mathbf{s})\frac{1}{\overline{\lambda}+1-\overline{h}\mathbf{s}} = \prod_{k=1}^{n} \mathcal{H}(s_k)\frac{1}{\lambda_k+1-h_k s_k}. \tag{35}$$

Then, from (18) and (35), we obtain the expression for the kernel k

$$\mathrm{k}(\mathbf{x}) = \prod_{k=1}^{n} \mathrm{k}(x_k) = \frac{1}{(2\pi i)^n}\prod_{k=1}^{n} \lim_{\mathrm{R}\to\infty}\int_{1-\nu_k-i\mathrm{R}}^{1-\nu_k+i\mathrm{R}} (\mathfrak{M}\mathrm{k})(s_k)x_k^{-s_k}ds_k$$

$$= \frac{1}{(2\pi i)^n}\prod_{k=1}^{n} \lim_{\mathrm{R}\to\infty}\int_{1-\nu_k-i\mathrm{R}}^{1-\nu_k+i\mathrm{R}} \mathcal{H}_k(s_k)\frac{1}{\lambda_k+1-h_k s_k}x_k^{-s_k}ds_k, \tag{36}$$

where the limits are taken in the topology of $\mathfrak{L}_{\nu,2}$.

According to (4) and (27), we have

$$\overline{\mathcal{H}}(\mathbf{s})\frac{1}{\overline{\lambda}+1-\overline{h}\mathbf{s}} = \overline{\mathcal{H}}(\mathbf{s})\frac{\Gamma(1-(-\overline{\lambda})-\overline{h}\mathbf{s})}{\Gamma(1-(-\overline{\lambda}-1)-\overline{h}\mathbf{s})}$$

$$= \overline{\mathcal{H}}^{\mathbf{m},\mathbf{n}+1}_{\mathbf{p}+1,\mathbf{q}+1}\left[\begin{matrix}(-\overline{\lambda},\overline{h}),(\mathbf{a}_i,\alpha_i)_{1,\mathbf{p}}\\(\mathbf{b}_j,\beta_j)_{1,\mathbf{q}},\ (-\overline{\lambda}-1,\overline{h})\end{matrix}\Big|\mathbf{s}\right]$$

$$= \prod_{k=1}^{n} \mathcal{H}^{m_k,n_k+1}_{p_k+1,q_k+1}\left[\begin{matrix}(-\lambda_k,h_k),(a_{i_k},\alpha_{i_k})_{1,p_k}\\(b_{j_k},\beta_{j_k})_{1,q_k},\ (-\lambda_k-1,h_k)\end{matrix}\Big|s_k\right]. \tag{37}$$

Denote by $\hat{\alpha}_k, \hat{\beta}_k$ $(k = 1, 2, \ldots, n)$ the constants $\widetilde{\alpha}_k, \widetilde{\beta}_k$ $(k = 1, 2, \ldots, n)$ in (24), respectively; by $\widetilde{a}_k^*$ $(k = 1, 2, \ldots, n)$, the constants a_k^* $(k = 1, 2, \ldots, n)$; and by $\widetilde{\Delta}_k$ $(k = 1, 2, \ldots, n)$, the constants Δ_k $(k = 1, 2, \ldots, n)$ in (25), respectively; and by $\widetilde{\mu}_k$ $(k = 1, 2, \ldots, n)$, the constants μ_k $(k = 1, 2, \ldots, n)$ in (26), respectively, for $\mathcal{H}^{m_k,n_k+1}_{p_k+1,q_k+1}$ $(k = 1, 2, \ldots, n)$ in (37). Then, $\hat{\alpha}_k = \widetilde{\alpha}_k$ $(k = 1, 2, \ldots, n)$; $\hat{\beta}_k = \min[\widetilde{\beta}_k, (1+\mathrm{Re}(\lambda_k))/h_k]$ $(k = 1, 2, \ldots, n)$; $\widetilde{a}_k^* = a_k^*$ $(k = 1, 2, \ldots, n)$; $\widetilde{\Delta}_k = \Delta_k$ $(k = 1, 2, \ldots, n)$; $\widetilde{\mu}_k = \mu_k - 1$ $(k = 1, 2, \ldots, n)$. Thus, it follows that

(a') $\hat{\alpha}_k < 1-\nu_k < \hat{\beta}_i$ $(k = 1, 2, \ldots, n)$;
from $\mathrm{Re}(\overline{\lambda}) > (1-\overline{\nu})\overline{h} - 1$, and either of the conditions
(b') $\widetilde{a}_k^* > 0$ $(k = 1, 2, \ldots, n)$;
(c') $\widetilde{a}_k^* = 0$ $(k = 1, 2, \ldots, n)$; or
$\widetilde{\Delta}_k(1-\nu_k) + \mathrm{Re}(\widetilde{\mu}_k) = \Delta_k(1-\nu_k) + \mathrm{Re}(\mu_k) - 1 \leq -1$
$(k = 1, 2, \ldots, n)$ holds. Applying Theorem 1 for $\mathbf{x} > 0$, then the equality

$$\mathrm{H}^{\mathbf{m},\mathbf{n}+1}_{\mathbf{p}+1,\mathbf{q}+1}\left[\mathbf{xt}\left|\begin{array}{c}(-\overline{\lambda},\overline{h}),(\mathbf{a}_i,\alpha_i)_{1,\mathbf{p}}\\(\mathbf{b}_j,\beta_j)_{1,\mathbf{q}},(-\overline{\lambda}-1,\overline{h})\end{array}\right.\right]$$
$$=\prod_{k=1}^{n}\mathrm{H}^{m_k,\overline{n}_k+1}_{p_k+1,q_k+1}\left[x_k\left|\begin{array}{c}(-\lambda_k,h_k),(a_{i_k},\alpha_{i_k})_{1,p_k}\\(b_{j_k},\beta_{j_k})_{1,q_k},(-\lambda_k-1,h_k)\end{array}\right.\right]$$
$$=\frac{1}{(2\pi i)^n}\prod_{k=1}^{n}\lim_{\mathrm{R}\to\infty}\int_{1-\nu_k-i\mathrm{R}}^{1-\nu_k+i\mathrm{R}}\mathcal{H}_k(s_k)\frac{1}{\lambda_k+1-h_ks_k}x_k^{-s_k}ds_k \tag{38}$$

holds almost everywhere. Then, (36) and (38) lead to the fact that the kernel k is given by

$$\mathrm{k}(\mathbf{x})=\mathrm{H}^{\mathbf{m},\mathbf{n}+1}_{\mathbf{p}+1,\mathbf{q}+1}\left[\mathbf{x}\left|\begin{array}{c}(-\overline{\lambda},\overline{h}),(\mathbf{a}_i,\alpha_i)_{1,\mathbf{p}}\\(\mathbf{b}_j,\beta_j)_{1,\mathbf{q}},(-\overline{\lambda}-1,\overline{h})\end{array}\right.\right],$$

and (33) is proven.

The representation (34) is proven similarly to (33). We use the equality

$$\overline{\mathcal{H}}(\mathbf{s})\frac{1}{\overline{\lambda}+1-\overline{h}\mathbf{s}}=-\overline{\mathcal{H}}(\mathbf{s})\frac{\Gamma(\overline{h}\mathbf{s}-\overline{\lambda}-1)}{\Gamma(\overline{h}\mathbf{s}-\overline{\lambda})}$$
$$=-\overline{\mathcal{H}}^{\mathbf{m}+1,\mathbf{n}}_{\mathbf{p}+1,\mathbf{q}+1}\left[\begin{array}{c}(\mathbf{a}_i,\alpha_i)_{1,\mathbf{p}},(-\overline{\lambda},\overline{h})\\(-\overline{\lambda}-1,\overline{h}),(\mathbf{b}_j,\beta_j)_{1,\mathbf{q}}\end{array}\middle|\mathbf{s}\right]$$
$$=-\prod_{k=1}^{n}\mathcal{H}^{m_k+1,\overline{n}_k}_{p_k+1,q_k+1}\left[\begin{array}{c}(a_{i_k},\alpha_{i_k})_{1,p_k},(-\lambda_k,h_k)\\(-\lambda_k-1,h_k),(b_{j_k},\beta_{j_k})_{1,q_k}\end{array}\middle|s_k\right]. \tag{39}$$

instead of (37). Thus, the statement (c) is proven. □

Let us prove (d). If $f\in\mathfrak{L}_{\overline{\nu},\overline{2}}\cap\mathfrak{L}_{\widetilde{\overline{\nu}},\overline{2}}$ and $\mathrm{Re}(\overline{\lambda})>\max[(1-\overline{\nu})\overline{h}-1,(1-\widetilde{\overline{\nu}})\overline{h}-1]$ or $\mathrm{Re}(\overline{\lambda})<\min[(1-\overline{\nu})\overline{h}-1,(1-\widetilde{\overline{\nu}})\overline{h}-1]$, then both transforms $\mathrm{H}f$ and $\widetilde{\mathrm{H}}f$ are given in (33) or (34), respectively, which shows that they are independent of $\overline{\nu}$.

Corollary 1. *Suppose that $\widetilde{\alpha}_k<\widetilde{\beta}_k$ $(k=1,2,\ldots,n)$, and that one of the following conditions holds:*
(a) $a_k^*>0$ $(k=1,2,\ldots,n)$;
(b) $a_k^*=0$ $(k=1,2,\ldots,n)$; $\Delta_k>0$ $(k=1,2,\ldots,n)$; and
$\widetilde{\alpha}_k<-\frac{\mathrm{Re}(\mu_k)}{\Delta_k}$ $(k=1,2,\ldots,n)$;
(c) $a_k^*=0$; $\Delta_k<0$ $(k=1,2,\ldots,n)$; and
$\widetilde{\beta}_k>-\frac{\mathrm{Re}(\mu_k)}{\Delta_k}$ $(k=1,2,\ldots,n)$;
(d) $a_k^*=0$ $(k=1,2,\ldots,n)$; $\Delta_k=0$, $(k=1,2,\ldots,n)$; and
$\mathrm{Re}(\mu_k)\leq0$ $(k=1,2,\ldots,n)$.
Then the H-transform (1) can be defined on $\mathfrak{L}_{\overline{\nu},\overline{2}}$ with
$\widetilde{\alpha}_k<\nu_k<\beta_k$ $(k=1,2,\ldots,n)$; $\nu_1=\nu_2=\cdots=\nu_n$.

Proof. When $1-\widetilde{\beta}_k<\nu_k<1-\widetilde{\alpha}_k$ $(k=1,2,\ldots,n)$, by Theorem 3, if either $a_k^*>0$ $(k=1,2,\ldots,n)$ or $a_k^*=0$ $(k=1,2,\ldots,n)$, $\Delta_k(1-\nu_k)\mathrm{Re}(\mu_k)\leq0$ $(k=1,2,\ldots,n)$ is satisfied, then the H-transform can be defined on $\mathfrak{L}_{\overline{\nu},\overline{2}}$, which is also valid when $\widetilde{\alpha}_k<\nu_k<\widetilde{\beta}_k$ $(k=1,2,\ldots,n)$. Hence, the corollary is clear in cases (a) and (d). When $\Delta_k>0$ and $\widetilde{\alpha}_k<-\frac{\mathrm{Re}(\mu_k)}{\Delta_k}$ $(k=1,2,\ldots,n)$, the assumption $\widetilde{\alpha}_k<\widetilde{\beta}_k$ $(k=1,2,\ldots,n)$ yields that there exists a vector $\overline{\nu}=(\nu_1,\nu_2,\ldots,\nu_n)$ such that $\widetilde{\alpha}_k<1-\nu_k\leq-\frac{\mathrm{Re}(\mu_k)}{\Delta_k}$ $(k=1,2,\ldots,n)$, and $\alpha_k<1-\nu_k\leq-\frac{\mathrm{Re}(\mu_k)}{\Delta_k}$ $(k=1,2,\ldots,n)$, which are required. For the case (c), the situation is similar, i.e., there exists $\overline{\nu}$ of the forms $\widetilde{\beta}_k>1-\nu_k\geq-\frac{\mathrm{Re}(\mu_k)}{\Delta_k}$ $(k=1,2,\ldots,n)$ and $\widetilde{\alpha}_k<1-\nu_k$ $(k=1,2,\ldots,n)$. Thus, the proof is completed. □

4. Conclusions

The multi-dimensional integral transformation with the Fox H-function is studied. Conditions are obtained for the boundedness and one-to-oneness of the operator of such a transformation from one Lebesgue-type weighted space of functions to another, and the analogues of the formula for integration by parts are proven. For the transformation under consideration, various integral representations are established. The results generalize those obtained earlier for the corresponding one-dimensional integral transform.

Due to the generality of the Fox H-function, many special integral transforms have the form studied in this paper, including operators with such kernels as generalized hypergeometric functions, classical hypergeometric functions, Bessel and modified Bessel functions and so on. Moreover, most important fractional integral operators, such as the Riemann–Liouville type, are covered by the class under consideration. The mapping properties in Lebesgue-weighted spaces, such as the boundedness, the range and the representations of the considered transformation, are established. In special cases, it is applied to the specific integral transforms mentioned above. We use a modern technique based on the extensive use of the Mellin transform and its properties. Moreover, we generalize our own previous results from the one-dimensional case to the multi-dimensional one. The multi-dimensional case is more complex and needs more delicate techniques.

Author Contributions: Investigation, S.S.; Writing—original draft, O.S. All authors have read and agreed to the published version of the manuscript.

Funding: This research received no external funding.

Data Availability Statement: No new data were created or analyzed in this study. Data sharing is not applicable to this article.

Conflicts of Interest: The authors declare no conflicts of interest.

References

1. Sitnik, S.M.; Skoromnik, O.V.; Shlapakov, S.A. Multi-dimensional generalized integral transform in the weighted spaces of summable functions. *Lobachevskii J. Math.* **2022**, *43*, 1170–1178. [CrossRef]
2. Samko, S.G.; Kilbas, A.A.; Marichev, O.I. *Fractional Integrals and Derivatives: Theory and Applications*; Gordon and Breach Science Publishers: London, UK, 1993.
3. Kilbas, A.A.; Srivastava, H.M.; Trujillo, J.J. *Theory and Applications of Fractional Differential Equations*; Elsevier: Amsterdam, The Netherlands, 2006.
4. Mathai, A.M.; Saxena, R.K. *The H-Function with Applications in Statistics and Other Disciplines*; Halsted Press: Ultimo, Australia; Wiley: New York, NY, USA, 1978.
5. Srivastava, H.M.; Gupta, K.C.; Goyal, S.L. *The H-Function of One and Two Variables with Applications*; South Asian Publishers: New Delhi, India, 1982.
6. Prudnikov, A.P.; Brychkov, Y.A.; Marichev, O.I. *Integrals and Series. More Special Functions*; Gordon and Breach: New York, NY, USA, 1990; Volume 3.
7. Kiryakova, V. *Generalized Fractional Calculus and Applications*; Wiley and Son: New York, NY, USA, 1994.
8. Kilbas, A.A.; Saigo, M. *H-Transforms. Theory and Applications*; Chapman and Hall: Boca Raton, FL, USA, 2004.
9. Shishkina, E.L.; Sitnik, S.M. *Transmutations, Singular and Fractional Differential Equations with Applications to Mathematical Physics*; Elsevier: Amsterdam, The Netherlands, 2020.
10. Prudnikov, A.P.; Brychkov, Y.A.; Marichev, O.I. Calculation of integrals and Mellin transformation. *Results Sci. Technol. Mat. Anal.* **1989**, *27*, 3–146.
11. Brychkov, Y.A.; Glaeske, H.Y.; Prudnikov, A.P.; Tuan, V.K. *Multidimensional Integral Transformations*; Gordon and Breach: Philadelphia, PA, USA, 1992.
12. Nikolski, S.M. *Approximation of Functions of Many Variables and Embedding Theorems*; Nauka: Moscow, Russia, 1975; 455p. (In Russian)

mathematics

MDPI

Article

Numerical Solutions of Inverse Nodal Problems for a Boundary Value Problem

Yong Tang [1], Haoze Ni [1], Fei Song [2,*] and Yuping Wang [2,*]

1 Information and Computational Sciences, College of Science, Nanjing Forestry University, Nanjing 210037, China
2 Department of Applied Mathematics, Nanjing Forestry University, Nanjing 210037, China
* Correspondence: songfei@njfu.edu.cn (F.S.); ypwang@njfu.com.cn (Y.W.)

Abstract: In this paper, we study inverse nodal problems for a boundary value problem. A uniqueness result for the potential function and a reconstruction method are obtained. By using the nodal points as input data, we compute the approximation solution of the potential function for the boundary value problem by the first kind Chebyshev wavelet method. Two numerical examples show that the first kind Chebyshev wavelet method for solving the inverse nodal problems for the boundary value problem is valid.

Keywords: inverse nodal problem; boundary value problem; potential function; Chebyshev wavelet

MSC: 34A55; 47E05

Citation: Tang, Y.; Ni, H.; Song, F.; Wang, Y. Numerical Solutions of Inverse Nodal Problems for a Boundary Value Problem. *Mathematics* **2022**, *10*, 4204. https://doi.org/10.3390/math10224204

Academic Editor: Maria Isabel Berenguer

Received: 9 October 2022
Accepted: 1 November 2022
Published: 10 November 2022

1. Introduction

We are concerned with the inverse nodal problem for the boundary value problem (BVP) $L := L(q, h, a)$ defined by

$$ly := -y'' + q(x)y = \rho^2 y, \quad 0 < x < 1, \tag{1}$$

associated with boundary conditions:

$$y'(0,\rho) - hy(0,\rho) = 0, \tag{2}$$
$$ay'(1,\rho) + \rho y(1,\rho) = 0, \tag{3}$$

where $a \neq 0, a, h \in \mathbb{R}$, ρ is the spectral parameter, $q(x)$ is a real-valued function and $q \in L^2[0,1]$.

Differential operators with boundary conditions having the spectral parameter frequently arise in nuclear physics, mathematics, quantum mechanics (see [1–9] and the references therein). In 2010, using the method of spectral mapping, Freiling and Yurko [1] studied three inverse problems for the Sturm–Liouville equation with boundary conditions polynomially dependent on the spectral parameter, and provided procedures to reconstruct this operator. In 1965, Li [2] showed that only one spectrum is sufficient to determine the potential function $q(x)$ of BVP $L(q, \infty, a)$ on $[0,1]$ by the quantum theory of scattering and presented an example to show that Li's theorem does not hold for $a = 0$.

The classical Sturm–Liouville operator $L_0 := L(q, h, H)$ is of the form (see [10]):

$$\begin{cases} ly := -y'' + q(x)y = \lambda y, & 0 < x < 1, \\ y'(0,\lambda) - hy(0,\lambda) = 0, \\ y'(1,\lambda) + Hy(1,\lambda) = 0, \end{cases}$$

where h and $q(x)$ are defined the above, λ is the spectral parameter, $H \in \mathbb{R}$. Borg [3] showed that two spectra with one common boundary condition and another differential boundary conditions are sufficient to determine the potential on $[0,1]$ together with the coefficients of the boundary conditions. Although there is only a little difference between operators L and L_0, many features are unlike, for example, Li's theorem for L and Borg's theorem for L_0. However, Hochstadt [4] studied the relationship between Li's theorem and Borg's theorem and improved the Li's theorem. He proved that one spectrum of operator L is equivalent to two spectra of operators $L_0(q,\infty,0)$ and $L_0(q,\infty,\infty)$.

The inverse nodal problem for differential operators is to recover the potential function and coefficients of boundary conditions by using its nodal data (see [11–13]), which was firstly studied by McLaughlin [11], Shen [12], Hald and McLaughlin [13]. Later on, there have been a lot of study of recovering the potential function by less nodal data. The uniqueness theorems and the reconstruction formulae were given by partial nodal data, for example, X.F. Yang [14]; Cheng, Law, and Tsay [15]; Guo and Wei [16]; C.-F. Yang [17]; Buterin and Shieh [18,19]; Wang and Yurko [20]; Wang, Shieh, and Wei [21]; and Wei, Miao, Ge, and Zhao [22], and the references therein). In particular, Chen, Cheng, and Law studied the stability of the inverse nodal problem for the Sturm–Liouville operator L_0 [23]. Since BVP $L(q,h,a)$ is not a special case of the operator in [1], Theorems 1 and 2 are also new results (see [9,20–22]).

In recent years, some numerical methods have been studied to determine the approximation solution of 1st type of Fredholm integral equation by Rashed [24,25]; Maleknejad, Saeedipoor, and Dehbozorgi [26]; Zhou and Xu [27], or other works. The approximation solutions of the inverse nodal problem for differential operators were studied by Akbarpoor, Koyunbakan, and Dabbaghian [28]; Gulsen, Yilmaz, and Akbarpoor [29]; Neamaty, Akbarpoor, and Yilmaz [30], respectively. In this study, we compute the approximation solution of the inverse nodal problem of BVP L by the first kind Chebyshev wavelet method and apply the first kind Chebyshev wavelet method for solving this problem.

In Section 2, we establish the uniqueness theorem for BVP L and give the reconstruction procedure. In Section 3, we find an approximation solution of the potential function $q(x)$ of BVP L from the first kind Chebyshev wavelet method. In Section 4, we present two numerical examples to show that the numerical method is valid.

2. Inverse Nodal Problem

In this section, we study the asymptotic formula of nodal points of the boundary value problem (1)–(3) and establish a uniqueness theorem for the inverse node problem with given nodal data.

Let $S(x,\rho), C(x,\rho), \varphi(x,\rho)$, and $\psi(x,\rho)$ be solutions of (1) with the initial conditions

$$S(0,\rho)=0, S'(0,\rho)=1, C(0,\rho)=1, C'(0,\rho)=0,$$

$$\varphi(0,\rho)=1, \varphi'(0,\rho)=h, \psi(1,\rho)=a, \psi'(1,\rho)=-\rho.$$

Denote $\tau=|\operatorname{Im}\rho|$, $\varphi_0=\arctan\frac{1}{a}$ and

$$q_1(x):=h+\frac{1}{2}\int_0^x q(t)\mathrm{d}t.$$

Then, the asymptotic formulae of $\varphi(x,\rho)$ and $\psi(x,\rho)$ are as follows:

$$\varphi(x,\rho) = \cos\rho x+q_1(x)\frac{\sin\rho x}{\rho}+\mathrm{o}\left(\frac{\mathrm{e}^{\tau x}}{|\rho|}\right), \quad 0\le x\le 1, \tag{4}$$

$$\varphi'(x,\rho) = -\rho\sin\rho x+q_1(x)\cos\rho x+\mathrm{o}(\mathrm{e}^{\tau x}), \quad 0\le x\le 1,$$

$$\psi(x,\rho) = \sqrt{1+a^2}\cos(\rho(1-x)-\varphi_0)+\mathrm{O}\left(\frac{\mathrm{e}^{\tau(1-x)}}{|\rho|}\right), \quad 0\le x\le 1, \tag{5}$$

$$\psi'(x,\rho) = \sqrt{1+a^2}\rho\sin(\rho(1-x)-\varphi_0)+\mathrm{O}\left(\mathrm{e}^{\tau(1-x)}\right), \quad 0\le x\le 1.$$

The characteristic function $\Delta(\rho)$ of L is defined by

$$\Delta(\rho) := \langle \psi, \varphi \rangle (x, \rho),$$

where $\langle \psi, \varphi \rangle (\rho, x) := \psi(x,\rho)\varphi'(x,\rho) - \psi'(x,\rho)\varphi(x,\rho)$, which is called Wronskian of ψ and φ. Clearly $\Delta(\rho)$ is independent of x (see [10]), and zeros of $\Delta(\rho)$ are called the eigenvalues of L. Denote the index set $\mathbf{A} := \{\pm 0, \pm 1, \pm 2, \cdots\}$ (For details, see [31]) and $\sigma(L) := \{\rho_n : \rho_n \in \mathbf{A}\}$ be the set of eigenvalues. Therefore, the asymptotic formula of $\Delta(\rho)$ is

$$\begin{aligned}\Delta(\rho) =& a\varphi'(x,\rho) + \rho\varphi(x,\rho)\\ =& \sqrt{1+a^2}[-\rho \sin(\rho - \varphi_0) + \omega \cos(\rho - \varphi_0) + \mathrm{o}(\mathrm{e}^{\tau})],\end{aligned}$$

where

$$\omega = h + \frac{1}{2}\int_0^1 q(t)\mathrm{d}t.$$

We have the asymptotic formulae of eigenvalue ρ_n:

$$\rho_n = n\pi + \varphi_0 + \frac{\omega}{n\pi + \varphi_0} + \frac{\kappa_n}{n}, \quad n \in \mathbf{A}, \quad |n| \gg 1, \tag{6}$$

where $\{\kappa_n\} \in l^2$. It follows from (6), all eigenvalues are real and simple for sufficiently large $|n|$. By direct calculation, we see that the n-th eigenfunction $\varphi(x, \rho_n)$ has exactly $|n|$ zeros $x_n^j \in (0,1)$, which satisfies the following formula:

$$\begin{aligned}&0 < x_n^1 < x_n^2 < \cdots < x_n^n < 1, \quad \text{if} \quad n > 0,\\ &0 < x_n^0 < x_n^{-1} < \cdots < x_n^{n+1} < 1, \text{ if} \quad n < 0,\end{aligned}$$

$$x_n^j = \frac{(j-1/2)\pi}{n\pi + \varphi_0} + \frac{2h + \int_0^{x_n^j} q(t)\mathrm{d}t}{2(n\pi + \varphi_0)^2} - \frac{(j-1/2)\pi\omega}{(n\pi + \varphi_0)^3} + \mathrm{o}\left(\frac{1}{n^2}\right). \tag{7}$$

for $n \gg 1$ uniformly with respect to j. Denote $l_n^j := x_n^{j+1} - x_n^j$,

$$X := X_+ \bigcup X_-, \quad X_+ := \bigcup_{n=1}^{\infty} \left\{x_n^j\right\}_{j=1}^{n-1} \quad \text{and} \quad X_- := \bigcup_{n=-\infty}^{-1} \left\{x_n^j\right\}_{j=n+1}^{0}.$$

It follows from (7)

$$l_n^j = \frac{\pi}{n\pi + \varphi_0} + \mathrm{o}\left(\frac{1}{n^2}\right) \tag{8}$$

and X is dense on $(0,1)$. We have

Theorem 1. *Given $X_0 \subseteq X$, where X_0 be dense on $(0,1)$. For each fixed $x \in [0,1]$, select a nodal sequence $\left\{x_{n_k}^{j_{n_k}}\right\} \subseteq X_0$ such that $\lim_{|n_k|\to\infty} x_{n_k}^{j_{n_k}} = x$, then*

$$\begin{aligned}\varphi_0 =& -\pi \lim_{|n_k|\to\infty} n_k\left(n_k l_{n_k}^{j_{n_k}} - 1\right),\\ f(x) :=& \lim_{|n_k|\to\infty} 2(n_k\pi + \varphi_0)^2\left(x_{n_k}^{j_{n_k}} - \frac{(j_{n_k} - \frac{1}{2})\pi}{n_k\pi + \varphi_0}\right)\\ =& \int_0^x q(t)\mathrm{d}t + 2h - 2\omega x.\end{aligned}$$

Proof. It follows from (8)

$$\varphi_0 = -\pi \lim_{|n_k|\to\infty} \left(n_k (n_k l_{n_k-1}^{j_{n_k}} - 1) + o(1) \right) = -\pi \lim_{|n_k|\to\infty} n_k \left(n_k l_{n_k-1}^{j_{n_k}} - 1 \right).$$

We reconstruct the coefficient a by

$$\frac{1}{a} = \tan \varphi_0. \tag{9}$$

It follows from (7)

$$\begin{aligned} f(x) &:= \lim_{|n_k|\to\infty} 2(n_k\pi + \varphi_0)^2 \left(x_{n_k}^{j_{n_k}} - \frac{(j_{n_k} - \frac{1}{2})\pi}{n_k\pi + \varphi_0} \right) \\ &= \lim_{|n_k|\to\infty} \left(\int_0^{\frac{(j_{n_k}-\frac{1}{2})\pi}{n_k\pi+\varphi_0}} q(t)\mathrm{d}t + 2h - 2\omega \frac{(j_{n_k} - \frac{1}{2})\pi}{n_k\pi + \varphi_0} + o(1) \right) \\ &= \int_0^x q(t)\mathrm{d}t + 2h - 2\omega x. \end{aligned} \tag{10}$$

In (10), let $x = 0$, we obtain

$$h = \frac{f(0)}{2}. \tag{11}$$

Taking the derivative with respect to x in (10), we get

$$f'(x) \overset{\text{a.e.}}{=} q(x) - 2\omega.$$

Then, we reconstruct the coefficient $q(x) - \int_0^1 q(t)\mathrm{d}t$ as follows

$$q(x) - \int_0^1 q(t)\mathrm{d}t \overset{\text{a.e.}}{=} f'(x) + 2h. \tag{12}$$

□

According to the Theorem 1, the coefficient of L is reconstructed from the nodal subset X_0 on $(0,1)$, and the reconstruction procedure is as follows:

Algorithm 1.

(1) *For each fixed* $x \in [0,1]$, *choose a sequence* $\left\{ x_{n_k}^{j_{n_k}} \right\} \subseteq X_0$ *such that* $\lim_{|n_k|\to\infty} x_{n_k}^{j_{n_k}} = x$;
(2) *The coefficient a is reconstructed by* (9);
(3) *Calculate the function* $f(x)$ *from* (10);
(4) *Reconstruct the coefficient h of the boundary conditions from* (11);
(5) *Recover the function* $q(x) - \int_0^1 q(t)\mathrm{d}t$ *by* (12).

From Algorithm 1, we obtain the uniqueness theorem

Theorem 2. $(q(x) - \int_0^1 q(t)\mathrm{d}t, h, a)$ *can be uniquely determined by the dense nodal subset* X_0 *on* $(0,1)$.

3. Numerical Solution of Inverse Nodal Problems

In this section, we study the following numerical solution of inverse node problems of L.

Numerical solution: for sufficiently large n, given the nodal points $x_n^j, j = \overline{1,n}$ and constants h, a, and $\int_0^1 q(t)\mathrm{d}t$, reconstruct the potential function $q(x)$. The solution $\varphi(x,\rho)$

of (1) can be written as follows (see [10]):

$$\varphi\left(x_n^j, \lambda_n\right) = \cos(\rho_n x_n^j) + q_1\left(x_n^j\right)\frac{\sin(\rho_n x_n^j)}{\rho_n} + \frac{1}{2\rho_n}\int_0^{x_n^j} q(t)\sin(\rho_n x_n^j)\cos(2\rho_n t)\,dt$$
$$-\frac{1}{2\rho_n}\int_0^{x_n^j} q(t)\cos(\rho_n x_n^j)\sin(2\rho_n t)\,dt + O\left(\frac{1}{\rho_n^2}\right) = 0. \quad (13)$$

This implies

$$\cos(\rho_n x_n^j) = O\left(\frac{1}{\rho_n}\right). \quad (14)$$

By virtue of (13) and (14), we obtain

$$\int_0^{x_n^j} q(t)\cos^2(\rho_n t)\,dt = -\rho_n \cot(\rho_n x_n^j) - h + O\left(\frac{1}{\rho_n}\right)$$
$$= -\rho_n^0 \cot(\rho_n^0 x_n^j) - h + O\left(\frac{1}{n}\right), \quad (15)$$

where $\rho_n^0 := n\pi + \varphi_0 + \frac{\omega}{n\pi+\varphi_0}$. Therefore, (15) implies

$$\int_0^{x_n^j} q(t)\cos^2(\rho_n^0 t)\,dt = -\rho_n^0 \cot(\rho_n^0 x_n^j) - h + O\left(\frac{1}{n}\right). \quad (16)$$

It follows from (16)

$$\int_0^{x_n^j} q(t)\cos^2(\rho_n^0 t)dt \cong -\rho_n^0 \cot(\rho_n^0 x_n^j) - h. \quad (17)$$

It is well known that Equation (17) is the first kind Fredholm integral equation. We convert the integral Equation (17) to a system of linear Equation (18) (see below). Then, the solution of the system of linear Equation (18) is an approximation solution of the potential function $q(t)$. We approximate the potential function $q(t)$ with the first kind Chebyshev polynomials as the basis functions.

Consider the Chebyshev wavelets on the interval $[0,1)$ (see [26])

$$\psi_{l,m}(t) = \begin{cases} 2^{k/2}\widetilde{T}_m\left(2^k t - 2l + 1\right), & \frac{l-1}{2^{k-1}} \le t < \frac{l}{2^{k-1}}, \\ 0, & \text{otherwise}, \end{cases}$$

where

$$\widetilde{T}_m(t) = \begin{cases} \frac{1}{\sqrt{\pi}}, & m = 0, \\ \sqrt{\frac{2}{\pi}}T_m(t), & m > 0, \end{cases}$$

where $l = 1, 2, \ldots, 2^{k-1}, m = 0, 1, \ldots, M-1$, k can be any positive integer, and $M \gg 1$. The functions $T_m(t)$ are the Chebyshev polynomials of degree m of the first kind on the interval $[-1,1]$, given by the following recursive formula:

$$T_0(t) = 1, \quad T_1(t) = t,$$
$$T_{m+1}(t) = 2tT_m(t) - T_{m-1}(t), \quad m = 1, 2, \ldots.$$

The function $f(t)$ on the interval $[0,1)$ is expressed as

$$f(t) = \sum_{l=1}^{2^{k-1}}\sum_{m=0}^{\infty} c_{l,m}\psi_{l,m}(t),$$

where

$$c_{l,m} =< q(t), \psi_{l,m}(t) >_{L^2_w[0,1)} = \int_0^1 q(t)\psi_{l,m}(t)w(2^k t - 2l + 1)dt,$$

$< \cdot, \cdot >_{L^2_w[0,1)}$ is the inner product on $L^2_w[0,1)$, $\|q\|_{2,w}$ is the norm of $q(x)$ on $L^2_w[0,1)$, and the weight function is $w(x) = \frac{1}{\sqrt{1-x^2}}$.

Using the Chebyshev wavelets to approximate the potential function $q(t)$, we have

$$q(t) \cong \sum_{l=1}^{2^{k-1}} \sum_{m=0}^{M-1} c_{l,m}\psi_{l,m}(t) = \mathbf{C}^{\mathrm{T}}\mathbf{\Psi}(\mathbf{t}), \tag{18}$$

where

$$\mathbf{C} = \left[c_{1,0}, \ldots, c_{1,M-1}, c_{2,0}, \ldots, c_{2,M-1}, \ldots, c_{2^{k-1},0}, \ldots, c_{2^{k-1},M-1}\right]^{\mathrm{T}},$$

$$\mathbf{\Psi}(\mathbf{t}) = \left[\psi_{1,0}(t), \ldots, \psi_{1,M-1}(t), \psi_{2,0}(t), \ldots, \psi_{2,M-1}(t), \ldots, \psi_{2^{k-1},0}(t), \ldots, \psi_{2^{k-1},M-1}(t)\right]^{\mathrm{T}}.$$

Substituting (18) into (17), we get

$$\sum_{l=1}^{2^{k-1}} \sum_{m=0}^{M-1} c_{l,m}\left(\int_0^{x_n^j} \psi_{l,m}(t)\cos^2(\rho_n^0 t)\mathrm{d}t\right) \cong -\rho_n^0 \cot(\rho_n^0 x_n^j) - h, \quad j = \overline{1,n}.$$

Next, we present the following theorem for the convergence of the given method. The readers refer to the references [25,26] for the convergence of the first kind Chebyshev wavelet method.

Theorem 3. *For each fixed $n = 2^{k-1}M$, $M \gg 1$, given n nodal points $\{x_n^j\}_{j=1}^n$ satisfying* (7) *together with the coefficients $(h, a, Q(1))$, then the potential function $q(x)$ can be written as an infinite sum of the first kind Chebyshev wavelets and this series converges to $q(x)$, that it is*

$$q(x) = \sum_{l=1}^{2^{k-1}} \sum_{m=0}^{\infty} c_{l,m}\psi_{l,m}(x)$$

and the approximation of potential function $q(x)$ is as follows:

$$q_{k,M}(x) = \sum_{l=1}^{2^{k-1}} \sum_{m=0}^{M-1} c_{l,m}\psi_{l,m}(x).$$

The solution of the inverse node problem can be computed by the following steps:

(1) Choose k, M. Set $n = 2^{k-1}M$.
(2) Calculate the unknown vector $\mathbf{C}$ by the following linear equation:

$$\mathbf{AC} = \mathbf{B},$$

where

$$\mathbf{A} = \begin{pmatrix} a_{1,0}^1 & a_{1,1}^1 & \cdots & a_{1,M-1}^1 & a_{2,0}^1 & \cdots & a_{2,M-1}^1 & \cdots & a_{2^{k-1},0}^1 & \cdots & a_{2^{k-1},M-1}^1 \\ a_{1,0}^2 & a_{1,1}^2 & \cdots & a_{1,M-1}^1 & a_{20}^2 & \cdots & a_{2,M-1}^2 & \cdots & a_{2^{k-1},0}^2 & \cdots & a_{2^{k-1},M-1}^2 \\ \vdots & & & & \vdots & & & & & \vdots & \\ a_{1,0}^{2^{k-1}M} & a_{1,1}^{2^{k-1}M} & \cdots & a_{1,M-1}^{2^{k-1}M} & a_{2,0}^{2^{k-1}M} & \cdots & a_{2,M-1}^{2^{k-1}M} & \cdots & a_{2^{k-1},0}^{2^{k-1}M} & \cdots & a_{2^{k-1},M-1}^{2^{k-1}M} \end{pmatrix},$$

where

$$a_{l,m}^j = \int_0^{x_n^j} \psi_{l,m}(t)\cos^2(\rho_n^0)t\mathrm{d}t,$$
$$l = \overline{1,2^{k-1}}, \quad m = \overline{0,M-1}, \quad n = 2^{k-1}M, \quad j = \overline{1,n},$$

$$\mathbf{B} = \begin{pmatrix} -\rho_n^0 \cot(\rho_n^0 x_n^1) - h \\ -\rho_n^0 \cot(\rho_n^0 x_n^2) - h \\ \cdot \\ \cdot \\ \cdot \\ -\rho_n^0 \cot(\rho_n^0 x_n^n) - h \end{pmatrix},$$

(3) To approximate $q(t_i), i = \overline{1, 2^{k-1}M}$, we use the following formula:

$$[q(t_i)] = \mathbf{C}^{\mathrm{T}}\mathbf{\Phi},$$

where

$$t_i = \frac{2i-1}{2^k M}, \quad i = 1, 2, \ldots, 2^{k-1}M,$$

and

$$\mathbf{\Phi} = \left[\mathbf{\Psi}\left(\frac{1}{2^k M}\right), \mathbf{\Psi}\left(\frac{3}{2^k M}\right), \ldots, \mathbf{\Psi}\left(\frac{2^k M - 1}{2^k M}\right)\right].$$

4. Numerical Examples

In this section, we use the first kind Chebyshev wavelet method to solve the inverse nodal problem for (1)–(3) and provide two numerical examples to demonstrate the accuracy of the numerical method by the Matlab software program.

Example 1. *Let the potential function $q(x) = \cos(4\pi x)$ and $h = 1, a = 1$. If $k = 4$ and $M = 5, 7, 9$ and x_n^j satisfy the Formula (7), respectively, we find three approximation solutions of the potential function $q(x)$.*

We use the first kind Chebyshev wavelet method to obtain an approximation of the potential function $q(x)$ as a solution of the inverse nodal problem for L. Numerical values of $q(x)$ and exact solutions of $q(x)$ with $k = 4$ and $M = 5, 7, 9$, respectively, we obtain three approximation solutions of the potential function $q(x)$ by the first kind Chebyshev wavelet method (see Figure 1).

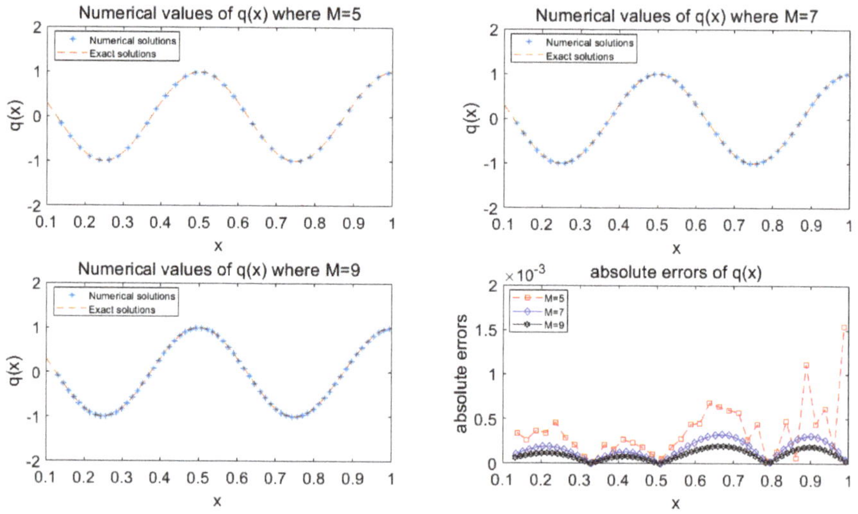

Figure 1. Numerical values of $q(x)$ and absolute errors between the approximation and exact solutions of $q(x)$ with $k = 4$ and $M = 5, 7, 9$ in Example 1.

Now, we give the second example for numerical values of $q(x)$ and exact solutions of $q(x)$ with $k = 4$ and $M = 5, 7, 9$ and $k = 5, 7$, $M = 5$ and $k = 6$, $M = 10$, respectively.

Example 2. *Let the potential function* $q(x) = 3x^2 - 1$ *and* $h = 1, a = 1$.

(1) *If* $k = 4$ *and* $M = 5, 7, 9$ *and* x_n^j *satisfy the Formula* (7), *we find three approximation solutions of the potential function* $q(x)$;

(2) *If* $k = 5, 7$, $M = 5$ *and* $k = 6$, $M = 10$ *and* x_n^j *satisfy the Formula* (7), *we find three approximation solutions of the potential function* $q(x)$.

We obtain approximation solutions of the potential function $q(x)$ by the first kind Chebyshev wavelet method (see Figures 2 and 3).

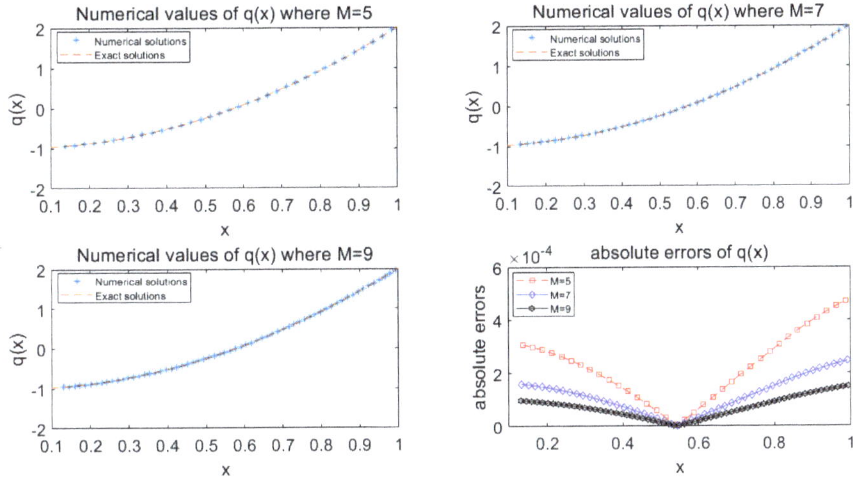

Figure 2. Numerical values of $q(x)$ and absolute errors between the approximation and exact solutions of $q(x)$ with $k = 4$ and $M = 5, 7, 9$ in Example 2.

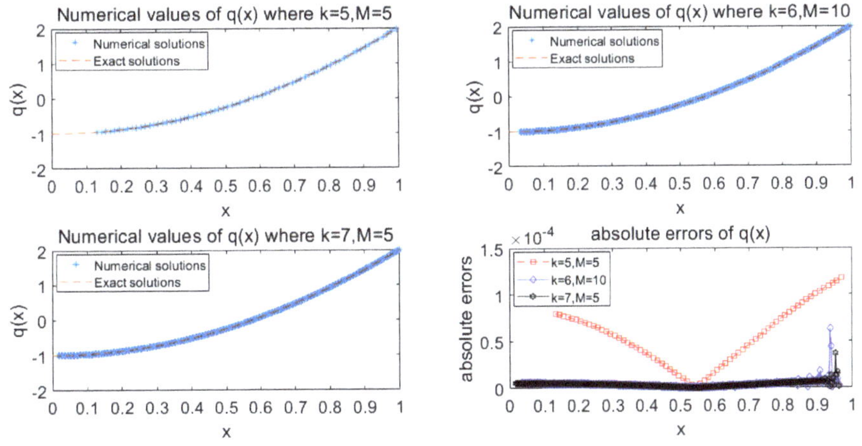

Figure 3. Numerical values of $q(x)$ and absolute errors between the approximation and exact solutions of $q(x)$ with $k = 5, 7$, $M = 5$, and $k = 6$, $M = 10$ in Example 2.

In the Figures 1–3, it can be seen that by increasing the values of n, the approximation solution of BVP L for the inverse nodal problem by the first kind Chebyshev wavelet method becomes more accurate and the error decreases. However, if n is not large, the errors near the boundary points are larger than others. If $n = 2^{7-1} \times 5$, the numerical solution is more effective.

5. Conclusions

In this work, we study the inverse nodal problem for Sturm–Liouville equation with one boundary condition having spectral parameter. The uniqueness theorem for BVP L and the reconstruction procedure are presented from the dense nodal set on the whole interval. By applying the first kind Chebyshev wavelet method, we compute three approximation solutions of BVP L for $k = 4$ and $M = 5,7,9$, respectively, in two examples. We still compute three approximation solutions of BVP L for $k = 5,7$, $M = 5$ and $k = 6$, $M = 10$, respectively, in Example 2. With increasing the values of n, the approximation solution of BVP L for the inverse nodal problem by the first kind Chebyshev wavelet method becomes more accurate and the error decreases. It is also proved that the first kind Chebyshev wavelet method for the approximation solution of BVP L for the inverse nodal problem is an effective method.

Author Contributions: Investigation, Y.T., H.N., F.S. and Y.W.; Methodology, F.S. and Y.W.; Software, Y.T. and H.N.; Writing—original draft, Y.T.; Writing—review & editing, F.S. and Y.W. All authors have read and agreed to the published version of the manuscript.

Funding: This research was funded by the innovation project of university students of Jiangsu Province (202110298138H) and the Natural Science Foundation of Jiangsu Province (Grants No. BK20190745).

Institutional Review Board Statement: Not applicable.

Informed Consent Statement: Not applicable.

Data Availability Statement: Not applicable.

Acknowledgments: The authors would like to thank the anonymous referees and Bondarenko, Department of Applied Mathematics and Physics, Samara National Research University, Samara, Russia, for valuable suggestions, which helped to improve the readability and quality of the paper. The first and second authors were supported in part by the innovation project of university students of Jiangsu Province (202110298138H). The third author was supported in part by the Natural Science Foundation of Jiangsu Province (Grants No. BK20190745).

Conflicts of Interest: The authors declare no conflict of interest.

References

1. Freiling, G.; Yurko, V.A. Inverse problems for Sturm-Liouville equations with boundary conditions polynomially dependent on the spectral parameter. *Inverse Probl.* **2010**, *26*, 055003. [CrossRef]
2. Li, Y.S. Inverse eigenvalue problem for a second-order differential equation with one boundary condition dependent on the spectral parameter. *Acta Math. Sin.* **1965**, *15*, 74–80.
3. Borg, G. Eine umkehrung der Sturm-Liouvilleschen eigenwertaufgabe. *Acta Math.* **1946**, *78*, 1–96. [CrossRef]
4. Hochstadt, H. On inverse problems associated with second-order differential operators. *Acta Math.* **1967**, *119*, 173–192. [CrossRef]
5. Fulton, C.T. Two-point boundary value problems with eigenvalue parameter contained in the boundary conditions. *Proc. R. Soc. Edinb.* **1977**, *77*, 293–308. [CrossRef]
6. Binding, P.A.; Browne, P.J.; Seddighi, K. Sturm-Liouville problems with eigenparameter dependent boundary conditions. *Proc. R. Soc. Edinb.* **1993**, *37*, 57–72. [CrossRef]
7. Browne, P.J.; Sleeman, B.D. Inverse nodal problem for Sturm-Liouville equation with eigenparameter dependent boundary conditions. *Inverse Probl.* **1996**, *12*, 377–381. [CrossRef]
8. Guliyev, N.J. A Riesz basis criterion for Schrödinger operators with boundary conditions dependent on the eigenvalue parameter. *Anal. Math. Phys.* **2020**, *10*, 2. [CrossRef]
9. Yang, C.-F.; Yang, X.P. Inverse nodal problems for the Sturm-Liouville equation with polynomially dependent on the eigenparameter. *Inverse Probl. Sci. Eng.* **2011**, *19*, 951–961. [CrossRef]
10. Freiling, G.; Yurko, V.A. *Inverse Sturm-Liouville Problems and Their Applications*; Nova Science Publishers: New York, NY, USA, 2001.
11. McLaughlin, J.R. Inverse spectral theory using nodal points as data—A uniqueness result. *J. Differ. Equ.* **1988**, *73*, 354–362. [CrossRef]
12. Shen, C.-L. On the nodal sets of the eigenfunctions of the string equations. *SIAM J. Math. Anal.* **1989**, *19*, 1419–1424. [CrossRef]
13. Hald, O.H.; McLaughlin, J.R. Solution of inverse nodal problems. *Inverse Probl.* **1989**, *5*, 307–377. [CrossRef]
14. Yang, X.F. A new inverse nodal problem. *J. Differ. Equ.* **2001**, *169*, 633–653. [CrossRef]
15. Cheng, Y.H.; Law, C.K.; Tsay, J. Remarks on a new inverse nodal problem. *J. Math. Anal. Appl.* **2000**, *248*, 145–155. [CrossRef]

16. Guo, Y.; Wei, G. Inverse problems: Dense nodal subset on an interior subinterval. *J. Differ. Equ.* **2013**, *255*, 2002–2017. [CrossRef]
17. Yang, C.-F. Solution to open problems of Yang concerning inverse nodal problems. *Isr. J. Math.* **2014**, *204*, 283–298. [CrossRef]
18. Buterin, S.A.; Shieh, C.-T. Inverse nodal problem for differential pencils. *Appl. Math. Lett.* **2009**, *22*, 1240–1247. [CrossRef]
19. Buterin, S.A.; Shieh, C.-T. Incomplete inverse spectral and nodal problems for differential pencils. *Results Math.* **2012**, *62*, 167–179. [CrossRef]
20. Wang, Y.P.; Yurko, V.A. On the inverse nodal problems for discontinuous Sturm-Liouville operators. *J. Differ. Equ.* **2016**, *260*, 4086–4109. [CrossRef]
21. Wang, Y.P.; Shieh, C.-T.; Wei, X. Partial inverse nodal problems for differential pencils on a star-shaped graph. *Math. Meth. Appl. Sci.* **2020**, *43*, 8841–8855. [CrossRef]
22. Wei, X.; Miao, H.; Ge, C.; Zhao, C. An inverse problem for Sturm-Liouville operators with nodal data on arbitrarily-half intervals. *Inverse Probl. Sci. Eng.* **2020**, *29*, 305–317. [CrossRef]
23. Chen, X.F.; Cheng, Y.H.; Law, C.K. Reconstructing potentials from zeros of one eigenfunction. *Trans. Am. Math. Soc.* **2011**, *363*, 4831–4851. [CrossRef]
24. Rashed, M.T. Numerical solution of a special type of integro-differential equations. *Appl. Math. Comput.* **2003**, *143*, 73–88. [CrossRef]
25. Rashed, M.T. Numerical solutions of the integral equations of the first kind. *Appl. Math. Comput.* **2003**, *145*, 413–420. [CrossRef]
26. Maleknejad, K.; Saeedipoor, E.; Dehbozorgi, R. Legendre wavelets direct method for the numerical solution of Fredholm integral equation of the first kind. In Proceedings of the World Congress on Engineering, London, UK, 29 June–1 July 2016; Volume 1.
27. Zhou, F.Y.; Xu, X.Y. Numerical solution of the convection diffusion equations by the second kind Chebyshev wavelets. *Appl. Math. Comput.* **2014**, *247*, 353–367. [CrossRef]
28. Akbarpoor, S.; Koyunbakan, H.; Dabbaghian, A. Solving inverse nodal problem with spectral parameter in boundary conditions. *Inverse Probl. Sci. Eng.* **2019**, *27*, 1790–1801. [CrossRef]
29. Gulsen, T.; Yilmaz, E.; Akbarpoor, S. Numerical investigation of the inverse nodal problem by Chebyshev interpolation method. *Therm. Sci.* **2018**, *22*, S123–S136. [CrossRef]
30. Neamaty, A.; Akbarpoor, S.; Yilmaz, E. Solving inverse Sturm-Liouville problem with separated boundary conditions by using two different input data. *Int. J. Comput. Math.* **2018**, *95*, 1992–2010. [CrossRef]
31. Buterin, S.A. On half inverse problem for differential pencils with the spectral parameter in boundary conditions. *Tamkang J. Math.* **2011**, *42*, 355–364. [CrossRef]

mathematics

MDPI

Article

Higher Monotonicity Properties for Zeros of Certain Sturm-Liouville Functions

Tzong-Mo Tsai

General Education Center, Ming Chi University of Technology, New Taipei City 24301, Taiwan; tsaitm@mail.mcut.edu.tw

Abstract: In this paper, we consider the differential equation $y'' + \omega^2\rho(x)y = 0$, where ω is a positive parameter. The principal concern here is to find conditions on the function $\rho^{-1/2}(x)$ which ensure that the consecutive differences of sequences constructed from the zeros of a nontrivial solution of the equation are regular in sign for sufficiently large ω. In particular, if $c_{\nu k}(\alpha)$ denotes the kth positive zero of the general Bessel (cylinder) function $C_\nu(x;\alpha) = J_\nu(x)\cos\alpha - Y_\nu(x)\sin\alpha$ of order ν and if $|\nu| < 1/2$, we prove that $(-1)^m\Delta^{m+2}c_{\nu k}(\alpha) > 0 \quad (m = 0,1,2,\ldots;k = 1,2,\ldots)$, where $\Delta a_k = a_{k+1} - a_k$. This type of inequalities was conjectured by Lorch and Szego in 1963. In addition, we show that the differences of the zeros of various orthogonal polynomials with higher degrees possess sign regularity.

Keywords: Sturm–Liouville equations; differences; zeros; completely monotonic functions; Bessel functions; orthogonal polynomials

MSC: 34B24; 33C10

1. Introduction

We consider the differential equation

$$y'' + \omega^2\rho(x)y = 0, \quad a \le x \le b, \tag{1}$$

where ω is a positive parameter and $\rho(x)$ is a positive C^∞-function on the interval $[a,b]$. By a Sturm–Liouville function, we mean a nontrivial real solution of (1). Let $\{x_k(\omega)\}$ denote the ascending sequence of the zeros of a Sturm–Liouville function in the interval $[a,b]$. The Sturm comparison theorem (see, e.g., p. 314 of [1] or p. 56 of [2]) states that the second differences of the sequence $\{x_k(\omega)\}$ are all positive if $\rho'(x) < 0$ and are all negative if $\rho'(x) > 0$. Our main purpose here is to move beyond the second differences and to show that higher consecutive differences of sequences constructed from $\{x_k(\omega)\}$ are regular in sign. Lorch and Szego [2] initiated the study of the sign regularity of higher differences of the sequences associated with Sturm–Liouville functions. In particular, if $c_{\nu k}(\alpha)$ denotes the kth positive zero of the general Bessel (cylinder) function

$$C_\nu(x;\alpha) = J_\nu(x)\cos\alpha - Y_\nu(x)\sin\alpha,$$

they proved that

$$(-1)^m\Delta^{m+1}c_{\nu k}(\alpha) > 0 \quad (m = 0,1,2,\ldots;k = 1,2,\ldots), \tag{2}$$

for $|\nu| > 1/2$, and conjectured (p. 71 of [2]) that, on the basis of numerical evidence

$$(-1)^m\Delta^{m+2}c_{\nu k}(\alpha) > 0 \quad (m = 0,1,2,\ldots;k = 1,2,\ldots). \tag{3}$$

for $|\nu| < 1/2$.

Citation: Tsai, T.-M. Higher Monotonicity Properties for Zeros of Certain Sturm-Liouville Functions. *Mathematics* **2023**, *11*, 2787. https://doi.org/10.3390/math11122787

Academic Editor: Natalia Bondarenko

Received: 21 May 2023
Revised: 15 June 2023
Accepted: 15 June 2023
Published: 20 June 2023

The symbol $\Delta^m a_k$ means, as usual, the mth (forward) difference of the sequence $\{a_k\}$:

$$\Delta^0 a_k = a_k, \quad \Delta^m a_k = \Delta^{m-1} a_{k+1} - \Delta^{m-1} a_k \quad (m = 1, 2, \ldots; k = 1, 2, \ldots).$$

Note that $C_\nu(x;\alpha)$ is a solution of the equation

$$y'' + q(x)y = 0, \quad x \in (0, +\infty), \tag{4}$$

with $q(x) = 1 - (\nu^2 - (1/4))x^{-2}$. Because $q'(x) = 2(\nu^2 - (1/4))x^{-3}$, we can see that the Sturm comparison theorem provides the results (2) for $m = 1$ and (3) for $m = 0$. It is mentioned in [2] that the signs of the first M differences of zeros of a Sturm–Liouville function of (4) could be inferred from the signs of $q^{(m)}(x)$, $m = 1, 2, \ldots, M$. Muldoon [3] made progress in (3), proving that (3) holds when $1/3 \le |\nu| < 1/2$ ([3], Corollary 4.2).

Our approach here is based on the ideas and results of [4], where the string equation $y'' + \lambda\rho(x)y = 0$ with $y(0) = y(1) = 0$ was considered. Using the eigenvalues and the nodal points, we constructed a sequence of piecewise continuous linear functions which converges to $\rho^{-1/2}(x)$ uniformly on $[0,1]$. Moreover, we obtained a formula for derivatives of $\rho^{-1/2}(x)$ in terms of the eigenvalues and the differences of the nodal points.

The rest of this paper is organized as follows. In Section 2, we use the zeros $x_k(\omega)$ of a Sturm–Liouville function as nodes to obtain a difference-derivative theorem (Lemma 1). In addition, we provide asymptotic estimates for $\rho^{-1/2}(x_k(\omega))$ as $\omega \to +\infty$ (Lemma 3). Then, we are able to express the higher differences $\Delta^{m+1} x_k(\omega)$ in terms of the derivatives of $\rho^{-1/2}(x)$ at those zeros. Moreover, the expression can be used to determine the regular manner of these differences (Theorems 1 and 2). In addition, we construct sequences from $x_k(\omega)$, where all the mth differences have the same sign (Corollary 1). The proofs of Lemmas 1 and 3 rely on a system of interlaced inductions, which is presented in Section 5. In Section 3, we use an approximation process for the zeros of the general Bessel function to prove the conjecture of Lorch and Szego (Theorem 3). In Section 4, the zeros of various orthogonal polynomials with higher degrees are shown to share similar sign regularity (Theorems 4 and 5).

The notation used throughout is standard. A function $\varphi(x)$ is said to be M-monotonic (resp., absolutely M-monotonic) on an interval I if

$$(-1)^m \varphi^{(m)}(x) \ge 0 \quad (resp., \varphi^{(m)}(x) \ge 0), \quad (x \in I; m = 0, 1, \ldots, M). \tag{5}$$

If (5) holds for $M = \infty$, then $\varphi(x)$ is said to be completely (resp., absolutely) monotonic on I. A sequence $\{a_k(\omega)\}$ depending on a positive parameter ω is said to be asymptotically M-monotonic (resp., asymptotically absolutely M-monotonic) if

$$(-1)^m \Delta^m a_k(\omega) \ge 0 \quad (resp., \Delta^m a_k(\omega) \ge 0), \quad (m = 0, 1, 2, \ldots, M; k = 1, 2, \ldots)$$

for sufficiently large ω.

Here, we should mention a number of recent studies related to this paper. In the proofs of Lemmas 1 and 3, we use the standard Taylor expansion of a function at the nodes. In fact, there have many different types of Taylor expansion; many interesting applications can be found in [5,6] and the references therein. The continuity of the coefficient function $\rho(x)$ ensures that the zeros of the solution of (1) have a regular asymptotic distribution. Readers interested in uniform distribution sequences can refer to [7]. Completely monotonic functions and sequences have specific representations, and arise in many research areas, such as moment problems and harmonic mappings. Interested readers can refer to [8–10] and the references therein.

2. Main Results

In this section, we consider the differential equation

$$y'' + \omega^2 \rho(x)y = 0, \quad a \le x \le b, \tag{6}$$

where ω is a positive parameter. We assume throughout that $\rho(x)$ is a positive C^∞-function on the interval $[a,b]$. The notation $f(x)$ is reserved for the function $\rho^{-1/2}(x)$. Let $y(x;\omega)$ be a nontrival real solution of (6) and let $x_1(\omega) < x_2(\omega) < \cdots$ be the zeros of $y(x;\omega)$ in the interval $[a,b]$. For $a \le x < b$, we denote by $k(x;\omega)$ the smallest positive integer k such that $x \le x_k(\omega)$. It is well known (see, e.g., [4,11]) that

$$\min_{[x_k(\omega),x_{k+1}(\omega)]} f \le \frac{\omega}{\pi}\Delta x_k(\omega) \le \max_{[x_k(\omega),x_{k+1}(\omega)]} f. \tag{7}$$

It follows that $\pi \min_{[a,b]} f \le \omega \Delta x_k(\omega) \le \pi \max_{[a,b]} f$. In particular, we have

$$\Delta x_k(\omega) = O(\omega^{-1}) \quad as \quad \omega \to +\infty. \tag{8}$$

Thus, by (7) and the continuity of f, we obtain $f(x) = \lim_{\omega\to+\infty} \frac{\omega}{\pi}\Delta x_{k(x;\omega)}(\omega)$, and for any fixed l,

$$\lim_{\omega\to+\infty} \frac{\Delta x_{k(x;\omega)+l}(\omega)}{\Delta x_{k(x;\omega)}(\omega)} = 1. \tag{9}$$

Note that (9) means that, because $\omega \to +\infty$, the sequence $x_k(\omega)$ behaves as if equally distributed.

If φ is m-times differentiable in $(t, t+md)$ and the lower derivatives of φ are continuous on $[t, t+md]$, a mean-value theorem ([12] p. 52, no. 98) for differences and derivatives states that there exists a δ such that

$$\Delta_d^m \varphi(t) = d^m \varphi^{(m)}(t + \delta m d),$$

where $\Delta_d \varphi(t) = \varphi(t+d) - \varphi(t)$. It is interesting to look for a difference-derivative theorem which can express the differences of a smooth function on the sequence $\{x_k(\omega)\}$ in terms of its derivatives at this sequence. The following lemma provides such a result.

Lemma 1. *Let $x_k = x_k(\omega)$. If φ is a C^∞-function on $[a,b]$, then for $m = 1,2,\ldots,$*

$$\Delta^m \varphi(x_k) = O(\omega^{-m}). \tag{10}$$

Moreover,

$$\Delta^m \varphi(x_k) = \sum_{q=1}^{m} A_{q,k}^{(m)} \varphi^{(q)}(x_{k+m-q}) + O(\omega^{-m-1}), \tag{11}$$

where the coefficients $A_{q,k}^{(m)}$ satisfy the recurrence relation

$$A_{1,k}^{(m)} = \Delta^m x_k, \quad A_{q,k}^{(m)} = \sum_{r=q-1}^{m-1} \binom{m-1}{r} A_{q-1,k+m-1-r}^{(r)} \Delta^{m-r} x_k, \tag{12}$$

for $q = 2,3,\ldots,m$.

To prove Lemma 1, a more detailed investigation into the behaviour of $x_k(\omega)$ is required. We use the Prüfer method to achieve this purpose. For each nontrivial solution $y(x;\omega)$ of (6), we define the Prüfer angle $\theta(x;\omega)$ as follows:

$$\omega\rho^{1/2}(x)\cot\theta(x;\omega) = \frac{y'(x;\omega)}{y(x;\omega)}.$$

Then, $\theta(x;\omega)$ satisfies the differential equation

$$\theta'(x;\omega) = \omega\rho^{1/2}(x) + \frac{\rho'(x)}{4\rho(x)}\sin 2\theta(x;\omega). \tag{13}$$

If we specify the initial condition for $\theta(x;\omega)$ to be $\theta(a;\omega) = \theta_a(\omega)$ with $0 \le \theta_a(\omega) < \pi$, then, by the standard results (see, e.g., [1] p. 315), we have

$$\theta(x_k(\omega);\omega) = k\pi, \tag{14}$$

and $k\pi \le \theta(x;\omega) \le (k+1)\pi$, $x \in [x_k(\omega), x_{k+1}(\omega)]$. Let $x_k = x_k(\omega)$. When integrating both sides of (13) from x_k to x_{k+1} and using (14), we find that

$$\pi = \omega \int_{x_k}^{x_{k+1}} \rho^{1/2}(x)dx + \int_{x_k}^{x_{k+1}} \frac{\rho'(x)}{4\rho(x)} \sin 2\theta(x;\omega)dx. \tag{15}$$

Taking the Taylor expansion of $(1/f)(x)$ at x_k and using (8), we obtain

$$\int_{x_k}^{x_{k+1}} \rho^{1/2}(x)dx = \sum_{r=0}^{m} \frac{(1/f)^{(r)}(x_k)}{(r+1)!}(\Delta x_k)^{r+1} + O(\omega^{-m-2}). \tag{16}$$

The estimate of the second integral in (15) is stated as the following lemma. Its proof consists of a reducible system of integrals which is provided in Appendix A.

Lemma 2. *Let $x_k = x_k(\omega)$. Then, for $m = 2, 3, \ldots$, we have*

$$\int_{x_k}^{x_{k+1}} \frac{\rho'(x)}{4\rho(x)} \sin 2\theta(x;\omega)dx = \sum_{r=0}^{m-2} \Delta F_r(x_k)\omega^{-r-1} + R_{m-2}(x_k), \tag{17}$$

where the functions F_r depend on $f = \rho^{-1/2}$ and

$$R_{m-2}(x_k) = O(\omega^{-m-1}). \tag{18}$$

Note that the first two functions F_r appearing in (17) are of the forms

$$F_0 = \frac{f'}{4} - \int \frac{(f')^2}{8f}dx \quad \text{and} \quad F_1 = 0. \tag{19}$$

For $m = 2, 3, \ldots$, using the estimates (16), (17) and (18), and multiplying (15) by $f(x_k)/\pi$, we find the estimate for $f(x_k)$:

$$f(x_k) = \frac{\omega}{\pi} \sum_{r=0}^{m} \frac{g_r(x_k)}{(r+1)!}(\Delta x_k)^{r+1} + \frac{1}{\pi} \sum_{r=0}^{m-2} (f\Delta F_r)(x_k)\omega^{-r-1} + O(\omega^{-m-1}), \tag{20}$$

where the functions $g_r = f(1/f)^{(r)}$ and $r = 0, 1, 2, \ldots, m$. Note that $g_0 = 1$. Moreover, if we apply the mth order difference operator to (20), we can find the estimates for differences of the function $f(x)$ at those zeros. Indeed, we have the following lemma.

Lemma 3. *Let $f(x)$ and $x_k = x_k(\omega)$ be the same as above. Then, for $m = 1, 2, 3, \ldots$, we have*

$$\Delta^m x_k = O(\omega^{-m}). \tag{21}$$

Moreover,

$$\Delta^m f(x_k) = \frac{\omega}{\pi}\Delta^{m+1} x_k + O(\omega^{-m-1}). \tag{22}$$

The proofs of Lemmas 1 and 3 are provided in Section 5.

Now, if we apply Lemma 1 to the function $f(x)$, then by (22), we have the estimate for the higher differences of $x_k = x_k(\omega)$:

$$\frac{\omega}{\pi}\Delta^{m+1} x_k = \sum_{q=1}^{m} A_{q,k}^{(m)} f^{(q)}(x_{k+m-q}) + O(\omega^{-m-1}). \tag{23}$$

Moreover, by using (8) and (12), and iterating (23) for m from 1 to M, then choosing a sufficiently large ω, we can ensure the monotonicity of the sequence $\{\Delta x_k(\omega)\}$ by f.

Theorem 1. *Let $x_k = x_k(\omega)$ and $f(x) = \rho^{-1/2}(x)$ be the same as above. If $f(x)$ is M-monotonic on the interval $[a,b]$, then the sequence $\{\Delta x_k(\omega)\}$ is asymptotically M-monotonic.*

Proof. Because

$$(-1)^m f^{(m)}(x) \geq 0 \quad (x \in [a,b]; m = 0,1,2,\ldots,M), \tag{24}$$

it suffices to show that

$$(-1)^{m-q} A_{q,k}^{(m)} \geq 0 \quad (q = 1,2,\ldots,m; m = 1,2,\ldots,M), \tag{25}$$

as $\omega \to +\infty$ to conclude that

$$(-1)^m \Delta^{m+1} x_k(\omega) \geq 0, \quad (m = 0,1,2,\ldots,M). \tag{26}$$

We prove (25) by induction on M. When $M = 1$, (25) reduces to $A_{1,k}^{(1)} \geq 0$, which is true because $A_{1,k}^{(1)} = \Delta x_k$, by (12). Now, suppose that (25) is true for N, with $1 \leq N < M$. By (23) for $m = N$, we have

$$\frac{\omega}{\pi}(-1)^N \Delta^{N+1} x_k = \sum_{q=1}^{N} [(-1)^{N-q} A_{q,k}^{(N)}][(-1)^q f^{(q)}(x_{k+N-q})] + O(\omega^{-N-1}),$$

which is nonnegative, as $\omega \to +\infty$ by the induction hypothesis, (24) and (21) for $m = N+1$. Thus, by (12) for $m = N+1$, $(-1)^N A_{1,k}^{(N+1)} = (-1)^N \Delta^{N+1} x_k \geq 0$ and for $q = 1,2,\ldots,N+1$,

$$(-1)^{N+1-q} A_{q,k}^{(N+1)} = \sum_{r=q-1}^{N} \binom{N}{r} [(-1)^{r-q+1} A_{q-1,k+N-r}^{(r)}][(-1)^{N-r} \Delta^{N+1-r} x_k] \geq 0,$$

again following the induction hypothesis. This proves (25) for $N+1$, and thereby proves the theorem. □

Note that if the factors $(-1)^m$ are deleted from the assumptions (24), followed by making the obvious changes in the above proof, conclusion (26) remains valid with amendation by eliminating the factors $(-1)^m$. Thus, we have the following theorem.

Theorem 2. *Let $x_k = x_k(\omega)$ and $f(x) = \rho^{-1/2}(x)$ be the same as mentioned above. If $f(x)$ is absolutely M-monotonic on the interval $[a,b]$, then the sequence $\{\Delta x_k(\omega)\}$ is asymptotically absolutely M-monotonic.*

As consequence of Lemma 1 and Theorems 1 and 2, we can use the zeros of a solution of (6) to construct sequences in which all mth differences have the same sign.

Corollary 1. *(a) Let $f(x)$ be M-monotonic on $[a,b]$. If $\varphi(x)$ is also M-monotonic on $[a,b]$, then the sequence $\{\varphi(x_k)\}$ is asymptotically M-monotonic.*

(b) Let $f(x)$ be absolutely M-monotonic on $[a,b]$. If $\varphi(x)$ is also absolutely M-monotonic on $[a,b]$, then the sequence $\{\varphi(x_k)\}$ is asymptotically absolutely M-monotonic.

Proof. Because $f(x)$ is M-monotonic on $[a,b]$, it can be seen from the proof of Theorem 1 that (25) holds. On the other hand, the M-monotonicity of $\varphi(x)$ on $[a,b]$ means that

$$(-1)^m \varphi^{(m)}(x) \geq 0 \quad (x \in [a,b]; m = 0,1,2,\ldots,M). \tag{27}$$

It now follows from (11), (25), (27) and (10) that

$$(-1)^m \Delta^m \varphi(x_k) = \sum_{q=1}^{m} [(-1)^{m-q} A_{q,k}^{(m)}][(-1)^q \varphi^{(q)}(x_{k+m-q})] + O(\omega^{-m-1}) \geq 0,$$

for all k and $m = 0, 1, 2, \ldots, M$, as $\omega \to +\infty$. The proof of part (b) is similar to that of part (a). □

Note that by the definition of the function $f(x) = \rho^{-1/2}(x)$, the conclusion of Theorem 1 (resp., Theorem 2) can be inferred directly from the assumptions on $\rho(x)$. In fact, $(-1)^m \rho^{(m+1)}(x) \geq 0$ (resp., $\rho^{(m+1)}(x) \leq 0$) on $[a, b]$ for $m = 0, 1, 2, \ldots, M-1$, implying $(-1)^m f^{(m)}(x) \geq 0$ (resp., $f^{(m)}(x) \geq 0$) on $[a, b]$ for $m = 1, 2, \ldots, M$. To examine these assertions, we can proceed by induction on M. For $M = 1$, per $f(x) = \rho^{-1/2}(x)$ and $f'(x) = (-1/2)\rho^{-3/2}(x)\rho'(x)$, the assertion is valid. For higher derivatives of $f(x)$, a general term of $f^{(m)}(x)$ would appear as

$$S_m = C[\rho]^{\alpha_0}[\rho']^{\alpha_1}[\rho'']^{\alpha_2} \cdots [\rho^{(m)}]^{\alpha_m}$$

with exponentials α_0 being a negative half-integer and $\alpha_1, \alpha_2, \ldots, \alpha_m$ all non-negative integers. The induction is carried through by differentiating S_m. We have

$$S_m' = C\alpha_0[\rho]^{\alpha_0-1}[\rho']^{\alpha_1+1}[\rho'']^{\alpha_2} \cdots [\rho^{(m)}]^{\alpha_m} + C\alpha_1[\rho]^{\alpha_0}[\rho']^{\alpha_1-1}[\rho'']^{\alpha_2+1} \cdots [\rho^{(m)}]^{\alpha_m}$$
$$+ \cdots + C\alpha_m[\rho]^{\alpha_0}[\rho']^{\alpha_1}[\rho'']^{\alpha_2} \cdots [\rho^{(m)}]^{\alpha_m-1}[\rho^{(m+1)}],$$

and under the conditions $(-1)^m \rho^{(m+1)}(x) \geq 0$ (resp., $\rho^{(m+1)}(x) \leq 0$) and the negative α_0, each term in the last sum has opposite sign (resp., the same sign) as S_m. Thus, $f^{(m)}(x)$ and $f^{(m+1)}(x)$ have alternating signs (resp., the same sign), completing the induction. Hence, we obtain the following corollary.

Corollary 2. *Let $x_k = x_k(\omega)$ be the same as above: (a) if $\rho'(x)$ is $(M-1)$-monotonic on $[a, b]$, then the sequence $\{\Delta x_k(\omega)\}$ is asymptotically M-monotonic, and*

(b) if $-\rho'(x)$ is absolutely $(M-1)$-monotonic on $[a, b]$, then the sequence $\{\Delta x_k(\omega)\}$ is asymptotically absolutely M-monotonic.

Although Corollary 2(a) is a partial result included in ([13], Theorem 3.3), the techniques employed in this section are independent of the methods in the series of papers [3,13,14] and the results of Hartman ([15], Theorems 18.1_n and 20.1_n). It provides the connection of the quantities between the differences of the zeros and the coefficient function $\rho(x)$, and might have some numerical interest.

One can find similar results concerned with the critical points of a Sturm–Liouville function of (6). In fact, by letting $x_k'(\omega)$ denote the kth critical point of a solution $y(x;\omega)$ of (6) in the interval $[a, b]$ and noting the definition of the Prüfer angle

$$\theta(x_k'(\omega); \omega) = (k - \frac{1}{2})\pi,$$

the procedures employed in this section are all valid. Thus, if we replace $\{x_k(\omega)\}$ in Theorems 1 and 2 and Corollaries 1 and 2 with $\{x_k'(\omega)\}$, the conclusions in these Theorems and Corollaries continue to hold.

3. Applications to Bessel Functions

Let $c_{\nu k}(\alpha)$ be the kth positive zero of the general Bessel (cylinder) function

$$C_\nu(x; \alpha) = J_\nu(x)\cos\alpha - Y_\nu(x)\sin\alpha,$$

where $J_\nu(x)$ and $Y_\nu(x)$ denote the Bessel functions with order ν of the first and second kind, respectively. The main results in this section are stated as follows.

Theorem 3. *(a) For $|\nu| < 1/2$, we have*

$$(-1)^m \Delta^{m+2} c_{\nu k}(\alpha) > 0 \quad (m = 0, 1, 2, \ldots; k = 1, 2, 3, \ldots).$$

(b) For $0 < |\nu| < 1/2$, we have

$$(-1)^m \Delta^{m+1} c_{\nu k}^{2|\nu|}(\alpha) > 0 \quad (m = 0, 1, 2, \ldots; k = 1, 2, 3, \ldots).$$

The Airy functions (see, e.g., [16] p. 18) satisfy the differential equation $y'' + \frac{x}{3} y = 0$. Here, we consider a broader class of functions, including the Airy functions, which satisfy the differential equation (see, e.g., [17] p. 97)

$$z'' + \omega^2 x^\gamma z = 0, \quad x \in (0, +\infty), \tag{28}$$

where $0 < \gamma < +\infty$. These functions are closely related to Bessel functions. Indeed,

$$z(x;\omega) = x^{1/2} C_\nu(2\nu\omega x^{1/(2\nu)}; \alpha), \quad \textit{where} \quad \nu = 1/(\gamma + 2),$$

is a nontrivial real solution of (28). Note that for each $\omega > 0$, the kth positive zeros $\xi_k(\omega)$ of $z(x;\omega)$ satisfies the identities

$$2\nu\omega(\xi_k(\omega))^{1/(2\nu)} = c_{\nu k}(\alpha) \quad \textit{and} \quad (2\nu\omega)^{2\nu} \xi_k(\omega) = c_{\nu k}^{2\nu}(\alpha).$$

Moreover, for each $\omega > 0$ and for $m = 0, 1, 2, \ldots$, we have

$$\Delta^{m+2} c_{\nu k}(\alpha) = 2\nu\omega \Delta^{m+2} (\xi_k(\omega))^{1/(2\nu)} \tag{29}$$

and

$$\Delta^{m+1} c_{\nu k}^{2\nu}(\alpha) = (2\nu\omega)^{2\nu} \Delta^{m+1} \xi_k(\omega). \tag{30}$$

Here, the identities (29) and (30) are the key to the regularity behaviour of the Bessel zeros.

To prove Theorem 3, we consider the family of differential equations

$$y'' + \omega^2 (x + a)^\gamma y = 0 \quad (a > 0; 0 < \gamma < +\infty), \tag{31}$$

on the interval $[0, b]$. Let $y_a(x;\omega)$ be a nontrivial real solution of (31) and let the sequence $\{x_k(\omega; a)\}$ be the zeros of $y_a(x;\omega)$ with ascending order in $[0, b]$. Following Theorem 1 with $f(x) = (x + a)^{-\gamma/2}$ and Corollary 1(a) with the function $\varphi(x) = (x + a)^{-1/(2\nu)}$, we have

$$(-1)^m \Delta^{m+1} x_k(\omega; a) \geq 0 \quad (m = 0, 1, 2, \ldots, M) \tag{32}$$

and

$$(-1)^m \Delta^m (x_k(\omega; a) + a)^{-1/(2\nu)} \geq 0 \quad (m = 0, 1, 2, \ldots, M), \tag{33}$$

as $\omega \to +\infty$. If we specify the initial conditions for the solution $y_a(x;\omega)$ of (31) to be

$$y_a(0;\omega) = z(a;\omega) \quad \textit{and} \quad y_a'(0;\omega) = z'(a;\omega),$$

then it is easy to verify that $y_a(x;\omega) = z(x + a;\omega)$ for $x \in [0, b]$; hence, for each k, $x_k(\omega; a) + a$ converges to ξ_k as $a \to 0^+$. Thus, for each $\omega > 0$, by (29) and (30) we have

$$\Delta^{m+2} c_{\nu k}(\alpha) = \lim_{a \to 0^+} 2\nu\omega \Delta^{m+2} (x_k(\omega; a))^{1/(2\nu)} \tag{34}$$

and

$$\Delta^{m+1}c_{\nu k}^{2\nu}(\alpha) = \lim_{a\to 0^+}(2\nu\omega)^{2\nu}\Delta^{m+1}x_k(\omega;a). \tag{35}$$

Recalling (15) and (17) with the function $\rho(x) = (x+a)^{\gamma}$ and denoting $x_k = x_k(\omega;a)$, we have

$$\omega\int_{x_k}^{x_{k+1}}(x+a)^{\gamma/2}dx = \pi - \sum_{r=0}^{m+1}\Delta F_r(x_k)\omega^{-r-1} - R_{m+1}(x_k). \tag{36}$$

Note that $\nu = 1/(\gamma+2)$ and $f(x) = (x+a)^{(2\nu-1)/(2\nu)}$. By (19), we have

$$\Delta F_0(x_k) = \frac{4\nu^2-1}{16\nu}\Delta(x_k+a)^{-1/(2\nu)}.$$

Thus, (36) becomes

$$2\nu\omega\Delta(x_k+a)^{1/(2\nu)} = \pi + \frac{1-4\nu^2}{16\nu\omega}\Delta(x_k+a)^{-1/(2\nu)} - \sum_{r=1}^{m+1}\Delta F_r(x_k)\omega^{-r-1} - R_{m+1}(x_k). \tag{37}$$

If we apply the difference operator Δ^{m+1} to (37), by (10) in the case $m+2$ and (18) in the case $m+3$, we can find

$$2\nu\omega\Delta^{m+2}(x_k+a)^{1/(2\nu)} = \frac{1-4\nu^2}{16\nu\omega}\Delta^{m+2}(x_k+a)^{-1/(2\nu)} + O(\omega^{-m-4}). \tag{38}$$

Moreover, multiplying (38) by $(-1)^m\omega^{m+3}$, we have

$$2\nu\omega^{m+4}(-1)^m\Delta^{m+2}(x_k+a)^{1/(2\nu)} = \frac{1-4\nu^2}{16\nu}\omega^{m+2}(-1)^m\Delta^{m+2}(x_k+a)^{-1/(2\nu)} + O(\omega^{-1}). \tag{39}$$

By (39), (33), (10) in the case $m+2$ and $0<\nu<1/2$, we have

$$(-1)^m\Delta^{m+2}(x_k+a)^{1/(2\nu)} \geq 0 \quad as \quad \omega\to+\infty. \tag{40}$$

Now, for each $a>0$, if we choose a sufficiently large $\omega = \omega(a)$ such that (40) and (32) hold, then by (34) and (35) we have

$$(-1)^m\Delta^{m+2}c_{\nu k}(\alpha) \geq 0 \quad (m=0,1,2,\ldots;k=1,2,3,\ldots), \tag{41}$$

and

$$(-1)^m\Delta^{m+1}c_{\nu k}^{2\nu}(\alpha) \geq 0 \quad (m=0,1,2,\ldots;k=1,2,3,\ldots). \tag{42}$$

Second, according to $Y_\nu(x) = (J_\nu(x)\cos\pi\nu - J_{-\nu}(x))/\sin\pi\nu$ (see, e.g., [17] p. 64), it is easy to verify that $C_{-\nu}(x;\alpha) = C_\nu(x;\alpha+\pi\nu)$; hence,

$$c_{-\nu k}(\alpha) = c_{\nu k}(\alpha+\pi\nu).$$

Thus, for $0<|\nu|<1/2$, (41) holds and (42) holds in the modified form:

$$(-1)^m\Delta^{m+1}c_{\nu k}^{2|\nu|}(\alpha) \geq 0 \quad (m=0,1,2,\ldots;k=1,2,3,\ldots). \tag{43}$$

Third, for $\nu=0$, any positive zero $c_{\nu k}(\alpha)$ of $C_\nu(x;\alpha)$ is definable as a continuously increasing function of the real variable ν (see, e.g., [17] p. 508), meaning that by an approximating process, (41) holds for all $|\nu|<1/2$.

Finally, because neither $\{\Delta^2 c_{\nu k}(\alpha)\}$ nor $\{\Delta c_{\nu k}^{2|\nu|}(\alpha)\}$ are constant sequences, the results of Lorch, Szego, and Muldoon for completely monotonic sequences ([2] p. 72 or [18] Theorem 2) guarantee the strict inequalities of (41) and (43). This completes the proof of Theorem 3.

4. Applications to Classical Orthogonal Polynomials

Several important classical orthogonal polynomials are related to Sturm–Liouville functions, such as the Hermite and Jacobi polynomials. In ([2] p. 71), Lorch, Szego, and their coworkers conjectured on the basis of numerical evidence that the θ-zeros of the Legendre polynomials, the special cases of Jacobi polynomials, and the positive zeros of the Hermite polynomials form sequences with mth differences having constant signs. In this section, we apply the results in Sections 2 and 3 to obtain partial answers for these conjectures.

4.1. Positive Zeros of Hermite Polynomials

Let $H_n(t)$ be the Hermite polynomial (see, e.g., [16] p. 105 (5.5.3)), defined by

$$H_n(t) = (-1)^n e^{t^2} (\frac{d}{dt})^n e^{-t^2}. \tag{44}$$

We consider the Hermite differential equation

$$H_n'' - 2tH_n' + 2nH_n = 0,$$

and the related equation

$$u'' + [(2n+1) - t^2]u = 0. \tag{45}$$

A simple calculation shows that (see, e.g., [16] p. 105 (5.5.2))

$$u_n(t) = e^{-t^2/2} H_n(t)$$

is a nontrivial solution of (45). From the general theory of orthogonal polynomials, we know that $H_n(t)$ has precisely n real zeros. By (44), we see that for even n it is the case that $H_n(t)$ is an even function of t, while for odd n $H_n(t)$ is an odd function of t. Accordingly, all zeros of $H_n(t)$ are placed symmetrically with respect to the origin, and the same phenomenon is clearly true for $u_n(t)$. For each n, the positive zeros of $H_n(t)$ are named by $h_1^{(n)} < h_2^{(n)} < \cdots < h_{[n/2]}^{(n)}$, where $[\cdot]$ is the greatest integer function.

The main result concerned with Hermite polynomials is as follows.

Theorem 4. *Let $h_k^{(n)}$ be as above. Then, for each k we have*

$$\Delta^m h_k^{(n)} \geq 0 \quad (m = 1, 2, \ldots, M), \tag{46}$$

for sufficiently large n.

Proof. For each n, by introducing the variable $x = t/\sqrt{2n+1}$ and letting $z_n(x) = u_n(t)$, Equation (45) is transformed into

$$z_n'' + (2n+1)^2(1 - x^2)z_n = 0.$$

We denote the kth positive zero of $z_n(x)$ by $\xi_k^{(n)}$, where $\xi_k^{(n)} = h_k^{(n)}/\sqrt{2n+1}$. Thus, we have

$$\Delta^m h_k^{(n)} = \sqrt{2n+1}\Delta^m \xi_k^{(n)}.$$

To prove (46), we consider the differential equation

$$y'' + (2n+1)^2(a - x^2)y = 0 \quad (a > 1; x \in [0,1]). \tag{47}$$

Let $\omega = 2n+1$, let $f(x) = (a-x^2)^{-1/2}$, let $y_n(x;a)$ be a nontrivial real solution of (47), and let $x_k^{(n)}(a)$ be the kth positive zero of $y_n(x;a)$. Then, from the following fact about $f^{(m)}(x)$

$$f^{(m)}(x) = \{a\, polynomial\, of\, x\, with\, nonnegative\, coefficients\}(a-x^2)^{-(2m+1)/2},$$

we know that $f^{(m)}(x) \geq 0$ on the interval $[0,1]$ for $m = 1,2,3,\ldots$. Thus, by Theorem 2, we obtain

$$\Delta^m x_k^{(n)}(a) \geq 0 \quad (m = 1,2,\ldots,M)$$

for sufficiently large n. If we specify the initial conditions for $y_n(x;a)$ to be

$$y_n(0;a) = z_n(0) \quad and \quad y_n'(0;a) = z_n'(0),$$

then it is easy to verify that $y_n(x;a)$ uniformly converges to $z_n(x)$ on the interval $[0,1]$ as $a \to 1^+$. Consequently, for $k = 1,2,\ldots,[\frac{n}{2}]$, the zero $x_k^{(n)}(a)$ converges to $\xi_k^{(n)}$ as $a \to 1^+$. Therefore, for fixed k,

$$\Delta^m \xi_k^{(n)} = \lim_{a\to 1^+} \Delta^m x_k^{(n)}(a) \geq 0,$$

and (46) holds. □

4.2. Zeros of Jacobi Polynomials

Considering $a > -1$ and $b > -1$, the Jacobi polynomial $P_n^{(a,b)}(x)$ (see, e.g., [16] p. 67 (4.3.1)) is defined by

$$(1-x)^a(1+x)^b P_n^{(a,b)}(x) = \frac{(-1)^n}{2^n n!}(\frac{d}{dx})^n\{(1-x)^{n+a}(1+x)^{n+b}\}.$$

Concerning the Jacobi polynomials $P_n^{(a,b)}(x)$ on the orthogonal interval $[-1,1]$, if we denote the zeros $x_k^{(n)} = x_k^{(n)}(a,b)$ of $P_n^{(a,b)}(x)$ with the descending order

$$1 > x_1^{(n)} > x_2^{(n)} > \cdots > x_n^{(n)} > -1,$$

then the θ-zeros $\theta_k^{(n)} = \theta_k^{(n)}(a,b)$ and $x_k^{(n)} = cos\theta_k^{(n)}$ of $P_n^{(a,b)}(cos\theta)$ behave as the order

$$0 < \theta_1^{(n)} < \theta_2^{(n)} < \cdots < \theta_n^{(n)} < \pi.$$

According to the uniform convergence theorem ([16] Theorem 8.1.1, p. 190)

$$\lim_{n\to+\infty} n^{-\alpha} P_n^{(a,b)}(cos\frac{x}{n}) = (\frac{x}{2})^{-a} J_a(x),$$

we know that

$$\lim_{n\to+\infty} n\theta_k^{(n)}(a,b) = j_{ak}.$$

Now, by Theorem 3(a), for $\nu = a$ and $\alpha = 0$, we have the following theorem.

Theorem 5. *For $|a| < 1/2$ and k fixed, we have*

$$(-1)^m \Delta^{m+2} \theta_k^{(n)}(a,b) \geq 0 \quad (m = 0,1,2,\ldots,M)$$

for sufficiently large n.

5. Proofs of Lemmas 1 and 3

In this section, we prove (10), (11), (12), (21), and (22) simultaneously by induction.

For $m = 1$, taking the Taylor expansion of φ at x_k

$$\varphi(x_{k+1}) = \varphi(x_k) + \varphi'(x_k)\Delta x_k + \varphi''(\xi_{k,2})\frac{(\Delta x_k)^2}{2}$$

where $x_k \le \xi_{k,2} \le x_{k+1}$ and using (8), we have

$$\Delta\varphi(x_k) = \varphi'(x_k)\Delta x_k + O(\omega^{-2}),$$

hence, (10), (11), and (12) are valid for $m = 1$. If we apply the first order difference operator to (20) and use (10) for $m = 1$ with $\varphi = F_0$, then we have

$$\Delta f(x_k) = \frac{\omega}{\pi}\Delta^2 x_k + \frac{\omega}{2!\pi}\Delta\{g_1(x_k)(\Delta x_k)^2\} + O(\omega^{-2}).$$

Because $\Delta\{\alpha_k\beta_k\} = \alpha_{k+1}(\Delta\beta_k) + (\Delta\alpha_k)\beta_k$, we have

$$\begin{aligned}\Delta f(x_k) &= \tfrac{\omega}{\pi}\Delta^2 x_k + \tfrac{\omega}{2!\pi}g_1(x_{k+1})\{\Delta x_{k+1}\Delta^2 x_k + \Delta^2 x_k \Delta x_k\} + O(\omega^{-2}) \\ &= \tfrac{\omega}{\pi}\Delta^2 x_k(1 + O(\omega^{-1})) + O(\omega^{-2}).\end{aligned}$$

Applying (10) for $m = 1$ again to the function $f(x)$, we find that $\Delta f(x_k) = O(\omega^{-1})$; now, we have

$$\Delta^2 x_k = O(\omega^{-2}),$$

hence,

$$\Delta f(x_k) = \frac{\omega}{\pi}\Delta^2 x_k + O(\omega^{-2}).$$

Thus, (21) for $m = 2$ and (22) for $m = 1$ are valid. The validity of (21) for $m = 2$ is the impetus of our induction argument.

Now, suppose that (10), (11), (12), (21), and (22) are fulfilled for $m = 1, 2, \ldots, N$. If we apply (10) for $m = N$ with $\varphi(x) = f(x)$ to (22) for $m = N$, then we have (21) for $m = N + 1$, that is,

$$\Delta^{N+1} x_k = O(\omega^{-N-1}).$$

Taking the Taylor expansion of φ at x_k

$$\varphi(x_{k+1}) = \varphi(x_k) + \sum_{p=1}^{N+1}\frac{\varphi^{(p)}(x_k)}{p!}(\Delta x_k)^p + \frac{\varphi^{(N+2)}(\xi_{k,N+2})}{(N+2)!}(\Delta x_k)^{N+2}, \tag{48}$$

where $x_k \le \xi_{k,N+2} \le x_{k+1}$, applying the Nth order difference operator to (48), and then using (21) for $m = 1$, we have

$$\Delta^{N+1}\varphi(x_k) = \sum_{p=1}^{N+1}\frac{1}{p!}\Delta^N\{\varphi^{(p)}(x_k)(\Delta x_k)^p\} + O(\omega^{-N-2}). \tag{49}$$

Following the product rule for higher differences, we know that

$$\Delta^N\{\varphi^{(p)}(x_k)(\Delta x_k)^p\} = \sum_{r=0}^{N}\binom{N}{r}\Delta^r\varphi^{(p)}(x_{k+N-r})\Delta^{N-r}(\Delta x_k)^p.$$

If we replace $\varphi(x_k)$ with $\varphi^{(p)}(x_{k+N-r})$ in (10) for $m = r$, $r = 1, 2, \ldots, N$ and use (21) for $m = 1, 2, \ldots, N + 1$, then we obtain

$$\Delta^N\{\varphi^{(p)}(x_k)(\Delta x_k)^p\} = O(\omega^{-N-p}) \quad (p = 1, 2, \ldots, N + 1). \tag{50}$$

Thus, (49) and (50) imply (10) for $m = N + 1$. Moreover, we have

$$\begin{aligned} \Delta^{N+1}\varphi(x_k) &= \Delta^N\{\varphi'(x_k)\Delta x_k\} + O(\omega^{-N-2}) \\ &= \sum_{r=0}^{N}\binom{N}{r}\Delta^r\varphi'(x_{k+N-r})\Delta^{N+1-r}x_k + O(\omega^{-N-2}). \end{aligned} \tag{51}$$

Applying (11) for $m = r$ to (51) with $\varphi'(x_{k+N-r})$ instead of $\varphi(x_k)$ for $r = 1, 2, \ldots, N$, we find

$$\begin{aligned} \Delta^{N+1}\varphi(x_k) = \; & \varphi'(x_{k+N})\Delta^{N+1}x_k \\ & + \sum_{r=1}^{N}\binom{N}{r}\{\sum_{q=1}^{r} A^{(r)}_{q,k+N-r}\varphi^{(q+1)}(x_{k+N-q})\}\Delta^{N+1-r}x_k + O(\omega^{-N-2}). \end{aligned} \tag{52}$$

If we change the order of the summation in (52) and shift the q index, then we can find

$$\begin{aligned} \Delta^{N+1}\varphi(x_k) = \; & \varphi'(x_{k+N})\Delta^{N+1}x_k \\ & + \sum_{q=2}^{N+1}\varphi^{(q)}(x_{k+N+1-q})\{\sum_{r=q-1}^{N}\binom{N}{r}A^{(r)}_{q-1,k+N-r}\Delta^{N+1-r}x_k\} + O(\omega^{-N-2}). \end{aligned}$$

Thus, (11) and (12) are valid for $m = N + 1$.

Finally, to prove (22) for $m = N + 1$, by applying the $(N + 1)$th order difference operator to (20) for $m = N + 1$, we have

$$\begin{aligned} \Delta^{N+1}f(x_k) = \; & \frac{\omega}{\pi}\sum_{r=0}^{N+1}\frac{1}{(r+1)!}\Delta^{N+1}\{g_r(x_k)(\Delta x_k)^{r+1}\} \\ & + \frac{1}{\pi}\sum_{r=0}^{N-1}\Delta^{N+1}\{f(x_k)\Delta F_r(x_k)\}\omega^{-r-1} + O(\omega^{-N-2}). \end{aligned} \tag{53}$$

Following the product rule for higher differences again, we have

$$\Delta^{N+1}\{g_r(x_k)(\Delta x_k)^{r+1}\} = \sum_{\beta=0}^{N+1}\binom{N+1}{\beta}\Delta^{\beta}g_r(x_{k+N+1-\beta})\Delta^{N+1-\beta}(\Delta x_k)^{r+1}.$$

Using (10) for $m = \beta$ with $g_r(x_{k+N+1-\beta})$ replacing $\varphi(x_k)$ for $\beta = 1, 2, \ldots, N + 1$ and using (21) for $m = 1, 2, \ldots, N + 1$, we obtain

$$\begin{aligned} & \Delta^{N+1}\{g_r(x_k)(\Delta x_k)^{r+1}\} \\ & = g_r(x_{k+N+1})\Delta^{N+1}(\Delta x_k)^{r+1} + \sum_{\beta=1}^{N+1}\binom{N+1}{\beta}\Delta^{\beta}g_r(x_{k+N+1-\beta})\Delta^{N+1-\beta}(\Delta x_k)^{r+1} \\ & = g_r(x_{k+N+1})(\Delta^{N+2}x_k)O(\omega^{-r}) + O(\omega^{-N-r-2}). \end{aligned} \tag{54}$$

On the other hand, applying (10) to the functions $f(x)$ and $F_r(x)$ for $m = 1, 2, \ldots, N + 1$, we have

$$\begin{aligned} & \Delta^{N+1}\{f(x_k)\Delta F_r(x_k)\} \\ & = f(x_{k+N+1})\Delta^{N+2}F_r(x_k) + \sum_{\beta=1}^{N+1}\binom{N+1}{\beta}\Delta^{\beta}f(x_{k+N+1-\beta})\Delta^{N+2-\beta}F_r(x_k) \\ & = O(\omega^{-N-1}) + O(\omega^{-N-2}). \end{aligned} \tag{55}$$

Applying the estimates (54) and (55) to (53), we obtain

$$\begin{aligned} \Delta^{N+1}f(x_k) &= \frac{\omega}{\pi}\Delta^{N+2}x_k + (\frac{\omega}{\pi}\Delta^{N+2}x_k)O(\omega^{-1}) + O(\omega^{-N-2}) \\ &= \frac{\omega}{\pi}\Delta^{N+2}x_k(1 + O(\omega^{-1})) + O(\omega^{-N-2}). \end{aligned} \tag{56}$$

If we replace $\varphi(x_k)$ with $f(x_k)$ in (10) for $m = N + 1$, then we have

$$\Delta^{N+1}f(x_k) = O(\omega^{-N-1}). \tag{57}$$

Note that (56) and (57) imply

$$\frac{\omega}{\pi}\Delta^{N+2}x_k = O(\omega^{-N-1}). \tag{58}$$

Then, by (56) and (58), we have (22) for $m = N+1$. This completes the proofs of Lemmas 1 and 3.

6. Conclusions

In this work, we consider the second-order differential equation $y'' + \omega^2\rho(x)y = 0$ on the interval $[a,b]$ associated with a positive parameter ω. When the function $\rho^{-1/2}(x)$ satisfies the (absolutely) M-monotonic condition on the interval $[a,b]$, we show that the difference of the zeros for a nontrivial solution of the equation satisfies the asymptotically (absolutely) M-monotonic property. As applications, we use an approximation process for the zeros of the Bessel function and prove the conjecture of Lorch and Szego. In addition, we show that the differences of the zeros of various orthogonal polynomials with higher degrees possess sign regularity.

On the basis of numerical evidence, Lorch, Szego, and their coworkers conjectured that the θ-zeros of the Legendre polynomials, the special cases of Jacobi polynomials, and the positive zeros of the Hermite polynomials are able to form absolutely monotonic sequences, that is, sequences in which all consecutive differences of the zeros are non-negative. In Theorem 5, the x-zeros of Jacobi polynomials are arranged in descending order, and hence the θ-zeros are arranged in increasing order, while the mth differences and $(m+1)$th differences of the θ-zeros of Jacobi polynomials are sign-alternating.

Funding: This research received no external funding.

Data Availability Statement: Not applicable.

Acknowledgments: The author would like to thank the anonymous reviewers who helped to improve the readability and quality of the paper. Thanks also to Professor Min-Jei Huang.

Conflicts of Interest: The author declares no conflict of interest.

Appendix A

Recalling $f(x) = \rho^{-1/2}(x)$ and the differential Equation (13) for the Prüfer angle $\theta(x;\omega)$, we have

$$\theta'(x;\omega) = \frac{\omega}{f(x)}\{1 - \frac{f'(x)}{2\omega}\sin 2\theta(x;\omega)\}. \tag{A1}$$

Then,

$$\{-\frac{f'}{2f}\sin 2\theta\}\frac{\theta'}{\{1-(f'/2\omega)\sin 2\theta\}\omega/f} = -\sum_{r=0}^{\infty}\{\frac{\omega^{-1}}{2}f'\sin 2\theta\}^{r+1}\theta', \tag{A2}$$

hence,

$$\int_{x_k}^{x_{k+1}}\frac{\rho'}{4\rho}\sin 2\theta dx = -\sum_{r=0}^{m-1}\frac{\omega^{-r-1}}{2^{r+1}}\int_{x_k}^{x_{k+1}}(f')^{r+1}(sin^{r+1}2\theta)\theta' dx + O(\omega^{-m-1}), \tag{A3}$$

where $\theta = \theta(x;\omega)$, $\theta' = \theta'(x;\omega)$ and $x_k = x_k(\omega)$.

To prove Lemma 2, we introduce the following integrals for a C^∞-function φ which is defined on $[a,b]$:

$$P_r[\varphi] = \int_{x_k}^{x_{k+1}}\varphi\cdot sin^r(2\theta)\cdot\theta' dx,$$

$$Q_r[\varphi] = \int_{x_k}^{x_{k+1}}\varphi\cdot sin^r(2\theta)\cdot cos(2\theta) dx,$$

and

$$R_r[\varphi] = \int_{x_k}^{x_{k+1}} \varphi \cdot sin^{r+1}(2\theta)dx,$$

where $r = 0, 1, 2, \ldots$. Now, (A3) can be written as

$$\int_{x_k}^{x_{k+1}} \frac{\rho'}{4\rho} \sin 2\theta dx = -\sum_{r=0}^{m-1} \frac{\omega^{-r-1}}{2^{r+1}} P_{r+1}[(f')^{r+1}] + O(\omega^{-m-1}). \tag{A4}$$

Via integration by parts, we have the following reduced formula for $P_{r+1}[\varphi]$:

$$P_{r+1}[\varphi] = \frac{-\varphi \cdot sin^r 2\theta \cdot cos2\theta}{2(r+1)} |_{x_k}^{x_{k+1}} + \frac{r}{r+1} P_{r-1}[\varphi] + \frac{1}{2(r+1)} Q_r[\varphi']. \tag{A5}$$

Introducing θ' in the same way as in (A2) and using integration by parts and (14), we have the following estimates for $Q_r[\varphi]$ and $R_r[\varphi]$:

$$Q_r[\varphi] = -\sum_{j=0}^{m-r-3} \frac{\omega^{-j-1}}{2^{j+1}(r+j+1)} R_{r+j}[(\varphi_j)'] + O(\omega^{-m+r}), \tag{A6}$$

and

$$R_r[\varphi] = \sum_{j=0}^{m-r-3} \frac{\omega^{-j-1}}{2^j} P_{r+j+1}[\varphi_j] + O(\omega^{-m+r+1}), \tag{A7}$$

where $\varphi_j = \varphi f(f')^j$. By applying the estimates (A6) and (A7) with suitable integrands to (A5) and then collecting the terms with the same order of ω in the sum together, we can find

$$\begin{aligned} P_{r+1}[\varphi] = & \tfrac{-\varphi \cdot sin^r 2\theta \cdot cos2\theta}{2(r+1)} |_{x_k}^{x_{k+1}} + \tfrac{r}{r+1} P_{r-1}[\varphi] \\ & - \textstyle\sum_{\beta=0}^{m-r-3} \tfrac{\omega^{-\beta-2}}{2^{\beta+2}(r+1)} \sum_{j=0}^{\beta} \tfrac{1}{r+j+1} P_{r+\beta+1}[(\varphi')_{j,\beta-j}] + O(\omega^{-m+r}), \end{aligned} \tag{A8}$$

where $\varphi_{j_1,j_2} = [(\varphi_{j_1})']_{j_2}$. By (A8) and (14), we have

$$P_1[\varphi] = \frac{-\Delta\varphi(x_k)}{2} - \sum_{\beta=0}^{m-3} \frac{\omega^{-\beta-2}}{2^{\beta+2}} \sum_{j=0}^{\beta} \frac{1}{j+1} P_{\beta+1}[(\varphi')_{j,\beta-j}] + O(\omega^{-m}), \tag{A9}$$

and

$$P_2[\varphi] = \frac{P_0[\varphi]}{2} - \sum_{\beta=0}^{m-4} \frac{\omega^{-\beta-2}}{2^{\beta+3}} \sum_{j=0}^{\beta} \frac{1}{j+2} P_{\beta+2}[(\varphi')_{j,\beta-j}] + O(\omega^{-m+1}). \tag{A10}$$

If we apply (A1) and (A7) to the integral $P_0[\varphi]$, then we have

$$P_0[\varphi] = \omega \int_{x_k}^{x_{k+1}} \frac{\varphi}{f} dx - \sum_{j=0}^{m-3} \frac{\omega^{-j-1}}{2^{j+1}} P_{j+1}[(\varphi f'/f)_j] + O(\omega^{-m+1}). \tag{A11}$$

Applying (A11) to (A10), we obtain

$$\begin{aligned} P_2[\varphi] = & \tfrac{\omega}{2} \textstyle\int_{x_k}^{x_{k+1}} \tfrac{\varphi}{f} dx - \sum_{j=0}^{m-3} \tfrac{\omega^{-j-1}}{2^{j+2}} P_{j+1}[(\varphi f'/f)_j] \\ & - \textstyle\sum_{\beta=0}^{m-4} \tfrac{\omega^{-\beta-2}}{2^{\beta+3}} \sum_{j=0}^{\beta} \tfrac{1}{j+2} P_{\beta+2}[(\varphi')_{j,\beta-j}] + O(\omega^{-m-1}). \end{aligned} \tag{A12}$$

In (A4), if we apply (A8) to the function $\varphi = (f')^{r+1}$ and use (A9) and (A12) to collect the reductions of those integrals $P_{r-1}[(f')^{r+1}]$ and $P_{r+\beta+1}[((f')^{r+1})'_{j,\beta-j}]$, then all reduction processes are stopped after a finite number of steps, while the remainders behave as $O(\omega^{-m-1})$. This completes the proof of Lemma 2.

References

1. Birkhoff, G.; Rota, G.-C. *Ordinary Differential Equations*, 4th ed.; Wiley: New York, NY, USA, 1989; p. 314.
2. Lorch, L.; Szego, P. Higher Monotonicity Properties of Certain Sturm-Liouville Functions. *Acta Math.* **1963**, *109*, 55–73. [CrossRef]
3. Muldoon, M.E. Higher monotonicity properties of certain Sturm-Liouville functions V. *Proc. R. Soc. Edinb. Sect. A* **1977**, *77*, 23–37. [CrossRef]
4. Shen, C.-L.; Tsai, T.-M. On a uniform approximation of the density function of a string equation using eigenvalues and nodal points and some related inverse nodal problems. *Inverse Probl.* **1995**, *11*, 1113–1123. [CrossRef]
5. Ali, A.H.; Páles, Z. Taylor-type expansions in terms of exponential polynomials. *Math. Inequal. Appl.* **2022**, *25*, 1123–1141. [CrossRef]
6. Kadum, Z.J.; Abdul-Hassan, N.Y. New Numerical Methods for Solving the Initial Value Problem Based on a Symmetrical Quadrature Integration Formula Using Hybrid Functions. *Symmetry* **2023**, *15*, 631. [CrossRef]
7. Kuipers, L.; Niederreiter, H. *Uniform Distribution of Sequences*; Dover Publications: Mineola, NY, USA, 2006.
8. Long, B.-Y.; Sugawa, T.; Wang, Q.-H. Completely monotone sequences and harmonic mappings. *Ann. Fenn. Math.* **2022**, *47*, 237–250. [CrossRef]
9. Wang, X.-F.; Ismail, M.E.H.; Batir, N.; Guo, S. A necessary and sufficient condition for sequences to be minimal completely monotonic. *Adv. Differ. Equ.* **2020**, *665*, 665. [CrossRef]
10. Aguech, R.; Jedidi, W. New characterizations of completely monotone functions and Bernstein functions, a converse to Hausdorff's moment characterization theorem. *Arab. J. Math. Sci.* **2019**, *23*, 57–82. [CrossRef]
11. Shen, C.-L. On the Barcilon formula for the string equation with a piecewise continuous density function. *Inverse Probl.* **2005**, *21*, 635–655. [CrossRef]
12. Pólya, G.; Szegö, G. *Problems and Theorems in Analysis*; Die Grundlehren der mathematischen Wissenschaften, Band 193; Springer: Berlin, Germany; New York, NY, USA, 1972; Volume II.
13. Lorch, L.; Muldoon, M.E.; Szego, P. Higher monotonicity properties of certain Sturm-Liouville functions III. *Can. J. Math.* **1970**, *22*, 1238–1265. [CrossRef]
14. Lorch, L.; Muldoon, M.E.; Szego, P. Higher monotonicity properties of certain Sturm-Liouville functions IV. *Can. J. Math.* **1972**, *24*, 349–368. [CrossRef]
15. Hartman, P. On Differential Equations and the Function $J_\mu^2 + Y_\mu^2$. *Am. J. Math.* **1961**, *83*, 154–188. [CrossRef]
16. Szegö, G. *Orthogonal Polynomials*, revised ed.; American Mathematical Society Colloquium Publications: New York, NY, USA, 1959; Volume 23.
17. Watson, G.N. *A Treatise on the Theory of Bessel Functions*, 2nd ed.; Cambridge University Press: Cambridge, UK, 1958.
18. Muldoon, M.E. Elementary remarks on multiply monotonic functions and sequences. *Can. Math. Bull.* **1971**, *14*, 69–72. [CrossRef]

mathematics

MDPI

Article

The Partial Inverse Spectral and Nodal Problems for Sturm–Liouville Operators on a Star-Shaped Graph

Xian-Biao Wei [1,*], Yan-Hsiou Cheng [2] and Yu-Ping Wang [3]

1 Department of Mathematics and Physics, Anhui Jianzhu University, Hefei 230601, China
2 Department of Mathematics and Information Education, National Taipei University of Education, Taipei City 106, Taiwan
3 Department of Applied Mathematics, Nanjing Forestry University, Nanjing 210037, China
* Correspondence: xbwei@ahjzu.edu.cn

Abstract: We firstly prove the Horváth-type theorem for Sturm–Liouville operators on a star-shaped graph and then solve a new partial inverse nodal problem for this operator. We give some algorithms to recover this operator from a dense nodal subset and prove uniqueness theorems from paired-dense nodal subsets in interior subintervals having a central vertex. In particular, we obtain some uniqueness theorems by replacing the information of nodal data on some fixed edge with part of the eigenvalues under some conditions.

Keywords: partial inverse spectral problem; partial inverse nodal problem; boundary value problem; graph; paired-dense nodal subset

MSC: 34A55; 34B09; 34L05; 47E05

1. Introduction

Consider the following boundary value problem $B := B(q,\alpha)$, $q(x) := \{q_l(x)\}_{l=1}^{p}$, $\alpha = \{\alpha_l\}_{l=1}^{p}$ on a star-shaped graph with p edges of identical length 1, defined as follows:

$$-y_l'' + q_l(x)y_l = \lambda y_l, \quad x \in (0,1), \quad l = \overline{1,p}, \tag{1}$$

associated with the separated boundary conditions at the pendant vertices 0

$$y_l(0,\lambda)\cos\alpha_l + y_l'(0,\lambda)\sin\alpha_l = 0, \quad l = \overline{1,p}, \tag{2}$$

and the standard matching conditions at the central vertex 1

$$y_1(1,\lambda) = y_l(1,\lambda), \quad l = \overline{2,p}, \quad \sum_{l=1}^{p} y_l'(1,\lambda) = 0, \tag{3}$$

where λ is the spectral parameter, $\alpha_l \in [0,\pi)$ and $q_l(x)$, and $l = \overline{1,p}$ is called the potential and is an integrable real-valued function on the l-th edge. The differential operators on quantum graphs have many applications in chemistry, mathematics, networks, spider webs, and so on (see [1–17] and the references therein).

The problem B is a natural extension of the classical Sturm–Liouville problem on the finite interval. The inverse nodal problems for the classical Sturm–Liouville operators are to recover the potential and boundary conditions by using its nodal data [18–23]. McLaughlin [22] firstly studied the inverse nodal problem for the classical Sturm–Liouville operator and showed that one set of nodal points can determine the Sturm–Liouville operators uniquely. The solution of the potential function to this problem was given by Hald and McLaughlin [19]. The uniqueness results show that the inverse nodal problem is overdetermined. Later on, there was much study focus on how to use less information of nodal

Citation: Wei, X.-B.; Cheng, Y.-H.; Wang, Y.-P. The Partial Inverse Spectral and Nodal Problems for Sturm–Liouville Operators on a Star-Shaped Graph. *Mathematics* **2022**, *10*, 3971. https://doi.org/10.3390/math10213971

Academic Editor: Carmen Chicone

Received: 28 September 2022
Accepted: 16 October 2022
Published: 26 October 2022

data to recover the potential. The uniqueness theorems and the reconstruction formulae are given by using twin-dense nodal subset [15,20,21,24–26], dense nodal subset [23,27,28], and partial nodal data [15,29–31]. Guo and Wei [30] presented a sharp condition on the nodal subset and proved the uniqueness for the classical Sturm–Liouville operator with a paired-dense nodal subset in interior subintervals under some conditions based on the Gesztesy–Simon theorem in [32]. In addition, the theory on dynamic Sturm–Liouville boundary value problems via variational methods was found in [33,34].

Beginning in 2002, Kuchment [5–8] studied quantum graphs and investigated the spectral properties of periodic boundary value problems for a carbon atom in graphene. In [12,13], Pivovarchik studied inverse spectral problems with Dirichlet boundary conditions for a star-shaped graph with p edges. He gave the asymptotic expansion of eigenvalues and showed that there are p sequences of eigenvalues where one sequence is simple while the others might not be. In particular, Law and Pivovarchik [35] discussed the multiplicity of the eigenvalues and interlacing properties between two spectral sets of the Sturm–Liouville problems defined on a tree. Recently, Luo, Jatulan and Law [36] gave a complete classification of Archimedean tilings for the periodic quantum graphs and investigated the sufficient conditions for point spectrum and continuous spectrum. Bondarenko [2] showed that if all components $q_l(x), l = \overline{1,p}$ but one on the graph are given a priori, the remaining component can be uniquely determined by two sequences of chosen eigenvalues and provided a constructive algorithm for the solution of the partial inverse problem. In [37], Wang and Shieh generalized Bondarenko's theorem by the methods in [38]. In this paper, we are going to solve the following partial inverse spectral problem for B:

IP1: (Inverse Problem 1) If $q_l(x) = \widetilde{q}_l(x)$ on $[0,1]$, α_l for $l \neq i_0$ and $q_{i_0}(x) = \widetilde{q}_{i_0}(x)$ on $[a_0,1]$ for some a_0, $a_0 \in (0,1]$ given a priori, recover $q_{i_0}(x)$ and α_{i_0} from part of the eigenvalues.

On the other hand, the inverse nodal problems on quantum graphs have been studied. In 2007, Currie and Watson [39] studied the inverse nodal problems on general graphs and showed that, for $q_i \in L^\infty$, one set of eigenvalues and nodal positions is sufficient to reconstruct the potentials $q_i's$. In 2008, Yurko [40] discussed the inverse nodal problem for B with $\alpha_l = 0, l = \overline{1,p}$ and proved that each component $q_l(x)$ can be uniquely determined up to a constant by a dense nodal set. Later on, Cheng [41] derived the asymptotics of eigenvalues of B with $q_l \in L^1, l = \overline{1,p}$ and presented direct and explicit formulae on recovering the potentials using a twin-dense nodal subset. Wang and Shieh [31] investigated the partial inverse nodal problem for B with Dirichlet boundary conditions from a twin-dense nodal subset in interior subintervals under some conditions. Therefore, we are going to solve the following partial inverse nodal problem for B with less nodal information:

IP2: (Inverse Problem 2) Recover the component $q_l(x), l = \overline{1,p}$ from given paired-dense nodal subsets on subintervals having a central vertex.

We firstly prove the Horváth-type theorem for B and extend Horváth's method in [38] for the classical Sturm–Liouville operator to B, which is also the theoretical basis for the solution of the partial inverse nodal problem for B. Then, we show that the components $\{q_l(x)\}_{l=1}^p$ for B can be uniquely determined up to a constant by a dense nodal subset corresponding to the first eigenvalue sequence in $[0,1]$; see Theorem 2. We also give algorithms to reconstruct $\{q_l(x)\}_{l=1}^p$ and $\{\alpha_l\}_{l=1}^p$ from a dense nodal subset. In Theorem 3, combined with the Horváth-type theorem for B, we show that if there is a paired-dense nodal subset corresponding to the first eigenvalue sequence in a interior subinterval, then, with a sufficiently large counting number corresponding to the first eigenvalue sequence, we can uniquely determine the components $\{q_l(x)\}_{l=1}^p$ up to a constant on the whole graph. Finally, in Theorem 4, without any nodal data on some i_0-th edge but with part of the eigenvalues, we can also uniquely determine components $\{q_l(x)\}_{l=1}^p$ up to a constant on the whole graph from a paired-dense nodal subset corresponding to the first eigenvalue sequence and sufficiently large counting numbers. We extend Guo-Wei's method in [30] for the classical Sturm–Liouville operator to B.

This article is organized as follows. In Section 2, we present preliminaries. We give the asymptotic formulae of nodal points. We present solutions to **IP1** in Section 3 and **IP2** in Section 4, respectively.

2. Preliminaries

Let $S_l(x,\lambda)$, $C_l(x,\lambda)$, and $\varphi_l(x,\lambda)$ be solutions of (1) for each $l=\overline{1,p}$ associated with the initial conditions:

$$S_l(0,\lambda)=0,\ S_l'(0,\lambda)=1,\ C_l(0,\lambda)=1,\ C_l'(0,\lambda)=0,$$
$$\varphi_l(0,\lambda)=\sin\alpha_l,\ \varphi_l'(0,\lambda)=-\cos\alpha_l.$$

Moreover, we have

$$\varphi_l(x,\lambda)=\sin\alpha_l C_l(x,\lambda)-\cos\alpha_l S_l(x,\lambda).$$

By the results in [42], we obtain the asymptotic formulae:

(a) If $\alpha_l=0$,

$$\begin{cases}\varphi_l(x,\lambda)=\dfrac{\sin\rho x}{\rho}-\dfrac{Q_l(x)\cos\rho x}{\rho^2}+o\left(\dfrac{e^{\tau x}}{\rho^2}\right), & 0<x<1,\\ \varphi_l'(x,\lambda)=\cos\rho x+\dfrac{Q_l(x)\sin\rho x}{\rho}+o\left(\dfrac{e^{\tau x}}{\rho}\right), & 0<x<1,\end{cases}\tag{4}$$

(b) If $0<\alpha_l<\pi$,

$$\begin{cases}\varphi_l(x,\lambda)=\sin\alpha_l\cos\rho x+(Q_l(x)\sin\alpha_l-\cos\alpha_l)\dfrac{\sin\rho x}{\rho}+o\left(\frac{e^{\tau x}}{\rho}\right), & 0<x<1,\\ \varphi_l'(x,\lambda)=-\rho\sin\alpha_l\sin\rho x+(Q_l(x)\sin\alpha_l-\cos\alpha_l)\cos\rho x+o(e^{\tau x}), & 0<x<1\end{cases}\tag{5}$$

for $|\lambda|\to\infty$, where $\rho=\sqrt{\lambda}$, $\tau=|\mathrm{Im}\rho|$, and

$$Q_l(x):=\frac{1}{2}\int_0^x q_l(t)\mathrm{d}t,\quad l=\overline{1,p}.$$

The characteristic function $\Delta(\lambda)$ of B is defined by

$$\Delta(\lambda):=\sum_{l=1}^{p}\varphi_l'(1,\lambda)\prod_{k=1,k\neq l}^{p}\varphi_k(1,\lambda),\tag{6}$$

which is an entire function in λ of order 1/2, where all zeros of $\Delta(\lambda)$ coincide with the eigenvalues of B. Denote $\sigma(B):=\cup_{m=1}^{p}M_m$ asthe eigenvalue set of B (counting with their multiplicities) where $M_m=\{\lambda_{m,n}\}_{n\in\mathbb{N}}$ and $\rho_{m,n}:=\sqrt{\lambda_{m,n}}$. We shall find the asymptotic formulae of nodal points separately corresponding to the three cases:

$$\begin{array}{l}\text{(I)}\ \ \alpha_l=0,\quad l=\overline{1,p};\\ \text{(II)}\ \ \alpha_l\in(0,\pi),\quad l=\overline{1,p};\\ \text{(III)}\ \alpha_l=0,\quad l=\overline{1,T},\quad\text{and}\quad \alpha_l\in(0,\pi),\quad l=\overline{T+1,p},\quad 1\le T\le p-1.\end{array}\tag{7}$$

By (Theorem 2.1 [41]), all eigenvalues are real. For the case I, there exist p sequences of eigenvalues $\lambda_{m,n}$ with the asymptotic formulae:

$$\begin{cases}\rho_{1,n}=\left(n-\dfrac{1}{2}\right)\pi+\dfrac{1}{2p(n-\frac{1}{2})\pi}\displaystyle\int_0^1(1-\cos((2n-1)\pi t))\left(\sum_{l=1}^{p}q_l(t)\right)\mathrm{d}t+O\left(\dfrac{1}{n^2}\right),\\ \rho_{m,n}=n\pi+\dfrac{\Lambda_{m,n,1}}{n\pi}+O\left(\dfrac{1}{n^2}\right),\quad m=\overline{2,p},\end{cases}\tag{8}$$

for $n \gg 1$, where $\Lambda_{m,n,1}$ is the $(m-1)$-th, $m = \overline{2,p}$ zero of the polynomial $p_1(\Lambda)$ of degree $(p-1)$

$$p_1(\Lambda) := \sum_{l=1}^{p} \prod_{i \neq l} \left[\Lambda - \frac{1}{2} \int_0^1 (1 - \cos((2n-1)\pi t)) q_i(t) \mathrm{d}t \right]. \tag{9}$$

For case II, there exist p sequences of eigenvalues $\lambda_{m,n}$ with the asymptotic formulae

$$\begin{cases} \rho_{1,n} = (n-1)\pi + \dfrac{1}{2(n-1)p\pi} \left(-2A_1 + \displaystyle\int_0^1 (1 + \cos(2(n-1)\pi t)) \left(\sum_{l=1}^{p} q_l(t) \right) \right) dt + O\left(\dfrac{1}{n^2}\right), \\ \rho_{m,n} = n\pi + \dfrac{\Lambda_{m,n,2}}{n\pi} + O\left(\dfrac{1}{n^2}\right), \; m = \overline{2,p}, \end{cases} \tag{10}$$

for $n \gg 1$, where

$$A_1 = \sum_{l=1}^{p} \cot \alpha_l, \tag{11}$$

and $\Lambda_{m,n,2}$ is the $(m-1)$-th, $m = \overline{2,p}$ zero of the polynomial $p_2(\Lambda)$ of degree $(p-1)$

$$p_2(\Lambda) := \sum_{l=1}^{p} \prod_{i \neq l} \left[\Lambda - \cot \alpha_i + \frac{1}{2} \int_0^1 (1 + \cos(2(n-1)\pi t)) q_i(t) \mathrm{d}t \right], \tag{12}$$

and for the case III, there exist p sequences of eigenvalues $\lambda_{m,n}$ with the asymptotic formulae

$$\begin{cases} \rho_{m,n} = n\pi + (-1)^m d_1 + \dfrac{\omega_1}{2n\pi} + o\left(\dfrac{1}{n}\right), \quad m = 1,2, \\ \rho_{m,n} = n\pi + \dfrac{\Lambda_{m,n,3}}{n\pi} + O\left(\dfrac{1}{n^2}\right), \quad m = \overline{3, T+1}, \\ \rho_{m,n} = \left(n - \frac{1}{2}\right)\pi + \dfrac{\Lambda_{m,n,4}}{\left(n - \frac{1}{2}\right)\pi} + O\left(\dfrac{1}{n^2}\right), \quad m = \overline{T+2, p}, \end{cases} \tag{13}$$

for $n \gg 1$, where

$$d_1 := \arcsin \sqrt{\frac{T}{p}}, \; A_2 := \sum_{l=T+1}^{p} \cot \alpha_l, \; \omega_1 := \frac{1}{p} \left((p-T) \sum_{l=1}^{T} Q_l(1) + T \sum_{l=T+1}^{p} Q_l(1) - TA_2 \right), \tag{14}$$

$\Lambda_{m,n,3}$ is the $(m-2)$-th, $m = \overline{3, T+1}$ root of the polynomial $p_3(\Lambda)$ of degree $(T-1)$

$$p_3(\Lambda) := \sum_{l=1}^{T} \prod_{i \neq l, i=1}^{T} \left(\Lambda - \frac{1}{2} \int_0^1 (1 - \cos(2n\pi t)) q_i(t) \mathrm{d}t \right), \tag{15}$$

and $\Lambda_{m,n,4}$ is the $(m-T-1)$-th, $m = \overline{T+2, p}$ root of the polynomial $p_4(\Lambda)$ of degree $(p-T-1)$

$$p_4(\Lambda) := \sum_{l=T+1}^{p} \prod_{i \neq l, i=T+1}^{p} \left(\Lambda - \cot \alpha_i + \frac{1}{2} \int_0^1 (1 + \cos((2n-1)\pi t)) q_i(t) \mathrm{d}t \right). \tag{16}$$

The function

$$m_l(x, \lambda) := -\frac{\varphi_l'(x, \lambda)}{\varphi_l(x, \lambda)}, \quad x \in (0, 1], \quad l = \overline{1, p},$$

is called the Weyl m-function of B_l, where the problem B_l is defined by by (1), (2) and $\varphi_l(1, \lambda) = 0$. Applying the same arguments as the proof of Marchenko's theorem in [43], one shows that the Weyl m-function $m_l(a, \lambda)$ uniquely determines $q_l(x)$ on $[0, a]$ with

$0 < a \leq 1$ and α_l. The eigenfunction $y(x, \lambda_{m,n})$ corresponding to the eigenvalue $\lambda_{m,n}$ of B is of the form:

$$y(x, \lambda_{m,n}) = \{c_l(\lambda_{m,n})\varphi_l(x, \lambda_{m,n})\}_{l=1}^{p},$$

where $c_l(\lambda_{m,n}), l = \overline{1,p}$ are constant, do not depend on x, and are not all zeros. The function $\varphi_l(x, \lambda_{1,n})$ is called the l-th component of $y(x, \lambda_{1,n})$. Let $x_{l,1,n}^{j}$ be the j-th nodal point of the l-th component $\varphi_l(x, \lambda_{1,n})$ corresponding to the eigenvalue $\lambda_{1,n}$, i.e., $\varphi_l(x_{l,1,n}^{j}, \lambda_{1,n}) = 0$, $l = \overline{1,p}$. The l-th component $\varphi_l(x, \lambda_{1,n})$ has exactly $n-1$ (simple) zeros inside the interval $(0,1)$, and

$$0 < x_{l,1,n}^{1} < x_{l,1,n}^{2} < \cdots < x_{l,1,n}^{j} < \cdots < x_{l,1,n}^{n-1} < 1.$$

For $l = \overline{1,p}$, let $X_{l,1} := \{x_{l,1,n}^{j}\}$ be the nodal set of the l-th component $\varphi_l(x, \lambda_{1,n})$ corresponding to M_1. Then, $X_{l,1}$ is dense on $[0,1]$ (see below for Lemma 1). Since we can only obtain the same nodal information from the same eigenvalues, we assume that $I_1 := \{n_{1,k}\}_{k=K_0}^{\infty}$ is a strictly increasing subsequence in $\mathbb{N}$ (where K_0 is defined in Lemma 2) such that

$$M_{1,0} := \left\{\lambda_{1,n_{1,k}} : \lambda_{1,n_{1,k_1}} < \lambda_{1,n_{1,k_2}} \text{ for any } n_{1,k_1} < n_{1,k_2},\ n_{1,k_1}, n_{1,k_2} \in I_1\right\}.$$

Next, we shall give the definition of a paired-dense nodal subset on a finite interval.

Definition 1. *For each $l = \overline{1,p}$, denote $W_{I_1}([a_l, b_l]) \subseteq X_{l,1} \bigcap [a_l, b_l]$ with $0 \leq a_l < b_l \leq 1$ on the l-th edge. The nodal subset $W_{I_1}([a_l, b_l])$ is called a paired-dense nodal subset on $[a_l, b_l]$ corresponding to I_1 if the following conditions hold:*

1. *For each $n_{1,k} \in I_1$, there exist some $j_k, r_k \geq 1$, $r_k \in \mathbb{N}$, such that $x_{l,1,n_{1,k}}^{j_k}, x_{l,1,n_{1,k}}^{j_k+r_k} \in W_{I_1}([a_l, b_l])$.*
2. *$\overline{W_{I_1}([a_l, b_l])} = [a_l, b_l]$.*

The definition of the paired-dense nodal subset was given in [30]. Clearly, the twin-dense nodal subset is a special case of the paired-dense nodal subset. Denote

$$\omega_0 = \frac{1}{p}\sum_{l=1}^{p} Q_l(1), \quad \alpha_n^j = \begin{cases} \dfrac{j}{n - \frac{1}{2}}, & \text{for I}, \\ \dfrac{j - \frac{1}{2}}{n-1}, & \text{for II}, \\ \dfrac{j}{n}, \quad l = \overline{1,T}, & \text{for III}, \\ \dfrac{j - \frac{1}{2}}{n}, \quad l = \overline{T+1,p}, & \text{for III}, \end{cases}$$

By the asymptotic behavior of $\lambda_{1,n}$ and $\varphi_l(x, \lambda_{1,n})$, one can easily obtain asymptotic behavior of nodal points. We omit the proof.

Lemma 1. *For three cases, the nodal points $x_{l,1,n}^{j}$ of the l-th component $\varphi(x, \lambda_{1,n})$ corresponding to the eigenvalue $\lambda_{1,n}$ have the asymptotic formulae:*

$$x_{l,1,n}^{j} = \alpha_n^j + \frac{1}{2(n-\frac{1}{2})^2\pi^2}\left(\int_0^{\alpha_n^j} q_l(t)\mathrm{d}t - 2\omega_0\alpha_n^j\right) + o\left(\frac{1}{n^2}\right), \quad \text{for I}, \tag{17}$$

$$x_{l,1,n}^{j} = \alpha_n^j - \frac{\cot \alpha_l}{(n-1)^2\pi^2} + \frac{1}{2(n-1)^2\pi^2}\int_0^{\alpha_n^j} q_l(t)\mathrm{d}t$$

$$+\frac{1}{2(n-1)^2\pi^2}\left(\frac{2A_1}{p}-\omega_0\right)\alpha_n^j+o\left(\frac{1}{n^2}\right), \text{ for II,} \tag{18}$$

$$x_{l,1,n}^j=\alpha_n^j+\frac{d_1\alpha_n^j}{n\pi}+\frac{1}{2(n\pi)^2}\int_0^{\alpha_n^j}q_l(t)\mathrm{d}t$$
$$+\frac{2d_1^2-\omega_1}{2(n\pi)^2}\alpha_n^j+o\left(\frac{1}{n^2}\right),\ l=\overline{1,T}, \quad \text{for III,} \tag{19}$$

$$x_{l,1,n}^j=\alpha_n^j+\frac{\alpha_n^j}{n\pi}-\frac{\cot\alpha_l}{n^2\pi^2}+\frac{1}{2(n\pi)^2}\int_0^{\alpha_n^j}q_l(t)\mathrm{d}t$$
$$+\frac{2d_1^2-\omega_1}{2(n\pi)^2}\alpha_n^j+o\left(\frac{1}{n^2}\right), \quad l=\overline{T+1,p}, \quad \text{for III} \tag{20}$$

for $n \gg 1$ uniformly in j, where ω_1, d_1, and A_1 are defined in (11) *and* (14).

3. Partial Inverse Spectral Problems

In this section, we shall study the partial inverse spectral problem for B. Let the boundary value problem $\widetilde{B}$ have the same form as B but with different coefficients. If a certain symbol γ denotes an object related to B, then the corresponding symbol $\widetilde{\gamma}$ with a tilde denotes the analogous object related to $\widetilde{B}$. Let $\widehat{\gamma}=\gamma-\widetilde{\gamma}$.

For $m=\overline{2,p}$, let $I_m:=\{n_{m,k}\}_{k=1}^{\infty}$ be a strictly increasing subsequence in $\mathbb{N}$, and denote $M_{m,0}:=\{\lambda_{m,n_{m,k}}: n_{m,k}\in I_m\}\subseteq M_m$. For each $m=\overline{1,p}$, the counting function corresponding to $M_{m,0}$ is defined by

$$N_{M_{m,0}}(t):=\sum_{\rho_{m,n_{m,k}}<t,\lambda_{m,n_{m,k}}\in M_{m,0}}1, \quad t\in\mathbb{R}^+.$$

By (6), this yields

$$\Delta(\lambda)=\gamma_1(\lambda)\varphi_{i_0}'(1,\lambda)+\gamma_2(\lambda)\varphi_{i_0}(1,\lambda), \tag{21}$$

where

$$\gamma_1(\lambda)=\prod_{l\neq i_0}\varphi_l(1,\lambda), \quad \text{and} \quad \gamma_2(\lambda)=\sum_{l\neq i_0}^{p}\varphi_l'(1,\lambda)\prod_{k\neq i_0,k\neq l}^{p}\varphi_k(1,\lambda).$$

Clearly the entire functions $\gamma_1(\lambda)$ and $\gamma_2(\lambda)$ in λ are only dependent on $q_l(x)$ and $\alpha_l, l\neq i_0$. If $\gamma_1(\lambda_{m,n})=\gamma_2(\lambda_{m,n})=0$, then we cannot obtain any information about the component $q_{i_0}(x)$ from the eigenvalue $\lambda_{m,n}$ by (21). Hence, we add the following Assumption 1:

Assumption 1. *For each $\lambda_{m,n_{m,k}}\in M_{m,0}$, such that*

$$\gamma_1^2(\lambda_{m,n_{m,k}})+\gamma_2^2(\lambda_{m,n_{m,k}})\neq 0, \quad m=\overline{1,p}.$$

We shall prove the following Horváth type-theorem for B, which is a solution to ***IP1****:*

Theorem 1. *Let $q_l(x)=\widetilde{q}_l(x)$ on $[0,1]$, $\alpha_l=\widetilde{\alpha}_l$ for $l\neq i_0$ and $q_{i_0}(x)=\widetilde{q}_{i_0}(x)$ on $[a_0,1]$ for some a_0, $a_0\in(0,1]$ be given a priori. If $M_{k,0}=\widetilde{M}_{k,0}$ with Assumption 1 satisfied, and there exist $t_0>0$, $0\leq\kappa_1\leq 1$, $\delta_1>0$ such that*

$$\sum_{m=1}^{p}N_{M_{m,0}}(t)\geq 2a_0\left\{\kappa_1\left[\frac{t}{\pi}+\frac{1}{2}\right]+(1-\kappa_1)\left(\left[\frac{t}{\pi}\right]+\frac{1}{2}\right)-1+\kappa_1+O(t^{-\delta_1})\right\}, \text{ if } \alpha_{i_0}=0; \tag{22}$$

$$\sum_{m=1}^{p}N_{M_{m,0}}(t)\geq 2a_0\left\{\kappa_1\left[\frac{t}{\pi}+\frac{1}{2}\right]+(1-\kappa_1)\left(\left[\frac{t}{\pi}\right]+\frac{1}{2}\right)+\kappa_1+O(t^{-\delta_1})\right\}, \text{ if } \alpha_{i_0}\neq 0 \tag{23}$$

for sufficiently large $t \geq t_0$, and

$$\lim_{t\to\infty} \frac{\sum_{m=1}^{p} N_{M_{m,0}}(t)}{t} = \frac{2a_0}{\pi},$$

where $[x]$ denotes the largest integer less than or equal to x, then

$$q_{i_0}(x) \overset{a.e.}{=} \widetilde{q}_{i_0}(x) \quad on \quad [0,1] \quad and \quad \alpha_{i_0} = \widetilde{\alpha}_{i_0}. \tag{24}$$

Proof. It follows from (1) for $l = i_0$

$$\int_0^1 \widehat{q}_{i_0}(x)\varphi_{i_0}(x,\lambda_{m,n_{m,k}})\widetilde{\varphi}_{i_0}(x,\lambda_{m,n_{m,k}})\mathrm{d}x = <\varphi_{i_0},\widetilde{\varphi}_{i_0}> (1,\lambda_{m,n_{m,k}}) - <\varphi_{i_0},\widetilde{\varphi}_{i_0}> (0,\lambda_{m,n_{m,k}}),$$

where $<\varphi_l,\widetilde{\varphi}_l> (x,\lambda) := \varphi_l(x,\lambda)\widetilde{\varphi}_l'(x,\lambda) - \varphi_l'(x,\lambda)\widetilde{\varphi}_l(x,\lambda), l = \overline{1,p}$, is called the Wronskian of $\varphi_l(x,\lambda)$ and $\widetilde{\varphi}_l(x,\lambda)$. This implies

$$<\varphi_{i_0},\widetilde{\varphi}_{i_0}> (1,\lambda_{m,n_{m,k}}) = \int_0^1 \widehat{q}_{i_0}(x)\varphi_{i_0}(x,\lambda_{m,n_{m,k}})\widetilde{\varphi}_{i_0}(x,\lambda_{m,n_{m,k}})\mathrm{d}x - \sin\widehat{\alpha}_{i_0}. \tag{25}$$

The assumption $q_l(x) = \widetilde{q}_l(x)$ on $[0,1]$ and $\alpha_l = \widetilde{\alpha}_l$ for all $l \neq i_0$ together with the initial conditions $\varphi_l(0,\lambda) = \widetilde{\varphi}_l(0,\lambda) = \sin\alpha_l,\ \varphi_l'(0,\lambda) = \widetilde{\varphi}_l'(0,\lambda) = -\cos\alpha_l$ show that

$$\varphi_l(1,\lambda_{m,n_{m,k}}) = \widetilde{\varphi}_l(1,\lambda_{m,n_{m,k}}) \quad \text{and} \quad \varphi_l'(1,\lambda_{m,n_{m,k}}) = \widetilde{\varphi}_l'(1,\lambda_{m,n_{m,k}}) \tag{26}$$

for all $m = \overline{1,p}$. From (21) and (26), it is clear that

$$\gamma_1(\lambda_{m,n_{m,k}}) = \widetilde{\gamma}_1(\lambda_{m,n_{m,k}}) \quad \text{and} \quad \gamma_2(\lambda_{m,n_{m,k}}) = \widetilde{\gamma}_2(\lambda_{m,n_{m,k}}) \tag{27}$$

By Assumption 1 and (21), we have

$$\begin{cases} \varphi_{i_0}'(1,\lambda_{m,n_{m,k}}) = -\dfrac{\gamma_2(\lambda_{m,n_{m,k}})}{\gamma_1(\lambda_{m,n_{m,k}})}\varphi_{i_0}(1,\lambda_{m,n_{m,k}}), \\ \widetilde{\varphi}_{i_0}'(1,\lambda_{m,n_{m,k}}) = -\dfrac{\gamma_2(\lambda_{m,n_{m,k}})}{\gamma_1(\lambda_{m,n_{m,k}})}\widetilde{\varphi}_{i_0}(1,\lambda_{m,n_{m,k}}), \end{cases} \quad \text{if} \quad \gamma_1(\lambda_{m,n_{m,k}}) \neq 0, \tag{28}$$

or

$$\begin{cases} \varphi_{i_0}(1,\lambda_{m,n_{m,k}}) = -\dfrac{\gamma_1(\lambda_{m,n_{m,k}})}{\gamma_2(\lambda_{m,n_{m,k}})}\varphi_{i_0}'(1,\lambda_{m,n_{m,k}}), \\ \widetilde{\varphi}_{i_0}(1,\lambda_{m,n_{m,k}}) = -\dfrac{\gamma_1(\lambda_{m,n_{m,k}})}{\gamma_2(\lambda_{m,n_{m,k}})}\widetilde{\varphi}_{i_0}'(1,\lambda_{m,n_{m,k}}), \end{cases} \quad \text{if} \quad \gamma_2(\lambda_{m,n_{m,k}}) \neq 0. \tag{29}$$

It follows from (25) and (27)–(29) that

$$<\varphi_{i_0},\widetilde{\varphi}_{i_0}> (1,\lambda_{m,n_{m,k}}) = 0, \quad \forall n_{m,k} \in I_m, \quad m = \overline{1,p}. \tag{30}$$

By $q_{i_0}(x) = \widetilde{q}_{i_0}(x)$ on $[a_0,1]$, this yields

$$<\varphi_{i_0},\widetilde{\varphi}_{i_0}> (a_0,\lambda_{m,n_{m,k}}) = 0, \quad \forall n_{m,k} \in I_m, \quad m = \overline{1,p}. \tag{31}$$

Define the function $K_{i_0}(\lambda)$ by

$$K_{i_0}(\lambda) := \frac{<\varphi_{i_0},\widetilde{\varphi}_{0,i_0}> (a_0,\lambda)}{\prod_{m=1}^{p} F_m(\lambda)}, \tag{32}$$

where

$$F_m(\lambda) := \prod_{\lambda_{m,n_{m,k}} \in M_{m,0}} \left(1 - \frac{\lambda}{\lambda_{m,n_{m,k}}}\right), \quad m = \overline{1,p}. \tag{33}$$

If $\lambda_{m,n_{m,k}} = 0$, we substitute $1 - \dfrac{\lambda}{\lambda_{m,n_{m,k}}}$ by λ in (33). If the eigenvalue $\lambda_{m,n_{m,k}} \in \cup_{m=1}^{p} M_{m0}$ is simple, then (31) guarantees that the function $K_{i_0}(\lambda)$ is analytical at $\lambda = \lambda_{m,n_{m,k}}$. By (8), (10) and (13), we see that the multiplicity of each eigenvalue can be only finite. Assume that the multiplicity of the eigenvalue $\lambda_{m,n_{m,k}} := \lambda_0 \in \cup_{m=1}^{p} M_{m,0}$ is $k_0, k_0 \geq 2$. Then,

$$\Delta(\lambda_0) = \frac{\mathrm{d}\Delta(\lambda)}{\mathrm{d}\lambda}\Big|_{\lambda=\lambda_0} = \cdots = \frac{\mathrm{d}^{k_0-1}\Delta(\lambda)}{\mathrm{d}\lambda^{k_0-1}}\Big|_{\lambda=\lambda_0} = 0. \tag{34}$$

Consequently, (21) and (34) show that

$$\begin{cases} \gamma_1(\lambda_0)\varphi'_{i_0}(1,\lambda_0) + \gamma_2(\lambda_0)\varphi_{i_0}(1,\lambda_0) = 0, \\ \gamma_1(\lambda_0)\frac{\mathrm{d}\varphi'_{i_0}(1,\lambda)}{\mathrm{d}\lambda}\big|_{\lambda=\lambda_0} + \gamma_2(\lambda_0)\frac{\mathrm{d}\varphi_{i_0}(1,\lambda)}{\mathrm{d}\lambda}\big|_{\lambda=\lambda_0} \\ +\frac{\mathrm{d}\gamma_1(\lambda)}{\mathrm{d}\lambda}\big|_{\lambda=\lambda_0}\varphi'_{i_0}(1,\lambda_0) + \frac{\mathrm{d}\gamma_2(\lambda)}{\mathrm{d}\lambda}\big|_{\lambda=\lambda_0}\varphi_{i_0}(1,\lambda_0) = 0, \\ \qquad \vdots \\ \sum_{k=0}^{k_0-1} C_{k_0-1}^{k}\frac{\mathrm{d}^k\gamma_1(\lambda)}{\mathrm{d}\lambda^k}\big|_{\lambda=\lambda_0}\frac{\mathrm{d}^{k_0-1-k}\varphi'_{i_0}(1,\lambda)}{\mathrm{d}\lambda^{k_0-1-k}}\big|_{\lambda=\lambda_0} \\ +\sum_{k=0}^{k_0-1} C_{k_0-1}^{k}\frac{\mathrm{d}^k\gamma_2(\lambda)}{\mathrm{d}\lambda^k}\big|_{\lambda=\lambda_0}\frac{\mathrm{d}^{k_0-1-k}\varphi_{i_0}(1,\lambda)}{\mathrm{d}\lambda^{k_0-1-k}}\big|_{\lambda=\lambda_0} = 0, \end{cases} \tag{35}$$

Similar to (35), we have

$$\begin{cases} \gamma_1(\lambda_0)\widetilde{\varphi}'_{i_0}(1,\lambda_0) + \gamma_2(\lambda_0)\widetilde{\varphi}_{i_0}(1,\lambda_0) = 0, \\ \gamma_1(\lambda_0)\frac{\mathrm{d}\widetilde{\varphi}'_{i_0}(1,\lambda)}{\mathrm{d}\lambda}\big|_{\lambda=\lambda_0} + \gamma_2(\lambda_0)\frac{\mathrm{d}\widetilde{\varphi}_{i_0}(1,\lambda)}{\mathrm{d}\lambda}\big|_{\lambda=\lambda_0} \\ +\frac{\mathrm{d}\gamma_1(\lambda)}{\mathrm{d}\lambda}\big|_{\lambda=\lambda_0}\widetilde{\varphi}'_{i_0}(1,\lambda_0) + \frac{\mathrm{d}\gamma_2(\lambda)}{\mathrm{d}\lambda}\big|_{\lambda=\lambda_0}\widetilde{\varphi}_{i_0}(1,\lambda_0) = 0, \\ \qquad \vdots \\ \sum_{k=0}^{k_0-1} C_{k_0-1}^{k}\frac{\mathrm{d}^k\gamma_1(\lambda)}{\mathrm{d}\lambda^k}\big|_{\lambda=\lambda_0}\frac{\mathrm{d}^{k_0-1-k}\widetilde{\varphi}'_{i_0}(1,\lambda)}{\mathrm{d}\lambda^{k_0-1-k}}\big|_{\lambda=\lambda_0} \\ +\sum_{k=0}^{k_0-1} C_{k_0-1}^{k}\frac{\mathrm{d}^k\gamma_2(\lambda)}{\mathrm{d}\lambda^k}\big|_{\lambda=\lambda_0}\frac{\mathrm{d}^{k_0-1-k}\widetilde{\varphi}_{i_0}(1,\lambda)}{\mathrm{d}\lambda^{k_0-1-k}}\big|_{\lambda=\lambda_0} = 0. \end{cases} \tag{36}$$

It follows from (35) and (36) that

$$< \varphi_{i_0}, \widetilde{\varphi}_{0,i_0} > (1,\lambda_0) = \frac{\mathrm{d} < \varphi_{i_0}, \widetilde{\varphi}_{0,i_0} > (1,\lambda)}{\mathrm{d}\lambda}\Big|_{\lambda=\lambda_0} = \cdots = \frac{\mathrm{d}^{k_0-1} < \varphi_{i_0}, \widetilde{\varphi}_{0,i_0} > (1,\lambda)}{\mathrm{d}\lambda^{k_0-1}}\Big|_{\lambda=\lambda_0} = 0.$$

This implies that the function $< \varphi_{i_0}, \widetilde{\varphi}_{0,i_0} > (1,\lambda)$ has zeros at λ_0 of at least k_0. Moreover, the function $< \varphi_{i_0}, \widetilde{\varphi}_{0,i_0} > (a_0,\lambda)$ has zeros at λ_0 of at least k_0. Thus, the function $K_{i_0}(\lambda)$ is analytical at $\lambda = \lambda_{m,n_{m,k}}$. Note that (4) and (5) show that

$$|< \varphi_{i_0}, \widetilde{\varphi}_{i_0} > (a_0,\lambda)| = \begin{cases} O\left(\frac{e^{2a_0\tau}}{|\rho|^2}\right), & \text{if} \quad \alpha_{i_0} = 0, \quad (37) \\ O(e^{2a_0\tau}), & \text{if} \quad \alpha_{i_0} \neq 0, \quad (38) \end{cases}$$

for $|\lambda| \to \infty$. By the results on the Weyl m-functions $m_l(x,\lambda)$ and $\widetilde{m}_l(x,\lambda)$ in [32], we have

$$\left|m_{i_0}(x,\lambda) - \widetilde{m}_{i_0}(x,\lambda)\right| = |i\rho + o(1) - (i\rho + o(1))| = o(1) \tag{39}$$

uniformly in $x \in [\delta, 1]$ for $|\lambda| \to \infty$ in any sector $\varepsilon_0 < \arg\lambda < \pi - \varepsilon_0$ for $\varepsilon_0 > 0$, where $\delta \in (0,1]$ (for details, see [32]). Consequently, it follows from (4), (5) and (39) that

$$|< \varphi_{i_0}, \widetilde{\varphi}_{i_0} > (a_0,\lambda)| = \begin{cases} o\left(\dfrac{e^{2a_0\tau}}{|\rho|^2}\right), & \text{if} \quad \alpha_l = 0, \quad (40) \\ o(e^{2a_0\tau}), & \text{if} \quad \alpha_l \neq 0, \quad (41) \end{cases}$$

for $|\lambda| \to \infty$ in any sector $\varepsilon_0 < \arg\lambda < \pi - \varepsilon_0$. By Levinson's estimate (see [44]), then the first formula of (8), or (10), or (13) and (22), or (23) imply that there exists a constant c_m such that

$$\frac{1}{\left|\prod_{m=1}^{p} F_m(\lambda)\right|} = O\left(e^{-2a_0\tau + \varepsilon\sqrt{|\lambda|}}\right), \quad \forall \lambda \in \bigcap_{m=1}^{p} D_{m,c_m} \tag{42}$$

for sufficiently large $|\lambda|$, where

$$D_{m,c_m} := \left\{\lambda : |\rho - \rho_{m,n_{m,k}}| \geq \frac{1}{8}c_m, \quad \lambda_{m,n_{m,k}} \in M_{m,0}\right\}.$$

Thus (37), (38), and (42) show that

$$|K_{i_0}(\lambda)| = O\left(e^{2\varepsilon\sqrt{|\lambda|}}\right), \quad \forall \lambda \in \bigcap_{m=1}^{p} D_{m,c_m} \tag{43}$$

for sufficiently large $|\lambda|$. Consequently, (43) and the maximum modulus principle show that the entire function $K_{i_0}(\lambda)$ is of the zero-exponential type, i.e., for arbitrary $\varepsilon > 0$, then

$$|K_{i_0}(\lambda)| \leq c e^{2\varepsilon\sqrt{|\lambda|}}, \quad \lambda \in \mathbb{C} \tag{44}$$

for sufficiently large $|\lambda|$, where c is constant. Noting that

$$\frac{\sin\sqrt{\lambda}}{\sqrt{\lambda}} = \prod_{n=1}^{\infty}\left(1 - \frac{\lambda}{n^2\pi^2}\right),$$

we obtain

$$\begin{aligned} \int_1^{\infty} \frac{\left[\frac{t}{\pi}\right]}{t}\frac{y^2}{y^2+t^2}dt &= \ln\left|\prod_{n=1}^{\infty}\left(1 - \frac{iy}{n^2\pi^2}\right)\right| + O(1) = \ln\left|\frac{\sin\sqrt{iy}}{\sqrt{iy}}\right| + O(1) \\ &= \sqrt{\frac{|y|}{2}} - \frac{1}{2}\ln|y| + O(1), \end{aligned} \tag{45}$$

$$\begin{aligned} \int_1^{\infty} \frac{\left[\frac{t}{\pi} + \frac{1}{2}\right]}{t}\frac{y^2}{y^2+t^2}dt &= \ln\left|\prod_{n=1}^{\infty}\left(1 - \frac{iy}{\left(n-\frac{1}{2}\right)^2\pi^2}\right)\right| + O(1) = \ln\left|\cos\sqrt{iy}\right| + O(1) \\ &= \sqrt{\frac{|y|}{2}} + O(1). \end{aligned} \tag{46}$$

$$\int_1^{\infty}\frac{1}{t}\frac{y^2}{y^2+t^2}\mathrm{d}t=\int_1^{\infty}\left(\frac{1}{t}-\frac{t}{y^2+t^2}\right)\mathrm{d}t=\frac{1}{2}\ln(y^2+1)=\ln|y|+O(1). \tag{47}$$

Next, we shall prove by two cases

$$|K_{i_0}(iy)|\le\begin{cases}\dfrac{c}{|y|^{1-a_0(1-\kappa_1)}}, & \text{if} \quad \alpha_{i_0}=0;\\ \dfrac{c}{|y|^{a_0\kappa_1}}, & \text{if} \quad \alpha_{i_0}\neq 0\end{cases} \tag{48}$$

for sufficiently large $y>0$. Here and below, we use the symbol c to represent a positive constant that may vary from one formula to another.

Case (1): All eigenvalues $\lambda_{m,n_{m,k}}\ge 1$. It follows from (45), (46), (47), (22) and (23) that

$$\begin{aligned}\ln\left|\prod_{m=1}^{p}F_m(iy)\right|&=\int_1^{\infty}\frac{\sum_{k=1}^{p}N_{k,0}(t)}{t}\frac{y^2}{y^2+t^2}\mathrm{d}t+O(1)\\ &\ge 2a_0\begin{cases}\int_1^{\infty}\frac{\kappa_1\left[\frac{t}{\pi}+\frac{1}{2}\right]+(1-\kappa_1)\left(\left[\frac{t}{\pi}\right]+\frac{1}{2}\right)-1+\kappa_1}{t}\frac{y^2}{y^2+t^2}\mathrm{d}t+O(1), & \text{if} \quad \alpha_{i_0}=0;\\ \int_1^{\infty}\frac{\kappa_1\left[\frac{t}{\pi}+\frac{1}{2}\right]+(1-\kappa_1)\left(\left[\frac{t}{\pi}\right]+\frac{1}{2}\right)+\kappa_1}{t}\frac{y^2}{y^2+t^2}\mathrm{d}t+O(1), & \text{if} \quad \alpha_{i_0}\neq 0\end{cases}\\ &=\begin{cases}2a_0\sqrt{\frac{|y|}{2}}-a_0(1-\kappa_1)\ln|y|+O(1), & \text{if} \quad \alpha_{i_0}=0;\\ 2a_0\sqrt{\frac{|y|}{2}}+a_0\kappa_1\ln|y|+O(1), & \text{if} \quad \alpha_{i_0}\neq 0.\end{cases}\end{aligned}$$

This implies

$$\begin{cases}\left|\prod_{m=1}^{p}F_m(iy)\right|\ge c\dfrac{e^{2a_0\sqrt{|y|/2}}}{|y|^{a_0(1+\kappa_1)}}, & \text{if} \quad \alpha_{i_0}=0;\\ \left|\prod_{m=1}^{p}F_m(iy)\right|\ge c|y|^{a_0(1-\kappa_1)}e^{2a_0\sqrt{|y|/2}}, & \text{if} \quad \alpha_{i_0}\neq 0\end{cases} \tag{49}$$

for a $y\in R^+$ that is sufficiently large. By (40), (41), and (49), we obtain (48).

Case (2): There exist $k_0\ge 1$ eigenvalues such that $\lambda_{m,n_{m,k_m}}<1$, $m=\overline{1,p}$, $k_m=\overline{1,k_{m,0}}$, where there may exist some $k_{m,0}$ such that $k_{m,0}=0$. If $k_{m,0}=0$, then $\lambda_{m,n_{m,k}}\ge 1$ for all $n_{m,k}$. Without loss of generality, we assume

$$\lambda_{m,n_{m,k_m}}<1,\quad m=\overline{1,p},\quad k_m=\overline{1,k_{m,0}},\quad \sum_{m=1}^{p}k_{m,0}=k_0.$$

Let

$$\mu_{m,n_{m,k_m}}=n_{m,k_m}\pi>1,\quad m=\overline{1,p},\quad S_{m,0}:=\{\lambda_{m,n_{m,k_m}}\}_{k_m=1}^{k_{m,0}},\quad k_m=\overline{1,k_{m,0}}$$

and

$$F_{1,m}(\lambda)=\prod_{k_m=1}^{k_{m,0}}\left(1-\frac{\lambda}{\mu_{m,n_{m,k_m}}}\right)\times\prod_{\lambda_{m,n_{m,k}}\in M_{m,0}\setminus S_{m,0}}\left(1-\frac{\lambda}{\lambda_{m,n_{m,k}}}\right),\quad m=\overline{1,p}. \tag{50}$$

Since

$$\lim_{y\to+\infty}\prod_{m=1}^{p}\prod_{k_m=1}^{k_{m,0}}\frac{1-\frac{iy}{\lambda_{m,n_{m,k_m}}}}{1-\frac{iy}{\mu_{m,n_{m,k_m}}}}=1,$$

then there exists a sufficiently large Y_0 such that

$$\left| \prod_{m=1}^{p} \prod_{k_m=1}^{k_{m,0}} \frac{1 - \frac{iy}{\lambda_{m,n_{m,k_m}}}}{1 - \frac{iy}{\mu_{m,n_{m,k_m}}}} \right| \geq \frac{1}{2} \tag{51}$$

for $y > Y_0$. Note that

$$\prod_{m=1}^{p} F_m(\lambda) = \prod_{m=1}^{p} \prod_{k_m=1}^{k_{m,0}} \frac{1 - \frac{\lambda}{\lambda_{m,n_{m,k_m}}}}{1 - \frac{\lambda}{\mu_{m,n_{m,k_m}}}} \times \prod_{m=1}^{p} F_{1,m}(\lambda).$$

By (49), (50) and (51), we also have (48). It follows from (48)

$$\lim_{y \to \infty} K_{i_0}(iy) = 0. \tag{52}$$

By the Phragmén-Lindelöf-type result in [20] together with (44) and (52), we obtain

$$K_{i_0}(\lambda) \equiv 0, \quad \lambda \in \mathbb{C}. \tag{53}$$

It follows from (53) that

$$< \varphi_{i_0}, \widetilde{\varphi}_{i_0} > (a_0, \lambda) = 0, \quad \forall \lambda \in \mathbb{C}, \quad l = \overline{1,p}.$$

Consequently,

$$m_{i_0}(a_0, \lambda) = \widetilde{m}_{i_0}(a_0, \lambda), \quad \forall \lambda \in \mathbb{C}, \quad l = \overline{1,p}. \tag{54}$$

By Marchenko's result in [21] together with (54), we have

$$\widehat{q}_{i_0}(x) \overset{a.e.}{=} 0 \quad \text{on} \quad [0, a_0], \quad \text{and} \quad \alpha_{i_0} = \widetilde{\alpha}_{i_0}. \tag{55}$$

The proof of Theorem 1 is completed. □

4. Partial Inverse Nodal Problems

In this section, we shall study the partial inverse nodal problem for B from a paired-dense nodal subset in an interior subinterval having a central vertex. For $l = \overline{1,p}$, we say $W_{I_1}([a_l, b_l]) = \widetilde{W}_{\widetilde{I}_1}([a_l, b_l])$ if for any $n_{1,k} \in I_1$ there exist $j_k, r_k, \widetilde{n}_{1,k}, \widetilde{j}_k \in \mathbb{N}$ such that $x_{l,1,n_{1,k}}^{j_k}, x_{l,1,n_{1,k}}^{j_k+r_k} \in W_{I_1}([a_l, b_l]), \widetilde{x}_{l,1,\widetilde{n}_{1,k}}^{\widetilde{j}_k}, \widetilde{x}_{l,1,\widetilde{n}_{1,k}}^{\widetilde{j}_k+r_k} \in \widetilde{W}_{\widetilde{I}_1}([a_l, b_l])$ and

$$x_{l,1,n_{1,k}}^{j_k} = \widetilde{x}_{l,1,\widetilde{n}_{1,k}}^{\widetilde{j}_k}, \quad x_{l,1,n_{1,k}}^{j_k+r_k} = \widetilde{x}_{l,1,\widetilde{n}_{1,k}}^{\widetilde{j}_k+r_k}.$$

We obtain the following three uniqueness theorems for B.

Theorem 2. *For each $l = \overline{1,p}$, let $X_{l,1,0} \subseteq X_{l,1}$ be a dense nodal subset on $[0,1]$; then*

$$q_l(x) - \widetilde{q}_l(x) \overset{a.e.}{=} 2\widehat{\omega}_0 \quad \text{on} \quad [0,1], \quad \alpha_l = \widetilde{\alpha}_l \quad \text{for} \quad l = \overline{1,p}, \quad \text{I} \quad \text{or} \quad \text{II}$$
$$q_l(x) - \widetilde{q}_l(x) \overset{a.e.}{=} \widehat{\omega}_1 \quad \text{on} \quad [0,1], \quad \alpha_l = \widetilde{\alpha}_l \quad \text{for} \quad l = \overline{1,p}, \quad \text{III}$$

Denote

$$C_1 = 2\widehat{\omega}_0, \quad \text{for} \quad \text{I}; \quad C_2 = 2\widehat{\omega}_0 - \frac{2\widehat{A}_1}{p}, \quad \text{for} \quad \text{II}; \quad C_3 = \widehat{\omega}_1, \quad \text{for} \quad \text{III}.$$

We need the following lemma to prove our main results in this paper.

Lemma 2. *Let* $0 \leq a_l < 1$ *for* $l = \overline{1,p}$. *If* $W_{I_1}([a_l,1]) = \widetilde{W}_{\tilde{I}_1}([a_l,1])$, *then there exists a large number* K_0 *such that*

$$\lambda_{1,n_{1,k}} - \widetilde{\lambda}_{1,\tilde{n}_{1,k}} = C_\nu, \quad \forall n_{1,k} \in \mathrm{I}_1, \quad \text{for} \quad \nu = 1,2,3, \tag{56}$$

By Theorems 1, 2 and Lemma 2, we prove Theorems 3 and 4, which are solutions to **IP2**.

Theorem 3. *Let* $0 \leq a_l < 1/2$ *for* $l = \overline{1,p}$ *and* $0 \leq \beta_1 = \max_{1\leq l\leq p}\{a_l\} < 1/2$. *If* $W_{I_1}([a_l,1]) = \widetilde{W}_{\tilde{I}_1}([a_l,1])$, *and there exist* $t_0 > 0$, $0 \leq \kappa_1 \leq 1$, *and* $\delta_1 > 0$ *such that*

$$N_{M_{1,0}}(t) \geq 2\beta_1\left\{\kappa_1\left[\frac{t}{\pi}+\frac{1}{2}\right] + (1-\kappa_1)\left(\left[\frac{t}{\pi}\right]+\frac{1}{2}\right)\right\} - 1 + \kappa_1 + O(t^{-\delta_1}) \text{ for I,} \tag{57}$$

$$N_{M_{1,0}}(t) \geq 2\beta_1\left\{\kappa_1\left[\frac{t}{\pi}+\frac{1}{2}\right] + (1-\kappa_1)\left(\left[\frac{t}{\pi}\right]+\frac{1}{2}\right) + \kappa_1\right\} + O(t^{-\delta_1}) \text{ for II, III} \tag{58}$$

for sufficiently large $t \geq t_0$, *and*

$$\lim_{t\to\infty} \frac{N_{M_{1,0}}(t)}{t} = \frac{2\beta_1}{\pi},$$

then

$$q_l(x) - \tilde{q}_l(x) \overset{a.e.}{=} C_\mu \quad \text{on} \quad [0,1], \quad \alpha_l = \tilde{\alpha}_l \quad \text{for} \quad l = \overline{1,p}, \quad \mu = 1,2,3, \tag{59}$$

Remark 1. *We can only study the partial inverse nodal problems for the cases* $0 \leq a_l < 1/2$, $l = \overline{1,p}$. *The general cases* $0 \leq a_l < 1$, $l = \overline{1,p}$ *require a separate investigation.*

Without any nodal data on the component $q_{i_0}(x)$, we have Theorem 4 from Theorems 3 and 1.

Theorem 4. *Let* $0 \leq a_l < 1/2$ *for* $l \neq i_0$ *and* $\beta_1 = \max_{l\neq i_0}\{a_l\}$. *Suppose that* $W_{I_1}([a_l,1]) = \widetilde{W}_{\tilde{I}_1}([a_l,1])$ *for* $l \neq i_0$, *and* $M_{m,0} = \widetilde{M}_{m,0}$ *for* $m \neq 1$, $M_{m,0}$ *for* $m = \overline{1,p}$ *satisfying the assumption* **(A)**, *and there exist* $t_0 > 0$, $0 \leq \kappa_\xi \leq 1$, $\delta_\xi > 0$, $\xi = 0,1$, *such that*

$$\begin{cases} N_{M_{1,0}}(t) \geq 2\beta_1\left\{\kappa_0\left[\frac{t}{\pi}+\frac{1}{2}\right] + (1-\kappa_0)\left(\left[\frac{t}{\pi}\right]+\frac{1}{2}\right) - 1 + \kappa_0 + O(t^{-\delta_0})\right\}, \\ \sum_{m=1}^{p} N_{M_{m,0}}(t) \geq 2\left\{\kappa_1\left[\frac{t}{\pi}+\frac{1}{2}\right] + (1-\kappa_1)\left(\left[\frac{t}{\pi}\right]+\frac{1}{2}\right) - 1 + \kappa_1 + O(t^{-\delta_1})\right\}, \end{cases} \text{for I;} \tag{60}$$

$$\begin{cases} N_{M_{1,0}}(t) \geq 2\beta_1\left\{\kappa_0\left[\frac{t}{\pi}+\frac{1}{2}\right] + (1-\kappa_0)\left(\left[\frac{t}{\pi}\right]+\frac{1}{2}\right) + \kappa_0 + O(t^{-\delta_0})\right\}, \\ \sum_{m=1}^{p} N_{M_{m,0}}(t) \geq 2\left\{\kappa_1\left[\frac{t}{\pi}+\frac{1}{2}\right] + (1-\kappa_1)\left(\left[\frac{t}{\pi}\right]+\frac{1}{2}\right) + \kappa_1 + O(t^{-\delta_1})\right\}, \end{cases} \text{for II, III} \tag{61}$$

for sufficiently large $t \geq t_0$;

$$\lim_{t\to\infty} \frac{N_{M_{1,0}}(t)}{t} = \frac{2\beta_1}{\pi}, \quad \lim_{t\to\infty} \frac{\sum_{m=1}^{p} N_{M_{m,0}}(t)}{t} = \frac{2}{\pi},$$

then (59) *holds.*

Next, we present proofs of Lemma 2 and Theorems 2–4.

Proof of Theorem 2. For each fixed $x \in [0,1]$ and $l = \overline{1,p}$, we choose $x^{j_k}_{l,1,n_{1,k}} \in X_{l,1,0}$ such that $\lim_{k\to\infty} x^{j_k}_{l,1,n_{1,k}} = x$. This implies $\lim_{k\to\infty} \alpha^{j_k}_{n_{1,k}} = x$. By (17)–(20) and the Riemann–Lebesgue lemma, we have

$$\begin{aligned}
f_{l,1,1}(x) :=& \lim_{k\to\infty} 2(n_{1,k} - \frac{1}{2})^2\pi^2\left(x^{j_k}_{l,1,n_{1,k}} - \alpha^{j_k}_{n_{1,k}}\right) \\
=& \lim_{k\to\infty}\left(\int_0^{\alpha^{j_k}_{n_{1,k}}} q_l(t)\mathrm{d}t - 2\omega_0\alpha^{j_k}_{n_{1,k}} + \mathrm{o}(1)\right) \\
=& \int_0^x q_l(t)\mathrm{d}t - 2\omega_0 x, \quad x \in [0,1] \quad \text{for} \quad \mathrm{I}; \qquad (62)\\
f_{l,1,2}(x) :=& \lim_{k\to\infty} 2(n_{1,k} - 1)^2\pi^2\left(x^{j_k}_{l,1,n_{1,k}} - \alpha^{j_k}_{n_{1,k}}\right) \\
=& \lim_{k\to\infty}\left(\int_0^{\alpha^{j_k}_{n_{1,k}}} q_l(t)\mathrm{d}t + \left(\frac{2A_1}{p} - 2\omega_0\right)\alpha^{j_k}_{n_{1,k}} + \mathrm{o}(1)\right) \\
=& -2\cot\alpha_l + \int_0^x q_l(t)\mathrm{d}t + \left(\frac{2A_1}{p} - 2\omega_0\right)x, \quad x \in [0,1] \quad \text{for} \quad \mathrm{II}; \qquad (63)\\
h_{l,1}(x) :=& \lim_{k\to\infty} (n_{1,k}\pi)\left(x^{j_k}_{l,1,n_{1,k}} - \alpha^{j_k}_{n_{1,k}}\right) \\
=& \lim_{k\to\infty}\left(d_1\alpha^{j_k}_{n_{1,k}} + \mathrm{o}(1)\right) \\
=& d_1 x, \quad x \in [0,1], \quad l = \overline{1,T}, \quad \text{for} \quad \mathrm{III}, \qquad (64)\\
f_{l,1,3}(x) :=& \lim_{k\to\infty} 2(n_{1,k}\pi)^2\left(x^{j_k}_{l,1,n_{1,k}} - \alpha^{j_k}_{n_{1,k}} - \frac{d_1\alpha^{j_k}_{n_{1,k}}}{n_{1,k}\pi}\right) \\
=& \lim_{k\to\infty}\left(\int_0^{\alpha^{j_k}_{n_{1,k}}} q_l(t)\mathrm{d}t - (\omega_1 - 2d_1^2)\alpha^{j_k}_{n_{1,k}} + \mathrm{o}(1)\right) \\
=& \int_0^x q_l(t)\mathrm{d}t - (\omega_1 - 2d_1^2)x, \quad x \in [0,1], \quad l = \overline{1,T}, \quad \text{for} \quad \mathrm{III}, \qquad (65)\\
f_{l,1,4}(x) :=& \lim_{k\to\infty} 2(n_{1,k}\pi)^2\left(x^{j_k}_{l,1,n_{1,k}} - \alpha^{j_k}_{n_{1,k}} - \frac{d_1\alpha^{j_k}_{n_{1,k}}}{n_{1,k}\pi}\right) \\
=& \lim_{k\to\infty}\left(-2\cot\alpha_l + \int_0^{\alpha^{j_k}_{n_{1,k}}} q_l(t)\mathrm{d}t - (\omega_1 - 2d_1^2)\alpha^{j_k}_{n_{1,k}} + \mathrm{o}(1)\right) \\
=& -2\cot\alpha_l + \int_0^x q_l(t)\mathrm{d}t - (\omega_1 - 2d_1^2)x, \quad x \in [0,1], \quad l = \overline{T+1,p}, \quad \text{for} \quad \mathrm{III}. \qquad (66)
\end{aligned}$$

By taking derivatives with respect to x in (62)–(66), we obtain

$$\begin{cases}
f'_{l,1,1}(x) \overset{a.e.}{=} q_l(x) - 2\omega_0, \quad x \in [0,1], \quad l = \overline{1,p}, & (67)\\
f'_{l,1,2}(x) \overset{a.e.}{=} q_l(x) + \left(\dfrac{2A_1}{p} - 2\omega_0\right), \quad x \in [0,1], \quad l = \overline{1,p}, & (68)\\
h'_{l,1}(x) = d_1 = \arcsin\sqrt{\dfrac{T}{p}}, & (69)\\
f'_{l,1,3}(x) = f'_{l,1,4}(x) \overset{a.e.}{=} q_l(x) - \omega_1 + 2d_1^2, \quad x \in [0,1], \quad l = \overline{1,p}. & (70)
\end{cases}$$

It follows from the assumption $X_{l,1,0} \overset{a.e.}{=} \widetilde{X}_{l,1,0}$ that

$$h_{l,1}(x) = \widetilde{h}_{l,1}(x) \quad \text{and} \quad f_{l,1,\nu}(x) = \widetilde{f}_{l,1,\nu}(x), \quad x \in [0,1], \quad l = \overline{1,p}, \quad \nu = \overline{1,4}. \qquad (71)$$

By (69), we find T by

$$T = p\sin^2 h'_{l,1}(x). \tag{72}$$

For cases II and III with $l = \overline{T+1,p}$, letting $x = 0$ in (63) and (66), we obtain

$$\alpha_l = \widetilde{\alpha}_l = \begin{cases} \operatorname{arccot}\dfrac{-f_{l,1,2}(0)}{2}, & l = \overline{1,p}, \quad \text{for} \quad \text{II}, \quad (73) \\ \operatorname{arccot}\dfrac{-f_{l,1,4}(0)}{2}, & l = \overline{T+1,p}, \quad \text{for} \quad \text{III}. \quad (74) \end{cases}$$

Furthermore, it follows from (71) that

$$f'_{l,1,\nu}(x) \overset{a.e.}{=} \widetilde{f}'_{l,1,\nu}(x), \quad x \in [0,1], \quad l = \overline{1,p}, \quad \nu = \overline{1,4}. \tag{75}$$

Consequently, (67)–(75) imply that

$$\widehat{q}_l(x) := q_l(x) - \widetilde{q}_l(x) \overset{a.e.}{=} C_\mu, \quad x \in [0,1], \quad l = \overline{1,p}, \quad \mu = 1,2,3. \tag{76}$$

This completes the proof of Theorem 2. □

The proof of Theorem 2 is constructive. We reconstruct the potential $q_l(x)$ up to a constant on the equilateral graph with the dense nodal subset $X_{l,1,0}$ on the l-th edge, $l = \overline{1,p}$, by the following algorithms:

Algorithm 1: For case I, reconstruct the potential $q_l(x)$ up to a constant by the following two steps:

(1) Find $f_{l,1,1}(x)$ by (62) for each $l = \overline{1,p}$.
(2) Reconstruct $q_l(x) - 2\omega_0$ on $(0,1)$ by (67).

Algorithm 2: For case II, reconstruct the potential $q_l(x)$ up to a constant by the following three steps:

(1) Find $f_{l,1,2}(x)$ by (63) for each $l = \overline{1,p}$.
(2) Reconstruct α_l for each $l = \overline{1,p}$ by (73), and then find A_1.
(3) Recover $q_l(x) - 2\omega_0$ on $(0,1)$ for each $l = \overline{1,p}$ by (68).

Algorithm 3: For case III, reconstruct the potential $q_l(x)$ up to a constant by the following four steps:

(1) Find $h_{l,1}(x)$ by (64) for each $l = \overline{1,p}$; reconstruct T by (72).
(2) Find $f_{l,1,3}(x)$ by (65) for each $l = \overline{1,T}$ and find $f_{l,1,4}(x)$ by (66) for each $l = \overline{T+1,p}$.
(3) Reconstruct α_l for each $l = \overline{T+1,p}$ by (74), and then find A_2.
(4) Recover $q_l(x) - \omega_1$ on $(0,1)$ for each $l = \overline{1,p}$ by (70).

Proof of Lemma 2. By suitably modifying the proof of Theorem 2, we obtain

$$q_l(x) - \widetilde{q}_l(x) \overset{a.e.}{=} C_\nu, \quad x \in [a_l, 1], \quad l = \overline{1,p}, \quad \nu = 1,2,3. \tag{77}$$

It follows from (17)–(20) that as $k \to \infty$

$$L_{l,1,n_{1,k}} := x_{l,1,n_{1,k}}^{j_k+r_k} - x_{l,1,n_{1,k}}^{j_k} = \begin{cases} \dfrac{r_k}{n_{1,k} - \dfrac{1}{2}} + o\left(\dfrac{1}{n_{1,k}^2}\right), & \text{for I}, \\ \dfrac{r_k}{n_{1,k} - 1} + o\left(\dfrac{1}{n_{1,k}^2}\right), & \text{for II}, \\ \dfrac{r_k}{n_{1,k}} + \dfrac{d_1 r_k}{n_{1,k}^2 \pi} + o\left(\dfrac{1}{n_{1,k}^2}\right), & \text{for III}, \ l = \overline{1,T}, \\ \dfrac{r_k}{n_{1,k}} + \dfrac{r_k}{n_{1,k}^2 \pi} + o\left(\dfrac{1}{n_{1,k}^2}\right), & \text{for III}, \ l = \overline{T+1,p}. \end{cases} \tag{78}$$

If the problems B and $\widetilde{B}$ belong to the same subcase in (7), say, case I, it follows from the first formula of (78) and the assumption that

$$L_{l,1,n_{1,k}} = \frac{r_k}{n_{1,k} - \frac{1}{2}} + o\left(\frac{1}{n_{1,k}^2}\right) = \frac{r_k}{\widetilde{n}_{1,k} - \frac{1}{2}} + o\left(\frac{1}{\widetilde{n}_{1,k}^2}\right) = \widetilde{L}_{l,1,n_{1,k}}, \quad \text{for} \quad k \gg 1.$$

Without loss of generality, we assume $\widetilde{n}_{1,k} \geq n_{1,k}$ here and below. This implies

$$\frac{r_k(n_{1,k} - \widetilde{n}_{1,k})}{\left(n_{1,k} - \frac{1}{2}\right)\left(\widetilde{n}_{1,k} - \frac{1}{2}\right)} = o\left(\frac{1}{n_{1,k}^2}\right) \quad \text{for} \quad k \gg 1. \tag{79}$$

It follows from (79) that

$$n_{1,k} = \widetilde{n}_{1,k} \quad \text{for} \quad k \gg 1. \tag{80}$$

If the problem B belongs to case II, while the problem $\widetilde{B}$ belongs to case III, then it follows from the second and third formulae of (78) and the assumption that

$$\frac{r_k}{n_{1,k} - 1} + o\left(\frac{1}{n_{1,k}^2}\right) = \frac{r_k}{\widetilde{n}_{1,k}} + \frac{d_1 r_k}{\widetilde{n}_{1,k}^2 \pi} + o\left(\frac{1}{\widetilde{n}_{1,k}^2}\right) \quad \text{for} \quad k \gg 1. \tag{81}$$

By virtue of (81), this yields

$$\frac{r_k}{n_{1,k} - 1} + o\left(\frac{1}{n_{1,k}^2}\right) = \frac{r_k}{\widetilde{n}_{1,k}} + \frac{d_1 r_k}{\widetilde{n}_{1,k}^2 \pi} + o\left(\frac{1}{\widetilde{n}_{1,k}^2}\right) \quad \text{for} \quad k \gg 1. \tag{82}$$

In particular, we have

$$\frac{r_k \widetilde{n}_{1,k}}{n_{1,k} - 1} - r_k - \frac{d_1 r_k}{\widetilde{n}_{1,k} \pi} = o\left(\frac{\widetilde{n}_{1,k}}{n_{1,k}^2}\right) + o\left(\frac{1}{\widetilde{n}_{1,k}}\right),$$

and hence

$$\lim_{k \to \infty} \frac{\widetilde{n}_{1,k}}{n_{1,k}} = 1.$$

By (82), we obtain

$$\frac{\widetilde{n}_{1,k} - n_{1,k} + 1 - \frac{d_1}{\pi}}{\widetilde{n}_{1,k}} = o\left(\frac{1}{n_{1,k}}\right) \quad \text{for} \quad k \gg 1.$$

This implies

$$\widetilde{n}_{1,k} - n_{1,k} + 1 - \frac{d_1}{\pi} = 0 \quad \text{for} \quad k \gg 1, \tag{83}$$

which is impossible by $0 < \frac{1}{\pi} \arcsin \sqrt{\frac{T}{p}} < \frac{1}{2}$ and $\widetilde{n}_{1,k} - n_{1,k} + 1 \geq 1$. Therefore, the problems B and $\widetilde{B}$ belong to the same subcase, and other cases can be treated similarly. This implies that (80) is valid for $k \gg 1$. Next, we only consider the problems B and $\widetilde{B}$ belonging to case III and $l \in \{T+1, \cdots, p\}$. For each $l = \overline{T+1, p}$, consider two Dirichlet boundary value problems defined on the interval $[x_{l,1,n_{1,k}}^{j_k}, x_{l,1,n_{1,k}}^{j_k + r_k}]$,

$$\begin{cases} -\varphi_l''(x,\lambda_{1,n_{1,k}}) + q_l(x)\varphi_l(x,\lambda_{1,n_{1,k}}) = \lambda_{1,n_{1,k}}\varphi_l(x,\lambda_{1,n_{1,k}}), & (84) \\ \varphi_l(x_{l,1,n_{1,k}}^{j_k},\lambda_{1,n_{1,k}}) = \varphi_l(x_{l,1,n_{1,k}}^{j_k+r_k},\lambda_{1,n_{1,k}}) = 0, & (85) \end{cases}$$

and

$$\begin{cases} -\widetilde{\varphi}_l''(x,\widetilde{\lambda}_{1,\widetilde{n}_{1,k}}) + \widetilde{q}_l(x)\widetilde{\varphi}_l(x,\widetilde{\lambda}_{1,\widetilde{n}_{1,k}}) = \widetilde{\lambda}_{1,\widetilde{n}_{1,k}}\widetilde{\varphi}_l(x,\widetilde{\lambda}_{1,\widetilde{n}_{1,k}}), & (86) \\ \widetilde{\varphi}_l(x_{l,1,n_{1,k}}^{j_k},\widetilde{\lambda}_{1,\widetilde{n}_{1,k}}) = \widetilde{\varphi}_l(x_{l,1,n_{1,k}}^{j_k+r_k},\widetilde{\lambda}_{1,\widetilde{n}_{1,k}}) = 0. & (87) \end{cases}$$

It follows from the first formula of (13) and (20) that

$$\rho_{1,n_{1,k}} x_{l,1,n_{1,k}}^{j_k} = \left(j_k - \frac{1}{2}\right)(d_1+\pi) + \frac{d_1}{2n_{1,k}\pi} + \mathrm{o}\left(\frac{1}{n_{1,k}}\right), \tag{88}$$

$$b_1 := -(2j_k-1)(d_1+\pi)d_1 - 2\pi\cot\alpha_l + \alpha_{n_{1,k}}^{j_k}(\omega_1+d_1^2) + \int_0^{\alpha_{n_{1,k}}^{j_k}} q_l(t)\mathrm{d}t;$$

$$\rho_{1,n_{1,k}} x_{l,1,n_{1,k}}^{j_k+r_k} = \left(j_k + r_k - \frac{1}{2}\right)(d_1+\pi) + \frac{d_2}{2n_{1,k}\pi} + \mathrm{o}\left(\frac{1}{n_{1,k}}\right), \tag{89}$$

$$b_2 := -(2j_k+r_k-1)(d_1+\pi)d_1 - 2\pi\cot\alpha_l + \alpha_{n_{1,k}}^{j_k}(\omega_1+d_1^2) + \int_0^{\alpha_{1,n_{1,k}}^{j_k+r_k}} q_l(t)\mathrm{d}t;$$

$$\widetilde{\rho}_{1,\widetilde{n}_{1,k}} x_{l,1,n_{1,k}}^{j_k} = \left(\widetilde{j}_k - \frac{1}{2}\right)(d_1+\pi) + \frac{\widetilde{d}_1}{2\widetilde{n}_{1,k}\pi} + \mathrm{o}\left(\frac{1}{n_{1,k}}\right), \tag{90}$$

$$\widetilde{b}_1 := -(2\widetilde{j}_k-1)(d_1+\pi)d_1 - 2\pi\cot\widetilde{\alpha}_l + \alpha_{n_{1,k}}^{j_k}(\widetilde{\omega}_1+d_1^2) + \int_0^{\alpha_{n_{1,k}}^{j_k}} \widetilde{q}_l(t)\mathrm{d}t;$$

$$\widetilde{\rho}_{1,n_{1,k}} x_{l,1,n_{1,k}}^{j_k+r_k} = \left(\widetilde{j}_k + r_k - \frac{1}{2}\right)(d_1+\pi) + \frac{\widetilde{d}_2}{2\widetilde{n}_{1,k}\pi} + \mathrm{o}\left(\frac{1}{n_{1,k}}\right), \tag{91}$$

$$\widetilde{b}_2 := -(2\widetilde{j}_k+r_k-1)(d_1+\pi)d_1 - 2\pi\cot\widetilde{\alpha}_l + \alpha_{n_{1,k}}^{j_k}(\widetilde{\omega}_1+d_1^2) + \int_0^{\alpha_{1,n_{1,k}}^{j_k+r_k}} \widetilde{q}_l(t)\mathrm{d}t.$$

By (84)–(87) and the integrations, we easily obtain

$$\int_{x_{l,1,n_{1,k}}^{j_k}}^{x_{l,1,n_{1,k}}^{j_k+r_k}} ((q_l(x)-\widetilde{q}_l(x)) - (\lambda_{1,n_{1,k}} - \widetilde{\lambda}_{1,\widetilde{n}_{1,k}}))\varphi_l(x,\lambda_{1,n_{1,k}})\widetilde{\varphi}_l(x,\widetilde{\lambda}_{1,\widetilde{n}_{1,k}})\mathrm{d}x = 0. \tag{92}$$

By virtue of (92) and $q_l(x) - \widetilde{q}_l(x) \overset{a.e.}{=} C_v$ on $[a_l, 1]$, we have

$$(C_v - (\lambda_{1,n_{1,k}} - \widetilde{\lambda}_{1,\widetilde{n}_{1,k}})) \int_{x_{l,1,n_{1,k}}^{j_k}}^{x_{l,1,n_{1,k}}^{j_k+r_k}} \varphi_l(x,\lambda_{1,n_{1,k}})\widetilde{\varphi}_l(x,\widetilde{\lambda}_{1,\widetilde{n}_{1,k}})\mathrm{d}x = 0. \tag{93}$$

On the other hand, it follows from (5)

$$\varphi_l(x,\lambda_{1,n_{1,k}})\widetilde{\varphi}_l(x,\widetilde{\lambda}_{1,\widetilde{n}_{1,k}}) = \sin\alpha_l \sin\widetilde{\alpha}_l \cos\rho_{1,n_{1,k}}x \cos\widetilde{\rho}_{1,\widetilde{n}_{1,k}}x + \mathrm{O}\left(\frac{1}{n_{1,k}}\right). \tag{94}$$

By virtue of (94), this yields

$$\int_{x_{l,1,n_{1,k}}^{j_k}}^{x_{l,1,n_{1,k}}^{j_k+r_k}} \varphi_l(x,\lambda_{1,n_{1,k}})\widetilde{\varphi}_l(x,\widetilde{\lambda}_{1,\widetilde{n}_{1,k}})\mathrm{d}x$$

$$
\begin{aligned}
&= \sin\alpha_l \sin\widetilde{\alpha}_l \int_{x_{l,1,n_{1,k}}^{j_k}}^{x_{l,1,n_{1,k}}^{j_k+r_k}} \cos\rho_{1,n_{1,k}} x \cos\widetilde{\rho}_{1,\widetilde{n}_{1,k}} x \mathrm{d}x + \mathrm{O}\left(\frac{1}{n_{1,k}^2}\right) \\
&= \frac{\sin\alpha_l \sin\widetilde{\alpha}_l}{2} \int_{x_{l,1,n_{1,k}}^{j_k}}^{x_{l,1,n_{1,k}}^{j_k+r_k}} \Big(\cos(\rho_{1,n_{1,k}} + \widetilde{\rho}_{1,\widetilde{n}_{1,k}})x + \cos(\rho_{1,n_{1,k}} - \widetilde{\rho}_{1,\widetilde{n}_{1,k}})x\Big) \mathrm{d}x + \mathrm{O}\left(\frac{1}{n_{1,k}^2}\right).
\end{aligned} \tag{95}
$$

For $k \gg 1$, (80) shows that

$$
\begin{aligned}
&\frac{1}{2L_{l,1,n_{1,k}}} \int_{x_{l,1,n_{1,k}}^{j_k}}^{x_{l,1,n_{1,k}}^{j_k+r_k}} \cos(\rho_{1,n_{1,k}} + \widetilde{\rho}_{1,\widetilde{n}_{1,k}}) x \mathrm{d}x \\
&= \frac{1}{2(\rho_{1,n_{1,k}} + \widetilde{\rho}_{1,n_{1,k}}) L_{l,1,n_{1,k}}} \Big(\sin(\rho_{1,n_{1,k}} + \widetilde{\rho}_{1,n_{1,k}}) x_{l,1,n_{1,k}}^{j_k+r_k} - \sin(\rho_{1,n_{1,k}} + \widetilde{\rho}_{1,n_{1,k}}) x_{l,1,n_{1,k}}^{j_k}\Big) \\
&= \frac{(-1)^{j_k+\widetilde{j}_k-1}}{2r_k(d_1+\pi)(1+\frac{1}{2n_{1,k}}) + \mathrm{o}\left(\frac{1}{n_{1,k}}\right)} \left(\sin\left((j_k+\widetilde{j}_k+2r_k-1)d_1 + \frac{b_1+\widetilde{b}_1}{2n_{1,k}\pi} + \mathrm{o}\left(\frac{1}{n_{1,k}}\right)\right)\right. \\
&\quad \left. - \sin\left((j_k+\widetilde{j}_k-1)d_1 + \frac{b_2+\widetilde{b}_2}{2n_{1,k}\pi} + \mathrm{o}\left(\frac{1}{n_{1,k}}\right)\right)\right) \\
&= \frac{(-1)^{j_k+\widetilde{j}_k-1}\Big(\sin(j_k+\widetilde{j}_k+2r_k-1)d_1 - \sin(j_k+\widetilde{j}_k-1)d_1\Big)}{2r_k(d_1+\pi)} + \mathrm{O}\left(\frac{1}{n_{1,k}}\right).
\end{aligned} \tag{96}
$$

It follows from the first formula of (13) and (80) that

$$
\begin{aligned}
&\frac{1}{2L_{l,1,n_{1,k}}} \int_{x_{l,1,n_{1,k}}^{j_k}}^{x_{l,1,n_{1,k}}^{j_k+r_k}} \cos(\rho_{1,n_{1,k}} - \widetilde{\rho}_{1,\widetilde{n}_{1,k}}) x \mathrm{d}x \\
&= \frac{1}{2L_{l,1,n_{1,k}}} \int_{x_{l,1,n_{1,k}}^{j_k}}^{x_{l,1,n_{1,k}}^{j_k+r_k}} \cos\left(\frac{\omega_1 x}{n_{1,k}} + \mathrm{o}\left(\frac{1}{n_{1,k}}\right)\right) \mathrm{d}x \\
&= \frac{1}{2L_{l,1,n_{1,k}}} \int_{x_{l,1,n_{1,k}}^{j_k}}^{x_{l,1,n_{1,k}}^{j_k+r_k}} \left[1 - 2\sin^2\left(\frac{\omega_1 x}{2n_{1,k}} + \mathrm{o}\left(\frac{1}{n_{1,k}}\right)\right)\right] \mathrm{d}x \\
&= \frac{1}{2} + \mathrm{O}\left(\frac{1}{n_{1,k}^2}\right).
\end{aligned} \tag{97}
$$

By (80), (96) and (97), there exists a sufficiently large constant K_0 such that

$$
\int_{x_{l,1,n_{1,k}}^{j_k}}^{x_{l,1,n_{1,k}}^{j_k+r_k}} \cos(\rho_{1,n_{1,k}} - \widetilde{\rho}_{1,\widetilde{n}_{1,k}}) x \mathrm{d}x > \left| \int_{x_{l,1,n_{1,k}}^{j_k}}^{x_{l,1,n_{1,k}}^{j_k+r_k}} \cos(\rho_{1,n_{1,k}} + \widetilde{\rho}_{1,\widetilde{n}_{1,k}}) x \mathrm{d}x \right| \tag{98}
$$

for all $k \geq K_0$. It follows from (92) and (98) that

$$
\begin{aligned}
&\left| \int_{x_{l,1,n_{1,k}}^{j_k}}^{x_{l,1,n_{1,k}}^{j_k+r_k}} \varphi(x,\lambda_{n_k}) \widetilde{\varphi}(x,\widetilde{\lambda}_{\widetilde{n}_k}) \mathrm{d}x \right| \\
&\geq \sin\alpha \sin\widetilde{\alpha} \left(\int_{x_{l,1,n_{1,k}}^{j_k}}^{x_{l,1,n_{1,k}}^{j_k+r_k}} \cos(\rho_{1,n_{1,k}} - \widetilde{\rho}_{1,\widetilde{n}_{1,k}}) x \mathrm{d}x - \left| \int_{x_{l,1,n_{1,k}}^{j_k}}^{x_{l,1,n_{1,k}}^{j_k+r_k}} \cos(\rho_{1,n_{1,k}} + \widetilde{\rho}_{1,\widetilde{n}_{1,k}}) x \mathrm{d}x \right| + \mathrm{O}\left(\frac{1}{n_{1,k}^2}\right)\right) \\
&> 0.
\end{aligned} \tag{99}
$$

Therefore, (93) and (99) imply that

$$
\widehat{\lambda}_{n_{1,k}} = \lambda_{n_{1,k}} - \widetilde{\lambda}_{\widetilde{n}_{1,k}} = C_\nu
$$

for all $n_{1,k} \in I_1$. This completes the proof of Lemma 2. $\square$

Next, we prove Theorem 3.

Proof of Theorem 3. By the assumption of Theorem 3 together with Lemma 2, we have

$$\begin{cases} q_l(x) - \widetilde{q}_l(x) \overset{a.e.}{=} C_\mu, \quad x \in [a_l, 1], \quad l = \overline{1,p}, \quad \mu = 1,2,3, & (100) \\ \lambda_{1,n_{1,k}} - \widetilde{\lambda}_{1,\widetilde{n}_{1,k}} = C_\mu, \quad \mu = 1,2,3, \quad \forall n_{1,k} \in \mathrm{I}_1. & (101) \end{cases}$$

Let $\widetilde{q}_{0,l}(x) := \widetilde{q}_l(x) + C_\mu$, $\widehat{q}_{0,l}(x) := q_l(x) - \widetilde{q}_{0,l}(x)$, and $\widetilde{\varphi}_{0,l}(x,\lambda)$ be the solution of

$$\begin{cases} u''(x,\lambda) + (\lambda - \widetilde{q}_{0,l}(x))u(x,\lambda) = 0, \quad 0 < x < 1, \\ u(0,\lambda) = \sin\alpha_l, \ u'(0,\lambda) = -\cos\alpha_l. \end{cases} \tag{102}$$

By a shift of the spectrum to the constant C_μ, then (100) and (101) imply

$$\begin{cases} \widehat{q}_{0,l}(x) \overset{a.e.}{=} 0 \quad \text{on} \quad [a_l, 1], \quad l = \overline{1,p}, & (103) \\ \lambda_{1,n_{1,k}} - \widetilde{\lambda}_{01,\widetilde{n}_{1,k}} = 0, \quad \forall n_{1,k} \in I_1. & (104) \end{cases}$$

Next, we prove

$$\widehat{q}_{0,l}(x) \overset{a.e.}{=} 0 \quad \text{on} \quad [0, a_l], \quad \text{and} \quad \alpha_l = \widetilde{\alpha}_l, \quad l = \overline{1,p}.$$

For each $\lambda_{1,n_{1,k}}$, (103) and (104) show that

$$< \varphi_l, \widetilde{\varphi}_{0,l} > (a_l, \lambda_{1,n_{1,k}}) = 0, \quad \forall\, n_{1,k} \in I_1. \tag{105}$$

It follows from (4) and (5) that

$$|< \varphi_l, \widetilde{\varphi}_{0,l} > (a_l, \lambda)| = \begin{cases} O\left(\dfrac{e^{2a_l\tau}}{|\rho|^2}\right), \quad \text{if} \quad \alpha_l = 0, & (106) \\ O(e^{2a_l\tau}), \quad \text{if} \quad \alpha_l \neq 0, & (107) \end{cases}$$

for $|\lambda| \to \infty$. Consequently, it follows from (4), (5) and (39) that

$$|< \varphi_l, \widetilde{\varphi}_{0,l} > (a_l, \lambda)| = \begin{cases} o\left(\dfrac{e^{2a_l\tau}}{|\rho|^2}\right), \quad \text{if} \quad \alpha_l = 0, & (108) \\ o(e^{2a_l\tau}), \quad \text{if} \quad \alpha_l \neq 0, & (109) \end{cases}$$

for $|\lambda| \to \infty$ in any sector $\varepsilon_0 < \arg\lambda < \pi - \varepsilon_0$. Define the function $K_{l,1}(\lambda)$ by

$$K_{l,1}(\lambda) := \frac{< \varphi_l, \widetilde{\varphi}_{0,l} > (a_l, \lambda)}{F_1(\lambda)}, \quad l = \overline{1,p},$$

Therefore, (105) together with the assumption on $M_{1,0}$ show that $K_{l,1}(\lambda)$ is an entire function in λ. By Levinson's estimate (see [44]), the first formula of (8), or (10), or (13) and (58) imply that there exists a constant c_1 such that

$$\frac{1}{|F_1(\lambda)|} = O\left(e^{-2\alpha_1\tau + \varepsilon\sqrt{|\lambda|}}\right), \quad \forall\lambda \in D_{1,c_1} \tag{110}$$

for sufficiently large $|\lambda|$. Thus (106), (107) and (110) for $m = 1$ show that

$$|K_{l,1}(\lambda)| = O\left(e^{-2(\alpha_1 - a_l)\tau + 2\varepsilon\sqrt{|\lambda|}}\right), \quad \forall\lambda \in D_{1,c_1} \tag{111}$$

for sufficiently large $|\lambda|$. Consequently, it follows from $a_l \leq \alpha_1 \leq \frac{1}{2}$, (111) and the maximum modulus principle that the entire function $K_{l,1}(\lambda)$ is of the zero-exponential type, and then for arbitrary $\varepsilon > 0$,

$$|K_{l,1}(\lambda)| \leq c e^{2\varepsilon\sqrt{|\lambda|}}, \quad \lambda \in \mathbb{C} \tag{112}$$

for sufficiently large $|\lambda|$. By calculations, we have

$$\begin{cases} |F_1(iy)| \geq c\dfrac{e^{2\alpha_1\sqrt{|y|/2}}}{|y|^{\alpha_1(1+\kappa_0)}}, & \text{for} \quad \text{I}; \\ |F_1(iy)| \geq c|y|^{\alpha_1(1-\kappa_0)}e^{2\alpha_1\sqrt{|y|/2}}, & \text{for} \quad \text{II}, \quad \text{III} \end{cases} \tag{113}$$

for a $y \in \mathrm{R}^+$ that is sufficiently large. It follows from (108), (109) and (113) that

$$|K_{l,1}(iy)| = o(1)$$

for a sufficiently large $y > 0$. This implies

$$\lim_{y \to \infty} K_{l,1}(iy) = 0. \tag{114}$$

By the Phragmén-Lindelöf-type result in [20] together with (112) and (114) again, we obtain

$$K_{l,1}(\lambda) \equiv 0, \quad \lambda \in \mathbb{C}. \tag{115}$$

It follows from (115) that

$$< \varphi_l, \widetilde{\varphi}_{0,l} > (a_l, \lambda) = 0, \quad \forall \lambda \in \mathbb{C}, \quad l = \overline{1,p}.$$

Consequently,

$$m_l(a_l, \lambda) = \widetilde{m}_{0,l}(a_l, \lambda), \quad \forall \lambda \in \mathbb{C}, \quad l = \overline{1,p}. \tag{116}$$

By (116) together with Marchenko's result in [21], we obtain

$$\widehat{q}_{0,l}(x) \overset{a.e.}{=} 0 \quad \text{on} \quad [0, a_l], \quad \text{and} \quad \alpha_l = \widetilde{\alpha}_l, \quad l = \overline{1,p}. \tag{117}$$

Hence, (103) and (117) show that (59) holds. The proof of Theorem 3 is completed. □

Proof of Theorem 4. We use the same symbols as these in Theorem 3. Applying the same arguments as the proof of Theorem 3, we have

$$\begin{cases} \widehat{q}_{0,l}(x) \overset{a.e.}{=} 0 \quad \text{on} \quad [0,1], \quad \text{and} \quad \alpha_l = \widetilde{\alpha}_l, \quad l \neq i_0, & (118) \\ \lambda_{1,n_{1,k}} - \widetilde{\lambda}_{01,\widetilde{n}_{1,k}} = 0, \quad \forall n_{1,k} \in I_1. & (119) \end{cases}$$

Next, we prove

$$\widehat{q}_{0,i_0}(x) \overset{a.e.}{=} 0 \quad \text{on} \quad [0,1], \quad \text{and} \quad \alpha_{i_0} = \widetilde{\alpha}_{i_0}. \tag{120}$$

It follows from (1) and (102) for $l = i_0$ that

$$< \varphi_{i_0}, \widetilde{\varphi}_{0,i_0} > (1, \lambda_{m,n_{m,k}}) = \int_0^1 \widehat{q}_{0,i_0}(x)\varphi_{i_0}(x, \lambda_{m,n_{m,k}})\widetilde{\varphi}_{0,i_0}(x, \lambda_{m,n_{m,k}})dx - \sin\widehat{\alpha}_{i_0}. \tag{121}$$

Similar to the argument as the proof of Theorem 1, one can complete the remaining proof of Theorem 4. □

Author Contributions: Conceptualization, X.-B.W.; Formal analysis, Y.-P.W.; Resources, Y.-H.C.; Writing—original draft, Y.-P.W.; Writing—review & editing, X.-B.W. All authors have read and agreed to the published version of the manuscript.

Funding: This research received no external funding.

Institutional Review Board Statement: Not applicable.

Informed Consent Statement: Not applicable.

Data Availability Statement: Not applicable.

Conflicts of Interest: The authors declare no conflict of interest.

References

1. Belishev, M.I. Boundary spectral inverse problem on a class of graphs(trees) by the BC-method. *Inverse Probl.* **2004**, *20*, 647–672. [CrossRef]
2. Bondarenko, N.P.; Shieh, C.-T. Partial inverse problems for Sturm-Liouville operators on trees. *Proc. R. Soc. Edinburg Sect. Math.* **2017**, *147*, 917–933. [CrossRef]
3. Bondarenko, N.P. A partial inverse problem for the Sturm-Liouville operator on a star-shaped graph. *Anal. Math. Phy.* **2018**, *8*, 155–168. [CrossRef]
4. Exner, P.; Keating, J.P.; Kuchment, P.; Sunada, T.; Teplyaev, A. Analysis on Graphs and Its Applications. In *Proceedings of Symposia in Pure Mathematics*; American Mathematical Soc.: Providence, RI, USA, 2008; Volume 77.
5. Kuchment, P. Graph models for waves in thin structures. *Waves Random Media* **2002**, *12*, 1–24. [CrossRef]
6. Kuchment, P. Quantum graphs I. Some basic structures. *Waves Random Media* **2004**, *14*, 107–128. [CrossRef]
7. Kuchment, P. Quantum graphs II. Some spectral properties of quantum and combinatorial graphs. *J. Phys. A* **2005**, *38*, 4887–4900. [CrossRef]
8. Kuchment, P.; Post, O. On the spectra of carbon nano-structures. *Comm. Math. Phys.* **2007**, *275*, 805–826. [CrossRef]
9. Kurasov, P.; Nowaczyk, M. Inverse spectral problem for quantum graphs. *J. Phys. A* **2005**, *38*, 4901–4915. [CrossRef]
10. Montrol, E. Quantum theory on a network. *J. Math. Phys.* **1970**, *11*, 635–648.
11. Nizhnik, L.P. Inverse eigenvalue problems for nonlocal Sturm-Liouville operators on a star graph. *Methods Funct. Anal. Topol.* **2012**, *18*, 68–78.
12. Pivovarchik, V. Inverse problem for the Sturm-Liouville equation on a simple graph. *SIAM J. Math. Anal.* **2000**, *32*, 801–819. [CrossRef]
13. Pivovarchik, V. Inverse problem for the Sturm-Liouville equation on a star-shaped graph. *Math. Nachr.* **2007**, *280*, 1595–1619. [CrossRef]
14. Wang, Y.P.; Bondarenko, N.; Shieh, C.-T. The inverse problem for differential pencils on a star-shaped graph with mixed spectral data. *Sci. China Math.* **2020**, *63*, 1559–1570. [CrossRef]
15. Wang, Y.P.; Yurko, V.A. On the inverse nodal problems for discontinuous Sturm-Liouville operators. *J. Differ. Equ.* **2016**, *260*, 4086–4109. [CrossRef]
16. Yang, C.F. Inverse spectral problems for the Sturm-Liouville operator on a d-star graph. *J. Math. Anal. Appl.* **2010**, *365*, 742–749. [CrossRef]
17. Yurko, V.A. Inverse spectral problems for Sturm-Liouville operators on graphs. *Inverse Probl.* **2005**, *21*, 1075–1086. [CrossRef]
18. Browne, P.J.; Sleeman, B.D. Inverse nodal problem for Sturm-Liouville equation with eigenparameter dependent boundary conditions. *Inverse Probl.* **1996**, *12*, 377–381. [CrossRef]
19. Hald, O.H.; McLaughlin, J.R. Solutions of inverse nodal problems. *Inverse Probl.* **1989**, *5*, 307–347. [CrossRef]
20. Law, C.K.; Shen, C.-L.; Yang, C.F. The inverse nodal problem on the smoothness of the potential function. *Inverse Probl.* **1999**, *15*, 253–263. [CrossRef]
21. Law, C.K.; Yang, C.F. Reconstructing the potential function and its derivatives using nodal data. *Inverse Probl.* **1998**, *14*, 299–312. [CrossRef]
22. McLaughlin, J.R. Inverse spectral theory using nodal points as data-a uniqueness result. *J. Differ. Equ.* **1988**, *73*, 354–362. [CrossRef]
23. Shieh, C.-T.; Yurko, V.A. Inverse nodal and inverse spectral problems for discontinuous boundary value problems. *J. Math. Anal. Appl.* **2008**, *347*, 266–272. [CrossRef]
24. Cheng, Y.H.; Law, C.K.; Tsay, J. Remarks on a new inverse nodal problem. *J. Math. Anal. Appl.* **2000**, *248*, 145–155. [CrossRef]
25. Yang, C.F. Solution to open problems of Yang concerning inverse nodal problems. *Isr. J. Math.* **2014**, *204*, 283–298. [CrossRef]
26. Yang, X.F. A new inverse nodal problem. *J. Differ. Equ.* **2001**, *169* 633–653. [CrossRef]
27. Keskin, B.; Ozkan, A.S. Inverse nodal problems for Dirac-type integro-differential operators. *J. Differ. Equ.* **2017**, *263*, 8838–8847. [CrossRef]
28. Kuryshova, Y.V.; Shieh, C.-T. An inverse nodal problem for integro-differential operators. *J. Inverse Ill-Posed Probl.* **2010**, *18*, 357–369. [CrossRef]
29. Guo, Y.; Wei, G. Inverse problems: Dense nodal subset on an interior subinterval. *J. Differ. Equ.* **2013**, *255*, 2002–2017. [CrossRef]

30. Guo, Y.; Wei, G. The sharp conditions of the uniqueness for inverse nodal problems. *J. Differ. Equ.* **2019**, *266*, 4432–4449. [CrossRef]
31. Wang, Y.P.; Shieh, C.-T. Inverse problems for Sturm-Liouville operators on a compact equilateral graph by partial nodal data. *Math. Meth. Appl. Sci.* **2021**, *44*, 693–704. [CrossRef]
32. Gesztesy, F.; Simon, B. Inverse spectral analysis with partial information on the potential II: The case of discrete spectrum. *Trans. Am. Math. Soc.* **2000**, *352*, 2765–2787. [CrossRef]
33. Graef, J.R.; Heidarkhani, S.; Kong, L. Nontrivial solutions for systems of Sturm-Liouville boundary value problems. *Differ. Equ. Appl.* **2014**, *6*, 255–265. [CrossRef]
34. Barilla, D.; Bohner, M.; Heidarkhani, S.; Moradi, S. Existence results for dynamic Sturm-Liouville boundary value problems via variational methods. *Appl. Math. Comput.* **2021**, *409*, 125614. [CrossRef]
35. Law, C.K.; Pivovarchik, V. Characteristic functions of quantum graphs. *J. Phys. A Math. Theor.* **2009**, *42*, 035302. [CrossRef]
36. Luo, Y.C.; Jatulan, E.; Law, C.K. Dispersion relations of periodic quantum graphs associated with Archimedean tilings (I). *J. Phys A Math. Theor.* **2019**, *52*, 165201. [CrossRef]
37. Wang, Y.P.; Shieh, C.-T. Inverse problems for Sturm-Liouville operators on a star-shaped graph with mixed spectral data. *Appl. Anal.* **2020**, *99*, 2371–2380. [CrossRef]
38. Horváth, M. On the inverse spectral theory of Schrödinger and Dirac operators. *Trans. Am. Math. Soc.* **2001**, *353*, 4155–4171. [CrossRef]
39. Currie, S.; Watson, B.A. Inverse nodal problems for Sturm-Liouville equations on graphs. *Inverse Probl.* **2007**, *23*, 2029–2040. [CrossRef]
40. Yurko, V.A. Inverse nodal problems for Sturm-Liouville operators on star-type graphs. *J. Inverse-Ill-Posed Probl.* **2008**, *16*, 715–722. [CrossRef]
41. Cheng, Y.H. Reconstruction of the Sturm-Liouville operator on a p-star graph with nodal data. *Rocky Mt. J. Math.* **2012**, *42*, 1431–1446. [CrossRef]
42. Freiling, G.; Yurko, V. *Inverse Sturm-Liouville Problems and Their Applications*; Nova Science Publishers: Huntington, NY, USA, 2001.
43. Marchenko, V.A. *Sturm-Liouville Operators and Their Applications*; Naukova Dumka, Kiev (1977) (Russian); English transl.; Birkhauser: Basel, Switzerland, 1986.
44. Levinson, N. *Gap and Density Theorems*; AMS Coll. Publ.: New York, NY, USA, 1940.

mathematics

Article

On the Asymptotic of Solutions of Odd-Order Two-Term Differential Equations

Yaudat T. Sultanaev [1,2,†], Nur F. Valeev [3,†] and Elvira A. Nazirova [4,*,†]

[1] Faculty of Physics and Mathematics, Bashkir State Pedagogical University n. a. M. Akmulla, Ufa 450008, Russia; sultanaevyt@gmail.com
[2] Center for Applied and Fundamental Mathematics of Moscow State University, Moscow 119991, Russia
[3] Institute of Mathematics with Computing Centre—Subdivision of the Ufa Federal Research Centre of the Russian Academy of Sciences, Ufa 450008, Russia; valeevnf@yandex.ru
[4] Institute of Informatics, Mathematics and Robotics, Ufa University of Science and Technology, Ufa 450074, Russia
* Correspondence: ellkid@gmail.com
† These authors contributed equally to this work.

Abstract: This work is devoted to the development of methods for constructing asymptotic formulas as $x \to \infty$ of a fundamental system of solutions of linear differential equations generated by a symmetric two-term differential expression of odd order. The coefficients of the differential expression belong to classes of functions that allow oscillation (for example, those that do not satisfy the classical Titchmarsh–Levitan regularity conditions). As a model equation, the fifth-order equation $\frac{i}{2}\left[(p(x)y''')'' + (p(x)y'')'''\right] + q(x)y = \lambda y$, along with various behaviors of coefficients $p(x)$, $q(x)$, is investigated. New asymptotic formulas are obtained for the case when the function $h(x) = -1 + p^{-1/2}(x) \notin L_1[1,\infty)$ significantly influences the asymptotics of solutions to the equation. The case when the equation contains a nontrivial bifurcation parameter is studied.

Keywords: asymptotic methods; oscillating coefficients; singular differential equations of odd order; Campbell's identity; quasi-derivatives; Shin–Zettl matrix

MSC: 34E10; 34E15; 34L20

Citation: Sultanaev, Y.T.; Valeev, N.F.; Nazirova, E.A. On the Asymptotic of Solutions of Odd-Order Two-Term Differential Equations. *Mathematics* **2024**, *12*, 213. https://doi.org/10.3390/math12020213

Academic Editor: Natalia Bondarenko
Received: 29 November 2023
Revised: 2 January 2024
Accepted: 5 January 2024
Published: 9 January 2024

1. Introduction

Analyzing the asymptotic behavior as $x \to \infty$ of a fundamental system of solutions of arbitrary-order singular differential equations, being of independent interest, is an effective method for studying qualitative spectral characteristics for corresponding differential operators [1–3]. As a rule, in these books, differential equations with regular coefficients with regular growth at infinity are investigated. Therefore, the study of the asymptotic behavior of solutions to equations with coefficients from other classes of functions is of particular interest. Such classes of functions were described by us in a previous paper [4]. Let us also note the works [5–7], where differential operators with distribution coefficients were studied.

For example, in the work [7], asymptotic formulas were obtained for the fundamental system of solutions of a two-term equation of even order:

$$(-1)^n(p(x)y^{(n)})^{(n)} + q(x)y = \lambda y, \quad x \in [1,\infty),$$

where the locally summable function p can be represented as $p(x) = (1 + r(x))^{-1}$, $r \in L_1[1,\infty)$ and q is a generalized function representable for some fixed k, $0 \le k \le n$, in the form $q = \sigma^{(k)}$, where $\sigma \in L_1[1,\infty)$ if $k < n$, $|\sigma|(1+|r|)(1+|\sigma|) \in L_1[1,\infty)$, if $k = n$.

Since 2014, we have been publishing a series of articles devoted to the study of the asymptotic behavior of solutions to singular ordinary differential equations with regularly oscillating coefficients [4,8–11]. In this case, a new approach was used for the study based on a sequence of matrix transformations and the use of Campbell's identity [12].

The use of this approach made it possible to obtain new asymptotic formulas in different cases. For example, in [4,11], new asymptotic formulas were obtained for solutions of the Sturm–Liouville equation

$$y'' + \left(\mu^2 + \frac{sin(x^\beta)}{x^\alpha}\right)y = 0, \;\; 0 < \alpha \leq 1, \; \beta > \frac{\alpha}{2} + 1$$

under some relations between α, β, μ. Note that μ has the meaning of a bifurcation parameter. By the way, this equation is one of the equations for testing new methods for constructing asymptotic formulas (see, for example, [13], p. 160).

Equations of odd order for irregular classes of coefficients (in the Titchmarsh–Levitan sense) have been studied less. In the works [7,10], the asymptotics of solutions of odd-order equations were studied in the case when the coefficient of the highest derivative is equal or equivalent to unity.

Here, we develop an approach that was proposed in [4,8–11] and can be implemented to study the asymptotic behavior as $x \to \infty$ of a fundamental system of solutions of two-term equations of arbitrary odd order of the form

$$ly = \frac{i}{2}\left[\left(p(x)y^{(n)}\right)^{(n+1)} + \left(p(x)y^{(n+1)}\right)^{(n)}\right] + q(x)y = \lambda y, \; x \geq 1 \tag{1}$$

for various behaviors of coefficients $p(x)$, $q(x)$.

This method allows us to significantly expand the classes of coefficients $p(x)$ and $q(x)$ for which we can write out the asymptotic behavior of solutions. In particular, new formulas are obtained in cases where $p(x)$ and $q(x)$ allow oscillations. Note that the new formulas obtained allow us to study the spectral properties of differential operators generated by the expression ly (1).

2. Transition to the Ordinary System of Differential Equations Using Quasi-Derivatives

Let us write Equation (1) in the form of a system of ordinary differential equations of the first order. To do this, we use the apparatus of quasi-derivatives (for more detail, see [14–16]). Let us define the functions $q_n(x) \in L_{1,loc}[1,\infty)$ so that

$$q_n^{(n)}(x) = q(x) \tag{2}$$

and introduce into consideration quasi-derivatives defined by the following formulas:

$$\begin{cases} z_1 = y, & z_{n+2} = \sqrt{p}z'_{n+1} - iq_n z_1 \\ z_2 = z'_1, & z_{n+3} = z'_{n+1} + iC_n^1 q_n z_2 \\ \dots & \dots \\ z_n = z'_{n-1}, & z_{2n} = z'_{2n-1} + i(-1)^{n-1}C_n^{n-2} q_n z_{n-1} \\ z_{n+1} = \sqrt{p}z'_n, & z_{2n+1} = z'_{2n} + i(-1)^n C_n^{n-1} q_n z_n. \end{cases} \tag{3}$$

Then, Equation (1) is equivalent to the relation

$$z'_{2n+1} = \lambda z_1 - i(-1)^{n+1}\frac{q_n}{\sqrt{p}}z_{n+1}.$$

Let us introduce the column vector $\mathbf{z} = column(z_1, z_2, ..., z_{2n+1})$ and write Equation (1) as a system of ordinary differential equations

$$\mathbf{z}' = S\mathbf{z},$$

where $S(x, \lambda)$ is the Shin–Zettl matrix [14].

$$S = \begin{pmatrix} 0 & 1 & 0 & \dots & 0 & 0 & 0 & 0 & \dots & 0 \\ 0 & 0 & 1 & \dots & 0 & 0 & 0 & 0 & \dots & 0 \\ & & & & & \dots\dots\dots & & & & \\ 0 & 0 & 0 & \dots & 1 & 0 & 0 & 0 & \dots & 0 \\ 0 & 0 & 0 & \dots & 0 & \frac{1}{\sqrt{p}} & 0 & 0 & \dots & 0 \\ \frac{iq_n}{\sqrt{p}} & 0 & 0 & \dots & 0 & 0 & \frac{1}{\sqrt{p}} & 0 & \dots & 0 \\ 0 & -inq_n & 0 & \dots & 0 & 0 & 0 & 1 & \dots & 0 \\ & & & & & \dots\dots\dots & & & & \\ 0 & 0 & 0 & \dots & 0 & (-1)^{n-1}inq_n & 0 & 0 & \dots & 1 \\ -i\lambda & 0 & 0 & \dots & 0 & 0 & \frac{(-1)^n iq_n}{\sqrt{p}} & 0 & \dots & 0 \end{pmatrix},$$

where the non-zero elements of the matrix $S(x, \lambda)$ are given by the formulas

$$s_{kj} = 1, \;\; j = 1 + k, \;\; k = \overline{1, n-1}, \;\; k = \overline{n+2, 2n}, \;\; s_{n,n+1} = s_{n+1,n+2} = \frac{1}{\sqrt{p}},$$

$$s_{n+1,1} = \frac{iq_n}{\sqrt{p}}, \; s_{n+k,k} = (-1)^{k-1} i C_n^{k-1} q_n, \; k = \overline{2, n},$$

$$s_{2n+1,2n-1} = \frac{(-1)^n iq_n}{\sqrt{p}}, \; s_{2n+1,1} = -i\lambda.$$

Note that from the relation $q_n^{(n)}(x) = q(x)$, the function $q_n(x)$ is determined up to a polynomial of order $n-1$. However, the fundamental system of solutions of Equation (1) does not depend on the choice of integration constants, which follows directly from Formula (3). Conditions for choosing the coefficients of the polynomial are formulated for each case under study.

Further, in order to avoid complicated formulas, we limit ourselves to considering the 5th-order two-term equation

$$ly = \frac{i}{2}\left[(p(x)y'')''' + (p(x)y''')''\right] + q(x)y = \lambda y, \; x \geq 1. \quad (4)$$

Using Formula (3), we introduce quasi-derivatives

$$\begin{cases} z_1 = y \\ z_2 = z_1' \\ z_3 = \sqrt{p} z_2' \\ z_4 = \sqrt{p} z_3' - iq_2 \, z_1 \\ z_5 = z_4' + 2iq_2 z_2. \end{cases}$$

Then, Equation (4) is equivalent to the relation

$$z_5' = -i\lambda z_1 + \frac{iq_2}{\sqrt{p}} z_3$$

and can be written as a system of ordinary differential equations:

$$\mathbf{z}' = \begin{pmatrix} 0 & 1 & 0 & 0 & 0 \\ 0 & 0 & 1/\sqrt{p} & 0 & 0 \\ iq_2/\sqrt{p} & 0 & 0 & 1/\sqrt{p} & 0 \\ 0 & -2iq_2 & 0 & 0 & 1 \\ -i\lambda & 0 & iq_2/\sqrt{p} & 0 & 0 \end{pmatrix} \mathbf{z},$$

where $\mathbf{z} = column(z_1, z_2, z_3, z_4, z_5)$.

Let the function $p(x)$ admit the representation

$$\frac{1}{\sqrt{p(x)}} = 1 + h(x),\ h(x) \in L_{1,loc}[1, \infty). \tag{5}$$

Let us write the last system of equations in the following form, taking into account (5):

$$\mathbf{z}' = (L_0 + h(x)L_1 + iq_2(x)D_0 + ih(x)q_2(x))D_1)\mathbf{z}, \tag{6}$$

$$L_0 = \begin{pmatrix} 0 & 1 & 0 & 0 & 0 \\ 0 & 0 & 1 & 0 & 0 \\ 0 & 0 & 0 & 1 & 0 \\ 0 & 0 & 0 & 0 & 1 \\ -i\lambda & 0 & 0 & 0 & 0 \end{pmatrix},\quad L_1 = \begin{pmatrix} 0 & 0 & 0 & 0 & 0 \\ 0 & 0 & 1 & 0 & 0 \\ 0 & 0 & 0 & 1 & 0 \\ 0 & 0 & 0 & 0 & 0 \\ 0 & 0 & 0 & 0 & 0 \end{pmatrix},$$

$$D_0 = \begin{pmatrix} 0 & 0 & 0 & 0 & 0 \\ 0 & 0 & 0 & 0 & 0 \\ 1 & 0 & 0 & 0 & 0 \\ 0 & -2 & 0 & 0 & 0 \\ 0 & 0 & 1 & 0 & 0 \end{pmatrix},\quad D_1 = \begin{pmatrix} 0 & 0 & 0 & 0 & 0 \\ 0 & 0 & 0 & 0 & 0 \\ 1 & 0 & 0 & 0 & 0 \\ 0 & 0 & 0 & 0 & 0 \\ 0 & 0 & 1 & 0 & 0 \end{pmatrix}.$$

3. Construction of Asymptotic Formulas

3.1. Case 1

Let us set

$$\tilde{h}(x) = q_2(x)h(x). \tag{7}$$

Let the following conditions be satisfied:

$$h(x),\ q_2(x) \in L_1[1, \infty),\ \tilde{h}(x) \in L_{1,loc}[1, \infty).$$

For example, these conditions are true for

$$h(x) = \frac{1}{x^{\gamma}},\ \gamma > 1;\quad q(x) = x^{\alpha} \sin x^{\beta},\ \ \alpha > 0,\ \ \beta > \frac{\alpha + 3}{2}.$$

Let the constant matrix T reduce the matrix L_0 to diagonal form. Let us make a replacement:

$$\mathbf{z} = T\mathbf{u},\ \ T^{-1}L_0T = \Lambda,\ \ \mu_k^5 = -i\lambda,\ k = \overline{1,5},$$

$$\Lambda = \begin{pmatrix} \mu_1 & 0 & 0 & 0 & 0 \\ 0 & \mu_2 & 0 & 0 & 0 \\ 0 & 0 & \mu_3 & 0 & 0 \\ 0 & 0 & 0 & \mu_3 & 0 \\ 0 & 0 & 0 & 0 & \mu_3 \end{pmatrix},\quad T = \begin{pmatrix} 1 & 1 & 1 & 1 & 1 \\ \mu_1 & \mu_2 & \mu_3 & \mu_4 & \mu_5 \\ \mu_1^2 & \mu_2^2 & \mu_3^2 & \mu_4^2 & \mu_5^2 \\ \mu_1^3 & \mu_2^3 & \mu_3^3 & \mu_4^3 & \mu_5^3 \\ \mu_1^4 & \mu_2^4 & \mu_3^4 & \mu_4^4 & \mu_5^4 \end{pmatrix}.$$

Then, system (5) takes the form

$$\mathbf{u}' = \Big(\Lambda + h(x)T^{-1}L_1T + iq_2(x)T^{-1}D_0T + iq_2(x)h(x)T^{-1}D_1T\Big)\mathbf{u}. \tag{8}$$

Obviously, because of the imposed conditions, System (8) satisfies the conditions of Lemma 1 in [3], p. 288, and is L-diagonal, which means we can write out asymptotic formulas as $x \to \infty$ for the fundamental system solutions of this system:

$$\mathbf{z}_k(x, \lambda) = T \cdot \mathbf{u}_k(x, \lambda) = e^{\mu_k x} \cdot T \cdot (\mathbf{e}_k + o(1)),\ \ k = \overline{1,5},$$

where $\mathbf{e}_k$ are unit basis vectors.

3.2. Case 2

Let the following conditions be satisfied:

$$q_2(x) \notin L_1[1,\infty), \;\; h(x),\, q_3(x),\, \tilde{h}(x) \in L_1[1,\infty). \tag{9}$$

These conditions are true for

$$h(x) = \frac{1}{x^\gamma}, \; \gamma > 1; \;\; q(x) = x^\alpha \sin x^\beta, \;\; \alpha > 0, \;\; \frac{\alpha+3}{2} \geq \beta > \frac{\alpha+4}{3}.$$

Following the approach outlined in the paper [8], we make a replacement in System (6):

$$\mathbf{z} = e^{iq_3(x)D_0}\mathbf{u}. \tag{10}$$

We obtain

$$\mathbf{u}' = e^{-iq_3(x)D_0}\left(L_0 + h(x)L_1 + i\tilde{h}(x)D_1\right)e^{iq_3(x)D_0}\mathbf{u}. \tag{11}$$

Let us apply Campbell's identity to transform the right-hand side of (11) to

$$e^{-iq_3(x)D_0}L_0e^{iq_2(x)D_0} = L_0 - iq_3(x)[D_0, L_0] + \frac{i^2q_3^2(x)}{2!}[D_0,[D_0,L_0]] -$$

$$-\frac{i^3q_3^3(x)}{3!}[D_0,[D_0,[D_0,L_0]]] + ...,$$

where $[A,B] = AB - BA$ is a matrix commutator.

Below, we use the following obvious consideration: if the matrix A is nilpotent, then a nonzero sequence of matrix commutators of the form $[A,[A,...,[A,B]]...]$ is finite.

Note that the matrix D_0 is nilpotent. By sequentially calculating the commutators on the right side of the last relation, we obtain that all terms, starting from the fourth, are equal to zero, and non-zero terms can be calculated:

$$[D_0, L_0] = \begin{pmatrix} 0 & 0 & 0 & 0 & 0 \\ -1 & 0 & 0 & 0 & 0 \\ 0 & 3 & 0 & 0 & 0 \\ 0 & 0 & -3 & 0 & 0 \\ 0 & 0 & 0 & 1 & 0 \end{pmatrix}, \quad [D_0,[D_0,L_0]] = \begin{pmatrix} 0 & 0 & 0 & 0 & 0 \\ 0 & 0 & 0 & 0 & 0 \\ 0 & 0 & 0 & 0 & 0 \\ 5 & 0 & 0 & 0 & 0 \\ 0 & 5 & 0 & 0 & 0 \end{pmatrix}.$$

Similar calculations can be carried out for the remaining terms on the right side of (11):

$$e^{-iq_3(x)D_0}h(x)L_1e^{iq_3(x)D_0} = h(x)L_1 - iq_3h(x)[D_0,L_1] + \frac{h(x)i^2q_3^2(x)}{2!}[D_0,[D_0,L_1]] + ...,$$

$$[D_0, L_1] = \begin{pmatrix} 0 & 0 & 0 & 0 & 0 \\ -1 & 0 & 0 & 0 & 0 \\ 0 & 2 & 0 & 0 & 0 \\ 0 & 0 & -2 & 0 & 0 \\ 0 & 0 & 0 & 1 & 0 \end{pmatrix}, \quad [D_0,[D_0,L_1]] = \begin{pmatrix} 0 & 0 & 0 & 0 & 0 \\ 0 & 0 & 0 & 0 & 0 \\ 0 & 0 & 0 & 0 & 0 \\ 4 & 0 & 0 & 0 & 0 \\ 0 & 4 & 0 & 0 & 0 \end{pmatrix}.$$

Because $[D_0, D_1] = 0$, the following representation is true:

$$e^{-iq_3(x)D_0}\tilde{h}(x)D_1e^{iq_3(x)D_0} = \tilde{h}(x)D_1.$$

Then, Equation (11) can be rewritten as

$$\mathbf{u}' = (L_0 + hL_1 + \tilde{h}D_1 - iq_3[D_0,L_0] + \frac{i^2q_3^2}{2!}[D_0,[D_0,L_0]] - iq_3h[D_0,L_1] + \frac{i^2hq_3^2}{2!}[D_0,[D_0,L_1]])\mathbf{u}. \tag{12}$$

Because of the imposed conditions on the functions $h(x),\ q(x)$, the last system can be written as

$$\mathbf{u}' = (L_0 + D(x))\mathbf{u},$$

where $D(x)$ is a matrix whose elements belong to $L_1[1,\infty)$. Just as in Case 1, let us make the replacement $\mathbf{u} = T\mathbf{v}$; then,

$$\mathbf{v}' = (\Lambda + T^{-1}D(x)T)\mathbf{v}. \tag{13}$$

System (13) satisfies the conditions of Lemma 1 in [3] and is L-diagonal, which means, taking into account (10), we can write asymptotic formulas for $x \to \infty$ for its fundamental system of solutions:

$$\mathbf{z}_k(x,\lambda) = e^{\mu_k x} \cdot e^{iq_3(x)D_0} \cdot T \cdot (\mathbf{e}_k + o(1)), \ \ k = \overline{1,5}.$$

where $\mathbf{e}_k$ are unit vectors.

Remark 1. *Let us note the importance of resulting Equation* (12)*. Imposing various conditions on the coefficients of this equation, $h(x)$, $q_3(x)h(x)$, $q_3^2(x)h(x), \tilde{h}(x)$, $q_3(x)$, and $q_2^3(x)$, different from the conditions in* (9) *, one can obtain different asymptotics of the fundamental system of solutions with nontrivial properties.*

3.3. Case 3

Let us define the function $h_1(x)$ so that

$$h_1'(x) = h(x). \tag{14}$$

Let us now consider the case when

$$h(x),\ q_2(x) \notin L_1[1,\infty),\ \ q_3(x),\ h_1(x),\ \tilde{h}(x) \in L_1[1,\infty).$$

For example, these conditions are true for

$$h(x) = \frac{1}{x^\gamma},\ 0 < \gamma < 1; \ \ q(x) = x^\alpha \sin x^\beta,\ \ 2 > \alpha > 0,\ \ \frac{\alpha+3}{2} \geq \beta > \frac{\alpha+3-\gamma}{2}.$$

Just as in Case 2, let us make a replacement in System (6)

$$\mathbf{z} = e^{h_1(x)L_1}\mathbf{u}. \tag{15}$$

Then, System (6) takes the form

$$\mathbf{u}' = e^{-h_1(x)L_1}\big(L_0 + iq_2(x)D_0 + i\tilde{h}(x)D_1\big)e^{h_1(x)L_1}\mathbf{u}. \tag{16}$$

Let us apply Campbell's identity to transform the right-hand side of (16):

$$e^{-h_1(x)L_1}L_0e^{h_1(x)L_1} = L_0 - h_1(x)[L_1,L_0] + \frac{h_1^2(x)}{2!}[L_1,[L_1,L_0]] - \frac{h_1^3(x)}{3!}[L_1,[L_1,[L_1,L_0]]] + \dots$$

Note that the matrix L_1 is nilpotent. By sequentially calculating the commutators on the right side of the last relation, we obtain that all terms, starting from the fourth, are equal to zero, and non-zero terms can be calculated:

$$[L_1,L_0] = \begin{pmatrix} 0 & 0 & -1 & 0 & 0 \\ 0 & 0 & 0 & 0 & 0 \\ 0 & 0 & 0 & 0 & 1 \\ 0 & 0 & 0 & 0 & 0 \\ 0 & 0 & 0 & 0 & 0 \end{pmatrix}, \quad [L_1,[L_1,L_0]] = \begin{pmatrix} 0 & 0 & 0 & 1 & 0 \\ 0 & 0 & 0 & 0 & 1 \\ 0 & 0 & 0 & 0 & 0 \\ 0 & 0 & 0 & 0 & 0 \\ 0 & 0 & 0 & 0 & 0 \end{pmatrix}.$$

Similar calculations can be carried out for the remaining terms on the right side of (16):

$$e^{-h_1(x)L_1} iq_2(x)D_0 e^{h_1(x)L_1} = iq_2(x)D_0 - iq_2(x)h_1(x)[L_1, D_0] + \frac{iq_2(x)h_1^2(x)}{2!}[L_1,[L_1,D_0]] -$$

$$-\frac{iq_2(x)h_1^3(x)}{3!}[L_1,[L_1,[L_1,D_0]]] + \frac{iq_2(x)h_1^4(x)}{4!}[L_1,[L_1,[L_1,[L_1,D_0]]]].$$

$$[L_1, D_0] = \begin{pmatrix} 0 & 0 & 0 & 0 & 0 \\ -1 & 0 & 0 & 0 & 0 \\ 0 & 2 & 0 & 0 & 0 \\ 0 & 0 & -2 & 0 & 0 \\ 0 & 0 & 0 & 1 & 0 \end{pmatrix}, \quad [L_1,[L_1,D_0]] = \begin{pmatrix} 0 & 0 & 0 & 0 & 0 \\ 0 & -2 & 0 & 0 & 0 \\ 0 & 0 & 4 & 0 & 0 \\ 0 & 0 & 0 & -2 & 0 \\ 0 & 0 & 0 & 0 & 0 \end{pmatrix},$$

$$[L_1,[L_1,[L_1,D_0]]] = \begin{pmatrix} 0 & 0 & 0 & 0 & 0 \\ 0 & 0 & 6 & 0 & 0 \\ 0 & 0 & 0 & -6 & 0 \\ 0 & 0 & 0 & 0 & 0 \\ 0 & 0 & 0 & 0 & 0 \end{pmatrix}, \quad [L_1,[L_1,[L_1,[L_1,D_0]]]] = \begin{pmatrix} 0 & 0 & 0 & 0 & 0 \\ 0 & 0 & 0 & -12 & 0 \\ 0 & 0 & 0 & 0 & 0 \\ 0 & 0 & 0 & 0 & 0 \\ 0 & 0 & 0 & 0 & 0 \end{pmatrix},$$

$$e^{-h_1(x)L_1} i\tilde{h}(x)D_1 e^{h_1(x)L_1} = i\tilde{h}(x)D_1 - ih_1(x)\tilde{h}(x)[L_1, D_1],$$

$$[L_1, D_1] = \begin{pmatrix} 0 & 0 & 0 & 0 & 0 \\ 1 & 0 & 0 & 0 & 0 \\ 0 & 0 & 0 & 0 & 0 \\ 0 & 0 & 0 & 0 & 0 \\ 0 & 0 & 0 & -1 & 0 \end{pmatrix}.$$

Because of the imposed conditions on the functions $h(x),\ q(x)$, the last system can be written as

$$\mathbf{u}' = (L_0 + iq_2(x)D_0 + D(x))\mathbf{u}.$$

where $D(x)$ is a matrix whose elements belong to $L_1[1,\infty)$. Unlike Case 2, the resulting system is not yet L-diagonal. Let us make one more transformation:

$$\mathbf{u} = e^{iq_3(x)D_0}\mathbf{v}. \tag{17}$$

Then, the last system will take the form

$$\mathbf{v}' = e^{-iq_3(x)D_0}(L_0 + iq_2(x)D_0 + D(x))e^{iq_3(x)D_0}\mathbf{v}. \tag{18}$$

Let us apply Campbell's identity to transform the right-hand side of (18). Just as in Case 2, taking into account the nilpotency of the matrix D_0 and sequentially calculating all the necessary matrix commutators, we obtain the following form of System (18):

$$\mathbf{v}' = (L_0 + \tilde{D}(x))\mathbf{v}.$$

Here, the matrix $\tilde{D}(x)$ is defined by the expression

$$\tilde{D}(x) = -iq_3(x)[D_0, L_0] + \frac{i^2 q_3^2}{2!}[D_0,[D_0,L_0]] + e^{-iq_3(x)D_0}D(x)e^{iq_3(x)D_0},$$

which because of the conditions imposed above on the functions $h(x),\ q(x)$, is obviously a matrix with elements summable over $[1,\infty)$.

Next, we make the replacement $\mathbf{v} = T\mathbf{s}$. Then,

$$\mathbf{s}' = (\Lambda + T^{-1}\tilde{D}(x)T)\mathbf{s}. \tag{19}$$

System (19) satisfies the conditions of Lemma 1 in [3] and is L-diagonal, which means, taking into account (15) and (17), we can write out asymptotic formulas as $x \to \infty$ for its fundamental system of solutions:

$$\mathbf{z}_k(x, \lambda) = e^{\mu_k x} \cdot e^{h_1(x)L_1} \cdot e^{iq_3(x)D_0} \cdot T \cdot (\mathbf{e}_k + o(1)), \ \ k = \overline{1,5},$$

where $\mathbf{e}_k$ are unit vectors.

Summarizing Cases 1–3, we find that we have proven the following theorem:

Theorem 1. *Let functions* $q_2(x),\ q_3(x),\ h(x),\ \tilde{h}(x),\ h_1(x)$ *be defined by Formulas* (2), (5), (7), *and* (14) *correspondingly and one of the following conditions be satisfied:*
(1) $h(x),\ q_2(x) \in L_1[1, \infty)$,
(2) $h(x),\ q_3(x),\ \tilde{h}(x) \in L_1[1, \infty)$,
(3) $q_3(x),\ h_1(x),\ \tilde{h}(x) \in L_1[1, \infty)$.
Then, the asymptotic formulas as $x \to \infty$ *for the fundamental system of solutions of Equation* (4) *are valid:*

$$y_j(x, \lambda) = e^{\mu_j x} \cdot (1 + o(1)), \quad j = \overline{1,5}.$$

In fact, we obtain the asymptotic formulas as $x \to \infty$ for vector function; we may also write down the asymptotic formulas for quasi-derivatives of solutions.

3.4. Counterexample

Let us show that the conditions of Theorem 1 are essential.

Let

$$h(x),\ q_2(x) \notin L_1[1, \infty),\ \ q_3(x),\ h_1(x) \in L_1[1, \infty),\ \ \tilde{h}(x) \in L_{1,loc}[1, \infty).$$

In the same way as in Case 3, we make sequential transformations

$$\mathbf{z} = e^{h_1(x)L_1}\mathbf{u}, \quad \mathbf{u} = e^{iq_3(x)D_0}\mathbf{w},$$

which brings Equation (6) to the form

$$\mathbf{w}' = (L_0 + ih(x)q_2(x))D_1)\mathbf{w} + F(x)\mathbf{w}, \tag{20}$$

where $F(x) \in L_1[1, \infty)$.

The last system of equations allows for a large variety in the asymptotic behavior as $x \to +\infty$ and can be the subject of a separate study.

We limit ourselves to considering a model example on which we demonstrate an unusual property of equations with oscillating coefficients, namely, the influence of the algebraic structure of the coefficients of the equation on the asymptotic behavior of the solutions.

Let

$$h(x) = a\sin(e^x),\, q_2(x) = \sin(ke^x),$$

from which

$$\tilde{h}(x) = h(x)q_2(x) = a\sin(e^x)\sin(ke^x) = \frac{1}{2}a[\cos((k-1)e^x) + \cos((k+1)e^x)].$$

Consider two cases: $k = \pm 1$ and $k \neq \pm 1$. Let $k \neq \pm 1$. Define the function $\tilde{h}_1(x)$ so that $\tilde{h}_1'(x) = \tilde{h}(x)$. Note that, in this case, $\tilde{h}_1(x) \in L_1[1, \infty)$.

In System (20), we set

$$\mathbf{w} = e^{i\tilde{h}_1(x)D_1}\mathbf{v};$$

then, for $\mathbf{v}$, we obtain the system

$$\mathbf{v}' = (L_0 + F_1(x))\mathbf{v}, \tag{21}$$

where $F_1(x) \in L_1[1,\infty)$ and which can easily be reduced to an L-diagonal system. Consequently, the main term of the asymptotics of the fundamental system of solutions (21), as above, are determined:

$$\mathbf{z}_j(x,\lambda) = e^{\mu_j x} \cdot T \cdot (\mathbf{e}_j + o(1)), \ \ j = \overline{1,5}.$$

Let $k = \pm 1$. Note that now

$$h(x)q_2(x) = a\sin(e^x)\sin(ke^x) =$$

$$\frac{1}{2}a[\cos((k-1)e^x) + \cos((k+1)e^x)] = \frac{1}{2}a(1+\cos(2e^x))$$

whence it follows that $h(x)q_2(x) \notin L_1[1,\infty)$. Let us denote $\sigma(x) = \frac{i}{2}a\cos(2e^x)$ and represent the system (21) in the following form:

$$\mathbf{w}' = \left(L_0 + \frac{i}{2}aD_1 + \sigma(x)D_1\right)\mathbf{w} + F(x)\mathbf{w}, \tag{22}$$

where $F(x) \in L_1[1,\infty)$, and the matrix $L_0 + \frac{i}{2}aD_1$ is constant. Considering

$$\sigma_1(x) = -\frac{i}{2}a\int_x^{+\infty} \cos(2e^\xi)d\xi \in L_1[1,\infty),$$

we make a replacement in System (22)

$$\mathbf{w} = e^{\sigma_1(x)D_1}\mathbf{v},$$

and again using the technique described above, we obtain the system

$$\mathbf{v}' = (L_0 + \frac{i}{2}aD_1)\mathbf{w} + F_1(x)\mathbf{w}. \tag{23}$$

Here, taking into account that $\sigma_1(x) \in L_1[1,\infty)$, we have $F_1(x) \in L_1[1,\infty)$. Let matrix $\hat{T}$ reduce matrix $L_0 + \frac{i}{2}aD_1$ to diagonal form, $\hat{\mu}_j\ , werej = \overline{1,5}$-eigenvalues of the matrix $L_0 + \frac{i}{2}aD_1$. Let us make a replacement:

$$\mathbf{v} = \hat{T}\mathbf{s}, \ \ \hat{T}^{-1}(L_0 + \frac{i}{2}aD_1)\hat{T} = \hat{\Lambda}.$$

Then, System (22) takes the form

$$\mathbf{s}' = (\hat{\Lambda} + \hat{T}^{-1}F_1(x)\hat{T})\mathbf{s}.$$

The resulting system is equivalent to the L-diagonal system. Then, as above, the fundamental system of solutions of Equation (6) can be represented as

$$\mathbf{z}_j(x,\lambda) = e^{\hat{\mu}_j x} \cdot \hat{T} \cdot (\mathbf{e}_j + o(1)), \ \ j = \overline{1,5}.$$

Thus, when the numerical coefficient k passes through the points $k = \pm 1$, the asymptotics of the fundamental system of solutions of Equation (4) undergoes a qualitative change. In other words, the points $k = \pm 1$ are bifurcation points for System (6) and corresponding Equation (4).

Such bifurcation points for differential equations and systems of equations with regularly oscillating coefficients are typical and were recently noted by us in works devoted to the study of the asymptotic behavior of the Sturm–Liouville equation with an oscillating potential [4,11].

4. Discussion

The results obtained have important applications in the spectral theory of differential operators generated by the left side of Equation (1). In particular, they make it possible to calculate the deficiency indices of the corresponding minimal differential operator.

The authors intend to investigate this issue in the future. In addition, we will be interested in the qualitative nature of the spectrum of such operators.

5. Conclusions

In this paper, we present a new approach to studying asymptotic behavior for $x \to \infty$ solutions of singular binomial differential equations of odd order for new classes of coefficients the corresponding differential expression.

This approach is based on the transition to a first-order system using quasi-derivatives and sequences of matrix transformations of this system. The key point of our approach is using the features of the algebraic matrix structure of the resulting system of ordinary differential equations.

This makes it possible to construct asymptotic formulas for solutions of new classes of equations, for example, for equations with oscillating coefficients. Note that similar results for such equations were not previously known.

Of particular interest is the counterexample we constructed, showing the phenomenon of "resonance" of oscillating coefficients: the existence of certain values of the numerical parameters of the coefficients of the equation at which the qualitative change asymptotic behavior of solutions. This property is typical, in general, for differential equations with oscillating coefficients of arbitrary order and was previously shown by us for the Sturm–Liouville equation.

Author Contributions: All three authors, on an equal level, discussed and posed the research questions in this paper. Y.T.S. helped prove the main results and type the manuscript. N.F.V. is the main author concerning the proofs of the main results. E.A.N. put the results into a more general frame and instructed the team on how to write the manuscript in this final form. All authors have read and agreed to the published version of the manuscript.

Funding: The studies of E.A.N. and Y.T.S. were funded by the Russian Science Foundation, project no. 23-21-00225.

Data Availability Statement: Data are contained within the article.

Conflicts of Interest: The authors declare no conflicts of interest.

References

1. Eastham, M.S.P. *The Asymptotic Solution of Linear Differential Systems, Applications of the Levinson Theorem*; Clarendon Press: Oxford, UK, 1989.
2. Fedoryuk, M.V. *Asymptotic Methods for Linear Ordinary Differential Equations*; Nauka: Moscow, Russia, 1983. (In Russian)
3. Naimark, M.A. *Linear Differential Operators*; Nauka: Moscow, Russia, 1969. (In Russian)
4. Nazirova, E.A.; Sultanaev, Y.T.; Valeeva, L.N. On a Method for Studying the Asymptotics of Solutions of Sturm–Liouville Differential Equations with Rapidly Oscillating Coefficients. *Math. Notes* **2022**, *112*, 1059–1064.
5. Konechnaja, N.N.; Mirzoev, K.A.; Shkalikov, A.A. On the Asymptotic Behavior of Solutions to Two-Term Differential Equations with Singular Coefficients. *Math. Notes* **2023**, *104*, 244–252. [CrossRef]
6. Mirzoev, K.A.; Konechnaja, N.N. Asymptotics of solutions to linear differential equations of odd order. *Vestnik Moskov. Univ. Ser. 1. Mat. Mekh.* **2020**, *75*, 22–26. [CrossRef]
7. Konechnaja, N.N.; Mirzoev, K.A.; Shkalikov, A.A. Asymptotics of Solutions of Two-Term Differential Equations. *Math. Notes* **2023**, *113*, 228–242. [CrossRef]
8. Nazirova, E.A.; Sultanaev, Y.T.; Valeev, N.F. On a new approach for studying asymptotic behavior of solutions to singular differential equations. *Trans. Ufa Math. J.* **2015**, 3, 9–14.

9. Myakinova, O.V.; Sultanaev, Y.T.; Valeev, N.F. On the Asymptotics of Solutions of a Singular nth-Order Differential Equation with Nonregular Coefficients. *Math. Notes* **2018**, *104*, 606–611.
10. Valeev, N.F.; Nazirova, E.A.; Sultanaev, Y.T. On a Method for Studying the Asymptotics of Solutions of Odd-Order Differential Equations with Oscillating Coefficients. [CrossRef] *Math. Notes* **2021**, *109*, 980–985. [CrossRef]
11. Nazirova, E.A.; Sultanaev, Y.T.; Valeev, N.F. The new asymptotics for solutions of the Sturm-Liouville equation. *Proc. Inst. Math. Mech. Acad. Sci. Azerbaijan* **2023**, *49*, 253–258.
12. Rossmann, W. *Lie Groups—An Introduction Through Linear Groups*; Oxford University Press: Oxford, UK, 2006.
13. Bellman, R. *Stability Theory of Differential Equations*; Mc-Graw-Hill: New York, NY, USA; Toronto, ON, Canada; London, UK, 1953.
14. Everitt, W.N.; Marcus, L. Boundary value problem and symplectic algebra for ordinary differential and quasi-differential operators. *AMS. Math. Surv. Monogr.* **1999**, *60*, 1–60.
15. Everitt, W.N.; Race, D. Some remarks on linear ordinary quasi-differential expressions. *Proc. Lond. Math. Soc.* **1987**, *3*, 300–320. [CrossRef]
16. Everitt, W.N.; Zettl, A. Differential operators generated by a countable number of quasi-differential expressions on the real line. *Proc. Lond. Math. Soc.* **1992**, *3*, 524–544. [CrossRef]

MDPI AG
Grosspeteranlage 5
4052 Basel
Switzerland
Tel.: +41 61 683 77 34

Mathematics Editorial Office
E-mail: mathematics@mdpi.com
www.mdpi.com/journal/mathematics

www.ingramcontent.com/pod-product-compliance
Lightning Source LLC
LaVergne TN
LVHW070245100526
838202LV00015B/2180